Numerical Methods in Physics with Python

Bringing together idiomatic Python programming, foundational numerical methods, and physics applications, this is an ideal standalone textbook for courses on computational physics. All the frequently used numerical methods in physics are explained, including foundational techniques and hidden gems on topics such as linear algebra, differential equations, root-finding, interpolation, and integration. The second edition of this introductory book features several new codes and 140 new problems (many on physics applications), as well as new sections on the singular-value decomposition, derivative-free optimization, Bayesian linear regression, neural networks, and partial differential equations. The last section in each chapter is an in-depth project, tackling physics problems that cannot be solved without the use of a computer. Written primarily for students studying computational physics, this textbook brings the non-specialist quickly up to speed with Python before looking in detail at the numerical methods often used in the subject.

Alex Gezerlis is Professor of Physics at the University of Guelph. Before moving to Canada, he worked in Germany, the United States, and Greece. He has received several research awards, grants, and allocations on supercomputing facilities. He has taught undergraduate and graduate courses on computational methods, as well as courses on quantum field theory, subatomic physics, and science communication.

Praise for the Second Edition

"Gezerlis' book *Numerical Methods in Physics* with Python is a beautiful example of how an established subject can be brought to the next level by making it very accessible and by introducing several insightful and interdisciplinary applications. This second edition considerably extends the set of exercises, resulting in an extremely useful resource for both students and teachers. Strongly recommended!"

Sonia Bacca, *Johannes Gutenberg-Universität Mainz*

"This new edition of *Numerical Methods*... is another great example of Gezerlis' passion for teaching and for doing so carefully and precisely. Especially welcome, in my view, are the addition of problems at the end of each chapter and the discussion of singular value decomposition (SVD) and Bayesian methods. The SVD is one of the crown jewels of linear algebra which modern students interested in machine learning will surely find beneficial. To physics, computer science, or engineering students mesmerized by the fast Fourier transform, Gezerlis' excellent explanation of it in Chapter 6 is likely to shed some light on the underlying divide-and-conquer algorithm, which is an essential classic."

Joaquin Drut, *University of North Carolina at Chapel Hill*

"A fantastic addition as an introductory textbook for computational physics. The book is timely, and the author made thoughtful and in my view many wise choices. The book is comprehensive and yet accessible to undergraduate students."

Shiwei Zhang, *Flatiron Institute and College of William & Mary*

Praise for the First Edition

"I enthusiastically recommend *Numerical Methods in Physics with Python* by Professor Gezerlis to any advanced undergraduate or graduate student who would like to acquire a solid understanding of the basic numerical methods used in physics. The methods are demonstrated with Python, a relatively compact, accessible computer language, allowing the reader to focus on understanding how the methods work rather than on how to program them. Each chapter offers a self-contained, clear, and engaging presentation of the relevant numerical methods, and captivates the reader with well-motivated physics examples and interesting physics projects. Written by a leading expert in computational physics, this outstanding textbook is unique in that it focuses on teaching basic numerical methods while also including a number of modern numerical techniques that are usually not covered in computational physics textbooks."

Yoram Alhassid, *Yale University*

"In *Numerical Methods in Physics with Python* by Gezerlis, one finds a resource that has been sorely missing! As the usage of Python has become widespread, it is too often the case that students take libraries, functions, and codes and apply them without a solid understanding of what is truly being done 'under the hood' and why. Gezerlis' book fills this gap with clarity and rigor by covering a broad number of topics relevant for physics, describing the underlying techniques and implementing them in detail. It should be an important resource for anyone applying numerical techniques to study physics."

Luis Lehner, *Perimeter Institute*

"Gezerlis' text takes a venerable subject – numerical techniques in physics – and brings it up to date and makes it accessible to modern undergraduate curricula through a popular, open-source programming language. Although the focus remains squarely on numerical techniques, each new lesson is motivated by topics commonly encountered in physics and concludes with a practical hands-on project to help cement the students' understanding. The net result is a textbook which fills an important and unique niche in pedagogy and scope, as well as a valuable reference for advanced students and practicing scientists."

Brian Metzger, *Columbia University*

Numerical Methods in Physics with Python

Second Edition

ALEX GEZERLIS
University of Guelph

Shaftesbury Road, Cambridge CB2 8EA, United Kingdom

One Liberty Plaza, 20th Floor, New York, NY 10006, USA

477 Williamstown Road, Port Melbourne, VIC 3207, Australia

314–321, 3rd Floor, Plot 3, Splendor Forum, Jasola District Centre, New Delhi – 110025, India

103 Penang Road, #05–06/07, Visioncrest Commercial, Singapore 238467

Cambridge University Press is part of Cambridge University Press & Assessment, a department of the University of Cambridge.

We share the University's mission to contribute to society through the pursuit of education, learning and research at the highest international levels of excellence.

www.cambridge.org
Information on this title: www.cambridge.org/9781009303859

DOI: 10.1017/9781009303897

First published 2020
Second edition published 2023

A catalogue record for this publication is available from the British Library.

ISBN 978-1-009-30385-9 Hardback
ISBN 978-1-009-30386-6 Paperback

Additional resources for this publication at www.cambridge.org/gezerlis2 and www.numphyspy.org

To Marcos, ψυχή βαθιά

My soul, rather than yearn for life immortal,
press into service every shift at your disposal.

Pindar

Contents

Preface

> The health of the eye seems to demand a horizon.
> We are never tired, so long as we can see far enough.
>
> Ralph Waldo Emerson

This is a textbook for advanced undergraduate (or beginning graduate) courses on Computational Physics. To explain what this means, I first go over what this book is *not*.

First, this is not a text that focuses mainly on physics applications and basic programming, only bringing up numerical methods as the need arises. It's true that such an approach would have the benefit of giving rise to beautiful visualizations and helping students gain confidence in using computers to study science. The disadvantage of this approach is that it tends to rely on external libraries, i.e., "black boxes". To make an analogy with non-computational physics, we teach students calculus before seeing how it helps us do physics. In other words, an instructor would not claim that derivatives are important but already well-studied, so we'll just employ a package that takes care of them. That being said, a physics-applications-first approach may be appropriate for a more introductory course (the type with a textbook that has the answers in the back) or perhaps as a computational addendum to an existing text on mechanics, electromagnetism, and so on.

Second, this is not a text addressing a small subset of modern computational methods. Depending on the instructor's interests and expertise, computational courses sometimes specialize on a single theme, such as: simulations (e.g., molecular dynamics or Monte Carlo), data analysis (e.g., uncertainty quantification), or partial differential equations (e.g., continuum dynamics). Such a targeted approach has the advantage of being intimately connected to research, at the cost of assuming students have picked up the necessary foundational material from elsewhere. To return to the analogy with non-computational physics, a first course on electromagnetism would never skip over things like basic electrostatics to get directly to, say, the Yang–Mills Lagrangian just because non-abelian gauge theory is more "current". Even so, an approach that focuses on modern computational technology is relevant to a more advanced course: once students have mastered the foundations, they can turn to state-of-the-art methods that tackle research problems.

The present text attempts to strike a happy medium: a broad spectrum of numerical methods is studied in detail and then applied to questions from undergraduate physics, via idiomatic implementations in the Python programming language. When selecting and discussing topics, I have prioritized pedagogy over novelty; this is reflected in the chapter titles, which are pretty standard. Of course, my views on what is pedagogically superior are mine alone, so the end result also happens to be original in some respects. Below, I touch upon some of the main features of this book, with a view to orienting the reader.

- **Idiomatic Python:** the book employs Python 3, which is a popular, open-source programming language. A pedagogical choice I have made is to start out with standard Python, use it for a few chapters, and only then turn to the NumPy library; I have found that this helps students who are new to programming in Python effectively distinguish between lists and NumPy arrays. The first chapter includes a discussion of modern programming idioms, which allow me to write shorter codes in the following chapters, thereby emphasizing the numerical method over programming details. This is somewhat counterintuitive: teaching more "advanced" programming than is usual in computational-physics books allows the programming to recede into the background. In other words, not having to fight with the programming language every step of the way makes it *easier* to focus on the physics (or the math).
- **Modern numerical-analysis techniques:** I devote an entire chapter to questions of numerical precision and roundoff error; I hope that the lessons learned there will pay off when studying the following chapters, which typically focus more on approximation-error themes. While this is not a volume on numerical analysis, it does contain a bit more on applied math than is typical: in addition to standard topics, this also includes modern techniques that haven't made it to computational-physics books before (e.g., automatic differentiation or interpolation at Chebyshev points). Similarly, the section on errors in linear algebra glances toward monographs on matrix perturbation theory. To paraphrase Forman Acton [2], the idea here is to ensure that the next generation does not think that an obligatory decimal point is slightly demeaning.
- **Methods "from scratch":** chapters typically start with a pedagogical discussion of a crude algorithm and then advance to more complex methods, in several cases also covering state-of-the-art techniques (when they do not require elaborate bookkeeping). Considerable effort is expended toward motivating and explaining each technique as it is being introduced. Similarly, the chapters are ordered in such a way that the presentation is cumulative. Thus, the book attempts to discuss things "from scratch", i.e., without referring to specialized background or more advanced references; physicists do not expect lemmas and theorems, but do expect to be convinced.[1] Throughout the text, the phrases "it can be shown"[2] and "stated without proof" are actively avoided, so this book may also be used in a flipped classroom, perhaps even for self-study. As part of this approach, I frequently cover things like convergence properties, operation counts, and the error scaling of different numerical methods. When space constraints made it impossible to reach for *simplex munditiis* in explaining a given method, I quietly omitted that method. This is intended as a "first book" on the subject, which should enable students to confidently move on to more advanced expositions.
- **Methods implemented:** while the equations and figures help explain why a method should work, the insight that can be gleaned from an existing implementation of a given algorithm is crucial. I have worked hard to ensure that these code listings are embedded in the main discussion, not tossed aside at the end of the chapter or in an online supplement. Even so, each implementation is typically given its own subsection, in order to

[1] *Nullius in verba*, the motto of the Royal Society, comes to mind. The idea, though not the wording, can clearly be traced to Heraclitus' fragment 50: "Listen, not to me, but to reason".

[2] An instance of *proof by omission*, but still better than "it can be easily shown" (*proof by intimidation*).

help instructors who are pressed for time in their selection of material. Since I wanted to keep the example programs easy to talk about, they are quite short, never longer than a page. In an attempt to avoid the use of black boxes, I list and discuss implementations of methods that are sometimes considered advanced (e.g., the QR eigenvalue method or the fast Fourier transform). While high-quality libraries like NumPy and SciPy contain implementations of such methods, the point of a book like this one is precisely to teach students how and why a given method works. The programs provided (whose filenames also appear in the book's index) can function as templates for further code development on the student's part, e.g., when solving the end-of-chapter problems.

- **Clear separation between numerical method and physics problem:** each chapter focuses on a given numerical theme. The first section always discusses physics scenarios that touch upon the relevant tools; these "motivational" topics are part of the standard undergrad physics curriculum, ranging from classical mechanics, through electromagnetism and statistical mechanics, to quantum mechanics. The bulk of the chapter then focuses on several numerical methods and their implementation, typically without bringing up physics examples. The last numbered section in each chapter is a Project: in addition to involving topics that were introduced in earlier sections (or chapters), these physics projects allow students to carry out calculations they wouldn't attempt without the help of a computer. These projects also provide a first taste of "programming-in-the-large". As a result of this design choice, the book may also be useful to beginning physics students or even students in other areas of science and engineering (with a more limited physics background). Even the primary audience may benefit from the structure of the text in the future, when tackling different physics questions. In the same spirit, the physics-oriented problems in each chapter's problem set are labelled with [$\mathcal{P}$]; these are placed near the end, presupposing the maturity developed while working on the earlier problems. (Since most problems involve some coding, the ones that are purely analytical are labelled with [$\mathcal{A}$], into the bargain.)
- **Second edition includes six new sections on:**

  - the singular-value decomposition (section 4.5),
  - derivative-free optimization (sections 5.5.2 and 5.6.5),
  - maximum-likelihood and Bayesian approaches to linear regression (section 6.6),
  - non-linear fitting via the Gauss–Newton method and neural networks (section 6.7),
  - finite-difference approaches to the diffusion equation (section 8.5.2).

  Six original codes are associated with these sections. Section 6.6 may be of special benefit to readers interested in experimental physics. I found that brief yet meaty introductions to these ideas are useful to physics students, at both the undergraduate and graduate levels. As always, the point was to avoid the dreaded phrase "it turns out that", i.e., the use of (analytical or programming) black boxes. In addition to the totally new material, using the book in a classroom setting has inspired a very large number of other modifications throughout the volume, ranging from minor tweaks (e.g., now explicitly citing problem numbers in the main text) to complete rewrites of selected first-edition sections. From start to finish, I have tried to navigate between Scylla (familiar notation obscuring conceptual subtleties) and Charybdis (too many strange-looking symbols).

- **Second edition includes 140 new problems on:** (a) extensions of techniques introduced in the main text, (b) topics that would otherwise take too many pages to discuss (e.g., problems 5.38, 5.39, and 5.40 on constrained minimization), and (c) a large number of physical applications: I have now included problems on standard themes (e.g., problem 7.65 on the Ising model in two dimensions, problem 8.58 on molecular dynamics for the Lennard–Jones potential, or problem 8.59 on the scattering of a wave packet from a barrier) as well as on topics that I have not encountered in other computational-physics textbooks (e.g., problem 6.67 on credible intervals for a relativistic particle's mass or problem 7.63 on the dimensional regularization of loop integrals). Sometimes a given physical theme carries over across chapters, for example: the Roche potential is visualized in problem 1.17, it is then extremized in problem 5.50 to find the Lagrange points, the volume of the Roche lobe is computed via quadrature in problem 7.56, and the Arenstorf orbit is arrived at by solving differential equations in problem 8.46.
  A word on solutions: standard practice is that computational-physics textbook authors either produce no solutions to the problems or provide solutions only to instructors teaching for-credit courses out of the textbook. I have followed the latter route, but I'm also providing (online) a subset of the solutions to all readers, as a self-study resource.
- **Topic sequence for different courses:** like many textbooks, this one contains more material than can be covered in a single semester. Here are two sample courses:
  - *Advanced undergraduate course:* sections 1.1–1.5, 2.1–2.4.3, 2.5.2, 3.1–3.3, 1.6, 4.1, 4.2.1–4.2.3, 4.3, 4.4.1, 5.1–5.2, 5.4, 5.5.2, 6.1–6.2.2, 6.5, 7.1–7.3, 7.5, 7.7.1–7.7.4, 8.1–8.3.1, 8.4.1, 8.5. Labs focus on Python programming; lectures mainly address numerical methods; physics content limited to motivation and homework assignments.
  - *Beginning graduate course:* appendix B, sections 2.1–2.5, 3.4, 4.1–4.6, 5.1, 5.3–5.6, 6.1, 6.2.2–6.2.3, 6.4–6.8, 7.1, 7.4–7.8, 8.1–8.4, 8.6. Python and an undergrad numerical course are prerequisites. Increased focus on analytical manipulations; programming limited to homework; lectures' physics content determined by a student poll.

Alas, adding 150 pages of new material for the second edition ran the risk of making this volume too expensive. With that in mind, I abridged sections 4.2 and 4.6, placing the original versions in the online supplement at `www.numphyspy.org`. This book continues to be dear to my heart; I hope the reader gets to share some of my excitement for the subject.

On the Epigraphs

I have translated 14 of the quotes appearing as epigraphs myself; in the remaining instances the original was in English. All 17 quotes are not protected by copyright. The sources are: DEDICATION: Pindar, *Pythian Odes*, 3.61–62 (~474 BCE), PREFACE: Ralph Waldo Emerson, *Nature*, Chapter III (1836 CE), CHAPTER 1: Immanuel Kant, *Lectures VI*, Philosophical Encyclopedia (~1780 CE), CHAPTER 2: Georg Wilhelm Friedrich Hegel, *The Phenomenology of Spirit*, Paragraph 74 (1807 CE), CHAPTER 3: Emily Dickinson, *Poem F372/J341* (1862 CE), CHAPTER 4: Vergil, *Georgics*, Book II, Line 412 (~29 BCE), CHAPTER 5: Karl

Kraus, *The Last Days of Mankind*, Act I, Scene 22 (~1918 CE), CHAPTER 6: Gabriel Lippmann, quoted in Henri Poincaré, *Calcul des probabilités*, Second edn, Section 108 (1912 CE), CHAPTER 7: Thucydides, *History of the Peloponnesian War*, Book IV, Paragraph 40 (~420 BCE), CHAPTER 8: Sophocles, *Oedipus Tyrannus*, Line 486 (~429 BCE), POSTSCRIPT: Socrates, quoted in Diogenes Laërtius, *Lives and Opinions of Eminent Philosophers*, Book 2 (~220 CE), APPENDIX A: Aristotle, *Metaphysics* Book III (B), 1001a1 (~330 BCE), APPENDIX B: Thomas Aquinas, *Commentary on Aristotle's Metaphysics*, Book IV (Γ), Lesson 1, Chapter 2, Commentary (1270 CE), APPENDIX C: Parmenides, *Fragment 5* (~475 BCE), (ONLINE) APPENDIX D: Callimachus, *Fragment 465* (~250 BCE), BIBLIOGRAPHY: Michel Eyquem de Montaigne, *Essay III.13*, On Experience (1588 CE), INDEX: James Joyce, *Ulysses*, Episode 16, Eumaeus (1922 CE).

Acknowledgments

My understanding of numerical methods has benefited tremendously from reading many books and papers. In the bibliography I mention only the works I consulted while writing.

I have learned a lot from my graduate students, my collaborators, as well as members of the wider nuclear physics, cold-atom, and astrophysics communities. This textbook is a pedagogical endeavor but it has unavoidably been influenced by my research, which is supported by the Natural Sciences and Engineering Research Council of Canada and the Canada Foundation for Innovation.

I would like to thank my Editor at Cambridge University Press, Vince Higgs, for his sagacious advice. Margaret Patterson did an excellent job copyediting both editions of the book, while Suresh Kumar helped with advanced LaTeX tricks. I am grateful to the anonymous reviewers for the positive feedback and suggestions for additions.

I wish to acknowledge the students taking the classes I taught; their occasional vacant looks resulted in my adding more explanatory material, whereas their (infrequent, even at 8:30 am) yawns made me shorten some sections. Eric Poisson made wide-ranging comments on the lecture notes that turned into the first edition. Aman Agarwal, Eli Bendersky, Liliana Caballero, Ryan Curry, Victoria Leaker, Benjamin Morling, Tristan Pitre and Sangeet-Pal Pannu spotted issues with individual sections or problems. Buried in this book are conceptual distinctions that attempt to preempt misconceptions which are both natural and widespread; I have enjoyed working with Martin Williams on spin-off journal articles.

"What makes us who we are is our choice of good or bad, not our opinion about it" (Aristotle, *Nicomachean Ethics*, 1112a3); Marcos Gezerlis made sure I approached technical subtleties with a similar outlook. Marcos was instrumental in expanding the new material into its current form, immediately picking up on the lacunae in early drafts; I knew I could stop writing when he reached the desired state of staring at nothing in particular while muttering "I understand now". I am indebted to my family (especially Myrsine and Ariadne), *inter multa alia*, for orienting me toward the subjective universal. While several people have helped me refine this book, any remaining poor choices are my responsibility.

1 Idiomatic Python

It's not always about speculating; at some point one must think about practice.
Immanuel Kant

This chapter is *not* intended as an introduction to programming in general or to programming with Python. A tutorial on the Python programming language can be found in the online supplement to this book; if you're still learning what variables, loops, and functions are, we recommend you go to our tutorial (see appendix A) before proceeding with the rest of this chapter. You might also want to have a look at the (official) Python Tutorial at `www.python.org`. Reference [41] is a readable book-length introduction to Python (also available online); Ref. [71] is another introduction, with nice material on visualization. Programming, like most other activities, is something you learn by doing. Thus, you should always try out programming-related material as you read it: *there is no royal road to programming*. Even if you have solid programming skills but no familiarity with Python, we recommend you work your way through one of the above resources, to familiarize yourself with the basic syntax. In what follows, we will take it for granted that you have worked through our tutorial and have modified the different examples to carry out further tasks. This includes solving many of the programming problems we pose there.

What this chapter *does* provide is a quick summary of Python features, with an emphasis on those which the reader is more likely not to have encountered in the past. In other words, even if you are already familiar with the Python programming language, you will most likely still benefit from reading this short chapter. Observe that the title at the top of this page is *Idiomatic* Python: this refers to coding in a *Pythonic* manner. The motive is not to proselytize but, rather, to let the reader work with the language (i.e., not against it); we aim to show how to write Python code that feels "natural". If this book was using, say, Julia or Rust instead of Python, we would still be making the same point: one should try to do the best job possible with the tools at one's disposal. As noted in the Preface, the use of idioms allows us to write shorter codes in the rest of the book, thereby emphasizing the numerical method over programming details; this is not merely an aesthetic concern but a question of cognitive consonance.

At a more mundane level, this chapter contains all the Python-related reference material we will need in this volume: reserved words, library functions, tables, and figures. Keeping the present chapter short is intended to help you when you're working through the following chapters and need to quickly look something up. Before summarizing Python features, we make some big-picture comments on the choice of language, as well as on programming in general.

1.1 Why Python?

Since computational physics is a fun subject, it is only appropriate that the programming involved should also be as pleasant as possible. In this book, we use Python 3, a popular, open-source programming language that has been described as "pseudocode that executes". Python is especially nice in that it doesn't require lots of boilerplate code; that, combined with the fact that one can use Python interactively, make it easy to write new programs. This is great from a pedagogical perspective, since it allows a beginner to start using the language without having to first study lengthy volumes. Importantly, Python's syntax is reasonably simple and leads to very readable code. Even so, Python is very expressive, allowing you to do more in a single line than is possible in many other languages. Furthermore, Python is cross-platform, providing a similar experience on Windows and Unix-like systems. Finally, Python comes with "batteries included": its standard library allows you to do a lot of useful work, without having to implement basic/unrelated things (e.g., sorting a list of numbers) yourself.

In addition to the functionality contained in core Python and in the standard library, Python is associated with a wider ecosystem, which includes libraries like Matplotlib, used to visualize data. Another member of the Python ecosystem, especially relevant to us, is the NumPy library (NumPy stands for "Numerical Python"); containing numerical arrays and several related functions, NumPy is one of the main reasons Python is so attractive for computational work. Another fundamental library is SciPy ("Scientific Python"), which provides many routines that carry out tasks like numerical integration and optimization in an efficient manner. A pedagogical choice we have made in this book is to start out with standard Python, use it for a few chapters, and only then turn to the `numpy` library; this is done in order to help students who are new to Python (or to programming in general) effectively distinguish between Python lists and `numpy` arrays. The latter are then used in the context of linear algebra (chapter 4), where they are indispensable, both in terms of expressiveness and in terms of efficiency.

Speaking of which, it's worth noting at the outset that, since our programs are intended to be easy to read, in some cases we have to sacrifice efficiency.[1] Our implementations are intended to be pedagogical, i.e., they are meant to teach you how and why a given numerical method works; thus, we almost never employ NumPy or SciPy functionality (other than `numpy` arrays), but produce our own functions, instead. We make some comments on alternative implementations here and there, but the general assumption is that you will be able to write your own codes using different approaches (or programming languages) once you've understood the underlying numerical method. If all you are interested in is a quick calculation, then Python along with its ecosystem is likely going to be your one-stop shop. As your work becomes more computationally challenging, you may need to switch to a compiled language; most work on supercomputers is carried out using languages like Fortran or C++ (or sometimes even C). Of course, even if you need to produce a hyperefficient code for your research, the insight you may gain from building a prototype in Python could

[1] Thus, we do not talk about things like Python's Global Interpreter Lock, cache misses, page faults, and so on.

be invaluable; similarly, you could write most of your code in Python and re-express a few performance-critical components using a compiled language. We hope that the lessons you pick up here (both on the numerical methods and on programming in general) will serve you well if you need to employ another environment in the future.

The decision to focus on Python (and NumPy) idioms is coupled to the aforementioned points on Python's expressiveness and readability: idiomatic code makes it easier to conquer the complexity that arises when developing software. (Of course, it does require you to first become comfortable with the idioms.) That being said, our presentation will be *selective*; Python has many other features that we will not go into. Most notably, we don't discuss how to define classes of your own or how to handle exceptions; the list of omitted features is actually very long.[2] While many features we leave out are very important, discussing them would interfere with the learning process for students who are still mastering the basics of programming. Even so, we do introduce topics that haven't often made it into computational-science texts (e.g., list comprehensions, dictionaries, for-else, array manipulation via slicing and `@`) and use them repeatedly in the rest of the book.

We sometimes point to further functionality in Python. For more, have a look at the bibliography and at The Python Language Reference (as well as The Python Standard Library Reference). Once you've mastered the basics of core Python, you may find books like Ref. [120] and Ref. [132] a worthwhile investment. On the wider theme of developing good programming skills, volumes like Ref. [103] can be enriching, as is also true of any book written by Brian Kernighan. Here we provide only the briefest of summaries.

1.2 Code Quality

We will not be too strict in this book about coding guidelines. Issues like code layout can be important, but most of the programs we will write are so short that this won't matter too much. If you'd like to learn more about this topic, your first point of reference should be PEP 8 – Style Guide for Python Code. Often more important than issues of code layout[3] are questions about how you write and check your programs. Here is some general advice:

- **Code readability matters** Make sure to target your program to humans, not the computer. This means that you should avoid using "clever" tricks. Thus, you should use good variable names and write comments that add value (instead of repeating the code). The human code reader that will benefit from this is first and foremost yourself, when you come back to your programs some months later.
- **Be careful, not swift, when coding** Debugging is typically more difficult than coding itself. Instead of spending two minutes writing a program that doesn't work and then requires you to spend two hours fixing it up, try to spend 10 minutes on designing the code and then carefully converting your ideas into program lines. It doesn't hurt to also use Python interactively (while building the program file) to test out components of the code one-by-one or to fuse different parts together.

[2] For example: decorators, coroutines, or type hints.

[3] Discussion of which, more often than not, sheds light on the narcissism of minor differences ("bikeshedding").

- **Untested code is wrong code** Make sure your program is working correctly. If you have an example where you already know the answer, make sure your code gives that answer. Manually step through a number of cases (i.e., mentally, or on paper, do the calculations the program is supposed to carry out). This, combined with judiciously placed printouts of intermediate variables, can go a long way toward ensuring that everything is as it should be. When modifying your program, ensure it still gives the original answer when you specialize the problem to the one you started with.
- **Write functions that do one thing well** Instead of carrying out a bunch of unrelated operations in sequence, you should structure your code so that it makes use of well-named (and well-thought-out) functions that do one thing and do it well. You should break down the tasks to be carried out and logically separate those into distinct functions. If you design these well, in the future you will be able to modify your programs to carry out much more challenging tasks, by only adding a few lines of new code (instead of having to change dozens of lines in an existing "spaghetti" code).
- **Use trusted libraries** In most of this book we are "reinventing the wheel", because we want to understand how things work (or don't work). Later in life, you should not have to always use "hand-made" versions of standard algorithms. As mentioned, there exist good (widely employed and tested) libraries like `numpy` that you should learn to make use of. The same thing holds, obviously, for the standard Python library: you should generally employ its features instead of "rolling your own".

One could add (much) more advice along these lines. Since our scope here is much more limited, we conclude by pointing out that in the Python ecosystem (or around it) there's extensive infrastructure [128] to carry out version control (e.g., `git`), testing (e.g., `doctest` and `unittest`), as well as debugging (e.g., `pdb`), program profiling and optimization, among other things. You should also have a look at the `pylint` tool.

1.3 Summary of Python Features

1.3.1 Basics

Python can be used interactively: this is when you see the Python prompt >>>, also known as a chevron. You don't need to use Python interactively: like other programming languages, the most common way of writing and running programs is to store the code in a file. Linear combinations of these two ways of using Python are also available, fusing interactive sessions and program files. In any case, your program is always executed by the Python interpreter. Appendix A points you in the direction of tools you could employ.

Like other languages (e.g., C or Fortran), Python employs variables, which can be integers, complex numbers, etc. Unlike those languages, Python is a dynamically typed language, so variables get their type from their value, e.g., `x = 0.5` creates a floating-point variable (a "float"). It may help you to think of Python values as being produced first and labels being attached to them after that. Numbers like `0.5` or strings like `"Hello"`, are known as *literals*. If you wish to print the value of a variable, you use the `print()` built-in function, i.e., `print(x)`. Further functionality is available in the form of standard-library modules, e.g., you can `import` the `sqrt` function that is to be found in the `math` module. Users can define their own modules: we will do so repeatedly. You can carry out arithmetic with variables, e.g., `x**y` raises `x` to the y-th power or `x//y` does "floor division". It's usually a good idea to group related operations using parentheses. Python also supports augmented assignment, e.g., `x += 1` or even multiple assignment, e.g., `x, y = 0.5, "Hello"`. This gives rise to a nifty way to swap two variables: `x, y = y, x`.

Comments are an important feature of programming languages: they are text that is ignored by the computer but can be very helpful to humans reading the code. That human may be yourself in a few months, at which point you may have forgotten the purpose or details of the code you're inspecting. Python allows you to write both single-line comments, via `#`, or docstrings (short for "documentation strings"), via the use of triple quotation marks. Crucially, we don't include explanatory comments in our code examples, since this is a book which explicitly discusses programming features in the main text. That being said, in your own codes (which are not embedded in a book discussing them) you should always include comments.

1.3.2 Control Flow

Control flow refers to programming constructs where not every line of code gets executed in order. A classic example is conditional execution via the `if` statement:

```
>>> if x!=0:
...         print("x is non-zero")
```

Indentation is important in Python: the line after `if` is indented, reflecting the fact that it belongs to the corresponding scenario. Similarly, the colon, :, at the end of the line containing the `if` is also syntactically important. If you wanted to take care of other possibilities, you could use another indented block starting with `else:` or `elif x==0:`. In the case of boolean variables, a common idiom is to write: `if flag:` instead of `if flag==True:`.

Another concept in control flow is the loop, i.e., the repetition of a code block. You can do this via `while`, which is typically used when you don't know ahead of time how many iterations you are going to need, e.g., `while x>0:`. Like conditional expressions, a `while` loop tests a condition; it then keeps repeating the body of the loop until the condition is no longer true, in which case the body of the block is jumped over and execution resumes from the following (non-indented) line. We sometimes like to be able to break out of a loop: if a condition in the middle of the loop body is met, then: (a) if we use `break` we will proceed to the first statement *after* the loop, or (b) if we use `continue` we skip not the entire loop, but the rest of the loop body *for the present iteration*.

A third control-flow construct is a `for` loop: this arises when you need to repeat a certain action a fixed number of times. For example, by saying `for i in range(3):` you will repeat whatever follows (and is indented) three times. Like C, *Python uses 0-based indexing* (which we will shorten to "0-indexing"), meaning that the indices go as 0, 1, 2 in this case. In general, range(n) gives the integers from 0 to n-1 and, similarly, range(m, n, i) gives the integers from m to n-1 in steps of i. Above, we mentioned how to use print() to produce output; this can be placed inside a loop to print out many numbers, each on a separate line; if you want to place all the output on the same line you do:

```
>>> for i in range(1,15,2):
...     print(0.01*i, end=" ")
```

that is, we've said end=" " after passing in the argument we wish to print. As we'll discuss in the following subsection, Python's for loop is incredibly versatile.

1.3.3 Data Structures

Python supports container entities, called data structures; we will mainly be using lists.

Lists A list is a container of elements; it can can grow when you need it to. Elements can have different types. You use square brackets and comma-separated elements when creating a list, e.g., zs = [5, 1+2j, -2.0]. You also use square brackets when indexing into a list, e.g., zs[0] is the first element and zs[-1] the last one. Lists are mutable sequences, meaning we can change their elements, e.g., zs[1] = 9, or introduce new elements, via append(). The combination of for loops and append() provides us with a powerful way to populate a list. For example:

```
>>> xs = []
>>> for i in range(20):
...     xs.append(0.1*i)
```

where we started with an empty list. In the following section, we'll see a more idiomatic way of accomplishing the same task. You can concatenate two lists via the addition operator, e.g., zs = xs + ys; the logical consequence of this is the idiom whereby a list can be populated with several (identical) elements using a one-liner, xs = 10*[0]. There are several built-in functions (applicable to lists) that often come in handy, most notably sum() and len().

Python supports a feature called slicing, which allows us to take a slice out of an existing list. Slicing, like indexing, uses square brackets: the difference is that slicing uses two integers, with a colon in between, e.g., ws[2:5] gives you the elements ws[2] up to (but not including) the element ws[5]. Slicing obeys convenient defaults, in that we can omit one of the integers in ws[m:n] without adverse consequences. Omitting the first index is interpreted as using a first index of 0, and omitting the second index is interpreted as using a second index equal to the number of elements. You can also include a third index: in ws[m:n:i] we go in steps of i. Note that list slicing uses colons, whereas the arguments of range() are comma-separated. Except for that, the pattern of start, end, stride is the same.

We are now in a position to discuss how copying works. In Python a new list, which is

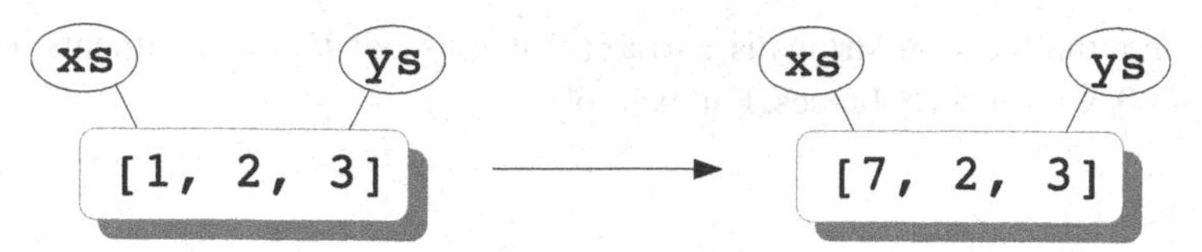

Fig. 1.1 Labelling and modifying a mutable object (in this case, a list)

assigned to be equal to an old list, is simply the old list by another name. This is illustrated in Fig. 1.1, which corresponds to the three steps `xs = [1,2,3]`, followed by `ys = xs`, and then `ys[0] = 7`. In other words, in Python we're not really dealing with variables, but with *labels* attached to values, since `xs` and `ys` are just different names for the same entity. When we type `ys[0] = 7` we are not creating a new value, simply modifying the underlying entity that both the `xs` and `ys` labels are attached to. Incidentally, things are different for simpler variables, e.g., `x=1; y=x; y=7; print(x)` prints `1` since `7` is a new value, not a modification of the value `x` is attached to. This is illustrated in Fig. 1.2, where we see that, while initially both variable names were labelling the same value, when we type `y=7` we create a new value (since the number 7 is a new entity, not a modification of the number 1) and then attach the `y` label to it.

Crucially, *when you slice you get a new list*, meaning that if you give a new name to a slice of a list and then modify that, then the original list is unaffected. For example, `xs = [1,2,3]`, followed by `ys = xs[1:]`, and then `ys[0] = 7` does not affect `xs`. This fact (namely, that slices don't provide views on the original list but can be manipulated separately) can be combined with another nice feature (namely, that when slicing one can actually omit both indices) to create a copy of the entire list, e.g., `ys = xs[:]`. This is a shallow copy, so if you need a deep copy, you should use the function `deepcopy()` from the standard module `copy`; the difference is immaterial here.

Tuples Tuples can be (somewhat unfairly) described as immutable lists. They are sequences that can neither change nor grow. They are defined using parentheses instead of square brackets, e.g., `xs = (1,2,3)`, but you can even omit the parentheses, `xs = 1,2,3`. Tuple elements are accessed the same way that list elements are, namely with square brackets, e.g., `xs[2]`.

Strings Strings can also be viewed as sequences, e.g., if `name = "Mary"` then `name[-1]` is the character `'y'`. Note that you can use either single or double quotation marks. Like tuples, strings are immutable. As with tuples, we can use + to concatenate two strings. A

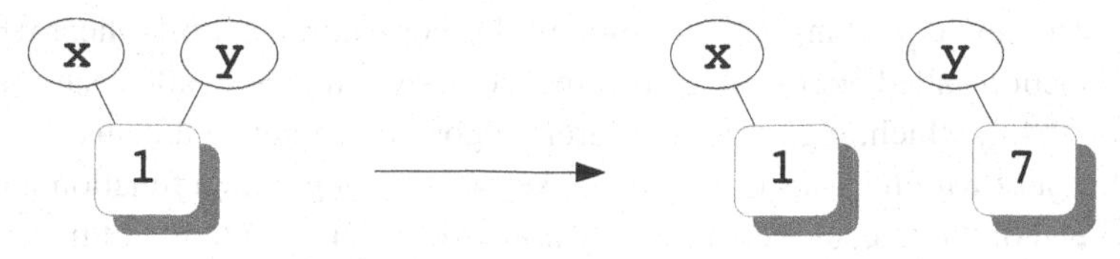

Fig. 1.2 Labelling immutable objects (in this case, integers)

useful function that acts on strings is `format()`: it uses *positional* arguments, numbered starting from 0, within curly braces. For example:

```
>>> x, y = 3.1, -2.5
>>> "{0} {1}".format(x, y)
'3.1 -2.5'
```

The overall structure is string-dot-format-arguments. This can lead to powerful ways of formatting strings, e.g.,

```
>>> "{0:1.15f} {1}".format(x, y)
'3.100000000000000 -2.5'
```

Here we also introduced a colon, this time followed by `1.15f`, where `1` gives the number of digits before the decimal point, `15` gives the number of digits after the decimal point, and `f` is a type specifier (that leads to the result shown for floats).

Dictionaries Python also supports dictionaries, which are called associative arrays in computer science (they're called maps in C++). You can think of dictionaries as being similar to lists or tuples, but instead of being limited to integer indices, with a dictionary you can use strings or floats as *keys*. In other words, dictionaries contain key and value pairs. The syntax for creating them involves curly braces (compare with square brackets for lists and parentheses for tuples), with the key-value pair being separated by a colon. For example, `htow = {1.41: 31.3, 1.45: 36.7, 1.48: 42.4}` is a dictionary associating heights to weights. In this case both the keys and the values are floats. We access a dictionary value (for a specific key) by using the name of the dictionary, square brackets, and the key we're interested in: this returns the value associated with that key, e.g., `htow[1.45]`. In other words, indexing uses square brackets for lists, tuples, strings, and dictionaries. If the specific key is not present, then we get an error. Note, however, that accessing a key that is not present *and then assigning* actually works: this is a standard way key:value pairs are introduced into a dictionary, e.g., `htow[1.43] = 32.9`.

1.3.4 User-Defined Functions

If our programs simply carried out a bunch of operations in sequence, inside several loops, their logic would soon become unwieldy. Instead, we are able to group together logically related operations and create what are called user-defined functions: just as in our earlier section on control flow, this refers to lines of code that are not necessarily executed in the order in which they appear inside the program file. For example, while the `math` module contains a function called `exp()`, we could create our own function called, say, `compexp()` as in section 2.4.4, which, e.g., uses a different algorithm to get to the answer. The way we introduce our own functions is via the `def` keyword, along with a function name and a colon at the end of the line, as well as the (by now expected) indentation of the code block that follows. Here's a function that computes the sum from 1 up to some integer:

```
>>> def sumofints(nmax):
...     val = sum(range(1,nmax+1))
...     return val
```

We are taking in the integer up to which we're summing as a parameter. We then ensure that `range()` goes up to (but not including) `nmax+1` (i.e., it includes `nmax`). We split the body of the function into two lines: first we define a new variable and then we *return* it, though we could have simply used a single line, `return sum(range(1,nmax+1))`. This function can be called (in the rest of the program) by saying `x = sumofints(42)`.

The function we just defined took in one parameter and returned one value. It could have, instead, taken in no parameters, e.g., summing the integers up to some constant; we would then call it by `x = sumofints()`. Similarly, it could have printed out the result, inside the function body, instead of returning it to the external world; in that case, where no `return` statement was used, the `x` in `x = sumofints(42)` would have the value `None`. Analogously, we could be dealing with several input parameters, or several return values, expressed by `def sumofints(nmin,nmax):`, or `return val1, val2`, respectively. The latter case is implicitly making use of a tuple.

We say that a variable that's either a parameter of a function or is defined inside the function is *local* to that function. If you're familiar with the terminology other languages use (pass-by-value or pass-by-reference), then note that Python employs *pass-by-assignment*, which for immutable objects behaves like pass-by-value (you *can't* change what's outside) and for mutable objects behaves like pass-by-reference (you *can* change what's outside), if you're not re-assigning. It's often a bad idea to change the external world from inside a function: it's best simply to return a value that contains what you need to communicate to the external world. This can become wasteful, but here we opt for conceptual clarity, always returning values without changing the external world. This is a style inspired by *functional programming*, which aims at avoiding *side effects*, i.e., changes that are not visible in the return value. (Unless you're a purist, input/output is fine.) Python also supports *nested functions* and *closures*. On a related note, Python contains the keywords `global` and `nonlocal` as well as function one-liners via `lambda`; some of these features are briefly touched upon in problem 1.4.

A related feature of Python is the ability to provide default parameter values:

```
>>> def cosder(x, h=0.01):
...     return (cos(x+h) - cos(x))/h
```

You can call this function with either `cosder(0.)` or `cosder(0., 0.001)`; in the former case, `h` has the value 0.01. Basically, the second argument here is *optional*. As a matter of good practice, you should make sure to always use immutable default parameter values. Finally, note that in Python one has the ability to define a function that deals with an indefinite number of positional or keyword arguments. The syntax for this is `*args` and `**kwargs`, but a detailed discussion would take us too far afield.

A pleasant feature of Python is that *functions are first-class objects*. As a result, we can pass them in as arguments to other functions; for example, instead of hard-coding the cosine as in our previous function, we could say:

```
>>> def der(f, x, h=0.01):
...     return (f(x+h) - f(x))/h
```

which is called by passing in as the first argument the function of your choice, e.g., `der(sin, 0., 0.05)`. Note how `f` is a regular parameter, but is used inside the function the same way we use functions (by passing arguments to them inside parentheses). We passed in the name of the function, `sin`, as the first argument and the `x` as the second argument.[4] As a rule of thumb, you should pass a function in as an argument if you foresee that you might be passing in another function in its place in the future. If you basically expect to always keep carrying out the same task, there's no need to add yet another parameter to your function definition. Incidentally, we really meant it when we said that in Python functions are first-class objects. You could even have a list whose elements are functions, e.g., `funcs = [sumofints, cos]`. Similarly, problem 1.2 explores a dictionary that contains functions as values (or keys).

1.4 Core-Python Idioms

We are now in a position to discuss Pythonic idioms: these are syntactic features that allow us to perform tasks more straightforwardly than would have been possible with the syntax introduced above. Using such alternative syntax to make the code more concise and expressive helps us write new programs, but also makes the lives of future readers easier. Of course, you do have to exercise your judgement.[5]

1.4.1 List Comprehensions

At the start of section 1.3.3, we saw how to populate a list: start with an empty one and use `append()` inside a `for` loop to add the elements you need. List comprehensions (often shortened to *listcomps*) provide us with another way of setting up lists. The earlier example can be replaced by `xs = [0.1*i for i in range(20)]`. This is much more compact (one line vs three). Note that when using a list comprehension the loop that steps through the elements of some other sequence (in this case, the result of stepping through `range()`) is placed *inside* the list we are creating! This particular syntax is at first sight a bit unusual, but very convenient and strongly recommended.

It's a worthwhile exercise to replace hand-rolled versions of code using listcomps. For example, if you need a new list whose elements are two times the value of each element in `xs`, you should *not* say `ys = 2*xs`: as mentioned earlier, this concatenates the two lists, which is not what we are after. Instead, what *does* work is `ys = [2*x for x in xs]`. More

[4] This means that we did *not* pass in `sin()` or `sin(x)`, as those wouldn't work.

[5] "A foolish consistency is the hobgoblin of little minds" (Ralph Waldo Emerson, *Self-Reliance*).

generally, if you need to apply a function to every element of a list, you could simply do so on the fly: `ws = [f(x) for x in xs]`. Another powerful feature lets you "prune" a list as you're creating it, e.g., `zs = [2*x for x in xs if x>0.3]`; this doubles an element only if that element is greater than 0.3 (otherwise it doesn't even introduce it).

1.4.2 Iterating Idiomatically

Our earlier example, `ys = [2*x for x in xs]`, is an instance of a significant syntactic feature: a `for` loop is not limited to iterating through a collection of integers in fixed steps (as in our earlier `for i in range(20)`) but can iterate through the list elements themselves *directly*.[6] This is a general aspect of iterating in Python, a topic we now turn to; the following advice applies to all loops (i.e., not only to listcomps).

One list Assuming the list `xs` already exists, you may be tempted to iterate through its elements via something like `for i in range(len(xs)):` then in the loop body you would get a specific element by indexing, i.e., `xs[i]`. The Pythonic alternative is to step through the elements themselves, i.e., `for x in xs:` and then simply use `x` in the loop body. Instead of iterating through an index, which is what the error-prone syntax `range(len(xs))` is doing, this uses Python's `in` to iterate directly through the list elements.

Sometimes, you need to iterate through the elements of a list `xs` in reverse: the old-school (C-influenced) way to do this is `for i in range(len(xs)-1, -1, -1):`, followed by indexing, i.e., `xs[i]`. This works, but all those `-1`'s don't make for light reading; instead, use Python's built-in `reversed()` function, saying `for x in reversed(xs):` and then using `x` directly. A final use case: you often need access to both the index showing an element's place in the list, and the element itself. The "traditional" solution would involve `for i in range(len(xs)):` followed by using `i` and `xs[i]` in the loop body. The Pythonic alternative is to use the built-in `enumerate()` function, `for i, x in enumerate(xs):` and then use `i` and `x` directly; this is more readable and less error-prone.

Two lists We sometimes want to iterate through two lists, `xs` and `ys`, in parallel. You should be getting the hang of things by now; the unPythonic way to do this would be `for i in range(len(xs)):`, followed by using `xs[i]` and `ys[i]` in the loop body. The Pythonic solution is to say `for x, y in zip(xs,ys):` and then use `x` and `y` directly. The `zip()` built-in function creates an iterable entity consisting of tuples, fusing the 0th element in `xs` with the 0th element in `ys`, the 1st element in `xs` with the 1st element[7] in `ys`, and so on.

Dictionaries We can use `for` to iterate Pythonically through more than just lists; in the tutorial you have learned that it also works for lines in a file. Similarly, we can iterate through the *keys* of a dictionary; for our earlier height-to-weight dictionary, you could say `for h in htow:` and then use `h` and `htow[h]` in the loop body. An even more Pythonic way of doing the same thing uses the `items()` method of dictionaries to produce all the key-value pairs: `for h,w in htow.items():` lets you use `h` and `w` inside the loop body.

[6] In other words, Python's `for` is similar to the `foreach` that some other languages have.
[7] In English we say "first", "second", etc. We'll use numbers, e.g., 0th, when using the 0-indexing convention.

Code 1.1 forelse.py

```
def look(target,names):
    for name in names:
        if name==target:
            val = name
            break
    else:
        val = None
    return val

names = ["Alice", "Bob", "Eve"]
print(look("Eve", names))
print(look("Jack", names))
```

For-else In this section we've spent some time discussing the line where `for` appears. We now turn to a way of using `for` loops that is different altogether: similarly to `if` statements, you can follow a `for` loop by an `else` (!). This is somewhat counterintuitive, but can be very helpful. The way this works is that the `for` loop is run as usual: if no `break` is encountered during execution of the `for` block, then control proceeds to the `else` block. If a `break` is encountered during execution of the `for` block, then the `else` block is not run. (Try re-reading the last two sentences after you study code 1.1.)

The `else` in a `for` loop is nice when we are looking for an item within a sequence and we need to do one thing if we find it and a different thing if we don't. An example is given in code 1.1, the first full program in the book. We list such codes in boxes with the filename at the top; we strongly recommend you download these Python codes and use them in parallel to reading the main text;[8] you would run this by typing `python forelse.py` or something along those lines; note that, unlike other languages you may have used in the past, there is no *compilation* stage. This specific code uses several of the Python features mentioned in this chapter: it starts with defining a function, `look()`, which we will discuss in more detail below. The main program uses a listcomp to produce a list of strings and then calls the `look()` function twice, passing in a different first argument each time; we don't even need a variable to hold the first argument(s), using string literal(s) directly. Since we are no longer using Python interactively, we employ `print()` to ensure that the output is printed on the screen (instead of being evaluated and then thrown away).

Turning to the `look()` function itself, it takes in two parameters: one is (supposed to be) a target string and the other one is a list of strings. The latter could be very long, e.g., the first names listed in a phone book. Note that there are three levels of indentation involved here: the function body gets indented, the `for` loop body gets indented, and then the conditional

[8] "One cannot so well grasp a thing and make it one's own, when it has been learned from someone else, as when one has discovered it oneself" (René Descartes, *Discourse on the Method*, part VI).

expression body also gets indented. Incidentally, our `for` loop is iterating through the list elements directly, as per our earlier admonition, instead of using error-prone indices. You should spend some time ensuring that you understand what's going on. Crucially, the `else` is indented at the level of the `for`, not at the level of the `if`. We also took the opportunity to employ another idiom mentioned above: `None` is used to denote the absence of a value. The two possibilities that are at play (target in sequence or target not in sequence) are probed by the two function calls in the main program. When the target is found, a `break` is executed, so the `else` is not run. When the target is not found, the `else` is run. It might help you to think of this `else` as being equivalent to `nobreak`: the code in that block is only executed when no `break` is encountered in the main part of the `for` loop. (Of course, this is only a mnemonic, since `nobreak` is not a reserved word in Python.) We will use the for-else idiom repeatedly in this volume (especially in chapter 5), whenever we are faced with an iterative task which may plausibly fail, in which case we wish to communicate that fact to the rest of the program. Since even some expert users are uncomfortable with the for-else idiom, problem 1.5 gives you a tour of the alternatives.

1.5 Basic Plotting with `matplotlib`

We will now visualize relationships between numbers via `matplotlib`, a plotting library (i.e., not part of core Python) which can produce quality figures: all the plots in this book were created using `matplotlib`. Inside the `matplotlib` package is the `matplotlib.pyplot` module, which is used to produce figures in a MATLAB-like environment.

A simple example is given in code 1.2. This starts by importing `matplotlib.pyplot` in the (standard) way which allows us to use it below without repeated typing of unnecessary characters. We then define a function, `plotex()`, that takes care of the plotting, whereas the main program simply introduces four list comprehensions and then calls our function. The listcomps also employ idiomatic iteration, in the spirit of applying what you learned in the previous section. If you're still a beginner, you may be wondering why we defined a Python function in this code. An important design principle in computer science goes by the name of *separation of concerns* (or sometimes *information hiding* or *encapsulation*): each aspect of the program should be handled separately. In our case, this means that each component of our task should be handled in a separate function.

Let's discuss this function in more detail. Its parameters are (meant to be) four lists, namely two pairs of x_i and y_i values. The function body starts by using `xlabel()` and `ylabel()` to provide labels for the x and y axes. It then creates individual curves/sets of points by using `matplotlib`'s function `plot()`, passing in the x-axis values as the first argument and the y-axis values as the second argument. The third positional argument to `plot()` is the *format string*: this corresponds to the color and point/line type. In the first case, we used `r` for red and `-` for a solid line. Of course, figures in this book are in black and white, but you can produce the color version using the corresponding Python code. In order to help you interpret this and other format strings, we list allowed colors and some

Code 1.2 plotex.py

```
import matplotlib.pyplot as plt

def plotex(cxs,cys,dxs,dys):
    plt.xlabel('x', fontsize=20)
    plt.ylabel('f(x)', fontsize=20)
    plt.plot(cxs, cys, 'r-', label='one function')
    plt.plot(dxs, dys, 'b--^', label='other function')
    plt.legend()
    plt.show()

cxs = [0.1*i for i in range(60)]
cys = [x**2 for x in cxs]
dxs = [i for i in range(7)]
dys = [x**1.8 - 0.5 for x in dxs]

plotex(cxs, cys, dxs, dys)
```

of the most important line styles/markers in table 1.1. The fourth argument to `plot()` is a keyword argument containing the label corresponding to the curve. In the second call to `plot()` we pass in a different format string and label (and, obviously, different lists); observe that we used two style options in the format string: `--` to denote a dashed line and `^` to denote the points with a triangle marker. The function concludes by calling `legend()`, which is responsible for making the legend appear, and `show()`, which makes the plot actually appear on our screen.

Table 1.1 Color, line styles, and markers in `matplotlib`

Character	Color	Character	Description
'b'	blue	'-'	solid line style
'g'	green	'--'	dashed line style
'r'	red	'-.'	dash-dot line style
'c'	cyan	':'	dotted line style
'm'	magenta	'o'	circle marker
'y'	yellow	's'	square marker
'k'	black	'D'	diamond marker
'w'	white	'^'	triangle-up marker

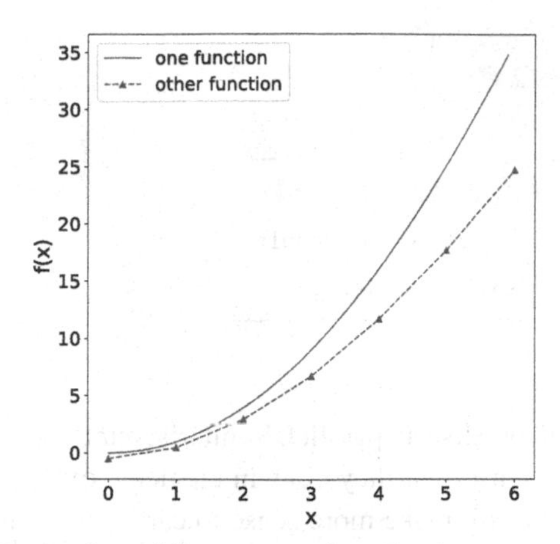

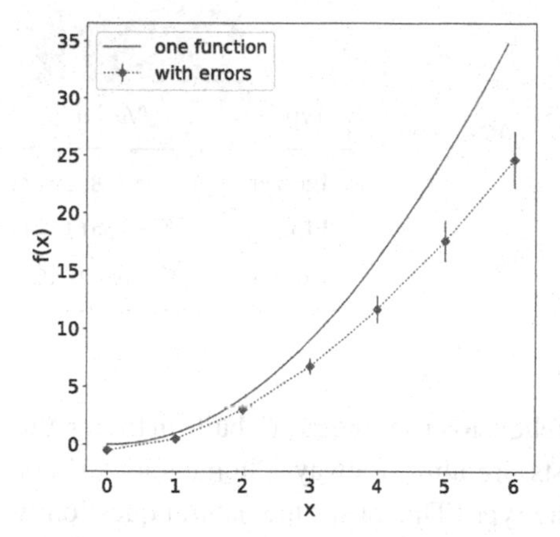

Fig. 1.3 Examples of figures produced using `matplotlib`

The result of running this program is in the left panel of Fig. 1.3. A scenario that pops up very often in practice involves plotting points with error bars:

```
dyerrs = [0.1*y for y in dys]
plt.errorbar(dxs, dys, dyerrs, fmt='b:D', label='with errors')
```

where we have called `errorbar()` to plot the points with error bars: the three positional arguments here are the x values, the y values, and the errors in the y values. After that, we pass in the format string using the keyword argument `fmt` and the label as usual. We thereby produce the right panel of Fig. 1.3.

We could fine-tune almost all aspects of our plots, including basic things like line width, font size, and so on. For example, we could get TeX-like equations by putting dollar signs inside our string, e.g., 'x_i' appears as x_i. We could control which values are displayed via `xlim()` and `ylim()`, we could employ a log-scale for one or both of the axes (using `xscale()` or `yscale()`), and much more. The online documentation can help you go beyond these basic features. Finally, we note that instead of providing `matplotlib` with Python lists as input, you could be using NumPy arrays; this is the topic we now turn to.

1.6 NumPy Idioms

NumPy arrays are not used in chapters 2 and 3 so, if you are still new to Python, you should focus on mastering Python lists: how to grow them, modify them, etc. Thus, you should *skip this section and the corresponding NumPy tutorial for now* and come back when you are about to start reading chapter 4. If you're feeling brave you can keep reading, but know that it is important to distinguish between Python lists and NumPy arrays.

In our list-based codes, we typically carry out the same operation over and over again, a fixed number of times, e.g., `contribs = [w*f(x) for w,x in zip(ws,xs)]`; this list com-

Table 1.2 Commonly used numpy data types

Type	Variants
Integer	`int8, int16, int32, int64`
Float	`float16, float32, float64, longdouble`
Complex	`complex64, complex128, complex256`

prehension uses the `zip()` built-in to step through two lists in parallel. Similarly, our Python lists are almost always "homogeneous", in the sense that they contain elements of only one type. This raises the natural question: wouldn't it make more sense to carry out such tasks using a homogeneous, fixed-length container? This is precisely what the Numerical Python (NumPy) array object does: as a result, it is fast and space-efficient. It also allows us to avoid, for the most part, having to write loops: even in our listcomps, which are more concise than standard loops, there is a need to explicitly step through each element one by one. Numerical Python arrays often obviate such syntax, letting us carry out mathematical operations on entire blocks of data in one step.[9] The standard convention is to import `numpy` with a shortened name: `import numpy as np`.

One-dimensional arrays One-dimensional (1d) arrays are direct replacements for lists. The easiest way to make an array is via the `array()` function, which takes in a sequence and returns an array containing the data that was passed in, e.g., `ys = np.array(contribs)`; the `array()` function is part of `numpy`, so we had to say `np.array()` to access it. There is both an `array()` function and an array object involved here: the former created the latter. Printing out an array, the commas are stripped, so you can focus on the numbers. If you want to see how many elements are in the array `ys`, use the `size` attribute, via `ys.size`. Remember: NumPy arrays are fixed-length, so the total number of elements cannot change. Another very useful attribute arrays have is `dtype`, namely the data type of the elements; table 1.2 collects several important data types. When creating an array, the data type can also be explicitly provided, e.g., `zs = np.array([5, 8], dtype=np.float32)`.

NumPy contains several handy functions that help you produce pre-populated arrays, e.g., `np.zeros(5)`, `np.ones(4)`, or `np.arange(6)`. The latter also works with float arguments, however, as we will learn in the following chapter, this invites trouble. For example, `np.arange(1.5,1.75,0.05)` and `np.arange(1.5,1.8,0.05)` behave quite differently. Instead, use the `linspace()` function to get a specified number of evenly spaced elements over a specified interval, e.g., `np.linspace(1.5,1.75,6)` or `np.linspace(1.5,1.8,7)`. There also exists a function called `logspace()`, which produces a logarithmically spaced grid of points.

Indexing for arrays works as for lists, namely with square brackets, e.g., `zs[2]`. Slicing appears, at first sight, to also be identical to how slicing of lists works. However, there is a crucial difference, namely that *array slices are views on the original array*; let's revisit our

[9] From here onward, we will not keep referring to these new containers as numpy arrays: they'll simply be called arrays, the same way the core-Python lists are simply called lists.

earlier example. For arrays, `xs = np.array([1,2,3])`, followed by `ys = xs[1:]`, and then `ys[0] = 7` *does* affect `xs`. NumPy arrays are efficient, eliminating the need to copy data: of course, one should always be mindful that different slices all refer to the same underlying array. For the few cases where you do need a true copy, you can use `np.copy()`, e.g., `ys = np.copy(xs[1:])` followed by `ys[0] = 7` does not affect `xs`. We could, just as well, copy the entire array over, without impacting the original. In what follows, we frequently make copies of arrays inside functions, in the spirit of impacting the external world only through the return value of a function.

Another difference between lists and arrays has to with *broadcasting*. Qualitatively, this means that NumPy often knows how to handle entities whose dimensionalities don't quite match. For example, `xs = np.zeros(5)` followed by `xs[:] = 7`, leads to an array of five sevens. Remember, since array slices are views on the original array, `xs[:]` cannot be used to create a copy of the array, but it *can* be used to broadcast one number onto many slots; this syntax is known as an "everything slice". Without it, `xs = 7` leads to `xs` becoming an integer (number) variable, not an array, which isn't what we want.

The real strength of arrays is that they let you carry out operations such as `xs + ys`: if these were lists, you would be concatenating them, but for arrays addition is interpreted as an elementwise operation (i.e., each pair of elements is added together).[10] If you wanted to do the same thing with lists you would need a listcomp and `zip()`. This ability to carry out such batch operations on all the elements of an array (or two) at once is often called *vectorization*. You could also use other operations, e.g., to sum the results of pairwise multiplication you simply say `np.sum(xs*ys)`. This is simply evaluating the scalar product of two vectors, in one short expression. Equally interesting is the ability NumPy has to also combine an array with a scalar: this follows from the aforementioned *broadcasting*, whereby NumPy knows how to interpret entities with different (but compatible) shapes. For example, `2*xs` doubles each array element; this is very different from what we saw in the case of lists. You can think of what's happening here as the value `2` being "stretched" into an array with the same size as `xs` and then an elementwise multiplication taking place. Needless to say, you could carry out several other operations with scalars, e.g., `1/xs`. Such combinations of broadcasting (whereby you can carry out operations between scalars and arrays) and vectorization (whereby you write one expression but the calculation is carried out for all elements) can be hard to grasp at first, but are very powerful (both expressive and efficient) once you get used to them.

Another very useful function, `np.where()`, helps you find specific indices where a condition is met, e.g., `np.where(2 == xs)` returns a tuple of arrays, so we would be interested in its 0th element. NumPy also contains several functions that look similar to corresponding `math` functions, e.g., `np.sqrt()`; these are are designed to take in entire arrays so they are almost always faster. NumPy also has functions that take in an array and return a scalar, like the `np.sum()` we encountered above. Another very helpful function is `np.argmax()`, which returns the index of the maximum value. You can also use NumPy's functionality to create your own functions that can handle arrays. For example, our earlier `contribs = [w*f(x) for w,x in zip(ws,xs)]` can be condensed into the much cleaner `contribs =`

[10] Conversely, if you wish to concatenate two arrays you cannot use addition; use `np.concatenate()`.

Table 1.3 Important attributes of numpy arrays

Attribute	Description
dtype	Data type of array elements
ndim	Number of dimensions of array
shape	Tuple with number of elements for each dimension
size	Total number of elements in array

ws*fa(xs), where fa() is designed to take in arrays (so, e.g., it uses np.sqrt() instead of math.sqrt()).

Two-dimensional arrays In core Python, matrices can be represented by lists of lists which are, however, quite clunky (as you'll further experience in the following section). In Python a list-of-lists is introduced by, e.g., LL = [[11,12], [13,14], [15,16]]. Just like for one-dimensional arrays, we can say A = np.array(LL) to produce a two-dimensional (2d) array that contains the elements in LL. If you type print(A) the Python interpreter knows how to strip the commas and split the output over three rows, making it easy to see that it's a two-dimensional entity, similar to a mathematical matrix.

Much of what we saw on creating one-dimensional arrays carries over to the two-dimensional case. For example, A.size is 6: this is the total number of elements, including both rows and columns. Another attribute is ndim, which tells us the dimensionality of the array: A.ndim is 2 for our example. The number of dimensions can be thought of as the number of distinct "axes" according to which we are listing elements. Incidentally, our terminology may be confusing to those coming from a linear-algebra background. In linear algebra, we say that a matrix $\mathbf{A}$ with m rows and n columns has "dimensions" m and n, or sometimes that it has dimensions $m \times n$. In contradistinction to this, the NumPy convention is to refer to an array like A as having "dimension 2", since it's made up of rows and columns (i.e., how many elements are in each row and in each column doesn't matter). There exists yet another attribute, shape, which returns a tuple containing the number of elements in each dimension, e.g., A.shape is (3, 2). Table 1.3 collects the attributes we'll need. You can also create two-dimensional arrays by passing a tuple with the desired shape to the appropriate function, e.g., np.zeros((3,2)) or np.ones((4,6)). Another function helps you make a square identity matrix via, e.g., np.identity(4). All of these functions produce floats, by default.

If you want to produce an array starting from a hand-rolled list, but wish to avoid the Python list-of-lists syntax (with double square brackets, as well as several commas), you can start from a one-dimensional list, which is then converted into a one-dimensional array, which in its turn is re-shaped into a two-dimensional array via the reshape() function, e.g., A = np.array([11,12,13,14,15,16]).reshape(3,2). Here's another example: A = np.arange(11,23).reshape(3,4). Observe how conveniently we've obviated the multitude of square brackets and commas of LL.

```
A[1,:]
11 12 13 14
15 16 17 18
19 20 21 22
```

```
A[:,2]
11 12 13 14
15 16 17 18
19 20 21 22
```

```
A[:2,:]
11 12 13 14
15 16 17 18
19 20 21 22
```

```
A[:,1:]
11 12 13 14
15 16 17 18
19 20 21 22
```

```
A[:2,1:]
11 12 13 14
15 16 17 18
19 20 21 22
```

```
A[:2,1:-1]
11 12 13 14
15 16 17 18
19 20 21 22
```

```
A[::2,:]
11 12 13 14
15 16 17 18
19 20 21 22
```

```
A[::2,:2]
11 12 13 14
15 16 17 18
19 20 21 22
```

```
A[::2,::2]
11 12 13 14
15 16 17 18
19 20 21 22
```

Fig. 1.4 Examples of slicing a 3×4 two-dimensional array

It is now time to see one of the nicer features of two-dimensional arrays in NumPy: the intuitive way to access elements. To access a specific element of `LL`, you have to say, e.g., `LL[2][1]`. Recall that we are using Python 0-indexing, so the rows are numbered 0th, 1st, 2nd, and analogously for the columns. This double pair of brackets, separating the numbers from each other, is quite cumbersome and also different from how matrix notation is usually done, i.e., A_{ij} or $A_{i,j}$. Thus, it comes as a relief to see that NumPy array elements can be indexed simply by using only one pair of square brackets and a comma-separated pair of numbers, e.g., `A[2, 1]`.

Slicing is equally intuitive, e.g., `A[1,:]` picks a specific row and uses an everything-slice for the columns, and therefore leads to that entire row. Figure 1.4 shows a few other examples: the highlighting shows the elements that are chosen by the slice shown at the top of each matrix. The most interesting of these is probably `A[:2,1:-1]`, which employs the Python convention of using `-1` to denote the last element, in order to slice the "middle" columns: `1:-1` avoids the first/0th column and the last column. We also show a few examples that employ a stride when slicing. To summarize, when we use two numbers to index (such as `A[2,1]`), we go from a two-dimensional array to a number. When we use one number to index/slice (such as `A[:,2]`), we go from a two-dimensional array to a one-dimensional array. When we use slices (such as `A[::2,:]`), we get collections of rows, columns, individual elements, etc. We can combine what we just learned about slicing two-dimensional arrays with what we already know about NumPy broadcasting. For example, `A[:,:] = 7` overwrites the entire matrix and you could do something analogous for selected rows, columns, etc.

Similarly to the one-dimensional case, *vectorized* operations for two-dimensional arrays allow us to, say, add together (elementwise) two square matrices, `A + B`. On the other hand, a simple multiplication is *also* carried out elementwise, i.e., each element in `A` is multiplied by the corresponding element in `B` when saying `A*B`. This may be unexpected behavior, if you were looking for a matrix multiplication. To repeat, *array operations are always carried out elementwise*, so when you multiply two two-dimensional arrays you get the *Hadamard product*, $(\mathbf{A} \odot \mathbf{B})_{ij} = A_{ij}B_{ij}$. The matrix multiplication that you know from linear algebra follows from a different formula, namely $(\mathbf{AB})_{ij} = \sum_k A_{ik}B_{kj}$, see Eq. (C.10). From Python 3.5 and onward[11] you can multiply two matrices using the `@` infix operator, i.e., `A@B`. In older versions you had to say `np.dot(A,B)`, which was not as intuitive as one would have liked. There are many other operations we could carry out, e.g., we can multiply together a two-dimensional and a one-dimensional matrix, `A@xs` or `xs@A`. It's easy to come up with more convoluted examples; here's another one: `xs@A@ys` is much easier to write (and read) than `np.dot(xs,np.dot(A,ys))`. Problem 1.8 studies the dot product of two one-dimensional arrays in detail: the main takeaway is that we compute it via `xs@ys`.[12] In a math course, to take the dot product (also known as the *inner product*) of two vectors you had to first take the transpose of the first vector (to have the dimensions match); conveniently, NumPy takes care of all of that for us, allowing us to simply say `xs@ys`.[13] Finally, *broadcasting* with two-dimensional arrays works just as you'd expect, e.g., `A/2` halves all the matrix elements.

NumPy contains several other important functions, with self-explanatory names. Since we just mentioned the (potential) need to take the transpose: NumPy has a function called `transpose()`; we prefer to access the transpose as an array attribute, i.e., `A.T`. You may recall our singing the praises of the `for` loop in Python; well, another handy idiom involves iterating over rows via `for row in A:` or over columns via `for column in A.T:`; marvel at how expressive both these options are. Iterating over columns will come in very handy in section 4.4.4 when we will be handling eigenvectors of a matrix.

In linear algebra the product of an $n \times 1$ column vector and a $1 \times n$ row vector produces an $n \times n$ matrix. If you try to do this "naively" in NumPy using 1d arrays, you will be disappointed: both `xs@ys` and `xs@(ys.T)` give a number (the same one). Inspect the `shape` attribute of `ys` and of `ys.T` to see what's going on. What you really need is the function `np.outer()`, which computes the outer product, e.g., `np.outer(xs, ys)`.

All the NumPy functions we mentioned earlier, such as `np.sqrt()`, also work on two-dimensional arrays, and the same holds for subtraction, powers, etc. Intriguingly, functions like `np.sum()`, `np.amin()`, and `np.amax()` also take in an (optional) argument `axis`: if you set `axis = 0` you operate across rows (column-wise) and `axis = 1` operates across columns (row-wise). Similarly, we can create our own user-defined functions that know how to handle an entire two-dimensional array/matrix at once. There are also several handy functions that are intuitively easier to grasp for the case of two-dimensional ar-

[11] Python 3.5 was already 8 years old when the present textbook came out.

[12] Make sure not to get confused by the fact that `A@B` produces a matrix but `xs@ys` produces a number.

[13] Similarly, when multiplying a vector by a matrix, `xs@A` is just as simple to write as `A@xs`.

rays, e.g., `np.diag()` to get the diagonal elements, or `np.tril()` to get the lower triangle of an array and `np.triu()` to get the upper triangle. In chapter 8 we will also need to use `np.meshgrid()`, which takes in two coordinate vectors and returns coordinate matrices. Another helpful function is `np.fill_diagonal()` which allows you to efficiently update the diagonal of a matrix you've already created. Finally, `np.trace()` often comes in handy.

The default storage-format for a NumPy array in memory is *row major*: as a beginner, you shouldn't worry about this too much, but it means that rows are stored in order, one after the other. This is the same format used by the C programming language. If you wish to use, instead, Fortran's column-major format, presumably in order to interoperate with code in that language, you simply pass in an appropriate keyword argument when creating the array. If you structure your code in the "natural" way, i.e., first looping through rows and then through columns, all should be well, so you can ignore this paragraph.

1.7 Project: Visualizing Electric Fields

Each chapter concludes with a physical application of techniques introduced up to that point. Since this is only the first chapter, we haven't covered any numerical methods yet. Even so, we can already start to look at some physics that is not so accessible without a computer by using `matplotlib` for more than just basic plotting. We will visualize a vector field, i.e., draw field lines for the electric field produced by several point charges.

1.7.1 Electric Field of a Distribution of Point Charges

Very briefly, let us recall *Coulomb's law*: the force on a test charge Q located at point P (at the position $\mathbf{r}$), coming from a single point charge q_0 located at $\mathbf{r}_0$ is given by:

$$\mathbf{F}_0 = k\frac{q_0 Q}{(\mathbf{r}-\mathbf{r}_0)^2}\frac{\mathbf{r}-\mathbf{r}_0}{|\mathbf{r}-\mathbf{r}_0|} \tag{1.1}$$

where Coulomb's constant is $k = 1/(4\pi\epsilon_0)$ in SI units (and ϵ_0 is the permittivity of free space). The force is proportional to the product of the two charges, inversely proportional to the square of the distance between the two charges, and points along the line from charge q_0 to charge Q. The electric field is then the ratio of the force $\mathbf{F}_0$ with the test charge Q in the limit where the magnitude of the test charge goes to zero. In practice, this gives us:

$$\mathbf{E}_0(\mathbf{r}) = kq_0\frac{\mathbf{r}-\mathbf{r}_0}{|\mathbf{r}-\mathbf{r}_0|^3} \tag{1.2}$$

where we cancelled out the Q and also took the opportunity to combine the two denominators. This is the electric field at the location $\mathbf{r}$ due to the point charge q_0 at $\mathbf{r}_0$.

If we were faced with more than one point charge, we could apply the *principle of superposition*: the total force on Q is made up of the vector sum of the individual forces acting on Q. As a result, if we were dealing with the n point charges $q_0, q_1, \ldots, q_{n-1}$ located at $\mathbf{r}_0, \mathbf{r}_1, \ldots, \mathbf{r}_{n-1}$ (respectively) then the situation would be that shown in Fig. 1.5. Our figure is in two dimensions for ease of viewing, but the formalism applies equally well to three dimensions. The total electric field at the location $\mathbf{r}$ is:

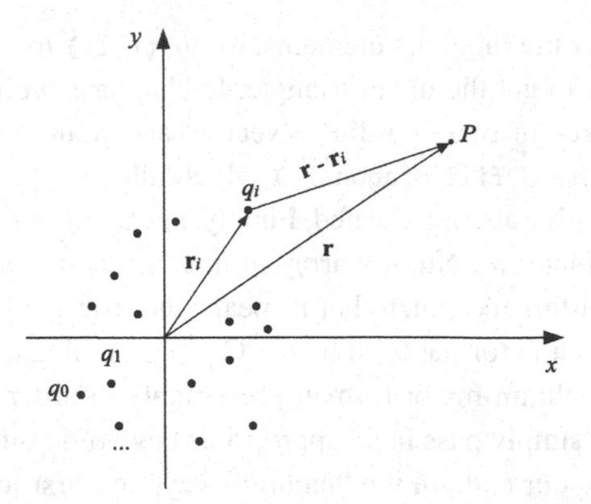

Fig. 1.5 Physical configuration made up of n point charges

$$\mathbf{E}(\mathbf{r}) = \sum_{i=0}^{n-1} \mathbf{E}_i(\mathbf{r}) = \sum_{i=0}^{n-1} kq_i \frac{\mathbf{r} - \mathbf{r}_i}{|\mathbf{r} - \mathbf{r}_i|^3} \tag{1.3}$$

namely, a sum of the individual electric field contributions, $\mathbf{E}_i(\mathbf{r})$. Note that you can consider this total electric field at any point in space, $\mathbf{r}$. Note, also, that the electric field is a vector quantity: at any point in space this $\mathbf{E}$ has a magnitude and a direction. One way of visualizing vector fields consists of drawing *field lines*, namely imaginary curves that help us keep track of the direction of the field. More specifically, the tangent of a field line at a given point gives us the direction of the electric field at that point. Field lines do not cross; they start at positive charges ("sources") and end at negative charges ("sinks").

1.7.2 Plotting Field Lines

We will plot the electric field lines in Python; while more sophisticated ways of visualizing a vector field exist (e.g., line integral convolution), what we describe below should be enough to give you a qualitative feel for things. While plotting functions (or even libraries) tend to change much faster than other aspects of the programming infrastructure, the principles discussed apply no matter what the specific implementation looks like.

We are faced with two tasks: first, we need to produce the electric field (vector) at several points near the charges as per Eq. (1.3) and, second, we need to plot the field lines in such a way that we can physically interpret what is happening. As in the previous code, we make a Python function for each task. For simplicity, we start from a problem with only two point charges (of equal magnitude and opposite sign). Also, we restrict ourselves to two dimensions (the Cartesian x and y).

Code 1.3 is a Python implementation, where Coulomb's constant is divided out for simplicity. We start by importing `numpy` and `matplotlib`, since the heavy lifting will be done

Code 1.3

vectorfield.py

```
import numpy as np
import matplotlib.pyplot as plt
from math import sqrt
from copy import deepcopy

def makefield(xs, ys):
    qtopos = {1: (-1,0), -1: (1,0)}
    n = len(xs)
    Exs = [[0. for k in range(n)] for j in range(n)]
    Eys = deepcopy(Exs)
    for j,x in enumerate(xs):
        for k,y in enumerate(ys):
            for q,pos in qtopos.items():
                posx, posy = pos
                R = sqrt((x - posx)**2 + (y - posy)**2)
                Exs[k][j] += q*(x - posx)/R**3
                Eys[k][j] += q*(y - posy)/R**3
    return Exs, Eys

def plotfield(boxl,n):
    xs = [-boxl + i*2*boxl/(n-1) for i in range(n)]
    ys = xs[:]
    Exs, Eys = makefield(xs, ys)
    xs=np.array(xs); ys=np.array(ys)
    Exs=np.array(Exs); Eys=np.array(Eys)
    plt.streamplot(xs, ys, Exs, Eys, density=1.5, color='m')
    plt.xlabel('$x$')
    plt.ylabel('$y$')
    plt.show()

plotfield(2.,20)
```

by the function `streamplot()`, which expects NumPy arrays as input. We also import the square root and the `deepcopy()` function, which can create a distinct list-of-lists.

The function `makefield()` takes in two lists, `xs` and `ys`, corresponding to the coordinates at which we wish to evaluate the electric field (x and y together make up $\mathbf{r}$). We also need some way of storing the $\mathbf{r}_i$ at which the point charges are located. We have opted to store these in a dictionary, which maps from charge q_i to position $\mathbf{r}_i$—take some time to consider alternative ("manual") implementations. For each position **r** we need to evaluate $\mathbf{E}(\mathbf{r})$: in two dimensions, this is made up of $E_x(\mathbf{r})$ and $E_y(\mathbf{r})$, namely the two Cartesian components of the total electric field. Focusing on only one of these for the moment, say $E_x(\mathbf{r})$, we realize that we need to store its value for any possible **r**, i.e., for any possible x and y values. We decide to use a list-of-lists, produced by a nested list comprehension. We then create another list-of-lists, for $E_y(\mathbf{r})$. We need to map out (i.e., store) the value of the x and y components of the total electric field, at all the desired values of the vector **r**, namely, on a two-dimensional grid made up of `xs` and `ys`. This entails computing the electric field (contribution from a given point charge q_i) at all possible y's for a given x, and then iterating over all possible x's. We also need to iterate over our point charges q_i and their locations $\mathbf{r}_i$ (i.e., the different terms in the sum of Eq. (1.3)); we do this by saying `for q, pos in qtopos.items():` at which point we unpack `pos` into `posx` and `posy`.

We thus end up with three nested loops: one over possible x values, one over possible y values, and one over i. All three of these are written idiomatically, employing `items()` and `enumerate()`. The latter was used to ensure that we won't only have access to the x and the y values, which are needed for the right-hand side of Eq. (1.3), but also to two indices (j and k) that will help us store the electric-field components in the appropriate list-of-lists entry, e.g., `Exs[k][j]`.[14] This storing is carried out after defining a helper variable to keep track of the vector magnitude that appears in the denominator in Eq. (1.3). You should think about the += a little bit: since the left-hand side is for given j and k, the summation is carried out only when we iterate over the q_i (and $\mathbf{r}_i$). Incidentally, our idiomatic iteration over the point charges means that we don't even need an explicit i index.

Our second function, `plotfield()`, is where we build our two-dimensional grid for the `xs` and `ys`.[15] We take in as parameters the length L and the number of points n we wish to use in each dimension and create our `xs` using a list comprehension; all we're doing is picking x's from $-L$ to L. We then create a copy of `xs` and name it `ys`. After this, we call our very own `makefield()` to produce the two lists-of-lists containing $E_x(\mathbf{r})$ and $E_y(\mathbf{r})$ for many different choices of **r**. The core of the present function consists of a call to `matplotlib`'s function `streamplot()`; this expects NumPy arrays instead of Python lists, so we convert everything over. If you skipped section 1.6, as you were instructed to do, you should relax: this call to `np.array()` is all you need to know for now (and until chapter 4). We also pass in to `streamplot()` two (optional) arguments, to ensure that we have a larger density of field lines and to choose the color. Most importantly, `streamplot()` knows how to take in `Exs` and `Eys` and output a plot containing curves with arrows, exactly like what we are trying to do. We also introduce x and y labels, using dollar signs to make the symbols look nicer.

The result of running this code is shown in the left panel of Fig. 1.6. Despite the fact that the charges are not represented by a symbol in this plot, you can easily tell that you are dealing with a positive charge on the left and a negative charge on the right. At this point, we realize that a proper graphic representation of field lines also has another feature: the density of field lines should correspond to the strength of the field (i.e., its magnitude). Our

[14] The confusing index order follows `streamplot()`'s documentation (see also page 630).
[15] This would all be much easier with two-dimensional arrays, but you skipped that section.

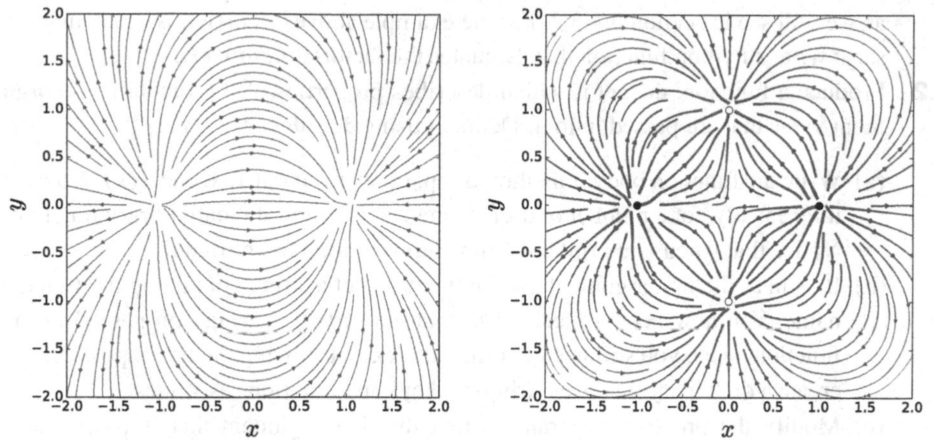

Fig. 1.6 Visualizing the electric fields resulting from two and four point charges

figure has discarded that information: the `density` argument we passed in had a constant value. This is a limitation of the `streamplot()` function.

There is a way to represent both the direction (as we already did) and the strength of the field using `streamplot()`, using the optional `linewidth` parameter. The argument passed in to `linewidth` can be a two-dimensional NumPy array, which keeps track of the strength at each point on the grid; it's probably better to pass in, instead, the logarithm of the magnitude at each point (possibly also using an offset). We show the result of extending our code to also include line width in the right panel of Fig. 1.6, where a stronger field is shown using a thicker line. This clearly shows that the field strength is larger near the charges (and in between them) than it is as you go far away from them. To make things interesting, this shows a different situation than the previous plot did: we are now dealing with four charges (two positive and two negative, all of equal magnitude). We also took the opportunity to employ symbols representing the position of the point charges themselves.

Problems

1.1 Study the following program, meant to evaluate the factorial of a positive integer:

```
def fact(n):
        return 1 if n==0 else n*fact(n-1)
print(fact(10))
```

This uses Python's version of the *ternary operator*. Crucially, it also uses *recursion*: we are writing the solution to our problem in terms of the solution of a smaller version of the same problem. The function then calls itself repeatedly. At some point we reach the *base case*, where the answer can be directly given. Recursion is helpful when the problem you are solving is amenable to a "divide-and-conquer" approach, as we will see in section 6.4.3.3. For the example above, recursion is quite unnecessary: write a Python function that evaluates the factorial *iterative*ly.

1.2 Produce a function, `descr()`, which describes properties of the function and argument value that are passed in to it. Define it using `def descr(f, x):`.

(a) Write the function body; this should separately print out `f`, `x`, and `f(x)`. Now call it repeatedly, for a number of user-defined and Python standard library functions. Notice that when you print out `f` the output is not human friendly.
(b) Pass in as the first argument not the function but a string containing the function name. You now need a mechanism that converts the string you passed in to a function (since you still need to print out `f(x)`). Knowing the list of possibilities, create a dictionary that uses strings as keys and functions as values.
(c) Modify the previous program so that the first argument that is passed in to `descr()` is a function, not a string. To produce the same output, you must also modify the body of the function since your dictionary will now be different.

1.3 Iterate through a list `xs` in reverse, printing out both the (decreasing) index and each element itself. Come up with both Pythonic and unPythonic solutions.

1.4 Imagine you have access to two Python functions, `fa()` and `fb()`, with one parameter each.[16] You now wish to apply the function `der()` that was defined in section 1.3.4 to a linear combination of `fa()` and `fb()`, namely `a*fa(x) + b*fb(x)`.

(a) Your initial reaction might be to define a third function, `fc()`; but what if you don't actually know the values of the `a` and `b` coefficients until you're already inside some other Python function, say, `caller()`? Based on what we've introduced in the main text, one idea is to define `fc()` to take three parameters (`x`, `a`, and `b`). This would then not interoperate with `der()`, which expects as its first argument a function taking in a single parameter. Try this.
(b) Your next instinct may be to modify the definition of `der()`: sometimes that's OK, but often it's just asking for trouble. What you're really after is a throwaway function that knows about the values of variables defined in the same scope. Do so via an anonymous function (via Python's `lambda` keyword) right where you need it. Look up the documentation and implement this solution.
(c) You may next be tempted to actually *name* your lambda function and then use it in `der()`. Try this. Unfortunately, this is explicitly discouraged in PEP 8.[17]
(d) Instead, define a *nested function* `func()` inside `caller()`, the latter calling `der()`.
(e) Use a *closure*: `func()` is nested inside `caller()`, the latter returning `func()`; recall that in Python you use the function name without the parentheses.

1.5 We will now elaborate on the for-else idiom introduced in `forelse.py`.

(a) First, we realize that our `look()` function, while helpful from a pedagogical perspective, is somewhat pointless: you didn't really need to write a function to

[16] The same idea applies to the two functions `sumofints()` and `cos()` that we encountered in the main text.

[17] It's easy to see why: if you feel the need to name your... anonymous function, then perhaps you shouldn't be using an anonymous function in the first place.

check for membership in a list. Directly use Python's `in` keyword to test whether or not a given string is to be found in a given list of strings.

(b) Now make `look()` slightly more useful by having it return not only the value (i.e., the string) but also the index where that value was found.

(c) Experiment with making `look()` shorter, by avoiding the for-else idiom altogether. First, do this the simplest way, i.e., by providing the default option `val = None` *before* you enter the loop. Second, avoid using `break` altogether by opting for two exit points in your function, i.e., two separate `return` statements.[18]

(d) The for-else idiom is actually better than the alternatives when exceptions (which we don't discuss elsewhere) are involved. Check the documentation to see how you can `raise` a `ValueError` in the `else` block of your `for` loop.

1.6 The following is implicitly defining a recurrence relation:

```
f0,f1 = 0,1
for i in range(n-1):
        f0,f1 = f1,f0+2*f1
```

We will now produce increasingly fancier versions of this code snippet.

(a) Define a function that takes in the cardinal number n and returns the corresponding latest value following the above recurrence relation. In other words, for $n = 0$ you should get 0, for $n = 1$ you should get 1, for $n = 2$ you should get 2, for $n = 3$ you should get 5, and so on.

(b) Define a *recursive* function taking in the cardinal number n and returning the corresponding latest value. The interface of the function will be identical to that of the previous part (the implementation will be different).

(c) Define a similar function that is more efficient. Outside the function, define a dictionary `ntoval = {0:0, 1:1}`. Inside the function, you should check to see if the n that was passed in exists as a key in `ntoval`: if it does, then simply return the corresponding value; if it doesn't, then carry out the necessary computation and augment the dictionary with a new key-value pair.

(d) If you take separation of concerns seriously, you may be feeling uncomfortable about accessing and modifying `ntoval` inside your function (since it is not being passed in as a parameter). Write a new function that looks like the one in the previous part, but takes in two parameters: `n` and `ntoval`.

(e) While part (d) respects separation of concerns, unfortunately it is not actually efficient. Write a similar function which uses a *mutable default parameter value*, i.e., it is defined by saying `def f5(n, ntoval = {0:0, 1:1}):`.

Test all five functions with $n = 8$: each of them should return 408. The functions in parts (c) and (e) should be efficient in the sense that if you now call them with, say, $n = 6$ they won't need to recompute the answer since they have already done so.

[18] Elsewhere in this book we always employ a single exit point, to make our codes easier to reason about.

1.7 This problem studies the quantity $(1 + 1/n)^n$ where $n = 10^1, 10^2, \ldots, 10^7$. Print out a nicely formatted table where the three columns are: (a) the value of n, (b) the quantity of interest computed with single-precision floating-point numbers, (c) the same quantity but now using doubles. You will need to use NumPy to get the singles to work. Keep in mind that the numbers shown in the second and third columns should be different (if they aren't, you're doing it wrong).

1.8 Investigate the relative efficiency of multiplying two one-dimensional NumPy arrays, `as` and `bs`; these should be large and with non-constant content. Do this in four distinct ways: (a) `sum(as*bs)`, (b) `np.sum(as*bs)`, (c) `np.dot(as,bs)`, and (d) `as@bs`. You may wish to use the `default_timer()` function from the `timeit` module. To produce meaningful timing results, repeat such calculations thousands of times (at least).

1.9 Take two matrices, **A** and **B**, which are $n \times m'$ and $m' \times m$, respectively. Implement matrix multiplication, without relying on `numpy`'s `@` or `dot()` as applied to matrices.

(a) Write the most obvious implementation you can think of, which takes in **A** and **B** and returns a new **C**. Use three loops.
(b) Write a function without the third loop by applying `@` to vectors.
(c) Write a third function that takes in **A** and **B** and returns **C**, but this time these are lists-of-lists, instead of arrays.
(d) Test the above three functions by employing specific examples for **A** and **B**, say 3×4 and 4×2, respectively.

1.10 Write your own functions that implement functionality similar to: (a) `np.argmax()`, (b) `np.where()`, and (c) `np.all()`, where the input will be a one-dimensional NumPy array. Note that `np.where()` is equivalent to `np.nonzero()` for this case.

1.11 Rewrite code 5.9, i.e., `action.py`, such that:

(a) The two lines involving `arr[1:-1]` now use an explicit loop and index.
(b) The calls to `np.fill_diagonal()` are replaced by explicit loop(s) and indices.

Compare the expressiveness (and total line-count) in your code vs that in `action.py`.

1.12 [$\mathcal{P}$] We now help you produce the right panel of Fig. 1.6:

(a) First try to produce the curves themselves. You'll need to appropriately place two positive and two negative charges and plot the resulting field lines. (Remember that dictionaries don't let you use duplicate keys, for good reason.)
(b) Introduce line width, by producing a list of lists containing the logarithm of the square root of the sum of the squares of the components of the electric field at that point, i.e., $\log\left[\sqrt{(E_x(\mathbf{r}))^2 + (E_y(\mathbf{r}))^2}\right]$ at each $\mathbf{r}$.
(c) Figure out how to add circles/dots (of color corresponding to the charge sign).
(d) Re-do this plot for the case of four positive (and equal) charges.

1.13 [$\mathcal{P}$] Problem 1.12 was made easier by the fact that you knew what the answer had to look like. You now need to produce a plot for the field lines (including line width) for the case of four alternating charges on a line. Place all four charges on the x axis, and give their x_i positions the values -1, -0.5, 0.5, and 1. The leftmost charge should be positive and the next one negative, and so on (they are all of equal magnitude).

1.14 [$\mathcal{P}$] Study methane *isotherms* with the van der Waals equation of state, Eq. (5.2).

(a) Plot 40 isotherms (i.e., constant-T curves, showing P vs v), where T goes from 162 to 210 K, v goes from $1.5b$ to $9b$ and the curves look smooth.

(b) If you solved the previous part correctly, you should barely be able to tell the different curves apart. Beautify your plot by employing an automated color map.

1.15 [$\mathcal{P}$] We address the problem of *wave interference*, by repurposing `vectorfield.py`. Work in two dimensions (just like in section 1.7) and assume that the combined effect of two objects dropped in a water basin can be modelled by:

$$W(\mathbf{r}) = A \sin(k|\mathbf{r} - \mathbf{r}_0|) + A \sin(k|\mathbf{r} - \mathbf{r}_1|) \tag{1.4}$$

You can think of this as an updated version of Eq. (1.3), where we are dealing with the linear addition of sinusoidal waves, instead of electric-field contributions. For simplicity, take $A = 1$, $k = 2\pi/0.3$, $\mathbf{r}_0 = (-1\ \ 0)^T$, and $\mathbf{r}_1 = (1\ \ 0)^T$. Since we are now faced with a density plot, you should employ (not `streamplot()` but) `imshow()` from Matplotlib; in each of x and y your plot should extend from -4 to $+4$.

1.16 [$\mathcal{P}$] We will now examine the simple (yet non-linear) pendulum on the *phase plane*. The total energy of a pendulum of mass m and length l is:

$$E = \frac{1}{2} m l^2 \dot{\theta}^2 + mgl(1 - \cos\theta) \tag{1.5}$$

where θ is the angle from the vertical. Your task is to create a plot of $\dot{\theta}$ vs θ, where you will show the contours of constant E using Matplotlib's `contour()`; feel free to repurpose `vectorfield.py`. It's best to vary θ from -2π to $+2\pi$; you may also have to play around with the `levels` parameter; take $m = 1$ kg, $l = 1$ m, and $g = 9.8$ m/s^2. Physically interpret the different regions you encounter; roughly speaking, you should be seeing open curves[19] (above and below) and eye-shaped areas (near the middle). The boundary between the two regions is known as a *separatrix*.

1.17 [$\mathcal{P}$] We will now discuss how to visualize *binary stars*. While stars in well-detached binaries will have spherical shapes, stars in close binaries will experience tidal distortions and will have nearly ellipsoidal shapes. The *Roche model* was introduced to account for either possibility: it studies the total gravitational potential for two masses that are in circular orbit (about their barycenter). Switching to a coordinate frame that eliminates the circular orbit of the two masses (a co-rotating/"synodic" frame), the (dimensionless) Roche potential at any point (x, y, z) can be written as:

$$\Phi = \frac{m_0}{\sqrt{(x - m_1)^2 + y^2 + z^2}} + \frac{m_1}{\sqrt{(x + m_0)^2 + y^2 + z^2}} + \frac{x^2 + y^2}{2} \tag{1.6}$$

The (dimensionless) masses can be written in terms of the mass ratio $q = m_0/m_1 \leq 1$ ($m_0 = q/(1 + q)$ and $m_1 = 1/(1 + q)$). We focus on the orbital plane ($z = 0$).

[19] The open curves have constant amplitude as θ is increased. We will see a different scenario in problem 8.43.

(a) Repurpose `vectorfield.py`, using Matplotlib's `contour()`, to visualize the curves of constant gravitational potential for the case of $q = 0.4$. Pass in an appropriate `levels` *list* to make your figure look good. (Note that a star's surface *is* an equipotential surface.) The characteristic ∞ shape you find in the middle is made up of two *Roche lobes*; a Roche lobe tells you the maximum volume that a star in a binary can occupy without losing gravitational control of its constituents.

(b) Another way you can visualize the Roche potential of Eq. (1.6) is via a surface plot. You can create such a figure via Matplotlib's `plot_surface()`. Go over its documentation, which will explain that you will first need to pass in `projection='3d'` when creating your axes. Make sure you can easily identify the different ridges and wells corresponding to the two stars as well as the area around them (e.g., via the use of `set_zlim()`).

1.18 [$\mathcal{P}$] The problem of a single particle in a one-dimensional lattice is a staple of introductory courses on solid-state physics; we will see how to tackle it for the case of a general potential in problem 4.56 (after learning about linear-algebra techniques). Here, we address a specific case, that of the *Kronig–Penney model*, involving an infinite periodic array of potential barriers (or, viewed from another perspective, of wells) of rectangular shape. Writing down the wave function within the barrier (of width $L - M$) and separately that within the well (of width M), and then imposing continuity of the wave function and of its derivative, one arrives at the condition:

$$\cos(KL) = \cos(k_0 M)\cosh[k_1(L-M)] + \frac{k_1^2 - k_0^2}{2k_0 k_1}\sin(k_0 M)\sinh[k_1(L-M)]$$
$$k_0 = \sqrt{2mE/\hbar^2}, \qquad k_1 = \sqrt{2m(V_0 - E)/\hbar^2} \tag{1.7}$$

V_0 is the barrier height, E the energy, and K the wave number in *Bloch's theorem*:

$$\psi(x + L) = e^{iKL}\psi(x) \tag{1.8}$$

This wave number obeys $-\pi < KL < \pi$. Equation (1.7) implicitly provides us with the energy-dispersion relation, i.e., $E(K)$. However, we don't know how to solve complicated equations yet (we tackle this challenge in problem 5.45, after learning about root-finding). Our approach here will be to: hand-pick a value of E, compute the right-hand side of the equation, and check to see if its absolute value is larger than unity; if so, then there is no solution for K. If, however, the absolute value of the right-hand side is smaller than unity, then we can find the value of K by taking the inverse cosine and dividing by L. (Since the cosine is an even function, you need to consider both K and $-K$ as equally acceptable solutions.) Take $L = 1, M = 0.4, \hbar^2/m = 1$ and study five thousand E values from (roughly) 0 to 110. Plot E vs K for the two cases of: (a) $V_0 = 0$ and (b) $V_0 = 35$. Do yourself a favor and employ the special functions contained in the complex-number-aware module `cmath` (i.e., not in `math`). Your plot for $V_0 = 0$ should be a parabola, which corresponds to a plane-wave dispersion; of course, your results will lie in $-\pi < KL < \pi$, so you will find a "folded" version of the parabola. Your plot for $V_0 = 35$ should exhibit *band gaps*, resulting from the fact that there are energy regions for which no solution exists.

2 Numbers

Should we not be concerned that this fear of erring is already the error itself?

Georg Wilhelm Friedrich Hegel

2.1 Motivation

We don't really have to provide a motivation regarding the importance of numbers in physics: both experimental measurements and theoretical calculations produce specific numbers. Preferably, one also estimates the uncertainty associated with these values.

We have, semi-arbitrarily, chosen to discuss a staple of undergrad physics education, the photoelectric effect. This arises when a metal surface is illuminated with electromagnetic radiation of a given frequency ν and electrons come out. In 1905 Einstein posited that quantization is a feature of the radiation itself. Thus, a light quantum (a *photon*) gives up its energy ($h\nu$, where h is now known as *Planck's constant*) to an electron, which has to break free of the material, coming out with a kinetic energy of:

$$T = h\nu - W \tag{2.1}$$

where W is the work function that relates to the specific material. The maximum kinetic energy T of the photoelectrons could be extracted from the potential energy of the electric field needed to stop them, via $T = eV_s$, where V_s is the stopping potential and e is the charge of the electron. Together, these two relations give us:

$$eV_s = h\nu - W \tag{2.2}$$

Thus, if one produces data relating V_s with ν, the slope would give us h/e.

In 1916, R. A. Millikan published a paper [107] titled "A direct photoelectric determination of Planck's h", where he did precisely that. Millikan's device included a remotely controlled knife that would shave a thin surface off the metal; this led to considerably enhanced photocurrents. It's easy to see why Millikan described his entire experimental setup as a "machine shop in vacuo". Results from this paper are shown in Fig. 2.1. Having extracted the slope h/e, the author then proceeded to compute h by "inserting my value of e".[1] The value Millikan extracted for Planck's constant was:

$$h = 6.56 \times 10^{-27} \text{erg s} \tag{2.3}$$

In his discussion of Fig. 2.1, Millikan stated that "it is a conservative estimate to place

[1] Recall that Millikan had measured e very accurately with his oil-drop experiment in 1913.

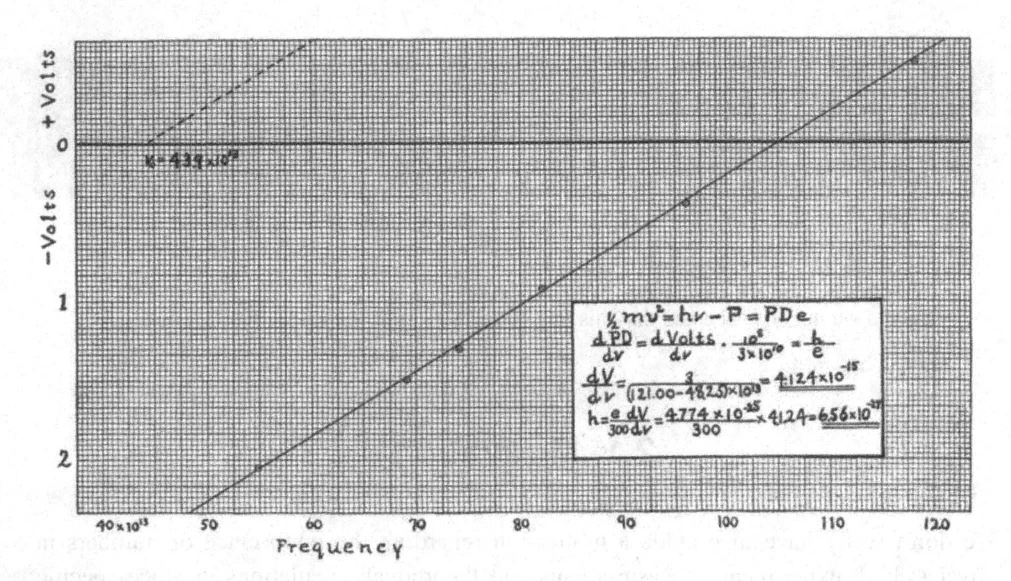

Fig. 2.1 Millikan's data on the photoelectric effect. Reprinted figure with permission from R. A. Millikan, *Phys. Rev.* 7, 355 (1916), Copyright 1916 by the American Physical Society.

the maximum uncertainty in the slope at about 1 in 200 or 0.5 per cent".Translating this to the above units, we find an uncertainty estimate of 0.03×10^{-27}erg s. Richardson and Compton's earlier experimental results [122] had a slope uncertainty that was larger than 60 percent. The above extraction is to be compared with the modern determination of $h = 6.626070040(81) \times 10^{-27}$erg s. Overall, Millikan's result was a huge improvement on earlier works and was important in the acceptance of Einstein's work and of light quanta.[2]

For the sake of completeness, we note that the aforementioned error estimate follows from *assuming* a linear relationship. It would have probably been best to start from Millikan's earlier comment that "the maximum possible error in locating any of the intercepts is say two hundredths of a volt" and then do a least-squares fit, as explained in problem 6.65. This isolated example already serves to highlight that even individual experimental measurements have associated uncertainties; nowadays, experimental data points are always given along with a corresponding error bar. Instead of multiplying the examples where experiment and theory interact fruitfully, we now turn to the main theme of this chapter, which is the presence of errors when storing and computing numbers.

2.2 Errors

In this text we use the word *accuracy* to describe the match of a value with the (possibly unknown) true value. On the other hand, we use the word *precision* to denote how many

[2] Intriguingly, Millikan himself was far from being convinced that quantum theory was relevant here, speaking of "the bold, not to say the reckless, hypothesis of an electro-magnetic light corpuscle".

digits we can use in a mathematical operation, whether these digits are correct or not. An inaccurate result arises when we have an error. This can happen for a variety of reasons, only one of which is limited precision.[3] Excluding "human error" and measurement uncertainty in the input data, there are typically two types of errors we have to deal with in numerical computing: approximation error and rounding error. In more detail:

- **Approximation errors** These are sometimes known as *truncation errors*. Here's an example. You are trying to approximate the exponential, e^x, using its Taylor series:

$$y = \sum_{n=0}^{n_{max}} \frac{x^n}{n!} \tag{2.4}$$

  Obviously, we are limiting the sum to the terms up to n_{max} (i.e., we are including the terms labelled $0, 1, \ldots, n_{max}$ and dropping the terms from $n_{max} + 1$ to ∞). As a result, it's fairly obvious that the value of y for a given x may depend on n_{max}. In principle, at the mere cost of running one's calculation longer, one can get a better answer.[4]
- **Roundoff errors** These are also known as *rounding errors*. This type of error appears every time a calculation is carried out using floating-point numbers: since these don't have infinite precision, some information is lost. Here's an example: using real numbers, it is easy to see that $(\sqrt{2})^2 - 2 = 0$. However, when carrying out the same operation in Python we get a non-zero answer:

```
>>> (sqrt(2))**2 - 2
4.440892098500626e-16
```

  This is because $\sqrt{2}$ cannot be evaluated with infinitely many digits on the computer. Thus, the (slightly inaccurate) result for `sqrt(2)` is then used to carry out a second calculation, namely the squaring. Finally, the subtraction is yet another mathematical operation that can lead to rounding error.[5] Often, roundoff errors do not go away even if you run the calculation longer.

In the present chapter we will talk quite a bit about roundoff errors. In the next chapter we talk about the combined effect of approximation and roundoff errors. The chapters after that typically focus only on approximation errors, i.e., on estimating how well a specific method performs in principle. Before we get that far, however, let us first try to introduce some basic concepts, without limiting ourselves to any one kind of error.

2.2.1 Absolute and Relative Error

Assume we are studying a quantity whose exact value is x. If $\tilde{x}$ is an approximate value for it, then we can define the *absolute error* as follows:[6]

[3] The value 1.23456789123456 is precisely determined yet, viewed as an approximation of π, quite inaccurate.
[4] But keep reading, since roundoff error becomes important here, too.
[5] Again, this is discussed much more thoroughly in the rest of the chapter.
[6] Other authors employ another definition of the absolute error, which differs by a minus sign (see also page 459).

$$\Delta x = \tilde{x} - x \tag{2.5}$$

We don't specify at this point the source of this absolute error: it could be uncertainties in the input data, an inaccuracy introduced by our imperfect earlier calculation, or the result of roundoff error (possibly accumulated over several computations). For example:

$$x_0 = 1.000, \qquad \tilde{x}_0 = 0.999 \tag{2.6}$$

corresponds to an absolute error of $\Delta x_0 = -10^{-3}$. This also allows us to see that the absolute error, as defined, can be either positive or negative. If you need it to be positive (say, in order to take its logarithm), simply take the absolute value.

We are usually interested in defining an *error bound* of the form:

$$|\Delta x| \leq \epsilon \tag{2.7}$$

or, equivalently:

$$|\tilde{x} - x| \leq \epsilon \tag{2.8}$$

where we hope that ϵ is "small". Having access to such an error bound means that we can state something very specific regarding the (unknown) exact value x:

$$\tilde{x} - \epsilon \leq x \leq \tilde{x} + \epsilon \tag{2.9}$$

This means that, even though we don't know the exact value x, we do know that it could be at most $\tilde{x} + \epsilon$ and at the least $\tilde{x} - \epsilon$. Keep in mind that if you know the actual absolute error, as in our $\Delta x_0 = -10^{-3}$ example above, then, from Eq. (2.5), you know that:

$$x = \tilde{x} - \Delta x \tag{2.10}$$

and there's no need for inequalities. The inequalities come into the picture when you don't know the actual value of the absolute error and only know a bound for the magnitude of the error. The error bound notation $|\Delta x| \leq \epsilon$ is sometimes rewritten in the form $x = \tilde{x} \pm \epsilon$, though you should be careful: this employs our definition of *maximal error* (i.e., the worst-case scenario) as above, not the usual *standard error* (i.e., the statistical concept you may have encountered in a lab course).

Of course, even at this early stage, one should think about exactly what we mean by "small". Our earlier case of $\Delta x_0 = -10^{-3}$ probably fits the bill. But what about:

$$x_1 = 1\,000\,000\,000.0, \qquad \tilde{x}_1 = 999\,999\,999.0 \tag{2.11}$$

which corresponds to an absolute error of $\Delta x_1 = -1$? Obviously, this absolute error is larger (in magnitude) than $\Delta x_0 = -10^{-3}$. On the other hand, it's not too far-fetched to say that there's something wrong with this comparison: x_1 is much larger (in magnitude) than x_0, so even though our approximate value $\tilde{x}_1$ is off by a unit, it "feels" closer to the corresponding exact value than $\tilde{x}_0$ was.

This is resolved by introducing a new definition. As before, we are interested in a quantity whose exact value is x and an approximate value for it is $\tilde{x}$. Assuming $x \neq 0$, we can define the *relative error* as follows:

$$\delta x = \frac{\Delta x}{x} = \frac{\tilde{x} - x}{x} \quad (2.12)$$

Obviously, this is simply the absolute error Δx divided by the exact value x. As before, we are not specifying the source of this relative error (input data uncertainties, roundoff, etc.). Another way to express the relative error is:

$$\tilde{x} = x(1 + \delta x) \quad (2.13)$$

You should convince yourself that this directly follows from Eq. (2.12). We will use this formulation repeatedly in what follows.[7]

Let's apply our definition of the relative error to the earlier examples:

$$\delta x_0 = \frac{0.999 - 1.000}{1.000} = -10^{-3}, \qquad \delta x_1 = \frac{999\,999\,999.0 - 1\,000\,000\,000.0}{1\,000\,000\,000.0} = -10^{-9} \quad (2.14)$$

The definition of the relative error is consistent with our intuition: $\tilde{x}_1$ is, indeed, a much better estimate of x_1 than $\tilde{x}_0$ is of x_0. Quite frequently, the relative error is given as a percentage: δx_0 is a relative error of -0.1% whereas δx_1 is a relative error of $-10^{-7}\%$.

In physics the values of an observable can vary by several orders of magnitude (according to density, temperature, and so on), so it is wise to employ the scale-independent concept of the relative error, when possible.[8] Just like for the case of the absolute error, we can also introduce a *bound for the relative error*:

$$|\delta x| = \left|\frac{\Delta x}{x}\right| \leq \epsilon \quad (2.15)$$

where now the phrase "ϵ is small" is unambiguous (since it doesn't depend on whether or not x is large). Finally, we note that the definiton of the relative error in Eq. (2.12) involves x in the denominator. If we have access to the exact value (as in our examples with x_0 and x_1 above), all is well. However, if we don't actually know the exact value, it is sometimes more convenient to use, instead, the approximate value $\tilde{x}$ in the denominator. Problem 2.2 discusses this alternative definition and its connection with what we discussed above.

2.2.2 Error Propagation

So far, we have examined the concepts of the absolute error and of the relative error (as well as the corresponding error bounds); no details were provided regarding the mathematical

[7] You can also expand the parentheses and identify $x\delta x = \Delta x$, to see that this leads to $\tilde{x} = x + \Delta x$.

[8] Once again, if you need the relative error to be positive, simply take the absolute value.

operations carried out using these values. We will now discuss the elementary operations (addition, subtraction, multiplication, division), to give you some insights into combining approximate values together. One of our goals is to see what happens when we put together the error bounds for two numbers a and b to produce an error bound for a third number x, i.e., we will study error propagation. We will also study more general scenarios, in which one is faced with more complicated mathematical operations (and possibly more than two numbers being operated on). In what follows, it's important to keep in mind that these are *maximal errors*, so our results will be different (and likely more pessimistic) than what you may have encountered in a standard course on experimental measurements.[9]

2.2.2.1 Addition or Subtraction

We are faced with two real numbers a and b, and wish to take their difference:

$$x = a - b \tag{2.16}$$

As usual, we don't know the exact values, but only the approximate values $\tilde{a}$ and $\tilde{b}$, so what we form instead is the difference of these:

$$\tilde{x} = \tilde{a} - \tilde{b} \tag{2.17}$$

Let us now apply Eq. (2.10) twice:

$$\tilde{a} = a + \Delta a, \qquad \tilde{b} = b + \Delta b \tag{2.18}$$

Plugging the last four equations into the definition of the absolute error, Eq. (2.5), we have:

$$\Delta x = \tilde{x} - x = (a + \Delta a) - (b + \Delta b) - (a - b) = \Delta a - \Delta b \tag{2.19}$$

In the third equality we cancelled what we could.

We now recall that we are interested in finding relations between error bounds. Thus, we take the absolute value and then use the triangle inequality to find:

$$|\Delta x| \leq |\Delta a| + |\Delta b| \tag{2.20}$$

You should convince yourself that a fully analogous derivation leads to exactly the same result for the case of addition of the two numbers a and b. Thus, our main conclusion so far is that *in addition and subtraction adding together the bounds for the absolute errors in the two numbers gives us the bound for the absolute error in the result.*

Let's look at an example. Assume that we have:

$$|4.56 - a| \leq 0.14, \qquad |1.23 - b| \leq 0.03 \tag{2.21}$$

(If you are in any way confused by this notation, look up Eq. (2.7) or Eq. (2.8).) Our finding in Eq. (2.20) implies that the following relation will hold, when $x = a - b$:

$$|3.33 - x| \leq 0.17 \tag{2.22}$$

[9] "It is probable that many quite improbable things should happen" (Aristotle, *Poetics*, 1456a25).

It's easy to see that this error bound, simply the sum of the two error bounds we started with, is larger than either of them. If we didn't have access to Eq. (2.20), we could have arrived at the same result the long way: (a) when a has the greatest possible value (4.70) and b has the smallest possible value (1.20), we get the greatest possible value for $a - b$, namely 3.50, and (b) when a has the smallest possible value (4.42) and b has the greatest possible value (1.26), we get the smallest possible value for $a - b$, namely 3.16.

As our main result in Eq. (2.20), applied just now in a specific example, shows, we simply add up the absolute error bounds. Again, this is different from what you do when you are faced with a "standard error" (i.e., the standard deviation of the sampling distribution of a statistic): in that case, the absolute errors add "in quadrature"; we will address this scenario in due course, see Eq. (6.129). We repeat that our result is more pessimistic (i.e., tries to account for the worst-case scenario). We won't keep repeating this warning below.

2.2.2.2 Catastrophic Cancellation

Let us examine the most interesting special case: $a \approx b$ (for which case $x = a - b$ is small). Dividing our result in Eq. (2.20) by x gives us the relative error (bound) in x:

$$|\delta x| = \left|\frac{\Delta x}{x}\right| \leq \frac{|\Delta a| + |\Delta b|}{|a - b|} \tag{2.23}$$

Now, express Δa and Δb in terms of the corresponding relative error: $\Delta a = a\delta a$ and $\Delta b = b\delta b$. Since $a \approx b$, you can factor $|a|$ out:

$$|\delta x| \leq (|\delta a| + |\delta b|)\frac{|a|}{|a - b|} \tag{2.24}$$

It's easy to see that if $a \approx b$ then $|a - b|$ will be much smaller than $|a|$ so, since the fraction will be large, the relative errors δa and δb will be magnified.[10]

Let's look at an example. Assume that we have:

$$|1.25 - a| \leq 0.03, \qquad |1.20 - b| \leq 0.03 \tag{2.25}$$

that is, a relative error (bound) $\delta a \approx 0.03/1.25 = 0.024$ or roughly 2.4% (this is approximate, because we divided by $\tilde{a}$, not by the, unknown, a). Similarly, the other relative error (bound) is $\delta b \approx 0.03/1.20 = 0.025$ or roughly 2.5%. From Eq. (2.24) we see that the relative error for the difference will obey:

$$|\delta x| \leq (0.024 + 0.025)\frac{1.25}{0.05} = 1.225 \tag{2.26}$$

where the right-hand side is an approximation (using $\tilde{a}$ and $\tilde{x}$). This shows us that two numbers with roughly 2.5% relative errors were subtracted and the result has a relative error which is more than one hundred percent! This is sometimes known as *subtractive or catastrophic cancellation*. For the purists, we note that *catastrophic cancellation* refers to the case where the two numbers we are subtracting are themselves subject to errors, as above.

[10] This specific issue doesn't arise in the case of addition, since there the denominator doesn't have to be tiny.

There also exists the scenario of *benign cancellation*, which shows up when you subtract quantities that are exactly known (though that is rarely the case in practice) or when the result of the subtraction does not need to be too accurate for what follows. The distinction between catastrophic and benign cancellation is further explained in problems 2.6 and 2.7 (including the classic example of a simple quadratic equation) and in section 2.4 below.

2.2.2.3 Multiplication or Division

We are faced with two real numbers a and b, and wish to take their product:

$$x = ab \tag{2.27}$$

As usual, we don't know the exact values, but only the approximate values $\tilde{a}$ and $\tilde{b}$, so what we form instead is the product of these:

$$\tilde{x} = \tilde{a}\tilde{b} \tag{2.28}$$

With some foresight, we will now apply Eq. (2.13) twice:

$$\tilde{a} = a(1 + \delta a), \qquad \tilde{b} = b(1 + \delta b) \tag{2.29}$$

Plugging the last four equations into the definition of the relative error, Eq. (2.12), we have:

$$\delta x = \frac{\tilde{x} - x}{x} = \frac{\tilde{a}\tilde{b} - ab}{ab} = \frac{a(1 + \delta a)b(1 + \delta b) - ab}{ab} = 1 + \delta a + \delta b + \delta a \delta b - 1 = \delta a + \delta b \tag{2.30}$$

In the fourth equality we cancelled the ab. In the fifth equality we cancelled the unit and we dropped $\delta a \delta b$ since this is a higher-order term (it is the product of two small terms).

We now recall that we are interested in finding relations between error bounds. Thus, we take the absolute value and then use the triangle inequality to find:

$$|\delta x| \leq |\delta a| + |\delta b| \tag{2.31}$$

In problem 2.1 you will carry out the analogous derivation for the case of division of the two numbers a and b, finding exactly the same result. Thus, our new conclusion is that *in multiplication and division adding together the bounds for the relative errors in the two numbers gives us the bound for the relative error in the result.* Observe that for addition or subtraction we were summing *absolute* error bounds, whereas here we are summing *relative* error bounds. Thus, typically multiplication and division don't cause too much trouble, whereas addition and (especially) subtraction can cause headaches.

Let's look at an example. Assume that we have the same numbers as in Eq. (2.25):

$$|1.25 - a| \leq 0.03, \qquad |1.20 - b| \leq 0.03 \tag{2.32}$$

that is, a relative error $\delta a \approx 0.03/1.25 = 0.024$ of roughly 2.4% and a relative error $\delta b \approx 0.03/1.20 = 0.025$ of roughly 2.5%. Our finding in Eq. (2.31) implies that:

$$\frac{|1.5 - x|}{|x|} \leq 0.049 \tag{2.33}$$

namely a relative error bound of roughly 4.9%. It's easy to see that this error bound, while larger than either of the relative error bounds we started with, is nowhere near as dramatic as what we found when we subtracted the same two numbers.

2.2.2.4 General Error Propagation: One Variable

We have studied (maximal) error propagation for a couple of simple cases of combining two numbers (subtraction and multiplication). But what about the error when you do something more complicated to a single number, e.g., take its square root or its logarithm?

Let us go over some notation. As per Eq. (2.5), the absolute error in a variable x is:

$$\Delta x = \tilde{x} - x \tag{2.34}$$

We now turn to a more involved quantity, namely $y = f(x)$. We wish to calculate:

$$\Delta y = \tilde{y} - y = f(\tilde{x}) - f(x) \tag{2.35}$$

What we'll do is to Taylor expand $f(\tilde{x})$ around x. This gives us:

$$\Delta y = f(x + \Delta x) - f(x) = f(x) + \frac{df(x)}{dx}\Delta x + \frac{1}{2}\frac{d^2 f(x)}{dx^2}(\Delta x)^2 + \cdots - f(x) \approx \frac{df(x)}{dx}\Delta x \tag{2.36}$$

In the third step we cancelled the $f(x)$ and disregarded the $(\Delta x)^2$ term and higher-order contributions: assuming Δx is small, this is legitimate. We have thereby shown that:

$$\Delta y \approx \frac{df(x)}{dx}\Delta x \tag{2.37}$$

In other words, the absolute *condition number* is $df(x)/dx$: this determines how strongly the absolute error in x will affect the absolute error in y. If you were faced with, say, $y = x^3$, you could estimate the absolute error in y by taking a derivative: $\Delta y \approx 3x^2\Delta x$. Obviously, when x is large the absolute error Δx gets amplified due to the presence of $3x^2$ in front.

A formal aside: let's define the *forward error* to be $\tilde{f}(x) - f(x)$; here x is the exact problem, $f(x)$ is the exact solution to the exact problem, and $\tilde{f}(x)$ is an approximate solution to the exact problem. Then, to take $f(\tilde{x}) = \tilde{f}(x)$ $(= \tilde{y})$ is to search for the approximate problem $\tilde{x}$ for which the exact solution $(f(\tilde{x}))$ is equal to the approximate solution to the exact problem $(\tilde{f}(x))$. In that case, $\tilde{x} - x$ is known as the *backward error* and bounding it is known as *backward error analysis*. As an example, let's tackle $f(x) = \sqrt{x}$ at $x = 3$, implying $y = \sqrt{3}$. Take $\tilde{y} = \widetilde{\sqrt{3}} = 1.8$ to be the result of our own (poor) approximation to the square-root function. The forward error is $\tilde{f}(x) - f(x) = \widetilde{\sqrt{3}} - \sqrt{3} \approx 0.068$; we notice that $f(\tilde{x}) = \sqrt{3.24} = 1.8 = \widetilde{\sqrt{3}} = \tilde{f}(x)$, so the backward error is $\tilde{x} - x = 0.24$.

It is straightforward to use Eq. (2.37) to get an estimate for the relative error in y:

$$\delta y = \frac{\Delta y}{y} \approx \frac{1}{f(x)}\frac{df(x)}{dx}\Delta x \tag{2.38}$$

If we now multiply and divide by x (assuming, of course, that $x \neq 0$) we find:

$$\delta y \approx \frac{x}{f(x)} \frac{df(x)}{dx} \delta x \tag{2.39}$$

which is our desired relation connecting δy with δx. Analogously to what we saw for the case of the absolute error, our finding here shows us that the coefficient in front of δx determines how strongly the relative error in x will affect the relative error in y. For example, if you were faced with $y = x^4$, then you could estimate the relative error in y as follows:

$$\delta y \approx \frac{x}{x^4} 4x^3 \delta x = 4\delta x \tag{2.40}$$

that is, the relative error in y is worse than the relative error in x, but not dramatically so.[11]

2.2.2.5 General Error Propagation: Many Variables

For future reference, we oberve that it is reasonably straightforward to generalize our approach above to the case of a function of many variables, i.e., $y = f(x_0, x_1, \ldots, x_{n-1})$: the total error Δy would then have contributions from each Δx_i and each partial derivative:

$$\Delta y \approx \sum_{i=0}^{n-1} \frac{\partial f}{\partial x_i} \Delta x_i \tag{2.41}$$

This is a general formula, which can be applied to functions of varying complexity. As a trivial check, we consider the case:

$$y = x_0 - x_1 \tag{2.42}$$

which a moment's consideration will convince you is nothing other than the subtraction of two numbers, as per Eq. (2.16). Applying our new general result to this simple case:

$$\Delta y \approx \Delta x_0 - \Delta x_1 \tag{2.43}$$

which is precisely the result we found in Eq. (2.19).

Equation (2.41) can now be used to produce a relationship for the relative error:

$$\delta y \approx \sum_{i=0}^{n-1} \frac{x_i}{f(x_0, x_1, \ldots, x_{n-1})} \frac{\partial f}{\partial x_i} \delta x_i \tag{2.44}$$

You should be able to see that this formula can be applied to almost all possible scenarios.[12] An elementary test would be to take:

$$y = x_0 x_1 \tag{2.45}$$

[11] You will apply both Eq. (2.37) and Eq. (2.39) to other functions when you solve problem 2.3.

[12] Of course, in deriving our last result we assumed that $y \neq 0$ and that $x_i \neq 0$.

which is simply the multiplication of two numbers, as per Eq. (2.27). Applying our new general result for the relative error to this simple case, we find:

$$\delta y \approx \frac{x_0}{x_0 x_1} x_1 \delta x_0 + \frac{x_1}{x_0 x_1} x_0 \delta x_1 = \delta x_0 + \delta x_1 \tag{2.46}$$

which is precisely the result in Eq. (2.30).

You will benefit from trying out more complicated cases, e.g., $y = \sqrt{x_0} + x_1^3 \log x_2$. In your resulting expression for δy, the coefficient in front of each δx_i tells you by how much (or if) the corresponding relative error is amplified.

2.3 Representing Real Numbers

Our discussion so far has been general: the source of the error, leading to an absolute or relative error or error bound, has not been specified. At this point, we will turn our attention to roundoff errors. In order to do that, we first go over the representation of real numbers on the computer and then discuss simple examples of mathematical operations. For the rest of the chapter, we will work on roundoff error alone: this will result from the representation of a number (i.e., storing that number) or the representation of an operation (e.g., carrying out a subtraction).

2.3.1 Basics

Computers use electrical circuits, which communicate using signals. The simplest such signals are *on* and *off*. These two possibilities are encoded in what is known as a *binary digit or bit*: bits can take on only two possible values, by convention 0 or 1.[13] All types of numbers are stored in binary form, i.e., as collections of 0s and 1s.

Python integers actually have unlimited precision, so we won't have to worry about them too much. In this book, we mainly deal with real numbers, so let's briefly see how those are represented on the computer. Most commonly, real numbers are stored using *floating-point representation*. This has the general form:

$$\pm \text{ mantissa} \times 10^{\text{exponent}} \tag{2.47}$$

For example, the speed of light in scientific notation is $+2.997\,924\,58 \times 10^8$ m/s.

Computers only store a finite number of bits, so cannot store exactly all possible real numbers. In other words, there are "only" finitely many exact representations/*machine numbers*.[14] These come in two varieties: normal and subnormal numbers. There are three ways of losing precision, as shown qualitatively in Fig. 2.2: *underflow* for very small numbers, *overflow* for very large numbers, and *rounding* for decimal numbers whose value falls

[13] You can contrast this to decimal digits, which can take on 10 different values, from 0 to 9.

[14] That is, finitely many decimal numbers that can be stored exactly using a floating-point representation.

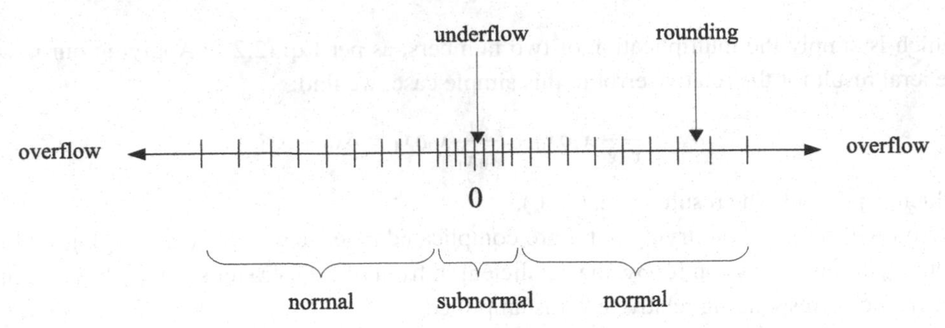

Fig. 2.2 Illustration of exactly representable floating-point numbers

between two exactly representable numbers. For more on these topics, you should look at appendix B; here we limit ourselves to simply quoting some results.

Python employs what are known as *double-precision floating point numbers*, also called *doubles*; their storage uses 64 bits in total. Doubles can represent:

$$\pm 4.9 \times 10^{-324} \leftrightarrow \pm 1.8 \times 10^{308} \tag{2.48}$$

This refers to the ability to store very large or very small numbers. Most of this ability is found in the term corresponding to the exponent. For doubles, if we try to represent a number that's larger than 1.8×10^{308} we get *overflow*. Similarly, if we try to represent a number that's smaller than 4.9×10^{-324} we get *underflow*. Keep in mind that being able to represent 4.9×10^{-324} does *not* mean that we are able to store 324 significant figures in a double. The number of significant figures (and the related concept of *precision*) is found in the coefficient in front (e.g., `1.8` or `1.234567`). For doubles, the precision is 1 part in $2^{52} \approx 2.2 \times 10^{-16}$, which amounts to 15 or 16 decimal digits.

2.3.2 Overflow

We can explore the above results programmatically. We will start from an appropriately large value, in order to shorten the number of output lines:

```
>>> large = 2.**1021
for i in range(3):
...     large *= 2
...     print(i, large)
0 4.49423283715579e+307
1 8.98846567431158e+307
2 inf
```

This is what we expected: from $2^{1024} \approx 1.7976 \times 10^{308}$ and onward we are no longer able to store the result in a double. You can check this by saying:

```
>>> 8.98846567431158e+307*1.999999999999999
1.797693134862315e+308
```

Multiplying by 2 would have led to `inf`, as above. Problem 2.4 investigates when underflow occurs.

2.3.3 Machine Precision

We already mentioned that the precision for doubles is limited to $\approx 2.2 \times 10^{-16}$. The precision is related to the distance between two vertical lines in the figure above for a given region of interest: as we noted, anything between the two lines gets rounded, either to the left or to the right. We now turn to the question of carrying out arithmetic operations using such numbers: this gives rise to the all-important question of *rounding*. This question arises every time we are trying to combine two floating-point numbers but the answer is not an exactly representable floating-point number. For example, 1 and 10 can be exactly represented as doubles, but $1/10$ cannot.

We first address an even simpler problem: five-digit decimal arithmetic. Let's assume we want to add together the two numbers 0.12345 and 1.2345. One could notice that $1.2345 = 0.12345 \times 10^1$ while $0.12345 = 0.12345 \times 10^0$, i.e., in an (imagined) system that used five-digit decimal arithmetic these two numbers would have the same mantissa and different exponents. However, that doesn't help us when adding: to add two mantissas we have to align them to use the same exponent (since that's how addition works). Adding them as real numbers (i.e., not five-digit decimal numbers) we get:

$$0.12345 + 1.2345 = 1.35795 \tag{2.49}$$

But our answer now contains six decimal digits, 1.35795. Since we're limited to five-digit decimal numbers, this leaves us with the option of *chopping* the result down to 1.3579 or *rounding* it up to 1.3580. Problems like this one also appear in other arithmetic operations and for other representational systems (like binary).

Turning back to the question of the machine representation of doubles, we try to make the concept of "precision" more specific. We define the *machine precision* ϵ_m as follows: it is the gap between the number 1.0, on the one hand, and the smallest possible number $\tilde{x}$ that is larger than 1.0 ($\tilde{x} > 1.0$), on the other hand. If you have read appendix B, you should know that (given the form of the mantissa for normal doubles) the smallest number we can represent obeying $\tilde{x} > 1.0$ is $1 + 2^{-52}$. The gap between this number and 1 is $2^{-52} \approx 2.2 \times 10^{-16}$. In other words:

$$\epsilon_m \approx 2.2 \times 10^{-16} \tag{2.50}$$

We will make use of this repeatedly in what follows.

Instead of delving into a study of binary arithmetic (analogous to what we did above for five-digit decimal arithmetic), let us investigate the definition of machine precision using

Python. We start with a small number and keep halving it, after which operation we add it on to `1.0`: at some point, this is going to give us back `1.0`: we then call the gap between `1.0` and the last number > 1 the machine precision. Explicitly:

```
>>> small = 1/2**50
>>> for i in range(3):
...         small /= 2
...         print(i, 1 + small, small)
0 1.0000000000000004 4.440892098500626e-16
1 1.0000000000000002 2.220446049250313e-16
2 1.0 1.1102230246251565e-16
```

As you can see, we started `small` at an appropriately small value, in order to shorten the number of output lines. We can further explore this topic interactively. At first sight, the results below might be confusing:

```
>>> 1. + 2.3e-16
1.0000000000000002
>>> 1. + 1.6e-16
1.0000000000000002
>>> 1. + 1.12e-16
1.0000000000000002
>>> 1. + 1.1e-16
1.0
```

We found that there exist numbers smaller than `2.2e-16` that when added to 1 lead to a result that is larger than 1. If you pay close attention, you will notice that `1. + 1.6e-16` or `1. + 1.12e-16` are rounded to the same number that `1. + 2.22e-16` corresponds to (namely, `1.0000000000000002`). However, below a certain point, we start rounding down to 1. In other words, for some values of `small` below the machine precision, we have a computed value of `1.+small` that is not 1, but corresponds to $1 + \epsilon_m$.[15]

Take a moment to appreciate that you can use a double to store a tiny number like 10^{-300}, but this doesn't mean that you can store $1 + 10^{-300}$: to do so, you would need 301 digits of precision (and all you have is 16).

2.3.4 Revisiting Subtraction

We discussed in an earlier section how bad catastrophic cancellation can be. At the time, we were investigating general errors, which could have come from several sources. Let us try to specifically investigate what happens in the case of subtraction when the *only* errors involved are those due to roundoff, i.e., let us assume that the relative error δx is a consequence of the fact that, generally speaking, x cannot be represented exactly on the

[15] Some authors call the smallest value of `small` for which `1.+small` doesn't round down to 1 the *unit roundoff* u: it is easy to see that this is related to the machine precision by $\epsilon_m = 2u$.

computer (i.e., it is typically not a machine number). Thus, when replacing x by its nearest double-precision floating-point number, we make a roundoff error: without getting into more detail, we will estimate the relative error's magnitude using the machine precision, which as we saw earlier is $\epsilon_m \approx 2.2 \times 10^{-16}$.[16] Obviously, this is only an estimate: for example, the error made when rounding to a subnormal number may be smaller (since subnormals are more closely spaced). Another example: if x is a machine number, then it can be exactly represented by $\tilde{x}$, so the relative error is actually 0. Our point in using ϵ_m is simply that one should typically not trust floating-point numbers for more than 15–16 (relative) decimal digits of precision.[17] If you think of ϵ_m as an error bound, then the fact that sometimes the error is smaller is OK.

Since we will estimate the relative error δx using ϵ_m, we see that the absolute error can be considerably larger: from Eq. (2.12) we know that $\Delta x = x\delta x$, so if x is very large then since we are taking $\epsilon_m \approx 2.2 \times 10^{-16}$ as being fixed at that value, it's obvious that Δx can become quite large. Let's take a specific example: given a relative error $\approx 10^{-16}$, a specific double of magnitude $\approx 10^{22}$ will have an absolute error in its last digit, of order 10^6.

Given that we use ϵ_m to find a general relative error in representing a real number via the use of a floating-point number, we can re-express our result for catastrophic cancellation from Eq. (2.24) as:

$$|\delta x| \leq \frac{|a|}{|a-b|} 2\epsilon_m \tag{2.51}$$

Due to $|a|/|a-b|$, even the *relative* error can be much larger than ϵ_m.

It might be worth making a distinction at this point between: (a) the loss of significant figures when subtracting two nearly equal numbers, and (b) the loss of even more digits when carrying out the subtraction using floating-point numbers (which have finite precision). Let's start from the first case. Subtract two nearly equal numbers, each of which has 20 significant figures:

$$1.2345678912345678912 - 1.2345678900000000000 = 0.0000000012345678912 \tag{2.52}$$

Note that here we subtracted real numbers (not floating-point representations) and wrote out the answer explicitly. It's easy to see that we started with 20 significant figures and ended up with 11 significant figures, even though we're dealing with real numbers/infinite precision. We now turn to the second case, which is the carrying out of this subtraction using floating-point numbers. We can use Python (doubles) to be explicit. We have:

```
>>> 1.2345678912345678912 - 1.2345678900000000000
1.234568003383174e-09
```

Comparing to the answer we had above (for real numbers), we see that we only match the first 6 (out of 11) significant figures. This is partly a result of the fact that each of our initial numbers is not represented on the computer using the full 20 significant figures, but only at most 16 digits. Explicitly:

[16] Note, however, that ϵ_m was defined in a related but slighty different context (for values around 1.0).

[17] Incidentally, though we loosely use the term significant figures here and elsewhere, it's better to keep in mind the relative error, instead, which is a more precise and base-independent measure.

```
>>> 1.2345678912345678912
1.234567891234568
>>> 1.23456789000000000
1.23456789
>>> 1.234567891234568 - 1.23456789
1.234568003383174e-09
```

This shows that we lose precision in the first number, which then has to lead to loss of precision in the result for the subtraction.

It's easy to see that things are even worse than that, though: using real numbers, the subtraction 1.234567891234568 – 1.23456789 would give us 0.000000001234568. Instead of that, we get `1.234568003383174e-09`. This is not hard to understand: a number like `1.234567891234568` typically has an absolute error in the last digit, i.e., of the order of 10^{-15}, so the result of the subtraction generally cannot be trusted beyond that absolute order (`1.234568003383174e-09` can be rewritten as `1234568.003383174e-15`). This is a result of the fact that in addition or subtraction the *absolute* error in the result comes from adding the absolute errors in the two numbers.

Just in case the conclusion is still not "clicking", let us try to discuss what's going on using relative errors. Here the exact number is $x = 0.0000000012345678912$ while the approximate value is $\tilde{x} = 0.000000001234568003383174$. Since this is one of those situations where we can actually evaluate the error, let us do so. The definition in Eq. (2.12) gives us $\delta x \approx 9.08684 \times 10^{-8}$. Again, observe that we started with two numbers with relative errors of order 10^{-16}, but subtracting them led to a relative error of order roughly 10^{-7}. Another way to look at this result is to say that we have explicitly evaluated the left-hand side of Eq. (2.51), δx. Now, let us evaluate the right-hand side of that equation. Here $a = 1.2345678912345678912$ and $a - b = 0.0000000012345678912$. The right-hand side comes out to be $|a|\, 2\epsilon_m/|a - b| \approx 2.2 \times 10^{-7}$. Thus, we find that the inequality is obeyed (of course), but the actual relative error is a factor of a few smaller than what the error bound would lead us to believe. This isn't too hard to explain: most obviously, the right-hand side of Eq. (2.51) contains a 2, stemming from the assumption that both $\tilde{a}$ and $\tilde{b}$ have a relative error of ϵ_m, but in our case b didn't change when we typed it into the Python interpreter.

You should repeat the above exercise (or something like it) for the cases of addition, multiplication, or division. It's easy to come up with examples where you add two numbers, one of which is poorly constrained, and then you get a large absolute error in the result (but still not as dramatic as in catastrophic cancellation). On the other hand, since in multiplication and division only the relative errors add up, you can convince yourself that since your starting numbers each have a relative error bound of roughly 10^{-16}, the error bound for the result is at worst twice that, which is typically not that bad.

2.3.5 Comparing Floats

Since only machine numbers can be represented exactly (other numbers are rounded to the nearest machine number), we have to be careful when comparing floating-point numbers.

We won't go into the question of how operations with floating-point numbers are actually implemented, but some examples may help explain the core issues.

Specifically, you should (almost) never compare two floating point variables $\tilde{x}$ and $\tilde{y}$ for equality: you might have an analytical expectation that the corresponding two real numbers x and y should be the same, but if the values of $\tilde{x}$ and $\tilde{y}$ are arrived at via different routes, their floating-point representations may well be different. A famous example:

```
>>> xt = 0.1 + 0.2
>>> yt = 0.3
>>> xt == yt
False
>>> xt
0.30000000000000004
>>> yt
0.3
```

The solution is to (almost) never compare two floating-point variables for equality: instead, take the absolute value of their difference, and check if that is smaller than some acceptable threshold, e.g., 10^{-10} or 10^{-12}. To apply this to our example above:

```
>>> abs(xt-yt)
5.551115123125783e-17
>>> abs(xt-yt) < 1.e-12
True
```

which behaves as one would expect.[18]

The above recipe (called an *absolute epsilon*) is fine when comparing natural-sized numbers. However, there are situations where it can lead us astray. For example:

```
>>> xt = 12345678912.345
>>> yt = 12345678912.346
>>> abs(xt-yt)
0.0010013580322265625
>>> abs(xt-yt) < 1.e-12
False
```

This makes it look like these two numbers are really different from each other, though it's plain to see that they aren't. The solution is to employ a *relative epsilon*: instead of comparing the two numbers to check whether they match up to a given small number, take into account the magnitude of the numbers themselves, thereby making the comparison relative. To do so, we first introduce a helper function:

[18] Well, almost: if you were carrying out the subtraction between 0.30000000000000004 and 0.3 using real numbers, you would expect their difference to be 0.00000000000000004. Instead, since they are floats, the answer turns out to be `5.551115123125783e-17`.

```
>>> def findmax(x,y):
...     return max(abs(x),abs(y))
```

This picks the largest magnitude, which we then use to carry out our comparison:

```
>>> xt = 12345678912.345
>>> yt = 12345678912.346
>>> abs(xt-yt)/findmax(xt,yt) < 1.e-12
True
```

Finally, note that there are situations where it's perfectly fine to compare two floats for equality, e.g., `while 1.+small != 1.:`. This specific comparison works: 1.0 is exactly representable in double-precision, so the only scenario where `1.+small` is equal to `1.0` is when the result gets rounded to `1.0`. Of course, this codicil to our rule (you can't compare two floats for equality, except when you can) in practice appears most often when comparing a variable to a literal: there's nothing wrong with saying `if xt == 10.0:` since 10.0 is a machine number and `xt` can plausibly round up or down to that value.

2.4 Rounding Errors in the Wild

Most of what we've had to say about roundoff error up to this point has focused on a single elementary mathematical operation (e.g., one subtraction or one addition). Of course, in actual applications one is faced with many more calculations (e.g., taking the square root, exponentiating), often carried out in sequence. It is often said that rounding error for the case where many iterations are involved leads to roundoff error buildup. This is not incorrect, but more often than not we are faced with one or two iterations that cause a problem, which can typically not be undone after that point. Thus, in the present section we turn to a study of more involved cases of rounding error.

2.4.1 Are Roundoff Errors Random?

At this point, many texts on computational science discuss a standard problem, that of roundoff error propagation when trying to evaluate the sum of n numbers x_0, x_1, . . ., x_{n-1}. One way to go about this is by applying our discussion from section 2.2.2.1. If each number has the same error bound ϵ, it's easy to convince yourself that since the absolute errors will add up, you will be left with a total error bound that is $n\epsilon$. This is most likely too pessimistic: it's hard to believe that the errors for n numbers would never cancel, i.e., they would all have the same sign and maximal magnitude.

What's frequently done, instead, is to assume that the errors in the different terms are independent, use the theory of random variables, and derive a result for the scaling with n of the absolute or relative error for the sum $\sum_{i=0}^{n-1} x_i$. We will address essentially the same

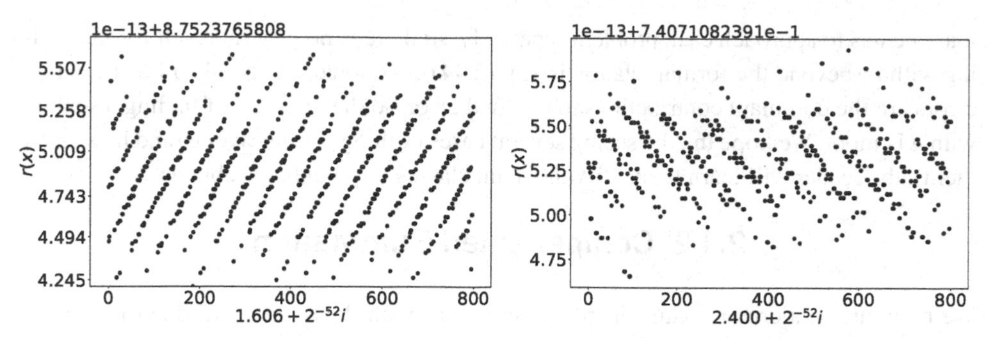

Fig. 2.3 Value of rational function discussed in the main text, at two different regions

problem when we introduce the central limit theorem in problem 6.47. There we will find that, if ϵ is the standard deviation for one number, then the standard error for the sum turns out to be $\sqrt{n}\epsilon$. For large n, it's clear that $\sqrt{n}$ grows much more slowly than n. Of course, this is comparing apples (maximal errors) with oranges (standard errors).

One has to pause, however, to think about the assumption that the errors of these x_i's are stochastically independent: many of the examples we will discuss in the rest of this chapter would have been impossible if the errors were independent. To take one case, the next contribution in a Taylor expansion is clearly correlated to the previous contribution. Rounding errors are not random: you get the same (predictable) answer every time you repeat the same computation. If you wish to study them probabilistically, you will need correlated, discrete (i.e., not continuous) random variables. These are points that, while known to experts (see, e.g., N. Higham's book [68]), are too often obscured in introductory textbooks. The confusion may arise from the fact that the roundoff error in the finite-precision *representation* of a real number may be modelled using a simple distribution; this is wholly different from the roundoff error in a *computation* involving floating-point numbers.

Perhaps it's best to consider a specific example.[19] Let's look at the rational function:

$$r(x) = \frac{4x^4 - 59x^3 + 324x^2 - 751x + 622}{x^4 - 14x^3 + 72x^2 - 151x + 112} \tag{2.53}$$

In problem 2.11 you will learn to code up polynomials using an efficient and accurate approach known as Horner's rule. In problem 2.12, you will apply what you've learned to this specific rational function. You will find what's shown in Fig. 2.3: this is clear evidence that our method of evaluating the function is sensitive to roundoff error: the typical error is $\approx 10^{-13}$. What's more, the roundoff error follows striking patterns: the left panel shows that the error is not uniformly random.[20] To highlight the fact that the pattern on the left panel is not simply a fluke, the right panel picks a different region and finds a different pattern (again, a mostly non-random one). In the aforementioned problem, you will not only reproduce these results, but also see how one could do better.

Where does this leave us? Since assuming totally independent standard errors is not warranted and assuming maximal errors is too pessimistic, how do we proceed? The answer is

[19] "When we perceive the whole at once, as in numerical computations, all agree in one judgment" (Samuel Johnson, *The History of Rasselas, Prince of Abyssinia*, chapter XXVIII).

[20] In which case we would be seeing "noise" instead of straight lines.

that one has to approach each problem separately, so there is no general result for the scaling with n (beyond the formal relation in Eq. (2.44)). As noted, often only a few rounding errors are the dominant contributions to the final error, so the question of finding a scaling with n is moot. We hope that by seeing several cases of things going wrong (and how to fix them), the reader will learn to identify the main classes of potential problems.

2.4.2 Compensated Summation

We now turn to a crucial issue regarding operations with floats; in short, due to roundoff errors, when you're dealing with floating-point numbers the associative law of algebra does not necessarily hold. You know that `0.1` added to `0.2` does not give `0.3`, but things are even worse than that: *the result of operations involving floating-point numbers may depend on the order in which these operations are carried out.* Here's a simple example:

```
>>> (0.7 + 0.1) + 0.3
1.0999999999999999
>>> 0.7 + (0.1 + 0.3)
1.1
```

Once again, operations that "should" give the same answers (i.e., that *do* give the same answer when dealing with real numbers) may not. This behavior is more than just a curiosity: it can have real-world consequences. In fact, here's an even more dramatic example:

```
>>> xt = 1.e20; yt = -1.e20; zt = 1.
>>> (xt + yt) + zt
1.0
>>> xt + (yt + zt)
0.0
```

In the first case, the two large numbers, `xt` and `yt`, cancel each other and we are left with the unit as the answer. In the second case, we face a situation similar to that in subsection 2.3.3: adding 1 to the (negative) large number `yt` simply rounds to `yt`; this is analogous to the `1. + small` we encountered on page 48, only this time we're faced with `large + 1.` and it is the unit that is dropped. Then, `xt` and `yt` cancel each other out (as before). If you're finding these examples a bit disconcerting, you are in good company.

Once you think about the problem more carefully, you might reach the conclusion that the issue that arose here is not too problematic: you were summing up numbers of wildly varying magnitudes, so you cannot trust the final answer too much. Unfortunately, as we'll see in section 2.4.4, sometimes you may not even be aware of the fact that the intermediate values in a calculation are large and of opposite signs (leading to cancellations), in which case you might not even know how much you should trust the final answer. A lesson that keeps recurring in this chapter is that you should get used to reasoning about your calculation, in contradistinction to blindly trusting whatever the computer produces.

We don't want to sound too pessimistic, so we will now see how to sum up numbers

kahansum.py

Code 2.1

```
def kahansum(xs):
    s = 0.; e = 0.
    for x in xs:
        temp = s
        y = x + e
        s = temp + y
        e = (temp - s) + y
    return s

if __name__ == '__main__':
    xs = [0.7, 0.1, 0.3]
    print(sum(xs), kahansum(xs))
```

very accurately. Our task is simply to sum up the elements of a list. In problem 2.15, we will see that often one can simply sort the numbers and then add them up starting with the smallest one. There are, however, scenarios where sorting the numbers to be summed is not only costly but goes against the task you need to carry out. Most notably, when solving initial-value problems in the study of ordinary differential equations (see chapter 8), the terms *have* to be added in the same order as that in which they are produced.

Here we will employ a nice trick, called *compensated summation* or *Kahan summation*. Qualitatively, what this does is to estimate the rounding error in each addition and then compensate for it with a correction term. More specifically, if you are adding together two numbers (a and b) and $\tilde{s}$ is your best floating-point representation for the sum, then if:

$$e = (a - \tilde{s}) + b \tag{2.54}$$

we can compute $\tilde{e}$ to get an estimate of the error $(a+b) - \tilde{s}$, namely the information that was lost when we evaluated $\tilde{s}$. While this doesn't help us when all we're doing is summing two numbers[21] it can really help when we are summing *many* numbers: add in this correction to the next term in your series, before adding that term to the partial sum.

Typically, compensated summation is more accurate when you are adding a large number of terms, but it can also be applied to the case we encountered at the start of the present section. Code 2.1 provides a Python implementation. This does what we described around Eq. (2.54): it estimates the error in the previous addition and then compensates for it. Note how our new function does not need any "length" arguments, since it simply steps through the elements of the list. We then encounter a major new syntactic feature of Python: the line `if __name__ == '__main__':` checks to see if we're running the present file as the main program (which we are). In this case, including this extra check is actually unnecessary: we could have just defined our function and then called it. We will see the importance of

[21] $\tilde{s} + \tilde{e}$ doesn't get you anywhere, since $\tilde{s}$ was already the best we could do!

this further check later, when we wish to call `kahansum()` without running the rest of the code. The output is `1.0999999999999999 1.1`. Thus, this simple function turns out to cure the problem we encountered earlier on. We don't want to spend too much time on the topic, but you should play around with compensated summation, trying to find cases where the direct sum does a poor job (here's another example: `xs = [123456789 + 0.01*i for i in range(10)]`).

Even our progress has its limitations: if you take `xs = [1.e20, 1., -1.e20]`, which is (a modified version of) the second example from the start of this section, you will see that compensated summation doesn't lead to improved accuracy. In general, if:

$$\sum_i |x_i| \gg \left| \sum_i x_i \right| \tag{2.55}$$

then compensated summation is not guaranteed to give a small relative error. On a different note, Kahan summation requires more computations than regular summation: this performance penalty won't matter to us in this book, but it may matter in real-world applications.

2.4.3 Naive vs Manipulated Expressions

We now go over a simple example showing how easy it is to lose accuracy if one is not careful; at the same time, we will see how straightforward it is to carry out an analytical manipulation that avoids the problem. The task at hand is to evaluate the function:

$$f(x) = \frac{1}{\sqrt{x^2+1} - x} \tag{2.56}$$

for large values of x. A Python implementation using list comprehensions is given in code 2.2. The output of running this code is:

```
10000 19999.99977764674
100000 200000.22333140278
1000000 1999984.77112922
10000000 19884107.85185185
```

The answer appears to be getting increasingly worse as the x is increased. Well, maybe: this all depends on what we expect the correct answer to be. On the other hand, the code/expression we are using is certainly not robust, as you can easily see by running the test case of $x = 10^8$. In Python this leads to a `ZeroDivisionError` since the terms in the denominator are evaluated as being equal. This is happening because for large values of x, we know that $x^2 + 1 \approx x^2$. We need to evaluate the square root very accurately if we want to be able to subtract a nearly equal number from it.

An easy way to avoid this problem consists of rewriting the starting expression:

$$f(x) = \frac{1}{\sqrt{x^2+1} - x} = \frac{\sqrt{x^2+1} + x}{(\sqrt{x^2+1} - x)(\sqrt{x^2+1} + x)} = \frac{\sqrt{x^2+1} + x}{x^2 + 1 - x^2} = \sqrt{x^2+1} + x \tag{2.57}$$

naiveval.py

Code 2.2

```
from math import sqrt

def naiveval(x):
        return 1/(sqrt(x**2 + 1) - x)


xs = [10**i for i in range(4,8)]
ys = [naiveval(x) for x in xs]
for x, y in zip(xs, ys):
        print(x, y)
```

In the second equality we multiplied numerator and denominator by the same expression. In the third equality we used a well-known identity in the denominator. In the fourth equality we cancelled terms in the denominator. Notice that this expression no longer requires a subtraction. If you implement the new expression, you will get the output:

```
10000 20000.000050000002
100000 200000.00000499998
1000000 2000000.0000005001
10000000 20000000.000000052
```

The errors now behave much better: for $x \gg 1$ we have $x^2 + 1 \approx x^2$, so we are essentially printing out $2x$. There are several other cases where a simple rewriting of the initial expression can avoid bad numerical accuracy issues (often by avoiding a subtraction).

2.4.4 Computing the Exponential Function

We now turn to an example where several calculations are carried out in sequence. Thus, if something goes wrong we must carefully sift through intermediate results to see what went wrong (and when). We focus on the task of computing the exponential function (assuming we have no access to a math library), by using the Taylor/Maclaurin series:

$$e^x = 1 + x + \frac{x^2}{2!} + \frac{x^3}{3!} + \cdots \tag{2.58}$$

We're clearly not going to sum infinitely many terms, so we approximate this expansion by keeping only the terms up to n_{max}:

$$e^x \approx \sum_{n=0}^{n_{max}} \frac{x^n}{n!} \tag{2.59}$$

A naive implementation of this algorithm would divide x raised to increasingly large powers by increasingly large factorials, summing the result of such divisions up to a specified point. This approach suffers from (at least) two problems: (a) calculating the power and the factorial is costly,[22] and (b) both x^n and $n!$ can become very large numbers (potentially overflowing) even though their ratio can be quite small.

Instead of calculating the power and factorial (only to divide them away), we take advantage of the fact that the n-th term in the expansion can be related to the $(n-1)$-th term:

$$\frac{x^n}{n!} = \frac{x}{n}\frac{x^{n-1}}{(n-1)!} \tag{2.60}$$

Thus, we can get a new term by multiplying the old term by x/n. This leads to a straightforward implementation that obviates the calculation of powers and factorials. Incidentally, it is easy to see that the magnitude of the terms grows, if $x > n$ holds; we will later examine the consequences of this fact.

Before turning to an implementation in Python, let us think about when to terminate our summation loop. There are generally two possibilities: (a) either we test for when the new term is "small", or (b) we test for when the running total has reached a desirable value. At first sight, it is difficult to implement (b), since we don't know the correct answer for the sum, so we turn to (a): this in its turn can be accomplished in (at least) two ways. First, we could terminate when the n-th term is a small fraction of the running total (say, less than 10^{-8}). This, however, seems needlessly restrictive, bringing us to: second, we could simply terminate when the n-th term underflows to zero. A moment's reflection, however, brings us back to point (b): at the end of the calculation, we're not really interested in the n-th term, but in the total sum. Thus, a better approach is to terminate the loop when it is determined that adding the n-th term to the running total doesn't change the sum.[23]

The above ideas are straightforwardly implemented in Python in code 2.3. Clearly, the test of terminating the loop when the latest term doesn't change the answer takes the form `while newsum != oldsum:` and the production of the latest term using Eq. (2.60) is given by `term *= x/n`. We also print a counter, the running sum, and the latest term; you can comment this line out later on. The loop in the main program calls our function for three different values of x, which we now discuss one at a time. The output for $x = 0.1$ is:

```
x, library exp(x): 0.1 1.1051709180756477
1 1.1 0.1
2 1.105 0.005000000000000001
3 1.1051666666666666 0.0001666666666666667
...
8 1.1051709180756446 2.480158730158731e-13
9 1.1051709180756473 2.75573192239859e-15
10 1.1051709180756473 2.75573192239859e-17
```

where we suppressed part of the output (as we will continue to do below). Note that if we limit ourselves to $n_{max} = 2$ then the answer we get is `1.105`. Using the language

[22] Not to mention that the latter would have to be coded up separately, if we're not using a math library.

[23] Take some time to understand this: we care about whether the latest term changes the answer or not, regardless of whether or not the last term on its own underflows to zero.

compexp.py

Code 2.3

```
from math import exp

def compexp(x):
    n = 0
    oldsum, newsum, term = 0., 1., 1.
    while newsum != oldsum:
        oldsum = newsum
        n += 1
        term *= x/n
        newsum += term
        print(n, newsum, term)
    return newsum

for x in (0.1, 20., -20.):
    print("x, library exp(x):", x, exp(x))
    val = compexp(x)
```

introduced around Eq. (2.4), this result suffers only from *approximation error*, not from *roundoff error* (and not from any roundoff error buildup): even if we were using real numbers (of infinite precision) to do the calculation for $n_{max} = 2$, we would have found 1.105.

In this case, since x is small, we observe that the value of `term` is decreasing with each new iteration. As advertised, the loop terminates when the value of `term` is small enough that it doesn't change the value of `newsum`. Comparing our final answer for $e^{0.1}$ with that provided by the `math` module's `exp()` function, we find agreement in 16 decimal digits, which is all one can hope for when dealing with double-precision floating-point numbers. We note, finally, that the convergence was achieved after only 10 steps.

We next turn to the output for $x = 20$:

```
x, library exp(x): 20.0 485165195.4097903
1 21.0 20.0
2 221.0 200.0
3 1554.3333333333335 1333.3333333333335
...
18 185052654.63711208 40944813.9157307
19 228152458.75893387 43099804.12182178
20 271252262.88075566 43099804.12182178
21 312299695.3777288 41047432.49697313
...
66 485165195.4097904 1.3555187344975148e-07
```

```
67 485165195.40979046 4.046324580589596e-08
68 485165195.40979046 1.1900954648792928e-08
```

We immediately observe that this case required considerably more iterations to reach convergence: here the final value of the sum is $\approx 5 \times 10^8$ and the smallest `term` is $\approx 1 \times 10^{-8}$.[24] In both cases, there are 16 or 17 orders of magnitude separating the final answer for the sum from the smallest `term`. Observe that the magnitude of `term` here first increases until a maximum, at which point it starts decreasing. We know from our discussion of Eq. (2.60) that the terms grow as long as $x > n$; since $x = 20$ here, we see that $n = 20$ is the turning point after which the terms start decreasing in magnitude.

Comparing our final answer for e^{20} with that provided by the `math` module's `exp()` function, we find agreement in 15 decimal digits. This is certainly not disappointing, given that we are dealing with doubles. We observe, for now, that the error in the final answer stems from the last digit in the 20th term; we will further elucidate this statement below.

Up to this point, we've seen very good agreement between the library function and our numerical sum of the Taylor expansion, for both small and large values of x. We now turn to the case of negative x. The output for $x = -20$ is:

```
x, library exp(x): -20.0 2.061153622438558e-09
1 -19.0 -20.0
2 181.0 200.0
3 -1152.3333333333335 -1333.3333333333335
...
18 21277210.34254431 40944813.9157307
19 -21822593.779277474 -43099804.12182178
20 21277210.34254431 43099804.12182178
21 -19770222.154428817 -41047432.49697313
...
93 6.147561828914624e-09 -8.56133790667976e-24
94 6.147561828914626e-09 1.8215612567403748e-24
95 6.147561828914626e-09 -3.8348658036639467e-25
```

In this case, too, we observe that the code required considerably more iterations to reach convergence (even more than what was needed for $x = 20$). The final answer for the sum here is much smaller than before ($\approx 6 \times 10^{-9}$), so we need to wait until `term` becomes roughly 16 orders of magnitude smaller than that ($\approx 4 \times 10^{-25}$). Again, observe that the magnitude of `term` here first increases until a maximum, at which point it starts decreasing. Just like in the previous case, the magnitude of `term` stops increasing after $n = 20$. As a matter of fact, the absolute value of every single `term` here is the same as it was for $x = 20$, as can be seen here by comparing the output for lines 1–3 and 18–21 to our earlier output.

Comparing our final answer for e^{-20} with that provided by the `math` module's `exp()`

[24] Compare with the case of $x = 0.1$, where the sum was of order 1 and the smallest `term` was $\approx 2 \times 10^{-17}$.

function, we find (the right order of magnitude, but) absolutely no decimal digits agreeing! In other words, for $x = -20$ our sum of the Taylor series is totally wrong.

Let us try to figure out what went wrong. We observe that the magnitude of `newsum` in the first several iterations is clearly smaller than the magnitude of `term`: this means that in addition to the absolute value of `term` growing, it's growing faster than the absolute value of `newsum`. This difference in the speed of growth becomes more dramatic in the 20th iteration, where `term` is more than two times bigger than the absolute value of `newsum`. What is the difference between the present case and that of $x = 20$? Clearly, it's related to the fact that the sign of `term` here oscillates from iteration to iteration. This is a result of the fact that $(-20)^n = -20^n$ for odd-n (take Eq. (2.59) and set $x = -20$). Since the `terms` have alternating signs, they cancel each other and at some point `newsum` (which itself also oscillated in sign for a while) starts to get smaller and smaller.

We've now seen why the negative x case is different: there is cancellation between numbers of comparable magnitude. We can do even better than this handwaving explanation, though. Each `term` is accurate to at most 16 decimal digits (since it's a double-precision floating-point number). Thus, the largest-magnitude `term` has an absolute error in its last digit, of order 10^{-8}. (Another way to say this is that every double has a relative error of roughly 10^{-16} and since this specific double has magnitude 10^8 it has an absolute error of roughly 10^{-8}.) Note that we are not here talking about the smallest `term` (adding which leaves `newsum` unchanged) but about the *largest* `term` (which has the largest error of all terms): as we keep adding more terms to `newsum`, this largest error is not reduced but is actually propagated over! Actually, things are even worse than that: the final answer for `newsum` has magnitude $\approx 6 \times 10^{-9}$ and is therefore even smaller than the error of $\approx 10^{-8}$ that we've been carrying along. Thus, the final answer has no correct significant digits! In the case of positive x (namely the $x = 20$ we studied above), the largest `term` also had an error of $\approx 10^{-8}$, but since in that case there were no cancellations, the final value of the sum was $\approx 5 \times 10^8$, leading to no error for the first 15 decimal digits.

We thus see that our algorithm for calculating e^x for negative x (via the Taylor expansion) is unstable because it introduces cancellation. As a matter of fact, in this specific case the cancellation was needless: we could have just as easily taken advantage of $e^x = 1/e^{-x}$ and then carried out the sum for a positive value of x (which would not have suffered from cancellation issues) and then proceeded to invert the answer at the end. In our example, we can estimate e^{-20} by summing the Taylor series for e^{20} and then inverting the answer:

```
>>> 1/485165195.40979046
2.061153622438557e-09
```

We have 15 digits of agreement! This shouldn't come as a surprise: the Taylor expansion for positive x doesn't contain any cancellations and leads to 15–16 correct significant digits. After that, dividing 1 by a large number with 15–16 correct significant digits leads to an answer with 15–16 correct significant digits. Of course, this quick fix does not necessarily apply to all Taylor expansions (e.g., the $\sin x$ of problem 2.14): some of them have the alternating signs built-in, though analytical manipulations can help there, too. Finally, the problem set addresses numerical misbehavior even when there's no subtraction in sight.

Code 2.4 recforw.py

```
from math import exp

def forward(nmax=22):
    oldint = 1 - exp(-1)
    for n in range(1,nmax):
        print(n-1, oldint)
        newint = n*oldint - exp(-1)
        oldint = newint

print("n = 20 answer is 0.0183504676972562")
print("n, f[n]")
forward()
```

2.4.5 An Even Worse Case: Recursion

In the previous subsection we examined a case where adding together several correlated numbers leads to an error (in the largest one) being propagated unchanged to the final answer, thereby making it inaccurate. In this subsection, we will examine the problem of recurrence/recursion, where even a tiny error in the starting expression can be multiplied by the factorial of a large number, thereby causing major headaches.

Our task is to evaluate integrals of the form:

$$f_n = \int_0^1 x^n e^{-x} dx \tag{2.61}$$

for different (integer) values of n, i.e., for $n = 0, 1, 2, \ldots$. To see how this becomes a recursive problem, we analytically evaluate the indefinite integral for the first few values of n:

$$\begin{aligned} &\int e^{-x} dx = -e^{-x}, && \int x e^{-x} dx = -e^{-x}(x+1), \\ &\int x^2 e^{-x} dx = -e^{-x}(x^2+2x+2), && \int x^3 e^{-x} dx = -e^{-x}(x^3+3x^2+6x+6) \end{aligned} \tag{2.62}$$

where we used integration by parts. This leads us to a way of relating the indefinite integral for the n-th power to the indefinite integral for the $(n-1)$-th power:

$$\int x^n e^{-x} dx = n \int x^{n-1} e^{-x} dx - x^n e^{-x} \tag{2.63}$$

It is now trivial to use this result in order to arrive at a recursive expression for the definite integral f_n in Eq. (2.61):

$$f_n = n f_{n-1} - e^{-1} \tag{2.64}$$

This works for $n = 1, 2, 3, \ldots$ and we already have the result $f_0 = 1 - e^{-1}$.

We code this up in the most straightforward way possible in code 2.4.[25] This clearly shows that we only need to keep track of two numbers at any point in time: the previous one and the current one. We start at the known result $f_0 = 1 - e^{-1}$ and then simply step through Eq. (2.64); note that our function is simply printing things out, not returning a final value. We start at $n = 0$ and the last value we print out is for $n = 20$; we're also providing the correct answer, arrived at via other means. We immediately notice the difference between the algorithm coded up in the previous subsection (where we were simply adding in an extra number) and what's going on here: even if we ignore any possible subtractive cancellation, Eq. (2.64) contains nf_{n-1} which means that any error in determining f_{n-1} is *multiplied* by n when producing the next number. If n is large, that can be a big problem. Since the expression we're dealing with is recursive/recurrent, even if we start with a tiny error, this is compounded by being multiplied by n every time through the loop. We get:

```
n = 20 answer is 0.0183504676972562
n, f[n]
0 0.6321205588285577
1 0.26424111765711533
2 0.16060279414278833
...
16 0.022201910404060943
17 0.009553035697593693
18 -0.19592479861475587
19 -4.090450614851804
20 82.17689173820752
```

Clearly, that escalated fast. Even though for the first 15 or so terms we see a gradual decline in magnitude, after that the pace picks up pretty fast. We are dealing with a numerical instability, which ends up giving us garbage for f_{20}. It's not hard to see why: since f_0 is stored as a double-precision floating-point number, it has an absolute error in its last digit, of order 10^{-16}. Focusing only on the nf_{n-1} term in Eq. (2.64), we see that by the time we get up to $n = 20$, our 10^{-16} error will have been multiplied by 20!: given that $20! \times 10^{-16} \approx 243$, this completely overwhelms our expected answer of ≈ 0.018.

The process we followed in the code above, starting at $n = 0$ and building up to a finite n, is called *forward recursion*. We will eliminate our headaches by a simple trick, namely the use of *backward recursion*: solve Eq. (2.64) for f_{n-1} in terms of f_n:

$$f_{n-1} = \frac{f_n + e^{-1}}{n} \tag{2.65}$$

The way to implement this new equation is to start at some large value of n, say $n = 30$, and then step *down* one integer at a time. We immediately realize that we don't actually know the value of f_{30}. However, the algorithm implied by Eq. (2.65) is much better behaved than what we were dealing with before: even if we start with a bad estimate of f_{30}, say with an error of 0.01, the error will be *reduced* with each iteration, since we are now

[25] We implement the recurrence ("recursive") relation Eq. (2.64), *without* a recursive Python function.

Code 2.5 recback.py

```
from math import exp

def backward(nmax=31):
        oldint = 0.01
        for n in reversed(range(20,nmax)):
                print(n, oldint)
                newint = (oldint + exp(-1))/n
                oldint = newint

print("n = 20 answer is 0.0183504676972562")
print("n, f[n]")
backward()
```

dividing by n. Thus, by the time we get down to $n = 20$, the error will have turned into $0.01/(30 \times 29 \times 28 \times \cdots \times 22 \times 21) \approx 9 \times 10^{-17} \approx 10^{-16}$, which happens to be quite good. Code 2.5 shows an implementation of backward recursion; running this, we get:

```
n = 20 answer is 0.0183504676972562
n, f[n]
30 0.01
29 0.012595981372381411
28 0.013119842156683579
...
22 0.016688929189184395
21 0.017480380470937 58
20 0.018350467697256186
```

We have agreement in the first 14 significant figures, which means that the fifteenth significant figure is off. This is a result of the aforementioned absolute error of $\approx 10^{-16}$; notice how the first significant figure is of order 10^{-2}.

2.4.6 When Rounding Errors Cancel

We now turn to a slightly more complicated example, which shows how function evaluations can be rather counterintuitive. The moral to be drawn from our discussion is that there is no substitute for thinking. As a matter of fact, this is a case where an approximation that seems to be bad at first sight ends up performing much better than we had expected.

Our goal is to examine the behavior of the function $f(x) = (e^x - 1)/x$ at small (or perhaps intermediate) x. We start from coding this up in Python in the obvious (naive) way, see the function `f()` in code 2.6. The output of running this code is:

```
1e-14 0.9992007221626409 1.000000000000005
1e-15 1.1102230246251565 1.0000000000000004
1e-16 0.0 1.0
-1e-15 0.9992007221626408 0.9999999999999994
-1e-16 1.1102230246251565 1.0
-1e-17 -0.0 1.0
```

Ignore the function `g()` and the last number in each row, for now. We see that for small x (whether negative or positive) the naive function gives not-very-accurate results, until at some point for even smaller x the answer is absolutely wrong.[26] It's obvious that our code is plagued by catastrophic cancellation.

One way to go about improving the solution would be to use the Maclaurin series for e^x:

$$\frac{e^x - 1}{x} \approx 1 + \frac{x}{2!} + \frac{x^2}{3!} + \cdots \tag{2.66}$$

The problem with this is that to get a desired accuracy we need to keep many terms (and the number of terms depends on the desired accuracy, see section 2.4.4). The trick in the function `g()`, instead, nicely makes use of standard rounding properties: it compares a float to a float-literal for equality. This is perfectly fine, as the relevant lines of code are there precisely to catch the cases where `w` is rounded to 1 or to 0. In the former case (e^x rounding to 1), it returns the analytical answer by construction. In the latter case (e^x rounding to 0, when x is large and negative), it plugs in the value for the rest of the expression. That leaves us with the crucial expression `(w-1)/log(w)`, which is equivalent to `(exp(x)-1)/log(exp(x))`, for all other cases.

Let us now look at the last number printed on each row, which corresponds to `g()`. It's obvious that the revised version is much better than the naive one. The reason this new function works so well is because it makes the exponential appear in both the numerator and the denominator: `(exp(x)-1)/log(exp(x))`. As a result, the roundoff error in the evaluation of the exponential `exp(x)` in the numerator is cancelled by the presence of the same roundoff error in the exponential `exp(x)` in the denominator.

This result is sufficiently striking that it should be repeated: our updated function works better because it plays off the error in `exp(x)` in the numerator, against the same error in `exp(x)` in the denominator. Let's study the case of $x = 9 \times 10^{-16}$ in detail. The algorithm in `f()` does the following calculation:

$$\frac{\texttt{e}^{\texttt{x}} - \texttt{1}}{\texttt{x}} = \frac{\texttt{8.881784197001252e} - \texttt{16}}{\texttt{9e} - \texttt{16}} = \texttt{0.9868649107779169} \tag{2.67}$$

where we are also showing the intermediate results that Python produces. Similarly, the algorithm in `g()` does the following calculation:

$$\frac{\texttt{e}^{\texttt{x}} - \texttt{1}}{\texttt{log}(\texttt{e}^{\texttt{x}})} = \frac{\texttt{8.881784197001252e} - \texttt{16}}{\texttt{8.881784197001248e} - \texttt{16}} = \texttt{1.0000000000000004} \tag{2.68}$$

Given the earlier outputs, you should not be surprised that the final answer is more accurate.

[26] In the limit $x \to 0$ we know the answer must be 1, from L'Hôpital's rule.

Code 2.6 cancel.py

```
from math import exp, log

def f(x):
    return (exp(x) - 1)/x

def g(x):
    w = exp(x)
    if w==0.:
        val = -1/x
    elif w==1.:
        val = 1.
    else:
        val = (w-1)/log(w)
    return val

xs = [10**(-i) for i in (14, 15, 16)]
xs += [-10**(-i) for i in (15, 16, 17)]
fvals = [f(x) for x in xs]
gvals = [g(x) for x in xs]
for x, fval, gval in zip(xs, fvals, gvals):
    print(x, fval, gval)
```

But look at the same calculation as that in `g()`, this time carried out using real numbers:

$$\frac{e^x - 1}{\log(e^x)} = \frac{9.000000000000004050\ldots \times 10^{-16}}{9.000000000000000000\ldots \times 10^{-16}} = 1.000000000000000450\ldots \quad (2.69)$$

The algorithm in `g()`, when employing floating-point numbers, produces an inaccurate numerator (`e`$^{\texttt{x}}$ `- 1`) and an inaccurate denominator (`log(e`$^{\texttt{x}}$`)`) that are divided to produce an accurate ratio. There is a lesson to learn here: we usually only look at intermediate results when something goes wrong. In the case under study, we found that the intermediate results are actually bad, but end up leading to desired behavior in the end.

The reasonably simple example we've been discussing here provides us with the opportunity to note a general lesson regarding numerical accuracy: it is quite common that mathematically elegant expressions, like `f()`, are numerically unreliable. On the other hand, numerically more accurate approaches, like `g()`, are often messier, containing several (ugly) cases. There's often no way around that.

2.5 Project: the Multipole Expansion in Electromagnetism

Our physics project involves a configuration of several point charges and the resulting electrostatic potential; we will then introduce what is known as the *multipole expansion*, whereby the complications in a problem are abstracted away and a handful of numbers are sufficient to give a good approximation to physical properties. We will be applying this to the Coulomb potential but, unsurprisingly, the same approach has also been very fruitful in the context of the Newton potential, i.e., the gravitational field of planets and stars.

Since this chapter has focused on numerics, so will our project: we will encounter series convergence as well as recurrence relations. We use as our starting point the electric-field-visualization machinery that we introduced in section 1.7. In our discussion of the multipole expansion we will introduce what are known as *Legendre polynomials*. In addition to helping us simplify the study of electrostatics in what follows, these will also make appearances in later chapters.

2.5.1 Potential of a Distribution of Point Charges

2.5.1.1 General Case

In the project at the end of the previous chapter we focused on the electric field, which is a vector quantity. In most of the present section, we will, instead, be studying the electrostatic potential. As you may recall from a course on electromagnetism, since the curl of the electric field $\mathbf{E}$ is zero, there exists a scalar function whose gradient is the electric field:

$$\mathbf{E}(\mathbf{r}) = -\nabla\phi(\mathbf{r}) \tag{2.70}$$

where $\phi(\mathbf{r})$ is called the *electrostatic potential.* In general there would also be a second term on the right-hand side, involving the vector potential $\mathbf{A}$, but here we are focusing only on the problem of static point charges.

The electrostatic potential at point P (located at $\mathbf{r}$) due to the point charge q_0 (which is located at $\mathbf{r}_0$) is simply:

$$\phi_0(\mathbf{r}) = k\frac{q_0}{|\mathbf{r} - \mathbf{r}_0|} \tag{2.71}$$

where, as before, Coulomb's constant is $k = 1/(4\pi\epsilon_0)$ in SI units (and ϵ_0 is the permittivity of free space). As you can verify by direct differentiation, this leads to the electric field in Eq. (1.2).[27] If we are faced with more than one point charge, we could apply the *principle of superposition* also to the electrostatic potential, similarly to what we did for the electric-field vector in Eq. (1.3). As a result, when dealing with the n point charges $q_0, q_1, \ldots, q_{n-1}$ located at $\mathbf{r}_0, \mathbf{r}_1, \ldots, \mathbf{r}_{n-1}$ (respectively), namely the configuration shown in Fig. 1.5, the total electrostatic potential at the location $\mathbf{r}$ is:

[27] We could have included an arbitrary offset here, but we haven't.

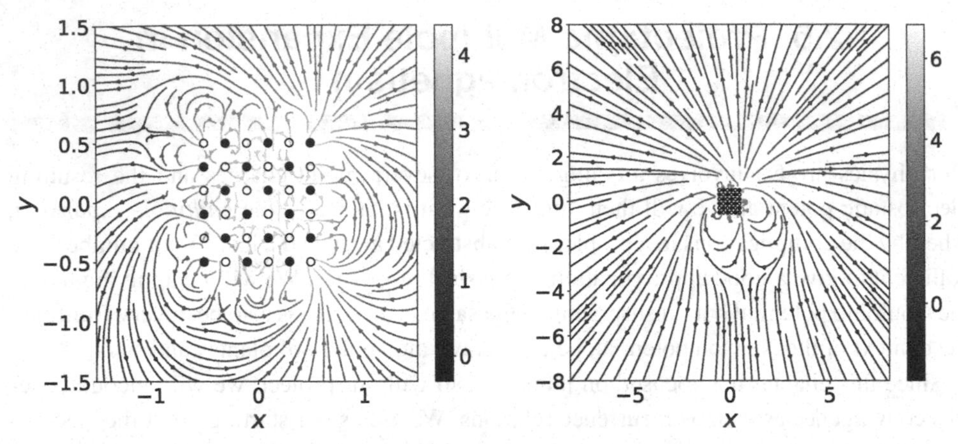

Fig. 2.4 Array of 36 charges of alternating sign and varying magnitudes

$$\phi(\mathbf{r}) = \sum_{i=0}^{n-1} \phi_i(\mathbf{r}) = \sum_{i=0}^{n-1} k \frac{q_i}{|\mathbf{r} - \mathbf{r}_i|} \tag{2.72}$$

i.e., a sum of the individual potential contributions. Obviously, this $\phi(\mathbf{r})$ is a scalar quantity. It contains $|\mathbf{r} - \mathbf{r}_i|$ in the denominator: this is the celebrated *Coulomb potential.*

2.5.1.2 Example: Array of 36 Charges

For the sake of concreteness, throughout this project we will study an arbitrarily chosen specific configuration: 36 charges placed in a square from −0.5 to 0.5 in both x and y directions. To keep things interesting, we will pick the q_i charges to have varying magnitudes and alternating signs. Using our code `vectorfield.py` from the previous chapter we are led to Fig. 2.4 for this case, where this time we are using a color map (in grayscale) instead of line width to denote the field strength.[28] This array of charges, distributed across six rows and six columns, is here accompanied by the electric field, similarly to what we saw in Fig. 1.6. In the present section we are interested in the electrostatic potential, not directly in the electric field, but you should be able to go from the former to the latter using Eq. (2.70). Our new figure shows many interesting features: looking at the left panel, the direction and magnitude of the electric field are quite complicated; the field appears to be strongest in between the charges. As we move farther away from the charge array, however, as shown in the right panel, we start noticing the big picture: the field is strongest inside the array, then we notice some intermediate-scale features at negative y, and then as we get farther from the configuration of charges the electric field appears reasonably simple. We recall that positive charges act as "sources": since at large distances all the arrows point outward, we see that, abstracting away all the details, "effectively" the charge array acts

[28] Here and below x and y are measured in meters.

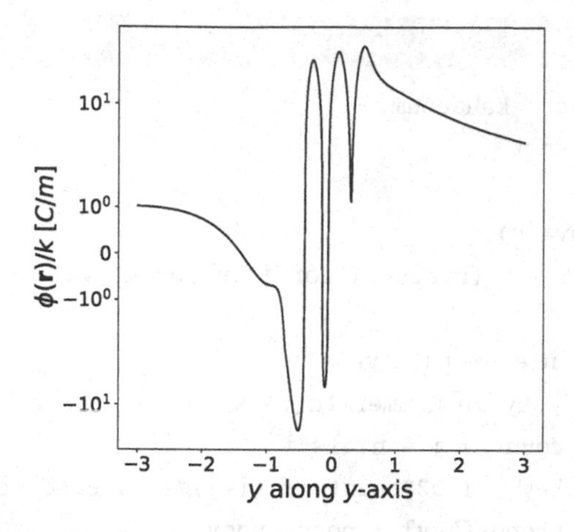

Fig. 2.5 Electrostatic potential for the 36-charge array, along the y axis

like an overall positive charge (though there may be more complicated residual effects not captured by this simple analogy).

As advertised, our main concern here is the electrostatic potential. The question then arises of how to visualize the potential: in principle, one could draw equipotential surfaces (curves), which would be orthogonal to the electric field lines of Fig. 2.4. Instead, with a view to what we will be studying in the following sections, here we opt for something simpler: we pick a specific direction on the x–y plane and study the potential along it. Since the right panel in Fig. 2.4 exhibits interesting behavior at negative y that is not present for positive y, we decide to study the y axis itself (i.e., set $x = 0$) for both positive and negative y. The result of evaluating $\phi(\mathbf{r})$ along the y axis as per Eq. (2.72) (for the given configuration of charges) is shown in Fig. 2.5, using a symmetrical-log scale.[29] The overall features are easy to grasp: we observe rapid oscillations at short distances, which is where the q_i charges are physically located, and then simpler behavior at larger distances. As we expected based on our electric-field visualization, the potential exhibits more structure (even changing sign) at intermediate distances along the negative y axis than it does along the positive y axis. Speaking of the sign, let's do a quick consistency check: at large distances along the positive y axis we see that the potential gets smaller as we increase y; since $\mathbf{E}(\mathbf{r}) = -\nabla\phi(\mathbf{r})$ we expect a positive electric field (pointing up). Similarly, along the negative y axis the potential gets larger as we increase the y magnitude at intermediate distances, so we expect the electric field, again, to be positive (i.e., point up). As hinted at in the leftmost part of our curve, as you keep increasing the y magnitude along the negative y axis the situation will actually change: at some point for $y \lesssim -3$ you will "curve down", implying a negative electric field (pointing down). All these findings are consistent with what is shown in Fig. 2.4: take a minute to inspect that figure more closely.

[29] Note that we've divided out k, so our values are given in units of C/m.

Code 2.7 chargearray.py

```
from kahansum import kahansum
from math import sqrt

def chargearray(nvals):
    vals = [-0.5 + i/(nvals-1) for i in range(nvals)]
    qtopos = {}
    for i,posx in enumerate(vals):
        for j,posy in enumerate(vals):
            count = j + nvals*i + 1
            key = 1.02*count if (i+j)%2==0 else -count
            qtopos[key] = posx, posy
    return qtopos

def vecmag(rs):
    sq = [r**2 for r in rs]
    return sqrt(kahansum(sq))

def fullpot(qtopos,rs):
    potvals = []
    for q,pos in qtopos.items():
        diffs = [r - po for r,po in zip(rs,pos)]
        R = vecmag(diffs)
        potvals.append(q/R)
    return kahansum(potvals)

if __name__ == '__main__':
    qtopos = chargearray(6)
    for y in 1,-1:
        rs = [0.,y]
        potval = fullpot(qtopos,rs)
        print(rs, potval)
```

2.5.1.3 Implementation

At this point, it's probably best to get a bit more specific: the results in these two figures correspond to a given choice (not only of charge placement and sign but also) of the magnitudes of the charges involved. Thinking about how to code up our 36-charge array, we could go about this task in several different ways, e.g., using distinct lists for charges, x coordinates, and y coordinates. However, since we already employed a Python dictionary in `vectorfield.py` in order to associate charges with their positions, it stands to reason that we should now generalize this approach: instead of filling in the keys and values by hand, we should do so programmatically. Code 2.7 provides a Python implementation. Overall, we see that this code produces the point-charge distribution, introduces a helper function, calculates $\phi(\mathbf{r})$ as per Eq. (2.72), and prints out the potential value at two points along the y axis. It should be easy to see that using these functions you can produce Fig. 2.5. Let's go over each aspect of this program in more detail.

The function `chargearray()` produces the dictionary `qtopos`. You may recall that this was accomplished in one line in the code implementing the project of the previous chapter: there, we were faced with only two (or at most four) charges, so we "hard-coded" the keys and values. To do things in a more systematic way, we first remember how to populate dictionaries: in section 1.3.3 we saw that the syntax is `htow[key] = value`. Producing a grid of 36 elements (6×6) is not too difficult: we could simply pick some values (as in `vals`) and iterate over x and y, adding in a new key/value pair each time (with the keys being the q_i and the values being the $\mathbf{r}_i$). Trying to be idiomatic, we make use of `enumerate()` so we can have access both to the value of the coordinate and to its index.[30]

Our task is complicated somewhat by the requirement that the charges be alternating: if on a given row the first charge is negative, the one to its right should be positive, then the next one negative and so on. This isn't too hard to implement, either: you could simply have an index and check whether or not that is even or odd. Unfortunately, in our case we want this to go on as you move to the next row: the last charge on the first row is positive, but the one immediately under it should be negative. In other words, we need to traverse our grid from left to right, then from right to left, and so on, just like in ancient Greek inscriptions.[31] This is accomplished in our test `if (i+j)%2==0`. To keep things interesting, we picked each q_i according to when we encountered it, via `count = j + nvals*i + 1`.[32] When constructing `key`, we artificially inflated the positive charge values: as you will discover when you play around with this code, the total charge (adding together all q_i) would have been zero had we not taken this step. Using Python's ternary operator, we fit everything on one line.

We then introduce an auxiliary function, which evaluates the magnitude of a vector using a list comprehension: observe how convenient Python's `for` is here, since we don't need to specify how many dimensions our vector has. We also employ our very own `kahansum()` from earlier in this chapter: while this is probably overkill for summing just a few numbers, it's a nice opportunity to employ functionality we've already developed. We are, finally, also in a position to see why we had included the line saying `if __name__ == '__main__':` all the way back in `kahansum.py`: since we're now not running that file as the main program, the lines that followed the `if` check are not executed this time. If we hadn't taken care of this check, the older file's output would be confusingly printed out every time we ran the present `chargearray.py` file.

The final function of this program implements Eq. (2.72): we sum the $\phi_i(\mathbf{r})$ contributions one by one. Using the `items()` method of dictionaries really pays off: once again, we

[30] We even kept things general and did not hard-code the value 6, instead passing it in as an argument.

[31] This pattern is called *boustrophedon*, namely as an ox turns with the plough at the end of the furrow.

[32] We add in a unit to make sure we don't include a charge of zero magnitude.

don't have to specify ahead of time how many charges we are faced with *or* how many dimensions our vectors have. We simply have access to one q_i and one $\mathbf{r}_i$ at a time. We first form the components of $\mathbf{r} - \mathbf{r}_i$ using a list comprehension as well as Python's `zip()`. We then store each $\phi_i(\mathbf{r})$ contribution and at the end use `kahansum()` to sum them all together. The main program simply picks two arbitrary points on the y axis and prints out the value of the potential $\phi(\mathbf{r})$ at those points; you can use these values as benchmarks later in this section, after we've developed the multipole expansion. Both the values and the signs are consistent (nay, identical) with what we saw in Fig. 2.5.

2.5.2 Expansion for One Point Charge

We now turn to our stated aim, which is to approximate a complicated electrostatic potential using only a few (or several) terms. To keep things manageable, we start from the simple case of a single point charge. This allows us to introduce Legendre polynomials without getting lost in a sea of indices; in the following subsection, we will apply what we've learned to the general case of a distribution of point charges.

2.5.2.1 First Few Terms

For one point charge q_0 (located at $\mathbf{r}_0$), the electrostatic potential at point P (at position $\mathbf{r}$) is given by Eq. (2.71):

$$\phi_0(\mathbf{r}) = k\frac{q_0}{|\mathbf{r} - \mathbf{r}_0|} \tag{2.73}$$

Roughly speaking, what we'll do is to massage the denominator such that it ends up containing only $\mathbf{r}$ (or $\mathbf{r}_0$) but not the difference between the two vectors.

With this in mind, let us examine the square of the denominator:

$$(\mathbf{r} - \mathbf{r}_0)^2 = r^2 + r_0^2 - 2rr_0\cos\theta_0 = r^2\left[1 + \left(\frac{r_0}{r}\right)^2 - 2\left(\frac{r_0}{r}\right)\cos\theta_0\right] \equiv r^2[1 + \alpha] \tag{2.74}$$

You may think of the first equality as the so-called law of cosines, or as simply expanding the square and expressing the dot product of the two vectors in terms of the angle θ_0 between them. In the second equality, we pulled out a factor of r^2. In the third equality we noticed the overall structure of the expression, by defining:

$$\alpha \equiv \left(\frac{r_0}{r}\right)\left(\frac{r_0}{r} - 2\cos\theta_0\right) \tag{2.75}$$

We can now use the binomial theorem, a standard Taylor-expansion result:

$$\frac{1}{(1+x)^m} = 1 - mx + \frac{m(m+1)}{2!}x^2 - \frac{m(m+1)(m+2)}{3!}x^3 + \cdots \tag{2.76}$$

We are in a position to expand the Coulomb potential from Eq. (2.73) as follows:

$$\frac{1}{|\mathbf{r} - \mathbf{r}_0|} = \frac{1}{r}\frac{1}{(1+\alpha)^{1/2}} = \frac{1}{r}\left(1 - \frac{1}{2}\alpha + \frac{3}{8}\alpha^2 - \frac{5}{16}\alpha^3 + \frac{35}{128}\alpha^4 - \cdots\right)$$

$$
\begin{aligned}
&= \frac{1}{r}\Bigg[1 - \frac{1}{2}\left(\frac{r_0}{r}\right)\left(\frac{r_0}{r} - 2\cos\theta_0\right) + \frac{3}{8}\left(\frac{r_0}{r}\right)^2\left(\frac{r_0}{r} - 2\cos\theta_0\right)^2 \\
&\quad - \frac{5}{16}\left(\frac{r_0}{r}\right)^3\left(\frac{r_0}{r} - 2\cos\theta_0\right)^3 + \frac{35}{128}\left(\frac{r_0}{r}\right)^4\left(\frac{r_0}{r} - 2\cos\theta_0\right)^4 - \cdots\Bigg] \\
&= \frac{1}{r}\Bigg[1 + \left(\frac{r_0}{r}\right)\cos\theta_0 + \left(\frac{r_0}{r}\right)^2\frac{1}{2}\left(3\cos^2\theta_0 - 1\right) + \left(\frac{r_0}{r}\right)^3\frac{1}{2}\left(5\cos^3\theta_0 - 3\cos\theta_0\right) \\
&\quad + \left(\frac{r_0}{r}\right)^4\frac{1}{8}\left(35\cos^4\theta_0 - 30\cos^2\theta_0 + 3\right) + \cdots\Bigg]
\end{aligned}
\tag{2.77}
$$

In the first equality we took one over the square root of Eq. (2.74). In the second equality we used the binomial theorem of Eq. (2.76), asssuming that $r > r_0$. In the third equality we plugged in our definition of α from Eq. (2.75). In the fourth equality we expanded out the parentheses and grouped terms according to the power of r_0/r. We then notice that each term inside the square brackets is given by a power of r_0/r times a *polynomial* of $\cos\theta_0$. As it so happens, this identification of the coefficients is precisely the way Legendre introduced in 1782 what are now known as *Legendre polynomials*.[33] In short, we have arrived at the following remarkable result:

$$
\frac{1}{|\mathbf{r} - \mathbf{r}_0|} = \frac{1}{r}\sum_{n=0}^{\infty}\left(\frac{r_0}{r}\right)^n P_n(\cos\theta_0) \tag{2.78}
$$

where the P_n are the Legendre polynomials. Specifically, putting the last two equations together we find the first few Legendre polynomials:

$$
\begin{aligned}
&P_0(x) = 1, \qquad P_1(x) = x, \qquad P_2(x) = \frac{1}{2}(3x^2 - 1), \\
&P_3(x) = \frac{1}{2}(5x^3 - 3x), \qquad P_4(x) = \frac{1}{8}(35x^4 - 30x^2 + 3)
\end{aligned}
\tag{2.79}
$$

For odd-n the polynomial is an odd function of x and, similarly, for even-n the polynomial is an even function of x. Higher-order polynomials can be derived analogously, by employing the binomial theorem, expanding the parentheses, and grouping terms.

Since this is a chapter on numerics, it's worth noticing that what we've accomplished with Eq. (2.78) is to trade a subtraction (which sometimes leads to catastrophic cancellation) on the left-hand side, for a sum of contributions on the right-hand side. Even if the signs oscillate, we are organizing our terms hierarchically.

A further point: in the derivation that led to our main result we assumed $r > r_0$. If we had been faced with a situation where $r < r_0$, we would have carried out an expansion in powers of r/r_0, instead, and the right-hand side would look similar to Eq. (2.78) but with the r and r_0 changing roles. In our specific example, we will be studying positions $\mathbf{r}$ away from the 36-charge array, so we will always be dealing with $r > r_0$.

[33] Legendre was studying the Newton potential, three years before Coulomb published his law.

2.5.2.2 Legendre Polynomials: from the Generating Function to Recurrence Relations

While you could, in principle, generalize the power-expansion approach above to higher orders (or even a general n-th order), it's fair to say that the manipulations become unwieldy after a while. In this subsection, we will take a different approach, one that also happens to be easier to implement programmatically. Let's take our main result of Eq. (2.78) and plug in the second equality of Eq. (2.74). We find:

$$\frac{1}{\sqrt{1-2\left(\frac{r_0}{r}\right)\cos\theta_0+\left(\frac{r_0}{r}\right)^2}}=\sum_{n=0}^{\infty}\left(\frac{r_0}{r}\right)^n P_n(\cos\theta_0) \tag{2.80}$$

If we now define $u \equiv r_0/r$ and $x \equiv \cos\theta_0$, our equation becomes:

$$\frac{1}{\sqrt{1-2xu+u^2}}=\sum_{n=0}^{\infty}u^n P_n(x) \tag{2.81}$$

This is merely a reformulation of our earlier result. In short, it says that the function on the left-hand side, when expanded in powers of u, has coefficients that are the Legendre polynomials. As a result, the function on the left-hand side is known as the *generating function* of Legendre polynomials.

At this point we haven't actually gained anything from our redefinitions in terms of x and u. To see the benefit, we differentiate Eq. (2.81) with respect to u and find:

$$\frac{x-u}{(1-2xu+u^2)^{3/2}}=\sum_{n=0}^{\infty}nP_n(x)u^{n-1} \tag{2.82}$$

If we now identify the $1/\sqrt{1-2xu+u^2}$ from Eq. (2.81) and move the remaining factor of $1/(1-2xu+u^2)$ to the numerator, we get:

$$(1-2xu+u^2)\sum_{n=0}^{\infty}nP_n(x)u^{n-1}+(u-x)\sum_{n=0}^{\infty}P_n(x)u^n=0 \tag{2.83}$$

where we also moved everything to the same side. We now expand the parentheses and end up with five separate summations:

$$\sum_{n=0}^{\infty}nP_n(x)u^{n-1}-\sum_{n=0}^{\infty}2nxP_n(x)u^n+\sum_{n=0}^{\infty}nP_n(x)u^{n+1}+\sum_{n=0}^{\infty}P_n(x)u^{n+1}-\sum_{n=0}^{\infty}xP_n(x)u^n=0 \tag{2.84}$$

We are faced with a power series in u (actually a sum of five power series in u) being equal to zero, regardless of the value of u. This implies that the coefficient of each power of u (separately) is equal to zero. Thus, if we take a given power to be j (e.g., $j = 17$) our five summations above give for the coefficient of u^j the following:

$$(j+1)P_{j+1}(x)-2jxP_j(x)+(j-1)P_{j-1}(x)+P_{j-1}(x)-xP_j(x)=0 \tag{2.85}$$

which, after some trivial re-arrangements, gives us:

$$P_{j+1}(x) = \frac{(2j+1)xP_j(x) - jP_{j-1}(x)}{j+1} \tag{2.86}$$

To step through this process, we start with the known first two polynomials, $P_0(x) = 1$ and $P_1(x) = x$, and calculate $P_n(x)$ by taking:

$$j = 1, 2, \ldots, n-1 \tag{2.87}$$

This is known as *Bonnet's recurrence relation*. It's similar in spirit to Eq. (2.64), but here we won't face as many headaches. We will implement our new relation in Python in the following section. For now, we carry out, instead, the simpler task of picking up where Eq. (2.79) had left off: we plug $j = 4$ into Eq. (2.86) to find:

$$P_5(x) = \frac{9xP_4(x) - 4P_3(x)}{5} = \frac{1}{8}\left(63x^5 - 70x^3 + 15x\right) \tag{2.88}$$

In principle, this approach can be followed even for large n values.

At this point, we could turn to deriving a relation for the first derivative of Legendre polynomials. It's pretty easy to derive a recurrence relation that does this: as you will find out in problem 2.20, instead of differentiating Eq. (2.81) with respect to u (as we did above), we could now differentiate that equation with respect to x. What this gives us is a recurrence relation for $P'_j(x)$. As the problem shows, we can do even better, deriving a formula that doesn't necessitate a separate recurrence process for the derivatives. This is:

$$P'_n(x) = \frac{nP_{n-1}(x) - nxP_n(x)}{1 - x^2} \tag{2.89}$$

This equation only needs access to the last two Legendre polynomials, $P_n(x)$ and $P_{n-1}(x)$, which we produced when we were stepping through Bonnet's formula. We are now ready to implement both Eq. (2.86) and Eq. (2.89) in Python.

2.5.2.3 Implementation

Code 2.8 is an implementation of our two main relations for the Legendre polynomials and their derivatives. The function `legendre()` basically implements Eq. (2.86). We observe, however, that the first polynomial that that equation produces is $P_2(x)$: to evaluate $P_{j+1}(x)$ you need $P_j(x)$ and $P_{j-1}(x)$. Thus, we have hard-coded two special cases that simply spit out $P_0(x)$ or $P_1(x)$ (as well as $P'_0(x)$ or $P'_1(x)$) if the input parameter `n` is very small. All other values of `n` are captured by the `else`, which contains a loop stepping through the values of j given in Eq. (2.87). Note a common idiom we've employed here: we don't refer to $P_{j-1}(x)$, $P_j(x)$, and $P_{j+1}(x)$ but to `val0`, `val1`, and `val2`. There's no need to store the earlier values of the polynomial, so we don't.[34] Note that each time through the loop

[34] Of course, we *do* calculate those, only to throw them away: another option would have been to also return all the $P_j(x)$'s to the user, since we get them "for free".

Code 2.8 legendre.py

```
import matplotlib.pyplot as plt

def legendre(n,x):
    if n==0:
        val2 = 1.
        dval2 = 0.
    elif n==1:
        val2 = x
        dval2 = 1.
    else:
        val0 = 1.; val1 = x
        for j in range(1,n):
            val2 = ((2*j+1)*x*val1 - j*val0)/(j+1)
            val0, val1 = val1, val2
        dval2 = n*(val0-x*val1)/(1.-x**2)
    return val2, dval2

def plotlegendre(der,nsteps):
    plt.xlabel('$x$', fontsize=20)

    dertostr = {0: "$P_n(x)$", 1: "$P_n'(x)$"}
    plt.ylabel(dertostr[der], fontsize=20)

    ntomarker = {1: 'k-', 2: 'r--', 3: 'b-.', 4: 'g:', 5: 'c^'}
    xs = [i/nsteps for i in range(-nsteps+1,nsteps)]
    for n,marker in ntomarker.items():
        ys = [legendre(n,x)[der] for x in xs]
        labstr = 'n={0}'.format(n)
        plt.plot(xs, ys, marker, label=labstr, linewidth=3)

    plt.ylim(-3*der-1, 3*der+1)
    plt.legend(loc=4)
    plt.show()

if __name__ == '__main__':
    nsteps = 200
    plotlegendre(0,nsteps)
    plotlegendre(1,nsteps)
```

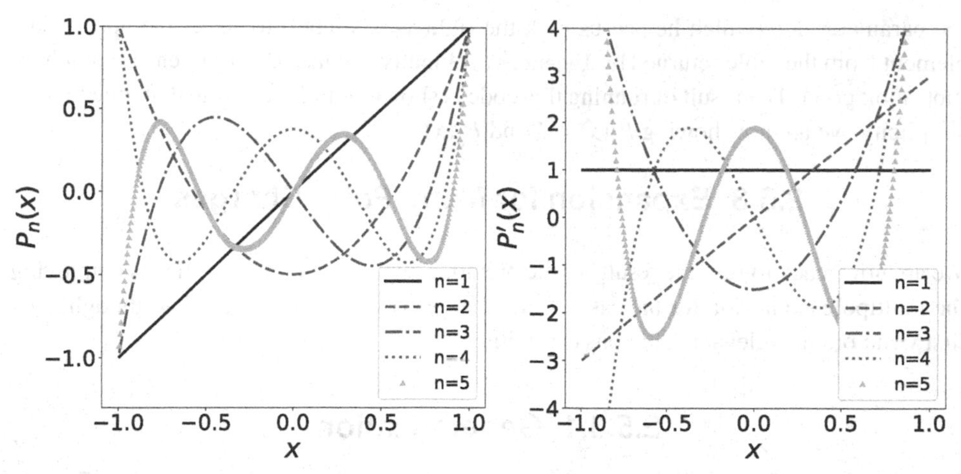

Fig. 2.6 Legendre polynomials (left panel) and their first derivatives (right panel)

we have to change the interpretation of `val0`, `val1`, and `val2`: we do this via an idiomatic swap of two variables. Once we're done iterating, we turn to the derivative. As discussed around Eq. (2.89), this is not a recurrence relation: it simply takes in the last two polynomials and outputs the value of the derivative of the last Legendre polynomial. Our function returns a tuple of values: for a given n and a given x, we output $P_n(x)$ and $P_n'(x)$.[35]

For most of this project, we provide the Python code that *could* be used to produce the figures we show, but we don't actually show the `matplotlib` calls explicitly. This is partly because it's straightforward to do so and partly because it would needlessly lengthen the size of this book. However, in the present case we've made an exception: as mentioned in the previous chapter, we follow the principle of "separation of concerns" and place all the plotting-related infrastructure in one function. The function `plotlegendre()` is used to plot five Legendre polynomials and their derivatives. These are 10 functions in total: as you can imagine, plotting so many different entities can become confusing if one is not careful. We have therefore taken the opportunity to show that Python dictionaries can also be helpful in tackling such mundane tasks as plotting. The first dictionary we employ determines the y-axis label: since we will be producing one plot for the Legendre polynomials and another plot for the Legendre-polynomial derivatives, we want these to have different labels.[36] We then use a second dictionary, to encapsulate the correspondence from n value to line or marker style: since we're plotting five different functions (each time), we need to be able to distinguish them from each other. Once again, a moment's thought would show that the alternative is to explicitly build the `ys` and then call `plot()` for each function separately.[37] In contradistinction to this, we employ a loop over the dictionary items, and use fancy-string formatting to ensure that each curve label employs the correct number. Also, we use

[35] You might need the derivative of Legendre polynomials if you wish to evaluate more involved electrostatic properties. We won't actually be doing that in what follows, but we *will* make use of the derivatives in section 7.5.2.2, when we discuss Gauss–Legendre integration.

[36] Take a moment to think of other possible solutions to this task: the one that probably comes to mind first is to copy and paste. Most of the time, this is a bad idea.

[37] But what happens if you want to add a couple more functions to the plot?

the parameter `der` (which helped us pick the right y-axis label) to select the appropriate element from the tuple returned by `legendre()`. Finally, we use `ylim()` to ensure that both plots look good. The result of running this code is shown in Fig. 2.6. Note that, due to their simplicity, we are not showing $P_0(x) = 1$ and $P_0'(x) = 0$.

2.5.3 Expansion for Many Point Charges

We are now ready to fuse the results of the previous two subsections: this means employing the multipole expansion for the case of our charge array. Before we do that, though, let's first write out the relevant equations explicitly.

2.5.3.1 Generalization

Earlier, we were faced with the potential coming from a single charge q_0, Eq. (2.73); we expanded the denominator and arrived at an equation in terms of Legendre polynomials, namely Eq. (2.78). In the present case, we would like to study the more general potential coming from n charges, as per Eq. (2.72):

$$\phi(\mathbf{r}) = \sum_{i=0}^{n-1} \phi_i(\mathbf{r}) = \sum_{i=0}^{n-1} k \frac{q_i}{|\mathbf{r} - \mathbf{r}_i|} \tag{2.90}$$

If we carry out an expansion like that in Eq. (2.78) for each of these denominators, we get:

$$\begin{aligned}\phi(\mathbf{r}) &= \sum_{i=0}^{n-1} \frac{1}{r} \sum_{n=0}^{\infty} k q_i \left(\frac{r_i}{r}\right)^n P_n(\cos\theta_i) = k \sum_{n=0}^{\infty} \frac{1}{r^{n+1}} \sum_{i=0}^{n-1} q_i r_i^n P_n(\cos\theta_i) \\ &= \frac{k}{r} \sum_{i=0}^{n-1} q_i + \frac{k}{r^2} \sum_{i=0}^{n-1} q_i r_i \cos\theta_i + \frac{k}{r^3} \sum_{i=0}^{n-1} q_i r_i^2 \frac{1}{2}\left(3\cos^2\theta_i - 1\right) + \cdots \end{aligned} \tag{2.91}$$

where θ_i is the angle between the vectors $\mathbf{r}$ and $\mathbf{r}_i$. In the second equality we interchanged the order of the summations and pulled out the denominators. In the third equality we wrote out the first few terms of the n summation and also plugged in the relevant $P_n(x)$ from Eq. (2.79).

At this point, we can spell out what was only implicit before: our approximation for the total potential at point $\mathbf{r}$ is given as a sum of terms of the form $1/r$, $1/r^2$, $1/r^3$, and so on:

$$\phi(\mathbf{r}) = \frac{k}{r} Q_0 + \frac{k}{r^2} Q_1 + \frac{k}{r^3} Q_2 + \cdots \tag{2.92}$$

with the coefficients Q_j being a (possibly complicated) combination of the q_i, r_i, and θ_i. Even if you haven't encountered this specific problem in a course on electromagnetism, the general principle should be familiar to you: the terms we are now dealing with have names such as *monopole*, *dipole*, *quadrupole*, etc. In general, what we're carrying out is known as the *multipole expansion*. The monopole coefficient Q_0 is simply a sum over all the charges; the corresponding contribution to the potential goes as $1/r$, just like in the

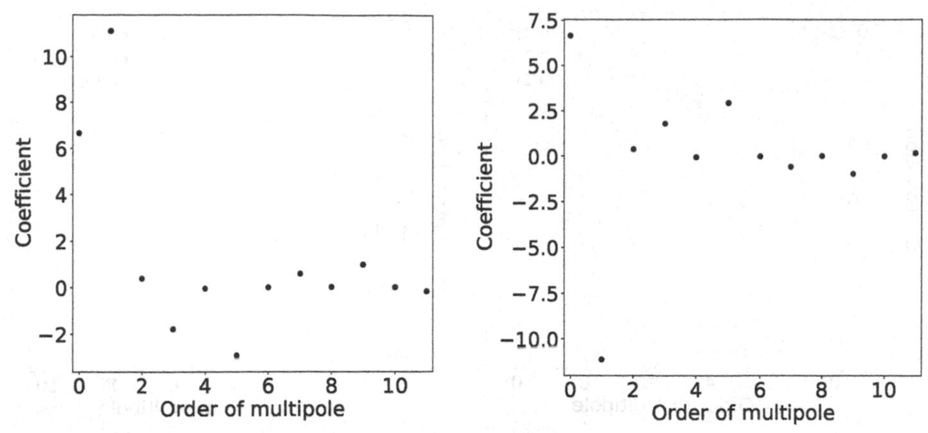

Fig. 2.7 Multipole coefficients along the positive (left) and negative (right) y axis

case of a single point charge. The dipole coefficient Q_1 is a sum over (q_i times) $r_i \cos\theta_i$, which can be re-expressed as $\hat{\mathbf{r}} \cdot \mathbf{r}_i$. Employing exactly the same argument, we see that the quadrupole coefficient Q_2 is a sum over q_i times $[3(\hat{\mathbf{r}} \cdot \mathbf{r}_i)^2 - r_i^2]/2$. Note that these coefficients may depend on the *direction* of $\mathbf{r}$, but not on its magnitude: all the dependence on the magnitude of $\mathbf{r}$ has been pulled out and is in the denominator.[38]

It's worth pausing for a moment to appreciate what we've accomplished in Eq. (2.92): by expanding in r_i/r (assuming, for now, that $r > r_i$) and interchanging the sums over n and i, we've expressed the full potential $\phi(\mathbf{r})$ (which we know from Eq. (2.72) is generally a complicated function of $\mathbf{r}$) as an expansion in powers of $1/r$, where the coefficients depend on the point charges, as well as the angles between $\mathbf{r}$ and $\mathbf{r}_i$. It goes without saying that increasingly large powers of $1/r$ have less and less of a role to play; of course, this also depends on the precise value of the coefficients. Thus, there are situations where the octupole, hexadecapole, and higher-order terms may need to be explicitly taken into account. Even so, what we've managed to do is to take a (possibly very complicated) distribution of point charges and encapsulate its effects on the total potential (along a given direction) into a few numbers, the Q_j.

Let's try to see the above insights applied to a specific case, that of our 36-charge array: Fig. 2.7 shows the coefficients along the y axis.[39] As already mentioned, the coefficients Q_j don't depend on the precise point on the y axis, only on whether $\mathbf{r}$ is pointing up or down. Focusing on the positive y axis for the moment (left panel), we find that for this specific scenario the coefficients exhibit interesting structure: the monopole coefficient has a large magnitude, but the dipole coefficient is even larger.[40] After that, the even-n coefficients seem to be pretty small, but the $n = 3, 5, 7, 9$ coefficients are sizable (and of both signs).

Still on the subject of our 36-charge array, let's now examine how the different coefficients are combined together. As you can see from Eq. (2.92), Q_j's of oscillating sign

[38] Incidentally, moving the origin of the coordinates may change the precise values of Q_j, but the overall interpretation remains the same.

[39] We picked the y axis for our $\mathbf{r}$'s in order to be consistent with what we showed in Fig. 2.5.

[40] Coefficients at different orders have different units.

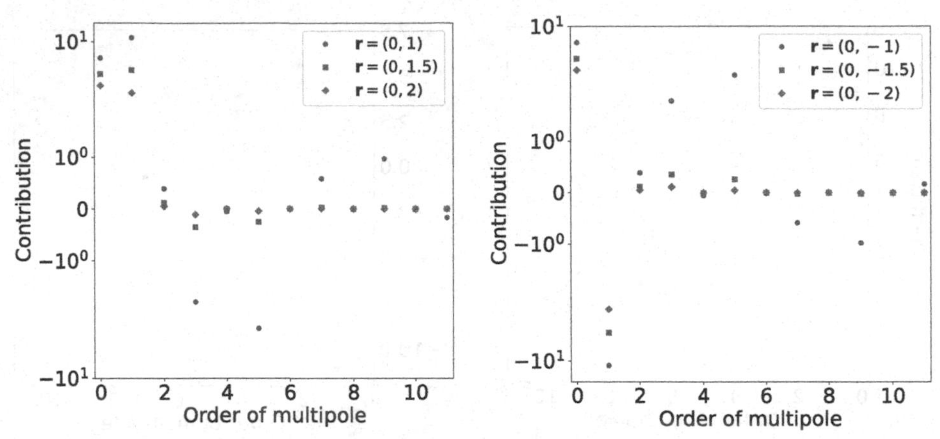

Fig. 2.8 Multipole contributions along the positive (left) and negative (right) y axis

and large magnitude end up giving positive and negative contributions: however, as noted in our discussion on page 69, these contributions are not of comparable magnitude, since they are multiplying powers of $1/r$. Thus, assuming r is reasonably large, the higher-order terms don't contribute too much. To reverse this argument, as you make r comparable to r_i you will need to keep track of an increasing number of terms in your multipole expansion. This is precisely what we find in the left panel of Fig. 2.8 (using a symmetrical-log scale): when you're close to the point charges the contributions mimic the size of the coefficients, but as you make r larger you need considerably fewer terms in the multipole expansion.[41]

The right panel of Fig. 2.7 shows the coefficients along the negative y axis: we notice that even-n multipole coefficients are unchanged, whereas odd-n multipole coefficients have flipped sign. Taking the most prominent example, the electric dipole term is the largest in magnitude for both cases, but is positive in the left panel and negative in the right panel (this is easy to understand: $\hat{\mathbf{r}} \cdot \mathbf{r}_i$ changes sign when you flip the direction of $\hat{\mathbf{r}}$). The contributions in the right panel of Fig. 2.8 behave as we would expect.

2.5.3.2 Implementation

In order to produce Fig. 2.7 on the coefficients and Fig. 2.8 on the different contributions in the multipole expansion, we needed to make use of a code implementing the formalism we've introduced thus far. Code 2.9 is a Python implementation, in which we have separated out the different aspects of the functionality: (a) decomposition into magnitudes and angles, (b) the coefficients, and (c) combining the coefficients together with the value of r to produce an approximation to the full potential. This is the first major project in this book. While the program doesn't look too long, you notice that it starts by `import`ing functionality from our earlier codes: `kahansum()` to carry out the compensated summation, `chargearray()` to produce our 36-charge array, `vecmag()` as a helper function, as well

[41] All contributions have the same units, C/m.

multipole.py

Code 2.9

```
from kahansum import kahansum
from chargearray import chargearray, vecmag
from legendre import legendre

def decomp(rs,ris):
    rmag = vecmag(rs); rimag = vecmag(ris)
    prs = [r*ri for r,ri in zip(rs,ris)]
    vecdot = kahansum(prs)
    costheta = vecdot/(rmag*rimag)
    return rmag, rimag, costheta

def multicoes(rs,qtopos,nmax=60):
    coes = [0. for n in range(nmax+1)]
    for n in range(nmax+1):
        for q,pos in qtopos.items():
            rmag, rimag, costheta = decomp(rs,pos)
            val = q*(rimag**n)*legendre(n,costheta)[0]
            coes[n] += val
    return coes

def multifullpot(rs,qtopos):
    coes = multicoes(rs,qtopos)
    rmag = vecmag(rs)
    contribs = [coe/rmag**(n+1) for n,coe in enumerate(coes)]
    return kahansum(contribs)

if __name__ == '__main__':
    qtopos = chargearray(6)
    for y in 1,-1:
        rs = [0.,y]
        potval = multifullpot(rs,qtopos); print(rs, potval)
```

as `legendre()` to compute the Legendre polynomials that we need.

Our first new function, `decomp()`, takes in two vectors ($\mathbf{r}$ and $\mathbf{r}_i$, given as lists containing their Cartesian components), and evaluates their magnitudes as well as the cosine of the angle between them. As is common with our programs, there are several edge cases that this function does *not* cover: there will be a division failing if either of $\mathbf{r}$ and $\mathbf{r}_i$ is placed at the origin. (While we're taking $r > r_i$, it's possible that our charge array could contain a charge at the coordinate origin.) Another issue with this function is that it is wasteful: it evaluates `rmag` each time it is called (i.e., 36 times in our case), even though the input `rs` hasn't changed: as usual, this is because we've opted in favor of code simplicity.

Our function, `multicoes()`, is a straightforward implementation of the second equality in Eq. (2.91). It takes in as parameters `qtopos` (the distribution of point charges), `rs` (the position **r** at which we are interested in evaluating/approximating the potential), and `nmax` (a parameter controlling how many terms we should keep in the multipole expansion). We provide a default parameter value for `nmax` which is presumably large enough that we won't have to worry about the quality of our approximation. Problem 2.24 asks you to investigate this in detail, for obvious reasons: why use a large number of terms in the multipole expansion when you don't need to? The structure of our function closely follows that of Eq. (2.91): the loop over n is outside and that over i inside (though, as earlier, there's no need to carry around an explicit i index). We use the `items()` method of Python dictionaries (which you should have gotten used to by now) to step through all the q_i's in our distribution. The only subtlety is that our call to `legendre()` ends with `[0]`: this results from the fact that `legendre()` returns a tuple of two numbers. While `multicoes()` is a pretty function, note that it (and the rest of the code) *assumes* that $r > r_i$ is true: problem 2.25 asks you to remove this assumption, rewriting the code appropriately.

Our function, `multifullpot()`, is a test: it computes the multipole-expansion approximation of Eq. (2.92) for the total potential $\phi(\mathbf{r})$, which can be compared to the complete calculation of it in Eq. (2.72). This sums the contribution from each multipole (with the coefficient for each multipole term coming from all the q_i charges), whereas the approach of `fullpot()` in our earlier code was to simply sum together the full contributions from each q_i, as per Eq. (2.72). This function is, as usual, idiomatic (employing a list comprehension, as well as `enumerate()`). At the end, it sums together (using `kahansum()`) all the contributions: this is different from Fig. 2.8, which shows each contribution separately.

The main program again picks two arbitrary points on the y axis and prints out the multipole-expansion approximation to the value of the potential $\phi(\mathbf{r})$ at those points. Comparing those numbers to the output of our earlier code `chargearray.py`, we see that they are in pretty good agreement with the full potential values. This isn't too surprising, since we picked `nmax=60` as our default parameter value. (Given that our points are pretty close to the charge array, this was probably a safe choice.) The results may feel somewhat underwhelming: this impression can be altered if you recall that this code can (be modified in order to) produce Fig. 2.7 on the coefficients and Fig. 2.8 on the different contributions in the multipole expansion. These figures can help you build insight about the specific geometry involved in our 36-charge array and on the role of the different multipole terms.

Problems

2.1 [$\mathcal{A}$] Study the propagation of the relative error, for the case of division, $x = a/b$, by analogy to what was done in the main text for the case of multiplication.

2.2 [$\mathcal{A}$] Using the notation of the main text, the absolute value of the relative error is $|\delta x| = |(\tilde{x}-x)/x|$. In practice, it is convenient to provide a bound for a distinct quantity, $|\tilde{\delta x}| = |(\tilde{x}-x)/\tilde{x}|$. Using these two definitions, try to find an inequality relating $|\tilde{\delta x}|$ and $|\delta x|$ (i.e., $A \leq |\tilde{\delta x}| \leq B$, where A and B contain $|\delta x|$); explain what this implies about the relative magnitudes of $|\tilde{\delta x}|$ and $|\delta x|$ in practical situations.

2.3 [$\mathcal{A}$] This problem studies error propagation for specific examples. Specifically, at $x = 2$ find the: (a) absolute error in $f(x) = \ln x$ for $\tilde{x} = 1.9$, (b) relative error in $f(x) = \sqrt{x}$ for $\tilde{x} = 1.95$, and (c) backward error for $f(x) = e^x$ and $\tilde{f}(2) = e^2 = 8$.

2.4 Without rounding error, the following code's output would be $f(x) = x$. (Note that this has nothing to do with catastrophic cancellation.) Determine what's going on, remembering our comment (section 2.4) on one or two iterations being the culprits.

```
from math import sqrt
def f(x,nmax=100):
        for i in range(nmax):
                x = sqrt(x)
        for i in range(nmax):
                x = x**2
        return x
for xin in (5., 0.5):
        xout = f(xin); print(xin, xout)
```

2.5 We now examine a case where plotting a function on the computer can seriously mislead us. The function we wish to plot is: $f(x) = x^6 + 0.1\log(|1 + 3(1 - x)|)$. Use 100 points from $x = 0.5$ to 1.5 to plot this function in `matplotlib`. Do you see a dip? Consider the function itself and reason about what you should be seeing. Then use a much finer grid and ensure that you capture the analytically expected behavior.

2.6 Take the standard quadratic equation: $ax^2 + bx + c = 0$, whose solutions are:

$$x_{\pm} = \frac{-b \pm \sqrt{b^2 - 4ac}}{2a} \tag{2.93}$$

Take $b > 0$ for concreteness. It is easy to see that when $b^2 \gg ac$ we don't get a catastrophic cancellation when evaluating $b^2 - 4ac$ (we may still get a "benign" cancellation). Furthermore, $\sqrt{b^2 - 4ac} \approx b$. However, this means that x_+ will involve catastrophic cancellation in the numerator. We will employ an analytical trick in order to help us preserve significant figures. Observe that the product of the two roots obeys the relation: $x_+x_- = c/a$. The answer now presents itself: use Eq. (2.93) to calculate x_-, for which no catastrophic cancellation takes place. Then, use $x_+x_- = c/a$ to calculate x_+. Notice that you ended up calculating x_+ via division only (i.e., without a catastrophic cancellation). Write a Python code that evaluates and prints out: (a) x_-, (b) x_+ using the "bad" formula, and (c) x_+ using the "good" formula. Take $a = 1, c = 1, b = 10^8$. Discuss the answers.[42]

[42] Keep in mind that if $b^2 \approx ac$, then $b^2 - 4ac$ would involve a catastrophic cancellation. Unfortunately, there is no analytical trick to help us get out of this problem.

2.7 We promised to return to the distinction between catastrophic and benign cancellation. Take $\tilde{x}$ and $\tilde{y}$ to be: $\tilde{x} = 1234567891234567.0$ and $\tilde{y} = 1234567891234566.0$. Now, if we try to evaluate $\tilde{x}^2 - \tilde{y}^2$ we will experience catastrophic cancellation: each of the squaring operations leads to a rounding error and then the subtraction exacerbates that dramatically. Write a Python code that does the following:

(a) Carries out the calculation $1234567891234567^2 - 1234567891234566^2$ using integers, i.e., exactly.
(b) Carries out the subtraction $1234567891234567.0^2 - 1234567891234566.0^2$ using floats, i.e., exhibiting catastrophic cancellation.
(c) Now, we will employ a trick: $x^2 - y^2$ can be re-expressed as $(x - y)(x + y)$. Try using this trick for the floats and see what happens. Does your answer match the integer answer or the catastrophic-cancellation answer? Why?

2.8 We will study the following function:

$$f(x) = \frac{1 - \cos x}{x^2} \tag{2.94}$$

(a) Start by plotting the function, using a grid of the form $x = 0.1 \times i$ for $i = 1, 2, \ldots, 100$. This should give you some idea of the values you should expect for $f(x)$ at small x.
(b) Verify your previous hunch by taking the limit $x \to 0$ and using L'Hôpital's rule.
(c) Now, see what value you find for $f(x)$ when $x = 1.2 \times 10^{-8}$. Does this make sense, even qualitatively?
(d) Use a trigonometric identity to avoid the cancellation. Evaluate the new function at $x = 1.2 \times 10^{-8}$ and compare with your analytical answer for $x \to 0$.

2.9 [$\mathscr{A}$] This problem focuses on analytical manipulations introduced in order to avoid a cancellation. Rewrite the following expressions in order to evaluate them for large x: (a) $\sqrt{x+1} - \sqrt{x}$, (b) $\frac{1}{x+1} - \frac{2}{x} + \frac{1}{x-1}$, and (c) $\frac{1}{\sqrt{x}} - \frac{1}{\sqrt{x+1}}$.

2.10 As a statistical warmup to chapters 6 and 7, evaluate the mean of the n values x_i: $\mu = \sum_{i=0}^{n-1} x_i/n$. You can evaluate the variance using a *two-pass* algorithm:

$$\sigma^2 = \frac{1}{n}\sum_{i=0}^{n-1}(x_i - \mu)^2 \tag{2.95}$$

This is called a two-pass algorithm because you need to evaluate the mean first, so you have to loop through the x_i once to get the mean and a second time time to get the variance. Many people prefer the following *one-pass* algorithm:

$$\sigma^2 = \left(\frac{1}{n}\sum_{i=0}^{n-1} x_i^2\right) - \mu^2 \tag{2.96}$$

You should be able to see that this formula allows you to keep running sums of the x_i and the x_i^2 values in parallel and then perform only one subtraction at the end. Naively, you might think that the two-pass algorithm will suffer from more roundoff error problems, since it involves n subtractions. On the other hand, if you solved problem 2.7 on $\tilde{x}^2 - \tilde{y}^2$, you might be more wary of subtracting the squares of two nearly equal numbers (which is what the one-pass algorithm does). Write two Python functions, one for each algorithm, and test them on the two cases below:

$$x_i = 0, 0.01, 0.02, \ldots, 0.09$$
$$x_i = 123456789.0, 123456789.01, 123456789.02, \ldots, 123456789.09 \quad (2.97)$$

2.11 This problem discusses error buildup when trying to evaluate polynomials without and with the use of *Horner's rule*. Take a polynomial of degree $n - 1$:

$$P(x) = p_0 + p_1 x + p_2 x^2 + p_3 x^3 + \cdots + p_{n-1} x^{n-1} \quad (2.98)$$

Write a function that takes in a list containing the p_i's, say `coeffs`, and the point x and evaluates the value of $P(x)$ in the naive way, i.e., from left to right.
Notice that this way of coding up the polynomial contains several (needless) multiplications. This is so because x^i is evaluated as $x \times x \times \cdots \times x$ (where there are $i - 1$ multiplications). Thus, this way of approaching the problem corresponds to:

$$1 + 2 + \cdots + n - 2 = \frac{(n-1)(n-2)}{2} \quad (2.99)$$

multiplications, from x^2 all the way up to x^{n-1}. If we rewrite the polynomial:

$$P(x) = p_0 + x(p_1 + x(p_2 + x(p_3 + \cdots + x(p_{n-2} + xp_{n-1})\cdots))) \quad (2.100)$$

then we can get away with only $n - 1$ multiplications (i.e., no powers are evaluated). This is obviously more efficient, but equally important is the fact that this way we can substantially limit the accumulation of rounding error (especially for polynomials of large degree). Write a function that takes in a list containing the p_i's, say `coeffs`, and the point x and evaluates the value of $P(x)$ in the new way, i.e., from right to left.
Apply the previous two functions to the (admittedly artificial) case of:

```
coeffs = [(-11)**i for i in reversed(range(8))]
```

and $x = 11.01$. Observe any discrepancy and discuss its origin.

2.12 This problem studies the rational function introduced in the main text, Eq. (2.53).

(a) Apply Horner's rule twice (once for the numerator and once for the denominator) to produce two plots, one for $x = 1.606 + 2^{-52}i$ and one for $x = 2.400 + 2^{-52}i$. Your results should look like Fig. 2.3.

(b) Create a new Python function that codes up the following expression:

$$s(x) = 4 - \frac{3(x-2)[(x-5)^2 + 4]}{x + (x-2)^2[(x-5)^2 + 3]} \quad (2.101)$$

which is a rewritten version of our rational function. Apply this new function for the previous two plots and compare the rounding error pattern, size, etc.

(c) Now introduce two more sets of results, this time for the starting expression for $r(x)$ produced using (not Horner's rule but) the naive implementation, using powers (i.e., the way you would have coded this up before doing problem 2.11). Interpret your findings.

2.13 This problem studies a new rational function:

$$t(x) = \frac{7x^4 - 101x^3 + 540x^2 - 1204x + 958}{x^4 - 14x^3 + 72x^2 - 151x + 112} \tag{2.102}$$

Notice that the denominator is the same as in problem 2.12.

(a) Plot $t(x)$, evaluated via Horner's rule, along with the following (equivalent) continued fraction, from $x = 0$ to $x = 4$:

$$u(x) = 7 - \frac{3}{x - 2 - \frac{1}{x-7+\frac{10}{x-2-\frac{2}{x-3}}}} \tag{2.103}$$

You may wish to know that:

$$u(1) = 10, \; u(2) = 7, \; u(3) = 4.6, \; u(4) = 5.5 \tag{2.104}$$

(b) Evaluate each of these functions for $x = 10^{77}$ (make sure to use floats in your code). Do you understand what is happening? Are you starting to prefer one formulation over the other? (What happens if you use integers instead of floats?)

(c) Plot the two Python functions ($t(x)$ and $u(x)$) for $x = 2.400 + 2^{-52}i$, where i goes from 0 to 800. Was your intuition (about which formulation is better) correct?

2.14 In `compexp.py` we employed the Taylor series for e^x. Do the same for $f(x) = \sin x$ at $x = 0.1$ and $x = 40$. Just like we did in the main text, you will first need to find a simple relation between the n-th term and the $(n-1)$-th term in the series. Then, use a trigonometric identity to study an equivalent (smaller) x instead of 40.

2.15 Here we study the *Basel problem*, namely the sum of the reciprocals of the squares of the positive integers. This also happens to be a specific value of the Riemann zeta function, $\zeta(2)$. The value can be calculated in a variety of ways and turns out to be:

$$\sum_{k=1}^{\infty} \frac{1}{k^2} = \frac{\pi^2}{6} = 1.644\,934\,066\,848\,226\,4\ldots \tag{2.105}$$

Here we will use Python to compute partial sums.

(a) Code this up by adding all the contributions from $k = 1, 2,$ up to some large integer; break out of the loop when the value of the sum stops changing. What is the value of the sum when this happens? Is there some meaning behind the value of the maximum integer (inverse squared) when this happens? Call this `nmaxd`.

(b) We were summing the contributions from largest to smallest, so by the time we got to the really tiny contributions they stopped mattering. The obvious fix here is to *reverse* the order in which we are carrying out the sum. This has the advantage of dealing with the tiniest contributions first (you may have encountered the term "subnormal numbers" in appendix B). The disadvantage is that our previous strategy regarding termination (break out when the sum stops changing) doesn't apply here: when summing in reverse the last few contributions are the largest in magnitude. Instead, simply pick a large maximum integer `nmaxr` from the start. Be warned that this will start to take a long time, depending on the CPU you are using. Do a few runs for `nmaxr` a multiple of `nmaxd` (4, 8, 16, 32). You should see the answer get (slowly) better. The beauty here is that the larger `nmaxr`, the better we can do. (Try increasing `nmaxd` for the direct method.)

(c) Employ compensated summation to carry out the sum. The Kahan sum function does well, despite the fact that it carries out the sum in direct order, i.e., by starting from the largest contribution. We don't have a nifty termination criterion, but increasing the number of terms can still make a difference.

2.16 In this problem we will learn how to *accelerate the convergence of a sequence*. We will accomplish this by transforming a sequence p_n obeying $\lim_{n\to\infty} p_n = \xi$ into another sequencc q_n that converges more quickly to ξ.

(a) Assume the sequence p_n converges to ξ like a geometric series with a factor C:

$$p_{n+1} - \xi = C(p_n - \xi), \qquad n = 1, 2, \ldots \tag{2.106}$$

Apply this equation a second time taking $n \to n+1$ and then eliminate C; you can then solve the resulting equation for ξ. This takes the form:

$$q_n = p_n - \frac{(p_{n+1} - p_n)^2}{p_n + p_{n+2} - 2p_{n+1}} \tag{2.107}$$

where we took the opportunity to write q_n on the left-hand side (instead of ξ) to emphasize that this is a new sequence (in practice, the starting relation Eq. (2.106) is likely to be only approximately valid). This technique is known as *Aitken extrapolation* or *Aitken's Δ^2 method.*[43]

(b) We will now apply Aitken extrapolation, Eq. (2.107), to the following problem:

$$\sum_{k=1}^{\infty} \frac{(-1)^{k+1}}{k} = \ln 2 = 0.693\,147\,180\,559\,945\,3\ldots \tag{2.108}$$

To do so, treat the partial sums as the terms of your original sequence:

$$p_n = \sum_{k=1}^{n} \frac{(-1)^{k+1}}{k} \tag{2.109}$$

You should find that when $n = 100$ or so the Aitken-extrapolated q_n's have converged on six significant figures of the true answer, whereas the (untransformed) partial sums p_n's are still struggling with the second significant figure.

2.17 In section 2.4.5 backward recursion avoided the problems of forward recursion. Another example is provided by *spherical Bessel functions* (of the first kind):

$$j_{n+1}(x) + j_{n-1}(x) = \frac{2n+1}{x} j_n(x) \tag{2.110}$$

which is analogous to Bonnet's recurrence relation, Eq. (2.86). Evaluate $j_8(0.5)$ in the "naive way", starting from the known functions:

$$j_0(x) = \frac{\sin x}{x}, \qquad j_1(x) = \frac{\sin x}{x^2} - \frac{\cos x}{x} \tag{2.111}$$

[43] A cousin of Richardson extrapolation, introduced in section 3.3.7 and re-encountered in later chapters.

and see what goes wrong. Then, use backward recursion starting from $n = 15$. Since your two starting guesses are arbitrary, you should normalize by computing:

$$j_8(0.5) = \tilde{j}_8(0.5)\frac{j_0(0.5)}{\tilde{j}_0(0.5)} \tag{2.112}$$

at the end, where $\tilde{j}_0(0.5)$ and $\tilde{j}_8(0.5)$ are the values you computed and $j_0(0.5)$ is the correctly normalized value from Eq. (2.111).

2.18 We will study the Fourier series of a periodic square wave. As you may already know, this gives rise to what is known as the *Gibbs phenomenon*, which is a "ringing" effect that appears at discontinuities. Specifically, from $-\pi$ to π we have:

$$f(x) = \begin{cases} \frac{1}{2}, & 0 < x < \pi \\ -\frac{1}{2}, & -\pi < x < 0 \end{cases} \tag{2.113}$$

The Fourier series of this function is:

$$f(x) = \frac{2}{\pi} \sum_{n=1,3,5,\ldots} \frac{\sin(nx)}{n} \tag{2.114}$$

(a) Create two Python functions, one for the square wave and another for its Fourier expansion. The latter should take in as an argument the maximum n up to which you wish the sum to go.
(b) Plot the square wave and the Fourier-expansion results for $n_{max} = 1, 3, 5, 7, 9$ (six curves in total). Where are the oscillation amplitudes largest?
(c) Note that this issue arises not from roundoff error, but from the nature of the Fourier series itself. To convince yourself that this is, indeed, the case, take the maximum n value to be 21, 51, 101, and so on. What do you find?

2.19 Compare the output of `legendre.py` with that of the recurrence relation for $n = 1000$:

$$P_n(x) = 2xP_{n-1}(x) - P_{n-2}(x) - \frac{xP_{n-1}(x) - P_{n-2}(x)}{n} \tag{2.115}$$

2.20 [$\mathcal{A}$] We now derive Eq. (2.89) for the first derivative of Legendre polynomials.

(a) Differentiate Eq. (2.81) with respect to x. Then, split the resulting equation into four separate sums, noticing that our power series in u is equal to zero, regardless of the value of u. Thus, the coefficient of each power of u (separately) is equal to zero. This will be a recurrence relation for $P_j'(x)$.
(b) Differentiate Eq. (2.86) with respect to x.
(c) Multiply the result of part (b) by 2 and add to that part (a) times $2j + 1$.
(d) Produce two equations (S and D, respectively), one being the sum of the results of part (a) and part (c) (divided by 2) and the other being the difference of the results of part (a) and part (c) (divided by 2). Your D should be of the form:

$$xP_n'(x) = P_{n-1}'(x) + nP_n(x) \tag{2.116}$$

(e) Take $j \to j+1$ in S and subtract x times D. The result is equivalent to Eq. (2.89).

2.21 This problem studies the evaluation of Bernoulli numbers and polynomials.

(a) In section 2.5.2.2 we saw how to go from a generating function to a recurrence relation for Legendre polynomials. We will now do something analogous for *Bernoulli numbers*. Start from the first equation and show the second:

$$\frac{u}{e^u - 1} = \sum_{j=0}^{\infty} \frac{B_j u^j}{j!}, \quad n - 1 = \sum_{j=1}^{n-1} B_{2j} \binom{2n}{2j} \tag{2.117}$$

You should use the Taylor series of the exponential and the Cauchy product.

(b) Compute the first 15 even Bernoulli numbers from the recurrence relation.

(c) Slightly generalizing the above generating function, we get to a corresponding relation for *Bernoulli polynomials*:

$$\frac{ue^{ut}}{e^u - 1} = \sum_{j=0}^{\infty} B_j(t) \frac{u^j}{j!} \tag{2.118}$$

Use this equation to derive the following properties:

$$B_j(0) = B_j, \qquad \frac{d}{dt} B_j(t) = jB_{j-1}(t), \qquad B_j(1) = (-1)^j B_j(0) \tag{2.119}$$

(d) We will now use the previous three properties to derive the celebrated *Euler–Maclaurin summation formula*. Since $B_0(t) = B_0 = 1$, we can write:

$$\int_0^1 g(t)dt = \int_0^1 g(t)B_0(t)dt \tag{2.120}$$

You can now use the second property and replace $B_0(t)$ with $B_1'(t)$. Integrate by parts and keep repeating the entire exercise until you find:

$$\int_0^1 g(t)dt = \frac{g(0)}{2} + \frac{g(1)}{2} - \sum_{j=1}^{m} \frac{1}{(2j)!} B_{2j} \left[g^{(2j-1)}(1) - g^{(2j-1)}(0) \right]$$
$$+ \frac{1}{(2m)!} \int_0^1 g^{(2m)}(t) B_{2m}(t) dt \tag{2.121}$$

Crucially, only even Bernoulli numbers appear here.

(e) To make the integral and derivatives easier, take $g(t) = e^t$. Print out the value of the left-hand side in Eq. (2.121) as well as the value of the right-hand side as you take $m = 1, 2, 3, \ldots, 10$ (always dropping the remainder term).

2.22 This problem studies the evaluation of *Chebyshev polynomials* and their extrema, known as *Chebyshev points*. First, implement the following recurrence relation:

$$T_{n+1}(x) = 2xT_n(x) - T_{n-1}(x) \tag{2.122}$$

starting from the known functions $T_0(x) = 1$ and $T_1(x) = x$. Plot the first few Chebyshev polynomials from $x = -1$ to $x = +1$. Second, use trigonometric identities to show (analytically) that the representation:

$$T_n(x) = \cos\left(n \cos^{-1} x\right) \tag{2.123}$$

is equivalent to that in Eq. (2.122). Finally, use Eq. (2.123) to show that the n extrema of $T_{n-1}(x)$ are:

$$x_j = -\cos\left(\frac{j\pi}{n-1}\right), \qquad j = 0, 1, \ldots, n-1 \tag{2.124}$$

This result will play an important role in chapter 6.

2.23 [$\mathcal{P}$] Using `vectorfield.py` and `chargearray.py`, produce panels analogous to those in Fig. 2.4, made up of 36 charges (along six rows and six columns). In this problem, in addition to changing the magnitude of the charges (to either +1 or −1) you should also change their placement (i.e., they no longer need to be alternating as in our figure). Separately investigate the cases where the field at large distances behaves like: (a) a dipole, and (b) a quadrupole.

2.24 [$\mathcal{P}$] For the 36-charge array of section 2.5 determine, as a function of position along the y axis, the minimum number of terms you need to keep in the multipole expansion in order to ensure your relative error in the total potential is less than 10^{-6}.

2.25 [$\mathcal{P}$] Generalize our code in `multipole.py` so that it works regardless of whether or not $r > r_i$. This necessitates changes to all three functions in that program.

2.26 [$\mathcal{P}$] *Spherical harmonics* are very important in gravity, electromagnetism, and quantum mechanics (among several other topics). Here we will visualize them, starting from scratch; to do so, we first have to produce some scaffolding.

(a) The generalization of Bonnet's recurrence relation, Eq. (2.86), for *associated Legendre polynomials*, takes the form:

$$(n-m)P_n^m(x) = x(2n-1)P_{n-1}^m(x) - (n+m-1)P_{n-2}^m(x) \tag{2.125}$$

To get going, you will appreciate having access to:

$$P_m^m(x) = (-1)^m\,(2m-1)!!\,(1-x^2)^{m/2} \tag{2.126}$$

which makes use of the double factorial (the product of *odd* integers). Another helpful starting expression is:

$$P_{m+1}^m(x) = x(2m+1)P_m^m(x) \tag{2.127}$$

Compute $P_n^m(x)$ by stepping through Eq. (2.125), making use of Eq. (2.126) and Eq. (2.127) as necessary. Use your function to plot $P_4^2(x)$.
The above approach works for $0 \le m \le n$. To deal with negative m values, use:

$$P_n^{-m}(x) = (-1)^m \frac{(n-m)!}{(n+m)!} P_n^m(x) \tag{2.128}$$

Plot $P_4^{-2}(x)$; if you want, you can compare against `scipy.special.lpmv()`.

(b) Write a Python function that computes the *real spherical harmonics*:

$$Y_{nm}(\theta,\phi) = \begin{cases} (-1)^m\sqrt{2}\sqrt{\frac{2n+1}{4\pi}\frac{(n-|m|)!}{(n+|m|)!}}\,P_n^{|m|}(\cos\theta)\sin(|m|\phi), & \text{if } m<0 \\ \sqrt{\frac{2n+1}{4\pi}}\,P_n^0(\cos\theta), & \text{if } m=0 \\ (-1)^m\sqrt{2}\sqrt{\frac{2n+1}{4\pi}\frac{(n-m)!}{(n+m)!}}\,P_n^m(\cos\theta)\cos(m\phi), & \text{if } m>0 \end{cases} \tag{2.129}$$

(c) Produce a grid for θ (from 0 to π) and a grid for ϕ (from 0 to 2π). Visualize $|Y_{42}|$ using a 3d polar plot, which is conveniently provided by `Axes3D.plot_surface()`. Just like in `vectorfield.py`, you should use nested lists to produce the relevant quantities and switch to NumPy arrays at the last second. Our problem is three-dimensional in Cartesian coordinates, so you will need to pass in as arguments the appropriately converted forms of $|Y_{42}|$, i.e., $|Y_{42}|\sin\theta\cos\phi$, $|Y_{42}|\sin\theta\sin\phi$, and $|Y_{42}|\cos\theta$ (i.e., $|Y_{42}|$ plays the role of the radial component, r).

2.27 [$\mathcal{P}$] In quantum field theory, a physically rich toy model is that of the partition function in the zero-dimensional scalar self-interacting case:

$$Z(g) = \frac{1}{\sqrt{2\pi}} \int_{-\infty}^{+\infty} dx e^{-x^2/2 - gx^4/4!} \tag{2.130}$$

We haven't seen how to numerically evaluate integrals yet (that will have to wait until chapter 7), so we evaluate this integral analytically to give:

$$Z(g) = \sqrt{\frac{3}{2\pi g}} e^{3/(4g)} K_{1/4}\left(\frac{3}{4g}\right) \tag{2.131}$$

where $K_\nu(z)$ is the modified Bessel function of the second kind (of real order ν), whose values can be computed via `scipy.special.kv()`.

(a) Expand the $e^{-gx^4/4!}$ term in Eq. (2.130), under the assumption $|g| \ll 1$ and interchange the order of the integration and the summation, and then use a standard identity for the resulting Gaussian-like integral. Your final result should be:

$$\mathcal{A}(g) = \sum_{m=0}^{\infty} c_m g^m, \qquad c_m = \left(-\frac{1}{4!}\right)^m \frac{(4m)!}{4^m (2m)! m!} \tag{2.132}$$

Note that $Z(g)$ is convergent for $g > 0$ and divergent for $g < 0$; as a result, the series expansion in Eq. (2.132) has zero radius of convergence. In other words, our perturbative expansion is divergent; it can still be useful, however, since it is an asymptotic expansion.[44] Behind all the complexity buried in the c_m's, the main point is that they grow factorially; also important is the fact that they have alternating signs, similarly to what we encountered in section 2.4.4 when computing the exponential function for negative arguments.

(b) Call the partial sum for $\mathcal{A}(g)$ (summing from 0 up to and including n) $\mathcal{A}_n(g)$. Plot $\mathcal{A}_n(g)$ vs g (from 0.01 to 0.45), for $n = 5$ and $n = 10$; also show the exact answer from Eq. (2.131). Do you understand which n is doing a better job?

(c) Pick $g = 0.1$ and plot the difference between $\mathcal{A}_n(g)$ and the exact answer from Eq. (2.131), this time vs n (i.e., $|\mathcal{A}_n(g) - Z(g)|$ vs n). The idea behind *optimal truncation* is to truncate right before you reach the smallest term, i.e., to pick the n that this plot tells you will be the best. Figure out the appropriate range of n's and how to present the results visually. Then, produce a new plot for $g = 0.3$.

[44] In order to summarize things intuitively, rewrite Eq. (C.2) as $f(x) = \sum_{m=0}^{n-1} c_m (x - x_0)^m + R_n(x)$. We are dealing with a *convergent series* when, for x fixed, $|R_n(x)| \to 0$ as $n \to \infty$. We have an *asymptotic series* when, for n fixed, $|R_n(x)| \ll |x - x_0|^{n-1}$ as $x \to x_0$. The term "asymptotic" is usually reserved for the divergent case.

(d) We now introduce the *Borel transform* of the series in Eq. (2.132):

$$\mathcal{B}_{\mathcal{A}}(g) = \sum_{m=0}^{\infty} \frac{c_m g^m}{m!} \tag{2.133}$$

which divides every term in the expansion $\mathcal{A}(g)$ by $m!$. Remarkably, in many cases this produces a convergent series (i.e., with finite radius of convergence). Take $g = 0.1$ and plot the difference between the partial sums corresponding to $\mathcal{B}_{\mathcal{A}}(g)$ (for $n = 1, 2, \ldots, 20$), on the one hand, and the exact answer from Eq. (2.131), on the other, vs n (i.e., $|\mathcal{B}_{\mathcal{A},n}(g) - Z(g)|$ vs n). Add new sets of points for $g = 0.3, 1.4, 1.6$ and interpret your results in terms of series convergence.

(e) In the last plot, it's a little difficult to say what's going on for $g = 1.4$ and (especially) for $g = 1.6$. This results from using a small maximum n: employing floats, you will soon run into overflow issues if you try to make n much larger. Write a new function that employs NumPy's long doubles to compute the partial sum corresponding to $\mathcal{B}_{\mathcal{A}}(g)$ up to $n = 150$.[45] Motivated by problem 2.15, you could also sum the terms in reverse. Plot the results for $g = 1.4$ and $g = 1.6$ to determine what's going on.

(f) While it is intriguing that $\mathcal{B}_{\mathcal{A}}(g)$ is sometimes a convergent expansion, the match between the Borel-transform partial sums and the exact answer is not always great. Also, you may be thinking that it was arbitrary to divide by the factorial in Eq. (2.133). Both these concerns can be addressed by defining the *Borel sum*:

$$B(g) = \int_0^{\infty} e^{-t} \mathcal{B}_{\mathcal{A}}(tg) dt \tag{2.134}$$

Take $t \to t/g$, interchange the order of the summation and integration, and use $m! = \int_0^{\infty} e^{-t} t^m dt$, to show that asymptotically the behavior of $B(g)$ is the same as that of $\mathcal{A}(g)$. We now have a formulation with the same asymptotics, but $B(g)$ is an analytic function at $g = 0$, so we can reconstruct the exact non-perturbative answer, $Z(g)$.[46] To summarize, we can take a divergent perturbative series, carry out a Borel transform, proceed to the Borel sum, and thereby recover the true integral whose behavior the divergent series was meant to capture.

2.28 [$\mathcal{P}$] When implementing Hartree–Fock theory for Coulomb interactions, one option is to employ Gaussian-type orbitals; the (nuclear-attraction) integrals involved in this approach—$f_n(x) = \int_0^1 dt \exp(-xt^2) t^{2n}$—obey the following recurrence relation:

$$f_n(x) = \frac{(2n-1) f_{n-1}(x) - e^{-x}}{2x} \tag{2.135}$$

which is closer to Eq. (2.86) than to Eq. (2.64), since this involves an x dependence. Start with $f_{30}(x) = 1$ and employ backward recursion to plot $f_{10}(x)$.

[45] Of course, NumPy's long doubles are platform-dependent, so they may not be enough to help you go all the way to $n = 150$. Check `np.finfo(np.longdouble)` to see what you are dealing with.

[46] Note that $\mathcal{B}_{\mathcal{A}}(g)$ has a finite radius of convergence, whereas the integral in Eq. (2.134) needs the Borel transform evaluated at all arguments. Thus, one has to analytically continue $\mathcal{B}_{\mathcal{A}}(g)$; to do so, one needs to know *all* the coefficients of the series. In realistic applications, one goes beyond the partial sum of the Borel transform (which is what we used above), employing a tool like the *Padé approximant* (which we don't go into).

3 Derivatives

After great pain, a formal feeling comes.

Emily Dickinson

3.1 Motivation

In this chapter and in all the following ones we start out with a section titled "Motivation", which has a dual purpose: (a) to provide examples of the chapter's theme drawn from the study of physics, and (b) to give a mathematical statement of the problem(s) that will be tackled in later sections.

3.1.1 Examples from Physics

Here are three examples of the use of derivatives in physics:

1. In classical mechanics, the definition of the *velocity* of a single particle is:

$$\mathbf{v} = \frac{d\mathbf{r}}{dt} \tag{3.1}$$

   where $\mathbf{r}$ is the position of the particle in a given reference frame. This definition tells us that the velocity is the time derivative of the position.
2. In classical electromagnetism, the equation connecting the electric field $\mathbf{E}$ with the vector potential $\mathbf{A}$ and scalar potential ϕ is:

$$\mathbf{E} = -\nabla\phi - \frac{\partial \mathbf{A}}{\partial t} \tag{3.2}$$

   The right-hand side contains both spatial derivatives and a time derivative.
3. The Lagrangian density for the vibrations of a continuous rod can be expressed as:

$$\mathcal{L} = \frac{1}{2}\left[\frac{1}{c^2}\left(\frac{\partial\varphi}{\partial t}\right)^2 - \left(\frac{\partial\varphi}{\partial x}\right)^2\right] \tag{3.3}$$

   Here c is the velocity of longitudinal elastic waves. The dynamical variable is $\varphi(x, t)$. As you can see, the spatial and temporal derivatives of this quantity together make up the Lagrangian density. Something very similar to this expression appears in state-of-the-art lattice field theory computations.

3.1.2 The Problem to Be Solved

More generally, our task is to evaluate the derivative of $f(x)$ at a specific point, $f'(x)$. If we have an analytical expression for the x-dependence of the function $f(x)$, then this problem is in principle trivial , even though humans are error-prone when dealing with complicated expressions. However, in many cases we, instead, have a set of n discrete data points (i.e., a table) of the form $(x_i, f(x_i))$ for $i = 0, 1, \ldots, n-1$.[1] This often happens when we are dealing with a computationally demanding task: while we could produce more-and-more sets of points, in practice, we are limited by the time it takes to produce those.

One could approach this problem in a number of ways. First, we can use interpolation or data fitting (see chapter 6) to produce a new function that approximates the data reasonably well, and then apply *analytical differentiation* to that function. This is especially helpful when we are dealing with noisy data. On the other hand, just because a function approximates a set of data points quite well, doesn't mean that it also captures information on the derivative (or even higher-order derivatives) of the function. In other words, interpolating/fitting and then taking the derivative doesn't provide much guidance regarding the error involved when we're interested in the derivative (do we know we're using the right analytical form?). The second class of approach is helpful in systematizing our ignorance: it is called the *finite-difference* approach, also known simply as numerical differentiation. In a nutshell, it makes use of the Taylor series expansion of the function we are interested in differentiating, around the specific point where we wish to evaluate the derivative. The third class of approach is known as *automatic differentiation*: this is as accurate as analytical differentiation, but it deals with numbers instead of mathematical expressions.

3.2 Analytical Differentiation

Derivatives like those discussed in the previous section are defined in the usual way:

$$\frac{df(x)}{dx} = \lim_{h \to 0} \frac{f(x+h) - f(x)}{h} \tag{3.4}$$

This gives us the derivative of a function f at the point x in terms of a limit. In practice, we typically don't apply this definition, but use, instead, standard differentiation rules for powers, quotients, products, as well as the well-known expressions for the derivative of several special functions. As you learned a long time ago, the application of these rules can help you differentiate any function you wish to, e.g.:

$$\frac{d}{dx} e^{\sin(2x)} = 2\cos(2x)\, e^{\sin(2x)} \tag{3.5}$$

You could similarly evaluate the second derivative, the third derivative, and so on.

[1] Note that when labelling our n points we start at 0 and end at $n-1$: this is consistent with Python's 0-indexing.

In most of this book, we are interested in *computing*, namely plugging actual numbers into mathematical expressions. Even so, there are times when carrying out a numerical evaluation might obscure underlying simplicities. Other times, carrying analytical calculations out by hand can get quite tedious. The solution for these scenarios is to use a computer algebra package. Such packages allow one to carry out symbolic manipulations on the computer. `Sage` is a mathematical software system with a "Python-like" syntax that was intended to be an alternative to solutions like `Maple`, `Mathematica`, or `Matlab`. In problem 3.1, we focus on a much more lightweight solution, namely `SymPy` (actually included as part of `Sage`): this is a Python module (that can be used like any other Python module) to carry out symbolic manipulations using Python idioms whenever possible.

3.3 Finite Differences

Let us return to the definition of the derivative given in Eq. (3.4):

$$\left.\frac{df(x)}{dx}\right|_{\tilde{x}} = \lim_{h\to 0}\frac{f(\tilde{x}+h)-f(\tilde{x})}{h} \tag{3.6}$$

where we are slightly changing our notation to show that this definition allows us to evaluate the derivative of $f(x)$ at the point $\tilde{x}$.[2] The obvious thing that comes to mind is to try to use this formula but, instead of taking the limit $h \to 0$, simply take h to be "small". This *ad hoc* approach probably has you wondering: (a) do we have any understanding of the errors involved? (b) do we know what "small" means? (c) do we realize that, typically, the smaller h becomes, the smaller the numerator becomes? Focusing on the last point: as we're making h smaller and smaller, we're producing a smaller and smaller numerator $f(\tilde{x}+h)-f(\tilde{x})$, since we're evaluating the function f at two points that are just next to each other: as if that wasn't enough, then we're simply dividing by that tiny number h (since it's in the denominator), magnifying any mistakes we made in the evaluation of the numerator.

In the rest of this section, we will try to remedy the problems of this *ad hoc* approach, in order to be more systematic. As mentioned above, we will make repeated use of the Taylor series expansion of the function we are interested in differentiating.

3.3.1 Non-Central-Difference Approximations

3.3.1.1 Forward Difference

With the aforementioned disclaimer that we don't distinguish between x and $\tilde{x}$, let us start from the all-important Taylor expansion of $f(x+h)$ around x, see Eq. (C.1):

$$f(x+h) = f(x) + hf'(x) + \frac{h^2}{2}f''(x) + \frac{h^3}{6}f'''(x) + \frac{h^4}{24}f^{(4)}(x) + \cdots \tag{3.7}$$

This can be trivially re-arranged to give:

$$f'(x) = \frac{f(x+h)-f(x)}{h} - \frac{h}{2}f''(x) + \cdots \tag{3.8}$$

[2] Informally, this distinction between a general point x and a specific point $\tilde{x}$ is often passed over.

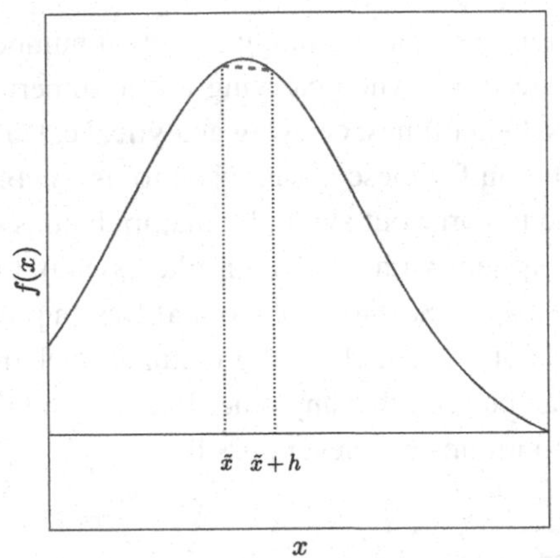

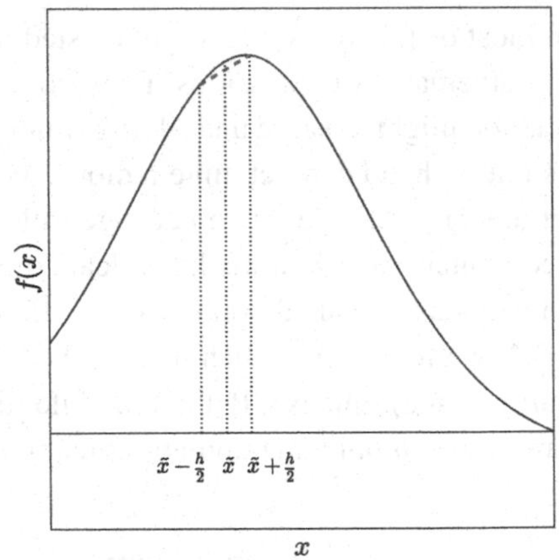

Fig. 3.1 First approximation of a first derivative: forward (left), central (right)

This naturally leads to an approximation for the value of the derivative of $f(x)$ at x:

$$f'(x) = \frac{f(x+h) - f(x)}{h} + O(h) \quad (3.9)$$

known as the *(first) forward-difference approximation*. To be clear, this approximation consists in using only the fraction on the right-hand side to evaluate the derivative: as the second term on the right-hand side shows, this suffers from an error $O(h)$: this is possibly not too bad when h is small. This recipe is called a *forward* difference because it starts at x and then moves in the forward/positive direction to $x + h$. Graphically, this is represented in the left panel of Fig. 3.1. Clearly, the forward difference is nothing other than the slope of the line segment connecting $f(x)$ and $f(x + h)$. This is not always a great approximation: for the example chosen, we would expect the slope of $f(x)$ to be positive.

Note that the formula we arrived at for the forward difference happens to be identical to the formula that we qualitatively discussed after Eq. (3.6): the difference here is that, due to the derivation starting from the Taylor series, we have a handle on the error this approximation corresponds to. It may still suffer from the issues mentioned above (tiny numerator, tiny denominator), but at least now we have some guidance on how well we're doing: if the h is not too small (so that we're still away from major roundoff issues), halving the h should double the quality of the approximation (in absolute terms).

Incidentally, we used in Eq. (3.9) the O symbol, which you may have not encountered before (see also appendix C.1). This is known as big-O notation. For our purposes, it is simply a concise way of encapsulating only the dependence on the most crucial parameter, without having to worry about constants, prefactors, etc. Thus, $O(h)$ means that the leading error is of order h. Since h will always be taken to be "small", we see that an error $O(h)$ is much larger/worse than, say, an error $O(h^6)$. Though there could, in principle, also be prefactors that complicate such a comparison, the essential feature is captured by the power-law (or other) dependence attached to the O symbol.[3]

3.3.1.2 Backward Difference

At this point, we observe that we started our derivation in Eq. (3.7) with $f(x+h)$ and then ended up with the prescription Eq. (3.9) for the forward difference. Obviously, we could just as well have started with the Taylor expansion of $f(x-h)$:

$$f(x-h) = f(x) - hf'(x) + \frac{h^2}{2}f''(x) - \frac{h^3}{6}f'''(x) + \frac{h^4}{24}f^{(4)}(x) + \cdots \quad (3.10)$$

Again, this can be trivially re-arranged to give:

$$f'(x) = \frac{f(x) - f(x-h)}{h} + \frac{h}{2}f''(x) - \cdots \quad (3.11)$$

This naturally leads to an approximation for the value of the derivative of $f(x)$ at x:

$$f'(x) = \frac{f(x) - f(x-h)}{h} + O(h) \quad (3.12)$$

known as the *(first) backward-difference approximation*. This is called a *backward* difference because it starts at x and then moves in the negative direction to $x-h$. It is nothing other than the slope of the line segment connecting $f(x-h)$ and $f(x)$. The forward and backward differences are very similar, so we focus on the former, for the sake of concreteness. These are collectively known as *non-central differences*, as used in the title of this subsection. "Non-central" is employed in contradistinction to "central", which we discuss in the following subsection. Before we do that, though, we will spend some time discussing the errors of the forward-difference approximation in more detail.

3.3.1.3 Error Analysis for the Forward Difference

We've already seen above that we want the h to be small[4] but, on the other hand, we don't want to make the h too small, as that will give rise to roundoff errors. Clearly, the best-possible choice of h will be somewhere in the middle.

We're here dealing with the combination of an approximation error $\mathcal{E}_{app}$, coming from the truncation of the Taylor series, and a roundoff error $\mathcal{E}_{ro}$, coming from the subtraction and division involved in the definition of the forward difference. In chapter 2 we defined the absolute error as "approximate minus exact", see Eq. (2.5); here we're using a new symbol, $\mathcal{E}$, to denote the magnitude of the absolute error. The magnitude of the approximation error

[3] Here we are interested in small h, in which case $O(h^2)$ is "better" than $O(h)$. In later chapters we will examine the dependence on the required number of function evaluations n (where n is large) so, there, $O(n^2)$ will be "worse" than $O(n)$. In both cases, $O(h)$ as $h \to 0$ or $O(n)$ as $n \to \infty$, we are dealing with *asymptotic notation*.

[4] So that the term $O(h)$ we're chopping off doesn't matter very much.

is immediately obvious, given Eq. (3.8):

$$\mathcal{E}_{app} = \frac{h}{2}|f''(x)| \tag{3.13}$$

Actually this result is not totally trivial: from what we know about the Lagrange remainder in a Taylor series, this expression should involve $|f''(\xi)|$, where ξ is a point between x and $x + h$. However, since h is small, it is a reasonable approximation to take $|f''(\xi)| \approx |f''(x)|$; we will make a similar approximation (more than once) below.

Turning to $\mathcal{E}_{ro}$, we remember that we are interested in evaluating $(f(x + h) - f(x))/h$. We focus on the numerator: this is subtracting two numbers that are very close to each other, bringing to mind the discussion in section 2.2.2.1. As per Eq. (2.20), to find the absolute error of the subtraction $f(x + h) - f(x)$ we add together the absolute errors in $f(x + h)$ and in $f(x)$. As per Eq. (2.51), we see that the absolute error in $f(x + h) - f(x)$ is $f(x)2\epsilon_m$, where we assumed $f(x + h) \approx f(x)$ and that the relative error for each function evaluation is approximated by the machine error. If we now ignore the error incurred by the division by h (which we know is a reasonable thing to do), we see that the absolute error in evaluating the forward difference, $(f(x + h) - f(x))/h$, will simply be the absolute error in the numerator divided by h:

$$\mathcal{E}_{ro} = \frac{2|f(x)|\epsilon_m}{h} \tag{3.14}$$

We can now explicitly see that for the case under study the approximation error decreases as h gets smaller, whereas the roundoff error increases as h gets smaller. Adding the approximation error together with the roundoff error gives us:

$$\mathcal{E} = \mathcal{E}_{app} + \mathcal{E}_{ro} = \frac{h}{2}|f''(x)| + \frac{2|f(x)|\epsilon_m}{h} \tag{3.15}$$

To minimize this total error, we set the derivative with respect to h equal to 0:

$$\frac{1}{2}|f''(x)| - \frac{2|f(x)|\epsilon_m}{h_{opt}^2} = 0 \tag{3.16}$$

where we called the h value that minimizes the total absolute error h_{opt}. For future reference, we note that Eq. (3.16) can be manipulated to give:

$$\frac{h_{opt}}{2}|f''(x)| = \frac{2|f(x)|\epsilon_m}{h_{opt}} \tag{3.17}$$

Our equation in Eq. (3.16) can be solved for the optimal value of h giving us:

$$h_{opt} = \sqrt{4\epsilon_m \left|\frac{f(x)}{f''(x)}\right|} \tag{3.18}$$

These results can be plugged into Eq. (3.15) to give the smallest possible error in the forward difference. We do this in two steps. First, we use Eq. (3.17) to find:

$$\mathcal{E}_{opt} = h_{opt}|f''(x)| \tag{3.19}$$

Second, we plug Eq. (3.18) into our latest result to find:

$$\mathcal{E}_{opt} = \sqrt{4\epsilon_m |f(x)f''(x)|} \tag{3.20}$$

For concreteness, assume that $f(x)$ and $f''(x)$ are of order 1. Our result in Eq. (3.18) then tells us that we should pick $h_{opt} = \sqrt{4\epsilon_m} \approx 3 \times 10^{-8}$. In that case, the (optimal/minimum) absolute error is also $\mathcal{E}_{opt} = \sqrt{4\epsilon_m} \approx 3 \times 10^{-8}$. These numbers might, of course, look quite different if the value of the function (or its second derivative) is much greater than or less than 1. Note that an error of 10^{-8} is not that impressive: in section 3.2 on analytical differentiation we had *no* differentiation error: the only error involved was the function evaluation, which is typically a significantly less important problem.

Keep in mind that statements like "the error in this approach is $O(h)$" (referring to the approximation error) are only true for reasonably well-behaved functions. If the corresponding derivative is infinite or doesn't exist, then you cannot rely on the straightforward results on scaling we discuss here.

3.3.2 Central-Difference Approximation

We will now try to find a way to approximate the derivative using a finite difference, but more accurately than with the forward/backward difference. As before, we start from the Taylor expansion of our function around x, but this time we choose to make a step of size $h/2$ (as opposed to the h in the previous section):

$$f\left(x+\frac{h}{2}\right) = f(x) + \frac{h}{2}f'(x) + \frac{h^2}{8}f''(x) + \frac{h^3}{48}f'''(x) + \frac{h^4}{384}f^{(4)}(x) + \cdots \tag{3.21}$$

We can also write down a similar Taylor expansion for a move in the negative direction:

$$f\left(x-\frac{h}{2}\right) = f(x) - \frac{h}{2}f'(x) + \frac{h^2}{8}f''(x) - \frac{h^3}{48}f'''(x) + \frac{h^4}{384}f^{(4)}(x) - \cdots \tag{3.22}$$

Comparing the last two equations, we immediately see that adding them or subtracting them can lead to useful patterns: the sum contains only even derivatives whereas the difference contains only odd derivatives. In the present case, we subtract the second equation from the first: all even derivatives along with the $f(x)$ term cancel. We can now re-arrange the result, solving for $f'(x)$:

$$f'(x) = \frac{f\left(x+\frac{h}{2}\right) - f\left(x-\frac{h}{2}\right)}{h} - \frac{h^2}{24}f'''(x) + \cdots \tag{3.23}$$

This naturally leads to an approximation for the value of the derivative of $f(x)$ at x:

$$f'(x) = \frac{f\left(x+\frac{h}{2}\right) - f\left(x-\frac{h}{2}\right)}{h} + O(h^2) \tag{3.24}$$

known as the *(first) central-difference approximation*. As the second term on the right-hand side shows, this suffers from an error $O(h^2)$: since h is generally small, h^2 is even smaller.

Just like for the forward-difference approximation in Eq. (3.9), computing the central difference requires only two function evaluations. This is called the *central* difference because these two evaluations are at $x - h/2$ and at $x + h/2$, i.e., they are centered at x: just like for the forward difference, these two points are h apart. Graphically, this is represented in the right panel of Fig. 3.1. Clearly, the central difference is nothing other than the slope of the line segment connecting $f(x - h/2)$ and $f(x + h/2)$. As the figure shows, for the example chosen this is a much better approximation than the forward difference in the left panel. We will discuss this approximation's error budget below: for now, note that if the h is not too small (so that we're still away from major roundoff issues), halving the h should quadruple the quality of the approximation (in absolute terms).

There exists one (very common in practice) situation where a central-difference approximation is simply not usable: if we have a set of n discrete data points (i.e., a table) of the form $(x_i, f(x_i))$ for $i = 0, 1, \ldots, n-1$ we will not be able to use a central difference to approximate the derivative at x_0 or at x_{n-1}: for any of the "middle" points we could always use two evaluations (one on the left, one on the right), but for the two endpoints we simply don't have points available "on the other side", so a forward/backward difference is necessary there. (We return to the question of points on a grid in section 3.3.6.)

3.3.2.1 Error Analysis for the Central Difference

We wish the h to be small,[5] but we should be cautious about not making the h too small, lest that give rise to roundoff errors. Once again, the best-possible choice of h will be somewhere in the middle. As you'll show in problem 3.2, for this case adding the approximation error together with the roundoff error gives us:

$$\mathcal{E} = \mathcal{E}_{app} + \mathcal{E}_{ro} = \frac{h^2}{24}|f'''(x)| + \frac{2|f(x)|\epsilon_m}{h} \tag{3.25}$$

You will also show that minimizing this total error leads to:

$$h_{opt} = \left(24\epsilon_m \left|\frac{f(x)}{f'''(x)}\right|\right)^{1/3}, \quad \mathcal{E}_{opt} = \left(\frac{9}{8}\epsilon_m^2 [f(x)]^2 |f'''(x)|\right)^{1/3} \tag{3.26}$$

For concreteness, assume that $f(x)$ and $f'''(x)$ are of order 1. Our result in Eq. (3.26) then tells us that we should pick $h_{opt} = (24\epsilon_m)^{1/3} \approx 2\times10^{-5}$. In that case, the (optimal/minimum) absolute error is $\mathcal{E}_{opt} = (9\epsilon_m^2/8)^{1/3} \approx 4\times10^{-11}$. These numbers might, of course, look quite different if the value of the function (or its third derivative) is much greater than or less than 1. This time our error of 10^{-11} is much better than that for the forward difference, though it's still likely worse than what we got in section 3.2 using analytical differentiation.

We just observed that the error for the central difference (10^{-11}) is considerably smaller than that for the forward difference (10^{-8}). Intriguingly, the optimal step size for the central difference (10^{-5}) is orders of magnitude *larger* than the optimal step size for the forward difference (10^{-8}): the better algorithm allows us to "get away with" a larger h.

[5] As before, so that the term $O(h^2)$ we're chopping off doesn't matter very much.

3.3.3 Implementation

We now turn to an example: we will employ the same function, $f(x) = e^{\sin(2x)}$, that we analytically differentiated in section 3.2. For concreteness, we study the derivative at a fixed x, namely $x = 0.5$. We will examine the absolute errors in approximating the derivative using forward and central differences in code 3.1. We first define one Python function, `f()`, corresponding to the mathematical function whose derivative we're interested in calculating, and then another Python function, `fprime()`, corresponding to the derivative itself: this will be used as an analytical benchmark, with which to compare the difference formula results. We then define functions giving the forward and central difference results, which take in as an argument the function we wish to differentiate (which could be changed later). We then re-encounter the `if __name__ == '__main__':` idiom. This will come in handy in a later section, when we will employ the functions defined in this program without running the rest of the code.

We use a few list comprehensions to store the step sizes and the difference-formula results corresponding to them. In order to make the output more legible than it would be by default, we employ a format string as discussed in chapter 1. The only new feature here is that we store the format string into the variable `rowf`, in order to make the line containing the `print()` itself easier to understand. We print things out this way because we will be faced with a cascade of zeros. The output of running this code is:

```
h     abs. error in fd   abs. error in cd
1e-01 0.3077044583376249 0.0134656094697689
1e-02 0.0260359156901186 0.0001350472492652
1e-03 0.0025550421497806 0.0000013505116288
1e-04 0.0002550180941236 0.0000000135077878
1e-05 0.0000254969542519 0.0000000001051754
1e-06 0.0000025492660578 0.0000000002500959
1e-07 0.0000002564334673 0.0000000011382744
1e-08 0.0000000255070782 0.0000000189018428
1e-09 0.0000000699159992 0.0000000699159992
1e-10 0.0000021505300500 0.0000021505300500
1e-11 0.0000332367747395 0.0000111721462455
```

Let's first look at the forward-difference results: we see that as h gets smaller by an order of magnitude, the absolute error also gets smaller by an order of magnitude: recall that for this approach the approximation error is $O(h)$. The minimum absolute error turns out to be $\approx 2.5 \times 10^{-8}$, which appears for the step size $h = 10^{-8}$. This is qualitatively consistent with what we saw in the section on the error analysis above; this time around we don't have guarantees that the function (or derivative) values are actually 1. Actually, the absolute error we found here is slightly smaller than expected, due to a cancellation of errors that we couldn't generally assume was going to be present (e.g., for another value of x). Beyond this point, as we keep reducing h, we see that the absolute error starts increasing: the roundoff error is now dominating, leading to increasingly poor results.

Code 3.1 finitediff.py

```
from math import exp, sin, cos

def f(x):
    return exp(sin(2*x))

def fprime(x):
    return 2*exp(sin(2*x))*cos(2*x)

def calc_fd(f,x,h):
    fd = (f(x+h) - f(x))/h
    return fd

def calc_cd(f,x,h):
    cd = (f(x+h/2) - f(x-h/2))/h
    return cd

if __name__ == '__main__':
    x = 0.5
    an = fprime(x)

    hs = [10**(-i) for i in range(1,12)]
    fds = [abs(calc_fd(f,x,h) - an) for h in hs]
    cds = [abs(calc_cd(f,x,h) - an) for h in hs]

    rowf = "{0:1.0e} {1:1.16f} {2:1.16f}"
    print("h     abs. error in fd       abs. error in cd")
    for h,fd,cd in zip(hs,fds,cds):
        print(rowf.format(h,fd,cd))
```

The results for the central-difference formula are completely analogous: we see that as h gets smaller by an order of magnitude, the absolute error gets smaller by two orders of magnitude: recall that for this approach the approximation error is $O(h^2)$. This process reaches a minimum at $h = 10^{-5}$: the absolute error there is $\approx 10^{-10}$, similarly to what was seen in our earlier discussion of the error budget. As before, reducing the h further gives disappointing results, since the roundoff error dominates beyond that point.

After you solve problem 3.3, you will produce Fig. 3.2. The step size h gets smaller as we go to the left, whereas the absolute value of the absolute error gets smaller as we go to the bottom of the plot. The behavior encountered in this plot (for a specific function at a

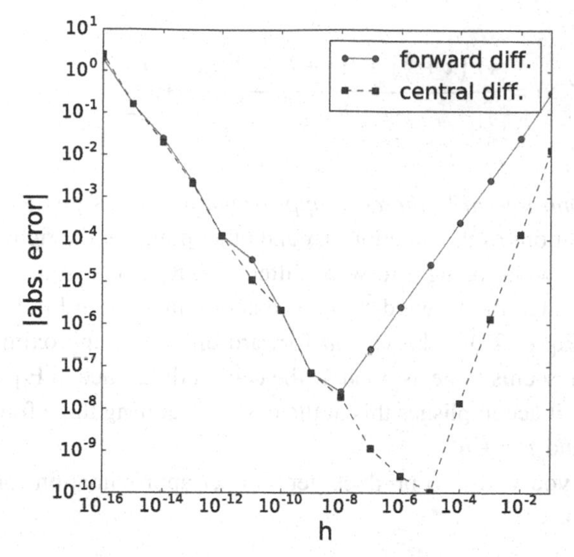

Log-log plot using forward-difference and central-difference formulas

Fig. 3.2

specific point) is pretty generic. On the right part of the plot we are dominated by the truncation/approximation error: there, the central difference is clearly superior, as the absolute error is consistently smaller. In this region, we can easily see our earlier statements in action: as h changes by an order of magnitude, the error of the forward difference changes by an order of magnitude, whereas the error of the central difference changes by two orders of magnitude. As we keep moving to the left, beyond a certain point roundoff error starts dominating and the forward difference is as good (or as bad) as the central difference. We stop plotting when the error is so large that we are completely failing at evaluating the derivative.

It's worth pointing out that the source of all the roundoff problems that *all* finite-difference formulas suffer from is that in each case *the sum of all the function evaluation coefficients is zero!* This is pretty obvious at this point, since we've only seen two finite-difference formulas—Eq. (3.9) which contained $+f(x+h) - f(x)$ and Eq. (3.24) which contained $+f(x+h/2) - f(x-h/2)$—but you should keep it in mind as you proceed. Having terms that nearly cancel each other invites trouble.

3.3.4 More Accurate Finite Differences

Up to this point, we've seen non-central and central finite differences; the forward difference had an error $O(h)$ and the central difference an error $O(h^2)$. It's important to note that both the forward difference, Eq. (3.9), and the central difference, Eq. (3.24), require two function evaluations to calculate a finite-difference ratio.

As you will show in problem 3.6, one can produce the following approximation for the value of the derivative of $f(x)$ at x:

$$f'(x) = \frac{4f\left(x+\frac{h}{2}\right) - f(x+h) - 3f(x)}{h} + \frac{h^2}{12}f'''(x) + \cdots \quad (3.27)$$

known as the *second forward-difference approximation*. It has this name because: (a) it also requires evaluations of the function at x and other points to the right (so it is a forward difference), and (b) while being a forward difference, it suffers from an error $O(h^2)$, i.e., it is more accurate than the forward-difference approximation in Eq. (3.9), which has an error of $O(h)$ (so Eq. (3.27) is the *second* forward-difference approximation). While this new approximation seems to be as good as the central difference in Eq. (3.24) which also has an error $O(h^2)$, it accomplishes this at the cost of requiring three function evaluations: $f(x)$, $f(x+h/2)$, and $f(x+h)$.

In problem 3.6 you will also produce yet another approximation for the value of the derivative of $f(x)$ at x:

$$f'(x) = \frac{27f\left(x+\frac{h}{2}\right) + f\left(x-\frac{3}{2}h\right) - 27f\left(x-\frac{h}{2}\right) - f\left(x+\frac{3}{2}h\right)}{24h} + \frac{3}{640}h^4 f^{(5)}(x) + \cdots \quad (3.28)$$

known as the *second central-difference approximation*, for obvious reasons. This approximation has an error $O(h^4)$, making it the most accurate finite-difference formula (for the first derivative) that we've encountered. However, it accomplishes this at the cost of requiring four function evaluations: $f(x+3h/2)$, $f(x+h/2)$, $f(x-h/2)$, and $f(x-3h/2)$. It's worth emphasizing what we noted above: the sum of all the function evaluation coefficients is zero for this formula, too $(27+1-27-1=0)$.

It should be clear by now that one can keep including more points, taking Taylor expansions, multiplying and adding/subtracting those together, to arrive at formulas (whether central or non-central) that are of even higher order in accuracy. As we've already noted, the problem with that approach is that one needs increasingly many function evaluations: in practice, evaluating $f(x)$ at different points is a costly task, so it turns out that more accurate expressions like those we've seen in this section are not too commonly encountered "in the wild". (You should also try to come up with a $O(h^4)$ method by using $f(x+h/2)-f(x-h/2)$ together with $f(x+h)-f(x-h)$.)

3.3.5 Second Derivative

In practice, we also need higher-order derivatives: the second derivative is incredibly important in all areas of physics. It should come as no surprise that one can set up forward, backward, and central difference expressions (of increasing sophistication) that approximate the second derivative. It is common to derive the simplest possible finite-difference formula for the second derivative by saying that the second derivative is the first derivative of the first derivative, which symbolically translates to:

$$f''(x) \approx \frac{f'\left(x+\frac{h}{2}\right)-f'\left(x-\frac{h}{2}\right)}{h} \tag{3.29}$$

This is nothing other than the central-difference formula, Eq. (3.24), applied once. Applying it twice more (on the right-hand side) would give us an explicit approximation formula for the second derivative in terms of function evaluations only.

Instead of pursuing that avenue, however, we will here follow the spirit of the previous sections, where we repeatedly used Taylor expansions, manipulating them to eliminate terms of our choosing. Specifically, we look, once again, at Eq. (3.21) and Eq. (3.22). At this point, we note that subtracting one of these equations from the other leads to odd derivatives alone. It is just as easy to see that summing these two Taylor expansions together:

$$f\left(x+\frac{h}{2}\right)+f\left(x-\frac{h}{2}\right) = 2f(x) + \frac{h^2}{4}f''(x) + \frac{h^4}{192}f^{(4)}(x) + \cdots \tag{3.30}$$

leads to even derivatives alone. Since we're interested in approximating the second derivative, we realize we're on the right track. Our result can be trivially re-arranged to give:

$$f''(x) = 4\frac{f\left(x+\frac{h}{2}\right)+f\left(x-\frac{h}{2}\right)-2f(x)}{h^2} - \frac{h^2}{48}f^{(4)}(x) - \cdots \tag{3.31}$$

This leads to an approximation for the value of the second derivative of $f(x)$ at x:

$$f''(x) = 4\frac{f\left(x+\frac{h}{2}\right)+f\left(x-\frac{h}{2}\right)-2f(x)}{h^2} + O(h^2) \tag{3.32}$$

known as the *(first) central-difference approximation* to the second derivative.

While it's not too common in physics to need derivatives beyond the second derivative, it should be straightforward to see how one would go about calculating higher-order derivatives. We've seen that sums of Taylor series give us even derivatives and differences of the Taylor series give us odd derivatives: we simply need to combine sufficiently many sums or differences, to cancel all unwanted terms.[6] For example, $f^{(4)}(x)$ can be approximated using $f(x+h/2)+f(x-h/2)$ together with $f(x+h)+f(x-h)$.

3.3.6 Points on a Grid

So far, we've been quite cavalier in our use of different step sizes h and the placement of different points at x, $x+h/2$, $x+h$, and so on. In other words, we were taking it for granted that f was at our disposal, meaning that we could evaluate the function at any point of our choosing. This does happen sometimes in practice: for example, in section 3.5 we will evaluate a kinetic energy by numerically taking the second derivative of a wave function,

[6] More systematically, one sets up a linear system of equations involving a *Vandermonde matrix*. A similar problem arises when computing the coefficients of, say, the first derivative using an unequally spaced grid.

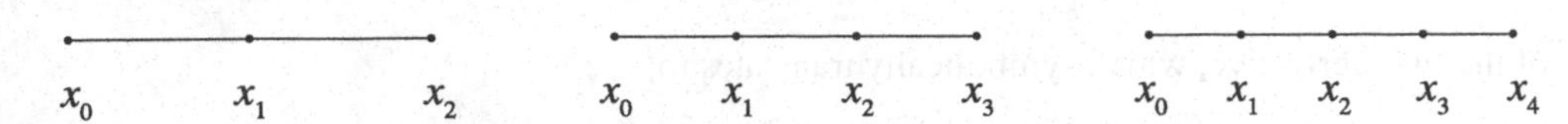

Fig. 3.3 Illustration of points and subintervals/panels

with an h that is up to us. However, as we noted in the first section of the present chapter, sometimes we don't control the function and the points at which it is evaluated but, instead, have access only to a set of n discrete data points (i.e., a table) of the form $(x_i, f(x_i))$ for $i = 0, 1, \ldots, n-1$. In this case, the function is known only at fixed points, x_i, not of our choosing.

3.3.6.1 Avoiding Error Creep

A very common use case is when the points x_i are on an equally spaced grid (also known as a *mesh*), from a to b. The n points are then given by the following relation:

$$x_i = a + ih \tag{3.33}$$

where, as usual, $i = 0, 1, \ldots, n-1$. The h is clearly given by:

$$h = \frac{b-a}{n-1} \tag{3.34}$$

Thus, even if this h is small (i.e., we have many points n), it is not of our choosing.

Since we will make heavy use of the last two formulas in what follows, let us take some time to interpret them. First, note that if you use all x_i's, then you are dealing with a *closed* approach: x_i ranges from a (for $i = 0$) to b (for $i = n-1$). In other words, in that case the endpoints of our interval would be included in our set of points. Second, we are using notation that will easily translate to Python's (or any other C-based language's) 0-indexing: we start counting at 0 and stop counting at $n-1$. To be explicit: we have n points in total (two of which are the endpoints). Third, we observe that the formula for the h, Eq. (3.34), simply takes the interval from a to b and splits it up into smaller pieces. When we are dealing with n points in total, we are faced with $n-1$ subintervals from a to b. In what follows, we will be using the terms *subinterval* and *panel* interchangeably.[7] This is illustrated in Fig. 3.3; note that the overall interval (from a to b) always stays the same.

Observe that Eq. (3.33) always starts at $x = a$ and then automatically transports you to your desired x_i, e.g., if $i = 17$ it brings you up to $x_{17} = a + 17h$. You can store all of these x_i

[7] You might choose to introduce a new variable containing the number of panels, $N = n-1$, in which case $h = (b-a)/N$. In that case, the i in x_i would go from 0 to N, since we have $N+1 = n$ points in total. This is perfectly legitimate, but does introduce the cognitive load of having to keep track of two different variables, n and N. We'll stick to n alone in most of what follows.

into a Python list by saying `xs = [a+i*h for i in range(n)]`. Note how `range` automatically ensures that `i` goes from `0` to `n-1`. While this may appear to be rather unremarkable, it's worth observing that there are those[8] who would prefer to use an alternative formula:

$$x_{i+1} = x_i + h \tag{3.35}$$

You may now be able to see where this is going: instead of storing all the x_i's so as to use them only once, you might be tempted to simply use a "running" `x` variable, which you can keep updating inside a loop by saying `x += h`. Unfortunately, this would be a mistake: h is almost certainly bound to be evaluated with limited precision. That means that each time we are adding h to the previous result we are committing another addition error. If you are dealing with thousands of points (or more) it's easy to see how an initial error of ϵ_m in evaluating x can be exacerbated. Since we're using these points on a grid in order to evaluate a finite difference, we are setting ourselves up for failure: our `x`'s are becoming progressively worse, therefore the ordinates (i.e., the `f(x)`) will also be wrong; this is a case of "systematic creep". It's easy to see that our starting expression in Eq. (3.33) is much better: it only involves one multiplication and one addition, and is therefore always to be preferred. If you read and understood chapter 2, this point should be quite straightforward. Of course, you may choose to avoid both the systematic creep (via the use of $x_i = a + ih$) and the storing of all the x_i's, simply by reapplying $x_i = a + ih$ each time through the loop.

3.3.6.2 Finite-Difference Formulas

Now, let us assume our task is to calculate the first derivative of $f(x)$ *at the same points* x_i. To be fully explicit, let's assume that the x_i's are 101 points from 0 to 5. This leads to a step size of $h = 0.05$: these points are $0, 0.05, 0.1, 0.15, 0.2, 0.25, 0.3, \ldots, 4.9, 4.95, 5.0$. For example, we are interested in $f'(3.7)$ but the neighboring values at our disposal are only $f(3.65)$, $f(3.7)$, and $f(3.75)$. The forward-difference formula in Eq. (3.9):

$$f'(x_i) = \frac{f(x_{i+1}) - f(x_i)}{h} + O(h) \tag{3.36}$$

shows that if we take $x_i = 3.7$ we can estimate $f'(3.7)$ using $f(3.7)$ and $f(3.75)$ with $h = 0.05$. You can even think of the denominator in Eq. (3.36) as $x_{i+1} - x_i$. However, things are not so simple for the case of the central-difference formula in Eq. (3.24):

$$f'(x_i) = \frac{f(x_{i+1/2}) - f(x_{i-1/2})}{h} + O(h^2) \tag{3.37}$$

This involves quantities like $f(x_{i+1/2}) \equiv f(x_i + h/2)$ that lie off-grid. Taking $x_i = 3.7$ and $h = 0.05$, this formula requires the function evaluations $f(3.675)$ and $f(3.725)$. But, as we already noted, we have access to $f(3.65)$, $f(3.7)$, and $f(3.75)$, not to $f(3.675)$ and $f(3.725)$. What we *can* do, is take the h in the central-difference formula in Eq. (3.24) to be twice as large, i.e., take $h = 0.1$ (obviously, this is not the same as the h that was used to produce the grid of points in Eq. (3.33)). For $h = 0.1$, the central-difference formula requires $f(3.65)$ and $f(3.75)$ in order to approximate $f'(3.7)$ (and is therefore still distinct from the forward-difference formula).

[8] Especially programmers who are averse to storing quantities that they won't make much use of later on.

To reiterate, the central-difference formula in Eq. (3.37) assumes that we have knowledge of $f(x_{i+1/2})$ and $f(x_{i-1/2})$: if all we have access to is a grid of points like that in Eq. (3.33), then we will not be able to use the central-difference formula in this form. We *will* be able to use it if we double the step size in Eq. (3.24), taking $h \to 2h$:

$$f'(x_i) = \frac{f(x_{i+1}) - f(x_{i-1})}{2h} + O(h^2) \tag{3.38}$$

where you can think of the denominator as $x_{i+1} - x_{i-1}$. In other words, by doubling the step size $h \to 2h$ used in the equation giving us the central difference, we are able to plug in the same value of h in the formula Eq. (3.38) as that used in the forward-difference Eq. (3.36), e.g., $h = 0.05$. This happens to be a general result: all the expressions we gave above that require a function evaluation at a midpoint could be recast in a form that can use function evaluations on a grid if we simply take $h \to 2h$. For example, our equation for the central-difference approximation to the second derivative from Eq. (3.32) becomes:

$$f''(x_i) = \frac{f(x_{i+1}) + f(x_{i-1}) - 2f(x_i)}{h^2} + O(h^2) \tag{3.39}$$

with $(x_{i+1} - x_i)^2$ in the denominator. This is probably easier to memorize, anyway.

Let's summarize how to evaluate the first derivative at points on a grid x_i when all we have are the function values $f(x_i)$. We've seen that we can approximate $f'(x_i)$ using: (a) Eq. (3.36) which corresponds to the forward-difference formula with step size h, and (b) Eq. (3.38) which corresponds to the central-difference formula with step size $2h$. The more accurate method (central difference) here uses a larger step size $2h$ and we know that a larger step size leads to less accuracy. Now, the question naturally arises: could it be that the inaccuracy stemming from the larger step size overpowers the accuracy coming from the fact that cental difference is a higher-order method?

In order to answer this question, we go back to our error analyses. For the forward difference, we already know that the total error is given by Eq. (3.15):

$$\mathcal{E} = \mathcal{E}_{app} + \mathcal{E}_{ro} = \frac{h}{2}|f''(x_i)| + \frac{2|f(x_i)|\epsilon_m}{h} \tag{3.40}$$

We will now find the corresponding expression for the new central difference. This will not be identical to Eq. (3.25), since we're now dealing with a step size of $2h$, as the formula we're analyzing is Eq. (3.38). One can analyze the new approximation from scratch, or simply replace $h \to 2h$ in Eq. (3.25) to get:

$$\mathcal{E} = \mathcal{E}_{app} + \mathcal{E}_{ro} = \frac{h^2}{6}|f'''(x_i)| + \frac{|f(x_i)|\epsilon_m}{h} \tag{3.41}$$

The trend exhibited by Eq. (3.41) appears to still be generally better than that of Eq. (3.40): the approximation term is still quadratic,[9] whereas the roundoff term is now half as big.

[9] And with a 6 in the denominator compared to the forward difference's 2 in the denominator.

One could always concoct artificial scenarios where $|f'''(x_i)|$ is dramatically larger than $|f''(x_i)|$, but for most purposes it's safe to say that the central-difference formula (even with twice the step size) is better than the forward-difference one.

As an aside, we note that if we had been less strict in our requirements for the derivative of $f(x)$, e.g., if we were OK with the possibility of getting the derivative at points other than the x_i, then we would have been able to use the unmodified central-difference formula of Eq. (3.24). To be explicit, taking $x = 3.725$ and $h = 0.05$, this formula requires the function evaluations $f(3.7)$ and $f(3.75)$ which we *do* have access to! The price to be paid, of course, is that this way we calculate $f'(3.725)$, not $f'(3.7)$ (i.e., $f'(x_{i+1/2})$ instead of $f'(x_i)$).

3.3.7 Richardson Extrapolation

We now turn to a nice technique, called Richardson extrapolation, which can be used to impove the accuracy of a numerical algorithm. Here we apply it to finite-difference formulas, but its applicability is much wider, so we will return to this tool when we study integrals in chapter 7 and differential equations in chapter 8.

3.3.7.1 General Formulation

Assume your task is to evaluate the quantity G, i.e., G is the exact answer of the calculation that you are currently trying to carry out approximately. Your approximate answer $g(h)$ depends on a parameter h, which is typically an increment, or step size. We write:

$$G = g(h) + \mathcal{E}_{app}(h) \tag{3.42}$$

where we explicitly noted that the approximation error also depends on h. The procedure we are about to introduce works only on reducing the approximation error (so we didn't also include a roundoff term $\mathcal{E}_{ro}$).

We will now make the (pretty widely applicable) assumption that the error term can be written as a sum of powers of h:

$$G = g(h) + Ah^p + Bh^{p+q} + Ch^{p+2q} + \cdots \tag{3.43}$$

where A, B, C are constants, p denotes the order of the leading error term and q is the increment in the order for the error terms after that. The idea behind Richardson extrapolation is to apply Eq. (3.43) twice, once for a step size h and once for a step size $h/2$:

$$\begin{aligned} G &= g(h) + Ah^p + O(h^{p+q}) \\ G &= g(h/2) + A\left(\frac{h}{2}\right)^p + O(h^{p+q}) \end{aligned} \tag{3.44}$$

Equating the two right-hand-sides gives us:

$$g(h) + Ah^p = g(h/2) + A\left(\frac{h}{2}\right)^p + O(h^{p+q}) \tag{3.45}$$

This equation can now be solved for Ah^p:

$$Ah^p = \frac{2^p}{2^p - 1}\left[g(h/2) - g(h)\right] + O(h^{p+q}) \tag{3.46}$$

This result, in turn, can be plugged back into the first relation in Eq. (3.44) to give:

$$G = \frac{2^p g(h/2) - g(h)}{2^p - 1} + O(h^{p+q}) \tag{3.47}$$

This is sometimes called an *extrapolated value*. This is a good time to lay all your possible worries to rest: Eq. (3.47) contains a subtraction on the right-hand-side (so it's possible some roundoff error may creep in) but, despite the fact that $g(h)$ and $g(h/2)$ presumably have similar values, no catastrophic cancellation is present, because of the 2^p coefficient.

To summarize, we started in Eq. (3.43) from a formula that has a leading error of $O(h^p)$; by using two calculations (one for a step size h and one for a step size $h/2$), we managed to eliminate the error $O(h^p)$ and are left with a formula that has error $O(h^{p+q})$. It's easy to see how this process could be repeated: by starting with two calculations each of which has error $O(h^{p+q})$, we can arrive at an answer with error $O(h^{p+2q})$, and so on.

3.3.7.2 Finite Differences

We now apply our general result in Eq. (3.47) to finite-difference formulas. In order to bring out the connections with other approaches touched upon in this chapter, we will carry out this task twice, once for the forward difference and once for the central difference.

Forward Difference

We will use our first forward-difference formula, Eq. (3.9):

$$D_{fd}(h) = \frac{f(x+h) - f(x)}{h} \tag{3.48}$$

where we also took the opportunity to employ new notation. Clearly, the leading error term in Eq. (3.48) is $O(h)$ (as we explicitly derived in section 3.3), meaning that the exponent in h^p is $p = 1$ for this case. Richardson extrapolation will eliminate this leading error term, leaving us with the next contribution, which is $O(h^2)$, as we know from Eq. (3.7).

Applying Eq. (3.47) with the present notation for $p = 1$ gives us:

$$\begin{aligned} R_{fd} &= 2D_{fd}(h/2) - D_{fd}(h) + O(h^2) = 2\frac{f\left(x+\frac{h}{2}\right) - f(x)}{h/2} - \frac{f(x+h) - f(x)}{h} + O(h^2) \\ &= \frac{4f\left(x+\frac{h}{2}\right) - f(x+h) - 3f(x)}{h} + O(h^2) \end{aligned} \tag{3.49}$$

In the second equality we plugged in our definition from Eq. (3.48) twice. In the third equality we grouped terms together. Our result is *identical* to the second forward-difference formula from Eq. (3.27): while there we had to explicitly derive things in terms of Taylor series, here we simply carried out one step of a Richardson extrapolation process.[10] To

[10] We could now carry out two calculations with errors $O(h^2)$, set $p = 2$ in Eq. (3.47) and we would get a result with error $O(h^3)$.

summarize, we employed a two-point formula (Eq. (3.48)) twice and ended up with a three-point formula (Eq. (3.49)) which is more accurate.

Central Difference

This time we will use our first central-difference formula, Eq. (3.24):

$$D_{cd}(h) = \frac{f\left(x+\frac{h}{2}\right) - f\left(x-\frac{h}{2}\right)}{h} \tag{3.50}$$

where we used the notation $D_{cd}(h)$ notation on the left-hand side. The leading error term in Eq. (3.50) is $O(h^2)$ (as we explicitly derived in section 3.3), meaning that the exponent in h^p is $p = 2$ for this case. Richardson extrapolation will eliminate this leading error term, leaving us with the next contribution, which is $O(h^4)$, as we know from Eq. (3.28).

Applying Eq. (3.47) with the present notation for $p = 2$ gives us:

$$\begin{aligned} R_{cd} &= \frac{4}{3}D_{cd}(h/2) - \frac{1}{3}D_{cd}(h) + O(h^4) \\ &= \frac{4}{3}\frac{f\left(x+\frac{h}{4}\right) - f\left(x-\frac{h}{4}\right)}{h/2} - \frac{1}{3}\frac{f\left(x+\frac{h}{2}\right) - f\left(x-\frac{h}{2}\right)}{h} + O(h^4) \\ &= \frac{8f\left(x+\frac{h}{4}\right) + f\left(x-\frac{h}{2}\right) - f\left(x+\frac{h}{2}\right) - 8f\left(x-\frac{h}{4}\right)}{3h} + O(h^4) \end{aligned} \tag{3.51}$$

In the second line we plugged in our definition from Eq. (3.50) twice. In the third line we grouped terms together. Our result is *identical* to the second central-difference formula that you were asked to derive after Eq. (3.28): while there we had to explicitly derive things in terms of Taylor series, here we simply carried out one step of a Richardson extrapolation process. If you're not seeing that the two results are identical, take $h \to 2h$ in the present result. To summarize, we employed a two-point formula (Eq. (3.50)) twice and ended up with a four-point formula (Eq. (3.51)) which is more accurate.[11]

Implementation

Code 3.2 is an implementation of Richardson extrapolation: crucially, this works by combining the outputs of our earlier forward- and central-difference functions. The main new feature in this code is that we have opted against copying and pasting functions from an earlier file: that would be needlessly error-prone. Instead, what we've done here is to import specific functions from `finitediff.py` which take care of the function itself, its analytical derivative, the forward difference, and the central difference. In addition to making the present code shorter, this has the added advantage that if we ever update those functions, we only need to change them in one location, namely in the file where they are defined.[12] The output of running this code is:

[11] Again, we could now carry out two calculations with errors $O(h^4)$, set $p = 4$ in Eq. (3.47) and we would get a result with error $O(h^6)$.

[12] As long as we don't break the interface, i.e., if we keep the same input and output conventions.

Code 3.2 richardsondiff.py

```
from finitediff import f, fprime, calc_fd, calc_cd

x = 0.5
an = fprime(x)

hs = [10**(-i) for i in range(1,7)]

rowf = "{0:1.0e} {1:1.16f} {2:1.16f}"
print("h     abs. err. rich fd      abs. err. rich cd")
for h in hs:
    fdrich = 2*calc_fd(f,x,h/2) - calc_fd(f,x,h)
    fd = abs(fdrich-an)
    cdrich = (4*calc_cd(f,x,h/2) - calc_cd(f,x,h))/3
    cd = abs(cdrich-an)
    print(rowf.format(h,fd,cd))
```

```
h     abs. err. rich fd  abs. err. rich cd
1e-01 0.0259686059827384 0.0000098728371007
1e-02 0.0002695720500450 0.0000000009897567
1e-03 0.0000027005434182 0.0000000000009619
1e-04 0.0000000270109117 0.0000000000043667
1e-05 0.0000000003389138 0.0000000000132485
1e-06 0.0000000006941852 0.0000000002500959
```

Starting from the forward-difference Richardson-extrapolated results: every time we reduce the h by an order of magnitude, the absolute error is reduced by two orders of magnitude, consistent with a method that has an approximation error of $O(h^2)$, see Eq. (3.49). This process reaches a minimum at $h = 10^{-5}$: the absolute error there is $\approx 3 \times 10^{-10}$. Note that this behaviour is very similar to the output of `finitediff.py` *for the central-difference column*, which also resulted from a method with an error of $O(h^2)$. The last line in the present output merely serves to show that roundoff has started dominating the forward-difference case, so we can no longer rely on the Richardson extrapolation process.

The central-difference Richardson-extrapolated results are analogous. The first time we reduce h by an order of magnitude we reduce the absolute error by a whopping *four* orders of magnitude, consistent with a method that has an approximation error of $O(h^4)$, see Eq. (3.51). This cannot continue indefinitely: in the next step we improve by three orders of magnitude, reaching a minimum at $h = 10^{-3}$: the absolute error there is $\approx 10^{-12}$, the best we've seen so far using a finite-difference(-related) scheme. As the h is further reduced roundoff error starts dominating, so the extrapolation process is no longer reliable.

Observe that Richardson extrapolation doesn't tell you what the leading error in the new theory will look like (in the language of Eq. (3.47), what value q takes): you need to determine that separately. Of course, you could always use the Richardson approach to produce a better theory (as we did above) and then carry out the needed Taylor expansions after the fact (i.e., knowing what form your better theory takes) to determine the approximation error. This is much easier than manually having to combine together several different Taylor expansions in the dark, i.e., without knowing what the answer should look like.[13]

3.4 Automatic Differentiation

We have seen two ways of taking derivatives on a computer: first, *analytically* using a symbolic algebra package or library and, second, using *finite-difference* formulas of varying sophistication. We now turn to a third option, known as *automatic differentiation* or, sometimes, *algorithmic differentiation*. (Problem 3.4 introduces a charming surrogate.)

Automatic differentiation is equivalent to analytical differentiation of elementary functions along with propagation using the chain rule. Crucially, this is *not* accomplished via the manipulation of expressions (as is done in symbolic differentiation, which is quite inefficient in practice) but using specific numbers. Thus, we never have access to an analytical expression providing the derivative of a function, but do get a result with nearly machine precision for the derivative at any one point (i.e., way better than finite differences). Thus, automatic differentiation is as accurate as symbolic differentiation, but it doesn't need to produce a general expression (which is, anyway, only going to be used at specific points).

Using Python, one has access to several implementations of automatic differentiation, e.g., JAX: the details are strongly time dependent so we will, instead, focus on the main idea. (Problem 3.18 asks you to experiment with `jax`, whereas problem 3.19 guides you toward building a bare-bones automatic differentiator yourself.) In the past, computational physics books did not mention automatic differentiation but, given its benefits and conceptual simplicity, it deserves a wider audience.

3.4.1 Dual Numbers

Extend any number a by also adding a second component:

$$\mathbf{a} = a + a'd \tag{3.52}$$

These are called *dual numbers*. This d is simply a placeholder telling us what the second component, a', is. This is analogous to complex numbers, where we go from x to $x + yi$ via the use of the imaginary unit, $i = \sqrt{-1}$. Here, we take $d^2 = 0$ (compare with $i^2 = -1$). If you're uncomfortable with the possibility of a number being non-zero but giving zero when squared (despite having read section 2.2.2), look up the term *Grassmann variable*.

[13] "When the sun has set, all beasts are black" (François Rabelais, *Pantagruel*, chapter 12).

It is easy to see how arithmetic works for dual numbers. For example, for addition:

$$\mathbf{a} + \mathbf{b} = (a + a'd) + (b + b'd) = a + b + (a' + b')d \tag{3.53}$$

Similarly, for multiplication we have:

$$\mathbf{a} \times \mathbf{b} = (a + a'd) \times (b + b'd) = ab + ab'd + a'bd + a'b'd^2 = ab + (ab' + a'b)d \tag{3.54}$$

To go to the third equality we made use of the fact that $d^2 = 0$. You can see how subtraction and division will turn out. What is emerging here is an arithmetic where the first component behaves as real numbers do, whereas the second component follows well-known rules of differentiation (for the sum, the product, etc.). At this point, we can drop the use of d entirely, by rewriting dual numbers using ordered pairs of real numbers:[14]

$$\mathbf{a} = (a, a') \tag{3.55}$$

This definition using an ordered pair is completely analogous to complex numbers, which are also given as $z = (x, y)$. As was implicit above, a gives us the value of a function at a specific point and a' the value of the derivative of that function at the same point. We can now re-express (and augment) the basic arithmetic rules as follows:

$$\begin{aligned}
\mathbf{a} + \mathbf{b} &= (a + b, a' + b') \\
\mathbf{a} - \mathbf{b} &= (a - b, a' - b') \\
\mathbf{a} \times \mathbf{b} &= (ab, ab' + a'b) \\
\mathbf{a} \div \mathbf{b} &= \left(\frac{a}{b}, \frac{a'b - ab'}{b^2}\right)
\end{aligned} \tag{3.56}$$

Thus, dual numbers give us a way to differentiate elementary mathematical operations; observe that we're using special symbols for all these operations (e.g., $\mathbf{+}$ instead of $+$) when applied to ordered pairs. In actual calculations, we'll need to also know how to treat constants, c, and independent variables, x. Simply define:

$$\begin{aligned}
\mathbf{c} &= (c, 0) \\
\mathbf{x} &= (x, 1)
\end{aligned} \tag{3.57}$$

Both of these are plausible, since the derivative of a constant is 0 and the derivative of x with respect to x is 1. In practice, we'll be faced with a function $f(x)$ for which we would like to produce the derivative. To accomplish this, we replace all occurrences of x with $\mathbf{x}$, all occurrences of constants c with $\mathbf{c}$, and also employ the basic arithmetic rules in Eq. (3.56). Thus, we arrive at a new function $\mathbf{f}(\mathbf{x})$: evaluating this at the specific point $(x_0, 1)$ we get the ordered pair $(f(x_0), f'(x_0))$, thereby achieving what we set out to do.

[14] In this book, bold lowercase symbols denote vectors (see appendix C.2); dual numbers basically fit the bill.

3.4.2 An Example

Let's apply all of the above rules to a specific example. Take:

$$f(x) = \frac{(x+2)(x-3)}{x-4} \tag{3.58}$$

We are interested in computing $f(6)$ and $f'(6)$. We can immediately see (by plugging in) that $f(6) = 12$. However, the value of $f'(6)$ is not as obvious. In order to find it, we employ the above prescription, promoting all numbers to dual numbers and interpreting the arithmetic operations appropriately. We have:

$$\begin{aligned}\mathbf{f(x)} &= \mathbf{(x+2)\times(x-3)\div(x-4)}\\ &= [(x,1)+(2,0)]\times[(x,1)-(3,0)]\div[(x,1)-(4,0)]\end{aligned} \tag{3.59}$$

Now, here's the beauty of it all. Without needing to find an expression for f', we will calculate the derivative at the specific point $\mathbf{x} = (6,1)$ by plugging in and using the rules:

$$\begin{aligned}\mathbf{f}((6,1)) &= [(6,1)+(2,0)]\times[(6,1)-(3,0)]\div[(6,1)-(4,0)]\\ &= (8,1)\times(3,1)\div(2,1) = (24,11)\div(2,1) = \left(12,-\frac{1}{2}\right)\end{aligned} \tag{3.60}$$

In the second equality we did the addition/subtractions. In the third equality we did the multiplication. In the fourth equality we carried out the division. This gives us $f(6) = 12$ (which we had already figured out) and $f'(6) = -1/2$ (which we hadn't). As advertised, we have arrived at the derivative of our function at a specific point simply by plugging in values and following the rules above, without having to produce an expression for f'.

3.4.3 Special Functions

With a view to extending this formalism, let's start with a polynomial:

$$p(a) = c_0 + c_1 a + c_2 a^2 + c_3 a^3 + \cdots + c_{n-1} a^{n-1} \tag{3.61}$$

We now wish to promote a to $\mathbf{a}$: we'll need to know how to handle powers of $\mathbf{a}$. These are easy to calculate via repeated use of the multiplication rule from Eq. (3.56):

$$\mathbf{a}^2 = \mathbf{a}\times\mathbf{a} = (a^2, 2aa') \tag{3.62}$$

Similarly:

$$\mathbf{a}^3 = \mathbf{a}^2\times\mathbf{a} = (a^3, 3a^2a') \tag{3.63}$$

and so on. Thus, promoting the a to (a, a') and the constants c_i to $(c_i, 0)$ gives us:

$$\begin{aligned}\mathbf{p(a)} &= (c_0,0)+(c_1,0)\times\mathbf{a}+(c_2,0)\times\mathbf{a}^2+(c_3,0)\times\mathbf{a}^3+\cdots+(c_{n-1},0)\times\mathbf{a}^{n-1}\\ &= (c_0,0)+(c_1a,c_1a')+(c_2a^2,c_22aa')+\cdots+(c_{n-1}a^{n-1},c_{n-1}(n-1)a^{n-2}a')\\ &= (c_0+c_1a+c_2a^2+\cdots+c_{n-1}a^{n-1},c_1a'+c_22aa'+\cdots+c_{n-1}(n-1)a^{n-2}a')\\ &= (p(a),a'p'(a))\end{aligned} \tag{3.64}$$

In the second line we carried out all the multiplications. In the third line we did the summations. In the fourth line we identified in the second component the presence of p', namely the derivative of the polynomial. This result, which we explicitly proved for polynomials, leads us to think about other functions, since we will also need to know how to take, say, the cosine of a dual number. We define the relevant (chain) rule as:

$$\mathbf{g}(\mathbf{a}) = \mathbf{g}\left((a, a')\right) = (g(a), a'g'(a)) \tag{3.65}$$

which looks completely analogous to our result for polynomials in Eq. (3.64) above. Note that the $'$ on the right-hand side of this equation has two distinct (though related) meanings: a' is the second component of $\mathbf{a}$, whereas g' refers to the derivative of the function g, which is arrived at through some other means. A few examples:

$$\begin{aligned}
\mathbf{sin}(\mathbf{a}) &= \mathbf{sin}\left((a, a')\right) = (\sin a, a' \cos a)\\
\mathbf{cos}(\mathbf{a}) &= \mathbf{cos}\left((a, a')\right) = (\cos a, -a' \sin a)\\
\mathbf{e}^{\mathbf{a}} &= \mathbf{e}^{(a,a')} = (e^a, a'e^a)\\
\mathbf{ln}(\mathbf{a}) &= \mathbf{ln}\left((a, a')\right) = \left(\ln a, \frac{a'}{a}\right)\\
\sqrt{\mathbf{a}} &= \sqrt{(a, a')} = \left(\sqrt{a}, \frac{a'}{2\sqrt{a}}\right)
\end{aligned} \tag{3.66}$$

We are using new symbols on the left-hand sides, in the spirit in which above we promoted the function f to $\mathbf{f}$ (or the function g to $\mathbf{g}$). These can either be seen as results of Eq. (3.65) or could be explicitly derived via the relevant Taylor expansion each time.

A software system that implements automatic differentiation knows how to apply all these formulas, even if it needs to apply more than one to the same expression. To make this crystal clear, we now turn to our usual example, $f(x) = e^{\sin(2x)}$ at $x = 0.5$. Following the rules of promotion, as above, we have:

$$\mathbf{f}(\mathbf{x}) = \mathbf{e}^{\mathbf{sin}((2,0)\times(x,1))} \tag{3.67}$$

We now plug in $x = 0.5$ to find:

$$\mathbf{f}\left((0.5, 1)\right) = \mathbf{e}^{\mathbf{sin}((2,0)\times(0.5,1))} = \mathbf{e}^{\mathbf{sin}((1,2))} = \mathbf{e}^{(\sin(1), 2\cos(1))} = \left(e^{\sin(1)}, 2\cos(1)e^{\sin(1)}\right) \tag{3.68}$$

In the second equality we carried out the multiplication as per Eq. (3.56). In the third equality we used the chain rule in Eq. (3.66) for the sine. In the fourth equality we used the chain rule in Eq. (3.66) for the exponential. Thus, we have arrived at the output $f(0.5) = e^{\sin(1)} \approx 2.319\,776\,824\,715\,853$ and $f'(0.5) = 2\cos(1)e^{\sin(1)} \approx 2.506\,761\,534\,986\,894$. These results agree (to within machine precision) with the answer we get by analytically evaluating the derivative and then plugging in $x = 0.5$.

It's important to understand that the procedure we followed was not actually that of taking a derivative. We were merely dealing with numbers, to which we applied the promotions to ordered pairs as per Eq. (3.57), the basic arithmetic rules in Eq. (3.56), and the chain rule in Eq. (3.65): the result turns out to be a derivative, even though the elementary operations involved simply the additions, divisions, etc. of real numbers.

Before concluding, it's worth underlining that the procedure we followed for automatic differentiation doesn't exactly address one of our main tasks, namely the case where we have access only to a table of points $(x_i, f(x_i))$ for $i = 0, 1, \ldots, n-1$.[15] Instead, automatic differentiation can be applied to the piece of code producing a function evaluation at one point: it then promotes all numbers to ordered pairs and follows simple rules to evaluate the function's derivative at the same point, without any major overhead.

3.5 Project: Local Kinetic Energy in Quantum Mechanics

We will now see how taking derivatives is an essential part of evaluating the kinetic energy in quantum mechanics (in the position basis). If you haven't studied quantum mechanics, you can skim through most of section 3.5.1. Even if you don't know what a wave function is, there are some lessons to be learned implementation-wise: we will show how one Python function can handle different physical scenarios, as long as we are careful to respect some general conventions about function-interface design.

Physics-wise, our coverage will revolve around one non-relativistic particle; we will study two different interaction terms (harmonic oscillator and free particle). To refresh your memory, the *time-independent Schrödinger equation* involves the Hamiltonian $\hat{H}$, the wave function ψ, and the energy E:

$$\hat{H}\psi = E\psi \tag{3.69}$$

This is an *eigenvalue problem*, about which we'll have a lot more to say in the rest of this volume, especially in chapter 4 (where we discuss linear algebra) and chapter 8 (where we study differential equations). In general, the Hamiltonian operator is made up of the kinetic-energy operator and the potential-energy operator: $\hat{H} = \hat{T} + \hat{V}$.

3.5.1 Single-Particle Wave Functions in One Dimension

We now go over some standard results, assuming that you've encountered this material before, so we will not repeat the explicit derivations.

[15] For that matter, symbolic differentiation, as discussed in section 3.2 doesn't either, except if one has first interpolated and then analytically differentiates the interpolating function.

3.5.1.1 Quantum Harmonic Oscillator

An important problem in one-dimensional single-particle quantum mechanics is the harmonic oscillator; for this case, the time-independent Schrödinger equation takes the form:

$$-\frac{\hbar^2}{2m}\frac{d^2\psi(x)}{dx^2} + \frac{1}{2}m\omega^2x^2\psi(x) = E\psi(x) \tag{3.70}$$

where $\hbar$ is Planck's constant (divided by 2π) and the particle's mass is m. The left-hand side is made up of the kinetic energy (involving a second derivative) and the potential energy (involving a term quadratic in x). The ω is the (classical) angular frequency.

Every quantum mechanics textbook discusses how to solve this problem. This means how to determine the energy eigenvalues:

$$E_n = \left(n + \frac{1}{2}\right)\hbar\omega \tag{3.71}$$

and the (normalized) eigenfunctions:

$$\psi_n(x) = \frac{1}{\sqrt{2^n n!}}\left(\frac{m\omega}{\pi\hbar}\right)^{1/4} H_n\left(\sqrt{\frac{m\omega}{\hbar}}x\right) e^{-m\omega x^2/(2\hbar)} \tag{3.72}$$

Note that both the eigenvalues and the eigenfunctions are labelled by a discrete index n (a non-negative integer): this goes to the heart of *quantization*. Notoriously, the ground state of the harmonic oscillator is $n = 0$, which is characterized by a finite energy as per Eq. (3.71); this is known as *zero-point motion*.

For a given n, the harmonic-oscillator eigenfunction is equal to (some prefactors times) a Gaussian term and another term involving an *Hermite polynomial*, $H_n(x)$. Hermite polynomials can be evaluated via a process similar to that introduced in section 2.5 for Legendre polynomials. They obey the *recurrence relation*:

$$H_{j+1}(x) = 2xH_j(x) - 2jH_{j-1}(x) \tag{3.73}$$

To step through this process, one starts with the known first two polynomials, $H_0(x) = 1$ and $H_1(x) = 2x$, and calculates $H_n(x)$ by taking:

$$j = 1, 2, \dots, n-1 \tag{3.74}$$

The first few Hermite polynomials are shown in the left panel of Fig. 3.4, which is analogous to the left panel of Fig. 2.6. Similarly, for the derivative one can use the relation:

$$H_n'(x) = 2nH_{n-1}(x) \tag{3.75}$$

though we don't need to evaluate these derivatives here (we will later on). To summarize, for a specified n the wave function is given by Eq. (3.72), which you can immediately compute if you know how to evaluate Hermite polynomials numerically.

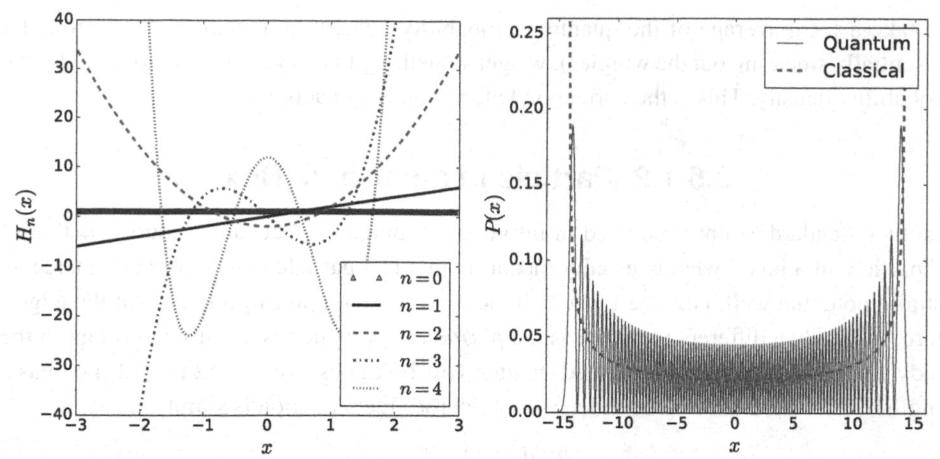

Fig. 3.4 First few Hermite polynomials (left) and harmonic oscillator probability density (right)

Correspondence Principle

Let us recall the harmonic oscillator in *classical* mechanics. If the amplitude of the oscillation is x_0, then the total energy E, which is a constant of the motion, is:

$$E = \frac{1}{2}m\omega^2 x_0^2 \tag{3.76}$$

This describes the extreme case where the kinetic energy is zero; there are two such *turning points*, $\pm x_0$. As you may recall, the particle oscillates between them, $-x_0 \leq x \leq x_0$. The regions $x^2 > x_0^2$ are *classically forbidden*, since they correspond to negative kinetic energy. (Things are different in quantum mechanics, since there the wave function may "leak" into the classically forbidden region.) Problem 3.20 guides you toward deriving the following equation for the *classical probability density* for the oscillator problem:

$$P_c(x) = \frac{1}{\pi\sqrt{x_0^2 - x^2}} \tag{3.77}$$

This is the probability that we will find the particle in an interval dx around the position x; it is larger near the turning points: that's where the speed of the particle is smallest.

A fascinating question arises regarding how we go from quantum to classical mechanics; this is described by what is known as the *correspondence principle*: as the quantum number n goes to infinity, we should recover classical behavior. In order to see this, the right panel of Fig. 3.4 plots results for the quantum harmonic oscillator for the case of $n = 100$, which is reasonably large. The solid curve, exhibiting many wiggles, is the probability density for the quantum case; as per the *Born rule*, this is the square of the modulus of the wave function, i.e., $P(x) = |\psi(x)|^2$. To compare apples to apples, we also study a classical problem (dashed curve) with an amplitude of:

$$x_0 = \sqrt{\frac{\hbar}{m\omega}(2n+1)} \tag{3.78}$$

which we found by equating Eq. (3.76) with Eq. (3.71). As the figure clearly shows, if we take a local average of the quantum probability density in a small interval around x (essentially smearing out the wiggles), we get something that is very similar to the classical probability density. This is the correspondence principle in action.

3.5.1.2 Particle in a Periodic Box

Another standard example covered in introductory quantum-mechanics courses is that of a "particle in a box"; what is usually meant by this is a particle that is confined inside an infinite potential well, i.e., one for which the wave function must go to zero at the edges. Here, we study a different scenario, i.e., a *periodic* box, which is usually important in the study of extended (i.e., non-confined) matter. Our box goes from $-L/2$ to $L/2$ (i.e., has a length of L); inside it, the time-independent Schrödinger equation is simply:

$$-\frac{\hbar^2}{2m}\frac{d^2\psi(x)}{dx^2} = E\psi(x) \tag{3.79}$$

since we're assuming there is no external potential inside the box. The eigenfunctions are simply plane waves, i.e., complex exponentials:

$$\psi_k(x) = \frac{1}{\sqrt{L}}e^{ikx} \tag{3.80}$$

The prefactor is such that our eigenfunction is normalized. We have labelled our eigenfunctions with the wave number k; it is related to the momentum via the relation $p = \hbar k$.

In order to find out more about this wave number/momentum, let us investigate the boundary conditions. We mentioned above that we are dealing with a *periodic* box; mathematically, this is expressed by saying $\psi_k(x) = \psi_k(x+L)$, which is precisely the relation that allowed us to stay inside a single box in the first place ($-L/2 \le x \le L/2$): the wave function simply repeats itself beyond that point. If we now combine this periodicity condition with our plane waves of Eq. (3.80), we find:

$$k = \frac{2\pi}{L}n \tag{3.81}$$

where n is an integer which could be positive, negative, or zero. This is already different from the case of the harmonic oscillator above. If we know the value of this quantum number n, i.e., we know the value of the wave number for a given state, then we are also able to calculate the energy corresponding to that state; to see this, plug the eigenfunction Eq. (3.80) into the eigenvalue equation Eq. (3.79) to find:

$$E = \frac{\hbar^2 k^2}{2m} = \frac{\hbar^2}{2m}\left(\frac{2\pi}{L}\right)^2 n^2 \tag{3.82}$$

In the second equality we used our result in Eq. (3.81). As expected for this problem (for which there is no attraction and therefore no bound state), the energy cannot be negative.

3.5.1.3 Implementation

Let us see how to implement our two possibilities for the wave function. The task at hand is pretty mundane; in order to make it a bit interesting, we have decided to write two

Code 3.3

psis.py

```
from math import sqrt, pi, factorial, exp
import cmath

def hermite(n,x):
    val0 = 1.; val1 = 2*x
    for j in range(1,n):
        val2 = 2*x*val1 - 2*j*val0
        val0, val1 = val1, val2
    dval2 = 2*n*val0
    return val2, dval2

def psiqho(x,nametoval):
    n = nametoval["n"]
    momohbar = nametoval["momohbar"]
    al = nametoval["al"]
    psival = momohbar**0.25*exp(-0.5*al*momohbar * x**2)
    psival *= hermite(n,sqrt(momohbar)*x)[0]
    psival /= sqrt(2**n * factorial(n) * sqrt(pi))
    return psival

def psibox(x,nametoval):
    n = nametoval["n"]
    boxl = nametoval["boxl"]
    return cmath.exp(2*pi*n*x*1j/boxl) / sqrt(boxl)

if __name__ == '__main__':
    x = 1.
    nametoval = {"n": 100, "momohbar": 1., "al": 1.}
    psiA = psiqho(x, nametoval)
    nametoval = {"n": -2, "boxl": 2*pi}
    psiB = psibox(x, nametoval)
    print(psiA, psiB)
```

functions *with the same interface*; that means that the external world will treat wave function code the same way, without caring about details of the implementation.

Both of our functions will take in two parameters: first, the position of the particle and, second, a dictionary bundling together any wave-function-specific parameters. This is not idle programming ("code golf"): as you will discover in the following subsection, a common interface makes a world of difference when you need to pass these two functions into a *third* function. We implement a general solution, similar to Python's `**kwargs`, to help students who are still becoming comfortable with dictionaries in Python.

Code 3.3 starts with import statements, as usual. We will need an exponential that handles real numbers (for the harmonic oscillator) and another one that can handle complex numbers (for the plane wave of the box problem). To avoid a name clash, we import `exp` explicitly and then import the `cmath` module, from which we intend to use the complex exponential `cmath.exp()`. Our code then defines a function that evaluates Hermite polynomial values and derivatives. This is very similar to our earlier `legendre()`, only this time we don't cater to the possibilities $n = 0$ or $n = 1$, in order to shorten the code.

Our first wave function, `psiqho()`, takes in a float and a dictionary, as advertised. This implements Eq. (3.72): in addition to the position of the particle, we also need to pass in the quantum number n, as well as the value of $m\omega/\hbar$. To spice things up, we also allow for the possibility of including an extra variational parameter, α, in the exponent of the Gaussian. That brings the total of extra parameters needed up to three. The dictionary `nametoval` is aptly named: it maps from a string (the name) to a value; thus, by passing in the appropriately initialized (in the external world) dictionary, we manage to bundle together all the necessary extra parameters, while still keeping our interface clean. The rest of the function is pretty straightforward.

Our second wave function, `psibox()`, also takes in a float and a dictionary. As before, the float is for the position of the particle and the dictionary for any other parameters needed to evaluate the wave function for this case. To implement Eq. (3.80), we need to pass in, first, the quantum number and, second, the length of the box side, L. Note how different the requirements of our two wave functions are: `psiqho()` takes in a position and then three more numbers, whereas `psibox()` accepts a position and then two more numbers; despite having different parameter needs, the two functions have the same interface, due to our decision to bundle the "extra" parameters into a dictionary. Of course, inside `psibox()` we immediately "unpack" this dictionary to get the values we need. We then use `cmath.exp()`.

The main program explicitly defines a `nametoval` dictionary that bundles the extra parameters for the oscillator problem, before passing it to `psiqho()`. The test code calling `psibox()` is analogous: first, a `nametoval` dictionary is defined; note that, while, the two dictionaries look quite different, the parts of the code *calling* `psiqho()` and `psibox()` are identical. The output of this code is not very important here; the main point is the common interface, which will help us below.

3.5.2 Second Derivative

You may have realized that the code we just discussed evaluates the wave functions *only*, i.e., it has nothing to do with energies. We will now see how to use that code to evaluate the kinetic energy, by implementing the second derivative. This will emphasize the aforementioned point on interface design, but will also be of wider applicability.

Before we discuss how to code things up, we first see *why* someone might want to evaluate the kinetic energy. Didn't we already solve this problem by finding the eigenenergies in Eq. (3.71) and Eq. (3.82)? Notice that here we're not trying to find the *total energy*, only the *kinetic energy*; note, further, that we're not trying to evaluate the *expectation value* of the kinetic energy but the value of the kinetic energy *for a given position*. Motivated by Eq. (3.69), $\hat{H}\psi = E\psi$, we define the following quantity, known as the *local kinetic energy*:

$$T_L = \frac{\hat{T}\psi}{\psi} \tag{3.83}$$

In general, the value of T_L will depend on x. The normalization of the wave function doesn't matter: it appears in both numerator and denominator, and cancels out.

Of course, for the simple cases studied here we've actually solved the problem of evaluating T_L, too: first, for the periodic box, there is no potential energy term, so the kinetic energy *is* the total energy; also, as per Eq. (3.82), for this specific case the energy does not depend on the position of the particle. Second, for the one-dimensional harmonic oscillator problem, given that $\hat{T}\psi + \hat{V}\psi = E\psi$, and the eigenenergy is given by Eq. (3.71), we find:

$$T_L = \frac{\hat{T}\psi}{\psi} = \left(n + \frac{1}{2}\right)\hbar\omega - \frac{1}{2}m\omega^2 x^2 \tag{3.84}$$

We can analytically determine T_L once we know all the quantities on the right-hand side. Obviously, we can find a local kinetic energy that is *negative*: just pick x to be large enough in magnitude; this is a situation that was not allowed in classical mechanics.

Even though the problem here is already solved ahead of time, the tools we will develop will be much more general, and therefore applicable also to situations where the wave function ψ is, say, not analytically known. Similarly, in the examples above we were studying *eigenfunctions*, for which the corresponding eigenvalues were also known; the approach that implements the local kinetic energy (i.e., basically, the second derivative) is general enough not to care whether or not your ψ is an eigenfunction. Finally, as we'll see in the Project of chapter 7, the local kinetic energy also arises in quantum Monte Carlo calculations for *many*-particle systems, so what we discuss here is part of the scaffolding for that, more involved, problem.

From the Schrödinger equations for our problems, Eq. (3.70) and Eq. (3.79), we see that they involve the second derivative (times $-\hbar^2/(2m)$). We now turn to the main theme of this chapter, namely the evaluation of derivatives via finite differences as per Eq. (3.39):

$$\frac{d^2\psi(x)}{dx^2} \approx \frac{\psi(x+h) + \psi(x-h) - 2\psi(x)}{h^2} \tag{3.85}$$

3.5.2.1 Implementation

Code 3.4 is an implementation of the local kinetic energy. We start by importing the two wave functions from code 3.3. The core of the new program is the function `kinetic()`. Its parameters are a wave function, a position, the infamous dictionary for extra parameters, and then an optional step size h for the finite differencing. This kinetic function interoperates with any wave function that respects our common interface (described above). We

Code 3.4 kinetic.py

```
from psis import psiqho, psibox
from math import pi

def kinetic(psi,x,nametoval,h=0.005):
    hom = 1.
    psiold = psi(x,nametoval)
    psip = psi(x+h,nametoval)
    psim = psi(x-h,nametoval)

    lapl = (psip + psim - 2.*psiold)/h**2
    kin = -0.5*hom*lapl/psiold
    return kin

def test_kinetic():
    x = 1.
    hs = [10**(-i) for i in range(1,6)]
    nametoval = {"n": 100, "momohbar": 1., "al": 1.}
    qhos = [kinetic(psiqho,x,nametoval,h) for h in hs]
    nametoval = {"n": -2, "boxl": 2*pi}
    boxs = [kinetic(psibox,x,nametoval,h) for h in hs]

    rowf = "{0:1.0e} {1:1.16f} {2:1.16f}"
    print("h        qho          box")
    for h,qho,box in zip(hs,qhos,boxs):
        print(rowf.format(h,qho,box))

if __name__ == '__main__':
    test_kinetic()
```

start by setting $\hbar^2/m = 1$ for simplicity. We then proceed to carry out the three function evaluations shown in Eq. (3.85). Note that we are not changing the value of x, but are employing "shifted" positions directly, wherever we need them. If you, instead, say something like `x += h`, you may be in for a surprise:[16] as you learned in chapter 2, steps like $x + h$ followed by $x - 2h$ followed by $x + h$ don't always give back x, as there may be roundoff error accumulation. The rest of the function is quite straightforward.

Next, we define a *test function*: this contains the type of code you may have been putting

[16] "[N]ot to recount great incidents, but to make small ones interesting" (Arthur Schopenhauer, *Parerga and Paralipomena*, second volume, paragraph 228).

in the main program so far (as did we, say, in code 3.3). This is an example of standard programming practice: by encapsulating your test code here, you are free to later define *other* test functions that probe different features. The body of the function itself is not too exciting: we define a list of h's for which we wish to evaluate the (finite-difference approximation to the) second derivative, and then we call `kinetic()`. Each time, we define the appropriate extra-parameter dictionary, and then pass in the appropriate wave function to `kinetic()`. This is followed by a nicely formatted print-out of the values we computed. The main program then does nothing other than call our one test function (which in its turn calls `kinetic()`, which goes on to call the wave functions repeatedly). The output is:

```
h     qho                  box
1e-01 83.7347487381334759  1.9933422158758454+0.0000000000000081j
1e-02 99.8252144561287480  1.9999333342235321+0.0000000000010316j
1e-03 99.9982508976935520  1.9999993332343806-0.0000000000642852j
1e-04 99.9999829515395646  1.9999999969522586-0.0000000035077577j
1e-05 99.9997983578354876  2.0000003679325493+0.0000008242844295j
```

In order to interpret this output, we first note that, for our chosen input parameters, we analytically expect the oscillator answer (see Eq. (3.84)) to be 100, while the box answer should be 2 (see Eq. (3.82)). As the step size gets smaller, we get an increasingly good approximation to the second derivative of the wave function; of course, as we keep reducing h at some point roundoff error catches up to us, so further decreases are no longer helpful. For the oscillator, from the right panel of Fig. 3.4 we immediately see that if we wish to have any hope of capturing the behavior of the wave function for $n = 100$, then we had better ensure that the step size is much smaller than the wiggles. Next, we notice that the box answer comes with an imaginary part, which Eq. (3.82) tells us should be zero: as the real part becomes better, the imaginary part becomes larger, so one would have to balance these two properties (and drop the imaginary part).

Problems

3.1 Make sure `SymPy` is installed on your system; use it to analytically differentiate the function $f(x) = e^{\sin(2x)}$. Then, evaluate the derivative at a few points from $x = 0$ to $x = 0.5$. Compare the latter values to the output of `fprime()` from code 3.1.

3.2 [$\mathcal{A}$] Justify Eq. (3.25) and then use it to show Eq. (3.26).

3.3 Create a function that takes in three lists and produces the output in Fig. 3.2. Feel free to use the `xscale()` and `yscale()` functions.

3.4 We turn to the fascinating approach of *complex-step differentiation*. First, Taylor expand $f(x+ih)$ around x, where f is an analytic function; take the imaginary part of both sides and divide by h to show that $f'(x) = \text{Im}[f(x+ih)/h] + O(h^2)$ holds. Then, augment Fig. 3.2 to include results from this new method; you'll need to employ complex arithmetic. Unmistakably, circumventing catastrophic cancellation pays off.

3.5 Modify `finitediff.py` to also plot the second forward-difference and the second-central difference results. Comment on whether the plot agrees with the analytical expectation coming from the approximation-error dependence on h.

3.6 [$\mathcal{A}$] Show Eq. (3.27) by combining the Taylor series for $f(x+h)$ and $f(x+h/2)$. Then, show Eq. (3.28) by subtracting two pairs of Taylor series (pairwise): $f(x+h/2)$ and $f(x-h/2)$, on the one hand, $f(x+3h/2)$ and $f(x-3h/2)$, on the other.

3.7 We turn to the first central-difference approximation to the second derivative.

(a) Start with the error analysis, including both approximation and roundoff error. Derive expressions for the h_{opt} and the $\mathcal{E}_{opt}$. Then, produce numerical estimates for h_{opt} and the $\mathcal{E}_{opt}$. Compare these results to those for the first derivative.

(b) Now code this problem up in Python (for the function $f(x) = e^{\sin(2x)}$ at $x = 0.5$) to produce both a table of numbers and a plot for the absolute error, with h taking on the values $10^{-1}, 10^{-2}, 10^{-3}, \ldots, 10^{-10}$.

3.8 This problem studies the second derivative of the following function:

$$f(x) = \frac{1 - \cos x}{x^2} \tag{3.86}$$

which you encountered in problem 2.8. We take $x = 0.004$.

(a) Start by analytically evaluating the second derivative, and plugging in $x = 0.004$.

(b) Give h the values $10^{-1}, 10^{-2}, \ldots, 10^{-6}$ and produce a log-log plot of the absolute error in the first central-difference approximation to the second derivative.

(c) Introduce a new set of points to the previous plot, this time evaluating the first central-difference approximation to the second derivative not of $f(x)$, but of the analytically rewritten version you arrived at in problem 2.8 (recall: you had used a trigonometric identity which enabled you to avoid the cancellation).

3.9 In section 2.5 we introduced Legendre polynomials via Eq. (2.81), the generating function which led to a recurrence relation, Eq. (2.86). Here we discuss another representation of Legendre polynomials, that produced by *Rodrigues' formula*:

$$P_n(x) = \frac{1}{2^n n!} \frac{d^n}{dx^n} \left[(x^2 - 1)^n \right] \tag{3.87}$$

(a) Analytically derive the leading coefficient of the Legendre polynomial:

$$a_n = \frac{(2n)!}{2^n (n!)^2} \tag{3.88}$$

This a_n is the coefficient multiplying x^n when you expand $P_n(x)$ out.

(b) For $h = 0.01$ employ Eq. (3.24) and Rodrigues' formula to compute $P_1(x)$. Then, compute the central-difference approximation to the second derivative (as the central difference of the central difference) to compute $P_2(x)$. Keep calling the earlier central-difference function(s) and go up to $P_8(x)$. Each time, you should plot your function against the output of `legendre.py`.

(c) What happens if you change h to 10^{-3}?

3.10 We will see how to implement the gradient of a scalar function using Python lists:

(a) Write a Python function to compute the gradient $\nabla\phi(\mathbf{x})$ via a forward-difference approximation, as per Eq. (5.108). This should work for any scalar function; test it on the $\phi(\mathbf{x})$ of Eq. (5.106).

(b) Use your new function to plot $\frac{\partial\phi}{\partial x}\frac{\partial\phi}{\partial y}$ in a format similar to that in Fig. 5.16.

3.11 We turn to the relationship between Horner's rule, Eq. (2.100), and the evaluation of the first derivative of a polynomial. To bridge the two, introduce *synthetic division*:

$$p(x) = (x - x_0)q(x) + p(x_0) \tag{3.89}$$

where $p(x)$ is of degree $n-1$ and $q(x)$ of degree $n-2$. Show that $p'(x_0) = q(x_0)$. Then, figure out the pattern obeyed by the coefficients of the different monomials in $q(x)$; use it to implement a version of Horner's rule that also computes $p'(x_0)$.

3.12 This problem deals with our analytical expectations for the total error, $\mathcal{E} = \mathcal{E}_{app} + \mathcal{E}_{ro}$, for Eq. (3.40) and Eq. (3.41). Let h take on the values $10^{-1}, 10^{-2}, 10^{-3}, 10^{-4}$ and plot the total error for the two cases. Assume $f(x) = 1$, $f''(x) = 0.1$, and $f'''(x) = 100$. Do you understand these results?

3.13 Produce a table of x_i and $e^{\sin(2x_i)}$ values, where x_i goes from 0 to 1.6 in steps of 0.08.

(a) Plot the forward-difference and central-difference results (for the first derivative) given these values. (Hint: if you cannot produce a result for a specific x, don't.) Then, introduce a curve for the analytical derivative.

(b) Use Richardson extrapolation for the forward difference for points on a grid and add an extra set of points to the plot. You can use:

$$R_{fd} = 2D_{fd}(h) - D_{fd}(2h) + O(h^2) \tag{3.90}$$

3.14 [$\mathcal{A}$] In the main text we used Richardson extrapolation in order to rederive a second forward-difference formula for the first derivative, see Eq. (3.49). Do the same (i.e., don't use Taylor series) for the third forward difference for the first derivative.

3.15 [$\mathcal{A}$] In the main text we derived a second central-difference formula for the first derivative in two distinct ways, using Taylor series and using Richardson extrapolation. Correspondingly derive a second central-difference formula for the second derivative.

3.16 We will now discuss an application of Richardson extrapolation which does *not* involve finite differences. Inspired by Archimedes' approach to approximating the value of π (namely inscribing a regular polygon in a circle with unit diameter), we write down the perimeter of a polygon with 2^n angles $g_n = 2^n \sin\left(\frac{\pi}{2^n}\right)$.

(a) Set $h = 2^{-n}$ and then Taylor/Maclaurin expand the sine so as to produce a systematic theory for the dependence of g on h, i.e., $g(h)$.

(b) In the previous part you should have found that, to begin with, in the language of Eq. (3.47) you have:

$$G = \frac{4g(h/2) - g(h)}{3} + O(h^4) \tag{3.91}$$

Take $n = 2, \dots, 10$ and apply $g_n = 2^n \sin\left(\frac{\pi}{2^n}\right)$. Then, use those values as input to Eq. (3.91) to produce extrapolated estimates.

(c) You should now see a pattern following from the above Taylor expansion. In the previous part you cancelled the error $O(h^2)$, so you were left with a theory having $p = 4$ and $q = 2$. Apply Eq. (3.47) to the results you produced in the previous part to carry out a further round of Richardson extrapolation. (This is quite similar to what we will encounter in Eq. (7.87).)

3.17 [$\mathcal{A}$] Explicitly show how automatic differentiation works on the following function:

$$f(x) = \frac{(x-5)(x-6)\sqrt{x}}{x-7} + \ln(8x) \tag{3.92}$$

to evaluate $f(4)$ and $f'(4)$.

3.18 For this problem you will need to make sure JAX is installed on your system. Use the `grad()` function to automatically differentiate $f(x) = e^{\sin(2x)}$ at $x = 0.5$, comparing to the answer mentioned in the main text. Be sure to enable double precision.

3.19 Write a basic automatic differentiator in Python, implementing the operations discussed in section 3.4.1. Test your code on example functions like that in Eq. (3.58). Probably the easiest way you can accomplish this task is via object orientation: define a new `class` called `dual`; you should implement the special methods `__add__`, `__sub__`, and so on, according to the rules given in Eq. (3.56). If you don't want each of these to start with a test using `isinstance()`, you should have your `__init__` convert constants as per Eq. (3.57).

3.20 [$\mathcal{P}$] This problem studies the classical and quantum harmonic oscillators.

(a) For the classical oscillator, take the solution $x = x_0 \sin(\omega t)$, which assumes that the particle is at $x = 0$ at $t = 0$. Calculate $P_c(x)$ by determining what fraction of the total time the particle will spend in an interval dx around x. In other words, use the relation $P_c(x)dx = dt/T$, where T is the period of oscillation.[17]

(b) Plot the squares of the quantum harmonic oscillator eigenfunctions for $n = 3, 10, 20, 150$ and compare with the classical solution(s).

3.21 [$\mathcal{P}$] Implement the three-dimensional harmonic oscillator, allowing different n in each Cartesian component. Test your function using `kinetic()` and `psiqho()` (i.e., the three-dimensional energy should be the sum of three one-dimensional energies).

3.22 [$\mathcal{P}$] Rewrite `psibox()` such that it applies to the three-dimensional particle in a box. Your wave vector should take the form $\mathbf{k} = 2\pi(n_x, n_y, n_z)/L$. In practice, you may wish to only input the "cardinal number" labelling the eigenstate. To do so, order the triples n_x, n_y, n_z according to the magnitude of the sum of their squares. Rewrite `psibox()` such that it takes in only the cardinal number (and the box size L) and evaluates the correct wave function. (You will probably want to first create another function which produces a dictionary mapping cardinal numbers to triples n_x, n_y, n_z.)

3.23 [$\mathcal{P}$] Rewrite `kinetic()` to use the second central-difference formula for the second derivative (from problem 3.15). Compare your answers to what you get in `kinetic.py` for both the harmonic-oscillator and the periodic-boundary cases.

[17] If you are unfamiliar with differential forms, you can treat the derivative as a fraction (however, see page 520).

3.24 [$\mathcal{P}$] This problem applies our function `kinetic()` to a physical setting that is different from what we encountered in our Project. Specifically, we wish to study a particle in one dimension, impinging on a simple-step barrier, i.e., our potential is:

$$V(x) = \begin{cases} 0, & x < 0 \\ V_0, & x > 0 \end{cases} \tag{3.93}$$

and we take the energy of the particle to be $E < V_0$. As you may have seen in a course on quantum mechanics, the wave function for this problem takes the form:

$$\psi(x) = \begin{cases} Ae^{ikx} + Be^{-ikx}, & x < 0 \quad (k = \sqrt{2mE}/\hbar) \\ Ce^{-\kappa x}, & x > 0 \quad (\kappa = \sqrt{2m(V_0 - E)}/\hbar) \end{cases} \tag{3.94}$$

and continuity of the wave function and its derivative gives us:

$$\frac{C}{A} = \frac{2}{1 + i\kappa/k}, \qquad \frac{B}{A} = \frac{1 - i\kappa/k}{1 + i\kappa/k} \tag{3.95}$$

Take $k = 2$ and $\kappa = 4$ and determine (both analytically and using Python) what the kinetic energy should be if the particle is located at $x < 0$ (or at $x > 0$).

3.25 [$\mathcal{P}$] Examine the derivative of a *noisy function* (e.g., from imperfectly measuring a particle's velocity v_i at different time slices t_i). Since we haven't talked about random numbers yet, we will model the effect of the noise by superimposing a highly oscillatory behavior on a slowly varying function. To make things concrete, let us study:

$$f(x) = 2 + 5\sin x + 0.1\sin(30x) \tag{3.96}$$

The last term has a small amplitude, but contains strong oscillations which have a dramatic effect on the function's derivative. Take 128 points placed on an equally spaced grid from 0 to 2π and produce a table of values, $(x_i, f(x_i))$.

(a) Plot $f(x)$ together with $g(x)$, where $g(x)$ does not contain the third (highly oscillatory) term. Observe that the two curves basically lie on top of each other.

(b) Create a plot that contains: (i) the analytically computed $f'(x)$, (ii) the analytically computed $g'(x)$, and (iii) the forward-difference approximation to $f'(x)$, using adjacent points on your grid. Observe that both the analytical $f'(x)$ and the forward-difference approximation to it are highly oscillatory and therefore quite different from the "underlying" behavior of $g'(x)$.

(c) Introduce a new set of points into your latest plot: (iv) the forward-difference approximation to $f'(x)$, using grid points which are twice removed (e.g., to estimate the derivative at x_{10} you use the function values at x_{12} and x_{10}). Observe that this set of points exhibits "noise" with smaller amplitude; this is because we have employed a larger step size h and have thereby smoothed out some of the "unphysical" oscillations due to the third term. (Parts (b) and (c) are employing the forward-difference approximation, i.e., we were able to do a better job without turning to a better finite-difference approximation.)

4 Matrices

> Praise large estates, but cultivate a small one.
>
> Vergil

4.1 Motivation

Linear algebra pops up almost everywhere in physics, so the matrix-related techniques developed below will be used repeatedly in later chapters. As a result, the present chapter is one of the longest ones in the book and in some ways constitutes its backbone. With this in mind, we will take the time to introduce several numerical techniques in detail. As in previous chapters, our concluding section addresses a physics setting where the numerical tools we have developed become necessary.

4.1.1 Examples from Physics

In contradistinction to this, in the current subsection we will discuss some elementary examples from undergrad physics, which don't involve heavy-duty computation, but do involve the same concepts.

1. **Rotations in two dimensions**
   Consider a two-dimensional Cartesian coordinate system. A point $\mathbf{r} = (x\ y)^T$ can be rotated counter-clockwise through an angle θ about the origin, producing a new point $\mathbf{r}' = (x'\ y')^T$. The two points' coordinates are related as follows:

$$\begin{pmatrix} \cos\theta & -\sin\theta \\ \sin\theta & \cos\theta \end{pmatrix} \begin{pmatrix} x \\ y \end{pmatrix} = \begin{pmatrix} x' \\ y' \end{pmatrix} \tag{4.1}$$

   The 2×2 matrix appearing here is an example of a *rotation matrix* in Euclidean space. If you know $\mathbf{r}'$ and wish to calculate $\mathbf{r}$, you need to solve this system of two linear equations. Observe that our rotation matrix is *not* symmetric (the two off-diagonal matrix elements are not equal). While symmetric matrices are omnipresent in physics, there are several scenarios where the matrix involved is not symmetric. Thus, we will study general matrices in most of what follows.
2. **Electrostatic potentials**
   Assume you have n electric charges q_j (which are unknown) held at the positions $\mathbf{R}_j$ (which are known). Further assume that you know the electric potential $\phi(\mathbf{r}_i)$ at the n

known positions $\mathbf{r}_i$. From the definition of the potential (as well as the fact that the potential obeys the principle of superposition), we see that:

$$\phi(\mathbf{r}_i) = \sum_{j=0}^{n-1} \left(\frac{k}{|\mathbf{r}_i - \mathbf{R}_j|} \right) q_j \tag{4.2}$$

where $i = 0, 1, \ldots, n-1$. If you assume you have four charges, the above relation turns into the following 4×4 linear system of equations:

$$\begin{pmatrix} k/|\mathbf{r}_0 - \mathbf{R}_0| & k/|\mathbf{r}_0 - \mathbf{R}_1| & k/|\mathbf{r}_0 - \mathbf{R}_2| & k/|\mathbf{r}_0 - \mathbf{R}_3| \\ k/|\mathbf{r}_1 - \mathbf{R}_0| & k/|\mathbf{r}_1 - \mathbf{R}_1| & k/|\mathbf{r}_1 - \mathbf{R}_2| & k/|\mathbf{r}_1 - \mathbf{R}_3| \\ k/|\mathbf{r}_2 - \mathbf{R}_0| & k/|\mathbf{r}_2 - \mathbf{R}_1| & k/|\mathbf{r}_2 - \mathbf{R}_2| & k/|\mathbf{r}_2 - \mathbf{R}_3| \\ k/|\mathbf{r}_3 - \mathbf{R}_0| & k/|\mathbf{r}_3 - \mathbf{R}_1| & k/|\mathbf{r}_3 - \mathbf{R}_2| & k/|\mathbf{r}_3 - \mathbf{R}_3| \end{pmatrix} \begin{pmatrix} q_0 \\ q_1 \\ q_2 \\ q_3 \end{pmatrix} = \begin{pmatrix} \phi(\mathbf{r}_0) \\ \phi(\mathbf{r}_1) \\ \phi(\mathbf{r}_2) \\ \phi(\mathbf{r}_3) \end{pmatrix} \tag{4.3}$$

which needs to be solved for the 4 unknowns q_0, q_1, q_2, and q_3. (As an aside, note that in this case, too, the matrix involved is *not* symmetric.)

3. **Principal moments of inertia**

Let's look at coordinate systems again. Specifically, in the study of the rotation of a rigid body about an arbitrary axis in three dimensions you may have encountered the *moment of inertia tensor*:

$$I_{\alpha\beta} = \int \rho(\mathbf{r}) \left(\delta_{\alpha\beta} r^2 - \mathbf{r}_\alpha \mathbf{r}_\beta \right) d^3 r \tag{4.4}$$

where $\rho(\mathbf{r})$ is the mass density, α and β denote Cartesian components, and $\delta_{\alpha\beta}$ is a Kronecker delta. The moment of inertia tensor is represented by a 3×3 matrix:

$$\mathbf{I} = \begin{pmatrix} I_{xx} & I_{xy} & I_{xz} \\ I_{yx} & I_{yy} & I_{yz} \\ I_{zx} & I_{zy} & I_{zz} \end{pmatrix} \tag{4.5}$$

The diagonal elements are moments of inertia and the off-diagonal elements are known as products of inertia. Given their definition, this matrix is symmetric (e.g., $I_{xy} = I_{yx}$). It is possible to employ a coordinate system for which the products of inertia vanish. The axes of this coordinate system are known as the *principal axes* for the body at the point O. In this case, the moment of inertia tensor is represented by an especially simple (diagonal) matrix:

$$\mathbf{I}_P = \begin{pmatrix} I_0 & 0 & 0 \\ 0 & I_1 & 0 \\ 0 & 0 & I_2 \end{pmatrix} \tag{4.6}$$

where I_0, I_1, and I_2 are known as the *principal moments* of the rigid body at the point O. In short, finding the principal axes is equivalent to diagonalizing a 3×3 matrix: this is an instance of the "eigenvalue problem", about which we'll hear much more below.

4.1.2 The Problems to Be Solved

Appendix C.2 summarizes some of the linear-algebra material you should already be familiar with. As far as the notation is concerned, you need to keep in mind that we employ indices that go from 0 to $n-1$, in order to be consistent with the rest of the book (and with Python). Also, we use bold uppercase symbols to denote matrices (e.g., **A**) and bold lowercase symbols to denote vectors (e.g., **x**). In this chapter, after we do some preliminary error analysis in section 4.2, we will be solving two large classes of problems. Both are very easy to write down, but advanced monographs have been written on the techniques employed to solve them in practice.

First, we look at the problem where we have n unknowns x_i, along with $n \times n$ coefficients A_{ij} and n constants b_i:

$$\begin{pmatrix} A_{00} & A_{01} & \dots & A_{0,n-1} \\ A_{10} & A_{11} & \dots & A_{1,n-1} \\ \vdots & \vdots & \ddots & \vdots \\ A_{n-1,0} & A_{n-1,1} & \dots & A_{n-1,n-1} \end{pmatrix} \begin{pmatrix} x_0 \\ x_1 \\ \vdots \\ x_{n-1} \end{pmatrix} = \begin{pmatrix} b_0 \\ b_1 \\ \vdots \\ b_{n-1} \end{pmatrix} \tag{4.7}$$

where we used a comma to separate two indices (e.g., $A_{1,n-1}$) when this was necessary to avoid confusion. These are n equations linear in n unknowns. In compact matrix form, this problem is written:

$$\mathbf{A}\mathbf{x} = \mathbf{b} \tag{4.8}$$

where **A** is sometimes called the *coefficient matrix*. (We will see below how to actually solve this problem, but for now we limit ourselves to saying that $|\mathbf{A}| \neq 0$, i.e., the matrix is non-singular, so it's made up of linearly independent columns.) Even though it looks very simple, this is a problem that we will spend considerable time solving in the present chapter. We will be doing this mainly by using the *augmented coefficient matrix* which places together the elements of **A** and **b**, i.e.:

$$(\mathbf{A}|\mathbf{b}) = \left(\begin{array}{cccc|c} A_{00} & A_{01} & \dots & A_{0,n-1} & b_0 \\ A_{10} & A_{11} & \dots & A_{1,n-1} & b_1 \\ \vdots & \vdots & \ddots & \vdots & \vdots \\ A_{n-1,0} & A_{n-1,1} & \dots & A_{n-1,n-1} & b_{n-1} \end{array}\right) \tag{4.9}$$

Note that this way we don't have to explicitly write out the elements of **x**.

In a course on linear algebra you will have seen examples of legitimate operations one can carry out while solving the system of linear equations. Such operations change the elements of **A** and **b**, but leave the solution vector **x** unchanged. More generally, we are allowed to carry out the following *elementary row operations*:

- *Scaling*: each row/equation may be multiplied by a constant (multiplies $|\mathbf{A}|$ by the same constant).
- *Pivoting*: two rows/equations may be interchanged (changes sign of $|\mathbf{A}|$).
- *Addition/Elimination*: a row/equation may be replaced by the sum of that row/equation with a multiple of any other row/equation (doesn't change $|\mathbf{A}|$).

In addition to providing the name and an explanation, we also mention parenthetically the effect of each operation on the determinant of the matrix $\mathbf{A}$. Keep in mind that these are operations that are carried out on the augmented coefficient matrix $(\mathbf{A}|\mathbf{b})$ so, for example, when you interchange two rows of $\mathbf{A}$ you should, obviously, also interchange the corresponding two elements of $\mathbf{b}$.

Second, we wish to tackle the standard form of the matrix eigenvalue problem:

$$\mathbf{A}\mathbf{v} = \lambda\mathbf{v} \tag{4.10}$$

Once again, in this form the problem seems quite simple, but a tremendous amount of work has gone toward solving this equation. Explicitly, our problem is equivalent to:

$$\begin{pmatrix} A_{00} & A_{01} & \dots & A_{0,n-1} \\ A_{10} & A_{11} & \dots & A_{1,n-1} \\ \vdots & \vdots & \ddots & \vdots \\ A_{n-1,0} & A_{n-1,1} & \dots & A_{n-1,n-1} \end{pmatrix} \begin{pmatrix} v_0 \\ v_1 \\ \vdots \\ v_{n-1} \end{pmatrix} = \lambda \begin{pmatrix} v_0 \\ v_1 \\ \vdots \\ v_{n-1} \end{pmatrix} \tag{4.11}$$

A crucial difference from the system we encountered in Eq. (4.8) is that here both the scalar/number λ and the column vector $\mathbf{v}$ are unknown. This λ is called an *eigenvalue* and $\mathbf{v}$ is called an *eigenvector*.

Let's sketch one possible approach to solving this problem. If we move everything to the left-hand side, Eq. (4.10) becomes:

$$(\mathbf{A} - \lambda\mathcal{I})\mathbf{v} = \mathbf{0} \tag{4.12}$$

where $\mathcal{I}$ is the $n \times n$ identity matrix and $\mathbf{0}$ is an $n \times 1$ column vector made up of 0s. It is easy to see that we are faced with a system of n linear equations: the coefficient matrix here is $\mathbf{A} - \lambda\mathcal{I}$ and the constant vector on the right-hand side is all 0s. Since we have $n + 1$ unknowns in total (λ, v_0, v_1, ..., v_{n-1}), it is clear that we will not be able to find unique solutions for all the unknowns.

A trivial solution is $\mathbf{v} = \mathbf{0}$. In order for a non-trivial solution to exist, we must be dealing with a coefficient matrix whose determinant vanishes, namely:

$$|\mathbf{A} - \lambda\mathcal{I}| = 0 \tag{4.13}$$

where the right-hand side contains the *number* zero. In other words, we are dealing with a non-trivial situation only when the matrix $\mathbf{A} - \lambda\mathcal{I}$ is singular. Expanding out the determinant gives us a polynomial equation which is known as the *characteristic equation*:

$$(-1)^n\lambda^n + c_{n-1}\lambda^{n-1} + \dots + c_1\lambda + c_0 = 0 \tag{4.14}$$

Thus, an $n \times n$ matrix has at most n distinct eigenvalues, which are the zeros of the characteristic polynomial.[1] When a zero occurs, say, twice, we say that that zero has *multiplicity* 2 (more properly, *algebraic multiplicity* 2). If a zero occurs only once, in other words if it has multiplicity 1, we are dealing with a *simple* eigenvalue.[2]

Having calculated the eigenvalues, one way to evaluate the eigenvectors is simply by using Eq. (4.10) again. Specifically, for a given/known eigenvalue, λ_i, one tries to solve the system of linear equations $(\mathbf{A} - \lambda_i \mathcal{I})\mathbf{v}_i = \mathbf{0}$ for $\mathbf{v}_i$. For each value λ_i, we will not be able to determine unique values of $\mathbf{v}_i$, so we will limit ourselves to computing the relative values of the components of $\mathbf{v}_i$. (You may recall that things can be even worse when you are dealing with repeated eigenvalues.) Incidentally, it's important to keep the notation straight: for example, $\mathbf{v}_0$ is a column vector, therefore made up of n elements. Using our notation above, the elements of $\mathbf{v}$ are v_0 and v_1, therefore the elements of $\mathbf{v}_0$ have to be called something like $(\mathbf{v}_0)_0$, $(\mathbf{v}_0)_1$, and so on (the first index tells us which eigenvector we're dealing with, the second index which component of that eigenvector).

4.2 Error Analysis

We now turn to a discussion of error estimation in work with matrices. In the spirit of chapter 2, this will entail us finding worst-case (pessimistic) error bounds. Note that this will not amount to a detailed error analysis of specific methods, say, for the solution of linear systems of equations. Instead, we will provide some general derivations of when a problem is "well-conditioned", typically by using matrix perturbation theory (i.e., by checking what happens if there are uncertainties in the input data). An explicit analysis of specific methods (like the pioneering work by Wilkinson in the 1960s on Gaussian elimination) shows that rounding errors are equivalent to perturbations of the input data, so in essence this is precisely what we will be probing.[3]

In what follows, after some preliminary definitions, we will investigate how linear systems, eigenvalues, and eigenvectors depend on the input data. The present section focuses on mathematical derivations which apply generally; if you're looking for more motivation, you can find a large number of examples in appendix D.1 (which is at `www.numphyspy.org`).

4.2.1 From *a posteriori* to *a priori* Estimates

Let's study a specific 2×2 linear system, namely $\mathbf{Ax} = \mathbf{b}$ for the case where:

$$(\mathbf{A}|\mathbf{b}) = \left(\begin{array}{cc|c} 0.2161 & 0.1441 & 0.1440 \\ 1.2969 & 0.8648 & 0.8642 \end{array}\right) \tag{4.15}$$

This problem was introduced by W. Kahan [79] (who was also responsible for many of the examples we studied in chapter 2). We are stating from the outset that this example is

[1] A fascinating aside: the matrix $\mathbf{A}$ satisfies its own characteristic equation (Cayley–Hamilton theorem).

[2] You will show in problem 4.1 that *the product of the eigenvalues is equal to the determinant* of the matrix.

[3] In other words, in linear algebra it is customary to engage in backward error analysis (recall page 39).

contrived. That being said, the misbehavior we are about to witness is not a phenomenon that happens only to experts who are looking for it. It is merely a more pronounced case of problematic behavior that does appear in the real world.

Simply put, there are two options on how to analyze errors: (a) an *a priori* analysis, in which case we try to see how easy/hard the problem is to solve before we begin solving it, and (b) an *a posteriori* analysis, where we have produced a solution, and attempt to see how good it is. Let's start with the latter option, namely an *a posteriori* approach.

Say you are provided with the following approximate solution to the problem in Eq. (4.15):

$$\tilde{\mathbf{x}}^T = \begin{pmatrix} 0.9911 & -0.4870 \end{pmatrix} \tag{4.16}$$

We are showing the transpose to save space on the page; we will keep doing this below. One way of testing how good a solution this is, is to evaluate the *residual vector*:

$$\mathbf{r} = \mathbf{b} - \mathbf{A}\tilde{\mathbf{x}} \tag{4.17}$$

Qualitatively, you can immediately grasp this vector's meaning: since the "true" solution $\mathbf{x}$ satisfies $\mathbf{A}\mathbf{x} = \mathbf{b}$, an approximate solution $\tilde{\mathbf{x}}$ should "almost" satisfy the same equation. Plugging in the matrices gives us for this case:

$$\mathbf{r}^T = \begin{pmatrix} -10^{-8} & 10^{-8} \end{pmatrix} \tag{4.18}$$

which might naturally lead you to the conclusion that our approximate solution $\tilde{\mathbf{x}}$ is pretty good, i.e., it may suffer from minor rounding-error issues (say, in the last digit) but other than that it's a done deal. Here's the thing, though: the exact solution to our problem is actually:

$$\mathbf{x}^T = \begin{pmatrix} 2 & -2 \end{pmatrix} \tag{4.19}$$

as you can easily see by substituting in the starting equation, $\mathbf{A}\mathbf{x} = \mathbf{b}$ (in other words, we get a zero residual vector for the exact solution). Thus, far from being only slightly off, our approximate "solution" $\tilde{\mathbf{x}}$ doesn't contain even a single correct significant figure.

With the disclaimer that there's much more that could be said at the *a posteriori* level, we now drop this line of attack and turn to an *a priori* analysis: could we have realized that solving the problem in Eq. (4.15) was difficult? How could we know that there's something pathological about it? One way of doing this is to make small perturbations to the input data. Imagine we didn't know the values of, say, the coefficients in $\mathbf{A}$ all that precisely. Would anything change then? Employing notation that is analogous to that in chapter 2, we are talking about the effect an absolute perturbation on the coefficient matrix, $\Delta\mathbf{A}$, would have on the solution vector. Such perturbations may result from rounding error: in physics applications, the matrices $\mathbf{A}$ and $\mathbf{b}$ are often themselves the result of earlier calculations, i.e., not set in stone. They may also result from uncertainty in the input data.

For the matrix in Eq. (4.15), changing only one element of $\mathbf{A}$ by less than 0.1% will have a dramatic impact on the solution to our problem. What is making this problem behave poorly (i.e., be very sensitive to tiny perturbations)? One criterion that is sometimes mentioned in this context is: since (as we see in appendix C.2) a non-invertible/singular matrix has determinant 0, it is *prima facie* plausible that matrices that have determinants that are "close to 0" are close to being singular. This immediately raises the question of what "small determinant" means. Intuitively, it makes sense to think of a "small determinant" as having something to do with the magnitude of the relevant matrix elements.

4.2.2 Norm Definitions and Properties

Let us provide our intuitions with quantitative backing. We will introduce the *matrix norm*, which measures the magnitude of **A**. There are several possible definitions of a norm, but here we will employ one of two possibilities. First, we have the *Frobenius norm*:

$$\|\mathbf{A}\|_F = \sqrt{\sum_{i=0}^{n-1}\sum_{j=0}^{n-1}|A_{ij}|^2} \tag{4.20}$$

Note that double vertical lines are used to denote the norm. This is different from single vertical lines, used to denote the determinant of a matrix or the absolute value of a real number or the modulus of a complex number. Another popular definition is that of the *infinity norm*:

$$\|\mathbf{A}\|_\infty = \max_{0\le i\le n-1}\sum_{j=0}^{n-1}|A_{ij}| \tag{4.21}$$

also known as the *maximum (absolute) row-sum norm*. Both of these definitions try to measure the magnitude of the various matrix elements. Other definitions choose different ways to accomplish this (e.g., maximum column sum)—see Eq. (4.208) for another example.

Regardless of the specific definition employed, all matrix norms for square matrices obey the following properties:

$$\begin{aligned}
\|\mathbf{A}\| &\ge 0\\
\|\mathbf{A}\| &= 0 \text{ if and only if all } A_{ij} = 0\\
\|k\mathbf{A}\| &= |k|\,\|\mathbf{A}\|\\
\|\mathbf{A}+\mathbf{B}\| &\le \|\mathbf{A}\| + \|\mathbf{B}\|\\
\|\mathbf{A}\mathbf{B}\| &\le \|\mathbf{A}\|\,\|\mathbf{B}\|
\end{aligned} \tag{4.22}$$

Notice that a matrix norm is a number, not a matrix.[4] The fourth of these relations is known as the *triangle inequality* and should be familiar to you from other contexts.

We can now return to the question of when the determinant is "small". A reasonable definition would be $|\det(\mathbf{A})| \ll \|\mathbf{A}\|$, where we took the absolute value on the left-hand side and used the det notation to avoid any confusion. This alleged criterion has the advantage that it takes into account the magnitude of the matrix elements. (Don't stop reading here, though.)

[4] This is analogous to a matrix determinant, which quantifies an entire matrix but is not a matrix itself.

In what follows, we'll also make use of norms of column *vectors*, so we briefly go over two such definitions:

$$||\mathbf{x}||_E = \sqrt{\sum_{i=0}^{n-1} |x_i|^2}, \quad ||\mathbf{x}||_\infty = \max_{0\le i\le n-1} |x_i| \tag{4.23}$$

These are the *Euclidean norm* and the *infinity norm*, respectively. The latter is also known as the maximum-magnitude norm.

4.2.3 Condition Number for Linear Systems

Appendix D.1 unambiguously shows that the criterion $|\det(\mathbf{A})| \ll ||\mathbf{A}||$ is flawed (its appearance in textbooks notwithstanding).[5] So where does this leave us? Just because we encountered a faulty criterion does not mean that a good one cannot be arrived at. We will now carry out an informal derivation that will point us toward a quantitative measure of ill-conditioning. (Spoiler alert: it does not involve the determinant.) This measure of the sensitivity of our problem to small changes in its elements will be called the *condition number*.

Let us start with the unperturbed problem:

$$\mathbf{A}\mathbf{x} = \mathbf{b} \tag{4.24}$$

and combine that with the case where $\mathbf{A}$ is slightly changed, with $\mathbf{b}$ being held constant. Obviously, this will also impact the solution vector:[6]

$$(\mathbf{A} + \Delta\mathbf{A})(\mathbf{x} + \Delta\mathbf{x}) = \mathbf{b} \tag{4.25}$$

Of course, you could have chosen to also perturb the elements of the constant vector $\mathbf{b}$ (either at the same time or separately). These scenarios are explored in problems 4.3 and 4.4.

Expanding out the parentheses in Eq. (4.25) and plugging in Eq. (4.24), we find:

$$\mathbf{A}\Delta\mathbf{x} = -\Delta\mathbf{A}(\mathbf{x} + \Delta\mathbf{x}) \tag{4.26}$$

Assuming $\mathbf{A}$ is non-singular (so you can invert it), you get:

$$\Delta\mathbf{x} = -\mathbf{A}^{-1}\Delta\mathbf{A}(\mathbf{x} + \Delta\mathbf{x}) \tag{4.27}$$

Taking the norm on both sides we find:

$$||\Delta\mathbf{x}|| = ||\mathbf{A}^{-1}\Delta\mathbf{A}(\mathbf{x} + \Delta\mathbf{x})|| \le ||\mathbf{A}^{-1}||\ ||\Delta\mathbf{A}||\ ||\mathbf{x} + \Delta\mathbf{x}|| \le ||\mathbf{A}^{-1}||\ ||\Delta\mathbf{A}||\ ||\mathbf{x}|| \tag{4.28}$$

[5] "Yet malice never was his aim;
He lash'd the vice, but spar'd the name."
(Jonathan Swift, *Verses on the Death of Dr. Swift*)

[6] As per Eq. (2.5), the absolute error is "approximate minus exact", so Eq. (4.25) is a way of writing $\tilde{\mathbf{A}}\tilde{\mathbf{x}} = \mathbf{b}$.

In the first step all we did was to take the absolute value of −1 (third property in Eq. (4.22)) and in the second step we simply applied the fifth property in Eq. (4.22), twice. In the third step we used the triangle inequality and dropped the second-order term. Using the non-negativity of norms (first property in Eq. (4.22)), we get:

$$\frac{||\Delta \mathbf{x}||}{||\mathbf{x}||} \leq ||\mathbf{A}^{-1}||\,||\Delta \mathbf{A}|| \tag{4.29}$$

Multiplying and dividing by a constant on the right-hand side gives us:

$$\frac{||\Delta \mathbf{x}||}{||\mathbf{x}||} \leq ||\mathbf{A}||\,||\mathbf{A}^{-1}||\,\frac{||\Delta \mathbf{A}||}{||\mathbf{A}||} \tag{4.30}$$

In other words, if you know an error bound on $||\Delta \mathbf{A}||/||\mathbf{A}||$ then that translates to an error bound on $||\Delta \mathbf{x}||/||\mathbf{x}||$. The coefficient in front of $||\Delta \mathbf{A}||/||\mathbf{A}||$ determines if a small perturbation gets magnified when solving for $\mathbf{x}$ or not.

This derivation naturally leads us to the introduction of the following *condition number*:

$$\kappa(\mathbf{A}) = ||\mathbf{A}||\,||\mathbf{A}^{-1}|| \tag{4.31}$$

A large condition number can lead to an amplification of a small perturbation: we say we are dealing with an *ill-conditioned* problem. If the condition number is of order unity, then a small perturbation is not amplified, so we are dealing with a *well-conditioned* problem (the condition number is bounded below by unity). Basically, the condition number manages to encapsulate the sensitivity to perturbations *ahead of time*: you can tell that you will be sensitive to small perturbations even before you start solving the linear system of equations. Qualitatively, this condition number tells us both how well- or ill-conditioned the solution of the linear problem $\mathbf{A}\mathbf{x} = \mathbf{b}$ is, as well as how well- or ill-conditioned the inversion of matrix $\mathbf{A}$ is. This dual role is not surprising: conceptually (though not in practice) solving a linear system is equivalent to inverting the matrix on the left-hand side. Obviously, the precise value of the condition number depends on which norm you are using.

To evaluate the condition number $\kappa(\mathbf{A})$ from Eq. (4.31) we need to first compute the matrix inverse $\mathbf{A}^{-1}$. This raises two issues: (a) wouldn't the computation of the inverse necessitate use of the same methods (such as those discussed in section 4.3 below) whose appropriateness we are trying to establish in the first place? and (b) computing the inverse requires $O(n^3)$ operations (as we will discuss below) where the coefficient in front is actually more costly than the task we were faced with (solving a system of equations). But then it hardly seems reasonable to spend so much effort only to determine ahead of time how well you may end up doing at solving your problem. Both of these concerns are addressed by using an alternative method to *estimate* the condition number (within a factor of 10 or so) with only $O(n^2)$ operations (see also problem 4.42).

4.2.4 Condition Number for Simple Eigenvalues

We are now interested in solving not the linear system of equations $\mathbf{A}\mathbf{x} = \mathbf{b}$ from Eq. (4.8), but the eigenvalue problem $\mathbf{A}\mathbf{v} = \lambda\mathbf{v}$ from Eq. (4.10). In appendix D.1 we explore examples showing that the same condition number as for the linear system problem, $\mathbf{Ax} = \mathbf{b}$, leads to disappointing results in connection with the eigenvalue problem. We now turn to an informal derivation that will point us toward a quantitative measure of conditioning *eigenvalues*. (Spoiler alert: it is not $\kappa(\mathbf{A})$.) This quantitative measure of the sensitivity of our problem to small changes in the input matrix elements will be called the *condition number for simple eigenvalues*; "simple" means we don't have repeated eigenvalues.

Let us start with the unperturbed problem:

$$\mathbf{A}\mathbf{v}_i = \lambda_i \mathbf{v}_i \tag{4.32}$$

This is Eq. (4.10) but with explicit indices, so we can keep track of the different eigenvalues and the corresponding eigenvectors. By complete analogy to what we did in our derivation for the linear system in Eq. (4.25) above, the relevant perturbed equation now is:

$$(\mathbf{A} + \Delta\mathbf{A})(\mathbf{v}_i + \Delta\mathbf{v}_i) = (\lambda_i + \Delta\lambda_i)(\mathbf{v}_i + \Delta\mathbf{v}_i) \tag{4.33}$$

Here we are carrying out an (absolute) perturbation of the matrix $\mathbf{A}$ and checking to see its impact on λ_i and on $\mathbf{v}_i$.[7]

So far (and for most of this chapter) we are keeping things general, i.e., we have not made an assumption that $\mathbf{A}$ is symmetric (which would have simplified things considerably). This is as it should be, since physics applications are not limited to the symmetric case (see Eq. (8.94) for an example). We now realize that we've been calling the column vectors $\mathbf{v}_i$ that appear in Eq. (4.32) "eigenvectors" though, properly speaking, they should be called *right eigenvectors*. If we have access to "right eigenvectors", then it stands to reason that we can also introduce the *left eigenvectors* $\mathbf{u}_i$ as follows:

$$\mathbf{u}_i^T \mathbf{A} = \lambda_i \mathbf{u}_i^T \tag{4.34}$$

where more generally we should have been taking the Hermitian conjugate/conjugate-transpose, $\dagger$, but this distinction won't matter to us, since in all our applications everything will be real-valued. Notice how this works: $\mathbf{A}$ is an $n \times n$ matrix whereas $\mathbf{v}_i$ and $\mathbf{u}_i$ are $n \times 1$ column vectors (so $\mathbf{u}_i^T$ is a $1 \times n$ row vector). Notice also a very simple way of evaluating left-eigenvectors if you already have a method to produce right eigenvectors: simply take the transpose of Eq. (4.34) to find:

$$\mathbf{A}^T \mathbf{u}_i = \lambda_i \mathbf{u}_i \tag{4.35}$$

Thus, the right eigenvectors of the transpose of a matrix give you the left eigenvectors of the same matrix (remarkably, corresponding to the *same* eigenvalues, as you will show in

[7] Actually, right now we are only interested in the impact on λ_i, so we'll try to eliminate $\Delta\mathbf{v}_i$ here: we return to the sensitivity of eigenvectors in the following subsection.

problem 4.7). Since we are not assuming that we're dealing with a symmetric matrix, in general $\mathbf{A} \neq \mathbf{A}^T$, so the left eigenvectors $\mathbf{u}_i$ are different from the right eigenvectors $\mathbf{v}_i$.

We will now use the last three boxed equations to derive an error bound on the magnitude of the change of an eigenvalue, $\Delta\lambda_i$. Start with Eq. (4.33) and expand the parentheses out. If you also take second-order changes (of the form $\Delta \times \Delta$) as being negligible, you find:

$$\mathbf{A}\Delta\mathbf{v}_i + \Delta\mathbf{A}\mathbf{v}_i = \lambda_i\Delta\mathbf{v}_i + \Delta\lambda_i\mathbf{v}_i \tag{4.36}$$

where we also made use of Eq. (4.32) in order to cancel two terms. Note that dropping higher-order terms is legitimate under the assumption we are dealing with small perturbations and changes, and simply determines the validity of our results (i.e., they are valid "to first order"). Multiplying the last equation by $\mathbf{u}_i^T$ on the left, we get:

$$\mathbf{u}_i^T\mathbf{A}\Delta\mathbf{v}_i + \mathbf{u}_i^T\Delta\mathbf{A}\mathbf{v}_i = \lambda_i\mathbf{u}_i^T\Delta\mathbf{v}_i + \Delta\lambda_i\mathbf{u}_i^T\mathbf{v}_i \tag{4.37}$$

But two of these terms cancel, as per our definition in Eq. (4.34), so we are left with:

$$\mathbf{u}_i^T\Delta\mathbf{A}\mathbf{v}_i = \Delta\lambda_i\mathbf{u}_i^T\mathbf{v}_i \tag{4.38}$$

Taking the absolute value of both sides and solving for $|\Delta\lambda_i|$, we have:

$$|\Delta\lambda_i| = \frac{|\mathbf{u}_i^T\Delta\mathbf{A}\mathbf{v}_i|}{|\mathbf{u}_i^T\mathbf{v}_i|} \tag{4.39}$$

We realize that we can apply the Cauchy–Schwarz inequality to the numerator:

$$|\mathbf{u}_i^T\Delta\mathbf{A}\mathbf{v}_i| \leq ||\mathbf{u}_i||\ ||\Delta\mathbf{A}||\ ||\mathbf{v}_i|| \tag{4.40}$$

This is very similar to the fifth property in Eq. (4.22), but here we're faced with an absolute value of a number on the left-hand side, not the norm of a matrix. We can now take the eigenvectors to be normalized such that $||\mathbf{u}_i|| = ||\mathbf{v}_i|| = 1$ (as is commonly done in standard libraries and we will also do in section 4.4 below).

This means that we have managed to produce an error bound on $|\Delta\lambda_i|$, as desired:

$$|\Delta\lambda_i| \leq \frac{1}{|\mathbf{u}_i^T\mathbf{v}_i|}\ ||\Delta\mathbf{A}|| \tag{4.41}$$

But this is fully analogous to what we had found for the perturbations in the case of the linear system of equations. The coefficient in front of $||\Delta\mathbf{A}||$ determines whether or not a small perturbation gets amplified in a specific case. Thus, we are led to introduce a new *condition number for simple eigenvalues*, as promised:

$$\kappa_{ev}^{\lambda_i}(\mathbf{A}) = \frac{1}{|\mathbf{u}_i^T\mathbf{v}_i|} \tag{4.42}$$

where the subscript is there to remind us that this is a condition number for a specific problem: for the evaluation of eigenvalues. The superscript keeps track of which specific eigenvalue's sensitivity we are referring to. To calculate the condition number for a given eigenvalue you first have to calculate the product of the corresponding left- and right-eigenvectors. An eigenvalue condition number that is close to 1 corresponds to an eigenvalue that is well-conditioned and an eigenvalue condition number that is much larger than 1 is ill-conditioned; as usual, the demarcation between the two cases is somewhat arbitrary.

Observe that we chose to study an *absolute* error bound, that is, a bound on $|\Delta\lambda_i|$ instead of one on $|\Delta\lambda_i|/|\lambda_i|$: this is reasonable, since an eigenvalue is zero if you're dealing with non-invertible matrices.[8] Take a moment to appreciate the fact that our derivation leading up to Eq. (4.42) did not employ the matrix inverse, $\mathbf{A}^{-1}$.

4.2.5 Sensitivity of Eigenvectors

Having studied the sensitivity to small perturbations of a linear system solution and of eigenvalue evaluations, the obvious next step is to do the same for the eigenvectors. You might be forgiven for thinking that this problem has already been solved (didn't we just produce a new condition number for eigenvalues in the previous subsection?), but things are not that simple. To reiterate, we are interested in probing the sensitivity of the eigenvalue problem:

$$\mathbf{A}\mathbf{v}_i = \lambda_i \mathbf{v}_i \tag{4.43}$$

to small perturbations. This time, we focus on the effect of the perturbations on the evaluation of the (right) eigenvectors, $\mathbf{v}_i$.

In appendix D.1 we explore examples showing that the same condition number as for the evaluation of the eigenvalues leads to disappointing results in connection with the eigenvector problem. We will now carry out an informal derivation that will point us toward a quantitative measure of conditioning *eigenvectors*. (Spoiler alert: it is not $\kappa_{ev}^{\lambda_i}(\mathbf{A})$.) This quantitative measure of the sensitivity of our problem to small changes in the input matrix elements will provide guidance regarding how to approach things *a priori*.

For simplicity, we are assuming we are dealing with distinct eigenvalues/linearly independent eigenvectors. Perturbing leads to Eq. (4.33):

$$(\mathbf{A} + \Delta\mathbf{A})(\mathbf{v}_i + \Delta\mathbf{v}_i) = (\lambda_i + \Delta\lambda_i)(\mathbf{v}_i + \Delta\mathbf{v}_i) \tag{4.44}$$

As you may recall, after a few manipulations we arrived at Eq. (4.36):

$$\mathbf{A}\Delta\mathbf{v}_i + \Delta\mathbf{A}\mathbf{v}_i = \lambda_i\Delta\mathbf{v}_i + \Delta\lambda_i\mathbf{v}_i \tag{4.45}$$

We now expand the perturbation in the eigenvector in terms of the other eigenvectors:

$$\Delta\mathbf{v}_i = \sum_{j\neq i} t_{ji}\mathbf{v}_j \tag{4.46}$$

where the coefficients t_{ji} are to be determined. We are employing here the linear independence of the eigenvectors. Note that this sum does not include a $j = i$ term: you can

[8] You can see this for yourself: the determinant of a matrix is equal to the product of its eigenvalues; when one of these is 0, the determinant is 0, so the matrix is non-invertible.

assume that if there existed a t_{ii} it could have been absorbed into our definition of what a perturbation for this eigenvector is.[9]

If we plug the last equation into the penultimate equation, we find:

$$\sum_{j\neq i}(\lambda_j - \lambda_i)t_{ji}\mathbf{v}_j + \Delta\mathbf{A}\mathbf{v}_i = \Delta\lambda_i\mathbf{v}_i \tag{4.47}$$

The λ_j arose because we also used our defining relation Eq. (4.43). We will now multiply our equation by the left eigenvector $\mathbf{u}_k^T$, keeping in mind that left and right eigenvectors for distinct eigenvalues are orthogonal to each other, $\mathbf{u}_k^T\mathbf{v}_i = 0$ for $k \neq i$. We find:

$$(\lambda_k - \lambda_i)t_{ki}\mathbf{u}_k^T\mathbf{v}_k + \mathbf{u}_k^T\Delta\mathbf{A}\mathbf{v}_i = 0 \tag{4.48}$$

We can solve this relation for t_{ki} and then plug the result into Eq. (4.46), thereby getting:

$$\Delta\mathbf{v}_i = \sum_{j\neq i}\frac{\mathbf{u}_j^T\Delta\mathbf{A}\mathbf{v}_i}{(\lambda_i - \lambda_j)\mathbf{u}_j^T\mathbf{v}_j}\mathbf{v}_j \tag{4.49}$$

This is our main result. Let's unpack it a little bit. First, we notice that (unlike our earlier results in condition-number derivations), the right-hand side contains a sum: the perturbation in one eigenvector contains contributions that are proportional to each of the other eigenvectors. Second, we observe that the numerator contains the perturbation in the input matrix, $\Delta\mathbf{A}$.[10] Third, and most significant, we see that our denominator contains two distinct contributions: (a) a $\mathbf{u}_j^T\mathbf{v}_j$ term, which is the same thing that appeared in our definition of the condition number for a simple eigenvalue in Eq. (4.42), and (b) a $\lambda_i - \lambda_j$ term, which encapsulates the separation between the eigenvalue λ_i and all other eigenvalues.

Thus, we have found that a perturbation in the input matrix will get amplified, first, if $\mathbf{u}_j^T\mathbf{v}_j$ is small or, second, *if any two eigenvalues are close*! In other words, the problem of evaluating eigen*vectors* may be ill-conditioned either because the eigen*value* problem for any of the eigenvalues is ill-conditioned, or because two (or more) eigenvalues are closely spaced. Intuitively, we already know that if two eigenvalues coincide then we cannot uniquely determine the eigenvectors, so our result can be thought of as a generalization of this to the case where two eigenvalues are close to each other.

4.3 Solving Systems of Linear Equations

In the previous section (and in appendix D.1) we studied in detail the conditioning of linear-algebra problems. We didn't explain how we produced the different answers for, e.g., the norm, the eigenvectors, etc. In practice, we used functions like the ones we will be introducing in the present and following sections. There is no cart-before-the-horse here,

[9] $\mathbf{v}_i$ becomes $\mathbf{v}_i + \Delta\mathbf{v}_i$, so any term in $\Delta\mathbf{v}_i$ that is proportional to $\mathbf{v}_i$ simply adjusts the coefficient in front of $\mathbf{v}_i$.

[10] If we wanted to take the norm, the $\mathbf{u}_j^T$ and $\mathbf{v}_i$ in the numerator would disappear, since $||\mathbf{u}_i|| = ||\mathbf{v}_i|| = 1$.

though: given that the problems were small, it would have been straightforward to carry out such calculations "by hand". Similarly, we won't spend more time talking about whether or not a problem is well- or ill-conditioned in what follows. We will take it for granted that that's an *a priori* analysis you can carry out on your own.

It's now time to discuss how to solve simultaneous linear equations on the computer:

$$\mathbf{A}\mathbf{x} = \mathbf{b} \tag{4.50}$$

We will structure our discussion in an algorithm-friendly way, i.e., we will write the equations in such a way as to enable a step-by-step implementation in Python later on.

In the spirit of this book, we always start with a more inefficient method, which is easier to explain. In the case of linear algebra, the more inefficient method turns out to be more *general* than other, specialized, methods which have been tailored to study problems with specific symmetries. Note that we will be focusing mainly on *direct methods*, namely approaches that transform the original problem into a form that is more easily solved. Indirect methods, briefly discussed in section 4.3.5, start with a guess for the solution and then refine it until convergence is reached. They are mainly useful when you have to deal with *sparse* problems, where many of the matrix elements are zero.

4.3.1 Triangular Matrices

We start with the simplest case possible, that of triangular matrices (for which all elements either above or below the diagonal are zero). This is not simply a toy problem: many of the fancier methods for the solution of simultaneous equations, like those we discuss below, manipulate the starting problem so that it ends up containing one or two triangular matrices at the end. Thus, we are here also providing the scaffolding for other methods.

We immediately note that in the real world one rarely stores a triangular matrix in its entirety, as that would be wasteful: it would entail storing a large number of 0s, which don't add any new information; if we know that a matrix is triangular, the 0s above or below the diagonal are implied. In state-of-the-art libraries, it is common to use a single matrix to store together an upper-triangular and a lower-triangular matrix (with some convention about the diagonal, since each of those triangular matrices also generally has non-zero elements there). Here, since our goal is pedagogical clarity, we will opt for coding up each algorithm "naively", namely by carrying around several 0s for triangular matrices. Once you get the hang of things, you will be able to modify our codes to make them more efficient (as problem 4.13 asks you to do).

4.3.1.1 Forward Substitution

Start with a *lower-triangular* matrix $\mathbf{L}$. The problem we are interested in solving is:

$$\mathbf{L}\mathbf{x} = \mathbf{b} \tag{4.51}$$

This trivial point might get obscured later, so we immediately point out that by **b** we simply mean the constant vector on the right-hand side (which we could have also named **c**, **d**, and so on. Similarly, **x** is merely the solution vector, which we could have also called, e.g., **y**. Our task is to find the solution vector. For concreteness, let us study a 3×3 problem:

$$\begin{pmatrix} L_{00} & 0 & 0 \\ L_{10} & L_{11} & 0 \\ L_{20} & L_{21} & L_{22} \end{pmatrix} \begin{pmatrix} x_0 \\ x_1 \\ x_2 \end{pmatrix} = \begin{pmatrix} b_0 \\ b_1 \\ b_2 \end{pmatrix} \tag{4.52}$$

This can be expanded into equation form:

$$\begin{aligned} L_{00}x_0 &= b_0 \\ L_{10}x_0 + L_{11}x_1 &= b_1 \\ L_{20}x_0 + L_{21}x_1 + L_{22}x_2 &= b_2 \end{aligned} \tag{4.53}$$

The way to find the solution-vector components should be fairly obvious: start with the first equation and solve it for x_0. Then, plug that answer into the second equation and solve for x_1. Finally, plug x_0 and x_1 into the third equation and solve for x_2. We have:

$$x_0 = \frac{b_0}{L_{00}}, \qquad x_1 = \frac{b_1 - L_{10}x_0}{L_{11}}, \qquad x_2 = \frac{b_2 - L_{20}x_0 - L_{21}x_1}{L_{22}} \tag{4.54}$$

This process is known as *forward substitution*, since we solve for the unknowns by starting with the first equation and moving forward from there. It's easy to see how to generalize this approach to the $n \times n$ case:

$$x_i = \left(b_i - \sum_{j=0}^{i-1} L_{ij}x_j \right) \frac{1}{L_{ii}}, \qquad i = 0, 1, \ldots, n-1 \tag{4.55}$$

with the understanding that, on the right-hand side, the sum corresponds to zero terms if $i = 0$, one term if $i = 1$, and so on.

4.3.1.2 Back Substitution

You can also start with an *upper-triangular* matrix **U**:

$$\mathbf{Ux} = \mathbf{b} \tag{4.56}$$

As above, for concreteness, we first study a 3×3 problem:

$$\begin{pmatrix} U_{00} & U_{01} & U_{02} \\ 0 & U_{11} & U_{12} \\ 0 & 0 & U_{22} \end{pmatrix} \begin{pmatrix} x_0 \\ x_1 \\ x_2 \end{pmatrix} = \begin{pmatrix} b_0 \\ b_1 \\ b_2 \end{pmatrix} \tag{4.57}$$

This can be expanded into equation form:

$$\begin{aligned} U_{00}x_0 + U_{01}x_1 + U_{02}x_2 &= b_0 \\ U_{11}x_1 + U_{12}x_2 &= b_1 \\ U_{22}x_2 &= b_2 \end{aligned} \tag{4.58}$$

For this case, too, the way to find the solution-vector components should be fairly obvious: start with the last equation and solve it for x_2. Then, plug that answer into the second equation and solve for x_1. Finally, plug x_2 and x_1 into the first equation and solve for x_0:

$$x_2 = \frac{b_2}{U_{22}}, \quad x_1 = \frac{b_1 - U_{12}x_2}{U_{11}}, \quad x_0 = \frac{b_0 - U_{01}x_1 - U_{02}x_2}{U_{00}} \tag{4.59}$$

This process is known as *back substitution*, since we solve for the unknowns by starting with the last equation and moving backward from there. It's easy to see how to generalize this approach to the $n \times n$ case:

$$x_i = \left(b_i - \sum_{j=i+1}^{n-1} U_{ij}x_j\right)\frac{1}{U_{ii}}, \quad i = n-1, n-2, \dots, 1, 0 \tag{4.60}$$

with the understanding that, on the right-hand side, the sum corresponds to zero terms if $i = n - 1$, one term if $i = n - 2$, and so on.

4.3.1.3 Implementation

Given the form in which we have expressed forward substitution, Eq. (4.55), and back substitution, Eq. (4.60), the Python implementation of code 4.1 essentially writes itself. A few observations are in order.

We used NumPy functionality throughout this code, though we could have employed Python lists, instead. However, as already mentioned, in this chapter we will be using NumPy arrays in order to help you become comfortable with them. (If you haven't learned how to use arrays, now is the time to do so.) Being familiar with NumPy arrays will also help you employ a wealth of other Python libraries after you are done reading this book: arrays are omnipresent in numerical computing, so this is a good opportunity to see them used in action to implement major linear-algebra algorithms.

In addition to extracting the size of the problem from the constant vector, our code has the following feature: the `numpy array` by the name of `xs` (to be returned by the function) is first created to contain zeros. (We could have employed `numpy.empty()` instead.) Then, each element of `xs` is evaluated and stored in turn. (In other words, we are not creating an empty list and then appending elements one at a time.) Also, keep in mind that we are storing the solution vector **x** (which mathematically is a column vector) in a one-dimensional NumPy array: this will be very convenient when we print it out (since it will all show up as one row of text).

Crucially, the lines of code that implement the aforementioned equations are a joy to read and write, given NumPy's syntax: the code looks essentially identical to the equation. Note how expressive something like `L[i,:i]` is: there's no need for double square brackets, while the slicing takes only the elements that should be used. More importantly, there's no need for a loop within a loop (the first loop going over i's and the second loop going over j's): `numpy` knows how to take the dot product of two vectors correctly when you use `@`.[11] It's nice to see that Python/NumPy also handles the cases of $i = 0$ (no products) and $i = 1$ (only one product) correctly. Note finally that `backsub()` uses Python's `reversed()`, as per our discussion in chapter 1.

We then encounter the `testcreate()` function, which essentially pulls a test matrix out of a hat: the crucial point here is that (in the spirit of the separation of concerns principle) we create the matrix **A** inside a function (which can then be re-used in code we write in the future). As we'll see below, our choice is not totally arbitrary; for now, merely note that this matrix is not symmetric. The code that creates the constant vector **b** is arbitrary, basically showcasing some of NumPy's functionality, as described in section 1.6. Analogously, we have created a function that runs the substitution functions and compares with the output of `numpy.linalg.solve()`, the standard `numpy` choice for solving linear coupled equations. In the spirit of all our codes so far, we pass in as an argument, `f`, our hand-rolled function which solves the problem (in what follows we will keep using this test function).

In the main program, after creating **A** and **b**, we use `numpy.tril()` to create a lower-triangular matrix starting from `A` (and `numpy.triu()` to create an upper-triangular matrix). This is only done in the spirit of pedagogy: the substitution function calls work just as well (giving the same answers) even if you don't first take the lower- or upper-triangular elements. Running this code, you see that there is no visible difference between the output of our substitution functions and that of `numpy.linalg.solve()`. Given the simplicity of the problem(s) we are solving here, this comes as no surprise.

4.3.1.4 Digression: Operation Counts

We will now engage in an activity which is very common in linear algebra: we will count how many floating-point operations it takes to carry out a specific calculation (these are also called "flops"). If a given computation scales poorly (e.g., exponentially) with the size of the problem, then it will be hard to increase the size much more than what is currently possible. If the scaling is not "too bad" (e.g., polynomial with a small power), then one can keep solving bigger problems without needing to employ dramatically new hardware.

In such studies, you will encounter the O symbol which, as you may recall from page 92, is known as big-O notation: a method that scales as $O(n^3)$ is better than another method that scales as $O(n^4)$ (for the same problem), since the power dominates over any prefactor when n is large. When one explicitly counts the number of additions/subtractions and multiplications/divisions, one is sometimes interested in the prefactor, e.g., $2n^3$ is better than $4n^3$, since the former requires only half as many operations. On the other hand, lower powers don't impact the scaling seriously so you may sometimes encounter expressions such as $2n^3 - 7n^2 + 5n$ written as $\sim 2n^3$, simply dropping the lower-degree terms.

[11] As discussed in problem 1.8, we could have used `numpy.dot(as,bs)` or `numpy.sum(as*bs)`, instead.

triang.py **Code 4.1**

```
import numpy as np

def forsub(L,bs):
    n = bs.size
    xs = np.zeros(n)
    for i in range(n):
        xs[i] = (bs[i] - L[i,:i]@xs[:i])/L[i,i]
    return xs

def backsub(U,bs):
    n = bs.size
    xs = np.zeros(n)
    for i in reversed(range(n)):
        xs[i] = (bs[i] - U[i,i+1:]@xs[i+1:])/U[i,i]
    return xs

def testcreate(n,val):
    A = np.arange(val,val+n*n).reshape(n,n)
    A = np.sqrt(A)
    bs = (A[0,:])**2.1
    return A, bs

def testsolve(f,A,bs):
    xs = f(A,bs); print(xs)
    xs = np.linalg.solve(A,bs); print(xs)

if __name__ == '__main__':
    A, bs = testcreate(4,21)
    L = np.tril(A)
    testsolve(forsub,L,bs)
    print(" ")
    U = np.triu(A)
    testsolve(backsub,U,bs)
```

Let's study the case of matrix–vector multiplication explicitly, see Eq. (C.9):

$$\mathbf{y} = \mathbf{A}\mathbf{x}, \quad y_i = \sum_{j=0}^{n-1} A_{ij} x_j \tag{4.61}$$

We see that for a given y_i we need, on the right-hand side, n multiplications and $n-1$ additions. Thus, since we have n terms for the y_i's in total, we are faced with $n \times n$ multiplications and $n \times (n-1)$ additions in total. If we add both of these results up, we find that matrix–vector multiplication requires precisely $2n^2 - n$ floating-point operations. As above, you will frequently see this re-expressed as $\sim 2n^2$ or even as $O(n^2)$.

In problem 4.34 you are asked to calculate the operation counts for vector–vector multiplication and matrix–matrix multiplication. These turn out to be $O(n)$ and $O(n^3)$, respectively, but you will also need to evaluate the exact prefactors (and lower-degree terms).

4.3.1.5 Operation Count for Forward Substitution

For concreteness, we will examine only forward substitution (but the answer turns out to be the same for back substitution). We copy here Eq. (4.55) for your convenience:

$$x_i = \left(b_i - \sum_{j=0}^{i-1} L_{ij} x_j\right) \frac{1}{L_{ii}}, \qquad i = 0, 1, \ldots, n-1 \tag{4.62}$$

It is easy to see that each of the x_i requires one division, so n divisions in total. It's equally easy to see that for a given x_i, we need to carry out i multiplications and i subtractions (check this for a few values of i if it's not immediately obvious). Thus, we can group the required operations into two categories. First, we require:

$$\sum_{i=0}^{n-1} i = \frac{(n-1)n}{2} = \frac{n^2 - n}{2} \tag{4.63}$$

additions/subtractions. Second, we require:

$$n + \sum_{i=0}^{n-1} i = n + \frac{(n-1)n}{2} = \frac{n^2 + n}{2} \tag{4.64}$$

multiplications/divisions. If we add both of these results up, we find that forward substitution requires precisely n^2 floating-point operations. This could be expressed as $O(n^2)$, but the latter form is less informative: in our explicit calculation we have found that the prefactor is exactly 1.

4.3.2 Gaussian Elimination

We now turn to the problem of solving linear simultaneous equations for the general case, i.e., when we are not dealing with a triangular matrix. We will solve:

$$\mathbf{A}\mathbf{x} = \mathbf{b} \tag{4.65}$$

for a general matrix $\mathbf{A}$. We start with the method known as *Gaussian elimination* (though it was used in China two thousand years earlier and by Newton more than a century before Gauss). In essence, this method employs the third elementary row operation we introduced in section 4.1.2: a row/equation may be replaced by the sum of that row/equation with a multiple of any other row/equation. After doing this repeatedly (in what is known as the *elimination phase*), we end up with an upper-triangular matrix, at which point we are at the *back substitution phase* which, as we just saw, is easy to carry out.

4.3.2.1 Example

Before we discuss the general algorithm and its implementation, it may be helpful to first show how to solve a specific example "by hand". Let us study the following 3×3 problem:

$$\begin{cases} 2x_0 + x_1 + x_2 = 8 \\ x_0 + x_1 - 2x_2 = -2 \\ 5x_0 + 10x_1 + 5x_2 = 10 \end{cases} \qquad \begin{pmatrix} 2 & 1 & 1 \\ 1 & 1 & -2 \\ 5 & 10 & 5 \end{pmatrix} \begin{pmatrix} x_0 \\ x_1 \\ x_2 \end{pmatrix} = \begin{pmatrix} 8 \\ -2 \\ 10 \end{pmatrix} \tag{4.66}$$

given in two equivalent forms. Again, we've seen that one way to compactly write down the crucial elements at play is to use the augmented matrix:

$$\left(\begin{array}{ccc|c} 2 & 1 & 1 & 8 \\ 1 & 1 & -2 & -2 \\ 5 & 10 & 5 & 10 \end{array}\right) \tag{4.67}$$

This combines the coefficient matrix $\mathbf{A}$ with the constant vector $\mathbf{b}$, the solution vector $\mathbf{x}$ being implied. You should spend some time looking at this augmented matrix, since that's what we are going to use below (i.e., the equations themselves will be implicitly given). Note that in this specific example the coefficient matrix $\mathbf{A}$ is not symmetric.

As already mentioned, Gaussian elimination employs the third elementary row operation, i.e., we will replace a row with that same row plus another row (times a coefficient). Specifically, we first pick a specific row, called the *pivot row*, which we multiply by a number and then subtract from the row we are transforming. Let's use j as the index that keeps track of the pivot row and i for the index corresponding to the row we are currently transforming (as usual, for a 3×3 problem our indices can have the values 0, 1, or 2). The operation we are carrying out is:

$$\text{New row } i = \text{row } i - \text{coefficient} \times \text{row } j \tag{4.68}$$

The coefficient is selected such that after the transformation the leading number in row i is a 0. Perhaps this will become more clear once you see the algorithm in action.

We begin with $j = 0$, taking the first equation as the pivot row. We then take $i = 1$ (the second row) as the row to be transformed: our goal is to eliminate the element in the first column (i.e., the term corresponding to x_0). To do this, we will replace the second row with the second row minus the first row times 0.5 (since $1 - 0.5 \times 2 = 0$). Obviously, we have to carry out this calculation for the entire row, giving us:

$$\left(\begin{array}{ccc|c} 2 & 1 & 1 & 8 \\ 0 & 0.5 & -2.5 & -6 \\ 5 & 10 & 5 & 10 \end{array}\right) \tag{4.69}$$

Next, for the same pivot row ($j = 0$), we will transform the third row ($i = 2$), by multiplying the pivot row by 2.5 and subtracting (since $5 - 2.5 \times 2 = 0$). We get:

$$\left(\begin{array}{ccc|c} 2 & 1 & 1 & 8 \\ 0 & 0.5 & -2.5 & -6 \\ 0 & 7.5 & 2.5 & -10 \end{array}\right) \tag{4.70}$$

We now see that our work with $j = 0$ as our pivot row is done: all the rows below it have been transformed such that the 0th (first) column contains a zero.

We now take $j = 1$, i.e., use the second equation as our pivot row. (We always use the latest version of the matrix, so our $j = 1$ pivot row will be the result of our earlier transformation.) The rows to be transformed always lie below the pivot row, so in this case there's only one row to change, $i = 2$ (the third row). We multiply the pivot row by 15 and subtract from the third row, in order to eliminate the element in the second column (i.e., the term corresponding to x_1), since $7.5 - 15 \times 0.5 = 0$. This gives us:

$$\left(\begin{array}{ccc|c} 2 & 1 & 1 & 8 \\ 0 & 0.5 & -2.5 & -6 \\ 0 & 0 & 40 & 80 \end{array}\right) \tag{4.71}$$

Our coefficient matrix is now in triangular form, so the elimination phase is done.

We now use Eq. (4.60) from the back substitution section:

$$\begin{aligned} x_2 &= \frac{80}{40} = 2 \\ x_1 &= \frac{-6 - (-2.5) \times x_2}{0.5} = \frac{-6 - (-2.5) \times 2}{0.5} = -2 \\ x_0 &= \frac{8 - 1 \times x_1 - 1 \times x_2}{2} = \frac{8 - 1 \times (-2) - 1 \times 2}{2} = 4 \end{aligned} \tag{4.72}$$

Thus, we have accomplished our task: we have solved our three simultaneous equations for the three unknowns, through a combination of elimination and back substitution.

4.3.2.2 General Case

Having gone through an explicit case step by step, it should be relatively straightforward to generalize this to the $n \times n$ problem. First, we repeat the augmented matrix of Eq. (4.9):

$$(\mathbf{A}|\mathbf{b}) = \left(\begin{array}{cccc|c} A_{00} & A_{01} & \dots & A_{0,n-1} & b_0 \\ A_{10} & A_{11} & \dots & A_{1,n-1} & b_1 \\ \vdots & \vdots & \ddots & \vdots & \vdots \\ A_{n-1,0} & A_{n-1,1} & \dots & A_{n-1,n-1} & b_{n-1} \end{array}\right) \tag{4.73}$$

As we saw in the previous subsection, Gaussian elimination modifies the coefficient matrix and the constant vector until the former becomes triangular. It is standard to do this by modifying the matrix elements of $\mathbf{A}$ and $\mathbf{b}$ (so, if you need the original values, you should make sure you've made a copy of them ahead of time). In other words, at some intermediate point in time the augmented matrix will look like this:

$$\left(\begin{array}{ccccccccc|c}
A_{00} & A_{01} & A_{02} & \dots & A_{0,j-1} & A_{0j} & A_{0,j+1} & \dots & A_{0,n-1} & b_0 \\
0 & A_{11} & A_{12} & \dots & A_{1,j-1} & A_{1j} & A_{1,j+1} & \dots & A_{1,n-1} & b_1 \\
0 & 0 & A_{22} & \dots & A_{2,j-1} & A_{2j} & A_{2,j+1} & \dots & A_{2,n-1} & b_2 \\
\vdots & \vdots & \vdots & \ddots & \vdots & \vdots & \vdots & \ddots & \vdots & \vdots \\
0 & 0 & 0 & \dots & A_{j-1,j-1} & A_{j-1,j} & A_{j-1,j+1} & \dots & A_{j-1,n-1} & b_{j-1} \\
0 & 0 & 0 & \dots & 0 & A_{jj} & A_{j,j+1} & \dots & A_{j,n-1} & b_j \\
0 & 0 & 0 & \dots & 0 & A_{j+1,j} & A_{j+1,j+1} & \dots & A_{j+1,n-1} & b_{j+1} \\
\vdots & \vdots & \vdots & \ddots & \vdots & \vdots & \vdots & \ddots & \vdots & \vdots \\
0 & 0 & 0 & \dots & 0 & A_{ij} & A_{i,j+1} & \dots & A_{i,n-1} & b_i \\
\vdots & \vdots & \vdots & \ddots & \vdots & \vdots & \vdots & \ddots & \vdots & \vdots \\
0 & 0 & 0 & \dots & 0 & A_{n-1,j} & A_{n-1,j+1} & \dots & A_{n-1,n-1} & b_{n-1}
\end{array}\right) \tag{4.74}$$

The snapshot we are showing corresponds to the case where j just became the pivot row, meaning that all the rows up to it have already been transformed, whereas all the rows below still have to be transformed. As already noted, the values of the **A** and **b** matrix elements shown here are the current ones, i.e., have already been transformed (so far) from the original values. (The first row is always left unchanged.)

You should convince yourself that j can take on the values:

$$j = 0, 1, 2, \dots, n-2 \tag{4.75}$$

The first possibility for the pivot row is the first row, meaning all other rows have to be transformed. The last row to be transformed is the last one, so the final pivot row is the penultimate row. Using the same notation as above, we call i the index corresponding to the row that is being transformed. Obviously, i has to be greater than j. Given what we just discussed, i takes on the values:

$$i = j+1, j+2, \dots, n-1 \tag{4.76}$$

As we saw in Eq. (4.68) and in the 3×3 case, Gaussian elimination works by multiplying the pivot row j by a coefficient and subtracting the result from row i which is currently being transformed (and storing the result back in row i). The coefficient is chosen such that (after the transformation) row i starts with a 0. Thus, looking at the snapshot in our augmented matrix, where the leading non-zero element of row i is A_{ij} and the leading non-zero element of row j is A_{jj}, we see that the coefficient has to be A_{ij}/A_{jj}—given that $A_{ij} - (A_{ij}/A_{jj})A_{jj} = 0$. In equation form:

$$\text{coefficient} = \frac{A_{ij}}{A_{jj}} \tag{4.77}$$

Incidentally, A_{jj} is sometimes called the *pivot element* since it is used (divided out) in order to eliminate the leading elements in the following rows. Obviously, the other elements in row i will end up having some new values (most likely non-zero ones).

At the end of this process, the matrix **A** contained in our augmented matrix will be upper triangular, so it will be straightforward to then apply Eq. (4.60) from the back substitution section to solve for all the unknowns.

4.3.2.3 Implementation

We have structured Eq. (4.74) and the surrounding discussion in such a way as to simplify producing a Python code that implements Gaussian elimination. We will structure our code in a modular way, such that it can apply to any input matrices **A** and **b**, impacting the external world only via its return value. Code 4.2 is a straightforward implementation of the afore-discussed algorithm.

We start out by importing some of the routines we created in our earlier code on triangular matrices. As already mentioned, we make copies of the input matrices, so we may update them at will, without impacting the rest of the program in an unexpected way (this is inefficient, but pedagogically superior).

The core of our `gauelim()` function consists of two loops: one over `j` which keeps track of the current pivot row and one over `i` which keeps track of which row we are currently updating, see Eq. (4.75) and Eq. (4.76). Thus, our choice in this chapter of modifying linear-algebra notation so that all indices start from 0 is starting to pay off.

In the inner loop, we always start from evaluating the coefficient A_{ij}/A_{jj} which will be used to subtract out the leading element in the row currently being updated. This elimination is carried out in the line `A[i,j:] -= coeff*A[j,j:]`, which employs NumPy functionality to carry out this modification for each column in row `i`. Notice how nice this is: we did *not* need to keep track of a third index (and therefore did not need to introduce a third loop). This reduces the cognitive load needed to keep track of what's going on: all the desired elements on one row are updated in one line. Actually, if we wanted to update the whole row, we would have said `A[i,:] -= coeff*A[j,:]`. This, too, employs NumPy functionality to our advantage, processing the entire row at one go. The earlier choice we made, however, to index using `A[i,j:]` instead of `A[i,:]` is better: not only is it not wasteful, processing only the non-zero elements, but it is also more transparent, since it clearly corresponds to Eq. (4.74). Actually, we *are* being slightly wasteful: the leading element in row `i` will end up being zero but we are carrying out the subtraction procedure for that column as well, instead of just assuming that it will vanish. (You might wish to implement this code with a third loop to appreciate our comments.) The code then makes the corresponding update to the **b** row element currently being processed.

Code 4.2

gauelim.py

```
from triang import backsub, testcreate, testsolve
import numpy as np

def gauelim(inA,inbs):
    A = np.copy(inA)
    bs = np.copy(inbs)
    n = bs.size

    for j in range(n-1):
        for i in range(j+1,n):
            coeff = A[i,j]/A[j,j]
            A[i,j:] -= coeff*A[j,j:]
            bs[i] -= coeff*bs[j]

    xs = backsub(A,bs)
    return xs

if __name__ == '__main__':
    A, bs = testcreate(4,21)
    testsolve(gauelim,A,bs)
```

A further point: if we were interested in saving some operations, we might introduce an `if` clause in the inner loop checking whether or not `A[i,j]` is already 0, before we divide it by the coefficient. But then you would be needlessly testing all the time for a condition which is rarely met.[12]

After we have used all possible pivot rows, our matrix will have been updated to be upper triangular. At this point, we simply call our earlier `backsub()` function. This seems to be a neat example of code re-usability. We allow ourselves to call functions inside our codes even if we haven't passed them in explicitly, as long as it's clear what's happening and how one would proceed to further update the code.

The main body of the code is quite straightforward: it creates our test matrix and calls the function that compares our new routine to the standard NumPy output. Observe how useful these earlier two functions turned out to be. We will employ them again several times in what follows. Running the code, we see that our simple Gaussian elimination code is already doing a good job matching the output of `numpy.linalg.solve()`. Note that you haven't seen this output vector before: in the earlier code we created a lower or upper-triangular matrix starting from `A`, whereas now we are using the entire matrix. We always get at least seven digits of agreement between the two solution vectors. This is good, but at

[12] If your matrix is sparse or banded and efficiency is a major concern, you should be using a different method.

this point it is a bit unclear why we're not doing a better job. We will return to this question in a later section.

4.3.2.4 Operation Count

Turning to a study of the operation count for Gaussian elimination, we keep in mind that we are interested in total floating-point operations. Thus, while it may help us to think of additions, multiplications, etc. separately, in the end we will add all of them up. It may be helpful to look at Eq. (4.74), which shows a snapshot of the Gaussian elimination algorithm at an intermediate time slice. We can separate the operations into two categories: (a) the conversion of **A** into an upper-triangular matrix together with the corresponding changes to **b** and (b) the back substitution of the resulting triangular problem. We already know from an earlier section that the back substitution phase requires n^2 floating-point operations, so we only have to address the first category.

From Eq. (4.68) we know that what we keep doing is the following operation:

$$\text{New row } i = \text{row } i - \text{coefficient} \times \text{row } j \tag{4.78}$$

We recall from Eq. (4.75) that the steps involved are organized by pivot row:

$$j = 0, 1, 2, \dots, n-2 \tag{4.79}$$

and are applied to the rows:

$$i = j+1, j+2, \dots, n-1 \tag{4.80}$$

in turn, as per Eq. (4.76).

In the first step, we have to modify the $n-1$ rows below the pivot row $j = 0$. To evaluate the coefficients for each of the $n-1$ rows we need $n-1$ divisions of the form A_{ij}/A_{jj}. For each of the $n-1$ distinct i's we will need to carry out n multiplications and n subtractions (one for each column in **A**) and one multiplication and one subtraction for **b**. Since there are going to be $n-1$ values that i takes on, we are led to a result of $(n+1)(n-1)$ multiplications and $(n+1)(n-1)$ subtractions. Putting these results together with the divisions, in this first step we are carrying out $(n-1) + 2(n+1)(n-1)$ floating-point operations.

Still on the subject of the conversion of **A**, we turn to the second step, namely using the pivot row $j = 1$. We have to modify the $n-2$ rows below that pivot row. That leads to $n-2$ divisions for the coefficients. Then, for each of the i's, we will need to carry out $n-1$ multiplications and $n-1$ subtractions (one for each remaining column in **A**) and 1 multiplication and 1 subtraction for **b**. Putting these results together, we are led to $(n-2) + 2n(n-2)$ floating-point operations in total for this step.

It's easy to see that the third step would lead to $(n-3) + 2(n-1)(n-3)$ floating-point operations. A pattern now emerges: for a given pivot row j, we have:

$$(n-1-j) + 2(n+1-j)(n-1-j) \tag{4.81}$$

floating-point operations. Thus, for all possible values of the pivot row j we will need:

$$N_{count} = \sum_{j=0}^{n-2} [(n-1-j) + 2(n+1-j)(n-1-j)] = \sum_{k=1}^{n-1} [k + 2(k+2)k] = \sum_{k=1}^{n-1} (2k^2 + 5k)$$
$$= 2\frac{n(n-1)(2n-1)}{6} + 5\frac{(n-1)n}{2} = \frac{2}{3}n^3 + \frac{3}{2}n^2 - \frac{13}{6}n \sim \frac{2}{3}n^3 \quad (4.82)$$

floating-point operations. In the second step we used a new dummy variable, noticing that $n-1-j$ goes from $n-1$ down to 1. In the fourth step we used the two standard results for $\sum_{k=0}^{n-1} k$ and $\sum_{k=0}^{n-1} k^2$.[13]

Since we already know that the back substitution stage has a cost of n^2, we see that the elimination stage is much slower and therefore dominates the cost of the calculation: symbolically, we have $2n^3/3 + n^2 \sim 2n^3/3$.

4.3.3 LU Method

While the Gaussian elimination method discussed in the previous section is a reasonably robust approach (see below), it does suffer from the following obvious problem: if you want to solve $\mathbf{Ax} = \mathbf{b}$ for the same matrix $\mathbf{A}$ but a different right-hand-side vector $\mathbf{b}$, you would have to waste all the calculations you have already carried out and call a function like `gauelim()` again, performing the Gaussian elimination on $\mathbf{A}$ from scratch. You may wonder if this situation actually arises in practice. The answer is: yes, it does. We will see an example in section 4.4 below, when we discuss methods that evaluate eigenvalues.

Thus, we are effectively motivating a new method (the *LU method*) by stressing the need to somehow "store" the result of the Gaussian elimination process for future use. As a matter of fact, several flavors of the LU method exist, but we are going to be discussing the simplest one, which is a short step away from what we've already seen so far.

4.3.3.1 LU Decomposition

Let us assume we are dealing with a non-singular matrix $\mathbf{A}$ which can be expressed as the product of a lower-triangular matrix $\mathbf{L}$ and an upper-triangular matrix $\mathbf{U}$:

$$\mathbf{A} = \mathbf{LU} \quad (4.83)$$

For obvious reasons, this is known as the *LU decomposition* (or sometimes as the *LU factorization*) of matrix $\mathbf{A}$. We will see in the following section (when we discuss pivoting, see also problem 4.16) that the story is a bit more complicated than this, but for now simply assume that you can carry out such a decomposition.

The LU decomposition as described above is not unique. Here and in what follows, we will make an extra assumption, namely that the matrix $\mathbf{L}$ is *unit* lower triangular, namely

[13] If you have not seen the latter result before, it is worth solving problem 4.9: while this has little to do with linear algebra, it's always gratifying when a series telescopes.

that it is a lower-triangular matrix with 1's on the main diagonal, i.e., $L_{ii} = 1$ for $i = 0, 1, \dots, n-1$. This is known as the *Doolittle decomposition*.[14]

It may be easiest to get a feel for how one goes about constructing this (Doolittle) LU decomposition via an example. As above, we will start from the simple 3×3 case. Instead of studying a specific matrix, as we did earlier, let us look at the general 3×3 problem and assume that we are dealing with:

$$\mathbf{L} = \begin{pmatrix} 1 & 0 & 0 \\ L_{10} & 1 & 0 \\ L_{20} & L_{21} & 1 \end{pmatrix}, \qquad \mathbf{U} = \begin{pmatrix} U_{00} & U_{01} & U_{02} \\ 0 & U_{11} & U_{12} \\ 0 & 0 & U_{22} \end{pmatrix} \tag{4.84}$$

By assumption, we can simply multiply them together (as per $\mathbf{A} = \mathbf{LU}$) to get the original (undecomposed) matrix $\mathbf{A}$:

$$\mathbf{A} = \begin{pmatrix} U_{00} & U_{01} & U_{02} \\ U_{00}L_{10} & U_{01}L_{10} + U_{11} & U_{02}L_{10} + U_{12} \\ U_{00}L_{20} & U_{01}L_{20} + U_{11}L_{21} & U_{02}L_{20} + U_{12}L_{21} + U_{22} \end{pmatrix} \tag{4.85}$$

What we'll now do is to apply Gaussian elimination to the matrix $\mathbf{A}$ only (notice there's no $\mathbf{b}$ anywhere in sight). This is in keeping with our declared intention to find a way of "storing" the results of the Gaussian elimination process for future use.

As we did in our earlier 3×3 example, we will apply Eq. (4.68) repeatedly. We start from the pivot row $j = 0$. We first take $i = 1$: the coefficient will be such that the leading column in row $i = 1$ becomes a 0, i.e., L_{10}. Thus:

$$\text{New row 1} = \text{row 1} - L_{10} \times \text{row 0} \tag{4.86}$$

This gives:

$$\mathbf{A} = \begin{pmatrix} U_{00} & U_{01} & U_{02} \\ 0 & U_{11} & U_{12} \\ U_{00}L_{20} & U_{01}L_{20} + U_{11}L_{21} & U_{02}L_{20} + U_{12}L_{21} + U_{22} \end{pmatrix} \tag{4.87}$$

Similarly, still for the pivot row $j = 0$, we now take $i = 2$:

$$\text{New row 2} = \text{row 2} - L_{20} \times \text{row 0} \tag{4.88}$$

gives us:

$$\mathbf{A} = \begin{pmatrix} U_{00} & U_{01} & U_{02} \\ 0 & U_{11} & U_{12} \\ 0 & U_{11}L_{21} & U_{12}L_{21} + U_{22} \end{pmatrix} \tag{4.89}$$

Remember, we always use the latest/updated version of the matrix $\mathbf{A}$. We will now take the pivot row to be $j = 1$ and the row to be updated as $i = 2$:

$$\text{New row 2} = \text{row 2} - L_{21} \times \text{row 1} \tag{4.90}$$

[14] The Doolittle decomposition is to be contrasted with other algorithms, like the Crout decomposition and the Cholesky decomposition, which we will not address here.

gives us:

$$\mathbf{A} = \begin{pmatrix} U_{00} & U_{01} & U_{02} \\ 0 & U_{11} & U_{12} \\ 0 & 0 & U_{22} \end{pmatrix} \tag{4.91}$$

We have reached a very interesting situation. Inverting this line of reasoning leads us to conclude that to LU decompose a matrix you simply carry out the process of Gaussian elimination: the final version of the matrix **A** will be **U**. Similarly, the off-diagonal matrix elements of the matrix **L** will simply be the coefficients used in the Gaussian elimination process (L_{10}, L_{20}, and L_{21} above).

For the purposes of illustration, we assumed that we could write down the LU decomposition and drew our conclusions about how that came about in the first place. If this is still not ringing a bell, you might want to see the reverse being done: starting with **A**, we wish to calculate the matrix elements of **L** and **U**. For concreteness, let us study the matrix given in Eq. (4.67):

$$\begin{pmatrix} 2 & 1 & 1 \\ 1 & 1 & -2 \\ 5 & 10 & 5 \end{pmatrix} \tag{4.92}$$

where we took the liberty of dropping the elements of the constant vector **b**. As just stated, **U** will be the end result of the Gaussian elimination process, namely Eq. (4.71), and, similarly, **L** will collect the coefficients we used in order to bring our matrix **A** to upper-triangular form:

$$\mathbf{U} = \begin{pmatrix} 2 & 1 & 1 \\ 0 & 0.5 & -2.5 \\ 0 & 0 & 40 \end{pmatrix}, \qquad \mathbf{L} = \begin{pmatrix} 1 & 0 & 0 \\ 0.5 & 1 & 0 \\ 2.5 & 15 & 1 \end{pmatrix} \tag{4.93}$$

All these matrix elements made their appearance as part of the Gaussian elimination process in the previous section. What we've done here is to collect them. See how effortless this all was. (If you're still not convinced, try multiplying **L** and **U** together.)

Notice that storing two matrices (one upper triangular and one lower triangular) is quite wasteful, especially given that we already know that the diagonal elements of **L** are always 1. It is very common in practice to store both **L** and **U** in a single matrix (with the diagonal elements of **L** being implied). Problem 4.13 asks you to implement this solution, but in the main text we opt for pedagogical clarity and use separate matrices.

In contradistinction to what we did in our earlier section on Gaussian elimination, here we do not proceed to discuss the general $n \times n$ case. The reason why should be transparent at this point: the (Doolittle) LU decomposition process is identical to the Gaussian elimination process. We simply have to store the end result and the intermediately used coefficients.

4.3.3.2 Solving a System Using LU Decomposition

It is now time to see how the above decomposition can help in solving the standard problem we've been addressing, namely the solution of:

$$\mathbf{Ax} = \mathbf{b} \tag{4.94}$$

for a general matrix **A**. Since we assume that we've been able to carry out the LU decomposition, we know that $\mathbf{A} = \mathbf{LU}$, so our problem is recast as:

$$\mathbf{LUx} = \mathbf{b} \tag{4.95}$$

It should become crystal clear how to solve this problem once we rewrite this equation as:

$$\mathbf{L}(\mathbf{Ux}) = \mathbf{b} \tag{4.96}$$

Thus:

$$\begin{aligned} \mathbf{Ly} &= \mathbf{b} \\ \mathbf{Ux} &= \mathbf{y} \end{aligned} \tag{4.97}$$

We can solve the first of these equations by forward substitution (since we're dealing with a lower-triangular problem). The result can then be used to pose and solve the second equation by back substitution (since it involves an upper-triangular matrix).

4.3.3.3 Implementation

Before showing any code, let us summarize our accomplishments: we have been able to LU decompose a given matrix, which can then be combined with a constant vector and solve the corresponding system of linear equations. Code 4.3 accomplishes both tasks.

The main workhorse of this code is the straightforward function `ludec()`; this is identical to our earlier `gauelim()`, but:

- we don't have to worry about the constant vector **b**, since at this stage all we're doing is LU-decomposing a given matrix **A**.
- we are calling the array variable that keeps getting updated `U`, instead of `A`.
- we store all the coefficients we calculated in `L` (while placing 1's in its diagonal – this could have been done at the end, instead of starting from an identity matrix).

We then create a function that solves $\mathbf{Ax} = \mathbf{b}$ as per Eq. (4.97), namely by first forward substituting and then back substituting. Our previously advertised strive for modularity has paid off: the main body of `lusolve()` consists of three calls to other functions.

The main program is also very simple, calling functions to create the relevant matrices and evaluate the solution to the system of equations. Running this code, we find at least seven digits of agreement for the solution vectors. You may or may not be intrigued to see that the LU-method solution vector turns out to be identical to the Gaussian elimination vector we saw above.

Code 4.3

ludec.py

```
from triang import forsub, backsub, testcreate, testsolve
import numpy as np

def ludec(A):
        n = A.shape[0]
        U = np.copy(A)
        L = np.identity(n)

        for j in range(n-1):
                for i in range(j+1,n):
                        coeff = U[i,j]/U[j,j]
                        U[i,j:] -= coeff*U[j,j:]
                        L[i,j] = coeff
        return L, U

def lusolve(A,bs):
        L, U = ludec(A)
        ys = forsub(L,bs)
        xs = backsub(U,ys)
        return xs

if __name__ == '__main__':
        A, bs = testcreate(4,21)
        testsolve(lusolve,A,bs)
```

4.3.3.4 Operation Count

You are asked to carry out a detailed study of the operation count for LU decomposition in problem 4.34. Here, we focus on qualitative features. The calculation in Eq. (4.82) contained several components: divisions, multiplications, and subtractions. These last two were added up for the elements of both **A** and **b** which were being updated. Obviously, while carrying out the LU decomposition, we do not have to modify **b**. This changes the operation count slightly, but the dominant term in this case, too, is $\sim 2n^3/3$. One should also take into account that for Gaussian elimination (after counting the multiplications and subtractions for the elements of **b**) we only needed a back substitution at the end. As Eq. (4.97) clearly shows, at the end of the LU method we will have to carry out both a forward and a back substitution.

You should observe that in this operation-count estimate (as in all others) we are only interested in the floating-point operations. Keep in mind that this only tells part of the story:

one should, in principle, also keep track of storage requirements. For example, in the way we chose to implement LU decomposition, it requires two $n \times n$ matrices, whereas Gaussian elimination needed only one.

4.3.3.5 Matrix Inverse

Note that we chose not to implement a function that inverts a given matrix **A**, since one can generally avoid matrix inversion in practice. Even so, as mentioned earlier, you may wish to produce a matrix-inversion Python function in order to calculate the condition number for the problem $\mathbf{Ax} = \mathbf{b}$.[15]

We know from Eq. (C.17) that the matrix inverse $\mathbf{A}^{-1}$ satisfies:

$$\mathbf{A}\mathbf{A}^{-1} = \mathcal{I} \tag{4.98}$$

This motivates the following trick: think of the identity $\mathcal{I}$ as a matrix composed of n column vectors, $\mathbf{e}_i$. (Each of these vectors contains only one non-zero element, whose placement depends on i.) Thus, instead of tackling the full relation $\mathbf{A}\mathbf{A}^{-1} = \mathcal{I}$ head on, we break the problem up into n problems, one for each column of the identity matrix:

$$\mathbf{A}\mathbf{x}_i = \mathbf{e}_i \tag{4.99}$$

where the $\mathbf{x}_i$ are the columns of the inverse matrix $\mathbf{A}^{-1}$. Explicitly:

$$\mathbf{A}^{-1} = \begin{pmatrix} \mathbf{x}_0 & \mathbf{x}_1 & \dots & \mathbf{x}_{n-1} \end{pmatrix}, \qquad \mathcal{I} = \begin{pmatrix} \mathbf{e}_0 & \mathbf{e}_1 & \dots & \mathbf{e}_{n-1} \end{pmatrix} \tag{4.100}$$

Keep in mind that bold-lower-case entities are column vectors.

We now see our first example of using the LU decomposition to solve linear systems for different constant vectors **b**: $\mathbf{A}\mathbf{x}_i = \mathbf{e}_i$ can be combined with $\mathbf{A} = \mathbf{LU}$:

$$\mathbf{L}(\mathbf{U}\mathbf{x}_i) = \mathbf{e}_i \tag{4.101}$$

Then, for each $\mathbf{x}_i$ (or each $\mathbf{e}_i$) we use the same LU decomposition and carry out first a forward substitution and then a back substitution (just like in Eq. (4.97) above). As you may recall, each forward/back substitution has a cost of n^2 which is much less costly than the $O(n^3)$ needed for LU decomposition. Thus, since we need to solve for n columns of the inverse, the total cost of this approach is $(n^2 + n^2) \times n = 2n^3$, in addition to the cost of the LU decomposition (which is of the same order of magnitude).

If we wanted to use Gaussian elimination to evaluate the inverse, we would incur a cost of $O(n^3)$ for each column of the inverse, leading to an overall cost of $O(n^4)$, which is generally more than we are willing to tolerate.

[15] We will encounter another application in section 6.5.3.3, when we introduce the covariance matrix.

4.3.3.6 Determinant

Recall from Eq. (C.16) that the determinant of a triangular matrix is simply the product of the diagonal elements. Given that $\mathbf{A} = \mathbf{LU}$:

$$\det(\mathbf{A}) = \det(\mathbf{L}) \times \det(\mathbf{U}) = \left(\prod_{i=0}^{n-1} 1\right) \times \left(\prod_{i=0}^{n-1} U_{ii}\right) = \prod_{i=0}^{n-1} U_{ii} \tag{4.102}$$

In the first equality we used a result from elementary linear algebra, namely that the determinant of a product of two matrices is equal to the product of the two matrix determinants. In the second equality we expressed the determinant of each triangular matrix as the product of the diagonal elements and used the fact that the diagonal elements of $\mathbf{L}$ are all 1. That led us, in the third equality, to the conclusion that the determinant of the matrix $\mathbf{A}$ can be evaluated by multiplying together the diagonal elements of the matrix $\mathbf{U}$.[16]

4.3.4 Pivoting

Even though this wasn't explicitly stated, so far we've limited ourselves to Gaussian elimination (and its cousin, Doolittle's LU method) for straightforward cases. In this section, we find out that things often get sticky. We go over a standard way to address the most glaring deficiencies and also provide some pointers to more complicated remedies.

4.3.4.1 Instability without Ill-Conditioning

We spent quite a bit of time in an earlier section doing an *a priori* analysis of linear-algebra problems. For example, we saw that we can quantify the ill-conditioning of the problem $\mathbf{Ax} = \mathbf{b}$ using a reasonably straightforward prescription. Ill-conditioning is a property that refers to the problem we are solving. However, as we now address, there are situations where one tries to solve a perfectly well-conditioned problem but the method of choice fails to give a satisfactory answer. Obviously, the fault in this case lies with the method, not with the problem. We will explore this by considering three examples.

First, we look at the following (very simple) 2×2 problem:

$$\left(\begin{array}{cc|c} 0 & -1 & 1 \\ 1 & 1 & 2 \end{array}\right) \tag{4.103}$$

This is a perfectly well-conditioned problem, with an easy-to-evaluate analytical solution of:

$$\mathbf{x}^T = \begin{pmatrix} 3 & -1 \end{pmatrix} \tag{4.104}$$

Consider how you would go about implementing Gaussian elimination in this case. You would start (and stop) with the pivot row $j = 0$ which would be used to update the row $i = 1$. The coefficient multiplying the pivot row is, as usual, A_{ij}/A_{jj}. This is already a major breakdown of our methodology: here $A_{jj} = 0$ so we cannot divide by it to evaluate the coefficient. If you try to use our `gauelim()` function, your output vector will be `[ nan  nan ]`,

[16] Observe that our result tells us that $\mathbf{A}$ is non-singular if and only if all the diagonal elements of $\mathbf{U}$ are non-zero.

where nan stands for “not a number” (as explained in appendix B). It is disheartening to see such a nice method fail at solving such a simple problem.

Second, in case you were thinking that this is not such a big deal, given that you could immediately identify the 0 in the first slot of the first row of the previous example, we now turn to a case where the problem is “hidden”. Examine the following augmented matrix:

$$\left(\begin{array}{ccc|c} 2 & 1 & 1 & 8 \\ 2 & 1 & -4 & -2 \\ 5 & 10 & 5 & 10 \end{array}\right) \tag{4.105}$$

This is basically the example from Eq. (4.67) where we’ve modified a couple of elements, while making sure that the solution vector is the same as in the original example:

$$\mathbf{x}^T = \begin{pmatrix} 4 & -2 & 2 \end{pmatrix} \tag{4.106}$$

as you can verify by substitution or via an analytical solution. Note that this matrix, too, is perfectly well-conditioned. Here’s the thing: if you wish to apply Gaussian elimination to this problem, you would (once again) start with the pivot row $j = 0$ and update the rows $i = 1$ and $i = 2$ in turn. After you update both rows you would be faced with:

$$\left(\begin{array}{ccc|c} 2 & 1 & 1 & 8 \\ 0 & 0 & -5 & -10 \\ 0 & 7.5 & 2.5 & -10 \end{array}\right) \tag{4.107}$$

We wish to turn to the pivot row $j = 1$ but notice that the pivot element $A_{jj} = 0$ so (as in the first example) we are not able to bring this matrix to an upper-triangular form. In other words, if we use our `gauelim()` function for this problem our output vector will be `[ nan  nan  nan ]`. As advertised, this problem was not (totally) obvious at first sight: some steps of the Gaussian elimination process had to be applied first.

Third, it turns out that Gaussian elimination has trouble providing good solutions not only when zeros appear as pivot elements in the starting (or updated) matrix, but even when the pivot element is very small. This is a general fact: when an algorithm cannot be applied in a given situation, it most likely canot be applied successfully in situations that are similar to it. Take our first example from this section, but instead of a 0 in the first element of the first row, assume you are dealing with a small number:

$$\left(\begin{array}{cc|c} 10^{-20} & -1 & 1 \\ 1 & 1 & 2 \end{array}\right) \tag{4.108}$$

This is still a perfectly well-conditioned problem, with an analytical solution:

$$\mathbf{x}^T = \begin{pmatrix} \dfrac{3}{1+10^{-20}} & \dfrac{-1+10^{-20}}{1+10^{-20}} \end{pmatrix} \tag{4.109}$$

Knowing that Python employs double-precision floating-point numbers (and that 10^{-20} is smaller than the machine precision), you would be tempted to expect that the numerical solution to this system would be: `[ 3  -1 ]`, namely the same solution as in our first example above. (As it so happens, `numpy.linalg.solve()` does give us this, correct, answer.)

Now, let us think of how to apply Gaussian elimination here. As in the first example, we only have one pivot row ($j = 0$) and we're updating only one row ($i = 1$). Here the pivot element, $A_{jj} = 10^{-20}$, is small but non-zero, so we can apply the Gaussian elimination prescription. After we do so, our augmented matrix will be:

$$\left(\begin{array}{cc|c} 10^{-20} & -1 & 1 \\ 0 & 1+10^{20} & 2-10^{20} \end{array}\right) \tag{4.110}$$

This situation is similar to what we encountered in section 2.4.2, when in `big + 1.` the unit was dropped. (Here we are faced with a 10^{20} because we have divided by the pivot element.) Despite the roundoff errors that emerge, nowhere are we dividing by zero, so the back substitution can proceed normally, but for the approximate matrix:

$$\left(\begin{array}{cc|c} 10^{-20} & -1 & 1 \\ 0 & 10^{20} & -10^{20} \end{array}\right) \tag{4.111}$$

Unfortunately, this leads to `[ 0  -1 ]`, which is very different from the correct solution.

4.3.4.2 Partial Pivoting

We now introduce a straightforward technique to handle problems like those discussed in the previous subsection. One way to remedy the situation is to eliminate the problem from the start. For example, in the case of our first matrix above, Eq. (4.103), if we were solving the fully equivalent problem:

$$\left(\begin{array}{cc|c} 1 & 1 & 2 \\ 0 & -1 & 1 \end{array}\right) \tag{4.112}$$

instead, no breakdown would have ever emerged. This matrix is already in upper triangular form, so we can proceed with back substitution. Of course, such an *ad hoc* approach (interchanging rows in the starting matrix because they look like they will misbehave) can only take you so far. What we are looking for is a general prescription.

Observe that the issue in all three of our earlier examples resulted from using a small number (which was, perhaps, even zero) as the pivot element.[17] This leads us to a general presciption that aims to avoid using small numbers as pivot elements, when there is a better alternative. To explain what we mean, let's look at the Gaussian-elimination intermediate time slice from Eq. (4.74) again. This snapshot describes the moment when we are about to use A_{jj} as the pivot element (so it could also be referring to our original matrix, if $j = 0$). As you may recall, in our process the coefficient (used to eliminate the leading element in the following rows) is A_{ij}/A_{jj}: if A_{jj} is very small, then that coefficient will be very large. This helps guide our strategy of what to do: instead of blindly using A_{jj} as the pivot element (since it might have a very small magnitude) look at all the elements in the same column as A_{jj} in the rows below it for the largest possible matrix-element magnitude. This would be located, say, in row k: then, simply interchange rows j and k. Now you have a larger-magnitude pivot element and the coefficient used in the elimination process is never

[17] In the second example, this occurred at an intermediate stage.

greater than 1 in absolute value. You can then proceed with the elimination process as usual. Symbolically, we search for the smallest-integer k that satisfies:

$$|A_{kj}| = \max_{j \leq m \leq n-1} |A_{mj}| \tag{4.113}$$

followed by the interchange of rows j and k. You should make sure to mentally distinguish the second index from the first index in this equation: we are always searching in the j column (which is why A_{kj} and A_{mj} have the same second index) for the element with the largest magnitude. This means we search all the rows from row j and below.

This prescription employs the second elementary row operation we introduced in section 4.1.2, namely *pivoting*: two rows/equations may be interchanged, as long as you remember to interchange the **A** and **b** elements appropriately.[18] More properly speaking, this is known as *partial pivoting*, in the sense that we are only interchanging rows (we could have been interchanging columns). By the way, there is some terminology overloading at play here: we use the word "pivot" to describe the pivot row or the pivot element, but we also say that "pivoting" is the interchange of two rows. While these two meanings are related (we interchange two rows if necessary before we use a pivot row/element to eliminate the leading element in the next rows) they are distinct and should not be confused. It might help you to think of "partial pivoting" as synonymous with "row interchanging".

4.3.4.3 Implementation

The above description should be enough for you to see how one goes about implementing partial pivoting in practice; the result is code 4.4. Note that this code is identical to that in `gauelim()`, the only change being that we've added four lines. Let's discuss them.

We employ `numpy.argmax()`—introduced in problem 1.10—to find the index of the element with the maximum value.[19] This search for `k` starts at row `j` and goes up to the bottom row, namely row `n-1`. We've employed NumPy's slicing to write this compactly: `A[j:,j]` contains all the elements in column `j` from row `j` until row `n-1`. Note however that, as a result of this slicing, `numpy.argmax()` will return an index that is "shifted down": its output will be `0` if the largest element is A_{jj}, `1` if the largest element is $A_{j+1,j}$, and so on. This is why when evaluating `k` we offset `numpy.argmax()`'s output by adding in `j`: now `k`'s possible values start at `j` and end at `n-1`.

After we've determined `k`, we check to see if it is different from `j`: we don't want to interchange a row with itself (since that doesn't accomplish anything). If `k` is different from `j`, then we interchange the corresponding rows of **A** and **b**. We do this using the standard swap idiom in Python, which employs multiple assignment (and does not require a temporary throwaway variable). The swapping of `bs` elements should be totally transparent to you. The swapping of the `A` rows is a little more complicated: as you may recall, NumPy

[18] Incidentally, we note that this operation changes the sign of det(**A**).

[19] Conveniently, if the maximum value appears more than once, this function returns the first occurrence.

gauelim_pivot.py

Code 4.4

```
from triang import backsub, testcreate, testsolve
import numpy as np

def gauelim_pivot(inA,inbs):
    A = np.copy(inA)
    bs = np.copy(inbs)
    n = bs.size

    for j in range(n-1):
        k = np.argmax(np.abs(A[j:,j])) + j
        if k != j:
            A[j,:], A[k,:] = A[k,:], A[j,:].copy()
            bs[j], bs[k] = bs[k], bs[j]

        for i in range(j+1,n):
            coeff = A[i,j]/A[j,j]
            A[i,j:] -= coeff*A[j,j:]
            bs[i] -= coeff*bs[j]

    xs = backsub(A,bs)
    return xs

if __name__ == '__main__':
    A, bs = testcreate(4,21)
    testsolve(gauelim_pivot,A,bs)
```

slicing rules imply that array slices are views on the original array. As a result, if you don't say `A[j,:].copy()` you will overwrite your elements instead of swapping rows.[20]

Running this code, we find that the results of our Gaussian elimination with partial pivoting function are closer to those of `numpy.linalg.solve()` than the results of our earlier `gauelim()` function. This should not come as a surprise, since `numpy.linalg.solve()` also uses (a LAPACK routine which employs) partial pivoting. Note that partial pivoting helps reduce roundoff error issues, even when these are not very dramatic (as here).

You may wish to test our function on the three earlier examples to see that it solves the problems we were facing before. In problem 4.16, you are asked to apply our partial pivoting prescription also to the LU-decomposition method: the only difference is that LU

[20] This is unrelated to whether or not you choose to swap using a temporary intermediate variable.

decomposition only updates the **A** matrix, so you should keep track of which rows you interchanged and later use that information to interchange the corresponding rows in **b**.

As a final implementation-related point, note that in our code we chose to actually swap the rows each time. One can envision an alternative strategy, whereby the rows are left in their initial order, so that the elements that end up being eliminated are not always below the pivot row. The end result of this process is still a triangular matrix, but this time in "hidden" form, since it will be a scrambled triangular matrix. You could then perform back substitution if you had kept track of the order in which rows were chosen as pivot rows. Obviously, this requires more bookkeeping but is likely more efficient than actually swapping the rows.

4.3.4.4 Beyond Partial Pivoting

We have now seen that the Gaussian elimination method (which can be unstable without partial pivoting, even for well-conditioned problems) can be stabilized by picking as the pivot element always the largest possible value (leading to coefficients that are never larger than 1 in absolute value). Unfortunately, even Gaussian elimination with partial pivoting can turn out to be unstable. To see this, take the third example from our earlier subsection and scale the second row by multiplying by 10^{-20}:

$$\left(\begin{array}{cc|c} 10^{-20} & -1 & 1 \\ 10^{-20} & 10^{-20} & 2\times 10^{-20} \end{array}\right) \tag{4.114}$$

This is in reality the same set of equations, so one should expect the same answer. Of course, that's not what happens in practice. Since now the first elements in the first and second rows have the same magnitude, no interchanges take place under partial pivoting. Thus, the same problem that we faced without partial pivoting will still be present now: as a result, both `gauelim_pivot()` and `numpy.linalg.solve()` give the wrong answer.

This new issue can be handled by changing our prescription from using the largest-possible pivot element to using as the pivot element the one that has the largest *relative* magnitude (i.e., in relation to other elements in its row). This is known as *scaled partial pivoting*. Symbolically, we first define for each row i a size/scale factor:

$$s_i = \max_{0\leq j\leq n-1} |A_{ij}| \tag{4.115}$$

Thus, s_i contains (the absolute value of) the largest element in row i. Then, instead of determining k as corresponding to the largest matrix element in column j (at or below A_{jj}), we determine k by finding the matrix element with the largest relative magnitude, i.e., the largest one in comparison to the other elements in its row. Thus, we search for the smallest-integer k that satisfies:

$$\frac{|A_{kj}|}{s_k} = \max_{j\leq m\leq n-1} \frac{|A_{mj}|}{s_m} \tag{4.116}$$

and then we interchange rows j and k and proceed as usual. You are asked to implement this prescription in problem 4.15.

Unfortunately, even scaled partial pivoting does not fully eradicate the potential for instability. An approach that *does* guarantee stability is *complete pivoting*, whereby both rows and columns are interchanged before proceeding with Gaussian elimination. That being said, it is extremely rare for (scaled) partial pivoting to fail in practice, so most libraries do not implement the (more costly) strategy of complete pivoting.

Since this subsection is called "beyond partial pivoting", it's worth observing that there exist situations where Gaussian elimination can proceed without pivoting. This happens for matrices that are positive definite[21] and symmetric. It also happens for diagonally dominant matrices (you might want to brush up on our definitions from appendix C.2). For example:

$$\left(\begin{array}{ccc|c} 2 & 1 & 1 & 8 \\ 5 & 10 & 5 & 10 \\ 1 & 1 & -2 & -2 \end{array}\right) \tag{4.117}$$

This matrix is diagonally dominant and no pivoting would (or would need to) take place.[22] We will re-encounter diagonal dominance as a criterion in the following section.

You might be wondering why we would be interested in avoiding pivoting. The first (and most obvious) reason is that it entails a computational overhead: evaluating k and swapping rows can become costly. Second, if the original coefficient matrix is symmetric and banded then pivoting would remove nice structural properties like these; if you were using a method that depended on (or was optimized for) that structure (unlike what we've been doing above) you would be in trouble.

4.3.5 Jacobi Iterative Method

We now briefly talk about iterative methods, also known as *relaxation methods*. As you may recall, *direct* methods (such as Gaussian elimination) construct a solution by carrying out a fixed number of operations: first you take a pivot row, then you eliminate the leading elements in the rows below it, etc. In contradistinction to this, *iterative* methods start with a guess for the solution $\mathbf{x}$ and then refine it until it stops changing; the number of iterations required is typically not known in advance. The number of iterations can depend on the structure of the matrix, which method is employed, which initial solution vector we guess, and which convergence criterion we use.

Actually, for a given set of equations an iterative method might not even converge at all. Putting all these factors together, it probably comes as no surprise to hear that iterative methods are typically slower than direct methods. On the other hand, these drawbacks are counterbalanced by the fact that iterative methods are generally more efficient when we are dealing with extremely sparse matrices, which may or may not be banded. In other words,

[21] A matrix $\mathbf{A}$ is *positive definite* if $\mathbf{x}^T\mathbf{A}\mathbf{x} > 0$ for all $\mathbf{x} \neq \mathbf{0}$. This definition is most commonly applied to Hermitian matrices, in which case it implies that the eigenvalues are all positive. For non-symmetric matrices, being positive definite implies that the real part of each eigenvalue is always positive.

[22] This happens to be our example from Eq. (4.67) with the second and third rows interchanged.

when most matrix elements are 0, such approaches really save computational time.[23] Not coincidentally, such iterative approaches are often used in the solution of boundary-value problems for partial differential equations; it is therefore perhaps beneficial to encounter them at the more fundamental level of linear algebra.

Here we only discuss the simplest possible iterative approach, namely Jacobi's method. We leave to problems 4.19 and 4.20 the study of the convergence for the Jacobi method as well as an extension to the Gauss–Seidel method. Our goal here is not to cover such approaches thoroughly. Rather, it is to give you a flavor of iterating to get the answer; we will re-encounter this strategy when we solve the eigenvalue problem below (section 4.4) and when we discuss root-finding for non-linear algebraic equations (chapter 5).

4.3.5.1 Algorithm

Let's write out the system of linear algebraic equations, $\mathbf{Ax} = \mathbf{b}$, using index notation:

$$\sum_{j=0}^{n-1} A_{ij}x_j = b_i \tag{4.118}$$

where, as usual, $i = 0, 1, 2, \ldots, n-1$. The Jacobi method is motivated by taking this equation and solving for the component x_i which corresponds to the diagonal element A_{ii}:

$$x_i = \left(b_i - \sum_{j=0}^{i-1} A_{ij}x_j - \sum_{j=i+1}^{n-1} A_{ij}x_j\right)\frac{1}{A_{ii}}, \quad i = 0, 1, \ldots, n-1 \tag{4.119}$$

Observe that the form of this equation is similar to what we had in Eq. (4.55) for forward substitution and in Eq. (4.60) for back substitution. Of course, the situation here is quite different: there we could solve for each of the x_i in turn using the other x_j's which we'd already evaluated. Here, however, we don't know any of the solution-vector components.

The Jacobi method starts by choosing an initial solution vector $\mathbf{x}^{(0)}$: the superscript in parentheses tells us the iteration number (in this case we are starting, so it's the 0th iteration). The method proceeds by plugging in $\mathbf{x}^{(0)}$ to the right-hand side of Eq. (4.119), to produce an improved solution vector, $\mathbf{x}^{(1)}$:

$$x_i^{(1)} = \left(b_i - \sum_{j=0}^{i-1} A_{ij}x_j^{(0)} - \sum_{j=i+1}^{n-1} A_{ij}x_j^{(0)}\right)\frac{1}{A_{ii}}, \quad i = 0, 1, \ldots, n-1 \tag{4.120}$$

This process is repeated: the components of $\mathbf{x}^{(1)}$ are plugged in to the right-hand side of Eq. (4.119), thereby producing a further improved solution vector, $\mathbf{x}^{(2)}$. For the general k-th iteration step, we now have our prescription for the Jacobi iteration method:

$$x_i^{(k)} = \left(b_i - \sum_{j=0}^{i-1} A_{ij}x_j^{(k-1)} - \sum_{j=i+1}^{n-1} A_{ij}x_j^{(k-1)}\right)\frac{1}{A_{ii}}, \quad i = 0, 1, \ldots, n-1 \tag{4.121}$$

[23] Another feature is that iterative methods are self-correcting, i.e., an iteration is independent of the roundoff error in the previous cycle.

If all is going well, the solution vector will asymptotically approach the exact solution. A natural question is when to end this process. The answer depends on our desired accuracy, giving rise to what is known as a *convergence criterion.*

Let us be a little careful here. We know from section 2.2.1 that the absolute error in evaluating a quantity is defined as "approximate value minus exact value". Of course, in an iterative method like the Jacobi algorithm, we don't know the exact value, but we can estimate how far along we are by seeing how much a given component changes from one iteration to the next, $x_i^{(k)} - x_i^{(k-1)}$ (more on this in problem 4.20). Of course, we are dealing with a solution *vector*, so we have to come up with a way of combining all the solution-vector component contributions into one estimate of how much/little the entire answer is changing. As we've done repeatedly in earlier chapters, we will opt for a *relative* error criterion (comparing $x_i^{(k)} - x_i^{(k-1)}$ with $x_i^{(k)}$). Specifically, for a given tolerance ϵ, we will use the following termination criterion:

$$\sum_{i=0}^{n-1} \left| \frac{x_i^{(k)} - x_i^{(k-1)}}{x_i^{(k)}} \right| \leq \epsilon \tag{4.122}$$

Several other choices can be made, e.g., adding the contributions $x_i^{(k)} - x_i^{(k-1)}$ in quadrature before taking the square root, or using the maximum $x_i^{(k)} - x_i^{(k-1)}$ difference.

For the sake of completeness, we point out that we could have been (a lot) more systematic in our choice of an error criterion. First, recall the residual vector from Eq. (4.17), $\mathbf{r} = \mathbf{b} - \mathbf{A}\tilde{\mathbf{x}}$. Employing this, a more robust stopping criterion could be:

$$||\mathbf{b} - \mathbf{A}\mathbf{x}^{(k)}|| \leq \epsilon \left(||\mathbf{A}|| ||\mathbf{x}^{(k)}|| + ||\mathbf{b}|| \right) \tag{4.123}$$

This has the advantage of checking directly against the expected behavior ($\mathbf{A}\mathbf{x} = \mathbf{b}$) as well as the relative magnitudes involved. In other words, it measures relative size but, unlike our *ad hoc* prescription in Eq. (4.122), here we use norms to combine all the components together in a consistent way.

A question that still remains open is how to pick the initial guess $\mathbf{x}^{(0)}$. As you can imagine, if you start with a guess that is orders of magnitude different from the exact solution vector, it may take a while for your iterative process to converge. In practice, one often takes $\mathbf{x}^{(0)} = \mathbf{0}$. This is one reason why we chose to divide in our relative error criterion by $x_i^{(k)}$ (instead of $x_i^{(k-1)}$). In any case, $x_i^{(k)}$ happens to be our best estimate so far, so it makes sense to use it in the denominator for our relative error.

Note that even if you pick a "good" initial guess, the Jacobi method is not always guaranteed to converge! As you will show in problem 4.20, a sufficient condition for convergence of the Jacobi iterative method is that we are dealing with diagonally dominant matrices, regardless of the initial guess.[24] Keep in mind that a system of equations might not be diagonally dominant, but can become so if one re-arranges the equations. Observe, finally, that even systems that are not (and cannot become) diagonally dominant can be solved with Jacobi's method for some choices of initial solution vectors.

[24] Diagonal dominance also appeared in the previous subsection on pivoting in Gaussian elimination.

Code 4.5 jacobi.py

```
from triang import testcreate, testsolve
import numpy as np

def termcrit(xolds,xnews):
    errs = np.abs((xnews - xolds)/xnews)
    return np.sum(errs)

def jacobi(A,bs,kmax=50,tol=1.e-6):
    n = bs.size
    xnews = np.zeros(n)

    for k in range(1,kmax):
        xs = np.copy(xnews)

        for i in range(n):
            slt = A[i,:i]@xs[:i]
            sgt = A[i,i+1:]@xs[i+1:]
            xnews[i] = (bs[i] - slt - sgt)/A[i,i]

        err = termcrit(xs, xnews)
        print(k, xnews, err)
        if err < tol:
            break
    else:
        xnews = None

    return xnews

if __name__ == '__main__':
    n = 4; val = 21
    A, bs = testcreate(n,val)
    A += val*np.identity(n)
    testsolve(jacobi,A,bs)
```

4.3.5.2 Implementation

Code 4.5 is a Python implementation of the Jacobi algorithm described above. We first define a short function `termcrit()` to encapsulate our termination criterion, Eq. (4.122).

We use numpy functionality to conveniently form and sum the ratios $[x_i^{(k)} - x_i^{(k-1)}]/x_i^{(k)}$, this time employing `numpy.sum()`. We then turn to the function `jacobi()`, which contains two loops, one to keep track of which Jacobi iteration we're currently at and one to go over the x_i in turn. Observe how convenient numpy's functionality is: we are carrying out the sums for j indices that are larger than or smaller than i, without having to employ a third loop (or even a third index: j is not needed in the code); we don't even have to employ `numpy.sum()`, since `@` takes care of everything for us.

Our program employs the `for-else` idiom that we introduced in section 1.4.2: as you may recall, the `else` is only executed if the main body of the `for` runs out of iterations without ever executing a `break` (which for us would mean that the error we calculated never became smaller than the pre-set error tolerance).

Observe that we have been careful to create a new copy of our vector each time through the outside loop, via `xs = np.copy(xnews)`. This is very important: if you try to use simple assignment (or even slicing) you will get in trouble, since you will only be making `xs` a synonym for `xnews`: this would imply that convergence is always (erroneously) reached the first time through the loop.

In the main program, we first create our test matrix `A`: in this case, we also modify it after the fact, to ensure "by hand" that it is diagonally dominant. Had we not, Jacobi's method would flail and return `None`. Observe that our definition of `jacobi()` uses default parameter values; this allows us to employ `testsolve()`, which assumes our system-solver only takes in two arguments, just like before. Running this code, we find that we converge in roughly three dozen iterations.[25] Of course, how fast we will converge will also depend on our initial guess for the solution vector (though we always take $\mathbf{x}^{(0)} = \mathbf{0}$ for simplicity). In problem 4.19, you are asked to modify this code, turning it into the Gauss–Seidel method: a tiny change in the algorithm makes it converge considerably faster.

4.4 Eigenproblems

We turn to the "second half" of linear algebra, namely the matrix eigenvalue problem:

$$\mathbf{A}\mathbf{v}_i = \lambda_i \mathbf{v}_i \tag{4.124}$$

The main trick we employed in the previous section is no longer applicable: subtracting a multiple of a row from another row (i.e., the elimination procedure) changes the eigenvalues of the matrix, so it's not an operation we'll be carrying out in what follows.

As usual, we will be selective and study the special case where our $n \times n$ matrix $\mathbf{A}$ has n eigenvalues λ_i that are all *distinct*. This simplifies things considerably, since it means that the n eigenvectors $\mathbf{v}_i$ are linearly independent. In this case, it is easy to show (as you will discover when you solve problem 4.21) that the following relation holds:

[25] You should comment out the call to `print()` once you are ready to use the code for other purposes.

$$\mathbf{V}^{-1}\mathbf{A}\mathbf{V} = \mathbf{\Lambda} \tag{4.125}$$

where $\mathbf{\Lambda}$ is the diagonal "eigenvalue matrix" made up of the eigenvalues λ_i:

$$\mathbf{\Lambda} = \begin{pmatrix} \lambda_0 & 0 & \dots & 0 \\ 0 & \lambda_1 & \dots & 0 \\ \vdots & \vdots & \ddots & \vdots \\ 0 & 0 & \dots & \lambda_{n-1} \end{pmatrix} \tag{4.126}$$

and $\mathbf{V}$ is the "eigenvector matrix", whose columns are the right eigenvectors $\mathbf{v}_i$:

$$\mathbf{V} = \begin{pmatrix} \mathbf{v}_0 & \mathbf{v}_1 & \dots & \mathbf{v}_{n-1} \end{pmatrix} \tag{4.127}$$

Equation (4.125) shows how we can *diagonalize* a matrix, $\mathbf{A}$; solving the eigenproblem (i.e., computing the eigenvalues and eigenvectors) is often called *diagonalizing a matrix*. Recast as $\mathbf{A} = \mathbf{V}\mathbf{\Lambda}\mathbf{V}^{-1}$, Eq. (4.125) is called the *eigendecomposition*.

While we study only diagonalizable matrices (i.e., we won't address *defective matrices*), these are *not* toy problems. Our approach will be quite general, i.e., we won't assume that our matrices are sparse or even symmetric; while many problems in physics lead to symmetric matrices, not all do.[26] That being said, the eigenvalue problem for non-symmetric matrices is messy, since the eigenvalues do not need to be real; here we study only non-symmetric matrices that have real (and *distinct*) eigenvalues—see, however, problem 4.37.

In section 4.1.2 we saw that writing out $\det(\mathbf{A} - \lambda\mathcal{I}) = 0$ leads to a characteristic equation (namely a polynomial set to 0). As you will learn in section 5.3.1, finding the zeros of polynomials is very often an ill-conditioned problem, even when the corresponding eigenvalue problem is perfectly well-conditioned. Thus, it's wiser, instead, to transform the matrix into a form where it's easy to read the eigenvalues off, while ensuring that the eigenvalues of the starting and final matrix are the same.

The methods we *do* employ to computationally solve the eigenvalue problem are *iterative*; this is different from the system-solving in the previous section, where some methods were direct and some were iterative. Your first encounter with such an iterative method (the Jacobi method) may have felt underwhelming: if the matrix involved was not diagonally dominant, that approach could not guarantee a solution. In contradistinction to this, here we will introduce the state-of-the-art QR method, currently the gold standard for the case where one requires all eigenvalues. Before we get to it, though, we will discuss the power and inverse-power methods: these help pedagogically, but will also turn out to be conceptually similar to the full-blown QR approach. By the end of this section, we will have produced a complete eigenproblem-solver (all eigenvalues and all eigenvectors); splitting the discussion into several constituent methods will allow us to introduce and test the necessary scaffolding, so that the final solver does not appear too forbidding.

Even though we will create from scratch a stable eigenproblem solver, we will not include all bells and whistles. Thus, we won't be discussing things like the Aitken extrapola-

[26] The literature on the eigenvalue problem is overwhelmingly focused on the case of symmetric matrices.

tion, the Householder transformation, the concept of deflation, and so on. In the real world, one often first carries out some "preprocessing", in order to perform the subsequent calculations more efficiently. In the spirit of the rest of the book, we are here more interested in grasping the essential concepts than in producing a hyperefficient library.[27]

4.4.1 Power Method

As already mentioned, we will start with the simplest possible method, which turns out to be intellectually related to more robust methods. The general problem we are trying to solve is $\mathbf{A}\mathbf{v}_i = \lambda_i \mathbf{v}_i$: λ_i are the true eigenvalues and $\mathbf{v}_i$ are the true eigenvectors (all of which are currently unknown). Since we're making the assumption that all n eigenvalues are distinct, we are free to sort them such that:

$$|\lambda_0| > |\lambda_1| > |\lambda_2| > \ldots > |\lambda_{n-1}| \tag{4.128}$$

The power method (in its simplest form) will give us access to only one eigenvalue and eigenvector pair. Specifically, it will allow us to evaluate the largest eigenvalue λ_0 (also known as the *dominant* eigenvalue) and the corresponding eigenvector $\mathbf{v}_0$.[28]

4.4.1.1 Algorithm: First Attempt

Let us immediately start with the power-method prescription and try to elucidate it after the fact. In short, we start from an *ad hoc* guess and then see how we can improve it, as is standard in iterative approaches. The method tells us to start from a vector $\mathbf{z}^{(0)}$ and simply multiply it by the matrix $\mathbf{A}$ to get the next vector in the sequence:

$$\mathbf{z}^{(k)} = \mathbf{A}\mathbf{z}^{(k-1)}, \quad k = 1, 2, \ldots \tag{4.129}$$

Note that, just like in section 4.3.5 on the Jacobi method, we are using superscripts in parentheses, (k), to denote the iteration count. Obviously, we have $\mathbf{z}^{(1)} = \mathbf{A}\mathbf{z}^{(0)}$, then $\mathbf{z}^{(2)} = \mathbf{A}\mathbf{z}^{(1)} = \mathbf{A}^2\mathbf{z}^{(0)}$, and so on, leading to:

$$\mathbf{z}^{(k)} = \mathbf{A}^k\mathbf{z}^{(0)}, \quad k = 1, 2, \ldots \tag{4.130}$$

This is the source of the name "power method": iterating through the steps of the algorithm, we see the powers of the original matrix $\mathbf{A}$ making their appearance.

To see what any of this has to do with calculating eigenvalues, we express our starting vector $\mathbf{z}^{(0)}$ as a linear combination of the (unknown) eigenvectors:

$$\mathbf{z}^{(0)} = \sum_{i=0}^{n-1} c_i \mathbf{v}_i \tag{4.131}$$

where the coefficients c_i are also unknown. Note that the i subscript here refers to which

[27] Even so, the problems discuss some of the more advanced aspects we left out.

[28] This time we are not starting from the most general method, but from the most specific method.

eigenvector $\mathbf{v}_i$ we are dealing with (so the m-th component of the i-th eigenvector would be denoted by $(\mathbf{v}_i)_m$). Now, putting the last two equations together, we have:

$$\mathbf{z}^{(k)} = \mathbf{A}^k \mathbf{z}^{(0)} = \sum_{i=0}^{n-1} c_i \mathbf{A}^k \mathbf{v}_i = \sum_{i=0}^{n-1} c_i \lambda_i^k \mathbf{v}_i = c_0 \lambda_0^k \mathbf{v}_0 + \lambda_0^k \sum_{i=1}^{n-1} c_i \left(\frac{\lambda_i}{\lambda_0}\right)^k \mathbf{v}_i, \quad k = 1, 2, \dots \tag{4.132}$$

In the second equality we pulled the $\mathbf{A}^k$ inside the sum. In the third equality we used our defining equation, $\mathbf{A}\mathbf{v}_i = \lambda_i \mathbf{v}_i$, repeatedly. In the fourth equality we separated out the first term in the sum and also took the opportunity to multiply and divide by λ_0^k in the second term. Looking at our result, we recall Eq. (4.128) telling us that λ_0 is the largest eigenvalue: this implies that in the second term, $(\lambda_i/\lambda_0)^k \to 0$ as $k \to \infty$. To invert this reasoning, the rate of convergence of this approach is determined by the ratio $|\lambda_1/\lambda_0|$: since we've already sorted the eigenvalues, we know that λ_1 is the eigenvalue with the second largest magnitude, so the $|\lambda_1/\lambda_0|$ ratio will be the largest contribution in the sum. Thus, if we further assume that $c_0 \neq 0$, i.e., that $\mathbf{z}^{(0)}$ has a component in the direction of $\mathbf{v}_0$, then we have reached the conclusion that as we progress with our iteration we tend toward a $\mathbf{z}^{(k)}$ in the direction of $\mathbf{v}_0$ (which is the eigenvector corresponding to the largest eigenvalue, λ_0).

We see from Eq. (4.132) that our $\mathbf{z}^{(k)}$ will tend to $c_0 \lambda_0^k \mathbf{v}_0$. Of course, we don't actually know any of these terms (c_0, λ_0, or $\mathbf{v}_0$). Even so, our conclusion is enough to allow us to evaluate the eigenvalue. To do so, introduce the *Rayleigh quotient* of a vector $\mathbf{x}$ as follows:

$$\mu(\mathbf{x}) = \frac{\mathbf{x}^T \mathbf{A} \mathbf{x}}{\mathbf{x}^T \mathbf{x}} \tag{4.133}$$

If $\mathbf{x}$ is an eigenvector, $\mu(\mathbf{x})$ obviously gives the eigenvalue; if $\mathbf{x}$ is not an eigenvector, $\mu(\mathbf{x})$ is the nearest substitute to an eigenvalue.[29] Thus, since Eq. (4.132) tells us that $\mathbf{z}^{(k)}$ will tend to be proportional to $\mathbf{v}_0$, we see that $\mu(\mathbf{z}^{(k)})$ will tend to λ_0 (since everything else will cancel out). We have therefore been able to calculate the dominant eigenvalue, λ_0. In other words, for k finite, $\mu(\mathbf{z}^{(k)})$ is our best estimate for λ_0.

4.4.1.2 Algorithm: Normalizing

At this point, we could also discuss how to get the dominant eigenvector, $\mathbf{v}_0$, from $\mathbf{z}^{(k)}$. Instead of doing that, however, we observe that we have ignored a problem in our earlier derivation: in Eq. (4.132) the λ_0^k will become unbounded (if $|\lambda_0| > 1$) or tend to 0 (if $|\lambda_0| < 1$). In order to remedy this, we decide to scale the sequence $\mathbf{z}^{(k)}$ between steps.

The simplest way to accomplish such a scaling is to introduce a new sequence $\mathbf{q}^{(k)}$ which has the convenient property that $||\mathbf{q}^{(k)}|| = 1$. In all that follows we are employing a Euclidian norm implicitly.[30] To do this, simply scale the $\mathbf{z}^{(k)}$ with its norm:

$$\mathbf{q}^{(k)} = \frac{\mathbf{z}^{(k)}}{||\mathbf{z}^{(k)}||} \tag{4.134}$$

[29] In the least-squares sense, introduced in section 6.5.

[30] Our choice to use these specific symbols, $\mathbf{q}$ and $\mathbf{z}$, will pay off in coming sections.

As you can see by taking the norm on both sides, we have $||\mathbf{q}^{(k)}|| = 1$, as desired. It's also easy to see that the Rayleigh quotient for our new vector $\mathbf{q}^{(k)}$ will be:

$$\mu(\mathbf{q}^{(k)}) = [\mathbf{q}^{(k)}]^T \mathbf{A}\mathbf{q}^{(k)} \tag{4.135}$$

since the denominator will give us 1.

Thus, our new normalized power-method algorithm can be summarized as the following sequence of steps for $k = 1, 2, \ldots$:

$$\begin{aligned} \mathbf{z}^{(k)} &= \mathbf{A}\mathbf{q}^{(k-1)} \\ \mathbf{q}^{(k)} &= \frac{\mathbf{z}^{(k)}}{||\mathbf{z}^{(k)}||} \\ \mu(\mathbf{q}^{(k)}) &= [\mathbf{q}^{(k)}]^T \mathbf{A}\mathbf{q}^{(k)} \end{aligned} \tag{4.136}$$

We get this process going by starting from a unit-norm initial vector $\mathbf{q}^{(0)}$. Similarly to what we had before, for this to work $\mathbf{q}^{(0)}$ should have a component in the direction of $\mathbf{v}_0$.

Problem 4.22 asks you to go through the steps of Eq. (4.132) for the case of Eq. (4.136). For the unscaled method we know that $\mathbf{z}^{(k)}$ is *equal* to $\mathbf{A}^k\mathbf{z}^{(0)}$, from Eq. (4.130). In the present, scaled, case, $\mathbf{q}^{(k)}$ is not equal to but *proportional* to $\mathbf{A}^k\mathbf{q}^{(0)}$: the multiplication by $\mathbf{A}$ keeps happening at every step, but now we are also rescaling the result of the multiplication so we end up with a unit norm each time.

Finally, observe that as part of our algorithm we have already produced not only our best estimate for the dominant eigenvalue, $\mu(\mathbf{q}^{(k)})$, but also the corresponding eigenvector, which is nothing other than $\mathbf{q}^{(k)}$, conveniently having unit norm. As before, the $\mathbf{z}^{(k)}$ will tend to be proportional to the dominant eigenvector $\mathbf{v}_0$; this time the scaling that leads to $\mathbf{q}^{(k)}$ simply removes all prefactors and we're left with $\mathbf{v}_0$.

4.4.1.3 Implementation

Code 4.6 is an implementation of Eq. (4.136) in Python. We create a function to evaluate the norm of vectors, as per Eq. (4.23), since we'll need it to do the rescaling after each matrix–vector multiplication; the square is produced by saying `np.sum(xs*xs)`, even though for these purposes we could have, instead, said `xs@xs`. The function `power()` takes in a matrix and an optional parameter of how many times to iterate. We start out with $\mathbf{q}^{(0)}$, a unit-norm vector that's pulled out of a hat. We then multiply $\mathbf{A}$ by our unit-norm vector, to produce a non-unit-norm vector, $\mathbf{z}^{(k)}$. We proceed to normalize $\mathbf{z}^{(k)}$ to get $\mathbf{q}^{(k)}$. This is then done again and again. Evaluating the Rayleigh quotient at each step would have been wasteful; instead, we wait until we're done iterating and then evaluate $[\mathbf{q}^{(k)}]^T \mathbf{A}\mathbf{q}^{(k)}$ only once.[31] We also wrote a function that tests one eigenvalue and eigenvector pair by comparing against `numpy.linalg.eig()`, a wrapper to state-of-the-art LAPACK functionality. The main program uses our old `testcreate()` function to produce a test matrix $\mathbf{A}$ (placing the unneeded $\mathbf{b}$ in a throwaway variable—an underscore). The final lines of the output are:

[31] NumPy knows how to interpret `qs@A@qs`, without us having to say `np.transpose(qs)@A@qs`.

Code 4.6 power.py

```
from triang import testcreate
import numpy as np

def mag(xs):
    return np.sqrt(np.sum(xs*xs))

def power(A,kmax=6):
    zs = np.ones(A.shape[0])
    qs = zs/mag(zs)
    for k in range(1,kmax):
        zs = A@qs
        qs = zs/mag(zs)
        print(k,qs)

    lam = qs@A@qs
    return lam, qs

def testeigone(f,A,indx=0):
    eigval, eigvec = f(A)
    print(" "); print(eigval); print(eigvec)
    npeigvals, npeigvecs = np.linalg.eig(A)
    print(" ")
    print(npeigvals[indx]); print(npeigvecs[:,indx])

if __name__ == '__main__':
    A, _ = testcreate(4,21)
    testeigone(power,A)
```

```
21.3166626635
[ 0.44439562 0.48218122 0.51720906 0.55000781]

21.3166626635
[ 0.44439562 0.48218122 0.51720906 0.55000781]
```

Check to see that the eigenvector stops changing after a few iterations. In the code we just discussed, we arbitrarily pick a given number of total iterations; it is in principle better to introduce a self-terminating criterion. Since our algorithm works by generating new vectors $\mathbf{q}^{(k)}$, the most obvious choice in this regard is Eq. (4.122):

$$\sum_{j=0}^{n-1} \left| \frac{q_j^{(k)} - q_j^{(k-1)}}{q_j^{(k)}} \right| \leq \epsilon \tag{4.137}$$

which is expressed in terms of the relative difference in the components. Problem 4.23 asks you to implement the power method with this criterion.

4.4.1.4 Operation Count

For iterative methods, the total operation count depends on the actual number of iterations required, which we generally cannot predict ahead of time (since it depends on the starting point, the error tolerance, the stopping criterion, etc.). For example, our 4×4 test matrix took four iterations, whereas if you use `testcreate()` for the 20×20 case you will need six total iterations. Thus, any operation count we encounter will also have to be multiplied by m, where m is the number of actual iterations needed. The value of m required is related to the separation between the eigenvalue we are after and the next closest one.

The bulk of the work of the power method is carried out by $\mathbf{z}^{(k)} = \mathbf{A}\mathbf{q}^{(k-1)}$: this is a matrix–vector multiplication. As we discussed in section 4.3.1.4, this costs $\sim 2n^2$ operations. The power method also requires the evaluation of the norm of $\mathbf{z}^{(k)}$. You will show in problem 4.34 that vector–vector multiplication costs $\sim 2n$. Thus, so far the cost is $\sim 2mn^2$. Finally, we also need the Rayleigh quotient, which is made up of a single matrix–vector multiplication, $\sim 2n^2$, and a single vector–vector multiplication, $\sim 2n$. In total, we have found that the operation count for the power method is $\sim 2(m + 1)n^2$.

4.4.2 Inverse-Power Method with Shifting

We will now discuss a variation of the power method, which will also allow us to evaluate one eigenvalue and eigenvector pair. This time, it will be for the eigenvalue with the smallest magnitude (in absolute value).

4.4.2.1 Algorithm

Let's start with our defining equation, $\mathbf{A}\mathbf{v}_i = \lambda_i \mathbf{v}_i$. Multiplying on the left by $\mathbf{A}^{-1}$ we find $\mathbf{v}_i = \lambda_i \mathbf{A}^{-1}\mathbf{v}_i$. Dividing both sides of this equation by λ_i, we get:

$$\mathbf{A}^{-1}\mathbf{v}_i = \lambda_i^{\text{inv}} \mathbf{v}_i \tag{4.138}$$

where λ_i^{inv} is an eigenvalue of the inverse matrix and we just showed that $\lambda_i^{\text{inv}} = 1/\lambda_i$.[32] We have proven that *the eigenvectors of* $\mathbf{A}^{-1}$ *are the same as the eigenvectors of* $\mathbf{A}$. Similarly, we've shown that *the eigenvalues of* $\mathbf{A}^{-1}$ *are the reciprocals of the eigenvalues of* $\mathbf{A}$.

[32] The special case $\lambda_i = 0$ is easy to handle: since the determinant of a matrix is equal to 0 when an eigenvalue is 0, such a matrix is not invertible, so we wouldn't be able to multiply by $\mathbf{A}^{-1}$ in the first place.

The basic idea is to apply the power method, Eq. (4.136), to the inverse matrix, $\mathbf{A}^{-1}$:

$$\mathbf{z}^{(k)} = \mathbf{A}^{-1}\mathbf{q}^{(k-1)} \tag{4.139}$$

If you now combine the fact that the power method determines the largest-magnitude eigenvalue with the result $\lambda_i^{\text{inv}} = 1/\lambda_i$, our new approach will allow us to evaluate the eigenvalue of the original matrix $\mathbf{A}$ with the smallest absolute value, namely λ_{n-1}, as per Eq. (4.128). Similarly, our scaled vector $\mathbf{q}^{(k)}$ will tend toward the corresponding eigenvector, $\mathbf{v}_{n-1}$. In addition to normalizing at every step, as in Eq. (4.136), we would also have to evaluate the Rayleigh quotient for the inverse matrix at the end: $[\mathbf{q}^{(k)}]^T\mathbf{A}^{-1}\mathbf{q}^{(k)}$. Of course, since we recently showed that $\mathbf{A}$ and $\mathbf{A}^{-1}$ have the same eigenvectors, we could just as easily evaluate the Rayleigh quotient for the original matrix $\mathbf{A}$:

$$\mu(\mathbf{q}^{(k)}) = [\mathbf{q}^{(k)}]^T\mathbf{A}\mathbf{q}^{(k)} \tag{4.140}$$

to evaluate λ_{n-1}. The results of using the two different Rayleigh quotients will not be the same, but related to each other (the first one will give $1/\lambda_{n-1}$ and the second one λ_{n-1}).

While conceptually this is all the "inverse power" method amounts to, in practice one can avoid the (costly) evaluation of the inverse matrix; multiply Eq. (4.139) by $\mathbf{A}$ from the left, to find:

$$\mathbf{A}\mathbf{z}^{(k)} = \mathbf{q}^{(k-1)} \tag{4.141}$$

This is a linear system of the form $\mathbf{A}\mathbf{x} = \mathbf{b}$ which can be solved to give us $\mathbf{z}^{(k)}$. Every step of the way we will be solving a system for the same matrix $\mathbf{A}$ but different right-hand sides: by saying $\mathbf{L}\mathbf{U}\mathbf{x} = \mathbf{b}$ we are then able to solve this system by forward and then back substitution, as per Eq. (4.97):

$$\begin{aligned} \mathbf{L}\mathbf{y} &= \mathbf{q}^{(k-1)} \\ \mathbf{U}\mathbf{z}^{(k)} &= \mathbf{y} \end{aligned} \tag{4.142}$$

for a different $\mathbf{q}^{(k-1)}$ each time. Thus, we LU-decompose once (this being the most costly step), and then we can step through Eq. (4.141) with a minimum of effort. To summarize, the inverse-power method consists of the following sequence of steps for $k = 1, 2, \ldots$:

$$\begin{aligned} \mathbf{A}\mathbf{z}^{(k)} &= \mathbf{q}^{(k-1)} \\ \mathbf{q}^{(k)} &= \frac{\mathbf{z}^{(k)}}{\|\mathbf{z}^{(k)}\|} \\ \mu(\mathbf{q}^{(k)}) &= [\mathbf{q}^{(k)}]^T\mathbf{A}\mathbf{q}^{(k)} \end{aligned} \tag{4.143}$$

As should be expected, we get this process going by starting from a unit-norm initial vector $\mathbf{q}^{(0)}$. Similarly to what we had before, for this to work $\mathbf{q}^{(0)}$ should have a component in the direction of $\mathbf{v}_{n-1}$. Observe how our procedure in Eq. (4.143) is almost identical to that of the power method in Eq. (4.136): the only (crucial) difference is that here we have $\mathbf{A}$ on the left-hand side so, instead of needing to carry out a matrix–vector multiplication at each step, we need to solve a linear system of equations at each step. Thus, the crucial part of this process is $\mathbf{A}\mathbf{z}^{(k)} = \mathbf{q}^{(k-1)}$, which is made easy by employing LU decomposition.

4.4.2.2 Eigenvalue Shifting

We could proceed at this point to implement the inverse-power method. Instead of doing that, however, we will first further refine it. As usual, let's start with our defining equation, $\mathbf{A}\mathbf{v}_i = \lambda_i \mathbf{v}_i$. Subtract from both sides of this equation $s\mathbf{v}_i$ where s is some scalar, known as the *shift*, for reasons that will soon become clear. The resulting equation is:

$$(\mathbf{A} - s\mathcal{I})\mathbf{v}_i = (\lambda_i - s)\mathbf{v}_i \tag{4.144}$$

which can be written as:

$$\mathbf{A}^*\mathbf{v}_i = \lambda_i^*\mathbf{v}_i, \quad \mathbf{A}^* = \mathbf{A} - s\mathcal{I}, \quad \lambda_i^* = \lambda_i - s \tag{4.145}$$

In words, if you can solve the eigenproblem for the matrix $\mathbf{A}^*$ you will have evaluated eigenvectors $\mathbf{v}_i$ which are identical to those of $\mathbf{A}$. Furthermore, you will have evaluated the eigenvalues λ_i^*, which are equal to the eigenvalues λ_i of matrix $\mathbf{A}$ shifted by s.

Consider applying the inverse-power method of the previous subsection to solve the first equation in Eq. (4.145). Conceptually, what we are suggesting to do is to apply the direct power method for the matrix $(\mathbf{A} - s\mathcal{I})^{-1}$ or, equivalently, the inverse-power method for the matrix $\mathbf{A} - s\mathcal{I}$.[33] To be explicit, we are choosing to follow this sequence of steps:

$$\begin{aligned} \mathbf{A}^*\mathbf{z}^{(k)} &= \mathbf{q}^{(k-1)} \\ \mathbf{q}^{(k)} &= \frac{\mathbf{z}^{(k)}}{||\mathbf{z}^{(k)}||} \\ \mu^*(\mathbf{q}^{(k)}) &= [\mathbf{q}^{(k)}]^T\mathbf{A}^*\mathbf{q}^{(k)} \end{aligned} \tag{4.146}$$

This will allow you to evaluate the smallest eigenvalue, λ_i^*, of the matrix $\mathbf{A}^*$. But, since $\lambda_i^* = \lambda_i - s$, finding the smallest λ_i^* is equivalent to having evaluated *that eigenvalue* λ_i *of the original matrix* $\mathbf{A}$ *which is closest to* s. This explains why we were interested in combining the inverse-power method with eigenvalue shifting: the "inverse-power" part allows us to find the smallest eigenvalue and the "shifting" part controls what "smallest" means (i.e., the one that minimizes $\lambda_i - s$).

As a matter of fact, given that $\mathbf{q}^{(k)}$ will be converging to $\mathbf{v}_i$ (which is an eigenvector of $\mathbf{A}^*$ and of $\mathbf{A}$) when evaluating the Rayleigh quotient we could use, instead, the same formula as before, namely $\mu(\mathbf{q}^{(k)}) = [\mathbf{q}^{(k)}]^T\mathbf{A}\mathbf{q}^{(k)}$: this would automatically give us λ_i.[34] In Eq. (4.146) we used $\mathbf{A}^*$ in the Rayleigh quotient for pedagogical clarity, but once you've understood what's going on there's no need to take that extra step.

Now, when carrying out the shift for the matrix $\mathbf{A}^*$, we can pick several different values for s and (one can imagine) evaluate all the eigenvalue and eigenvector pairs for matrix

[33] We will sometimes mention the *direct* power method, to distinguish it from the inverse-power method.

[34] If we used $\mu^*(\mathbf{q}^{(k)}) = [\mathbf{q}^{(k)}]^T\mathbf{A}^*\mathbf{q}^{(k)}$, instead, we would get λ_i^* and would still have to add s back in, in order to get λ_i, given that $\lambda_i^* = \lambda_i - s$.

A.[35] What we'll do, instead, in coming sections is to assume we have access to an *estimate* for a given eigenvalue: in that case, the inverse-power method with shifting allows you to refine that estimate. Significantly, our new method is also quite useful when you already know an eigenvalue (even very accurately) but don't know the corresponding eigenvector: by having s be (close to) the true eigenvalue λ_i, a few iterations will lead to $\mathbf{q}^{(k)}$ converging to $\mathbf{v}_i$. (This will come in very handy in section 4.4.4 below.)

Before turning to Python code, let us examine a final application of eigenvalue shifting. Recall that the power method's convergence is determined by the ratio $|\lambda_1/\lambda_0|$, where λ_1 is the eigenvalue with the second largest magnitude. Similarly, since the (unshifted) inverse-power method converges toward the smallest eigenvalue λ_{n-1}, the rate of convergence will depend on the ratio $|\lambda_{n-1}/\lambda_{n-2}|$: if λ_{n-2} is much larger than λ_{n-1} then we'll converge rapidly (and if it's not much larger, then we'll have trouble converging). One can envision, then, employing a shift s that makes the ratio $|\lambda^*_{n-1}/\lambda^*_{n-2}|$ as small as possible. As a matter of fact, taking s to be very close to λ_{n-1} should be enough to enforce this (since $\lambda^*_{n-1} = \lambda_{n-1} - s$). We won't be following this route in what follows, but it's worth knowing that eigenvalue shifting is frequently used to *accelerate convergence*.

4.4.2.3 Implementation

Code 4.7 is a Python implementation of the inverse-power method. Comparing our new function to `power()`, we notice three main differences. First, at the start we shift our original matrix to produce $\mathbf{A}^* = \mathbf{A} - s\mathcal{I}$. We then apply our method to this, shifted, matrix. At the end of the process, if successful, we evaluate the Rayleigh quotient for the *original* matrix, which allows us to evaluate the eigenvalue λ_i of the original matrix that is closest to the hand-picked shift s. Second, we implement a self-stopping criterion, since later we intend to fuse our inverse-power shifted function with another method, and we don't want to be wasteful (e.g., carrying out hundreds of iterations when one or two will do). Thus, we have been careful to create a new copy of our vector each time through the loop, via `qs = np.copy(qnews)`. This code employs the `for-else` idiom, just like `jacobi()` did: now that we know what we are testing against in order to succeed, we can also account for the possibility of failing to converge in the given number of iterations.[36] Importantly, we also check to see if we need to flip the sign of our vector: when the eigenvalue is negative, left on its own the sign of $\mathbf{q}^{(k)}$ would change from iteration to iteration.[37] Our new function would always think it's failing to converge, since a vector is very different from its opposite; we take care of this by checking for sign flips and adjusting accordingly. Third, `power()` is implementing the (direct) power method so the bulk of its work is carried out using `@`. Here, we have to solve a linear system of equations: to do that, we first LU-decompose the shifted matrix $\mathbf{A}^*$. This is done only once, *outside* the loop: then, inside the loop we only use the forward and back substitution functions, which are considerably less costly.

[35] If you have no prior input this is an inefficient way of finding all eigenvalues, since we don't know how to pick s, e.g., should we use a grid? Problem 4.36 introduces a more systematic approach: deflation.

[36] The relative criterion is suboptimal here, since individual components of eigenvectors can be zero.

[37] The function `power()` doesn't worry about this, since it blindly keeps iterating regardless of the sign: the easiest way to test this is to say `A = -A` in the main program after creating the test matrix.

invpowershift.py

Code 4.7

```
from triang import forsub, backsub, testcreate
from ludec import ludec
from jacobi import termcrit
from power import mag, testeigone
import numpy as np

def invpowershift(A,shift=20,kmax=200,tol=1.e-8):
    n = A.shape[0]
    znews = np.ones(n)
    qnews = znews/mag(znews)
    Astar = A - np.identity(n)*shift
    L, U = ludec(Astar)

    for k in range(1,kmax):
        qs = np.copy(qnews)
        ys = forsub(L,qs)
        znews = backsub(U,ys)
        qnews = znews/mag(znews)

        if qs@qnews<0:
            qnews = -qnews

        err = termcrit(qs, qnews)
        print(k, qnews, err)

        if err < tol:
            lam = qnews@A@qnews
            break

    else:
        lam = qnews = None

    return lam, qnews

if __name__ == '__main__':
    A, _ = testcreate(4,21)
    testeigone(invpowershift,A)
```

Running this code, the final lines of the output are:

```
21.3166626635
[ 0.44439562 0.48218122 0.51720906 0.55000781]

21.3166626635
[ 0.44439562 0.48218122 0.51720906 0.55000781]
```

Crucially, we did *not* need to carry out `kmax` iterations. We have picked $s = 20$, precisely because we wanted to find the eigenvalue that is the same as that given by the direct power method (i.e., the largest one). We encourage you to play around with `shift` and find all the eigenvalue and eigenvector pairs for this simple 4×4 example.

4.4.2.4 Operation Count

Just like in the case of the power method above, when estimating the operation count for the inverse-power method we don't really know ahead of time how fast it will converge. Thus, any operation count we encounter will also have to be multiplied by m, where m is the number of actual iterations required.

The bulk of the work of the inverse-power method is carried out by $\mathbf{A}^*\mathbf{z}^{(k)} = \mathbf{q}^{(k-1)}$: this is a linear system of equations that has to be solved again and again. As we discussed earlier, we carry out a single LU decomposition for $\mathbf{A}^*$. We recall that an LU decomposition costs $\sim 2n^3/3$ operations. (You might be wondering why we didn't bother counting how many operations were required to build up the matrix $\mathbf{A}^*$. The reason is that since the shift only affects the diagonal, this would be $O(n)$ so it doesn't matter.)

The inverse-power method also requires a forward substitution (n^2) and a back substitution (n^2), as well as a vector-norm evaluation ($\sim 2n$), each time through the loop. In total, these three contributions add up to $\sim 2mn^2$. Similarly, the Rayleigh quotient evaluation costs $\sim 2n^2$. In total, we have found that the operation count for the inverse-power method is $\sim 2(m+1)n^2 + 2n^3/3$. Thus, we see that while the power method cost involves n^2, the inverse-power method cost involves n^3. This change in order of magnitude costs arose solely due to the LU decomposition; said another way, if you have pre-LU-decomposed your matrix, then the inverse-power method costs as much as the direct power method. (Said yet another way: typically $m \approx n$, in which case both the direct and the inverse-power method scale as $O(n^3)$.)

4.4.3 QR Method

The (direct or inverse) power method that we've discussed so far gives us only one eigenvalue at a time (either the largest or the smallest). As we saw, you could combine the latter method with eigenvalue shifting and then try to step through all the eigenvalues of your matrix. In the present section, we will discuss a robust and scalable method used to evaluate *all* the eigenvalues of a matrix at one go. We will be introducing the *QR method*: this approach, developed by J. Francis in the early 1960s, takes its name from the *QR decomposition* (also known as the *QR factorization*) and then adds a clever trick that allows one to simply read off all the eigenvalues of our matrix. In order to better grasp this trick, we will also make a slight detour into similarity transformations and the related approach known as "simultaneous iteration". Make sure to keep the terminology straight: we use the QR *decomposition* in order to express a matrix as the product of two other matrices, while we use the QR *method* in order to evaluate all eigenvalues of a matrix.

4.4.3.1 QR Decomposition

We start from the concept of the QR decomposition, which turns out to be very significant: according to the authors of Ref. [148], this is the most important idea in numerical linear algebra. As usual, we will not cover all variations of the approach, nor will we see all possible applications. That being said, we *will* derive things explicitly and implement them in Python for a quite general case.

It's worth keeping in mind that (as mentioned above) all methods that compute eigenvalues are *iterative*, but the QR decomposition is a *direct* method: just like the LU decomposition we discussed in an earlier section, it goes through a fixed number of steps, until it has decomposed the starting matrix in the desired form.[38]

Let's be explicit. The QR decomposition starts with a matrix **A** and decomposes it into the product of an orthogonal matrix **Q** and an upper-triangular matrix **R**. (This upper-triangular matrix is called **R** and not **U**, as we did above, for historical reasons.) Symbolically, we say that any real square matrix can be factorized as:

$$\mathbf{A} = \mathbf{QR} \tag{4.147}$$

Recall from appendix C.2 that a matrix is called orthogonal if the transpose is equal to the inverse, $\mathbf{Q}^{-1} = \mathbf{Q}^T$. We can recast this definition as $\mathbf{Q}^T\mathbf{Q} = \mathcal{I}$, showing that an orthogonal matrix has orthonormal columns (i.e., columns that are orthogonal unit vectors) – you can see that a better name would have been "orthonormal matrix".

Now, a bit on strategy: we will provide what is known as a "constructive" proof. That means that we will explicitly show how to construct the orthogonal matrix **Q** starting from our original matrix **A**. When we are done with that process, we will show that it is easy to find (as a matter of fact, we will have already found) an upper-triangular matrix **R** that, when multiplied by **Q** on the right, allows us to reconstruct the original matrix **A**.

Evaluating Q: We start by constructing the orthogonal matrix **Q**. We will employ an old method which you may have encountered in a course on linear algebra: *Gram–Schmidt orthogonalization*. Let us write our starting matrix **A** in terms of its columns $\mathbf{a}_j$:

$$\mathbf{A} = \begin{pmatrix} \mathbf{a}_0 & \mathbf{a}_1 & \dots & \mathbf{a}_{n-1} \end{pmatrix} \tag{4.148}$$

[38] There's no contradiction here: QR decomposition may be direct, but the QR method, which actually calculates the eigenvalues, *is* iterative.

Our task now is to start from assuming that the column vectors $\mathbf{a}_j$ are linearly independent and try to produce an *orthonormal* set of column vectors $\mathbf{q}_j$. When we've accomplished that task, we will have already produced our orthogonal matrix $\mathbf{Q}$:

$$\mathbf{Q} = \begin{pmatrix} \mathbf{q}_0 & \mathbf{q}_1 & \dots & \mathbf{q}_{n-1} \end{pmatrix} \tag{4.149}$$

since $\mathbf{Q}$ will be made up of the orthonormal column vectors $\mathbf{q}_j$ we just constructed.

To ensure that everything is transparent, we will build these orthonormal $\mathbf{q}_j$ column vectors one at a time. The first vector, $\mathbf{q}_0$, is very easy to produce: simply pick it to be in the direction of $\mathbf{a}_0$ and scale it by the norm of $\mathbf{a}_0$ (to ensure that $\mathbf{q}_0$ is a unit vector):

$$\mathbf{q}_0 = \frac{\mathbf{a}_0}{\|\mathbf{a}_0\|} \tag{4.150}$$

Unfortunately, this trivial prescription (simply divide by the norm) is not enough to produce the next orthonormal vector, $\mathbf{q}_1$; this is because we need the $\mathbf{q}_j$ to be orthogonal to each other and we have no guarantee that $\mathbf{a}_1$ is orthogonal to $\mathbf{a}_0$ (actually, these two almost certainly are *not* orthogonal to each other).

Here's the Gram–Schmidt prescription: in order to find a vector that is orthogonal to $\mathbf{q}_0$, take the second vector, $\mathbf{a}_1$, and subtract out its component in the direction of $\mathbf{q}_0$:

$$\mathbf{a}'_1 = \mathbf{a}_1 - (\mathbf{q}_0^T \mathbf{a}_1)\mathbf{q}_0 \tag{4.151}$$

This is the part of $\mathbf{a}_1$ that does *not* point in the direction of $\mathbf{a}_0$ (or of $\mathbf{q}_0$). You should explicitly check that $\mathbf{a}'_1$ is perpendicular to $\mathbf{q}_0$. The only thing that's left in order to produce our second orthonormal vector $\mathbf{q}_1$ is to normalize $\mathbf{a}'_1$:

$$\mathbf{q}_1 = \frac{\mathbf{a}'_1}{\|\mathbf{a}'_1\|} \tag{4.152}$$

which gives us a unit vector.

You are probably starting to discern the pattern, but let's do one more step explicitly. We've already determined $\mathbf{q}_0$ and $\mathbf{q}_1$. To determine $\mathbf{q}_2$ we, again, start with $\mathbf{a}_2$; this time we want to subtract any component in the plane of $\mathbf{q}_0$ and $\mathbf{q}_1$. Thus, we take $\mathbf{a}_2$ and subtract its component in the direction of $\mathbf{q}_0$ as well as its component in the direction of $\mathbf{q}_1$:

$$\mathbf{a}'_2 = \mathbf{a}_2 - (\mathbf{q}_0^T \mathbf{a}_2)\mathbf{q}_0 - (\mathbf{q}_1^T \mathbf{a}_2)\mathbf{q}_1 \tag{4.153}$$

Having made $\mathbf{a}'_2$ orthogonal to both $\mathbf{q}_0$ and $\mathbf{q}_1$ (check this!), all that's left is to scale $\mathbf{a}'_2$:

$$\mathbf{q}_2 = \frac{\mathbf{a}'_2}{\|\mathbf{a}'_2\|} \tag{4.154}$$

Clearly, the general pattern is, for $j = 0, 1, \dots, n-1$:

$$\begin{aligned} \mathbf{a}'_j &= \mathbf{a}_j - \sum_{i=0}^{j-1} (\mathbf{q}_i^T \mathbf{a}_j)\mathbf{q}_i \\ \mathbf{q}_j &= \frac{\mathbf{a}'_j}{\|\mathbf{a}'_j\|} \end{aligned} \tag{4.155}$$

where we've assumed that you can also extend this definition naturally to $j = 0$ (for this case, there are no terms in the sum, so $\mathbf{a}'_0 = \mathbf{a}_0$ and we agree with Eq. (4.150)). We have therefore succeeded in constructing the orthonormal set of $\mathbf{q}_j$ vectors. In other words, we have produced the orthogonal matrix $\mathbf{Q}$, as desired.

Evaluating R: We now turn to the matrix $\mathbf{R}$. Let us, momentarily, *assume* that $\mathbf{A} = \mathbf{QR}$ holds, and try to see if there is an easy way to determine $\mathbf{R}$. Using the notation employing column vectors, $\mathbf{A} = \mathbf{QR}$ can be rewritten as:

$$\begin{pmatrix} \mathbf{a}_0 & \mathbf{a}_1 & \mathbf{a}_2 & \dots & \mathbf{a}_{n-1} \end{pmatrix} = \begin{pmatrix} \mathbf{q}_0 & \mathbf{q}_1 & \mathbf{q}_2 & \dots & \mathbf{q}_{n-1} \end{pmatrix} \begin{pmatrix} R_{00} & R_{01} & R_{02} & \dots & R_{0,n-1} \\ 0 & R_{11} & R_{12} & \dots & R_{1,n-1} \\ 0 & 0 & R_{22} & \dots & R_{2,n-1} \\ \vdots & \vdots & \vdots & \ddots & \vdots \\ 0 & 0 & 0 & \dots & R_{n-1,n-1} \end{pmatrix} \tag{4.156}$$

Let's explicitly carry out the matrix multiplication and identify matrix elements (which themselves are column vectors) on the left-hand and the right-hand side. We find:

$$\begin{aligned} \mathbf{a}_0 &= R_{00}\mathbf{q}_0 \\ \mathbf{a}_1 &= R_{01}\mathbf{q}_0 + R_{11}\mathbf{q}_1 \\ \mathbf{a}_2 &= R_{02}\mathbf{q}_0 + R_{12}\mathbf{q}_1 + R_{22}\mathbf{q}_2 \\ &\vdots \\ \mathbf{a}_j &= R_{0j}\mathbf{q}_0 + R_{1j}\mathbf{q}_1 + \dots + R_{jj}\mathbf{q}_j \\ &\vdots \\ \mathbf{a}_{n-1} &= R_{0,n-1}\mathbf{q}_0 + R_{1,n-1}\mathbf{q}_1 + \dots + R_{n-1,n-1}\mathbf{q}_{n-1} \end{aligned} \tag{4.157}$$

where we also showed an intermediate case, $\mathbf{a}_j$. These equations can be solved for $\mathbf{q}_j$:

$$\begin{aligned} &\mathbf{q}_0 = \frac{\mathbf{a}_0}{R_{00}}, \quad \mathbf{q}_1 = \frac{\mathbf{a}_1 - R_{01}\mathbf{q}_0}{R_{11}}, \quad \mathbf{q}_2 = \frac{\mathbf{a}_2 - R_{02}\mathbf{q}_0 - R_{12}\mathbf{q}_1}{R_{22}}, \quad \dots, \\ &\mathbf{q}_j = \frac{\mathbf{a}_j - \sum_{i=0}^{j-1} R_{ij}\mathbf{q}_i}{R_{jj}}, \quad \dots, \quad \mathbf{q}_{n-1} = \frac{\mathbf{a}_{n-1} - \sum_{i=0}^{n-2} R_{i,n-1}\mathbf{q}_i}{R_{n-1,n-1}} \end{aligned} \tag{4.158}$$

Comparing this general result for $\mathbf{q}_j$ with what we found in Eq. (4.155), we see that it is appropriate to identify the matrix elements of $\mathbf{R}$ as follows:

$$\begin{aligned} R_{ij} &= \mathbf{q}_i^T \mathbf{a}_j, \quad j = 0, 1, \dots, n-1, \quad i = 0, 1, \dots, j-1 \\ R_{jj} &= \|\mathbf{a}'_j\| = \|\mathbf{a}_j - \sum_{i=0}^{j-1} R_{ij}\mathbf{q}_i\|, \quad j = 0, 1, \dots, n-1 \end{aligned} \tag{4.159}$$

where, implicitly, $R_{ij} = 0$ for $i > j$. Note a somewhat subtle point: we have chosen the diagonal elements of **R** to be positive, $R_{jj} > 0$, in order to match our earlier definitions. This wasn't necessary; however, if you do make the assumption $R_{jj} > 0$, then the QR decomposition is *uniquely determined*.

Crucially, both $R_{ij} = \mathbf{q}_i^T \mathbf{a}_j$ and $R_{jj} = ||\mathbf{a}'_j||$ are quantities that we have already evaluated in the process of constructing the matrix **Q**. That means that we can carry out those computations in parallel, building up the matrices **Q** and **R** together. This will become clearer in the following subsection, when we provide Python code that implements this prescription.

To summarize, we have been able to produce an orthogonal matrix **Q** starting from the matrix **A**, as well as an upper-triangular matrix **R** which, when multiplied together, give us the original matrix: $\mathbf{A} = \mathbf{QR}$. We have therefore accomplished the task we set out to accomplish. There are other ways of producing a QR decomposition, but for our purposes the above constructive proof will suffice.

Backing up for a second, we realize that what Gram–Schmidt helped us accomplish was to go from a set of n linearly independent vectors (the columns of **A**) to a set of n orthonormal vectors (the columns of **Q**). These ideas can be generalized to the infinite-dimensional case: this is the *Hilbert space*, which you may have encountered in a course on quantum mechanics. Without getting into much detail, we note that it is possible to extend the concept of a vector such that it becomes continuous, namely a *function*! If we also appropriately extend the definition of the inner product (to be the integral of the product of two functions), then all these ideas about orthonormalization start to become applicable to functions. As a matter of fact (as you'll see in problem 4.27), if you start from the monomials $1, x, x^2, \ldots$ and apply the Gram–Schmidt orthogonalization procedure, you will end up with (multiples of) the *Legendre polynomials* which are, indeed, orthogonal to each other. This is neither our first nor our last encounter with these polynomials.

4.4.3.2 QR Decomposition: Implementation

When thinking about how to implement the QR decomposition procedure, we direct our attention to Eq. (4.155), which tells us how to construct the matrix **Q**, and Eq. (4.159), which tells us how to construct the matrix **R**. The obvious way to test any routine we produce is to see if the product **QR** really is equal to the original matrix **A**. Thus, what we can do is inspect the norm $||\mathbf{A} - \mathbf{QR}||$ to see how well we decomposed our matrix (this may bring to mind the residual vector $\mathbf{r} = \mathbf{b} - \mathbf{A}\tilde{\mathbf{x}}$ from section 4.2.1).

Another concern emerges at this point: we saw that the Gram–Schmidt procedure guarantees the orthogonality of the matrix **Q**. In other words, mathematically, we know by construction that $\mathbf{Q}^T\mathbf{Q} = \mathcal{I}$ should hold. However, since we are carrying out these calculations for floating-point numbers (i.e., in the presence of roundoff errors), it's worth explicitly investigating how well this orthogonality holds. To do that, we will also evaluate the norm $||\mathbf{Q}^T\mathbf{Q} - \mathcal{I}||$ explicitly. Deviations of this quantity from zero will measure how poorly we are actually doing in practice.

Code 4.8 is a Python implementation of QR decomposition, together with the aforementioned tests. We immediately see that this is *not* an iterative method: we don't have a `kmax` or something along those lines controlling how many times we'll repeat the entire process. Put differently, it is n, namely the size of the problem, that directly controls how

qrdec.py Code 4.8

```
from triang import testcreate
from power import mag
import numpy as np

def qrdec(A):
    n = A.shape[0]
    Ap = np.copy(A)
    Q = np.zeros((n,n))
    R = np.zeros((n,n))
    for j in range(n):
        for i in range(j):
            R[i,j] = Q[:,i]@A[:,j]
            Ap[:,j] -= R[i,j]*Q[:,i]

        R[j,j] = mag(Ap[:,j])
        Q[:,j] = Ap[:,j]/R[j,j]
    return Q, R

def testqrdec(A):
    n = A.shape[0]
    Q, R = qrdec(A)
    diffa = A - Q@R
    diffq = np.transpose(Q)@Q - np.identity(n)
    print(n, mag(diffa), mag(diffq))

if __name__ == '__main__':
    for n in range(4,10,2):
        A, _ = testcreate(n,21)
        testqrdec(A)
```

many iterations we'll need to carry out. The function `qrdec()` is a mostly straightforward implementation of the equations we developed above. Note how R_{ij} is evaluated by multiplying two one-dimensional NumPy arrays together (so there was no need to take the transpose). The code then proceeds to use the already-evaluated R_{ij} in order to build up $\mathbf{a}'_j$: note that we use columns of $n \times n$ matrices throughout, so we've also created `Ap`, which corresponds to a collection of the columns $\mathbf{a}'_j$ into a matrix $\mathbf{A}'$.[39] The values over which

[39] $\mathbf{A}'$ starts out as a copy of $\mathbf{A}$, because that's always the first step in the Gram–Schmidt prescription for $\mathbf{a}'_j$: start from $\mathbf{a}_j$ and subtract out what you need to.

`i` and `j` range are those given in Eq. (4.159). When $j = 0$, no iteration over i takes place. Once we're done with the inner loop, we evaluate the diagonal elements R_{jj} and use them to normalize the columns of $\mathbf{A}'$ thereby producing the matrix $\mathbf{Q}$.

We've also created a function that computes the norms $||\mathbf{A}-\mathbf{QR}||$ and $||\mathbf{Q}^T\mathbf{Q}-\mathcal{I}||$. These are easy to implement; note, however, that in this case the transposition in $\mathbf{Q}^T$ has to be explicitly carried out in our code. Another thing that might raise a red flag is our use of `mag()` from `power.py`. As you may recall, that function had been developed in order to compute the magnitude of vectors, not the norm of matrices. However, due to the pleasant nature of Python and NumPy, it works just as well when you pass in a matrix. This happens to be the reason we had opted to say `np.sum(xs*xs)` instead of `xs@xs`: if you pass in a matrix in the place of `xs`, `xs@xs` carries out a matrix multiplication, and that's not how the matrix norm is defined in Eq. (4.20).

In the main body of the code, we print out the test norms for a few problem dimensionalities. For the 4×4 problem both the reconstruction of $\mathbf{A}$ using $\mathbf{Q}$ and $\mathbf{R}$, on the one hand, and the orthogonality of $\mathbf{Q}$, on the other hand, perform reasonably well, even in the presence of roundoff errors. As a matter of fact, $||\mathbf{A} - \mathbf{QR}||$ is tiny, close to machine precision. We notice that something analogous holds for the 6×6 and 8×8 problems as well: the product $\mathbf{QR}$ is a great representation of the matrix $\mathbf{A}$. Unfortunately, the same cannot be said for the degree of orthogonality of the matrix $\mathbf{Q}$: clearly this gets quite bad even for moderate problem sizes.

Actually, the orthogonality (or lack thereof) is troubling even for the 4×4 case. Problem 4.26 asks you to use QR decomposition to solve $\mathbf{Ax} = \mathbf{b}$; you'll see that even for the 4×4 problem the solution is disappointing. Of course, this is a section on eigenvalue evaluation, not on linear-system solving: we are building QR decomposition as a step in our QR method (and it will turn out that this lack of orthogonality is not as troubling for our purposes). That being said, it's worth knowing that Gram–Schmidt orthogonalization, which we just discussed, is poorly behaved in the presence of roundoff errors. This is why the algorithm presented above is known as *classical Gram–Schmidt*, in contradistinction to what is known as *modified Gram–Schmidt*. This, as you'll find out when you solve problem 4.25, has better orthogonality properties.[40]

4.4.3.3 Similarity Transformations

We now make a quick detour, before picking up the main thread of connecting the QR decomposition with the general QR method. First, recall from Eq. (4.125) that when we diagonalize a matrix $\mathbf{A}$ we manage to find the matrices $\mathbf{V}$ and $\mathbf{\Lambda}$ such that:

$$\mathbf{V}^{-1}\mathbf{AV} = \mathbf{\Lambda} \tag{4.160}$$

where $\mathbf{\Lambda}$ contains the eigenvalues of $\mathbf{A}$ and $\mathbf{V}$ is made up of the eigenvectors of $\mathbf{A}$.

Assume there exists another (non-singular) matrix, $\mathbf{S}$, such that:

[40] Problem 4.28 introduces a distinct approach, which employs a sequence of Householder transformations.

$$\mathbf{A}' = \mathbf{S}^{-1}\mathbf{A}\mathbf{S} \tag{4.161}$$

It is obvious why we assumed that $\mathbf{S}$ is non-singular: we need to use its inverse. Crucially, we are *not* assuming here that $\mathbf{A}'$ is diagonal or that $\mathbf{S}$ is made up of the eigenvectors. We are simply carrying out a specific transformation (multiply the matrix $\mathbf{A}$ by $\mathbf{S}^{-1}$ on the left and by $\mathbf{S}$ on the right) for a given matrix $\mathbf{S}$. This is known as a *similarity transformation* (and we say that the matrices $\mathbf{A}$ and $\mathbf{A}'$ are *similar*).

You may ask yourself: since $\mathbf{S}$ does not (necessarily) manage to diagonalize our starting matrix $\mathbf{A}$, then why are we bothering with this similarity transformation? To grasp the answer, let us start with the matrix eigenvalue problem from Eq. (4.124), $\mathbf{A}\mathbf{v}_i = \lambda_i\mathbf{v}_i$. Now, plug in the expression for $\mathbf{A}$ in terms of $\mathbf{A}'$ that results from Eq. (4.161):

$$\mathbf{S}\mathbf{A}'\mathbf{S}^{-1}\mathbf{v}_i = \lambda_i\mathbf{v}_i \tag{4.162}$$

If you multiply on the left by $\mathbf{S}^{-1}$ you get:

$$\mathbf{A}'\mathbf{S}^{-1}\mathbf{v}_i = \lambda_i\mathbf{S}^{-1}\mathbf{v}_i \tag{4.163}$$

At this point, you can define:

$$\mathbf{v}'_i \equiv \mathbf{S}^{-1}\mathbf{v}_i \tag{4.164}$$

thereby getting:

$$\mathbf{A}'\mathbf{v}'_i = \lambda_i\mathbf{v}'_i \tag{4.165}$$

This is a quite remarkable result: we have found that our two *similar matrices have the same eigenvalues*! This correspondence does not hold for the eigenvectors, though: as Eq. (4.164) shows us, $\mathbf{v}'_i \equiv \mathbf{S}^{-1}\mathbf{v}_i$: the eigenvectors of $\mathbf{A}$ and $\mathbf{A}'$ are related to each other but not identical. Problem 4.29 asks you to show that $\mathbf{A}$ and $\mathbf{A}'$ also have the same determinant.

One more definition: if the matrix $\mathbf{S}$ is unitary, then we say that the matrices $\mathbf{A}$ and $\mathbf{A}'$ are *unitarily similar*.[41] Since in this chapter we are focusing on real matrices, we are more interested in the case where $\mathbf{S}$ is *orthogonal*. In keeping with our earlier convention, let's call such an orthogonal matrix $\mathbf{Q}$. Since for an orthogonal matrix we know that $\mathbf{Q}^{-1} = \mathbf{Q}^T$, we immediately see that our latest similarity transformation takes the form:

$$\mathbf{A}' = \mathbf{Q}^T\mathbf{A}\mathbf{Q} \tag{4.166}$$

Such a transformation is especially attractive, since it is trivial to come up with the inverse of an orthogonal matrix (i.e., one doesn't have to solve a linear system of equations n times in order to compute the inverse).

Let's now see how this all ties in to the theme of this section: we have introduced similarity transformations (including the just-mentioned orthogonal similarity transformation)

[41] When $\mathbf{A}^\dagger\mathbf{A} = \mathbf{A}\mathbf{A}^\dagger$ (i.e., $\mathbf{A}$ is a *normal matrix*), $\mathbf{A}'$ will be diagonal: we say that $\mathbf{A}$ is unitarily diagonalizable.

as a way of transforming a matrix (*without* necessarily diagonalizing it) while still preserving the same eigenvalues. This is worth doing because, if we pick $\mathbf{Q}$ in such a way as to make $\mathbf{A}'$ *triangular*,[42] then we can simply read off the eigenvalues of $\mathbf{A}'$ (which are also the eigenvalues of $\mathbf{A}$) from the diagonal. This *Schur decomposition* greatly simplifies the task of evaluating eigenvalues for our starting matrix $\mathbf{A}$. Of course, at this stage we haven't explained *how* to find this matrix $\mathbf{Q}$;[43] we have simply claimed that it is possible to do so.

At this point, an attentive reader may ask: since a similarity transformation is nothing other than a determinant (and eigenvalue) preserving transformation, why don't we simply use the LU decomposition instead of going through all this trouble with orthogonal matrices? After all, as we saw in Eq. (4.102), an LU decomposition allows us to straightforwardly evaluate the determinant of a matrix. Hadn't we said that triangular matrices have their eigenvalues on their diagonal (which is consistent with the fact that the product of the diagonal elements of $\mathbf{U}$ gives the determinant)? Unfortunately, this argument doesn't work. While we can, indeed, read off the eigenvalues of $\mathbf{U}$ from the diagonal, these are *not* the same as the eigenvalues of $\mathbf{A}$: the LU decomposition is constructed by saying $\mathbf{A} = \mathbf{LU}$ and this is *not* a similarity transformation (which would have preserved the eigenvalues).

4.4.3.4 Simultaneous Iteration: First Attempt

The method of *simultaneous iteration* is a generalization of the power method to more than one eigenvectors. This new approach *prima facie* doesn't have too much to do with similarity transformations. Near the end of this section, however, we will use some of the entities introduced earlier to carry out precisely such a similarity transformation.

As usual, we assume that our eigenvalues are distinct (so we can also take it for granted that they've been sorted). You may recall from section 4.4.1 that the power method starts from an *ad hoc* guess and then improves it. More specifically, in Eq. (4.129) we started from a vector $\mathbf{z}^{(0)}$ and then multiplied by the matrix $\mathbf{A}$ repeatedly, leading to:

$$\mathbf{z}^{(k)} = \mathbf{A}^k \mathbf{z}^{(0)}, \quad k = 1, 2, \ldots \tag{4.167}$$

which was the justification for the name "power method". We then proceeded to show the connection with calculating eigenvalues, by expressing our starting vector $\mathbf{z}^{(0)}$ as a linear combination of the (unknown) eigenvectors:

$$\mathbf{z}^{(0)} = \sum_{i=0}^{n-1} c_i \mathbf{v}_i \tag{4.168}$$

Combining the last two equations led to:

$$\mathbf{z}^{(k)} = \sum_{i=0}^{n-1} c_i \lambda_i^k \mathbf{v}_i \tag{4.169}$$

where we could then proceed to single out the eigenvalue we were interested in and have all other ratios decay—see Eq. (4.132).

42 Actually, *quasi-triangular*, i.e., block triangular (in problem 4.37 we use the term "almost" triangular).
43 What certainly *doesn't* work is to use the $\mathbf{Q}$ coming out of $\mathbf{A} = \mathbf{QR}$ (try this).

We now sketch the most straightforward generalization of the above approach to more eigenvectors. You could, in principle, address the case of 2, 3, ... eigenvectors; instead, we will attempt to apply the power method to n initial vectors in order to extract all n eigenvectors of the matrix $\mathbf{A}$. Since you need n starting vectors, your initial guess can be expressed as an $n \times n$ matrix with these guess vectors as columns:

$$\mathbf{Z}^{(0)} = \begin{pmatrix} \mathbf{z}_0^{(0)} & \mathbf{z}_1^{(0)} & \mathbf{z}_2^{(0)} & \dots & \mathbf{z}_{n-1}^{(0)} \end{pmatrix} \tag{4.170}$$

This is a direct generalization of the one-eigenvector case ($\mathbf{Z}^{(0)}$ instead of $\mathbf{z}^{(0)}$). As always, the superscript in parentheses tells us which iteration we're dealing with and the subscript corresponds to the column index. We want our starting guess to be made up of n linearly independent vectors: one way to accomplish this is to take $\mathbf{Z}^{(0)}$ to be the $n \times n$ identity matrix and use its columns one at a time, as we saw in Eq. (4.100):

$$\mathcal{I} = \begin{pmatrix} \mathbf{e}_0 & \mathbf{e}_1 & \dots & \mathbf{e}_{n-1} \end{pmatrix} \tag{4.171}$$

In complete analogy to the one-eigenvector case, Eq. (4.129), our tentative prescription for the simultaneous iteration algorithm is to get the next matrix in the sequence simply by multiplying by our matrix $\mathbf{A}$:

$$\mathbf{Z}^{(k)} = \mathbf{A}\mathbf{Z}^{(k-1)}, \quad k = 1, 2, \dots \tag{4.172}$$

After k applications of the matrix $\mathbf{A}$ we will have:

$$\mathbf{Z}^{(k)} = \mathbf{A}^k\mathbf{Z}^{(0)} = \begin{pmatrix} \mathbf{z}_0^{(k)} & \mathbf{z}_1^{(k)} & \mathbf{z}_2^{(k)} & \dots & \mathbf{z}_{n-1}^{(k)} \end{pmatrix} \tag{4.173}$$

Again, to see what this has to do with calculating eigenvalues, we first expand the initial guess column $\mathbf{z}_j^{(0)}$ in terms of the actual eigenvectors:

$$\mathbf{z}_j^{(0)} = \sum_{i=0}^{n-1} c_{ij}\mathbf{v}_i \tag{4.174}$$

where we observe that our expansion coefficients c_{ij} now have two indices, one for the dummy summation variable and one to keep track of which initial column vector we are referring to.[44] If we then act with the $\mathbf{A}$ matrix k times we correspondingly find:

$$\mathbf{z}_j^{(k)} = \sum_{i=0}^{n-1} c_{ij}\lambda_i^k\mathbf{v}_i \tag{4.175}$$

One might hope to pull out of each sum the term we're interested in each time and watch the other terms decay away. At this stage, you may recall from section 4.4.1 that the unnormalized power method suffered from a problem: the eigenvalues raised to the k-th power become unbounded or tend to zero. However, for not-too-large values of k this is not a huge problem: you can simply iterate, say, 5 or 50 times and then at the end of this process scale each column of $\mathbf{Z}^{(k)}$ with its own norm. Assuming your eigenvalues are not too huge or tiny, this should be enough to keep things manageable.

[44] We are assuming that these c_{ij} put together, as well as their leading principal minors, are non-singular.

Problem 4.33 asks you to implement both scenarios: (a) fully unnormalized simultaneous iteration and (b) unnormalized simultaneous iteration that is scaled at the end with the norm of each column. As you will discover there, your woes are not limited to the problem of unboundedness (or vanishing values). Unfortunately, as k increases (even if you normalize at the end) the vectors $\mathbf{z}_j^{(k)}$ all converge to the same eigenvector of $\mathbf{A}$, namely the dominant eigenvector, $\mathbf{v}_0$. This is disappointing: we generalized the power method to the case of n eigenvectors, only to end up with n copies of the same, dominant, eigenvector. Do not despair.

4.4.3.5 Simultaneous Iteration: Orthonormalizing

Upon closer inspection, we realize what's going on: since we are now dealing with more than one eigenvector, normalizing columns is not enough: what we need to do, instead, is to ensure that the dependence of one column on any of the other columns is projected out. That is, in addition to normalizing, we also need to *orthogonalize*. But that's precisely what the Gram–Schmidt orthogonalization prescription does! Thus, we can ensure that we are dealing with orthonormal vectors by carrying out a QR decomposition *at each step*.

Given the above motivation, we'll proceed to give the prescription for the simultaneous iteration method (also known as *orthogonal iteration*) and later explore some of its fascinating properties. This prescription is a direct generalization of the power method of Eq. (4.136). We carry out the following steps for $k = 1, 2, \dots$:

$$\begin{aligned} \mathbf{Z}^{(k)} &= \mathbf{A}\mathbb{Q}^{(k-1)} \\ \mathbf{Z}^{(k)} &= \mathbb{Q}^{(k)}\mathbf{R}^{(k)} \end{aligned} \tag{4.176}$$

where there's no third line containing a Rayleigh quotient, since at this stage we're only interested in getting the eigenvectors. The way to read this prescription is as follows: start with $\mathbb{Q}^{(0)} = \mathcal{I}$ (as discussed above), then multiply by $\mathbf{A}$ to get $\mathbf{Z}^{(1)}$, then QR-decompose $\mathbf{Z}^{(1)}$ to get the orthonormal set of columns $\mathbb{Q}^{(1)}$, at which point you multiply by $\mathbf{A}$ to get $\mathbf{Z}^{(2)}$, and so on and so forth. In problem 4.33, you will implement this algorithm. It's really nice to see how similar the code you will develop is to the `power()` function that we developed earlier (of course, this has a lot to do with how user-friendly the Python and NumPy syntax is). This is a result of generalizing Eq. (4.136) to deal with matrices instead of vectors (and, correspondingly, generalizing the normalization to an orthonormalization).

You may have noticed that, when QR-decomposing $\mathbf{Z}^{(k)}$, we made use of different-looking symbols for the Q and for the R. The reasons for this will become clearer in the following section but, for now, note that our strange-looking $\mathbb{Q}^{(k)}$ is simply the orthogonal matrix that comes out of the QR decomposition for $\mathbf{Z}^{(k)}$. Since the notation appears a bit lopsided ($\mathbb{Q}^{(k)}$ is multiplied by $\mathbf{R}^{(k)}$, not with $\mathbb{R}^{(k)}$) you may be happy to hear that we are now ready to *define* such an $\mathbb{R}^{(k)}$ matrix, as follows:

$$\mathbb{R}^{(k)} \equiv \mathbf{R}^{(k)}\mathbf{R}^{(k-1)}\dots\mathbf{R}^{(2)}\mathbf{R}^{(1)} \tag{4.177}$$

In words, $\mathbb{R}^{(k)}$ is simply the product (in reverse order) of all the $\mathbf{R}^{(i)}$ matrices.[45]

In problem 4.2 you are asked to show that the product of two upper-triangular matrices is another upper-triangular matrix. From this it follows that $\mathbb{R}^{(k)}$ is an upper-triangular matrix. This result makes the following expression look very impressive:

$$\mathbf{A}^k = \mathbf{Q}^{(k)}\mathbb{R}^{(k)} \tag{4.178}$$

Just to be clear, we haven't shown that this significant relation is actually true (but we soon will). In words, this is saying that the $\mathbf{Q}^{(k)}$ that appeared in the last step of the simultaneous-iteration prescription, multiplied by *all* the $\mathbf{R}^{(i)}$ matrices that have made their appearance up to that point (a product which is equal to $\mathbb{R}^{(k)}$), gives us a QR decomposition of the k-th power of the matrix $\mathbf{A}$.[46]

Let us now prove Eq. (4.178) by induction. The base case $k = 0$ is straightforward: $\mathbf{A}^0 = \mathcal{I}$, this being consistent with the fact that $\mathbb{R}^{(0)} = \mathcal{I}$ as per our earlier definition, as well as $\mathbf{Q}^{(0)} = \mathcal{I}$ which was our assumption for the starting point. Now, we assume that the relation holds for the $k-1$ case:

$$\mathbf{A}^{k-1} = \mathbf{Q}^{(k-1)}\mathbb{R}^{(k-1)} \tag{4.179}$$

and we will show that it will also hold for the k case in Eq. (4.178). We have:

$$\mathbf{A}^k = \mathbf{A}\mathbf{A}^{k-1} = \mathbf{A}\mathbf{Q}^{(k-1)}\mathbb{R}^{(k-1)} = \mathbf{Z}^{(k)}\mathbb{R}^{(k-1)} = \mathbf{Q}^{(k)}\mathbf{R}^{(k)}\mathbb{R}^{(k-1)} = \mathbf{Q}^{(k)}\mathbb{R}^{(k)} \tag{4.180}$$

In the first equality we simply separated out one of the $\mathbf{A}$ matrices. In the second equality we used our $k-1$ step from Eq. (4.179). In the third equality we used our expression from the first line in Eq. (4.176). In the fourth equality we used our expression from the second line in Eq. (4.176). In the fifth equality we grouped together the two R terms, as per the definition in Eq. (4.177). We have reached the desired conclusion that Eq. (4.178) is true.

Qualitatively, simultaneous-iteration entities like $\mathbf{Q}^{(k)}$ and $\mathbb{R}^{(k)}$ allow us to QR-decompose successive powers $\mathbf{A}^k$, i.e., they allow us to construct orthonormal bases for the k-th power of $\mathbf{A}$. For a symmetric $\mathbf{A}$, this amounts to having computed the eigenvectors (which are orthonormal). As you will show in problem 4.32, for a non-symmetric $\mathbf{A}$ (for which the true eigenvectors are linearly independent, but not orthonormal), $\mathbf{Q}^{(k)}$ converges toward the orthogonal "factor " of the eigenvector matrix from Eq. (4.127), i.e., toward $\tilde{\mathbf{Q}}$ in $\mathbf{V} = \tilde{\mathbf{Q}}\mathbf{U}$ (where $\mathbf{U}$ is an upper-triangular matrix and $\tilde{\mathbf{Q}}$ is orthogonal). In either case, $\mathbf{Q}^{(k)}$ is related to the eigenvectors of the matrix $\mathbf{A}$.

Next, we try to extract the eigenvalues. For the power method we saw the Rayleigh quotient for a normalized vector in Eq. (4.135):

$$\mu(\mathbf{q}^{(k)}) = [\mathbf{q}^{(k)}]^T\mathbf{A}\mathbf{q}^{(k)} \tag{4.181}$$

A straightforward generalization to the n-dimensional problem is to define:

$$\mathbf{A}^{(k)} \equiv [\mathbf{Q}^{(k)}]^T\mathbf{A}\mathbf{Q}^{(k)} \tag{4.182}$$

where we introduced (yet another piece of) new notation on the left-hand side. This is our

[45] Since there's no such thing as a $\mathbf{Z}^{(0)}$, there's no such thing as an $\mathbf{R}^{(0)}$. If you wanted to generalize the definition in Eq. (4.177) to the case of $k = 0$, then $\mathbb{R}^{(0)} = \mathcal{I}$ would be the most natural assumption.

[46] It's extremely important at this point to distinguish between exponents like $\mathbf{A}^k$ and iteration counts like $\mathbf{Q}^{(k)}$.

first *orthogonal similarity transformation* in the flesh! We will now show that $\mathbf{A}^{(k)}$ (which is always similar to the matrix $\mathbf{A}$, meaning it has the same eigenvalues) converges to an *upper-triangular* matrix for the case of general $\mathbf{A}$:

$$\mathbf{A}^{(k)} = \tilde{\mathbf{Q}}^T\mathbf{A}\tilde{\mathbf{Q}} = \tilde{\mathbf{Q}}^T\mathbf{A}\mathbf{V}\mathbf{U}^{-1} = \tilde{\mathbf{Q}}^T\mathbf{V}\mathbf{\Lambda}\mathbf{U}^{-1} = \tilde{\mathbf{Q}}^T\tilde{\mathbf{Q}}\mathbf{U}\mathbf{\Lambda}\mathbf{U}^{-1} = \mathbf{U}\mathbf{\Lambda}\mathbf{U}^{-1} \tag{4.183}$$

In the first equality we used the fact that $\mathbf{Q}^{(k)}$ converges toward $\tilde{\mathbf{Q}}$. In the second equality we used the earlier relation, $\mathbf{V} = \tilde{\mathbf{Q}}\mathbf{U}$, after solving for $\tilde{\mathbf{Q}}$. In the third equality we used Eq. (4.125), multiplied by $\mathbf{V}$ on the left. In the fourth equality we used $\mathbf{V} = \tilde{\mathbf{Q}}\mathbf{U}$ again. In the fifth equality we used the fact that $\tilde{\mathbf{Q}}$ is an orthogonal (i.e., orthonormal) matrix. Our final result is the product of three upper-triangular matrices so it, too, is an upper-triangular matrix (note that the inverse of an upper-triangular matrix is also upper-triangular). This derivation (of the Schur decomposition) was carried out for the general case of a non-symmetric matrix $\mathbf{A}$; for the (simpler) case of symmetric matrices, the $\mathbf{A}^{(k)}$ converges to a *diagonal* form. In either case, the eigenvalues can be simply read off the diagonal of $\mathbf{A}^{(k)}$.

To summarize, we started from the prescription in Eq. (4.176), which generalized the power method, and ended up with the following two important equations:

$$\begin{aligned} \mathbf{A}^k &= \mathbf{Q}^{(k)}\mathbf{R}^{(k)} \\ \mathbf{A}^{(k)} &= [\mathbf{Q}^{(k)}]^T\mathbf{A}\mathbf{Q}^{(k)} \end{aligned} \tag{4.184}$$

Here we are merely collecting our earlier results for ease of reference. The first equation is Eq. (4.178) which we proved by induction above: this is related to the eigenvectors, via $\mathbf{Q}^{(k)}$. The second equation is Eq. (4.182) which we essentially took as a definition (motivated by what it means to be a Rayleigh quotient): this is what allows us to compute eigenvalues, as the elements on the diagonal of $\mathbf{A}^{(k)}$.

Note that throughout this section we've been carrying out this prescription for $k = 1, 2, \ldots$: in other words, we didn't really mention a termination criterion. As in the single-eigenvector power method, we could either run this for a fixed number of iterations, or try to get fancier and check how much, say, $\mathbf{A}^{(k)}$ changes from one iteration to the next. When solving problem 4.33 you should opt for the, simpler, first option; once that is working, you may choose to go back and make things more robust.

4.4.3.6 QR Method: Algorithm

We are now in a position to tackle the full QR method, which is an ingenious eigenvalue-computing approach. Describing this algorithm is very simple: so simple, as a matter of fact, that stating it without proof gives the impression that there is something magical going on. As you will soon see, the QR method can be implemented very easily. Having introduced the simultaneous-iteration method above (which you are asked to implement in problem 4.33), the QR method itself will (it is hoped) appear much more transparent. As a matter of fact, we will show that it is fully equivalent to simultaneous iteration, but simply goes about calculating the relevant entities with different priorities.

Let's first go over the QR method's astoundingly simple prescription and then proceed to show the equivalence with the simultaneous-iteration approach. We start with the initial condition $\mathbf{A}^{(0)} = \mathbf{A}$ and then carry out the following steps for $k = 1, 2, \ldots$:

$$\begin{aligned} \mathbf{A}^{(k-1)} &= \mathbf{Q}^{(k)}\mathbf{R}^{(k)} \\ \mathbf{A}^{(k)} &= \mathbf{R}^{(k)}\mathbf{Q}^{(k)} \end{aligned} \tag{4.185}$$

In essence, we start from a matrix $\mathbf{A}^{(k-1)}$ and try to produce a new matrix $\mathbf{A}^{(k)}$. More specifically, we start with $\mathbf{A}^{(0)} = \mathbf{A}$ (as already mentioned), then QR decompose this matrix to produce $\mathbf{Q}^{(1)}$ and $\mathbf{R}^{(1)}$. (Note that these two symbols look alike: we are using $\mathbf{Q}$, not $\mathbb{Q}$.) Here comes the crucial part: in order to produce the next matrix in our sequence, $\mathbf{A}^{(1)}$, we take $\mathbf{Q}^{(1)}$ and $\mathbf{R}^{(1)}$ and *multiply them in reverse order*! Now that we have $\mathbf{A}^{(1)}$ we QR-decompose that and then multiply in reverse order again, and so on and so forth.

Qualitatively, our new algorithm is QR-decomposing and then multiplying the resulting matrices in reverse order (again and again). To show that this is equivalent to simultaneous iteration, we will prove that the QR method leads to precisely the same two equations as did simultaneous iteration, namely Eq. (4.184) repeated here for your convenience:

$$\begin{aligned} \mathbf{A}^{k} &= \mathbb{Q}^{(k)}\mathbb{R}^{(k)} \\ \mathbf{A}^{(k)} &= [\mathbb{Q}^{(k)}]^{T}\mathbf{A}\mathbb{Q}^{(k)} \end{aligned} \tag{4.186}$$

Our goal now is to show that these two relations hold: once we've shown this, the first of these two equations will allow us to compute (the orthogonal factor of the) eigenvectors and the second to compute eigenvalues. Before we launch into the formal proofs, we realize that there's something missing: Eq. (4.185) involves $\mathbf{Q}^{(k)}$ and $\mathbf{R}^{(k)}$ whereas Eq. (4.186) makes use of $\mathbb{Q}^{(k)}$ and $\mathbb{R}^{(k)}$. Defining the latter is not too difficult: we simply assume that our earlier definition from Eq. (4.177) will carry over:

$$\mathbb{R}^{(k)} \equiv \mathbf{R}^{(k)}\mathbf{R}^{(k-1)}\ldots\mathbf{R}^{(2)}\mathbf{R}^{(1)} \tag{4.187}$$

Motivated by this, we correspondingly define $\mathbb{Q}^{(k)}$ as follows:

$$\mathbb{Q}^{(k)} \equiv \mathbf{Q}^{(1)}\mathbf{Q}^{(2)}\ldots\mathbf{Q}^{(k-1)}\mathbf{Q}^{(k)} \tag{4.188}$$

Crucially, these $\mathbf{Q}$ matrices are multiplied in the opposite order. Problem 4.2 asks you to show that the product of two orthogonal matrices is an orthogonal matrix: once you've shown this, it is easy to see that $\mathbb{Q}^{(k)}$ is orthogonal.[47]

Let us now prove the two relations in Eq. (4.186). We begin from the second equation, which we didn't really have to prove for the case of simultaneous iteration, since there

[47] You were probably suspecting that this would be the case since, in the previous subsection on simultaneous iteration, $\mathbb{Q}^{(k)}$ came out of the QR decomposition of $\mathbf{A}^{k}$, as per the first relation in Eq. (4.184) – this argument doesn't quite work, though, since we haven't yet shown this equation to be true for the present case.

we took it to be the result of a definition (and proceeded to show that $\mathbf{A}^{(k)}$ becomes upper triangular). The base case $k = 0$ is:

$$\mathbf{A}^{(0)} = [\mathbf{Q}^{(0)}]^T \mathbf{A} \mathbf{Q}^{(0)} = \mathbf{A} \tag{4.189}$$

Since our definition of $\mathbf{Q}^{(k)}$ in Eq. (4.188) involves $k = 1$ and up, it makes sense to assume that $\mathbf{Q}^{(0)}$ is the identity: this is what leads to $\mathbf{A}^{(0)} = \mathbf{A}$, which we know is true since we took it to be our starting point when introducing the QR method in Eq. (4.185). We now assume that the relation we are trying to prove holds for the $k - 1$ case:

$$\mathbf{A}^{(k-1)} = [\mathbf{Q}^{(k-1)}]^T \mathbf{A} \mathbf{Q}^{(k-1)} \tag{4.190}$$

and will try to show that it will also hold for the k case. Before we do that, we need a bit more scaffolding. Turn to the first relation in Eq. (4.185) and multiply on the left by $[\mathbf{Q}^{(k)}]^T$. Since $\mathbf{Q}^{(k)}$ is orthogonal, this gives us:

$$\mathbf{R}^{(k)} = [\mathbf{Q}^{(k)}]^T \mathbf{A}^{(k-1)} \tag{4.191}$$

If you're wondering how we know that $\mathbf{Q}^{(k)}$ is orthogonal, remember that it is what comes out of the QR decomposition of the matrix $\mathbf{A}^{(k-1)}$. Now, combining our latest result with the second relation in Eq. (4.185) we find:

$$\mathbf{A}^{(k)} = [\mathbf{Q}^{(k)}]^T \mathbf{A}^{(k-1)} \mathbf{Q}^{(k)} \tag{4.192}$$

We can plug in our $k - 1$ hypothesis from Eq. (4.190) for $\mathbf{A}^{(k-1)}$ to find:

$$\mathbf{A}^{(k)} = [\mathbf{Q}^{(k)}]^T [\mathbf{Q}^{(k-1)}]^T \mathbf{A} \mathbf{Q}^{(k-1)} \mathbf{Q}^{(k)} \tag{4.193}$$

At this point we marvel at how appropriate our definition of $\mathbf{Q}^{(k)}$ in Eq. (4.188) was.[48] We are hence able to group the extra term on each side, giving:

$$\mathbf{A}^{(k)} = [\mathbf{Q}^{(k)}]^T \mathbf{A} \mathbf{Q}^{(k)} \tag{4.194}$$

which is what we had set out to prove.

We turn to the first equation in Eq. (4.186), which we will also prove by induction. As in the previous section, the base case $k = 0$ is straightforward: $\mathbf{A}^0 = \mathcal{I}$, this being consistent with the facts that $\mathbf{R}^{(0)} = \mathcal{I}$ and $\mathbf{Q}^{(0)} = \mathcal{I}$, which are fully analogous to each other. Now, we assume that the relation holds for the $k - 1$ case:

$$\mathbf{A}^{k-1} = \mathbf{Q}^{(k-1)} \mathbf{R}^{(k-1)} \tag{4.195}$$

and will try to show that this means it will also hold for the k case. Our derivation will be somewhat similar to that in Eq. (4.180), but the intermediate steps will only make use of relations corresponding to the QR method (not simultaneous iteration). We have:

$$\mathbf{A}^k = \mathbf{A}\mathbf{A}^{k-1} = \mathbf{A}\mathbf{Q}^{(k-1)}\mathbf{R}^{(k-1)} = \mathbf{Q}^{(k-1)}\mathbf{A}^{(k-1)}\mathbf{R}^{(k-1)} = \mathbf{Q}^{(k-1)}\mathbf{Q}^{(k)}\mathbf{R}^{(k)}\mathbf{R}^{(k-1)} = \mathbf{Q}^{(k)}\mathbf{R}^{(k)} \tag{4.196}$$

In the first equality we separated out one of the $\mathbf{A}$ matrices. In the second equality we used our $k - 1$ step from Eq. (4.195). In the third equality we used our $k - 1$ step from

[48] Recall that when taking the transpose of a product of matrices, you get the product of the transpose of each matrix in reverse order.

Eq. (4.190): this is no longer a hypothesis (since we've shown Eq. (4.194) to be true, we know it will also hold for the case of $k-1$). We first multiplied Eq. (4.190) by $\mathbf{Q}^{(k-1)}$ on the left. In the fourth equality we used the first relation in Eq. (4.185) to eliminate $\mathbf{A}^{(k-1)}$. In the fifth equality we once again marvelled at the appropriateness of our definitions of $\mathbf{Q}^{(k)}$ in Eq. (4.188) and of $\mathbf{R}^{(k)}$ in Eq. (4.187). We have thus proved what we had set out to.

Summarizing where things now stand, we see that if we start with the QR-method prescription of Eq. (4.185), then we can show that Eq. (4.186) holds. But those are precisely the same relations that were true for the case of simultaneous iteration. If this abstract conclusion does not satisfy you, then simply step through the simultaneous iteration and QR-method prescriptions: with the initial conditions that we chose to employ, the two methods give identical intermediate (step-by-step) and final results. In other words, they are the *same* method. This raises the natural question: why bother using the QR method, which required an analogy with the simultaneous-iteration method in order to be interpreted? Could we not just use the simultaneous-iteration method directly?

Even though the two methods are identical, simultaneous iteration spends its time dealing with the eigenvectors: $\mathbf{Q}^{(k)}$ appears in Eq. (4.176) organically, and this is what gives us the orthogonal factor of the eigenvectors, $\tilde{\mathbf{Q}}$. You can also go through the trouble of defining $\mathbf{A}^{(k)}$ in the simultaneous-iteration approach, as per Eq. (4.182), but that requires two matrix multiplications, and is not even necessary to keep going: in its bare-bones formulation, the simultaneous-iteration method keeps producing new $\mathbf{Z}^{(k)}$ and $\mathbf{Q}^{(k)}$. On the other hand, the QR method in its basic formulation of Eq. (4.185) keeps evaluating $\mathbf{A}^{(k)}$ which, since we now know Eq. (4.186) to be true, gives you the eigenvalues of the original matrix $\mathbf{A}$ in the diagonal. You can also go through the trouble of defining $\mathbf{Q}^{(k)}$ as per Eq. (4.188), but this is usually much more costly than you would like (even so, problem 4.35 asks you to implement this approach by modifying our code). Any hint of magic in the QR method's workings should have (automagically) vanished by now: the second relation in Eq. (4.186), $\mathbf{A}^{(k)} = [\mathbf{Q}^{(k)}]^T\mathbf{A}\mathbf{Q}^{(k)}$, clearly shows that the second relation in Eq. (4.185), $\mathbf{A}^{(k)} = \mathbf{R}^{(k)}\mathbf{Q}^{(k)}$, is simply carrying out an orthogonal similarity transformation/Schur decomposition.

Before we turn to the Python code, we emphasize that the QR method as shown here is quite inefficient: in production, eigenvalue shifting (employed earlier in the context of the inverse-power method) is typically combined with the QR-method trick. There are several other upgrades one can carry out, some of which are explored in the problem set.

4.4.3.7 QR Method: Implementation

After these somewhat lengthy derivations, we turn to a near-trivial implementation of the QR method. This will make use of the QR-decomposition function `qrdec()` we introduced above. In essence, as per Eq. (4.185), all code 4.9 is doing is a QR decomposition followed by multiplying the resulting matrices in reverse order. As was to be expected, we start from $\mathbf{A}^{(0)} = \mathbf{A}$ and then keep decomposing and multiplying over and over again. This is done for a fixed number of times (passed in as an argument) for simplicity. Once we're done iterating, we store and return the elements on the diagonal of $\mathbf{A}^{(k)}$. Note that, since we're not implementing a fancy termination criterion, we didn't need to keep track of the iteration count: instead of $\mathbf{A}^{(k-1)}$, $\mathbf{Q}^{(k)}$, $\mathbf{R}^{(k)}$, and $\mathbf{A}^{(k)}$, this code uses simply `A`, `Q`, and `R`.

Code 4.9 qrmet.py

```
from triang import testcreate
from qrdec import qrdec
import numpy as np

def qrmet(inA,kmax=100):
    A = np.copy(inA)
    for k in range(1,kmax):
        Q, R = qrdec(A)
        A = R@Q
        print(k, np.diag(A))

    qreigvals = np.diag(A)
    return qreigvals

if __name__ == '__main__':
    A, _ = testcreate(4,21)
    qreigvals = qrmet(A,6)
    print(" ")
    npeigvals, npeigvecs = np.linalg.eig(A); print(npeigvals)
```

In the current version of the code, when we call `qrmet()` we print out the step-by-step values of our Rayleigh quotients: as the iteration count increases, these diagonal elements converge to the eigenvalues of $\mathbf{A}$. The main program also outputs the result from `numpy.linalg.eig()`. Running this code, we see that each eigenvalue rapidly converges to a fixed value. As advertised, the QR method has allowed us to evaluate *all* eigenvalues at one go. You should spend some time playing with this code. For example, you should check to see that as the iteration count increases, $\mathbf{A}^{(k)}$ becomes upper triangular (also check that for a symmetric $\mathbf{A}$ we get a diagonal $\mathbf{A}^{(k)}$). As always, you should comment out the call to `print()` inside `qrmet()` once you are ready to use the code for other purposes. (We will assume this below.)

4.4.3.8 QR Method: Operation Count

To compute the operation count required for the full QR method, we need to first count the operation cost of the QR decomposition prescription and then of the QR method itself.

Problem 4.34 asks you to evaluate the cost of the QR decomposition in detail. Here we will look at the leading contribution. Focus on the innermost loop (which contains the operations carried out the largest number of times). We see from Eq. (4.155) that we need to evaluate the inner product $\mathbf{q}_i^T\mathbf{a}_j$ (also known as R_{ij}). Since each of $\mathbf{q}_i$ and $\mathbf{a}_i$ is an $n \times 1$

dimensional vector, we see that this inner product requires n multiplications and $n - 1$ additions. Still in the innermost loop, $\mathbf{a}_j - \sum_{i=0}^{j-1}(\mathbf{q}_i^T \mathbf{a}_j)\mathbf{q}_i$ (also known as $\mathbf{a}_j - \sum_{i=0}^{j-1} R_{ij}\mathbf{q}_i$) for a fixed j and for a given i contribution to the sum requires n multiplications and n subtractions (one for each of the components of our column vectors). Putting all the operations together, we find $4n - 1$ or $\sim 4n$. Recalling that this was for fixed i and j, we explicitly sum over all possible values of these indices, as given in Eq. (4.159):

$$\sum_{j=0}^{n-1}\sum_{i=0}^{j-1} 4n = \sum_{j=0}^{n-1} 4nj = 4n\frac{(n-1)n}{2} \sim 2n^3 \tag{4.197}$$

In the penultimate step we used Eq. (4.63) and in the last step we kept only the leading term. Thus, we have found that a QR decomposition requires $\sim 2n^3$ floating-point operations. This is three times larger than the cost of an LU decomposition.

As we saw in Eq. (4.185), in addition to a QR decomposition, the QR method carries out a matrix multiplication at each iteration. Since each matrix multiplication costs $\sim 2n^3$ (as you showed in problem 4.34), assuming we need m QR-method iterations, the total operation count will be $4mn^3$ (half of the cost coming from the QR decomposition and half from the matrix multiplication). Now this is where things get tricky: what is the size of m? This depends on how well- (or ill-) separated the eigenvalues are, but generally speaking it is not too far off to assume that m will be of the same order of magnitude as n. In that case, our QR method is looking like it has a scaling of $O(n^4)$: for most practical purposes, this is not good enough. Fortunately, some basic preprocessing of our starting matrix $\mathbf{A}$ (bringing it into a so-called Hessenberg form) improves things dramatically: this reduces the amount of work needed per iteration down from $O(n^3)$ to $O(n^2)$. If you need to implement such an approach, this should be enough to get your literature search going.[49]

4.4.4 All Eigenvalues and Eigenvectors

We are now at a pretty good place: in `qrmet()` we have a function that computes all the eigenvalues of a matrix. If we're interested in building our very own (bare-bones) general-purpose library to compute all eigenvalue and eigenvector pairs (similarly to what `numpy.linalg.eig()` does), we are halfway there. We could modify our code above to also evaluate $\mathbf{Q}^{(k)}$; for a non-symmetric matrix, this would give us only the orthogonal factor $\tilde{\mathbf{Q}}$ (see, however, problem 4.39). Wishing to find the actual eigenvectors, with the QR-method results for the eigenvalues in place, we will now, instead, employ the shifted inverse-power method to extract the eigenvectors. Once you have reasonably good estimates for the eigenvalues, the shifted inverse-power method converges very fast (typically in a couple of iterations).

Code 4.10 is a Python implementation of such an approach. Our `eig()` function returns a 1d NumPy array containing the eigenvalues and a 2d NumPy array containing the eigenvectors in its columns. This function does what we said it would: it calls `qrmet()` to evaluate (a first version of) the eigenvalues and then `invpowershift()` for each eigenvalue

[49] Bringing our matrix into Hessenberg form is actually carried out by employing a sequence of Householder transformations. As mentioned in footnote 40, the relevant concept is introduced in problem 4.28.

Code 4.10 eig.py

```
from triang import testcreate
from invpowershift import invpowershift
from qrmet import qrmet
import numpy as np

def eig(A,eps=1.e-12):
    n = A.shape[0]
    eigvals = np.zeros(n)
    eigvecs = np.zeros((n,n))
    qreigvals = qrmet(A)
    for i, qre in enumerate(qreigvals):
        eigvals[i], eigvecs[:,i] = invpowershift(A,qre+eps)
    return eigvals, eigvecs

def testeigall(f,A):
    eigvals, eigvecs = f(A)
    npeigvals, npeigvecs = np.linalg.eig(A)
    print(eigvals); print(npeigvals)
    print(" ")
    for eigvec, npeigvec in zip(eigvecs.T,npeigvecs.T):
        print(eigvec); print(npeigvec)
        print(" ")

if __name__ == '__main__':
    A, _ = testcreate(4,21)
    testeigall(eig,A)
```

separately, returning an eigenvalue and eigenvector pair at a time. Note that `qrmet()` in its turn calls `qrdec()` and similarly `invpowershift()` calls `ludec()`. Thus, `eig()` is so short only because we have already done the heavy lifting in earlier sections.

Turning to some more detailed features of this program, we observe that our code (in an attempt to be idiomatic) includes the expression `enumerate(qreigvals)`: this is mixing Python with NumPy functionality. Depending on your perspective, you might be impressed by how seamless this is, or might want a NumPy-centric solution (in which case, see page 539 on `numpy.ndenumerate()`). Another detailed feature: we choose *not* to simply pass the specific eigenvalue that we get each time from `enumerate(qreigvals)` into `invpowershift()`: instead, we pass in the QR-method eigenvalue shifted slightly (by adding in `eps`). You can see why this is necessary by setting `eps` to have the value `0`: the inverse-power method misbehaves if your shift is not very close to but identical to the eigenvalue that would have come out as output.[50] Implicit in all this is that our `eig()` function computes the QR-method eigenvalues, then *does not* print them out or return them: instead, they are used as first approximations for the eigenvalues (shifts) passed into the inverse-power method function. The eigenvalues returned were computed by the latter function (and in some cases therefore end up being slightly modified).

We then proceed to introduce another test function, `testeigall()`, which first compares our eigenvalues with those of `numpy.linalg.eig()` and then does the same for each eigenvector in turn. As an aside, while we are in favor of code re-use, many times we simply write a new test function on the spot: this is often better than going through contortions to make an earlier test function work for a more general case (or to make a new very general test function that you then backport to all earlier tests). A nice feature of this test function is that we iterate over columns of NumPy arrays by stepping through the transpose of the matrix. This is what we encountered as `for column in A.T` on page 20. This time we have two matrices that we want to step through, so we use `zip()`. This is another case of mixing Python and NumPy functionality.

Running this code, we see that (for the digits printed) we have near-perfect agreement for all eigenvalues and eigenvectors. (One of the eigenvectors has the opposite sign, but this is totally arbitrary.) As it turns out, the last/smallest eigenvalue we now get is slightly different from what the QR-method had given us. You should spend some time playing with this code for other input matrices, also taking the opportunity to tweak the input parameters in the functions that work together to produce the `eig()` output. As part of that, you may wish to uncomment the `print()` line in `invpowershift()` to explicitly check how many iterations of the inverse-power method were needed to evaluate each eigenvector, given that the QR method had already produced reasonably decent eigenvalues.

4.5 The Singular-Value Decomposition

In this chapter we have encountered: (a) the LU decomposition, which helped us efficiently solve linear systems of equations, (b) the QR decomposition, which helped us orthonormalize (and was used as a step in solving the eigenvalue problem), and (c) the eigendecomposition (arrived at via (d) the Schur decomposition), to which we return on page 199. We now turn to a fifth, equally important, factorization. The *singular-value decomposition* (SVD) is typically introduced in linear-algebra courses as a formal device, with the implementation being relegated to libraries; in the spirit of the rest of the book, here we will provide some qualitative insights and an existence proof, but also produce an implementation of this factorization, which will allow you to experiment with specific examples.

4.5.1 Definition and Significance

The singular-value decomposition amounts to factorizing an *arbitrary* matrix **A** as a product of a two orthogonal matrices and one diagonal matrix. In keeping with our main inter-

[50] It's a worthwhile exercise for you to figure out exactly which line in which program is the culprit.

ests, we here limit ourselves to square matrices (of dimensions $n \times n$), but it is important to note that every single matrix (whether square or not) has an SVD. Let us look at this decomposition and start to interpret each term:

$$\mathbf{A} = \mathbf{U}\boldsymbol{\Sigma}\mathbf{V}^T \tag{4.198}$$

All three of $\mathbf{U}$, $\boldsymbol{\Sigma}$, and $\mathbf{V}$ are $n \times n$ matrices for us (but see problem 4.45 for the general case). Here $\mathbf{U}$ is an orthogonal matrix, i.e., a matrix satisfying $\mathbf{U}^T\mathbf{U} = \mathcal{I}$. The columns $\mathbf{u}_i$ of $\mathbf{U}$ are called the *left-singular vectors*. The next matrix, $\boldsymbol{\Sigma}$, is diagonal and is made up of the σ_i which are the (non-negative) *singular values*; similarly to Eq. (4.128), we will assume that the singular values have been sorted such that:

$$\sigma_0 \geq \sigma_1 \geq \ldots \sigma_{n-1} \geq 0 \tag{4.199}$$

Given this, the singular values are *unique*. The third matrix $\mathbf{V}$ is, like $\mathbf{U}$, orthogonal; its columns $\mathbf{v}_i$ are called the *right-singular vectors*. The two orthogonal matrices involved here are related in a very specific way: multiplying Eq. (4.198) by $\mathbf{V}$ on the right, we find $\mathbf{AV} = \mathbf{U}\boldsymbol{\Sigma}$. Writing this out in terms of its columns, we have:

$$\mathbf{A}\mathbf{v}_i = \sigma_i\mathbf{u}_i \tag{4.200}$$

This looks *almost* like our standard eigenvalue problem from Eq. (4.124) but, of course, here the vectors on the left-hand side and the right-hand side are different.

Let us now begin to interpret the SVD. Qualitatively, this new decomposition allows us to say that any matrix is diagonal, if you pick the appropriate orthonormal bases for its domain and range. To understand what this means, consider our standard linear-system problem from Eq. (4.50):

$$\mathbf{Ax} = \mathbf{b} \tag{4.201}$$

As a reminder, Gaussian elimination (and LU decomposition) focused on triangularizing $\mathbf{A}$, so that the system was then easy to handle. We will now see that the SVD helps us *diagonalize* the same system of equations. Multiply Eq. (4.201) from the left by $\mathbf{U}^T$ and also plug in the SVD from Eq. (4.198):

$$\mathbf{U}^T\mathbf{U}\boldsymbol{\Sigma}\mathbf{V}^T\mathbf{x} = \mathbf{U}^T\mathbf{b} \tag{4.202}$$

We now use the orthogonality of $\mathbf{U}$; also, motivated by the analogous argument near Eq. (4.164), we realize that we can express the domain in the basis of the right-singular vectors of $\mathbf{A}$ (the columns of $\mathbf{V}$) and, similarly, we can express the range in the basis of the left-singular vectors of $\mathbf{A}$ (the columns of $\mathbf{U}$):

$$\mathbf{x}' = \mathbf{V}^T\mathbf{x}, \qquad \mathbf{b}' = \mathbf{U}^T\mathbf{b} \tag{4.203}$$

Plugging these two expansions into Eq. (4.202) leads to:

$$\boldsymbol{\Sigma}\mathbf{x}' = \mathbf{b}' \tag{4.204}$$

which, as advertised, shows that the problem is diagonal when you employ the appropriate orthogonal coordinates.

At this point, you may be thinking that we have already encountered another sort of diagonalization, that associated with solving the eigenproblem. Equation (4.125) can be rewritten as:[51]

$$\mathbf{A} = \mathbf{W}\mathbf{\Lambda}\mathbf{W}^{-1} \tag{4.205}$$

where, crucially, this $\mathbf{W}$ is the eigenvector matrix, whose columns are the right eigenvectors, as per Eq. (4.127). If we express both the domain and the range in the basis of the eigenvectors (cf. the argument around Eq. (4.164)):

$$\mathbf{x}' = \mathbf{W}^{-1}\mathbf{x}, \qquad \mathbf{b}' = \mathbf{W}^{-1}\mathbf{b} \tag{4.206}$$

then $\mathbf{Ax} = \mathbf{b}$ turns into:

$$\mathbf{\Lambda}\mathbf{x}' = \mathbf{b}' \tag{4.207}$$

Thus, it may appear that the SVD is roughly similar to the eigendecomposition. However, it is crucial to keep in mind the following fundamental differences:

- The SVD can be applied to any matrix (even a defective matrix or a non-square one). The eigendecomposition is different in that not even all square matrices can be decomposed this way (i.e., it only works for diagonalizable matrices).
- The SVD uses two bases, $\mathbf{U}$ and $\mathbf{V}$, whereas the eigendecomposition only a single one (denoted by $\mathbf{W}$ above).
- The bases employed by the SVD are orthonormal, while (as discussed on page 189) the eigenvectors of a non-symmetric $\mathbf{A}$ are linearly independent but not orthonormal.
- The singular values σ_i are non-negative (and real), whereas the eigenvalues λ_i can be either positive or negative (or complex).

A further connection between the singular-value decomposition and the eigendecomposition is that the SVD is typically computed by rewriting things in terms of an eigenvalue problem. Since this often leads to misunderstandings, we postpone discussing that aspect.[52]

The SVD has a multitude of applications in linear algebra; some of these are due to the efficient algorithms that can be used in computing it, whereas others have to do with the stability properties it gives rise to. One notable application relates to the connection between singular values and the condition number of a linear system of equations. Another nice example is that the SVD allows you to work with non-invertible matrices; for example, it can help you to solve a homogeneous linear system of equations. A large family of applications involves the concept of *matrix approximation*, i.e., the ability to use one matrix that encapsulates the essential features of another matrix. In problem 6.43 we will come across yet another application of the SVD, namely its use in implementing least-squares fitting of function parameters to data. Several of these applications are touched upon in the present chapter's problem set, together with other aspects like the generalization to rectangular matrices, as well as a sub-optimal way of computing the SVD.

[51] Since $\mathbf{V}$ is being used for a different purpose in the SVD, we now use $\mathbf{W}$ for the eigenvector matrix.
[52] "[I]t is merely the bridge, and one cannot live on a bridge" (Georg Simmel, *Money in Modern Culture*).

4.5.2 Proving that the SVD Exists

Speaking of computing the SVD, it's worth noting that we haven't actually said how this is accomplished; we discuss this in the following subsection. Even so, we will now follow the example of earlier sections in this chapter and prove by induction that the decomposition in Eq. (4.198) is legitimate; in other words, while above we took it for granted that an arbitrary matrix can be decomposed as per Eq. (4.198), we will now explicitly show this. As in most of this book (and above), we focus on the case of a square ($n \times n$) matrix but, crucially, we make *no* assumptions about **A** being symmetric, diagonalizable, etc. The base case is for $n = 1$: instead of matrices, we are faced with scalars, i.e., numbers. It is easy to see that $A = U\Sigma V^T$ holds if we take $U = A/|A|, \Sigma = |A|$, and $V = 1$.[53] Before we assume that the singular-value decomposition of Eq. (4.198) holds for the $(n-1)\times(n-1)$ case and use that to show that it also holds for the $n\times n$ case, we will need to set up some scaffolding.

Most notably, we first introduce the *spectral norm* for matrices (also called the *2-norm*), in terms of the Euclidean norm for vectors:

$$||\mathbf{A}||_2 = \max_{||\mathbf{x}||_E \neq 0} \frac{||\mathbf{A}\mathbf{x}||_E}{||\mathbf{x}||_E} = \max_{||\mathbf{x}||_E = 1} ||\mathbf{A}\mathbf{x}||_E \tag{4.208}$$

where we assume **A** to be an $n \times n$ matrix and **x** an $n \times 1$ vector; we have included two equivalent definitions.[54] We take **v** to be the $n \times 1$ unit vector at which the maximum is attained. Thus, $||\mathbf{v}||_E = 1$ and:

$$||\mathbf{A}||_2 = ||\mathbf{A}\mathbf{v}||_E \equiv \sigma > 0 \tag{4.209}$$

hold. We also pick another unit vector **u** as follows:

$$\mathbf{u} = \frac{\mathbf{A}\mathbf{v}}{||\mathbf{A}\mathbf{v}||_E} \tag{4.210}$$

We now extend **v** into an orthonormal basis: this means we build up the $n \times n$ orthogonal matrix $\mathbf{V} = (\mathbf{v}\ \tilde{\mathbf{V}})$. Here $\tilde{\mathbf{V}}$ is an $n \times (n-1)$ matrix. Similarly, we extend **u** into the $n \times n$ orthogonal matrix $\mathbf{U} = (\mathbf{u}\ \tilde{\mathbf{U}})$, where $\tilde{\mathbf{U}}$ is an $n \times (n-1)$ matrix. We now have:

$$\mathbf{U}^T\mathbf{A}\mathbf{V} = \begin{pmatrix} \mathbf{u}^T \\ \tilde{\mathbf{U}}^T \end{pmatrix} \mathbf{A}\,(\mathbf{v}\ \tilde{\mathbf{V}}) = \begin{pmatrix} \mathbf{u}^T\mathbf{A}\mathbf{v} & \mathbf{u}^T\mathbf{A}\tilde{\mathbf{V}} \\ \tilde{\mathbf{U}}^T\mathbf{A}\mathbf{v} & \tilde{\mathbf{U}}^T\mathbf{A}\tilde{\mathbf{V}} \end{pmatrix} \tag{4.211}$$

For the top-left element we have:

$$\mathbf{u}^T\mathbf{A}\mathbf{v} = \frac{(\mathbf{A}\mathbf{v})^T(\mathbf{A}\mathbf{v})}{||\mathbf{A}\mathbf{v}||_E} = ||\mathbf{A}\mathbf{v}||_E = \sigma \tag{4.212}$$

In the first equality we used Eq. (4.210). In the second equality we applied the definition

[53] Incidentally, the following existence proof carries over almost in its entirety if **A** has dimensions $m \times n$, i.e., is rectangular (with, say, $m > n$). The only substantive difference is precisely in the base case: **A** would then be a column vector and we would have to use a norm instead of the absolute value.

[54] We should more properly be using the supremum, but the distinction is immaterial here.

of the Euclidean norm (and cancelled one term). In the third equality we used our earlier notation from Eq. (4.209). For the bottom-left element in Eq. (4.211) we have:

$$\tilde{\mathbf{U}}^T\mathbf{A}\mathbf{v} = \tilde{\mathbf{U}}^T\mathbf{u}\|\mathbf{A}\mathbf{v}\|_E = \mathbf{0} \tag{4.213}$$

In the first equality we used Eq. (4.210) and in the second equality we took advantage of the fact that $\mathbf{U}$ is an orthogonal matrix; here $\mathbf{0}$ is an $(n-1)\times 1$ column vector. We can also define the $1\times(n-1)$ row vector corresponding to the top-right element in Eq. (4.211) to be $\mathbf{w}^T = \mathbf{u}^T\mathbf{A}\tilde{\mathbf{V}}$. Finally, we also denote the $(n-1)\times(n-1)$ matrix in the bottom-right element in Eq. (4.211) by a new symbol: $\tilde{\mathbf{A}} = \tilde{\mathbf{U}}^T\mathbf{A}\tilde{\mathbf{V}}$. Putting it all together, Eq. (4.211) has turned into:

$$\mathbf{U}^T\mathbf{A}\mathbf{V} = \begin{pmatrix} \sigma & \mathbf{w}^T \\ \mathbf{0} & \tilde{\mathbf{A}} \end{pmatrix} \tag{4.214}$$

We will now show that the top-right element is also zero, i.e., $\mathbf{w} = \mathbf{0}$. To see this, examine the following:

$$\left\|\begin{pmatrix} \sigma & \mathbf{w}^T \\ \mathbf{0} & \tilde{\mathbf{A}} \end{pmatrix}\begin{pmatrix} \sigma \\ \mathbf{w} \end{pmatrix}\right\|_E = \left\|\begin{pmatrix} \sigma^2 + \|\mathbf{w}\|_E^2 \\ \tilde{\mathbf{A}}\mathbf{w} \end{pmatrix}\right\|_E = \sqrt{\left(\sigma^2 + \|\mathbf{w}\|_E^2\right)^2 + \|\tilde{\mathbf{A}}\mathbf{w}\|_E^2} \geq \sigma^2 + \|\mathbf{w}\|_E^2 \tag{4.215}$$

In the first step we carried out the matrix–vector multiplication. In the second step we (again) applied the definition of the Euclidean norm for vectors, Eq. (4.23). The final step is an elementary inequality. Our result will be very useful in the following derivation:

$$\sigma = \|\mathbf{A}\|_2 = \|\mathbf{U}^T\mathbf{A}\mathbf{V}\|_2 = \max_{\|\mathbf{x}\|_E\neq 0} \frac{\|\mathbf{U}^T\mathbf{A}\mathbf{V}\mathbf{x}\|_E}{\|\mathbf{x}\|_E} \geq \frac{\left\|\mathbf{U}^T\mathbf{A}\mathbf{V}\begin{pmatrix} \sigma \\ \mathbf{w} \end{pmatrix}\right\|_E}{\left\|\begin{pmatrix} \sigma \\ \mathbf{w} \end{pmatrix}\right\|_E} = \sqrt{\sigma^2 + \|\mathbf{w}\|_E^2} \tag{4.216}$$

In the first step we reiterated our definition from Eq. (4.209). In the second step we used the fact that orthogonal matrices preserve the 2-norm (which you are guided toward showing in problem 4.40). In the third step we applied the 2-norm definition from (the first equality in) Eq. (4.208). In the fourth step we applied the definition of the maximum. In the fifth step, we used our result from Eq. (4.215) and cancelled with the denominator. Comparing the starting with the final expression, we conclude that $\mathbf{w} = \mathbf{0}$. This means that both off-diagonal elements in Eq. (4.214) are zero.

We are now (finally) ready to carry out the inductive step. We assume that the SVD holds for the $(n-1)\times(n-1)$ matrix $\tilde{\mathbf{A}}$, i.e., $\tilde{\mathbf{A}} = \mathbf{U}_0\mathbf{\Sigma}_0\mathbf{V}_0^T$ and plug into Eq. (4.214):

$$\mathbf{U}^T\mathbf{A}\mathbf{V} = \begin{pmatrix} \sigma & \mathbf{0}^T \\ \mathbf{0} & \mathbf{U}_0\mathbf{\Sigma}_0\mathbf{V}_0^T \end{pmatrix} = \begin{pmatrix} 1 & \mathbf{0}^T \\ \mathbf{0} & \mathbf{U}_0 \end{pmatrix}\begin{pmatrix} \sigma & \mathbf{0}^T \\ \mathbf{0} & \mathbf{\Sigma}_0 \end{pmatrix}\begin{pmatrix} 1 & \mathbf{0}^T \\ \mathbf{0} & \mathbf{V}_0 \end{pmatrix}^T \tag{4.217}$$

To convince yourself that the second step is valid you may simply carry out the matrix multiplications. Multiplying by $\mathbf{U}$ on the left and by $\mathbf{V}$ on the right, we get:

$$\mathbf{A} = \left[\mathbf{U}\begin{pmatrix} 1 & \mathbf{0}^T \\ \mathbf{0} & \mathbf{U}_0 \end{pmatrix}\right]\begin{pmatrix} \sigma & \mathbf{0}^T \\ \mathbf{0} & \mathbf{\Sigma}_0 \end{pmatrix}\left[\mathbf{V}\begin{pmatrix} 1 & \mathbf{0}^T \\ \mathbf{0} & \mathbf{V}_0 \end{pmatrix}\right]^T \tag{4.218}$$

We recognize this as a singular-value decomposition of the $n \times n$ matrix $\mathbf{A}$: think of multiplying each of the terms in square brackets by its transpose. This concludes the proof.

4.5.3 Computing the SVD

An interesting property that relates the SVD of the matrix $\mathbf{A}$ with an eigendecomposition involving the same matrix is that the eigenvalues of the symmetric matrix $\mathbf{A}^T\mathbf{A}$ are σ_i^2 (and the right-singular vectors are the corresponding orthonormal eigenvectors). This property is very commonly used to compute the SVD in books written by non-mathematicians. For the sake of completeness, we guide you toward showing this property and implementing this approach in problem 4.41. However, problem 4.42 shows that this avenue leads to an unstable algorithm, i.e., it ends up being much more sensitive to perturbations than the initial problem was; in other words, it loses in practicality the little that it has gained in intuitiveness.

Instead of following that approach, we will now set up the computation of the SVD in a manner that is equally straightforward to grasp, but is also more stable. As usual, we will not provide a hyperefficient algorithm like those implemented in state-of-the-art libraries (which also involve a reduction to bidiagonal form); however, we will employ the same essential idea as is implemented in those libraries, without making a detour to the (instability-causing) $\mathbf{A}^T\mathbf{A}$ matrix.

The main idea involved is to start, just like we did before Eq. (4.200), by multiplying Eq. (4.198) by $\mathbf{V}$ on the right:

$$\mathbf{AV} = \mathbf{U\Sigma} \tag{4.219}$$

Similarly, we can take the transpose of Eq. (4.198) and then multiply by $\mathbf{U}$ on the right to find:

$$\mathbf{A}^T\mathbf{U} = \mathbf{V\Sigma} \tag{4.220}$$

where we made use of the fact that $\mathbf{\Sigma}^T = \mathbf{\Sigma}$ (since $\mathbf{\Sigma}$ is diagonal). The trick now is to notice that the last two relations are fully equivalent to the following block 2×2 equation:

$$\begin{pmatrix} \mathbf{0} & \mathbf{A}^T \\ \mathbf{A} & \mathbf{0} \end{pmatrix} \frac{1}{\sqrt{2}} \begin{pmatrix} \mathbf{V} & \mathbf{V} \\ \mathbf{U} & -\mathbf{U} \end{pmatrix} = \frac{1}{\sqrt{2}} \begin{pmatrix} \mathbf{V} & \mathbf{V} \\ \mathbf{U} & -\mathbf{U} \end{pmatrix} \begin{pmatrix} \mathbf{\Sigma} & \mathbf{0} \\ \mathbf{0} & -\mathbf{\Sigma} \end{pmatrix} \tag{4.221}$$

Explicitly carry out the matrix multiplications on both sides to convince yourself that we are merely restating Eq. (4.219) and Eq. (4.220).

We now realize this is a diagonalization of the $2n \times 2n$ matrix:

$$\mathbf{H} = \begin{pmatrix} \mathbf{0} & \mathbf{A}^T \\ \mathbf{A} & \mathbf{0} \end{pmatrix} \tag{4.222}$$

That is to say, Eq. (4.221) takes the form of Eq. (4.205):

$$\mathbf{H} = \mathbf{W\Lambda W}^{-1} \tag{4.223}$$

where:

$$\mathbf{W} = \frac{1}{\sqrt{2}}\begin{pmatrix} \mathbf{V} & \mathbf{V} \\ \mathbf{U} & -\mathbf{U} \end{pmatrix}, \qquad \mathbf{\Lambda} = \begin{pmatrix} \mathbf{\Sigma} & \mathbf{0} \\ \mathbf{0} & -\mathbf{\Sigma} \end{pmatrix} \tag{4.224}$$

In words, half (i.e., n) of the eigenvalues of $\mathbf{H}$ are the σ_i (and the other n eigenvalues differ by an overall sign). Similarly, we can find $\mathbf{V}$ and $\mathbf{U}$ by looking at, say, the first n columns of the $2n \times 2n$ matrix $\mathbf{W}$: the first n rows give us $\mathbf{V}$ and the next n rows $\mathbf{U}$. (The $1/\sqrt{2}$, which cancels in Eq. (4.221), is there to ensure that the eigenvectors are normalized, as usual.) We have been able to transform the problem of computing the SVD of $\mathbf{A}$ into the eigenproblem of $\mathbf{H}$ (and $\mathbf{A}^T\mathbf{A}$ did not appear anywhere).

4.5.4 Implementation

To compute the SVD in Python, we take guidance from Eq. (4.221): first, we set up the $2n{\times}2n$ matrix $\mathbf{H}$ of Eq. (4.222) in a dedicated function, `makeH()`, see `svd.py`, i.e., code 4.11. We carry out the actual SVD in a separate function, which takes in $\mathbf{A}$, uses `makeH()` to produce $\mathbf{H}$ and then computes $\mathbf{H}$'s eigenvalues and eigenvectors. As per Eq. (4.224), the eigenvalues can be used to extract $\mathbf{A}$'s singular values and the eigenvectors give us the left- and right-singular vectors. Conceptually, that's all there is to it.

Practically speaking, the story is a little more complicated: as we already know from Eq. (4.224), the eigenvalues of $\mathbf{H}$ will come in pairs of $\pm\sigma_i$. However, this causes convergence problems (to put it mildly) if you try to use `qrmet()`, our go-to approach for eigenvalue computation. (Recall that the QR method, in the form in which we set it up, can handle only eigenvalues of distinct magnitudes, see Eq. (4.128); the problem set discusses extensions.) To address this issue, in `svd()` we employ the simplest device possible: an arbitrary shift of the origin, similar in spirit to that of Eq. (4.145): $\mathbf{H}^* = \mathbf{H} - s\mathcal{I}$. This separates out the members of each pair, allowing a straightforward application of `qrmet()`. Then, the arbitrary shift is added back in, again similarly to what we did in Eq. (4.145).

We then employ `np.nonzero()`, introduced in problem 1.10, to extract the indices of the non-negative eigenvalues. These are first employed to extract the singular values of $\mathbf{A}$, namely the σ_i. Our next task is to extract the $\mathbf{U}$ and $\mathbf{V}$ matrices from the eigenvectors; to get the latter, we employ `invpowershift()`, just like we did in code 4.10, i.e., our eigenproblem library in `eig.py`. For pedagogical clarity, we first compute the entire $2n \times 2n$ matrix $\mathbf{W}$ and then proceed to extract the singular vectors: the "second quadrant" of $\mathbf{W}$ should give us $\mathbf{V}$ and the "third quadrant" of $\mathbf{W}$ should provide us with $\mathbf{U}$. Of course, the actual eigenvectors that we are faced with can be (and are) scrambled, so we employ the indices that `np.nonzero()` gave us to keep things straight. Since we are interested in $\mathbf{U}$ and $\mathbf{V}$ matrices that have orthonormal columns, we multiply by the extra $\sqrt{2}$ factor, as per Eq. (4.224); each time, we stack together a sequence of one-dimensional arrays to produce a two-dimensional array (employing `np.column_stack()`, which was introduced in our NumPy tutorial). On the `return` line, we produce a diagonal matrix containing the singular values; crucially, we return not $\mathbf{V}$ but $\mathbf{V}^T$, as is standard in SVD libraries.

In the main program, we first make a test matrix $\mathbf{A}$ and then call our very own `svd()`, comparing with what comes out of `np.linalg.svd()`. To keep the output manageable, we

Code 4.11 svd.py

```
from triang import testcreate
from invpowershift import invpowershift
from qrmet import qrmet
import numpy as np

def makeH(A):
    n = A.shape[0]
    H = np.zeros((2*n,2*n))
    H[n:,:n] = A
    H[:n,n:] = A.T
    return H

def svd(A, solver=qrmet):
    H = makeH(A)
    shift = 1.
    Hstar = H - np.identity(H.shape[0])*shift
    Hvals = solver(Hstar) + shift
    indices = (Hvals>=0).nonzero()[0]
    S = Hvals[indices]

    n = A.shape[0]
    vals = np.zeros(2*n)
    vecs = np.zeros((2*n,2*n))
    for i, qre in enumerate(Hvals):
        vals[i], vecs[:,i] = invpowershift(H,qre+1.e-20,tol=1.e-8)
    Vs = [vecs[:n,i] for i in indices]
    V = np.sqrt(2)*np.column_stack(Vs)
    Us = [vecs[n:,i] for i in indices]
    U = np.sqrt(2)*np.column_stack(Us)
    return U, np.diag(S), V.T

if __name__ == '__main__':
    A, _ = testcreate(4,21)
    A -= np.identity(4)
    U, S, VT = svd(A)
    print(np.all(np.linalg.norm(A - U@S@VT)<1.e-6), S)
    npU, npS, npVT = np.linalg.svd(A)
    diffA = A - npU@np.diag(npS)@npVT
    print(np.all(np.linalg.norm(diffA)<1.e-6), npS)
```

only print out the singular values (i.e., not the singular vectors). However, we also carry out a non-trivial reconstruction test: we check to see if the starting matrix $\mathbf{A}$ contains the same matrix elements (within some tolerance) as the SVD product $\mathbf{U\Sigma V}^T$. For the example matrix we have chosen, we find that the right-hand side of Eq. (4.198) is very close to the starting matrix on the left-hand side.

If you experiment with the test matrix $\mathbf{A}$, you will soon find cases where the singular values that are produced by our `svd()` do not match those of `np.linalg.svd()` (and, therefore, the SVD reconstruction of $\mathbf{A}$ also fails to pass our check). This generally happens when the singular values in $\mathbf{\Sigma}$ span several orders of magnitude. This is due to our choice of eigenvalue solver, i.e., it is unrelated to our overall approach of setting up the $2n \times 2n$ matrix $\mathbf{H}$ of Eq. (4.222) and then diagonalizing it. To help you in these scenarios, probed in the problem set, we have set up `svd()` accordingly: it takes in a parameter `solver`, which in our code is `qrmet()`, but could be replaced by, say, `np.linalg.eigvals()`. When you do that, you will find that our `svd()` again matches the output of `np.linalg.svd()` very well, even for "problematic" matrices $\mathbf{A}$ (see problem 4.46 for another approach).

4.6 Project: the Schrödinger Eigenvalue Problem

The *time-independent Schrödinger equation* is a prototypical eigenvalue problem:

$$\hat{H}|\psi\rangle = E|\psi\rangle \tag{4.225}$$

where $\hat{H}$ is the Hamiltonian operator, $|\psi\rangle$ is a state vector (called a ket by Dirac) in a Hilbert space, and E is the energy. In a course on quantum mechanics you likely heard the terms "eigenstates" and "eigenenergies". At the time, it was pointed out to you that Eq. (4.225) is an eigenvalue equation: it contains the same state vector on the left-hand side as on the right-hand side and is therefore formally similar to Eq. (4.124).

In practice, when solving Eq. (4.225) for a given physical problem, we typically get a differential equation, as we will see in chapter 8. In the present section, we limit ourselves to the case of one or more particles with spin-half, where there are no orbital degrees of freedom.[55] As we will see below, our problem maps onto a (reasonably) straightforward matrix form, where you don't have to worry about non-matrix features; that is, the $\hat{H}$ doesn't have a kinetic energy in it. Thus, the problem of spin-half particles becomes a direct application of the eigenproblem machinery we built earlier.

4.6.1 Physics Setup

The theory of interacting spin-half particles is fascinating: even if you have taken a course on quantum mechanics (QM), you may not be familiar with the approach we take here. In contradistinction to the standard textbook treatment, here we are interested in the *uncoupled representation*, where we consider the two-particle, three-particle, etc., system as being made up of individual particles. This offloads the bulk of the work onto the computer,

[55] You should be able to generalize this to the case of spin-one once you've understood our approach.

making it easy to generalize to the case of four (or more) particles with a minimum of complications (as you will find out when solving problems 4.50 and 4.51). This topic is worth fleshing out in detail but, since it lies somewhat outside the main thread of this book, it has been placed in appendix D.2 (as part of the online supplement at `www.numphyspy.org`). Here we briefly summarize some of the main results from that appendix and spend the majority of our time on implementing things in Python (in section 4.6.2).

Regardless of how many particles one is faced with, in QM one starts with operators and state vectors living in a given Hilbert space. However, *the physics in quantum mechanics is contained in the inner products, or the matrix elements*. As a result, relations that in the Hilbert-space language involved actions on kets, turn into relations involving matrices and column vectors. For example, the Schrödinger equation of Eq. (4.225) turns into:

$$\mathbf{H}\boldsymbol{\psi} = E\boldsymbol{\psi} \tag{4.226}$$

Note that this is a *matrix* eigenvalue problem. It cannot be overemphasized that this is a *result*: many texts jump from Eq. (4.225) to Eq. (4.226) without any explanation, so you are likely to benefit from examining the detailed argument in appendix D.2.

For the case of a single spin-half particle, the $\mathbf{H}$ is a 2×2 matrix and $\boldsymbol{\psi}$ a 2×1 vector. The interaction arises due to the presence of an external magnetic field; $\mathbf{H}$ is proportional to the *Pauli spin matrix* $\boldsymbol{\sigma}_z$. As you may recall:

$$\boldsymbol{\sigma}_x = \begin{pmatrix} 0 & 1 \\ 1 & 0 \end{pmatrix}, \quad \boldsymbol{\sigma}_y = \begin{pmatrix} 0 & -i \\ i & 0 \end{pmatrix}, \quad \boldsymbol{\sigma}_z = \begin{pmatrix} 1 & 0 \\ 0 & -1 \end{pmatrix} \tag{4.227}$$

For two spin-half particles, one employs a *tensor product* (denoted by $\otimes$) to express two-particle operators and state vectors in terms of the corresponding one-particle entities. The passage to a matrix representation, along the lines of Eq. (4.226), is greatly simplified if one introduces the *Kronecker product* between two matrices $\mathbf{U}$ and $\mathbf{V}$:

$$W_{ab} = (\mathbf{U} \otimes \mathbf{V})_{ab} = U_{ik} V_{jl}$$
$$\text{where } a = pi + j, \;\; b = pk + l \tag{4.228}$$

If $\mathbf{U}$ is $n \times n$ and $\mathbf{V}$ is $p \times p$, the original four indices take on the values:

$$i = 0, 1, \ldots, n-1, \quad k = 0, 1, \ldots, n-1, \quad j = 0, 1, \ldots, p-1, \quad l = 0, 1, \ldots, p-1 \tag{4.229}$$

As a result, the new indices take on the values:

$$a = 0, 1, \ldots, np-1, \quad b = 0, 1, \ldots, np-1 \tag{4.230}$$

We can use this to produce two-particle (i.e., 4×4) matrices, for example:

$$\mathbf{S}_{Iz} = \mathbf{S}_z \otimes \mathcal{I} \tag{4.231}$$

For this problem the matrix version of the Schrödinger equation of Eq. (4.226) still holds, but this time around **H** is a 4 × 4 matrix:

$$\mathbf{H} = -\omega_{\mathrm{I}}\mathbf{S}_{\mathrm{I}z} - \omega_{\mathrm{II}}\mathbf{S}_{\mathrm{II}z} + \gamma\left(\mathbf{S}_{\mathrm{I}x}\mathbf{S}_{\mathrm{II}x} + \mathbf{S}_{\mathrm{I}y}\mathbf{S}_{\mathrm{II}y} + \mathbf{S}_{\mathrm{I}z}\mathbf{S}_{\mathrm{II}z}\right) \tag{4.232}$$

and $\boldsymbol{\psi}$ a 4×1 vector. As you can see in Eq. (4.232), each of the particles I and II is interacting with an external magnetic field **B**, but the particles are also interacting with each other.

The case of three spin-half particles is completely analogous. One employs the Kronecker product to build up three-particle (i.e., 8 × 8) matrices, for example:

$$\mathbf{S}_{\mathrm{II}y} = \mathcal{I} \otimes \mathbf{S}_y \otimes \mathcal{I} \tag{4.233}$$

The matrix version of the Schrödinger equation now involves the 8 × 8 matrix:

$$\mathbf{H} = -\omega_{\mathrm{I}}\mathbf{S}_{\mathrm{I}z} - \omega_{\mathrm{II}}\mathbf{S}_{\mathrm{II}z} - \omega_{\mathrm{III}}\mathbf{S}_{\mathrm{III}z} + \gamma\left(\mathbf{S}_{\mathrm{I}\cdot\mathrm{II}} + \mathbf{S}_{\mathrm{I}\cdot\mathrm{III}} + \mathbf{S}_{\mathrm{II}\cdot\mathrm{III}}\right) \tag{4.234}$$

and $\boldsymbol{\psi}$ an 8 × 1 vector. As above, the particles are interacting with the magnetic field and (pairwise) with each other—$\mathbf{S}_{\mathrm{I}\cdot\mathrm{II}}$, and so on, can be expressed in terms of Cartesian components, as in Eq. (4.232).

4.6.2 Implementation

To keep things manageable, we provide three codes, the first of which is the program that builds the infrastructure (namely the Pauli spin matrices and the Kronecker product). The second program then employs 4×4 matrices to diagonalize the two-spin Hamiltonian. Once we've discussed the two-spin code and its output, we turn to our main attraction, which is a separate program implementing a three-spin function. This is where the 8 × 8 matrices are built and used.

4.6.2.1 Kronecker Product

In code 4.12 we code up the Pauli spin matrices (for future use) and the Kronecker product. The function `paulimatrices()` simply "hard-codes" the Pauli matrices from Eq. (4.227). As usual, we have opted to create a 2d NumPy array using a Python list containing all the elements, followed by a call to `reshape()`, in order to avoid a list-of-lists. The `dtype` of our arrray elements is here inferred from the arguments we pass in: this means that one of our Pauli matrices contains complex numbers and the other two floats. Importantly, we have chosen to write `paulimatrices()` in such a way that it returns a tuple of three NumPy arrays. This is a pattern that will re-emerge: in an attempt to keep things easier to reason about, we have chosen to "mix" conventions, so we create Python entities (tuples or lists) that contain NumPy arrays. It's almost certainly more efficient to do everything in NumPy: in this case, that would imply returning a 3d NumPy array, where one of the indices would keep track of which Cartesian component is being referred to and the other two indices would be "Pauli indices", namely the indices that tell us which bra and which ket we are using to sandwich our operator(s). Finally, notice that these Pauli matrices are one-particle entities. We will be able to combine them with other matrices to produce two-particle entities using the Kronecker product.

Code 4.12 kron.py

```
import numpy as np

def paulimatrices():
    sigx = np.array([0.,1,1,0]).reshape(2,2)
    sigy = np.array([0.,-1j,1j,0]).reshape(2,2)
    sigz = np.array([1.,0,0,-1]).reshape(2,2)
    return sigx, sigy, sigz

def kron(U,V):
    n = U.shape[0]
    p = V.shape[0]
    W = np.zeros((n*p,n*p), dtype=np.complex64)
    for i in range(n):
        for k in range(n):
            for j in range(p):
                for l in range(p):
                    W[p*i+j,p*k+l] = U[i,k]*V[j,l]
    return W

if __name__ == '__main__':
    sigx, sigy, sigz = paulimatrices()
    allones = np.ones((3,3))
    kronprod = kron(sigx,allones); print(kronprod.real)
```

Speaking of which, the function `kron()` is reasonably straightforward.[56] After figuring out n and p, which determine the dimensions of our matrices, we create an $np \times np$ matrix where we will store the output: this is made to be complex, in order to be as general as possible (we already saw that the y Pauli matrix is imaginary, so we need to employ complex arithmetic). We then proceed to carry out an iteration over all the one-particle indices $(ikjl)$, in order to implement Eq. (4.228), namely $W_{ab} = U_{ik}V_{jl}$ where $a = pi + j$ and $b = pk + l$. This is the only time in the book that we use four nested loops (typically a slow operation for large problems).

The main program takes the Kronecker product of a given Pauli spin matrix (a one-body, i.e., 2×2 entity) and a 3×3 matrix of 1's. This is a mathematical test case, i.e., there is no physics being probed here. Run the code to see what you get. Then, to highlight the fact that the Kronecker product is not commutative, try putting the Pauli matrix in the second slot. Even though `kron()` returns a complex matrix, we take the real part just before printing

[56] We could have used `numpy.kron()` but, as usual, we prefer to roll our own.

things out, for aesthetic reasons. You should try to work out what the output should be before proceeding; this will be a 6×6 matrix. In order to interpret the results, it's probably best to look at Eq. (D.118). When the Pauli spin matrix comes first, we get 3×3 blocks, each of which repeats a single matrix element of $\boldsymbol{\sigma}_x$ from Eq. (4.227). When the matrix full of 1's comes first, we get 2×2 blocks each of which is a $\boldsymbol{\sigma}_x$.

4.6.2.2 Two Particles

Code 4.13 contains a Python implementation of the two-spin Hamiltonian. This program imports the needed functionality from our earlier code(s) and then defines our new physics function `twospins()`, which takes in as inputs the coefficients $\omega_{\rm I}$, $\omega_{\rm II}$, and γ. We first set $\hbar$ to 1, to keep things simple; while this is part of what are known as "natural units", it won't have any effect in what follows. We then call `paulimatrices()` and create a 2×2 identity matrix for later use. Our main design choice (hinted at above) was to produce and store our 4×4 matrices in a list. The first time we do this is when we employ a list comprehension to create `SIs`. Note how Pythonic this is: we use no indices and directly iterate through the Pauli-matrices tuple. This (together with repeated calls to our Kronecker product function, `kron()`) allows us to produce the 4×4 matrices $\mathbf{S}_{\rm Ix}$, $\mathbf{S}_{\rm Iy}$, and $\mathbf{S}_{\rm Iz}$ in turn, storing them together in a list of three elements.[57] As we saw in Eq. (4.231), when we're dealing with particle `I` we put the Pauli spin matrix before the identity (e.g., $\mathbf{S}_{\rm Iz} = \mathbf{S}_z \otimes \mathcal{I}$), whereas when we're dealing with particle `II` the identity comes first (e.g., $\mathbf{S}_{\rm IIx} = \mathcal{I} \otimes \mathbf{S}_x$). You may now start to realize why we labelled the particles using the Roman numerals `I` and `II`: we already have to deal with Cartesian-component indices and Pauli indices. If we had also chosen to label the particles using the numbers 1 and 2 (or, heaven forfend, 0 and 1) it would have been easier to make a mistake.

We then evaluate the interacting term as per Eq. (4.232). We have six matrices that we need to multiply pairwise and then add up the results (giving you one 4×4 matrix). We have, again, chosen to use a list comprehension: this produces the matrix multiplications and stores them in a list, whose elements (which are matrices) are then added together using Python's `sum()`.[58] Depending on your eagerness to be Pythonic at all costs, you may have opted to avoid using any indices whatsoever, saying instead:

```
SIdotII = sum([SI@SII for SI,SII in zip(SIs,SIIs)])
```

Of course, this is starting to strain the limits of legibility. If you're still not comfortable with list comprehensions, simply write out the three matrix multiplications:

```
SIs[0]@SIIs[0] + SIs[1]@SIIs[1] + SIs[2]@SIIs[2]
```

[57] Again, you might prefer to use a 3d NumPy array. In that case, you can implement the interacting-spins term much more concisely, if you know what you're doing. A further point: while `numpy.dot()` and `@` are for us fully equivalent, they behave differently for 3d, 4d, etc. arrays.

[58] In turn, when Python tries to sum these matrices together, it knows how to do that properly, since they are NumPy arrays – recall that summing lists together would have concatenated them and that's not what we're trying to do here. Using `np.sum()` is also wrong: do you understand why?

Code 4.13 twospins.py

```
from kron import paulimatrices, kron
from qrmet import qrmet
import numpy as np

def twospins(omI,omII,gam):
    hbar = 1.
    paulis = paulimatrices()
    iden = np.identity(2)

    SIs = [hbar*kron(pa,iden)/2 for pa in paulis]
    SIIs = [hbar*kron(iden,pa)/2 for pa in paulis]
    SIdotII = sum([SIs[i]@SIIs[i] for i in range(3)])

    H = -omI*SIs[2] - omII*SIIs[2] + gam*SIdotII
    H = H.real
    return H

if __name__ == '__main__':
    H = twospins(1.,2.,0.5)
    qreigvals = qrmet(H); print(qreigvals)
```

If you know that you will never have to deal with more (or fewer) than three Cartesian components, then it's OK to explicitly write out what you're doing.

We then proceed to construct the 4×4 Hamiltonian matrix $\mathbf{H}$ of Eq. (4.232), using several of the results we have already produced. So far we've been doing our calculations using complex numbers, as we should. When we took the dot product that produced $\mathbf{S}_{\mathrm{I}\cdot\mathrm{II}}$, however, everything became real again. We therefore keep only the real part of `H`, since our eigenvalue methods (developed earlier in this chapter) won't work with complex matrices.

The main program is incredibly short: we first call `twospins()`, with some arbitrary input parameters. Now that we have the matrix `H`, we call our very own `qrmet()` to evaluate the matrix's eigenvalues and then print out the result.[59] Running this code, we see that we get a non-diagonal 4×4 Hamiltonian matrix, which we then diagonalize and get 4 eigenvalues. Of course, a 4×4 matrix is not too large, so you could have tried to solve this problem the old-fashioned way, via $|\mathbf{A} - \lambda \mathcal{I}| = 0$. As the dimensionality keeps growing, this is, in general, less feasible. While the output of our code is short, you shouldn't let that mislead you: our program carries out a number of calculations that would have required some effort to do by hand. As always, we urge you to play around with this code, printing out intermediate results. You can start by printing out matrices like $\mathbf{S}_{\mathrm{I}z}$ or $\mathbf{S}_{\mathrm{II}x}$; you can

[59] Problem 4.49 explores why we're not using our `eig()` function instead.

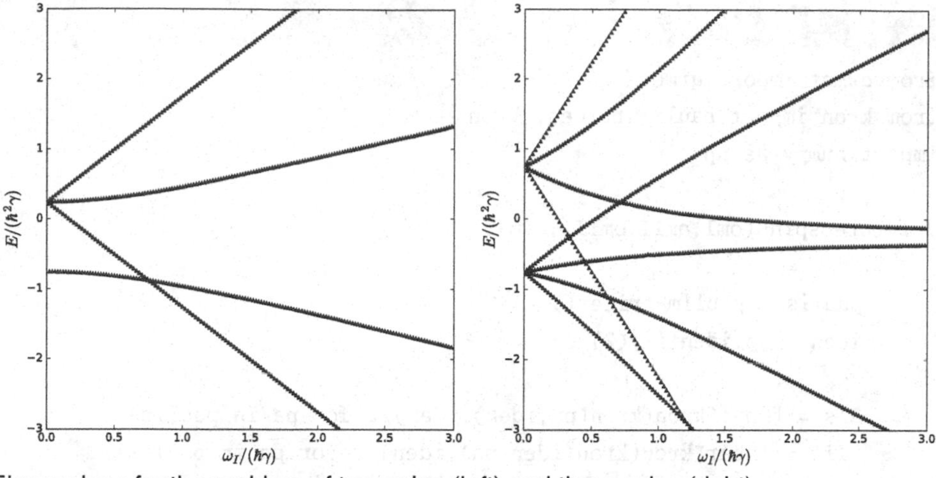

Fig. 4.1 Eigenvalues for the problem of two spins (left) and three spins (right)

compare with the matrices derived in appendix D.2. Another thing to try is to study the limits of no magnetic field or the opposite limit of no interaction between the spins. In one of these cases, our function `qrmet()` has trouble converging to the right answer. You will explore this in problem 4.48.

In the left panel of Fig. 4.1, we show the result of using `twospins()` to plot the total energy as a function of the ω_{I} (assuming $\omega_{\text{II}}/\omega_{\text{I}} = 2$, just like in our code). In both the axes, we have made everything dimensionless by dividing out an appropriate product of a power of $\hbar$ and the coupling constant γ. Physically, you can envision tuning the magnetic field B to tune the magnitude of ω_{I}. At vanishing magnetic field, we find three of the states becoming degenerate in energy; this is the spin-triplet that you find in the coupled representation (namely, three states with total spin 1). Another feature of this plot is that, at some point, there is a crossing between eigenenergies; you should spend some time trying to understand why this happens. It may help you to know that physically this situation is similar to the Zeeman effect in hydrogen (only there, $\omega_{\text{II}} \ll \omega_{\text{I}}$ holds).

4.6.2.3 Three Particles

Code 4.14 is a Python implementation for the case of three particles, building on our earlier work; in problem 4.50 you are asked to study the case of four spin-half particles.[60]

We first import the functions `paulimatrices()` and `kron()` from our earlier code. Our only new function, `threespins()`, is a modification of `twospins()`. While we, of course, have to introduce a new variable to keep track of the matrices corresponding to particle `III`, that's not the main change. The core modification is that, in the list comprehension that creates these matrices, we now have to call the `kron()` function twice. This is hardly surprising if you bring to mind equations like Eq. (4.233), $\mathbf{S}_{\text{II}y} = \mathcal{I} \otimes \mathbf{S}_y \otimes \mathcal{I}$. The Kronecker product is not commutative, so it matters whether the Pauli matrix is first, second, or third.

[60] You are free to try to implement the general case of any number of particles, but it's probably best to do that *after* you've done the four-particle case the quick and dirty way.

Code 4.14 threespins.py

```
from qrmet import qrmet
from kron import paulimatrices, kron
import numpy as np

def threespins(omI,omII,omIII,gam):
    hbar = 1.
    paulis = paulimatrices()
    iden = np.identity(2)

    SIs = [hbar*kron(kron(pa,iden),iden)/2 for pa in paulis]
    SIIs = [hbar*kron(kron(iden,pa),iden)/2 for pa in paulis]
    SIIIs = [hbar*kron(kron(iden,iden),pa)/2 for pa in paulis]

    SIdotII = sum([SIs[i]@SIIs[i] for i in range(3)])
    SIdotIII = sum([SIs[i]@SIIIs[i] for i in range(3)])
    SIIdotIII = sum([SIIs[i]@SIIIs[i] for i in range(3)])

    H = -omI*SIs[2] - omII*SIIs[2] - omIII*SIIIs[2]
    H += gam*(SIdotII+SIdotIII+SIIdotIII)
    H = H.real
    return H

if __name__ == '__main__':
    np.set_printoptions(precision=3)
    H = threespins(1.,2.,3.,0.5)
    qreigvals = qrmet(H); print(qreigvals)
```

A natural consequence of taking two Kronecker products is that at the end we will be dealing with 8×8 matrices. Turning to the dot-product terms like $\mathbf{S}_{\text{I}\cdot\text{II}}$, we observe that the relevant line, `SIdotII = sum([SIs[i]@SIIs[i] for i in range(3)])`, is identical to what we used in the two-particle code, despite the fact that the matrices being manipulated are now 8×8: Python and NumPy work together to accomplish exactly what we want. The only difference in this regard is that since we now have three particles, we have to form more pairs, also constructing $\mathbf{S}_{\text{I}\cdot\text{III}}$ and $\mathbf{S}_{\text{II}\cdot\text{III}}$. Our function then proceeds to create the Hamiltonian matrix, directly following Eq. (4.234).

The main program is very short: the only change in comparison to the two-particle case (in addition to having to pass in ω_{III} as an argument) is a beautifying fix ensuring that we don't print out too many digits of precision. Running this code, we realize that we are dealing with an 8×8 matrix that is not diagonal. That is where our function `qrmet()`

comes in handy and evaluates the eight eigenvalues. You could have tried using $|\mathbf{A}-\lambda\mathcal{I}| = 0$ to compute these eigenvalues, but you would be faced with a polynomial of eighth order: while some symmetries of the matrix might help you, you would get into trouble the second you modified our Hamiltonian to contain a new term, say $\hat{S}_{Ix}$.

Like in the two-spin case, we also used `threespins()` in order to build some physical intuition. The right panel of Fig. 4.1 shows the total energy for our eight states as a function of the ω_I (assuming $\omega_{II}/\omega_I = 2$ and $\omega_{III}/\omega_I = 3$, just like in our code). As before, we have ensured that all quantities plotted are dimensionless. At vanishing magnetic field, we see the states forming two quartets, with each of the four states in each quartet being degenerate in energy. As you will discover when you solve problem 4.48, the situation is actually a bit more complicated: we are here faced with a spin-quartet (with total spin 3/2) and two spin-doublets (each of which has total spin 1/2). Moving to other features of our plot, we see that this time around we find several more eigenvalue crossings.

Writing `threespins()` was not harder than writing `twospins()`, yet the problem we solved was much larger: as long as our infrastructure (here mainly the Kronecker product) is robust, we can address more demanding scenarios without much effort on our part (of course, the total operation cost for the computer can grow rather rapidly). Sometimes the knowledge that you've already built up your machinery is enough to convince you to tackle a problem that you may have shied away from if all you had at your disposal was paper and pencil. Concretely: you will benefit from experimenting with this code, printing out matrices like $\mathbf{S}_{IIIx}$ and comparing with what you derive (8×8 matrices) analytically. We hope that this process will convince you of the point we were making before in abstract terms, regarding how the computer can help. If that doesn't work, then solving problem 4.51 (asking you to repeat this exercise for the 10-particle case) should do the trick.

Problems

4.1 [$\mathcal{A}$] Show that the product of the eigenvalues of a matrix is equal to its determinant. Hint: consider the factorization of the characteristic polynomial.

4.2 [$\mathcal{A}$] Show that the product of two (a) upper-triangular matrices is an upper triangular matrix, and (b) orthogonal matrices is an orthogonal matrix.

4.3 [$\mathcal{A}$] In the main text we derived an error bound for $\|\Delta\mathbf{x}\|/\|\mathbf{x}\|$ for a linear system of equations, in Eq. (4.30), under the assumption that the vector $\mathbf{b}$ was held constant. Carry out an analogous derivation, assuming $\mathbf{b}$ is perturbed but $\mathbf{A}$ is held fixed.

4.4 [$\mathcal{A}$] We will now help you derive an error bound for the solution of a linear system of equations, $\|\Delta\mathbf{x}\|/\|\mathbf{x}\|$, keeping neither $\mathbf{A}$ nor $\mathbf{b}$ constant. We'll need some scaffolding involving a matrix $\mathbf{W}$ satisfying $\|\mathbf{W}\| < 1$.

(a) Prove by contradiction that $\mathcal{I} - \mathbf{W}$ is non-singular.

(b) Use $(\mathcal{I} - \mathbf{W})(\mathcal{I} - \mathbf{W})^{-1} = \mathcal{I}$ to show that:

$$\|(\mathcal{I} - \mathbf{W})^{-1}\| \leq \frac{1}{1 - \|\mathbf{W}\|} \tag{4.235}$$

(c) Combine $\mathbf{Ax} = \mathbf{b}$ with the generalization of Eq. (4.25), $(\mathbf{A}+\Delta\mathbf{A})(\mathbf{x}+\Delta\mathbf{x}) = \mathbf{b}+\Delta\mathbf{b}$, to show that:

$$\Delta\mathbf{x} = (\mathcal{I} + \mathbf{A}^{-1}\Delta\mathbf{A})^{-1}\mathbf{A}^{-1}(-\Delta\mathbf{A}\mathbf{x} + \Delta\mathbf{b}) \tag{4.236}$$

(d) Use the earlier parts to show that:

$$\frac{\|\Delta\mathbf{x}\|}{\|\mathbf{x}\|} \leq \frac{\kappa(\mathbf{A})}{1 - \kappa(\mathbf{A})\|\Delta\mathbf{A}\|/\|\mathbf{A}\|}\left(\frac{\|\Delta\mathbf{A}\|}{\|\mathbf{A}\|} + \frac{\|\Delta\mathbf{b}\|}{\|\mathbf{b}\|}\right) \tag{4.237}$$

4.5 For a matrix $\mathbf{A}$, implement: (a) finding the lower and upper triangular part, (b) the Frobenius norm, and (c) the infinity norm. You can use NumPy but, obviously, you shouldn't use the functions that have been designed to give the corresponding answer each time.

4.6 Rewrite `triang.forsub()` and `triang.backsub()` in less idiomatic form.

4.7 [$\mathcal{A}$] Prove that left and right eigenvalues are equivalent (something which does not hold for left and right eigenvectors).

4.8 Evaluate the norm and condition number $\kappa(\mathbf{A})$ for the 4×4, 10×10, and 20×20 matrices you get from `triang.testcreate()`; feel free to use NumPy functionality.

4.9 [$\mathcal{A}$] In this problem we will evaluate the sum $\sum_{k=0}^{n-1} k^2$ needed for Eq. (4.82). The trick we'll use is to start from the following sum:

$$\sum_{k=0}^{N}\left[(k+1)^3 - k^3\right] \tag{4.238}$$

First evaluate this sum without expanding the parentheses: you will notice that it involves a telescoping series and is therefore straightforward. Now that you know what this sum is equal to, expand the parentheses and solve for the sum we started with. Finally, it should be trivial to re-express the final result for the case $N = n - 1$.

4.10 Update `gauelim.py` such that it can handle multiple constant vectors. In other words, generalize `gauelim()` such that it takes in an $n \times n$ matrix $\mathbf{A}$ and an $n \times m$ matrix $\mathbf{B}$, where m is the number of distinct constant vectors you will be handling. The output should be an $n \times m$ matrix $\mathbf{X}$, collecting all the solution vectors:

$$\mathbf{X} = \begin{pmatrix}\mathbf{x}_0 & \mathbf{x}_1 & \dots & \mathbf{x}_{m-1}\end{pmatrix}, \qquad \boldsymbol{B} = \begin{pmatrix}\mathbf{b}_0 & \mathbf{b}_1 & \dots & \mathbf{b}_{m-1}\end{pmatrix} \tag{4.239}$$

Test out the case where $\mathbf{A}$ is a 4×4 matrix and $\mathbf{B}$ is 4×3.

4.11 Intriguingly, the LU decomposition of a matrix $\mathbf{A}$ preserves tridiagonality (meaning that $\mathbf{L}$ and $\mathbf{U}$ are tridiagonal if $\mathbf{A}$ is tridiagonal). Take the matrix whose main diagonal has elements d_0 to d_{n-1}, the diagonal below that has elements c_1 to c_{n-1}, and the diagonal above that has elements e_0 to e_{n-2}. There is no need to create and employ an $n \times n$ NumPy array: we need to store only the vectors $\mathbf{c}$, $\mathbf{d}$, and $\mathbf{e}$. Rewrite `ludec()` and `lusolve()` so that they apply to this problem. Before you start coding, you should first write out the equations involved.

4.12 Use LU decomposition to calculate the inverse of a matrix $\mathbf{A}$ as per Eq. (4.99) and the determinant as per Eq. (4.102). Test your answers by comparing to the output of `numpy.linalg.inv()` and `numpy.linalg.det()`.

4.13 Rewrite the `ludec()` function, where you now don't store the **L** and **U** matrices separately. Then, rewrite the `lusolve()` function to use your modified `ludec()`. Can you still use the `forsub()` and `backsub()` functions?

4.14 [$\mathcal{A}$] Prove (by contradiction) that Doolittle's LU decomposition is unique.

4.15 Implement scaled partial pivoting for the Gaussian elimination method, as per the discussion around Eq. (4.116):

$$\frac{|A_{kj}|}{s_k} = \max_{j \le m \le n-1} \frac{|A_{mj}|}{s_m} \tag{4.240}$$

Remember to also swap the scale factors when you are interchanging a row.

4.16 Implement pivoting (whether scaled or not) for the case of LU decomposition. While you're at it, also calculate the determinant for the case of pivoting (you have to keep track of how many row interchanges you carried out).

4.17 Use Gaussian elimination or LU decomposition (without and with pivoting) for:

```
A = np.array([4., 4, 8, 4, 4, 5, 3, 7, 8, \
              3, 9, 9, 4, 7, 9, 5]).reshape(4,4)
bs = np.array([1., 2, 3, 4])
```

Does pivoting help in any way? What are your conclusions about this matrix? You sometimes hear the advice that, since the problem arises from one of the U_{ii}'s being zero, you should replace it with a small number, say 10^{-20}. Does this work?

4.18 We will now employ the Jacobi iterative method to solve a linear system of equations, for the case where the coefficient matrix is sparse; we won't need to explicitly store **A**. Focus on the following *cyclic tridiagonal* matrix:

$$(\mathbf{A}|\mathbf{b}) = \left(\begin{array}{cccccccccc|c} 4 & -1 & 0 & 0 & \ldots & 0 & 0 & 0 & -1 & 1 \\ -1 & 4 & -1 & 0 & \ldots & 0 & 0 & 0 & 0 & 1 \\ \vdots & \vdots & \vdots & \vdots & \ddots & \vdots & \vdots & \vdots & \vdots & \vdots \\ 0 & 0 & 0 & 0 & \ldots & 0 & -1 & 4 & -1 & 1 \\ -1 & 0 & 0 & 0 & \ldots & 0 & 0 & -1 & 4 & 1 \end{array}\right) \tag{4.241}$$

Something similar appears when you study differential equations with periodic boundary conditions, as we will see in section 8.4.2. (Note that **A** is "almost" a tridiagonal matrix.) For this problem, the Jacobi iterative method of Eq. (4.121) turns into:

$$\begin{aligned} x_0^{(k)} &= \left(1 + x_1^{(k-1)} + x_{n-1}^{(k-1)}\right)\frac{1}{4} \\ x_i^{(k)} &= \left(1 + x_{i-1}^{(k-1)} + x_{i+1}^{(k-1)}\right)\frac{1}{4}, \quad i = 1, \ldots, n-2 \\ x_{n-1}^{(k)} &= \left(1 + x_0^{(k-1)} + x_{n-2}^{(k-1)}\right)\frac{1}{4} \end{aligned} \tag{4.242}$$

Implement the Jacobi iterative method for this set of equations. Your input parameters should be only the total number of iterations, the tolerance, as well as n (which determines the size of the problem), i.e., there are no longer any **A** and **b** to be passed in. Choose a tolerance of 10^{-5} and print out the solution vectors for $n = 10$ and for

$n = 20$, each time also comparing with the output of `numpy.linalg.solve()`. (To produce the latter, you will have to form **A** and **b** explicitly.)

4.19 A small modification to Eq. (4.121) leads to the *Gauss–Seidel method*:

$$x_i^{(k)} = \left(b_i - \sum_{j=0}^{i-1} A_{ij} x_j^{(k)} - \sum_{j=i+1}^{n-1} A_{ij} x_j^{(k-1)} \right) \frac{1}{A_{ii}}, \quad i = 0, 1, \ldots, n-1 \tag{4.243}$$

The Gauss–Seidel method uses the improved values as soon as they become available. Implement the Gauss–Seidel method in Python and compare its convergence with that of the Jacobi method, for the same problem as in `jacobi.py`.

4.20 [$\mathcal{A}$] This problem reformulates (in matrix form) the Jacobi iterative method for solving linear systems of equations and justifies its convergence criterion.

(a) Split the matrix **A** in terms of a diagonal matrix, a lower triangular matrix, the identity matrix, and an upper triangular matrix:

$$\mathbf{A} = \mathbf{D}(\mathbf{L} + \mathcal{I} + \mathbf{U}) \tag{4.244}$$

Now, if you are told that the following relation:

$$\mathbf{x}^{(k)} = \mathbf{B}\mathbf{x}^{(k-1)} + \mathbf{c} \tag{4.245}$$

is fully equivalent to the Jacobi method of Eq. (4.121), express **B** and **c** in terms of known quantities. (Unsurprisingly, a similar identification can be carried out for the Gauss–Seidel method of problem 4.19, as well.)

(b) We will now write the matrix relation Eq. (4.245) for the exact solution vector **x**:

$$\mathbf{x} = \mathbf{B}\mathbf{x} + \mathbf{c} \tag{4.246}$$

Combine this with Eq. (4.245) to relate $||\mathbf{x}^{(k)} - \mathbf{x}||$ with $||\mathbf{x}^{(k)} - \mathbf{x}^{(k-1)}||$. Discuss how small $||\mathbf{B}||$ has to be for $||\mathbf{x}^{(k)} - \mathbf{x}^{(k-1)}||$ to be a good correctness criterion.

(c) Examine the infinity norm $||\mathbf{B}||_\infty$ and see what conclusion you can reach about its magnitude for the case where **A** is diagonally dominant. (In practice, one tackles the problem $\mathbf{P}^{-1}\mathbf{A}\mathbf{x} = \mathbf{P}^{-1}\mathbf{b}$ instead, where **P** is known as a *preconditioner*.)

4.21 [$\mathcal{A}$] Prove Eq. (4.125), $\mathbf{V}^{-1}\mathbf{A}\mathbf{V} = \mathbf{\Lambda}$, for the 3×3 case. Hint: first prove $\mathbf{A}\mathbf{V} = \mathbf{V}\mathbf{\Lambda}$.

4.22 [$\mathcal{A}$] To see that the normalized (direct) power method of Eq. (4.136) finds $\mathbf{v}_0$, first show that $\mathbf{q}^{(k)} = \mathbf{A}^k\mathbf{q}^{(0)}/||\mathbf{A}^k\mathbf{q}^{(0)}||$ and then apply the argument of Eq. (4.132).

4.23 Implement the (direct) power method with the stopping criterion in Eq. (4.137). You will have to keep the $\mathbf{q}^{(k)}$ in the current iteration and in the previous one.

4.24 Apply Aitken extrapolation, Eq. (2.107), to the sequence of eigenvalue estimates produced by the (direct) power-method algorithm, Eq. (4.136). For $k = 1, 2, 3, 4, 5$ print out the Rayleigh-quotient values $\mu(\mathbf{q}^{(k)})$ as well as the accelerated sequence corresponding to them. Hint: you have to make some assumptions about the first two values of the untransformed sequence in order to start applying the Aitken formula.

4.25 This problem studies the *modified Gram–Schmidt* method, which is less sensitive to numerical errors than classical Gram–Schmidt. Conceptually, classical Gram–Schmidt produces new $\mathbf{a}'_j$ vectors by subtracting out any non-orthogonal components of the original vectors $\mathbf{a}_j$, as per Eq. (4.155). The problem with this is that if in the calculation of, say, $\mathbf{q}_3$ a numerical roundoff error is introduced, this error will then be propagated to the computation of $\mathbf{q}_4$, $\mathbf{q}_5$, and so on. In contradistinction to this, modified Gram–Schmidt tries to correct for the error in $\mathbf{q}_3$ when computing $\mathbf{q}_4$, $\mathbf{q}_5$, and so on.[61] In equation form, what modified Gram–Schmidt does is to replace the relations from Eq. (4.150) to Eq. (4.155) as follows:

$$\begin{aligned}
\mathbf{q}_0 &= \frac{\mathbf{a}_0}{||\mathbf{a}_0||}\\
\mathbf{a}_j'^{(0)} &= \mathbf{a}_j - (\mathbf{q}_0^T\mathbf{a}_j)\mathbf{q}_0, \qquad j = 1, 2, \ldots, n-1\\
\mathbf{q}_1 &= \frac{\mathbf{a}_1'^{(0)}}{||\mathbf{a}_1'^{(0)}||}
\end{aligned} \tag{4.247}$$

So far, there's nothing really "modified" going on. Then, the next step is:

$$\begin{aligned}
\mathbf{a}_j'^{(1)} &= \mathbf{a}_j'^{(0)} - (\mathbf{q}_1^T\mathbf{a}_j'^{(0)})\mathbf{q}_1, \qquad j = 2, 3, \ldots, n-1\\
\mathbf{q}_2 &= \frac{\mathbf{a}_2'^{(1)}}{||\mathbf{a}_2'^{(1)}||}
\end{aligned} \tag{4.248}$$

Note how the inner product is taken not with the original vector, but with the updated one (which has a superscript in parentheses). Thus, for the general case we have:

$$\begin{aligned}
\mathbf{q}_i &= \frac{\mathbf{a}_i'^{(i-1)}}{||\mathbf{a}_i'^{(i-1)}||}\\
\mathbf{a}_j'^{(i)} &= \mathbf{a}_j'^{(i-1)} - (\mathbf{q}_i^T\mathbf{a}_j'^{(i-1)})\mathbf{q}_i, \qquad j = i+1, \ldots, n-1
\end{aligned} \tag{4.249}$$

Implement QR decomposition in Python using the modified Gram–Schmidt approach. You should carefully think about how to structure your new code: the prescription above builds the required matrices row by row (whereas `qrdec()` works column by column), so it may help you to first restructure your classical Gram–Schmidt code to also work row by row. At that point, you will see that the only difference between the two methods is whether the inner products are computed using the original or the updated vectors. Then, evaluate $||\mathbf{Q}^T\mathbf{Q} - \mathcal{I}||$ for the `testcreate()` problem and compare with the orthogonality properties of classical Gram–Schmidt.

4.26 We will now use QR decomposition to solve a linear system of equations, $\mathbf{A}\mathbf{x} = \mathbf{b}$. This equation can be rewritten as: $\mathbf{QR}\mathbf{x} = \mathbf{b}$. We can take advantage of the orthogonality of $\mathbf{Q}$ to re-express this as: $\mathbf{R}\mathbf{x} = \mathbf{Q}^T\mathbf{b}$. But now the right-hand side contains only known quantities and the left-hand side has the upper-triangular matrix $\mathbf{R}$, so a back substitution is all that's needed. Implement this approach in Python for both classical and modified Gram–Schmidt (if you haven't solved problem 4.25, use only classical Gram–Schmidt).

4.27 [$\mathcal{A}$] We now generalize the Gram–Schmidt orthonormalization process—Eq. (4.150) to Eq. (4.155)—to work with functions. We realize that we are lacking the concept of

[61] This approach is somewhat similar in spirit to that of the Gauss–Seidel method (problem 4.19), in that updated values are used but, of course, we are comparing apples with oranges.

the "length" of a function (which would correspond to the norm used in the denominator to normalize) as well as the concept of the "inner product" of two functions (which would be needed to subtract out any component that's not orthogonal). Let's introduce the latter, namely the inner product of the function $f(x)$ with the function $g(x)$. We choose to work with the symmetrical interval $-1 \leq x \leq 1$ and define:

$$(f, g) \equiv \int_{-1}^{1} f(x)g(x)dx \tag{4.250}$$

assuming these are real functions. It is then straightforward to define the length of a function as simply $\sqrt{(f, f)}$. We are now in a position to start following the Gram–Schmidt steps. We will call q_j (non-bold) the result of orthonormalizing the monomials $a_0 = 1, a_1 = x, a_2 = x^2, a_3 = x^3, \ldots$. Taking them in order, we have:

$$q_0 = \frac{a_0}{\sqrt{(a_0, a_0)}} = \frac{1}{\sqrt{(1, 1)}} = \frac{1}{\sqrt{\int_{-1}^{1} 1^2 dx}} = \frac{1}{\sqrt{2}} \tag{4.251}$$

Next we have:

$$a'_1 = a_1 - (q_0, a_1)q_0 = x - \left(\frac{1}{\sqrt{2}}, x\right)\frac{1}{\sqrt{2}} = x - \frac{1}{\sqrt{2}}\int_{-1}^{1} \frac{1}{\sqrt{2}} x dx = x \tag{4.252}$$

which was particularly easy to calculate since the integral vanishes (1 and x were already orthogonal). Normalizing, we get:

$$q_1 = \frac{a'_1}{\sqrt{(a'_1, a'_1)}} = \frac{x}{\sqrt{(x, x)}} = \frac{x}{\sqrt{\int_{-1}^{1} x^2 dx}} = \sqrt{\frac{3}{2}}x \tag{4.253}$$

The next step is:

$$\begin{aligned} a'_2 = a_2 - (q_0, a_2)q_0 - (q_1, a_2)q_1 &= x^2 - \left(\frac{1}{\sqrt{2}}, x^2\right)\frac{1}{\sqrt{2}} - \left(\sqrt{\frac{3}{2}}x, x^2\right)\sqrt{\frac{3}{2}}x \\ &= x^2 - \frac{1}{\sqrt{2}}\int_{-1}^{1} \frac{1}{\sqrt{2}} x^2 dx - \sqrt{\frac{3}{2}}x \int_{-1}^{1} \sqrt{\frac{3}{2}} x^3 dx = x^2 - \frac{1}{3} \end{aligned} \tag{4.254}$$

In the last step the second integral vanished. Now, to normalize:

$$q_2 = \frac{a'_2}{\sqrt{(a'_2, a'_2)}} = \frac{x^2 - 1/3}{\sqrt{(x^2 - 1/3, x^2 - 1/3)}} = \frac{x^2 - 1/3}{\sqrt{\int_{-1}^{1} (x^2 - 1/3)^2 dx}} = \sqrt{\frac{5}{2}}\left(\frac{3}{2}x^2 - \frac{1}{2}\right) \tag{4.255}$$

(a) Carry out the calculation that leads to a'_3 and from there to q_3.

(b) Explicitly check the orthonormality of the q_j's, by seeing that:

$$(q_n, q_m) \equiv \int_{-1}^{1} q_n q_m dx = \delta_{nm} \tag{4.256}$$

holds for the first few orthonormal polynomials.

(c) Any polynomial of degree $n-1$ can be expressed as a linear combination of our orthonormal q_j's as follows:

$$r_{n-1}(x) = \sum_{j=0}^{n-1} c_j q_j \tag{4.257}$$

Use this expansion and the orthonormality property in Eq. (4.256) to show that:

$$(r_{n-1}, q_n) \equiv \int_{-1}^{1} r_{n-1} q_n dx = 0 \tag{4.258}$$

In words, q_n is orthogonal to all polynomials of a lower degree.

(d) The q_j we have been finding are multiples of the orthogonal *Legendre polynomials*. Compare the first few q_j's to Eq. (2.79). Also, use that equation to confirm that the normalization of the Legendre polynomials obeys:

$$(P_n(x), P_m(x)) \equiv \int_{-1}^{1} P_n(x) P_m(x) dx = \frac{2}{2n+1} \delta_{nm} \tag{4.259}$$

(e) Our constructive proof above is convincing when it comes to the orthogonality of Legendre polynomials of different degrees ($n \neq m$), since we are explicitly subtracting out any component that is not orthogonal. However, our checking of the normalization ($n = m$) for only the first few cases leaves something to be desired. In preparation for tackling the case of general n, first show that $P_n(1) = 1$ and $P_n(-1) = (-1)^n$; to see this, write out the left-hand side of Eq. (2.81) for $x = 1$ as a geometric series, consider the right-hand side of Eq. (2.81), and then use the uniqueness of the power series (finally, repeat the argument for $x = -1$). Armed with your latest results, as well as Eq. (2.116), explicitly compute $\int_{-1}^{1} [P_n(x)]^2 dx$ by carrying out an integration by parts.

4.28 We now use *Householder transformations* to carry out a QR decomposition.[62]

(a) We define a *Householder matrix* as $\mathbf{P} = \mathcal{I} - 2\mathbf{w}\mathbf{w}^T$, where $\mathbf{w}$ is a unit-norm vector. (Our definition involves the product of a column vector and a row vector; we will see something similar in Eq. (5.89).) Show that $\mathbf{P}$ is symmetric and orthogonal. You have thereby shown that $\mathbf{P}^2 = \mathcal{I}$.

(b) If two vectors $\mathbf{x}, \mathbf{y}$ satisfy $\mathbf{P}\mathbf{x} = \mathbf{y}$, then show that $\mathbf{w} = (\mathbf{x} - \mathbf{y})/\|\mathbf{x} - \mathbf{y}\|$.

(c) Specialize to $\mathbf{y} = \alpha\|\mathbf{x}\|\mathbf{e}_0$, where $\mathbf{e}_0$ is the first column of the identity matrix, as per Eq. (4.100). (Discuss why you should choose $\alpha = -\text{sign}(x_0)$.) Notice that $\mathbf{P}\mathbf{x}$ zeroes out all the elements of a general vector $\mathbf{x}$ below the leading element; this is reminiscent of the first step in Gaussian elimination.

(d) Produce a Householder matrix $\mathbf{P}^{(0)}$ that zeroes out the all the elements except the leading one in the first column of an $n \times n$ matrix $\mathbf{A}$. This $\mathbf{P}^{(0)}$ acting on the entire $\mathbf{A}$ thereby transforms the other elements (outside the first column). We now need to zero out the elements in $\mathbf{P}^{(0)}\mathbf{A}$ below its 1, 1 element. We form a new Householder matrix $\mathbf{P}^{(1)}$ based on the lower $n-1$ elements of the second

[62] A pithy (yet somewhat hard to memorize) summarization: the Gram–Schmidt technique carries out triangular orthogonalization, while the Householder method proceeds via orthogonal triangularization [148].

column of $\mathbf{P}^{(0)}\mathbf{A}$; note that this leads to $\mathbf{P}^{(1)}$ being $(n-1)\times(n-1)$. Similarly, $\mathbf{P}^{(2)}$ would be $(n-2)\times(n-2)$, and so on. We opt for the (wasteful) option of padding:

$$\mathbf{Q}^{(k)} = \begin{pmatrix} \mathcal{I} & \mathbf{0} \\ \mathbf{0} & \mathbf{P}^{(k)} \end{pmatrix} \tag{4.260}$$

where $\mathcal{I}$ is $k\times k$, $\mathbf{P}^{(k)}$ is $(n-k)\times(n-k)$ and, therefore, $\mathbf{Q}^{(k)}$ is $n\times n$. Explain why $\mathbf{R} = \mathbf{Q}^{(n-2)}\mathbf{Q}^{(n-3)}\ldots\mathbf{Q}^{(1)}\mathbf{Q}^{(0)}\mathbf{A}$ is upper triangular. Use the fact that these matrices are orthogonal to show that $\mathbf{Q} = \mathbf{Q}^{(0)}\mathbf{Q}^{(1)}\ldots\mathbf{Q}^{(n-3)}\mathbf{Q}^{(n-2)}$ is also orthogonal. Multiplying $\mathbf{Q}$ and $\mathbf{R}$ gives you the initial $\mathbf{A}$.

(e) Implement this prescription in Python to produce a QR decomposition.

4.29 [$\mathcal{A}$] Show that $\mathbf{A}'$ from Eq. (4.161), $\mathbf{A}' = \mathbf{S}^{-1}\mathbf{A}\mathbf{S}$, and $\mathbf{A}$ have the same determinant.

4.30 [$\mathcal{A}$] Show that a number z is a Rayleigh quotient of the matrix $\mathbf{A}$ if and only if it is a diagonal entry of $\mathbf{Q}^T\mathbf{A}\mathbf{Q}$, for an orthogonal matrix $\mathbf{Q}$. (Recall that "if and only if" means you have to prove this both ways.) The conclusion that follows from this is that Rayleigh quotients are simply diagonal entries of matrices (but you first have to orthogonally transform to the appropriate coordinate system).

4.31 Finding the eigenvectors of an upper triangular matrix $\mathbf{U}$ is (much) easier than finding the eigenvectors of a general matrix. Most obviously, the first column of the identity matrix, $\mathbf{e}_0$ as per Eq. (4.100), is an eigenvector of our triangular matrix. To see how to find the rest, first partition our $n \times n$ matrix as follows:

$$\mathbf{U} = \begin{pmatrix} \mathbf{U}_{00} & \mathbf{U}_{0i} & R \\ 0 & U_{ii} & S \\ 0 & 0 & T \end{pmatrix} \tag{4.261}$$

where, crucially, U_{ii} is a matrix *element* and $\mathbf{U}_{00}$ is a *submatrix*; this means that $\mathbf{U}_{0i}$ is a *vector* (R, S, and T won't matter in what follows).

First, write down the eigenvector of $\mathbf{U}$ with eigenvalue U_{ii} as $\begin{pmatrix}\mathbf{x}_0 & 1 & 0\end{pmatrix}^T$ and show that you can find $\mathbf{x}_0$ by solving the upper-triangular system of equations:

$$(\mathbf{U}_{00} - U_{ii}\mathcal{I})\mathbf{x}_0 = -\mathbf{U}_{0i} \tag{4.262}$$

Second, implement this approach programmatically to find all the eigenvectors of an upper-triangular matrix. (Return unit-norm versions of the eigenvectors.)

4.32 [$\mathcal{A}$] We now show that, for the algorithm of simultaneous iteration, $\mathbf{Q}^{(k)}$ converges toward $\tilde{\mathbf{Q}}$. First, use Eq. (4.125) to show that $\mathbf{A}^k = \mathbf{V}\mathbf{\Lambda}^k\mathbf{V}^{-1}$. Then, carry out two decompositions, $\mathbf{V} = \tilde{\mathbf{Q}}\mathbf{U}$ and $\mathbf{V}^{-1} = \mathbf{L}\mathbf{R}$, where $\tilde{\mathbf{Q}}$ is orthogonal, $\mathbf{L}$ is unit lower-triangular, with $\mathbf{U}$ and $\mathbf{R}$ being upper triangular. Combining these three equations, and introducing an identity expressed as $\mathbf{\Lambda}^{-k}\mathbf{\Lambda}^k$, a unit lower-triangular term of the form $\mathbf{\Lambda}^k\mathbf{L}\mathbf{\Lambda}^{-k}$ emerges: argue why its off-diagonal elements go to 0 as k is increased. You have thereby shown that $\mathbf{A}^k$ is equal to $\tilde{\mathbf{Q}}$ times an upper-triangular matrix. Comparing this to Eq. (4.178), since the factorization is unique, identify the large-k version of $\mathbf{Q}^{(k)}$ with $\tilde{\mathbf{Q}}$.

4.33 This problem implements simultaneous iteration step by step.

(a) Start with Eq. (4.172), namely $\mathbf{Z}^{(k)} = \mathbf{A}\mathbf{Z}^{(k-1)}$. Implement this and check how well you do on the eigenvectors for the 4×4 `testcreate()` example.
(b) Now repeat the previous prescription and simply normalize each column at the end (without any QR decomposition).
(c) Implement the full simultaneous iteration from Eq. (4.176) and compare $\mathbf{Q}^{(k)}$ with the (QR-decomposed) output of `numpy.linalg.eig()`.
(d) Implement $\mathbf{A}^{(k)}$ as per Eq. (4.182) and compare with the `numpy.linalg.eig()` eigenvalues.

4.34 [$\mathcal{A}$] Including all contributions (not just the leading term), evaluate the operation counts for (a) vector–vector multiplication, (b) matrix–matrix multiplication, (c) linear system solution via the LU method, and (d) QR decomposition.

4.35 Use the QR method to compute the orthogonal factor of the eigenvector matrix. If you followed the derivation in the main text, you'll know that this means you should write a Python code that modifies `qrmet()` to also compute $\mathbf{Q}^{(k)}$.

4.36 We will now see how to make the QR method more efficient, by introducing two interrelated ideas: *shifting and deflation*. Instead of Eq. (4.185), the prescription is:

$$\mathbf{A}^{(k-1)} - s^{(k)}\mathcal{I} = \mathbf{Q}^{(k)}\mathbf{R}^{(k)}, \qquad \mathbf{A}^{(k)} = \mathbf{R}^{(k)}\mathbf{Q}^{(k)} + s^{(k)}\mathcal{I} \tag{4.263}$$

where we also allow for the possibility of tuning the shift $s^{(k)}$ iteratively.

(a) Show that the second relation in Eq. (4.186) carries over, unchanged.
(b) Look at the bottom-right element of our matrix iterate:

$$A^{(k)}_{(n-1),(n-1)} = [\mathbf{e}_{n-1}]^T \mathbf{A}^{(k)} \mathbf{e}_{n-1} \tag{4.264}$$

where $\mathbf{e}_{n-1}$ is the last column of the identity matrix, as per Eq. (4.100). Show why this equation, combined with the second relation in Eq. (4.186), can be used to make a connection with the Rayleigh quotient. As a result, you can always pick $s^{(k)}$ to be the (latest) bottom-right element of $\mathbf{A}^{(k)}$.
(c) Assume that the shift $s^{(k)}$ is exactly equal to an eigenvalue. Explain why $\mathbf{R}^{(k)}$ is singular in that case and use that fact to show that:

$$\mathbf{A}^{(k)} = \begin{pmatrix} \mathbf{A}' & \mathbf{a} \\ 0 & s^{(k)} \end{pmatrix} \tag{4.265}$$

that is, the last row is full 0s, except for the last element (which is the eigenvalue). This is a smaller problem: $\mathbf{A}'$ has dimensions $(n-1)\times(n-1)$, so we can focus on it to find another eigenvalue; this is *deflation*. You can repeat this process, down to the point where you are left with a 1×1 (or 2×2) submatrix.
(d) Implement shifting and deflation for the QR method, as introduced in the previous parts of this problem. Employ some criterion on how small the elements on the last row (except for the last element) need to be each time, before you declare them to be "zero". Test your function on: (i) the $\mathbf{A}$ of `testcreate(4,21)` and (ii) a matrix of whose (distinct) eigenvalues two have the same magnitude (e.g., ± 2). Recall that the unshifted QR method struggles in the latter case.

4.37 Sometimes even real matrices have complex eigenvalues: the shifted QR method (of problem 4.36) would need to be re-implemented in complex arithmetic. We will now see an ingenious trick, employing a *double shift*, that manages to keep working with real numbers and still keep the same speed of convergence. Assume (for now) that your starting matrix is $\mathbf{A}^{(k-1)}$ and focus on its bottom-right 2×2 submatrix:

$$\mathbf{B} = \begin{pmatrix} A^{(k-1)}_{(n-2),(n-2)} & A^{(k-1)}_{(n-2),(n-1)} \\ A^{(k-1)}_{(n-1),(n-2)} & A^{(k-1)}_{(n-1),(n-1)} \end{pmatrix} \tag{4.266}$$

This submatrix has two eigenvalues, γ_0 and γ_1, which may be real or a complex conjugate pair. The idea is to decide to carry out not one, but two shifted QR steps:

$$\begin{aligned} \mathbf{A}^{(k-1)} - \gamma_0 \mathcal{I} &= \mathbf{Q}^{(k)}\mathbf{R}^{(k)}, \qquad & \mathbf{A}^{(k)} &= \mathbf{R}^{(k)}\mathbf{Q}^{(k)} + \gamma_0 \mathcal{I}, \\ \mathbf{A}^{(k)} - \gamma_1 \mathcal{I} &= \mathbf{Q}^{(k+1)}\mathbf{R}^{(k+1)}, \qquad & \mathbf{A}^{(k+1)} &= \mathbf{R}^{(k+1)}\mathbf{Q}^{(k+1)} + \gamma_1 \mathcal{I} \end{aligned} \tag{4.267}$$

Even if you start with a real matrix, a complex shift would require complex arithmetic; even assuming that the second complex shift *should* bring you back to the reals, this is not likely to happen, due to numerical error (recall chapter 2) .

(a) First, show that:

$$\left[\mathbf{Q}^{(k)}\mathbf{Q}^{(k+1)}\right]\left[\mathbf{R}^{(k+1)}\mathbf{R}^{(k)}\right] = \left[\mathbf{A}^{(k-1)}\right]^2 - (\gamma_0 + \gamma_1)\mathbf{A}^{(k-1)} + (\gamma_0\gamma_1)\mathcal{I} \tag{4.268}$$

Then, observe that the right-hand side is a real matrix, even if γ_0 and γ_1 are complex. To see this, recall the elementary properties of the eigenvalues of 2×2 matrices: $\gamma_0 + \gamma_1 = \text{trace}(\mathbf{B})$ and $\gamma_0\gamma_1 = \det(\mathbf{B})$.

(b) Interpret the left-hand side of Eq. (4.268) as the QR-decomposition of the real matrix on the right-hand side. Set $\mathbf{Z} = \mathbf{Q}^{(k)}\mathbf{Q}^{(k+1)}$ and show that:

$$\mathbf{A}^{(k+1)} = \mathbf{Z}^T \mathbf{A}^{(k-1)} \mathbf{Z} \tag{4.269}$$

starting from the second relation in Eq. (4.186) which, as you showed in problem 4.36, is still true (for a single shift). This result means that we can carry out the double shift without having to actually use Eq. (4.267): Eq. (4.269) takes us from $\mathbf{A}^{(k-1)}$ directly to $\mathbf{A}^{(k+1)}$.

(c) Use the above results to implement the double-shift QR method. To be explicit: first, form the real matrix on the right-hand side of Eq. (4.268). Second, QR decompose that. Third, use Eq. (4.269) to produce your updated matrix. You should repeat all of these steps several times. (No need to employ deflation.) It goes without saying that you should use floats (i.e., not complex numbers) throughout this process. The end result will be a matrix that is "almost" triangular: real eigenvalues will lie on the diagonal, whereas each complex-conjugate eigenvalue pair will correspond to a single sub-diagonal matrix element. In other words, a complex-conjugate eigenvalue pair will give rise to a 2×2 submatrix centered on the diagonal (from which you can extract the complex-conjugate eigenvalue pair analytically or otherwise). Test your Python function on the 5×5 matrix:

```
A[0,:] = 2; np.fill_diagonal(A[1:,:], 3)
```

4.38 We now use a single shift in the QR method (as per Eq. (4.263)), chosen in a way that makes it more efficient than the simple case of Eq. (4.264). We limit ourselves to real matrices with real eigenvalues. The new idea is known as the *Wilkinson shift*.

(a) Start with the 4×4 non-symmetric matrix made up of the following numbers $-10, +2, +3, +4, \ldots, +14, +15, -10$ distributed across rows. Wilkinson said to pick as the shift $s^{(k)}$ that eigenvalue of the matrix **B** (from Eq. (4.266)) which is closest to the bottom-right element, $A^{(k-1)}_{(n-1),(n-1)}$. In the case of a tie, arbitrarily pick one of the two eigenvalues. Implement this approach and compare the number of iterations required with that following from the shift of Eq. (4.264).

(b) Turn to the case of a symmetric matrix, for example the tridiagonal 4×4 matrix with 2's on the main diagonal and 1's on the two diagonals next to it. For a symmetric matrix, the two off-diagonal elements of **B** are equal to each other. In this case, a numerically stable expression for the Wilkinson shift is:

$$s^{(k)} = A^{(k-1)}_{(n-1),(n-1)} - \text{sign}(\delta)\frac{\left[A^{(k-1)}_{(n-2),(n-1)}\right]^2}{|\delta| + \sqrt{\delta^2 + \left[A^{(k-1)}_{(n-2),(n-1)}\right]^2}} \tag{4.270}$$

where $\delta = [A^{(k-1)}_{(n-2),(n-2)} - A^{(k-1)}_{(n-1),(n-1)}]/2$. If it happens that $\delta = 0$, you can take $\text{sign}(\delta)$ to be either $+1$ or -1. Implement this approach and compare the number of iterations required with that following from the shift of Eq. (4.264).

(c) We'll now turn to a case where even the symmetric-matrix Wilkinson shift of Eq. (4.270) gets into trouble. Take **A** to be the result of `testcreate(4,21)` and produce **H** as per Eq. (4.222); this **H** is symmetric, even when the **A** you started from is non-symmetric (as in our example). Run the code you developed in the previous part for **H** and see it fail. What is going wrong is that for such a problem, at the outset, all four elements of the submatrix **B** are zero, so Eq. (4.270) leads to NaN; similarly, the simpler approach of Eq. (4.264) leads to a vanishing shift (i.e., no shift). Modify the function you wrote for the symmetric-matrix Wilkinson shift in the previous part, such that it now checks to see if it's dealing with this scenario and, if so, it arbitrarily sets the shift to $+1$. Also update (the same way) the function you wrote that uses the shift of Eq. (4.264).

4.39 The conclusion of problems 4.32 and 4.35, namely that $\mathbf{Q}^{(k)}$ gives us the orthogonal factor of the eigenvector matrix, may have felt underwhelming. We will now see how triangularization helps not only with the computation of the eigenvalues (which are on the diagonal), but also with finding the eigenvectors.

(a) From the second equation in Eq. (4.186) we have:

$$\mathbf{A}^{(k)}[\mathbf{Q}^{(k)}]^T = [\mathbf{Q}^{(k)}]^T\mathbf{A} \tag{4.271}$$

and from Eq. (4.125) we have $\mathbf{AV} = \mathbf{V\Lambda}$. Combine these two relations to show that $[\mathbf{Q}^{(k)}]^T\mathbf{V}$ is the right-eigenvector matrix corresponding to $\mathbf{A}^{(k)}$.

(b) Using the approach of problem 4.31, you can compute eigenvectors of triangular matrices. Combine that with the code you developed for problem 4.35 (for the unshifted QR method discussed in the main text) to compute the eigenvectors of $\mathbf{A}^{(k)}$ and then compute the eigenvectors of a general matrix $\mathbf{A}$ as per part (a) of the present problem.

(c) Code up the same approach for the case of the shifted QR method (introduced in problem 4.36). Since the $\mathbf{Q}^{(k)}$'s now don't all have the same dimensions (due to deflation), you should "pad" them with an (appropriately dimensioned) identity matrix at the bottom right.

For the last two parts, you should first use `qrdec()` and then `np.linalg.qr()`. To understand why they behave differently, check $[\mathbf{Q}^{(k)}]^T\mathbf{Q}^{(k)}$ in both cases.

4.40 [$\mathcal{A}$] Show that an orthogonal matrix $\mathbf{Q}$ is norm-preserving (for the 2-norm), i.e., $\|\mathbf{QA}\|_2 = \|\mathbf{A}\|_2$ and $\|\mathbf{AQ}\|_2 = \|\mathbf{A}\|_2$ hold. Hint: work with the square of the norm.

4.41 We will examine how to compute the SVD in a manner that was described in the main text as being unstable: we will make use of the matrix $\mathbf{A}^T\mathbf{A}$. As in `svd.py`, we here limit ourselves to square matrices.

(a) Plug $\mathbf{A} = \mathbf{U\Sigma V}^T$ into $\mathbf{A}^T\mathbf{A}$ and show that a diagonalization of the latter can be used to find the (squares of the) singular values of $\mathbf{A}$ as well as the matrix $\mathbf{V}$ in its SVD. Motivated by this success, proceed to diagonalize $\mathbf{AA}^T$ to find the matrix $\mathbf{U}$. Implement this approach (using `np.linalg.eig()` for the diagonalizations) for the same example $\mathbf{A}$ as in `svd.py`. Form the product $\mathbf{U\Sigma V}^T$ and compare it against $\mathbf{A}$. Do you understand what went wrong? Feel free to benchmark against the output of `np.linalg.svd()`.

(b) Thinking of how to improve things, we realize that once we've computed $\mathbf{\Sigma}$ and $\mathbf{V}$, we don't actually need to resort to a diagonalization of $\mathbf{AA}^T$. Instead, rewrite $\mathbf{A} = \mathbf{U\Sigma V}^T$ as $\mathbf{AV} = \mathbf{U\Sigma}$ and solve for $\mathbf{U}$. Implement this approach, once again employing `np.linalg.eig()` for the diagonalization.

(c) Repeat part (b), this time using `qrmet()` and `invpowershift()` for the diagonalization. Compare $\mathbf{U}$, $\mathbf{\Sigma}$, and $\mathbf{V}$ to what you were getting before.

(d) Using the function you developed for part (b), turn to the example matrix that comes out of `testcreate(4,21)`. Check both the singular values and the SVD reconstruction of $\mathbf{A}$. Now, for the $\mathbf{A}$ of `testcreate(4,21)`, run our `svd()` from code 4.11; in order to compare apples with apples, use `np.linalg.eigvals()` as the `solver()`. What do you find?

4.42 This problem addresses the relationship between the SVD and the condition number for solving linear systems, $\kappa(\mathbf{A})$.

(a) First, show that our earlier definition of the spectral norm in Eq. (4.208) has the consequence that $\|\mathbf{A}\|_2 = \sigma_0$. In words: the 2-norm of a matrix $\mathbf{A}$ is the square root of the maximum eigenvalue of $\mathbf{A}^T\mathbf{A}$ (i.e., the dominant singular value of $\mathbf{A}$). Then, show that $\|\mathbf{A}^{-1}\|_2 = 1/\sigma_{n-1}$. Combine the last two results to find:

$$\kappa(\mathbf{A}) = \frac{\sigma_0}{\sigma_{n-1}} \tag{4.272}$$

This way you don't need to actually compute the inverse to compute $\mathbf{A}$'s condition number, consistently with what we noted at the end of section 4.2.3.

(b) Show that the condition number of the matrix $\mathbf{A}^T\mathbf{A}$ is equal to the square of the condition number of the matrix $\mathbf{A}$:

$$\kappa(\mathbf{A}^T\mathbf{A}) = \kappa(\mathbf{A})^2 \tag{4.273}$$

Thus, as was stated in the main text and experienced in problem 4.41, it is increasingly unstable to use $\mathbf{A}^T\mathbf{A}$ if what you're really after is the SVD of $\mathbf{A}$.[63]

(c) Use `np.linalg.cond()` as well as `np.linalg.svd()`—as per Eq. (4.272)—to compute the condition numbers of $\mathbf{A}$ and of $\mathbf{A}^T\mathbf{A}$, where $\mathbf{A}$ is the matrix that comes out of `testcreate(4,21)`; this shows how much worse the ill-conditioning becomes.

4.43 We will now examine the relationship between a matrix's singular values and its invertibility. Along the way, we will also learn how to use the SVD to solve a *homogeneous linear system* of equations, $\mathbf{Ax} = \mathbf{0}$.

(a) Show that the inverse of a matrix $\mathbf{A}$ (if it exists) satisfies:

$$\mathbf{A}^{-1} = \mathbf{V}\boldsymbol{\Sigma}^{-1}\mathbf{U}^T \tag{4.274}$$

Then, show that this holds programmatically (for an example $\mathbf{A}$).
The *rank* of a matrix (i.e., the number of linearly independent columns) is equal to the number of its non-zero singular values. For an $n \times n$ matrix $\mathbf{A}$, we say that $\mathbf{A}$ has *full rank* when the rank is n; otherwise, we say that $\mathbf{A}$ is *rank deficient*. Equation (4.274) tells you that you shouldn't be looking for the inverse if your matrix is rank deficient—however, see problem 6.43.

(b) Come up with a 5×5 matrix $\mathbf{A}$ for which you can find a non-trivial solution to $\mathbf{Ax} = \mathbf{0}$ (i.e., a solution other than $\mathbf{x} = \mathbf{0}$).

(c) Using both our `svd()` and `np.linalg.svd()` compute your $\mathbf{A}$'s singular values. Explain why our `svd()` cannot be used to compute $\mathbf{U}$ and $\mathbf{V}$.

(d) Find a non-trivial solution to $\mathbf{Ax} = \mathbf{0}$ for your $\mathbf{A}$ matrix, by employing the right-singular vector(s) corresponding to the vanishing singular value(s). The $\mathbf{x}$ you will find belongs to $\mathbf{A}$'s (right) *null space*.

4.44 This problem discusses what is known as a *low-rank approximation* to a matrix $\mathbf{A}$. This comes up in the study of image compression: instead of carrying a lot of unnecessary information around in a huge matrix, one only needs a handful of (judiciously chosen) quantities. The idea is to start from the SVD, $\mathbf{A} = \mathbf{U}\boldsymbol{\Sigma}\mathbf{V}^T$, and then produce a new matrix as follows:

$$\tilde{\mathbf{A}}^{(k)} = \sum_{i=0}^{k} \sigma_i \mathbf{u}_i \mathbf{v}_i^T \tag{4.275}$$

where $\mathbf{u}_i$ and $\mathbf{v}_i$ are the i-th column of $\mathbf{U}$ and $\mathbf{V}$, respectively. Note that this equation employs the outer product (producing a matrix); we will encounter something similar in section 5.4.3. As a specific example, use Matplotlib's `imshow()` to visualize the output (`Phi`) of code 8.8 (`poisson.py`). Then, use `np.linalg.svd()` and implement

[63] We will re-encounter this result in Eq. (6.155) but the dimensionality/ill-conditioning, there, will be small.

Eq. (4.275); plot $\tilde{\mathbf{A}}^{(0)}$, $\tilde{\mathbf{A}}^{(1)}$, $\tilde{\mathbf{A}}^{(2)}$, $\tilde{\mathbf{A}}^{(3)}$, and $\tilde{\mathbf{A}}^{(4)}$. To further understand what's going on, plot the singular values of `Phi` vs their cardinal number (i.e., σ_i vs i) and discuss what you see.

4.45 We now turn to the *SVD of a rectangular matrix* $\mathbf{A}$. Take $\mathbf{A}$'s dimensions to be $m \times n$, where we limit ourselves for concreteness to the case of $m \geq n$ (a tall/skinny matrix). The "full SVD" of this matrix, $\mathbf{A} = \mathbf{U}\boldsymbol{\Sigma}\mathbf{V}^T$, would involve a $\mathbf{U}$ that is $m \times m$, a $\boldsymbol{\Sigma}$ that is $m \times n$, and a $\mathbf{V}$ that is $n \times n$.
You should implement, instead, what is known as the *reduced SVD* (also known as the "compact SVD"): this takes into account the fact that the rightmost columns of $\mathbf{U}$ would only be acting on the bottom-most rows of $\boldsymbol{\Sigma}$ (which are full of zeros). Thus, the reduced SVD amounts to not bothering with these "silent" columns and rows and working with a $\mathbf{U}$ that is $m \times n$ and a $\boldsymbol{\Sigma}$ that is $n \times n$. Test your implementation against `np.linalg.svd()`'s output (passing in the optional argument `full_matrices=False`) for the matrix produced by the following call:

```
np.sqrt(np.arange(21,36).reshape(5,3))
```

Crucially, your $\boldsymbol{\Sigma}$ should contain only the non-zero singular values.

4.46 You may be disappointed to hear that our `svd.py` cannot handle the $\mathbf{A}$ matrix coming from `testcreate(4,21)`. To get this to work, first, you will need to combine the shifted QR method of problem 4.36 with the improved symmetric-matrix Wilkinson shift (introduced in part (c) of problem 4.38). This should work great for the singular values. However, the reconstruction of the $\mathbf{A}$ matrix fails, again. Find a couple of ways of tweaking the body of `svd()` to eliminate this issue.

4.47 [$\mathcal{P}$] In our study of differential equations, in chapter 8, we will learn how to tackle the one-particle time-independent Schrödinger equation for any potential term:

$$\frac{\hat{p}^2}{2m}|\psi\rangle + V(\hat{x})|\psi\rangle = E|\psi\rangle \tag{4.276}$$

Here, we focus on an important specific case, the quantum harmonic oscillator, $V(\hat{x}) = m\omega^2\hat{x}^2/2$. In the position basis this Schrödinger equation takes the form of Eq. (3.70). In a course on quantum mechanics you also learned about the energy basis, involving Dirac's trick with creation and annihilation operators. In terms of the energy eigenstates $|n\rangle$, the matrix elements of the position and momentum operators are:

$$\begin{aligned}\langle n'|\hat{x}|n\rangle &= \sqrt{\frac{\hbar}{2m\omega}}\left[\sqrt{n}\delta_{n',n-1} + \sqrt{n+1}\delta_{n',n+1}\right]\\ \langle n'|\hat{p}|n\rangle &= i\sqrt{\frac{\hbar m\omega}{2}}\left[\sqrt{n+1}\delta_{n',n+1} - \sqrt{n}\delta_{n',n-1}\right]\end{aligned} \tag{4.277}$$

These are the elements of infinite-dimensional matrices: n and n' start at 0, but they don't go up to a given maximum value. Keep in mind that the eigenvalues involved (the n that keeps track of the cardinal number of the eigenenergy) are *discrete*, even though we have not carried out any numerical discretization procedure ourselves; this is just the nature of the problem.
Truncate these matrices up to a maximum of, say, $n = 10$, $n' = 10$ and implement them. Then, by using matrix multiplications to carry out the necessary squarings,

verify that the Hamiltonian is diagonal in this basis; this is hardly surprising, but it is reassuring. Compare the eigenvalues to the (analytically known) answers of Eq. (3.71). Did you expect the value you found in the bottom element of the diagonal? To see what's going on, evaluate the trace of $\mathbf{XP} - \mathbf{PX}$, where $\mathbf{X} = \{\langle n'|\hat{x}|n\rangle\}$ and $\mathbf{P} = \{\langle n'|\hat{p}|n\rangle\}$, and compare with your expectations from Heisenberg's commutation relation.

4.48 [$\mathcal{P}$] The angular-momentum addition rule tells us that if we have a particle with j_0 and another one with j_1, the angular momentum of the pair, J, satisfies:

$$|j_0 - j_1| \leq J \leq j_0 + j_1 \tag{4.278}$$

You should apply this rule (twice) to the case of three electron spins. That means that you should first add two of the angular momenta, and then add to the resulting pair angular momentum the third angular momentum you had left out (this is known as *recoupling*). Employing this approach, you should find the total possible spin and the z projection of the spin for the 3-spin-half system. You should then employ `threespins()` to find the same answer. (If you try to use `twospins()` to evaluate the same limit – even though the Hamiltonian matrix is still diagonal – something breaks, as mentioned in the main text. Do you understand why?)

4.49 [$\mathcal{P}$] Employ our two and three spin codes, for the same input parameters as in the main text, to produce the eigenvectors/state vectors using (a version of) our `eig()` function. You first have to modify our implementation of the shifted inverse-power method so that it doesn't compare the last two iterations, but simply carries out a fixed total number of iterations. (You should discuss why this step is necessary here.) You can use `numpy.linalg.eig()` as a benchmark while you're developing your version.

4.50 [$\mathcal{P}$] Study the case of four spins. You should first rewrite the expression for the Hamiltonian: this will instruct you on how the spin operators and matrices should be produced for the present case. Finally, code this up in Python and test the eigenvalues to make sure that they reduce to the correct non-interacting limit. (It's OK to use `numpy.linalg.eig()` if you're having trouble converging in that limit.)

4.51 [$\mathcal{P}$] Generalize our codes to generate, say, the $\mathbf{S}_{\text{I}x}$ matrix for the case of five (or 10) particles. The number of particles should be kept general, passed in as an argument. The matrices won't fit on the screen, so print out selected matrix elements as a test. Your setup should be such that you could handle a different matrix, e.g., $\mathbf{S}_{\text{IV}y}$.

4.52 [$\mathcal{P}$] This problem and the following one study the vibrations of the *harmonic chain*, which can be viewed as a model of a linear polyatomic molecule. (It is also the discrete version of the continuous rod discussed around Eq. (3.3).) This is illustrated in Fig. 4.2, which shows a number of masses, m_i, connected by springs of constant k and equilibrium spacing a. The figure also shows a snapshot of horizontal vibration, marking each mass's displacement with respect to its rest position ($\eta_i = x_i - x_{0i}$). We allow for the possibility of applying a driving force on the first mass.

The figure shows the case of four masses and three springs, but it is easy to see how to generalize this to more masses/springs: the crucial point is that the first and last mass should be treated differently from "middle" masses. Using Newton's second

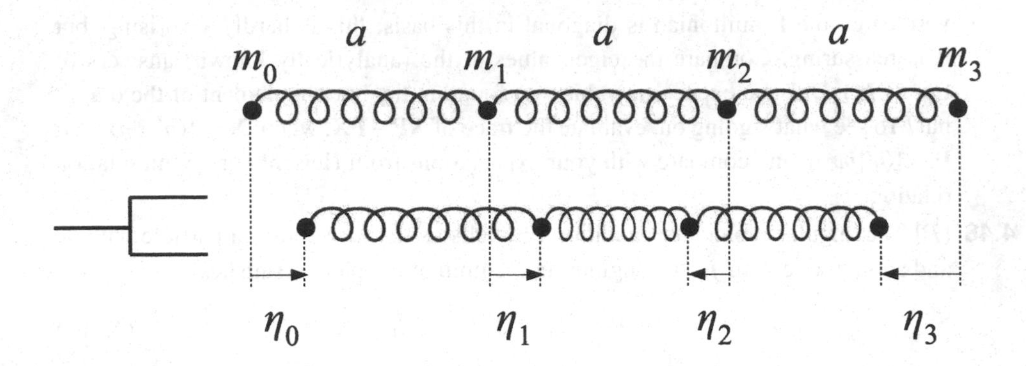

Fig. 4.2 Harmonic chain, made up of masses connected by springs, in and out of equilibrium

law we get the following equations of motion:

$$
\begin{aligned}
m_0 \frac{d^2\eta_0}{dt^2} &= k(\eta_1 - \eta_0) + F_d \\
m_i \frac{d^2\eta_i}{dt^2} &= k(\eta_{i+1} - \eta_i) + k(\eta_{i-1} - \eta_i), \qquad i = 1, 2, \ldots, n-2 \\
m_{n-1} \frac{d^2\eta_{n-1}}{dt^2} &= k(\eta_{n-2} - \eta_{n-1})
\end{aligned}
\tag{4.279}
$$

where we have a linear restoring force on each mass, in addition to allowing for a driving force F_d on the first mass. We need to treat the first and last masses differently, since they are connected to a single spring; each "middle" mass is connected to two springs. If we apply a harmonic driving force, $F_d = ce^{i\omega t}$, this will make all the masses oscillate with the same angular frequency, ω, i.e., $\eta_i = q_i e^{i\omega t}$. Plug this expression into Eq. (4.279) to produce a tridiagonal system of linear equations, $\mathbf{Ax} = \mathbf{b}$. Solve this system using `gauelim_pivot.py` for the case of $c = 2$, $k = 10$, $\omega = 3$; take $n = 50$. Plot the amplitudes q_i vs the index i and interpret the result for the cases where the individual masses are: (a) all equal to 1.0, and (b) $0.1, 0.2, \ldots, 5.0$.

4.53 [$\mathcal{P}$] Still on the subject of the harmonic chain, we now examine the scenario where there is no external driving force in Fig. 4.2 and Eq. (4.279), i.e., $F_d = 0$. Treat $\eta_i = q_i e^{i\omega t}$ as an *Ansatz*, i.e., plug it into Eq. (4.279) and look for normal-mode solutions. Show that the problem then takes the form:

$$\mathbf{Av} = \lambda \mathbf{Bv} \tag{4.280}$$

where $\lambda = \omega^2$ and $\mathbf{B}$ is the term containing the masses. Equation (4.280) is an example of what is known as the *generalized eigenvalue problem*. In the present case, since the mass-matrix is diagonal, we can trivially multiply both sides by $\mathbf{B}^{-1}$ to end up with a standard eigenvalue problem.

For $k = 10$ and $n = 50$, use `qrmet()` to plot the (sorted) possible ω values for the cases where the individual masses are: (a) all equal to 1.0, and (b) $0.1, 0.2, \ldots, 5.0$.

4.54 [$\mathcal{P}$] The discrete version of the moment of inertia tensor, Eq. (4.4) for n masses, is:

$$I_{\alpha\beta} = \sum_{k=0}^{n-1} m_k \left(\delta_{\alpha\beta} r_k^2 - (\mathbf{r}_k)_\alpha (\mathbf{r}_k)_\beta\right) \tag{4.281}$$

Note that α and β are Cartesian components, whereas k is labelling the masses. We will study the *ammonia molecule* NH_3, taking $m_0 = 14.0031$ and $m_{1-3} = 1.0079$ (both in atomic units). Take the positions relative to the center of mass to be:

$$\mathbf{r}_0 = (0, -5.9194916 \times 10^{-6}, 0.067766340)^T, \mathbf{r}_1 = (0, -0.93770592, -0.3138337)^T,$$
$$\mathbf{r}_2 = (0.8121, 0.46889408, -0.3138337)^T, \mathbf{r}_3 = (-0.8121, 0.46889408, -0.3138337)^T \tag{4.282}$$

Set up the matrix corresponding to the moment of inertia tensor and diagonalize it to compute the principal moments of the molecule, like in Eq. (4.6), After you sort the principal moments, you should find $I_0 = I_1 < I_2$, which means that you are dealing with an oblate symmetric top.

4.55 [$\mathcal{P}$] In the present problem we will see how to solve the time-independent Schrödinger equation, Eq. (3.69), for a general potential. As a reminder, for a one-particle problem in one dimension, this takes the form:

$$\left[-\frac{\hbar^2}{2m}\frac{d^2}{dx^2} + V(x)\right]\psi(x) = E\psi(x) \tag{4.283}$$

We will discuss differential equations in chapter 8. Here, we see how to recast this problem as a linear-algebra one; the general approach is sometimes known as *basis diagonalization*.

(a) Let us expand the (unknown) true wave function $\psi(x)$ in terms of a complete and orthonormal set of basis states, $\phi_n(x)$:

$$\psi(x) = \sum_{n=0}^{\infty} c_n \phi_n(x) \tag{4.284}$$

Show that if you multiply by $\phi_{n'}^*(x)$ and integrate over all of space, then the Schrödinger equation in Eq. (4.283) turns into the matrix eigenvalue problem $\mathbf{Hc} = E\mathbf{c}$, where the matrix $\mathbf{H}$ has the elements:

$$H_{n'n} = \int_{-\infty}^{+\infty} \phi_{n'}^*(x)\left[-\frac{\hbar^2}{2m}\frac{d^2}{dx^2} + V(x)\right]\phi_n(x) \tag{4.285}$$

The matrix-diagonalization techniques introduced in the present chapter will allow us to solve the one-body quantum-mechanical problem for a general potential $V(x)$—assuming we can somehow compute the needed derivatives and integrals in Eq. (4.285).

(b) Qualitatively, what we've done so far is to recast the ordinary differential equation in Eq. (4.283) as a matrix problem, using the (known) basis states, $\phi_n(x)$; the latter are chosen in order to make the calculations easier. As you can imagine, you could make a number of different choices for both the $V(x)$ (i.e., the specific problem you wish to solve) and the $\phi_n(x)$ (i.e., the tool you use to conveniently solve your problem).

In order to illustrate the principles involved, here we will combine the two problems introduced in our Project on derivatives from the previous chapter, section 3.5. We will compute the eigenvalues of the harmonic oscillator, $V(x) = m\omega^2 x^2/2$, pretending that we do not know about Eq. (3.71); to do so, we will use a plane-wave basis set, i.e., the set of eigenfunctions corresponding to a particle in a periodic box, Eq. (3.80). To employ our latest notation, the integral in Eq. (4.285) now goes from $-L/2$ to $+L/2$ and the basis states will be $\phi_n(x) = e^{i2\pi nx/L}/\sqrt{L}$. We will be truncating the expansion in Eq. (4.284) after a finite number of terms; thus, the matrix $\mathbf{H}$ will have dimensions $(2n_{max}+1)\times(2n_{max}+1)$ (recall Eq. (3.81), which tells us that *this* n can be either positive or negative). Employing the same approach as in Eq. (2.62), show that (up to a constant):

$$\int dx e^{i2\pi(n-n')x/L} x^2 = e^{i2\pi(n-n')x/L} \frac{L[iL^2 + 2(n-n')L\pi x - 2i(n-n')^2\pi^2 x^2]}{4(n-n')^3\pi^3} \tag{4.286}$$

where we assumed that $n' \neq n$. Distinguish between the two cases of $n' = n$ and $n' \neq n$ and use the above result to show that:

$$\int_{-L/2}^{L/2} dx e^{i2\pi(n-n')x/L} x^2 = (1-\delta_{n'n})(-1)^{n-n'} \frac{L^3}{2(n-n')^2\pi^2} + \delta_{n'n}\frac{L^3}{12} \tag{4.287}$$

(c) Combine Eq. (4.287) with the corresponding matrix element for the kinetic energy. Build up the full $\mathbf{H}$ matrix and then use `qrmet()` to compute the eigenvalues. Comparing against Eq. (3.71), see how large L and n_{max} need to be for you to get decent results for the first five eigenvalues.

Once you're done, you may be feeling a bit underwhelmed: we set up very general machinery, but then used it only to reproduce known behavior. Your impression will soon change: in chapters 6 and 7 we will learn techniques that can handle any integral of the form of Eq. (4.285), at which point we will be able to diagonalize an arbitrary single-particle potential. Furthermore, problem 4.56 shows us that the present formalism (after a slight tweak) is sufficiently robust to tackle a periodic array of oscillator potentials.

4.56 [$\mathcal{P}$] In problem 1.18 on the Kronig–Penney model, we briefly stated Bloch's theorem, Eq. (1.8), and observed that the wave number that appears in it obeys $-\pi < KL < \pi$; we didn't say so at the time, but this delimits the first *Brillouin zone*. Thus, in the presence of a periodic potential, the boundary condition is not $\phi_n(x+L) = \phi_n(x)$—as it was in problem 4.55 and in section 3.5—but Eq. (1.8). Instead of finding $k = 2\pi n/L$ (as in Eq. (3.81)), for a case of a periodic potential we get:

$$k = \frac{2\pi n}{L} + K \tag{4.288}$$

Put another way, the basis states are now not $\phi_n(x) = e^{i2\pi nx/L}/\sqrt{L}$, but:

$$\phi_n^K(x) = \frac{1}{\sqrt{L}} e^{iKx} e^{i2\pi nx/L} \tag{4.289}$$

The $(\phi_{n'}^{K}(x))^*$, which you need to write down the $\mathbf{H}$ that generalizes Eq. (4.285), is:

$$(\phi_{n'}^{K}(x))^* = \frac{1}{\sqrt{L}} e^{-iKx} e^{-i2\pi n' x/L} \tag{4.290}$$

This means that the K simply drops out of the potential part of $\mathbf{H}$; in other words, the only modification of the formalism in problem 4.55 that is required for the study of a periodic potential relates to the kinetic energy. That portion becomes:

$$\frac{\hbar^2}{2m}\left(\frac{2\pi n}{L}\right)^2 \to \frac{\hbar^2}{2m}\left(\frac{2\pi n}{L} + K\right)^2 \tag{4.291}$$

Modify the code you developed for problem 4.55, to cater to the present situation of a periodic potential: since your old $\mathbf{H}$ corresponded to a harmonic oscillator, your new one (with the modified kinetic term, as above) will correspond to a periodic array of harmonic potentials. Hand-pick many K values satisfying $-\pi < KL < \pi$ and produce a plot of the first five energy levels (E vs K). You have to diagonalize a new $\mathbf{H}$ matrix for each value of K. Just like in the (simpler) problem on the Kronig–Penney model, you should be finding band gaps, namely energy regions for which no solution exists. You have to be careful about the meaning/value of L, to ensure that you are probing an interesting regime.

4.57 [$\mathcal{P}$] We studied single-particle quantum mechanics in section 3.5, many interacting spins in section 4.6, and we will address many-particle quantum mechanics for spinless particles in section 7.8. What's missing here is the study of many-particle quantum mechanics for particles with spin.[64] Crucially, a many-particle wave function for the case where the particles obey, say, Fermi–Dirac statistics is *not* a simple product of single-particle wave functions.

Work in one spatial dimension and take the single-particle orbitals to be the set of eigenfunctions of a particle in a periodic box, Eq. (3.80), i.e., $\phi_j(x) = e^{i2\pi jx/L}/\sqrt{L}$. We are faced with n particles, with positions x_k for $k = 0, 1, \ldots, n-1$ and, correspondingly, n orbitals ϕ_j; the latter will be occupied in order of increasing energy—recall from Eq. (3.82) that $E_j = \hbar^2 4\pi^2 j^2/(2mL^2)$. Notice that (after $j = 0$) there is a degeneracy: the wave functions come in pairs, e.g., $j = \pm 3$ correspond to the same energy. The many-fermion wave function needs to be antisymmetric under the exchange of two particles; it can be conveniently written as a *Slater determinant* of the $n \times n$ matrix whose elements are $\phi_j(x_k)$. You can build up this $\mathbf{\Phi}$ by taking $\Phi_{kj} = \phi_j(x_k)$, as we will do on page 320. In the present setting, you could just as well have taken $\Phi_{jk} = \phi_j(x_k)$ but nothing would have changed, given that $\det(\mathbf{\Phi}^T) = \det(\mathbf{\Phi})$.

Note that $k = 0, 1, \ldots, n-1$ but j can be both positive and negative (and zero); specialize to the case of odd-n, for which you should observe that you will go up to a maximum j value appearing in $n = 2j_{max} + 1$ (make sure you understand this before proceeding). For $L = 1$ and nine particles with positions `np.arange(2,11)**0.4`, compute the Slater determinant (don't worry about the normalization); you should use Eq. (4.102) after you generalize `ludec.py` to work with complex numbers; your code should also work for, e.g., thirteen particles.

[64] This is touched upon in problem 7.51, for which the present problem provides scaffolding.

5 Zeros and Minima

But this age does not have the courage to cook up emblems for its own decline.

Karl Kraus

5.1 Motivation

5.1.1 Examples from Physics

Non-linear equations are omnipresent in physics. Here are a few examples:

1. **Van der Waals equation of state**
   As you may recall, the classical ideal gas "law" has the form:
   $$Pv = RT \tag{5.1}$$
   where P is the pressure, v is the molar volume, $R = 0.08206$ liter atm K^{-1} mole^{-1} is the gas constant, and T is the temperature. This ignores the finite size of the gas molecules, as well as the mutual attraction between molecules. Correcting for these features "by hand" (or via the formalism of statistical mechanics) leads to the so-called *van der Waals equation of state*:
   $$\left(P + \frac{a}{v^2}\right)(v - b) = RT \tag{5.2}$$
   where a and b are parameters that depend on the gas; for example, for methane $a = 2.253$ liter^2 atm mole^{-2} and $b = 0.04278$ liter mole^{-1}. Assume you know the pressure and the temperature and wish to evaluate the molar volume. As you can see by multiplying Eq. (5.2) by v^2, you get a cubic equation in v. Even though it's messy to do so, cubic equations can be solved analytically; however, van der Waals is not the final word on equations of state: you can keep adding more terms in an attempt to make your model more realistic (as is done in, e.g., the virial expansion), so the resulting equation becomes increasingly complicated.
2. **Quantum mechanics of a particle in a finite well**
   A standard part of any introduction to quantum mechanics is the study of a single particle in a one-dimensional finite square well potential. Assume that the potential is V_0 for $x < -a$ and $x > a$ and is 0 inside the well, i.e., for $-a < x < a$. Solving the one-dimensional time-independent Schrödinger equation leads to states of even and odd

parity. The even solutions obey the transcendental equation:

$$k \tan ka = \kappa \tag{5.3}$$

where k and κ are related to the mass m of the particle and to the energy of the state (and to V_0); k characterizes the wave function inside the well and κ that outside the well. These two quantities are related by:

$$k^2 + \kappa^2 = \frac{2mV_0}{\hbar^2} \tag{5.4}$$

In a first course on quantum mechanics you typically learn that these two equations must be solved for k and κ "graphically". But what does that really mean? And what would happen if the potential was slightly more complicated, leading to a more involved set of equations? It would be nice to have access to general principles and techniques.

3. **Gravitational force coming from two masses**

The previous examples involved one or two polynomial or transcendental equations. We now turn to a case where several quantities are unknown and the equations we need to solve are non-linear.

Consider two masses, M_R and M_S, located at $\mathbf{R}$ and $\mathbf{S}$, respectively. Crucially, while the masses are known, the positions where they are located are unknown. That means that we need to determine the coordinates of $\mathbf{R}$ and $\mathbf{S}$, i.e., $2 \times 3 = 6$ numbers. Thankfully, we are free to make force measurements on a test mass m at positions of our choosing. First, we carry out a measurement of the force on m at the position $\mathbf{r}_0$ (which we pick):[1]

$$\mathbf{F}_0 = -GM_R m \frac{\mathbf{r}_0 - \mathbf{R}}{|\mathbf{r}_0 - \mathbf{R}|^3} - GM_S m \frac{\mathbf{r}_0 - \mathbf{S}}{|\mathbf{r}_0 - \mathbf{S}|^3} \tag{5.5}$$

This is a vector relation so it corresponds to three scalar equations; however, we are dealing with six unknowns. To remedy the situation, we make another measurement of the force on m, this time at position $\mathbf{r}_1$:

$$\mathbf{F}_1 = -GM_R m \frac{\mathbf{r}_1 - \mathbf{R}}{|\mathbf{r}_1 - \mathbf{R}|^3} - GM_S m \frac{\mathbf{r}_1 - \mathbf{S}}{|\mathbf{r}_1 - \mathbf{S}|^3} \tag{5.6}$$

The last two vector relations put together give us six scalar equations in six unknowns. Due to the nature of Newton's gravitational force, these are coupled non-linear equations; this is not a problem you would enjoy approaching using paper and pencil.

5.1.2 The Problem(s) to Be Solved

At its most basic level, the problem we are faced with is as follows: you are given a function $f(x)$ and you need to find its *zeros*. In other words, you need to solve the equation $f(x) = 0$ (the solutions of this equation are known as *roots*).[2] Thus, our task is to solve:

$$f(x) = 0 \tag{5.7}$$

[1] Notice the formal similarity between this equation and Eq. (1.3).

[2] In practice, the words *root* and *zero* are often used interchangeably.

where $f(x)$ is generally non-linear. Just like our main problems in the chapter on linear algebra, our equation this time is very easy to write down. We are faced with one equation in one unknown: if the function $f(x)$ is linear, then the problem is trivial. If, however, it is non-linear then arriving at a solution may be non-trivial. The $f(x)$ involved may be a polynomial, a special function, or a complicated programming routine (which makes each function evaluation quite costly). In what follows, we will denote our root by x^*.

To give an example: $f(x) = x^2 - 5 = 0$ is very easy to write down. Anyone can "solve" it by saying $x^2 = 5$, followed by $x = \pm\sqrt{5}$. But what is the value of $\sqrt{5}$? Of course, you can use a calculator, or even `from math import sqrt` in Python. But how do *those* figure out which number gives 5 when squared? Roughly two thousand years ago, Hero of Alexandria came up with a specific prescription that addresses this question. In our case, we are not really interested in square roots, specifically, but in any non-linear equation. We will encounter several general-use methods in section 5.2.

A complication arises when you are faced with a single polynomial and need to evaluate its zeros. In equation form:

$$c_0 + c_1 x + c_2 x^2 + c_3 x^3 + \cdots + c_{n-1} x^{n-1} = 0 \tag{5.8}$$

This is a single equation, but it has $n - 1$ roots. As we will discover in section 5.3, this problem presents its own set of challenges.

A more complicated version of the earlier problem (one non-linear equation in one variable) is the case where we have n simultaneous non-linear equations in n unknowns:

$$\begin{cases} f_0(x_0, x_1, \ldots, x_{n-1}) & = 0 \\ f_1(x_0, x_1, \ldots, x_{n-1}) & = 0 \\ & \vdots \\ f_{n-1}(x_0, x_1, \ldots, x_{n-1}) & = 0 \end{cases} \tag{5.9}$$

which can be recast using our vector notation from chapter 4 as follows:

$$\mathbf{f}(\mathbf{x}) = \mathbf{0} \tag{5.10}$$

As we will discover in section 5.4, the crudest but most surefire method of solving a single non-linear equation cannot help us when we are faced with several simultaneous equations.

The final problem we attack in this chapter is related yet distinct: instead of trying to find the zeros of a function, find the points where the function attains a minimum (or, as we will see, a maximum):

$$\min \phi(\mathbf{x}) \tag{5.11}$$

where $\mathbf{x}$ bundles together the variables $x_0, x_1, \ldots, x_{n-1}$ but ϕ produces scalar values.[3] This problem is tackled in sections 5.5 and 5.6. As always, entire books have been written about each of these topics, so we will by necessity be selective in our exposition.

5.2 Non-linear Equation in One Variable

As noted in the previous section, our problem is very easy to write down: $f(x) = 0$. In order to make things concrete, in what follows we will be trying to solve a specific equation using a variety of methods. That equation is:

$$e^{x-\sqrt{x}} - x = 0 \tag{5.12}$$

Obviously, in our case $f(x) = e^{x-\sqrt{x}} - x$. Our problem involves both an exponential and a square root, so it is clearly non-linear.

As a general lesson, before we start discussing detailed root-finding methods, we first try to get some intuition on the specific problem we are facing. Most simply, we can do that by plotting the function; the result is shown in the left panel of Fig. 5.1. Just by looking at this plot, we see that our function has two zeros, one near $x^* \approx 1$ and another one near $x^* \approx 2.5$. As a matter of fact, we can immediately see by substituting that the first root is exactly $x^* = 1$. In the same spirit of trying to understand as much as possible regarding our problem, we also went ahead and plotted the first derivative of our function in the right panel of Fig. 5.1. We find that the derivative varies from negative to positive values, with the change from one to the other happening between our two roots. Similarly, we notice that for x small and x large the derivative is large in magnitude, but it has smaller values for intermediate x's.

Now, turning to ways of solving our equation: we'll need a method that can find both roots, preferably in comparable amounts of time. (While we already analytically know one of the roots, we'll keep things general.) The most naive approach, having already plotted the function, is to try out different values (or perhaps even a very fine grid) near where we expect the roots to be. This is not a great idea, for several reasons: first, it is very wasteful, requiring a large number of function evaluations. What do you do if the first grid spacing you chose was not fine enough? You pick a smaller grid spacing and start evaluating the function again. Second, as noted in the previous section, in some cases even evaluating the function a single time may be very costly, so you might not even be able to plot the function in the first place. Finally, it is easy to see that this brute-force approach will immediately fail when you are faced with a multidimensional problem (since you'd need a multidimensional grid, with costly function evaluations at each point on the grid).

At a big-picture level, we can divide root-finding methods into two large classes:

- **Bracketing methods**: these are approaches which start with two values of x, let us call

[3] Obviously, if $n = 1$ you need to minimize a function of a single variable.

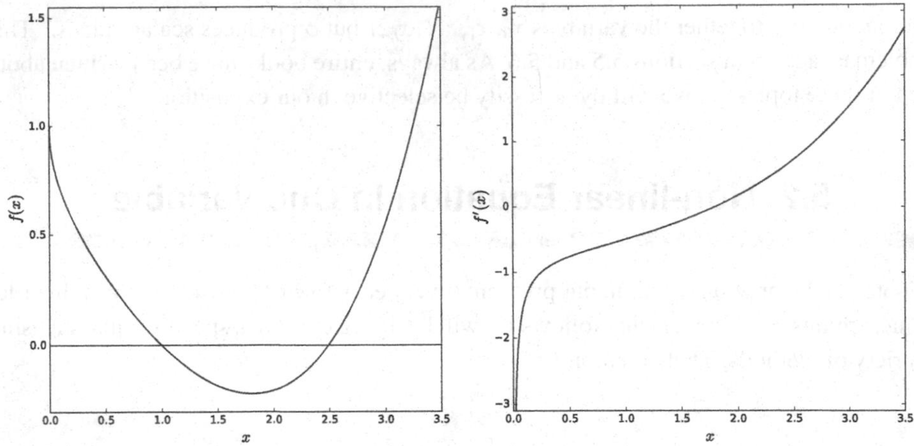

Fig. 5.1 Non-linear equation in one variable, for the case of $f(x) = e^{x-\sqrt{x}} - x$

them x_0 and x_1, which bracket the root (i.e., we already know that $f(x_0)$ and $f(x_1)$ have opposite signs). Of the methods we discuss below, bisection and Ridders' are bracketing methods.

- **Non-bracketing methods**, also known as open-domain methods: these approaches may need one or two starting estimates of the root, but the root itself does not need to be bracketed in the first place. Of the methods introduced below, fixed-point iteration, Newton's, and the secant methods are all non-bracketing.

While the distinction between bracketing and non-bracketing methods is very important and should always be borne in mind when thinking about a given approach, we will *not* introduce these methods in that order. Instead, we will employ a pedagogical approach, starting with simple methods first (which end up being very slow) and increasing the sophistication level as we go. Let us note at the outset that we will provide Python implementations for three out of the five methods, inviting you to code up the rest in the problems at the end of the chapter. Similarly, while we provide a detailed study of convergence properties in the first few subsections, we don't do so for all five methods.

5.2.1 Conditioning

In section 4.2 we encountered the question of how $\mathbf{x}$, in the problem $\mathbf{A}\mathbf{x} = \mathbf{b}$, is impacted when we slightly perturb $\mathbf{A}$. This led us to introduce the condition number $\kappa(\mathbf{A})$ in Eq. (4.31), as a measure of how much (or how little) a perturbation in the coefficient matrix, $||\Delta\mathbf{A}||/||\mathbf{A}||$, is amplified when evaluating the effect on the solution vector, $||\Delta\mathbf{x}||/||\mathbf{x}||$.

We will now do something analogous, this time for the case of absolute changes, just like we did for the eigenvalue problem in Eq. (4.41). The problem we are now solving is the non-linear equation $f(x) = 0$, so the perturbations in the system would be perturbations of the value of the function f. As it so happens, this is also closely related to what we studied in section 2.2.2.4 on error propagation for functions of one variable. To be specific, the exact root is here denoted by x^*, so our approximate value for it would be $\tilde{x}^*$. What

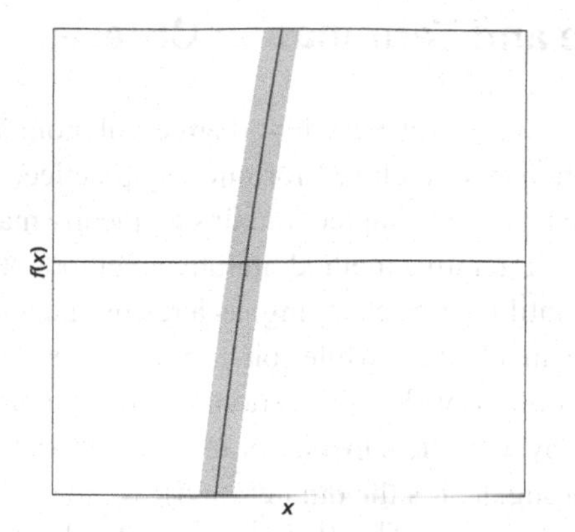

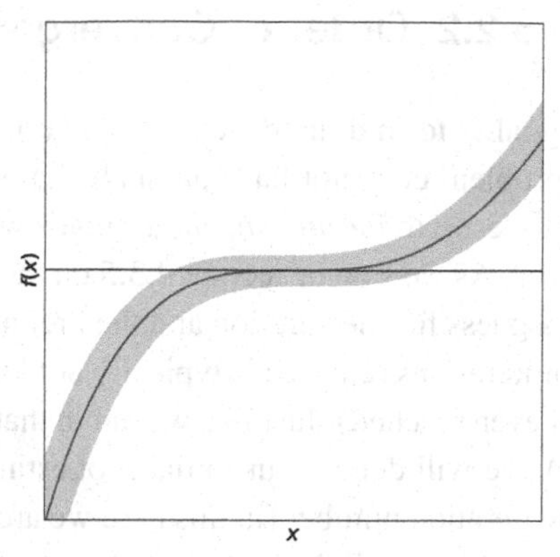

Fig. 5.2 Examples of conditioning for root-finding: well- (left) and ill- (right)

we'll do is to Taylor expand $f(\tilde{x}^*)$ around x^*. This gives us:

$$f(\tilde{x}^*) - f(x^*) = f(x^* + \Delta x^*) - f(x^*) = f(x^*) + f'(x^*)(\tilde{x}^* - x^*) + \cdots - f(x^*)$$
$$\approx f'(x^*)(\tilde{x}^* - x^*) \quad (5.13)$$

This is identical to the derivation we carried out in Eq. (2.36). Just like then, we made the assumption that $\tilde{x}^* - x^*$ is small, in order to drop higher-order terms.

We now remember that we are considering the question of how well- or ill-conditioned a given non-linear problem $f(x) = 0$ is; in other words, what we are after is a good estimate of x^*. To do that, we take our last result and solve it for $\tilde{x}^* - x^*$:

$$\tilde{x}^* - x^* \approx \frac{1}{f'(x^*)}[f(\tilde{x}^*) - f(x^*)] \quad (5.14)$$

allowing us to introduce a *condition number* for our problem:

$$\kappa_f = \frac{1}{f'(x^*)} \quad (5.15)$$

which you can immediately identify as the inverse of the (absolute) condition number we encountered for the problem of function evaluation in Eq. (2.37). Clearly, the magnitude of κ_f tells us whether or not a small change in the value of the function near the root is amplified when evaluating the root itself. Qualitatively, we see that a function $f(x)$ which crosses the x axis "rapidly" leads to large $f'(x^*)$ and therefore a small condition number κ_f; this is a *well-conditioned* problem. On the other hand, a problem for which $f(x)$ is rather flat near the root, crossing the x axis "slowly", will correspond to a small $f'(x^*)$ and therefore a large condition number κ_f; this is an *ill-conditioned* problem. This should be easy to grasp: if a function is quite flat near the root, it will be difficult to tell the root apart from its neighbors, see Fig. 5.2 (where the band reflects our uncertainty in the function value). As before, we need to (somewhat arbitrarily) decide how to distinguish between rapid and slow behavior.

5.2.2 Order of Convergence and Termination Criteria

Our goal is to find an x^* such that $f(x^*) = 0$. Except for very low-degree polynomials, this problem does not have an analytical solution with a closed formula. In practice, we have to resort to *iterative methods*; that's why the present chapter contains atypically many figures.[4] As we saw in section 4.3.5 on the Jacobi iterative method, iterative methods start with a guess for the solution and then refine it until it stops changing; as a result, the number of iterations required is typically not known in advance (while sometimes convergence is not even reached). Just like we did in that section (as well as in section 4.4 on eigenproblems), we will denote our initial root estimate by $x^{(0)}$: the superscript in parentheses tells us the iteration number (in this case we are starting, so it's the 0th iteration).

As noted, we will be computing a sequence of iterates, $x^{(0)}, x^{(1)}, x^{(2)}, \ldots$, and $x^{(k)}$, which (we hope) will be approaching x^*. For now, let us assume that we are dealing with a convergent method and that $x^{(k)}$ will be close to the root x^*; we'll discuss the meaning of "close" below. If there exist a constant $m \neq 0$ and a number p such that:

$$|x^{(k)} - x^*| \leq m|x^{(k-1)} - x^*|^p \tag{5.16}$$

for k sufficiently large, then m is called the *asymptotic error constant* and p is called the *order of convergence*. If $p = 1$ then we are dealing with a *linearly convergent* method, if $p = 2$ with a *quadratically convergent* method, and so on. For example, the fixed-point iteration method will turn out to be linearly convergent, whereas Newton's method will turn out to be quadratically convergent. It's worth noting that we can also find cases where p is not an integer: for example, the secant method turns out to have $p \approx 1.618$ (see problem 5.12). We say that such methods are *superlinearly convergent*.[5]

Generally speaking, the order of convergence is a formal result associated with a given method, so it doesn't always help us when we are faced with an actual iterative procedure. As we found out in chapter 2 on numerics, sometimes general error bounds are too pessimistic; a general convergence trend doesn't tell you what you should expect for the specific $f(x)$ you are dealing with. To make things concrete, the order of convergence discussed in the previous paragraph relates, on the one hand, the distance between the current iterate and the true value of the root ($|x^{(k)} - x^*|$) with, on the other, the distance between the previous iterate and the true value of the root ($|x^{(k-1)} - x^*|$). In practice, we don't actually know the true value of the root (x^*): that is what we are trying to approximate. In other words, we wish to somehow quantify $|x^{(k)} - x^*|$ when all we have at our disposal are the iterates themselves, $x^{(0)}, x^{(1)}, x^{(2)}, \ldots, x^{(k)}$.

Thinking about how to proceed, one option might be to check the absolute value of the function and compare it to a small tolerance:

$$|f(x^{(k)})| \leq \epsilon \tag{5.17}$$

[4] "To describe this without using visible models would be labor spent in vain" (Plato, *Timaeus*, 40d3).

[5] Strictly speaking, superlinear order means that $|x^{(k)} - x^*| \leq d_k|x^{(k-1)} - x^*|$ and $d_k \to 0$ hold; for the case of quadratic convergence we can take $d_k = m|x^{(k-1)} - x^*| \to 0$, so we see that quadratic order is also superlinear.

Unfortunately, this is an absolute criterion, which doesn't take into account the "typical" values of the function; it may be that the function is very steep near its root, in which case an ϵ like 10^{-8} would be too demanding (and therefore hard to satisfy). As another example, the function might have typical values of order 10^{15}: a test like our $\epsilon = 10^{-8}$ would simply be asking too much of our iterates. Conversely, the function might be very flat near its root (i.e., ill-conditioned) or involve tiny magnitudes, say 10^{-10}, even away from the root. You may be tempted, then, to somehow divide out a "typical" value of $f(x)$, in order to turn this criterion into a relative one, but it's not always clear what qualifies as typical.

In what follows, we will take another approach. Instead of checking to see how small the value of the function itself is, we will check the iterates themselves to see if they are converging to a given value. In other words, we will test the distance between the current iteration and the previous one, $|x^{(k)} - x^{(k-1)}|$.[6] We are already suspicious of using absolute criteria, so we come up with the following test:

$$\frac{|x^{(k)} - x^{(k-1)}|}{|x^{(k)}|} \leq \epsilon \tag{5.18}$$

Of course, if the numbers involved are tiny, even this criterion might lead us astray. We won't worry about this in our implementations below, but it may help you to know that one way of combining absolute and relative tolerance testing is as follows:

$$\frac{|x^{(k)} - x^{(k-1)}|}{1 + |x^{(k)}|} \leq \epsilon \tag{5.19}$$

If the iterates $x^{(k)}$ themselves are tiny and getting increasingly close to each other, here the numerator will be small but the denominator will be of order 1.

If you are feeling a bit uncomfortable about the transition from $|x^{(k)} - x^*|$ to (a possibly relativized) $|x^{(k)} - x^{(k-1)}|$ in our termination criterion, you should bring to mind problem 4.20 on the convergence properties of the Jacobi method. There, we directly proved a corresponding criterion for the case of that iterative method. If this is still not enough, you'll be pleased to find out that we will explicitly address the transition from one quantity to the other (for the case of fixed-point iteration) in section 5.2.3.3.

5.2.3 Fixed-Point Iteration

5.2.3.1 Algorithm

The first method we will introduce addresses a problem that is very similar, though not identical, to the $f(x) = 0$ problem that we started with. Specifically, this method solves:

$$x = g(x) \tag{5.20}$$

[6] As you may recall, in the case of linear algebra we were faced with an analogous but more complicated problem, since there the iterates themselves were vectors, so we had to somehow combine them together into a termination criterion; this led to Eq. (4.122), or the more robust Eq. (4.123).

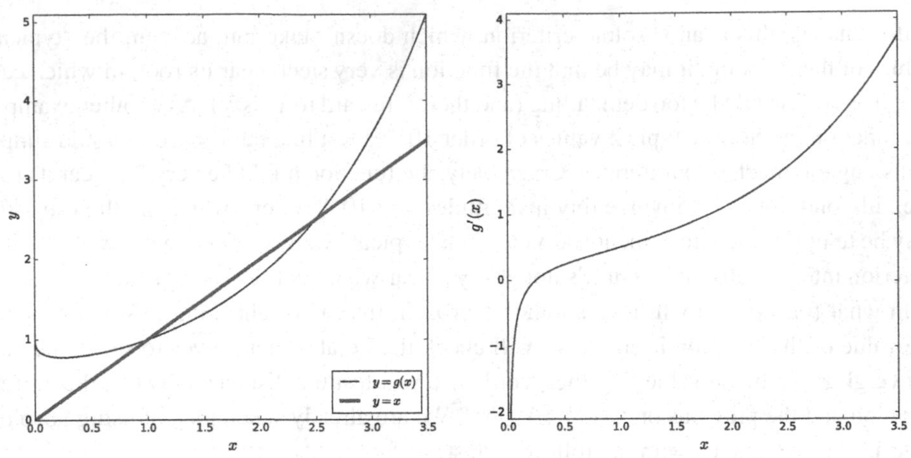

Fig. 5.3 Non-linear equation of the fixed-point form, for the case of $g(x) = e^{x-\sqrt{x}}$

As you may know, a point x^* satisfying this relation is called a *fixed point* of g. For example, the function $g(x) = x^2 - 6$ has two fixed points, since $g(3) = 3$ and $g(-2) = -2$. Similarly, the function $g(x) = x + 2$ has no fixed points, since x is never equal to $x + 2$.

We could always take $g(x) = x$ and rewrite it as $f(x) = g(x) - x = 0$ so it appears that the problem we are now solving is equivalent to the one we started out with. (Not quite. Keep reading.) Thus, let us see how to solve $g(x) = x$ on the computer. Again, for concreteness, we study our problem from Eq. (5.12): $e^{x-\sqrt{x}} - x = 0$ can be trivially re-arranged to give $g(x) = e^{x-\sqrt{x}}$. We have plotted $y = x$ and $y = g(x)$ in the left panel of Fig. 5.3, where you can immediately see that this example leads to two fixed points, namely two points where the curves meet. Unsurprisingly, these two fixed points are at $x^* \approx 1$ and $x^* \approx 2.5$, similarly to what we saw in our original problem.[7]

The simplest approach to solving Eq. (5.20) is basically to just iterate it, starting from an initial estimate $x^{(0)}$. Plug $x^{(0)}$ into the right-hand side and from there get $x^{(1)}$:

$$x^{(1)} = g(x^{(0)}) \tag{5.21}$$

As you may have guessed, this process is repeated: $x^{(1)}$ is then plugged into the right-hand side of Eq. (5.20), thereby producing a further improved solution $x^{(2)}$. For the general k-th iteration step, we now have our prescription for the *fixed-point iteration* method:

$$x^{(k)} = g(x^{(k-1)}), \quad k = 1, 2, \ldots \tag{5.22}$$

If all is going well, the solution vector will asymptotically approach the exact solution. This process of guessing a given value, evaluating the next one, and so on, is illustrated in Fig. 5.4.[8]

[7] It's not yet obvious why, but we have also plotted the first derivative of our $g(x)$ in the right panel of Fig. 5.3—this is the same as what we saw in the right panel of Fig. 5.1, but with an offset.

[8] Incidentally, this is strongly reminiscent of what we did in the section on the Jacobi iterative method, when faced with Eq. (4.119). Of course, there we had to find many values all at once and the problem was linear.

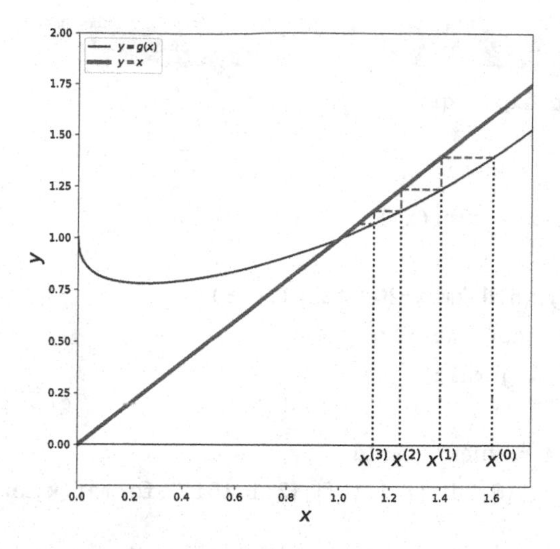

Fig. 5.4 Illustration of the fixed-point method for our example function

5.2.3.2 Implementation

Observe that we have been making the (hyperoptimistic) assumption that everything will "go well". We will make this question concrete soon, when we discuss the convergence properties of the fixed-point iteration method. For now, we turn to a Python implementation for the case of $g(x) = e^{x-\sqrt{x}}$, see code 5.1. This trial-and-error approach will emphasize what "going well" and "going wrong" mean in practice.

This program is reasonably straightforward; in essence, it is a simpler version of the Jacobi code we saw in the chapter on linear algebra (i.e., code 4.5). After defining our $g(x)$ function, we create a function that implements the fixed-point iteration method. The latter function takes in as arguments: (a) $g(x)$, since we might wish to solve a different problem later on, (b) the initial estimate for the root $x^{(0)}$, (c) the maximum number of iterations we should keep going for (in order to make sure the process terminates even if we've been unsuccessful), and (d) the relative tolerance which will be used to determine whether we should stop iterating. Since we don't plan on changing the last two parameters often, they are given default values.

The body of the function contains a single loop which employs the `for-else` idiom that we also encountered in earlier chapters: as you may recall, the `else` is only executed if the main body of the `for` runs out of iterations without ever executing a `break` (which for us would mean that the error we calculated never became smaller than the pre-set error tolerance).

In addition to evaluating the new $x^{(k)}$ each time, we also keep track of the difference between it and the previous iterate, so that we can form the ratio $|x^{(k)} - x^{(k-1)}|/|x^{(k)}|$. We take the opportunity to print out our iteration counter, as well as the latest iterate and the (absolute) change in the value of the iterate. Obviously, you should comment out the call to `print()` inside `fixedpoint()` once you are ready to use the code for other purposes. We then say `xold = xnew` in preparation for the next iteration (should it occur).

Code 5.1 fixedpoint.py

```
from math import exp, sqrt

def g(x):
    return exp(x - sqrt(x))

def fixedpoint(g,xold,kmax=200,tol=1.e-8):
    for k in range(1,kmax):
        xnew = g(xold)

        xdiff = xnew - xold
        print("{0:2d} {1:1.16f} {2:1.16f}".format(k,xnew,xdiff))

        if abs(xdiff/xnew) < tol:
            break

        xold = xnew
    else:
        xnew = None

    return xnew

if __name__ == '__main__':
    for xold in (0.99, 2.499):
        x = fixedpoint(g,xold)
        print(x)
```

In the main program, we simply pick a couple of values of $x^{(0)}$ and run our fixed-point iteration function using them. Wishing to reduce the number of required iterations, we pick our $x^{(0)}$ to be close to the actual root each time. Running this code we see that, for the first fixed point, we converge to the (analytically known) solution after 20 iterations. The absolute difference $|x^{(k)}-x^{(k-1)}|$ roughly gets halved at each iteration, so we eventually reach convergence. Obviously, the number of iterations required also depends on the starting value. For example, if we had picked $x^{(0)} = 1.2$ we would have needed 25 iterations.[9]

Turning to the second fixed point, things are more complicated. Even though we started very close to the fixed point (based on our knowledge of Fig. 5.3), we fail to converge. The absolute difference $|x^{(k)} - x^{(k-1)}|$ initially doubles at each iteration and then things get even worse, so we eventually crash. The `OverflowError` could have been avoided if we had

[9] Incidentally, if you're still uncomfortable with the for-else idiom, try giving `kmax` a value of, say, `10`.

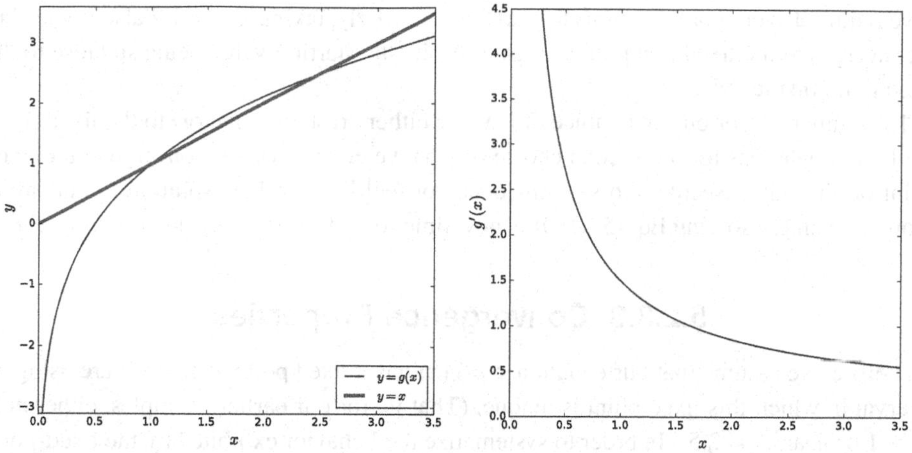

Fig. 5.5 Non-linear equation of the fixed-point form, for the case of $g(x) = \ln x + \sqrt{x}$

introduced another test of the form $|x^{(k)} - x^{(k-1)}| \geq |x^{(k-1)} - x^{(k-2)}|$, intended to check whether or not we are headed in the right direction. In any case, at this point we don't really know what's different for the case of one fixed point vs the other. Thinking that maybe we picked the "wrong" starting estimate, we try again with $x^{(0)} = 2.49$; this time the run doesn't crash: the absolute difference $|x^{(k)} - x^{(k-1)}|$ first keeps increasing in magnitude, but at some point it starts decreasing, leading to eventual convergence. Unfortunately, though, this converges to the wrong fixed point! The method is giving us the first fixed point, which we had already found.

We'll let you try out several other initial estimates $x^{(0)}$ (whether by hand or systematically). That exercise won't add anything new: we always either crash or converge to the fixed point on the left. Since we're stuck, we decide to try out a different approach. Recall that the fixed-point iteration method solves the problem $x = g(x)$. Since the equation we wished to solve was $e^{x-\sqrt{x}} - x = 0$, we acted as if the two problems were equivalent and then proceeded. Here's the thing, though: our choice of $g(x)$ was *not unique*! Specifically, we could take the natural logarithm of both sides in $e^{x-\sqrt{x}} = x$ to get $x - \sqrt{x} = \ln x$, which then has the form $x = g(x)$, but this time for $g(x) = \ln x + \sqrt{x}$. We have plotted $y = x$ and this new $y = g(x)$ in the left panel of Fig. 5.5; as before, we have two fixed points, namely two points where the curves meet.[10] Unsurprisingly, these two fixed points are at $x^* \approx 1$ and $x^* \approx 2.5$, just like in the previous version of our problem. We have also plotted the first derivative of our new $g(x)$ in the right panel of Fig. 5.5; this is considerably different from both the original problem's derivative, Fig. 5.1, and our previous fixed-point derivative, Fig. 5.3.

If we code up our new function we find that our iterative procedure now crashes for the case of the *first* fixed point (using the same initial estimate as before)! Using $x^{(0)} = 2.49$, we now *do* converge to the fixed point on the right, even if we take a bit longer to get there. For the sake of completeness, note that $x^{(0)} = 2.499$ would have also worked, but would

[10] Intriguingly, between the fixed points the two curves appear to be "closer" to each other than the corresponding two curves in Fig. 5.3 were.

have required even more iterations to converge. Similarly, taking $x^{(0)} = 1.2$ also works, but it converges to the fixed point on the right (despite the starting value being so close to the fixed point on the left).

To summarize, for our first choice of $g(x)$ we either crash or converge to the fixed point on the left, whereas for our second choice of $g(x)$ we either crash or converge to the fixed point on the right. Needless to say, this does not feel like a robust solution to our initial problem, namely solving Eq. (5.12). It is now time to understand why this is taking place.

5.2.3.3 Convergence Properties

As before, we assume that our equation $x = g(x)$ has a fixed point x^* and we are using an interval in which this fixed point is unique. (That is, for our earlier examples, either near $x^* \approx 1$ or near $x^* \approx 2.5$.) In order to systematize the behavior exhibited by the fixed-point iteration method, we start with our defining relationship, Eq. (5.22):

$$x^{(k)} = g(x^{(k-1)}) \tag{5.23}$$

Since x^* is a fixed point, we know that:

$$x^* = g(x^*) \tag{5.24}$$

holds. Subtracting the last two equations gives us:

$$x^{(k)} - x^* = g(x^{(k-1)}) - g(x^*) \tag{5.25}$$

Let us now employ a Taylor expansion. This is something that we've already done repeatedly in chapter 3 on finite differences and that we'll keep on doing.[11] Let us write down the Taylor expansion of $g(x^*)$ around $x^{(k-1)}$:

$$g(x^*) = g(x^{(k-1)}) + (x^* - x^{(k-1)})g'(x^{(k-1)}) + \cdots \tag{5.26}$$

If we stop at first order, we can rewrite this expression as:

$$g(x^*) = g(x^{(k-1)}) + (x^* - x^{(k-1)})g'(\xi) \tag{5.27}$$

where ξ is a point between x^* and $x^{(k-1)}$.[12]

If we now plug $g(x^*)$ from Eq. (5.27) into Eq. (5.25), we find:

$$x^{(k)} - x^* = g'(\xi)\left(x^{(k-1)} - x^*\right) \tag{5.28}$$

Note how on the right-hand side the order of x^* and $x^{(k-1)}$ has changed. If we now simply take the absolute value of both sides we find that our result is identical to that in Eq. (5.16) with $|g'(\xi)| \leq m$ and $p = 1$. We see that at each iteration the distance of our iterate from the true solution is multiplied by a factor of $|g'(\xi)|$; thus, if $|g'(\xi)| \leq m < 1$ we will be

[11] Most notably in chapter 7 on integration and in chapter 8 on differential equations.

[12] Taylor's theorem truncated to such a low order, as here, is known as the mean-value theorem.

converging to the solution. More explicitly, if we start with a $x^{(0)}$ sufficiently close to x^*, we will have:

$$|x^{(k)} - x^*| \leq m|x^{(k-1)} - x^*| \leq m^2|x^{(k-2)} - x^*| \leq \ldots \leq m^k|x^{(0)} - x^*| \tag{5.29}$$

and since $m < 1$ we see that $x^{(k)}$ converges to x^* *linearly* (since $p = 1$). Note that each application of the mean-value theorem needs its own ξ, but this doesn't matter as long as $|g'(\xi)| \leq m$ holds for each one of them. On the other hand, if $|g'(\xi)| > 1$ this process will diverge. Analogously, if $|g'(\xi)| < 1$ but close to 1, convergence will be quite slow. Finally, if $g'(x) = 0$ at or near the root, then the first-order term in the error vanishes; as a result, in this case the fixed-point iteration converges quadratically.[13]

Let us now apply these findings to our earlier examples. For the first $g(x)$ we have the derivative $g'(x)$ in the right panel of Fig. 5.3. We find that near the left fixed point $m < 1$ ($m = 0.5$ to be precise) and near the right fixed point $m > 1$. This is consistent with our findings using Python in the previous subsection: we cannot converge to the fixed point on the right but we do converge to the one on the left. Similarly, for the second $g(x)$ we have the derivative $g'(x)$ in the right panel of Fig. 5.5. We find that near the left fixed point $m > 1$ and near the right fixed point $m < 1$. Again, this is consistent with what we found using Python: we cannot converge to the fixed point on the left but we do converge to the one on the right. As it so happens, the latter case has $m \approx 0.72$, which explains why it took us a bit longer to converge in this case.

Before we conclude, let us return to the question of relating $|x^{(k)} - x^*|$ (which appears in the above derivation) to $|x^{(k)} - x^{(k-1)}|$ (which appears in our termination criterion). We can take Eq. (5.28) and add $g'(\xi)x^* - g'(\xi)x^{(k)}$ to both sides. After a trivial re-arrangement:

$$[1 - g'(\xi)]\left[x^{(k)} - x^*\right] = g'(\xi)\left[x^{(k-1)} - x^{(k)}\right] \tag{5.30}$$

If we now take the absolute value on both sides and assume $|g'(\xi)| \leq m < 1$, we find:

$$|x^{(k)} - x^*| \leq \frac{m}{1-m}|x^{(k)} - x^{(k-1)}| \tag{5.31}$$

Thus, as long as $m < 0.5$ (in which case $m/(1-m) < 1$) the "true convergence test" $|x^{(k)} - x^*|$ is even smaller than the quantity $|x^{(k)} - x^{(k-1)}|$ which we evaluate in our termination criterion. Conversely, if $m > 0.5$ it is possible that our termination criterion doesn't constrain true convergence very well; to take an extreme example, for $m = 0.99$ we have $m/(1 - m) = 99$ so the two quantities may be two orders of magnitude apart.[14]

Finally, note that Eq. (5.31) does not include roundoff errors. The derivation above may be straightforwardly generalized to include those. The main feature is that the final computational error will only depend on the roundoff error made in the *final* iteration. This is good news, as it implies that iterative methods are *self-correcting*: minor issues in the first few iterations will not impact the final accuracy.

[13] In problem 5.8 you will see that we can recast Newton's method as a fixed-point iterative process.

[14] Even so, our result is merely an error *bound*, so the actual error can be much smaller.

5.2.4 Bisection Method

Fixed-point iteration allowed us to study a root-finding method (including its convergence properties) in detail. Unfortunately, as we saw, the mapping from $f(x) = 0$ to $x = g(x)$ is not unique. Also, if $|g'(\xi)| > 1$ the fixed-point iteration method diverges. Thus, it would be nice to see a different approach, one that always succeeds. This is precisely what the *bisection* method accomplishes (well, almost—keep reading); it is a slow method, which doesn't really generalize to higher-dimensional problems, but it is safe and systematic.

5.2.4.1 Algorithm and Convergence

The bisection method assumes you have already bracketed the root; that means that you have found an x_0 and an x_1 for which $f(x_0)$ and $f(x_1)$ have opposite signs. We know from Bolzano's theorem that when a continuous function has values of opposite sign in a given interval, then the function has a root in that same interval (this is a corollary of the intermediate-value theorem). This works as follows:

- Take the interval (x_0, x_1) (for which we know $f(x_0)f(x_1) < 0$) and evaluate the midpoint:

$$x_2 = \frac{x_0 + x_1}{2} \tag{5.32}$$

- At this point, we will halve the original interval (x_0, x_1), thereby explaining why this approach is also known as the *internal halving* method. Thus, we are producing two subintervals: (x_0, x_2) and (x_2, x_1). The sign of $f(x_2)$ will determine in which of these two subintervals the root lies.
- If $f(x_0)f(x_2) < 0$, then the root is in (x_0, x_2). Thus, we could rename $x_1 \leftarrow x_2$ and repeat the entire process (which started with an x_0 and an x_1 that bracket the root).
- If, on the other hand, $f(x_0)f(x_2) > 0$, then the root is in (x_2, x_1). Thus, we could rename $x_0 \leftarrow x_2$ and repeat the entire process.
- There is also the extreme scenario where $f(x_0)f(x_2) = 0$ but, since we are dealing with floating-point numbers this is unlikely. Even so, if it does happen, it's merely informing you that you've already found the root.

Take a moment to realize that the notation we are using here is distinct from what we were doing above; x_0 and x_1 were never true "iterates", i.e., we never really thought x_1 was closer to the right answer than x_0. We could have just as easily called them a and b.[15] Actually, it may help you to think of the initial interval as (a, b); here a and b never change (since they simply correspond to the original bracket) whereas x_0 and x_1 are the numbers that keep getting updated[16] (and simply start out with the values $x_0 = a$ and $x_1 = b$).

The best estimate at a given iteration is x_2, given by Eq. (5.32). Thus, using our earlier notation $x^{(0)}$ is the first x_2, $x^{(1)}$ is the next x_2 (after we've halved the interval and found

[15] However, this notation would get really confusing below, since Ridders' method needs a, b, c, and d.
[16] "The ends of the earth were closing in" (G. K. Chesterton, *The Man Who Was Thursday*, chapter 6).

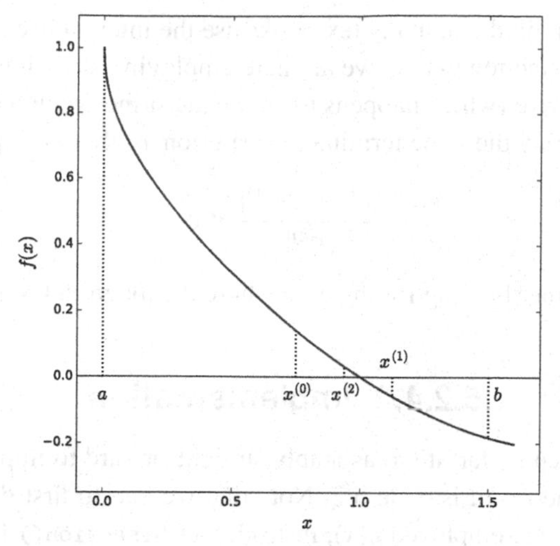

Fig. 5.6 Illustration of the bisection method for our example function

the next midpoint), $x^{(2)}$ is the next x_2 (once we've halved the interval again), and so on. In general, at the k-th iteration $x^{(k)}$ will be the latest midpoint. This is illustrated in Fig. 5.6: for this case, the initial midpoint is $x^{(0)}$, the midpoint between that and b is $x^{(1)}$, and so on.

Since we are always maintaining the bracketing of the root and the interval in which the root is bracketed always gets halved, we find that:

$$|x^{(k)} - x^*| \leq \frac{b-a}{2^{k+1}} \tag{5.33}$$

Remember that the original interval had length $b - a$. Instead of taking Eq. (5.33) for granted, let's quickly derive it. For $k = 0$ we have evaluated our first midpoint, but have not actually halved the interval (i.e., we haven't carried out a selection of one of the two subintervals). Since the first midpoint is at $(b + a)/2$, the largest possible error we've made in our determination of the root is either:

$$\frac{a+b}{2} - a = \frac{b-a}{2} \tag{5.34}$$

or:

$$b - \frac{a+b}{2} = \frac{b-a}{2} \tag{5.35}$$

which amounts to the same answer. Now, for $k = 1$ we have already picked one of the subintervals (say, the one on the left) and evaluated the midpoint of *that*: this is $[a+(a+b)/2]/2$. The largest possible error in *that* determination is the distance between the latest midpoint and either of the latest endpoints; this comes out to be $(b - a)/4$. The general case follows the same pattern. In the end, Eq. (5.33) tells us that as k becomes sufficiently large we converge to x^*. Since there's always a factor of $1/2$ relating $|x^{(k)} - x^*|$ with $|x^{(k-1)} - x^*|$, we see from our general result on convergence, Eq. (5.16), that here we have $m = 1/2$ and $p = 1$, i.e., *linear* convergence.

You should keep in mind that many textbooks use the interval itself in their termination criterion. In contradistinction to this, we are here employing at each iteration only a given number, $x^{(k)}$, as an iterate (which happens to be the latest incarnation of the midpoint, x_2). This allows us to employ the same termination criterion as always, Eq. (5.18):

$$\frac{|x^{(k)} - x^{(k-1)}|}{|x^{(k)}|} \leq \epsilon \tag{5.36}$$

with the same disclaimer holding for the case where the magnitudes involved are tiny.

5.2.4.2 Implementation

Given what we've seen so far, it's reasonably straightforward to implement the bisection method in Python; the result is code 5.2. Note that we had to first define $f(x)$, since the fixed-point iteration code employed $g(x)$, instead. Our `bisection()` function employs the interface you'd expect, first taking in $f(x)$. Then, it accepts the points determining the starting interval, x_0 and x_1.[17] The `kmax` and `tol` parameters are identical to what we had before.

The core of our new function implements Eq. (5.32) and the bullet points that follow. That implies an evaluation of a midpoint, then a halving of the interval, followed by a comparison between old and new midpoints. Note that we are trying not to be wasteful in our function evaluations, calling `f()` only once per iteration; to do that, we introduce the new variables `f0` and `f2` which get updated appropriately. We *are* being wasteful in evaluating `x2` at the start of the loop, since that will always be the same as the midpoint we had computed the last time through the loop (of course, this is a simple operation involving floats and therefore less costly than a function evaluation). When printing out values (which you can later comment out, as usual) we decided this time to also print out the (absolute) value of the function itself; even though our disclaimers above about how this could lead one astray are still valid, it's nice to have extra information in our output. Observe that the termination criterion is identical to what we had before; as per Eq. (5.18), we evaluate $|x^{(k)} - x^{(k-1)}|/|x^{(k)}|$, only this time the $x^{(k)}$ are the midpoints.

In the main program, we run our function twice, once for each root. The bracketing in this case is chosen to be quite wide: we simply look at the left panel of Fig. 5.1 and try to be conservative. Running this code leads us to finding both roots; already, things are much better than with the fixed-point iteration method. In each case we need more than a couple dozen iterations (so our approach to the final answer was somewhat slow), but success was guaranteed. Looking at the ouput in a bit more detail, we recall that the first column contains the iteration counter, the second column has the latest midpoint, the third column is the (absolute) difference in the last two midpoints, and the final column contains the absolute value of the function itself. Notice that the difference in the third columns gets halved (exactly) from one iteration to the next. As far as the function value itself goes, we generally tend to reduce its magnitude when iterating, though for the case of the root on the right the value of the function first grows a little.

[17] The fixed-point iteration method was different, taking in only a single starting value, $x^{(0)}$.

bisection.py Code 5.2

```
from math import exp, sqrt

def f(x):
    return exp(x - sqrt(x)) - x

def bisection(f,x0,x1,kmax=200,tol=1.e-8):
    f0 = f(x0)
    for k in range(1,kmax):
        x2 = (x0+x1)/2
        f2 = f(x2)

        if f0*f2 < 0:
            x1 = x2
        else:
            x0, f0 = x2, f2

        x2new = (x0+x1)/2
        xdiff = abs(x2new-x2)
        rowf = "{0:2d} {1:1.16f} {2:1.16f} {3:1.16f}"
        print(rowf.format(k,x2new,xdiff,abs(f(x2new))))

        if abs(xdiff/x2new) < tol:
            break
    else:
        x2new = None

    return x2new

if __name__ == '__main__':
    root = bisection(f,0.,1.5)
    print(root); print("")
    root = bisection(f,1.5,3.)
    print(root)
```

5.2.5 Newton's Method

We now turn to *Newton's method* (sometimes also called the *Newton–Raphson method*): this is the simplest fast method used for root-finding. It also happens to generalize to larger-dimensional problems in a reasonably straightforward manner. At a big-picture level, Newton's method requires more input than the approaches we saw earlier: in addition to being able to evaluate the function $f(x)$, one must also be able to evaluate its first derivative $f'(x)$. This is obviously trivial for our example above, where $f(x)$ is analytically known, but may not be so easy to access for the case where $f(x)$ is an externally provided (costly) routine. Furthermore, to give the conclusion ahead of time: there are many situations in which Newton's method can get in trouble, so it always pays to think about your specific problem instead of blindly trusting a canned routine. Even so, if you already have a reasonable estimate of where the root may lie, Newton's method is usually a fast and reliable solution.

5.2.5.1 Algorithm and Interpretation

We will assume that $f(x)$ has continuous first and second derivatives. Also, take $x^{(k-1)}$ to be the last iterate we've produced (or just an initial guess). Similarly to what we did in Eq. (5.26) above, we will now write down a Taylor expansion of $f(x)$ around $x^{(k-1)}$; this time we go up to one order higher:

$$f(x) = f(x^{(k-1)}) + \left(x - x^{(k-1)}\right) f'(x^{(k-1)}) + \frac{1}{2}\left(x - x^{(k-1)}\right)^2 f''(\xi) \tag{5.37}$$

where ξ is a point between x and $x^{(k-1)}$. If we now take $x = x^*$ then we have $f(x^*) = 0$. If we further assume that $f(x)$ is linear (in which case $f''(\xi) = 0$), we get:

$$0 = f(x^{(k-1)}) + \left(x^* - x^{(k-1)}\right) f'(x^{(k-1)}) \tag{5.38}$$

which can be re-arranged to give:

$$x^* = x^{(k-1)} - \frac{f(x^{(k-1)})}{f'(x^{(k-1)})} \tag{5.39}$$

In words: for a linear function, an initial guess can be combined with the values of the function and the first derivative (at that initial guess) to locate the root.

This motivates Newton's method: if $f(x)$ is non-linear, we still use the same formula, Eq. (5.39), this time in order to evaluate not the root but our next iterate (which, we hope, brings us closer to the root):

$$x^{(k)} = x^{(k-1)} - \frac{f(x^{(k-1)})}{f'(x^{(k-1)})}, \quad k = 1, 2, \ldots \tag{5.40}$$

As we did in the previous paragraph, we are here neglecting the second derivative term in the Taylor expansion. However if we are, indeed, converging, then $(x^{(k)} - x^{(k-1)})^2$ in Eq. (5.37) will actually be smaller than $x^{(k)} - x^{(k-1)}$, so all will be well.

Another way to view the prescription in Eq. (5.40) is *geometrically*: we approximate $f(x)$ with its tangent at the point $(x^{(k-1)}, f(x^{(k-1)}))$; then, the next iterate, $x^{(k)}$, is the point where that tangent intercepts the x axis. If this is not clear to you, look at Eq. (5.38) again. Figure 5.7 shows this step being applied repeatedly for our example function from Eq. (5.12). We make an initial guess, $x^{(0)}$, and then evaluate the tangent at the point $(x^{(0)}, f(x^{(0)}))$. We

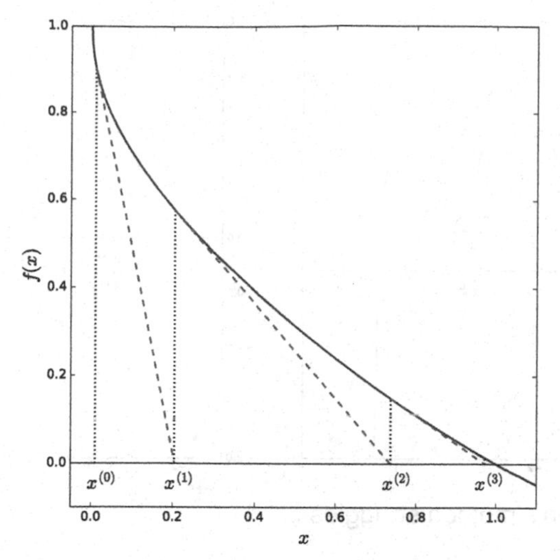

Fig. 5.7 Illustration of Newton's method for our example function

call $x^{(1)}$ the point where that tangent intercepts the x axis and repeat. For our example, this process brings us very close to the root in just a few steps.

5.2.5.2 Convergence Properties

We now turn to the convergence properties of Newton's method. To orient the reader, what we'll try to do is to relate $x^{(k)} - x^*$ to $x^{(k-1)} - x^*$, as per Eq. (5.16). We will employ our earlier Taylor expansion, Eq. (5.37), and take $x = x^*$, but this time without assuming that the second derivative vanishes. Furthermore, we assume that we are dealing with a simple root x^*, for which we therefore have $f'(x^*) \neq 0$; that means we can also assume $f'(x) \neq 0$ in the vicinity of the root. We have:

$$0 = f(x^{(k-1)}) + \left(x^* - x^{(k-1)}\right) f'(x^{(k-1)}) + \frac{1}{2}\left(x^* - x^{(k-1)}\right)^2 f''(\xi) \tag{5.41}$$

where the left-hand side is the result of $f(x^*) = 0$.

Dividing by $f'(x^{(k-1)})$ and re-arranging, we find:

$$-\frac{f(x^{(k-1)})}{f'(x^{(k-1)})} - x^* + x^{(k-1)} = \frac{\left(x^* - x^{(k-1)}\right)^2 f''(\xi)}{2f'(x^{(k-1)})} \tag{5.42}$$

The first and third terms on the left-hand side can be combined together to give $x^{(k)}$, as per Newton's prescription in Eq. (5.40). This leads to:

$$x^{(k)} - x^* = \left[\frac{f''(\xi)}{2f'(x^{(k-1)})}\right]\left(x^{(k-1)} - x^*\right)^2 \tag{5.43}$$

This is our desired result, relating $x^{(k)} - x^*$ to $x^{(k-1)} - x^*$. Taking the absolute value of both

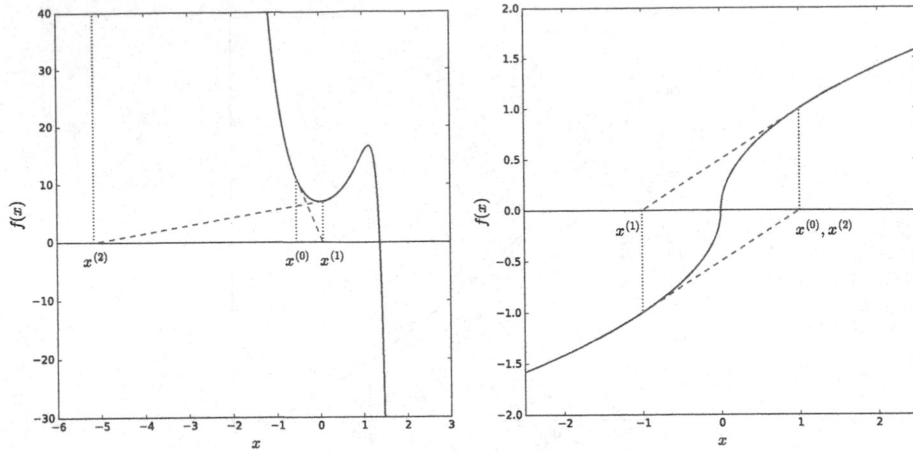

Fig. 5.8 Cases where Newton's method struggles

sides, our result is identical to that in Eq. (5.16), under the assumption that:

$$\frac{f''(\xi)}{2f'(x^{(k-1)})} \leq m \tag{5.44}$$

Since the right-hand side of Eq. (5.43) contains a square, we find that, using the notation of Eq. (5.16), $p = 2$, so if $m < 1$ holds then *Newton's method is quadratically convergent*, sufficiently close to the root.[18] This explains why our iterates approached the root so rapidly in Fig. 5.7, even though we intentionally picked a poor starting point.

5.2.5.3 Multiple Roots and Other Issues

Of course, as already hinted at above, there are situations where Newton's method can misbehave. As it so happens, our starting point in Fig. 5.7 was *near* 0, but not actually 0. For our example function $f(x) = e^{x-\sqrt{x}} - x$, the first derivative has a $\sqrt{x}$ in the denominator, so picking $x^{(0)} = 0$ would have gotten us in trouble. This can easily be avoided by picking another starting point. There are other problems that relate not to our initial guess, but to the behavior of $f(x)$ itself.

An example is given in the left panel of Fig. 5.8. Our initial guess $x^{(0)}$ is perfectly normal and does not suffer from discontinuities, or other problems. However, by taking the tangent and finding the intercept with the x axis, our $x^{(1)}$ happens to be near a local extremum (minimum in this case); since $f'(x^{(k-1)})$ appears in the denominator in Newton's prescription in Eq. (5.40), a small derivative leads to a large step, considerably away from the root. Note that, for this misbehavior to arise, our previous iterate doesn't even need to be "at" the extremum, only in its neighborhood. It's worth observing that, for this case, the root $x^* \approx 1.4$ actually *is* found by Newton's method after a few dozen iterations: after slowly creeping back toward the minimum, one iterate ends up overshooting to the right this time (past the root), after which point the solution is slowly but surely reached.

[18] Again, problem 5.8 recasts Newton's method as a special case of fixed-point iteration to find the same result.

Things are even worse in the case of the right panel of Fig. 5.8, where our procedure enters an infinite cycle. Here we start with a positive $x^{(0)}$, then the tangent brings us to a negative iterate (which happens to be the exact opposite of where we started, $x^{(1)} = -x^{(0)}$), and then taking the tangent and finding the intercept brings us back to where we started, $x^{(2)} = x^{(0)}$. After that, the entire cycle keeps repeating, without ever coming any closer to the root which we can see is at $x^* = 0$. Of course, this example is somewhat artificial; in a real-world application, the function would likely not be perfectly symmetric.

Another problem arises when we are faced with a root that is *not* simple. As you may recall, in our study of the convergence properties for Newton's method above, we assumed that $f'(x^*) \neq 0$. This is not the case when you are faced with a multiple root.[19] This is a problem not only for the convergence study but also for the prescription itself: Eq. (5.40) contains the derivative in the denominator, so $f'(x^*) = 0$ can obviously lead to trouble.

As an example, in Fig. 5.9 we are plotting the function:

$$f(x) = x^4 - 9x^3 + 25x^2 - 24x + 4 \tag{5.45}$$

which has one root at $x^* \approx 0.21$, another one at $x^* \approx 4.79$, as well as a *double root* at $x^* = 2$. This is easier to see if we rewrite our function as:

$$f(x) = (x - 2)^2 \, (x^2 - 5x + 1) \tag{5.46}$$

As it turns out, Newton's method can find the double root, but it takes a bit longer than you'd expect to do so. Intriguingly, this is an example where the bisection method (and *all* bracketing methods) cannot help us: the function does not change sign at the root $x^* = 2$ (i.e., the root is not bracketed). As you will find out when you solve problem 5.9, Newton's method for this case converges linearly! If you would like to have it converge quadratically (as is usually the case), you need to implement Newton's prescription in a slightly modified form, namely:

$$x^{(k)} = x^{(k-1)} - 2\frac{f(x^{(k-1)})}{f'(x^{(k-1)})}, \quad k = 1, 2, \ldots \tag{5.47}$$

as opposed to Eq. (5.40). Problem 5.9 discusses the source of this mysterious 2 prefactor.

Unfortunately, in practice we don't know ahead of time that we are dealing with a double root (or a triple root, etc.), so we cannot pick such a prefactor that guarantees quadratic convergence. As you will discover in problem 5.10, a useful trick (which applies regardless of your knowledge of the multiplicity of the root) is to apply Newton's prescription, but this time not on the function $f(x)$ but on the function:

$$w(x) = \frac{f(x)}{f'(x)} \tag{5.48}$$

As you will show in that problem, $w(x)$ has a simple root at x^*, regardless of the multiplicity of the root of $f(x)$ at x^*. Of course, there's no free lunch, and you end up introducing cancellations (which may cause headaches) when evaluating the derivative $w'(x)$, which you'll need to do in order to apply Eq. (5.40) to $w(x)$.

[19] Applying our definition in Eq. (5.15), we see that for multiple roots, where $f'(x^*) = 0$, the root-finding problem is always ill-conditioned. This is challenging for all approaches, not only for Newton's method.

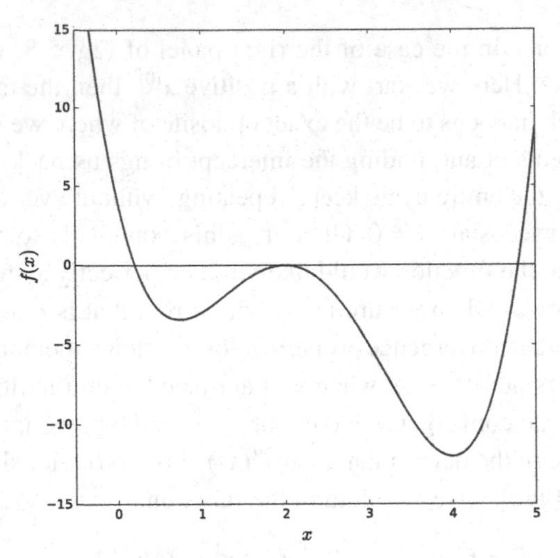

Fig. 5.9 A case where the equation has a double root

Note, finally, that there are several other modifications to the plain Newton algorithm that one could carry out (e.g., what is known as *backtracking*, where a "bad" step is discarded). Some of these improvements are explored in the problems at the end of the chapter. Many such tricks happen to be of wider applicability, meaning that we can employ them even when we're not using Newton's method.

5.2.6 Secant Method

Given the simplicity of Newton's method, as well as the fact that it generalizes to other problems, it would be natural to provide an implementation at this point. However, since (modified) versions of Newton's method are implemented below, we don't give one now for the basic version of the algorithm. Instead, we proceed to discuss other approaches, starting from the *secant method*, which is quite similar to Newton's method. As you may recall, the biggest new requirement that the latter introduced was that you needed to evaluate not only the function $f(x)$, but also the derivative $f'(x)$. In some cases that is too costly, or even impossible. You might be tempted to approximate the derivative using the finite-difference techniques we introduced in chapter 3. However, those would necessitate (at least) two function evaluations per step. The method we are about to introduce is better than that, so will be our go-to solution for the cases where we wish to avoid derivatives.

5.2.6.1 Algorithm and Interpretation

As already hinted at, the secant method replaces the evaluation of the derivative, needed for Newton's method, by a single function evaluation. Although a naive finite-difference calculation would be bad (how do you pick h?), the secant method is inspired by such an

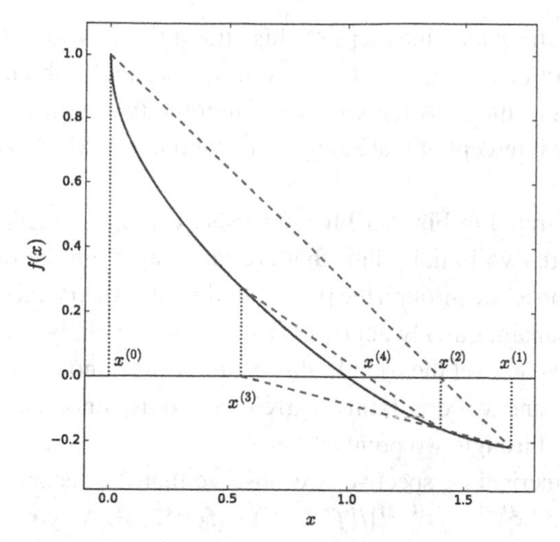

Fig. 5.10 Illustration of the secant method for our example function

approach. Specifically, this new method starts from Newton's Eq. (5.40):

$$x^{(k)} = x^{(k-1)} - \frac{f(x^{(k-1)})}{f'(x^{(k-1)})}, \quad k = 1, 2, \ldots \tag{5.49}$$

and approximates the derivative $f'(x^{(k-1)})$ using the latest two points and function values:

$$f'(x^{(k-1)}) \approx \frac{f(x^{(k-1)}) - f(x^{(k-2)})}{x^{(k-1)} - x^{(k-2)}} \tag{5.50}$$

which looks like a finite difference, where the spacing h is picked not by being on a grid or by hand, but by whatever values our latest iterates have; this implies that the spacing changes at every iteration.

Combining the last two equations, our prescription for the secant method is:

$$x^{(k)} = x^{(k-1)} - f(x^{(k-1)}) \frac{x^{(k-1)} - x^{(k-2)}}{f(x^{(k-1)}) - f(x^{(k-2)})}, \quad k = 2, 3, \ldots \tag{5.51}$$

where we start the iteration at $k = 2$, since we need *two* initial guesses as starting points. Once we get going, though, we need to evaluate the function only *once* per iteration. Note how this prescription doesn't need the derivative $f'(x)$ anywhere.

You may be wondering where the name "secant" comes from. This word refers to a straight line that cuts a curve in two or more points.[20] You can understand why the method is so named if you think about the geometric interpretation of our prescription in Eq. (5.51). From elementary geometry, you may recall that the line that goes through the two points (x_0, y_0) and (x_1, y_1) is:

$$y - y_0 = \frac{y_1 - y_0}{x_1 - x_0}(x - x_0) \tag{5.52}$$

[20] Secant comes from the Latin *secare*, "to cut" or "to divide".

If you now compute the x axis intercept of this straight line, you will find that it is precisely of the same form as in Eq. (5.51). Thus, in the secant method, at a given step, we are producing a straight line[21] as the secant through the two points $(x^{(k-1)}, f(x^{(k-1)}))$ and $(x^{(k-2)}, f(x^{(k-2)}))$. The intercept of that straight line with the x axis gives us the next iterate, $x^{(k)}$, and so on.

This is nicely illustrated in Fig. 5.10 for our usual example function. Even though we (purposely) begin with two initial points that are far away from the root, we see that this process of constructing a line through the points, finding the x axis intercept, and then using the latest two points again, quickly approaches the root. Crucially, we do *not* require that our two initial guesses bracket the root. In this example, $x^{(0)}$ and $x^{(1)}$ do happen to bracket the root, but then $x^{(1)}$ and $x^{(2)}$ don't; this is irrelevant to us, since we're simply drawing a straight line that goes through two points.

From a purely numerical perspective, we observe that the secant method requires the evaluation of the ratio $[x^{(k-1)} - x^{(k-2)}]/[f(x^{(k-1)}) - f(x^{(k-2)})]$. As you may have suspected, typically both numerator and denominator will be small, so this ratio will be evaluated with poor relative accuracy. Fortunately, it's the coefficient in front, $f(x^{(k-1)})$, that matters more. In problem 5.12 you will show that the order of convergence of the secant method is the golden ratio, $p \approx 1.618$. This is superlinear convergence: not as fast as Newton's method, but the fact that we don't need to evaluate the derivative may make this trade-off worthwhile.

5.2.6.2 Implementation

Code 5.3 is a Python program that implements our prescription from Eq. (5.51). Observe that the parameter list is identical to that of our `bisection()` function from a previous section.[22] The rest of the function is pretty straightforward: we keep forming the ratio of the difference between abscissas and function values, use it to produce a new iterate, and evaluate the difference between our best current estimate and the previous one. This part of the code happens to be easier to read (and write) than `bisection()` was, since there we needed notation for the old vs the new midpoint. As usual, we print out a table of values for informative purposes and then check to see if we should break out of the loop. It's worth observing that we are trying to avoid being wasteful when computing `ratio` and `x2`: similarly to what we did in `bisection()`, we introduce intermediate variables `f0` and `f1` to ensure that there is only one function evaluation per iteration.

The main program uses initial guesses for each of the roots. For the first root, we pick $x^{(0)}$ and $x^{(1)}$ to match what we saw in Fig. 5.10. For the second root, we pick two values, close to where we expect the root to lie, as is commonly done in practice. Running this code, we see that not too many iterations are required to reach the desired tolerance target. This is much faster than the fixed-point iteration; even more importantly, the secant method allows us to evaluate both roots, without having to carry out any analytical manipulations

[21] This time not as the tangent at the point $(x^{(k-1)}, f(x^{(k-1)}))$, as we did in Newton's method.

[22] Of course, there the `x0` and `x1` were required to bracket the root, whereas here they don't have to. While we're on the subject, note that in our earlier program `x0`, `x1`, and `x2` stood for x_0, x_1, and x_2, whereas now `x0`, `x1`, and `x2` stand for $x^{(k-2)}$, $x^{(k-1)}$, and $x^{(k)}$.

secant.py Code 5.3

```
from bisection import f

def secant(f,x0,x1,kmax=200,tol=1.e-8):
    f0 = f(x0)
    for k in range(1,kmax):
        f1 = f(x1)
        ratio = (x1 - x0)/(f1 - f0)
        x2 = x1 - f1*ratio

        xdiff = abs(x2-x1)
        x0, x1 = x1, x2
        f0 = f1

        rowf = "{0:2d} {1:1.16f} {2:1.16f} {3:1.16f}"
        print(rowf.format(k,x2,xdiff,abs(f(x2))))

        if abs(xdiff/x2) < tol:
            break
    else:
        x2 = None

    return x2

if __name__ == '__main__':
    root = secant(f,0.,1.7); print(root); print("")
    root = secant(f,2.,2.1); print(root)
```

first. Of course, there was some guesswork involved when picking the two pairs of initial guesses. The secant method also took considerably fewer iterations than the bisection method and barely more than Newton's method. All in all, with basically no bookkeeping and no derivative-evaluations, it has allowed us to finds both roots fast.

5.2.7 Ridders' Method

We now turn to *Ridders'* method: this approach is "only" 40 years old, making it one of the most modern topics in this volume.[23] We back up for a moment to realize that the last

[23] And, in all likelihood, in your undergraduate education.

two methods we've encountered (Newton's, secant) produced the next iterate by finding the x axis intercept of a straight line; they only differ in how that straight line is produced. Ridders' method follows the same general idea, of finding the x axis intercept of a straight line, but chooses this line not by the value of the function $f(x)$ or its derivative $f'(x)$, but through a clever trick. While our derivation below will be somewhat long, the prescription we arrive at is quite simple, requiring very little bookkeeping.

Ridders' is a bracketing method, so it assumes $f(x_0)f(x_1) < 0$. The main idea is to multiply $f(x)$ by the unique exponential function which turns it into a straight line:

$$R(x) = f(x)\, e^{Qx} \tag{5.53}$$

That is, regardless of the shape of $f(x)$, the new $R(x)$ will be linear, $R(x) = c_0 + c_1 x$. Since we have three undetermined parameters (c_0, c_1, and Q), we will use three points to pin them down. We take these to be our initial bracket's endpoints, x_0 and x_1, as well as their midpoint, $x_2 = (x_0 + x_1)/2$. A moment's thought will convince you that, since $R(x)$ is a straight line, we will have:

$$R_2 = \frac{R_0 + R_1}{2} \tag{5.54}$$

where we are using the notation $R_i \equiv R(x_i)$; we will soon also employ the corresponding $f_i \equiv f(x_i)$. Let us further define:

$$d = x_2 - x_0 = x_1 - x_2 \tag{5.55}$$

Using Eq. (5.53) and Eq. (5.54), we find:

$$f_0\, e^{Qx_0} + f_1\, e^{Qx_1} - 2f_2\, e^{Qx_2} = 0 \tag{5.56}$$

Factoring out e^{Qx_0} leads to:

$$f_1\, e^{2Qd} - 2f_2\, e^{Qd} + f_0 = 0 \tag{5.57}$$

which you can see is a quadratic equation in e^{Qd}. Solving it leads to:

$$e^{Qd} = \frac{f_2 - \text{sign}(f_0)\sqrt{f_2^2 - f_0 f_1}}{f_1} \tag{5.58}$$

Since we are starting with the assumption that $f_0 f_1 < 0$, we see that $f_2^2 - f_0 f_1 > 0$, as it should be. For the same reason, the square root of $f_2^2 - f_0 f_1$ is larger in magnitude than f_2.

The $-\text{sign}(f_0)$ is deduced from the fact that $e^{Qd} > 0$. To see this, we distinguish between two possible scenarios, (a) $f_0 < 0$ and $f_1 > 0$, (b) $f_0 > 0$ and $f_1 < 0$. Taking the first scenario, we see that the denominator in Eq. (5.58) is positive, so the numerator should also be positive. For this scenario, there are two possibilities regarding the sign of f_2. If $f_2 > 0$, then the square root term has to come in with a plus sign in order to get a positive numerator. Similarly, if $f_2 < 0$ then the first term in the numerator is negative so, again, the square root term (which is larger in magnitude) has to come in with a plus sign to ensure

$e^{Qd} > 0$. An analogous line of reasoning leads to a minus sign in front of the square root for the second scenario, $f_0 > 0$ and $f_1 < 0$. Overall, we need $-\text{sign}(f_0)$, as in Eq. (5.58).

Looking at Eq. (5.58), we see that we have determined Q in terms of the known values d, f_0, f_1, and f_2. That means that we have fully determined $R(x)$, as per our starting Eq. (5.53). Given that $R(x)$ is a straight line, we can play the same game as in the secant method, namely we can find the x axis intercept. To be clear, that means that we will produce a new point x_3 as the x axis intercept of the line going through (x_1, R_1) and (x_2, R_2), which of course also goes through (x_0, R_0); we apply Eq. (5.51):

$$x_3 = x_2 - R_2\frac{x_2 - x_1}{R_2 - R_1} = x_2 - \frac{d}{R_1/R_2 - 1} = x_2 + \text{sign}(f_0)\frac{f_2 d}{\sqrt{f_2^2 - f_0 f_1}} \tag{5.59}$$

In the second equality we brought the R_2 to the denominator and identified d in the numerator (along with changing the sign of both numerator and denominator). In the third equality we employed the fact that:

$$\frac{R_1}{R_2} = \frac{f_1\, e^{Qx_1}}{f_2\, e^{Qx_2}} = \frac{f_1}{f_2}e^{Qd} \tag{5.60}$$

together with our result in Eq. (5.58) for e^{Qd}. To reiterate, we found the x axis intercept of $R(x)$; we then re-expressed everything in terms of the function $f(x)$, which was how our initial problem was formulated.

Our last result in Eq. (5.59) is something you could immediately implement programmatically. It only requires $f(x)$ evaluations to produce a root estimate, x_3. We'll follow Ridders' original work in making an extra step, in order to remove the factor $\text{sign}(f_0)$.[24] Divide both the numerator and denominator by f_0 to find:

$$x_3 = x_2 + (x_1 - x_2)\frac{f_2/f_0}{\sqrt{(f_2/f_0)^2 - f_1/f_0}} \tag{5.61}$$

where we also took the oportunity to re-express d as per Eq. (5.55).[25] We now notice a nice property of Ridders' prescription: since $f_0 f_1 < 0$, we see that the denominator is larger in magnitude than the numerator. That shows that x_3 will certainly stay within the original bracket between x_0 and x_1; thus, Ridders' method is guaranteed to converge.

So far, we've discussed only the procedure that allows us to produce our first root estimate, x_3, i.e., we haven't seen how to iterate further. Even so, it may be instructive at this point to see Ridders' method in action; in the left panel of Fig. 5.11 we show our example function, as usual. We have picked our initial bracket x_0 and x_1 in such a way as to highlight that the function $f(x)$ and the straight line $R(x)$ are totally different entities. As a reminder, in Ridders' method we start from x_0 and x_1, we evaluate the midpoint x_2, and

[24] In the literature after Ridders' paper, that factor is given as $\text{sign}(f_0 - f_1)$. However, since we already know that $f_0 f_1 < 0$, it's simpler to use $\text{sign}(f_0)$.

[25] This equation for x_3 might make you a bit uncomfortable from a numerical perspective: f_0, f_1, and f_2 are expected to become very small in magnitude as we approach the root, so perhaps we shouldn't be dividing these small numbers by each other and then subtracting the ratios. Recall our comments on page 256.

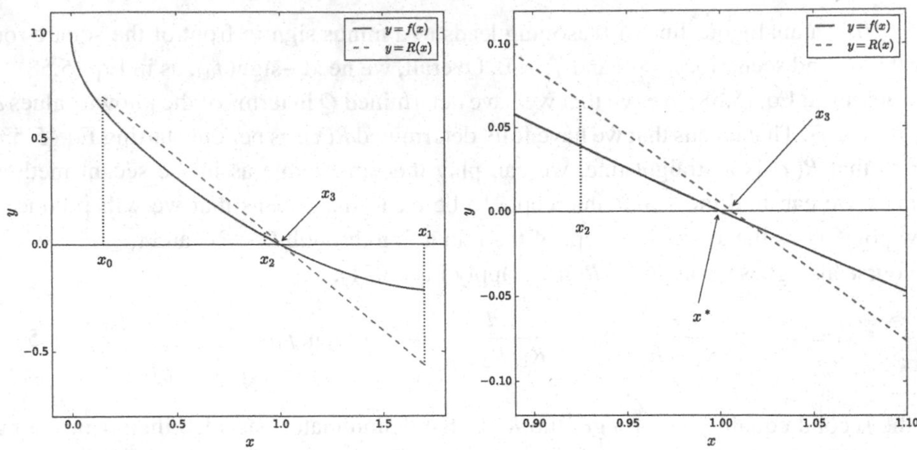

Fig. 5.11 Illustration of Ridders' method for our example function

then use the function values f_0, f_1, and f_2 as per Eq. (5.58) and Eq. (5.53) to produce the line $R(x)$. Our root estimate x_3 is the point where $R(x)$ crosses the x axis. As you can see in Fig. 5.11, for this problem just one iteration of this algorithm was enough to get very close to the true root x^*. In order to clarify that the x axis intercepts of $f(x)$ and of this first $R(x)$ are not necessarily the same, we provide a zoomed-in version of the plot in the right panel of Fig. 5.11.

After we've evaluated our first estimate for the root, in the way just discussed, Ridders' method goes on to form a new bracket, made up of x_3 and one of the existing x_i values. Specifically, we check if $f_3 f_i < 0$ for $i = 0, 1, 2$ to ensure that the root is still bracketed between x_3 and a specific x_i. We then rename the two endpoints appropriately, i.e., to x_0 and x_1, and repeat the entire process.

Overall, Ridders' method is a reliable bracketing method. As we've seen in our example case, it converges more rapidly than the bisection method, typically quadratically. Of course, this is slightly misleading, since each iteration of Ridders' method requires two function evaluations, one for the midpoint and one for an endpoint (which was an x axis intercept in the previous iteration). Even so, this is still superlinear and therefore faster than the bisection method.

5.2.8 Summary of One-Dimensional Methods

We've already encountered five different methods that solve non-linear equations. Here's a quick summary of their main features:

- *Fixed-point method*: a non-bracketing method; simple; doesn't always converge.
- *Bisection method*: a bracketing method; simple; always converges; linear convergence.
- *Newton's method*: a non-bracketing method; requires the function derivative; usually quadratic convergence; sometimes slow or worse.[26]

[26] Leading some people to say that Newton's method should be avoided; this is too strong a statement.

- *Secant method*: a non-bracketing method; similar to Newton's method but doesn't require the function derivative; usually superlinear convergence; sometimes slow or worse.
- *Ridders' method*: a bracketing method; superlinear convergence; rarely slow.

There are even more speedy and reliable methods out there. Most notably, *Brent's method* is a bracketing approach which is as reliable as the bisection method, with a speed similar to that of the Newton or secant methods (for non-pathological functions). However, the bookkeeping it requires is more complicated, so in the spirit of the rest of this volume we omit it from our presentation. Note that for all bracketing methods our earlier disclaimer applies: if you have a root that is not bracketed, like in Fig. 5.9, you will not be able to use such an approach.

Combining all of the above pros and cons, in what follows we will generally reach for Newton's method (especially when generalizing to multiple dimensions below) or the secant method (when we don't have any information on the derivative, as in section 8.3.1.1). When you need the speed of these methods and the reliability of bracketing methods, you can employ a "hybrid" approach, as touched upon in problem 5.7.

5.3 Zeros of Polynomials

As noted at the start of this chapter, a special case of solving a non-linear equation in one variable has to do with polynomial equations, of the form:

$$c_0 + c_1 x + c_2 x^2 + c_3 x^3 + \cdots + c_{n-1} x^{n-1} = 0 \tag{5.62}$$

The reason this gets its own section is as follows: we know from the fundamental theorem of algebra that a polynomial of degree $n - 1$ has $n - 1$ roots, so this problem is already different from what we were dealing with above, where only one or two roots needed to be evaluated.

5.3.1 Challenges

Let us start out by discussing the special challenges that emerge in polynomial root-finding. We've already encountered one of these in section 5.2.5.3; as you may recall, for *multiple roots* we have $f'(x^*) = 0$ so, using our definition from Eq. (5.15), we saw that the root-finding problem in this case is always ill-conditioned. We already learned that bracketing methods may be inapplicable in such scenarios, whereas methods (like Newton's) which are usually lightning fast, slow down.

Based on the intuition you developed when studying error analysis in linear algebra, in section 4.2, you probably understand what being ill-conditioned means: if you change the coefficients of your problem slightly, the solution is impacted by a lot. Let's look at a

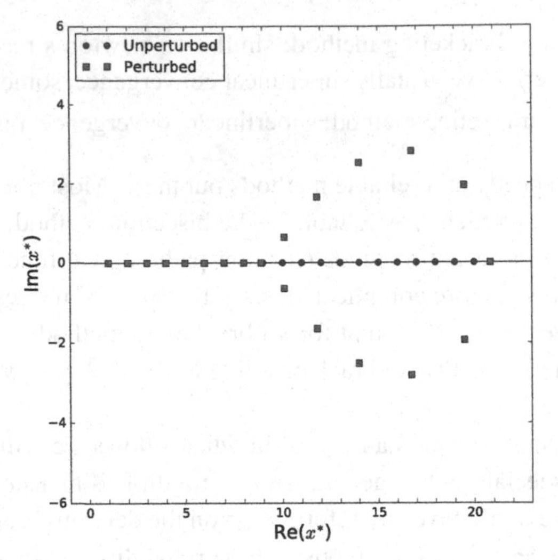

Fig. 5.12 Roots of the Wilkinson polynomial on the complex plane

specific example, motivated by Eq. (5.46):

$$x^2 - 4x + 4 = 0 \tag{5.63}$$

This has a double root at $x^* = 2$, as you can easily see by gathering the terms together into $(x-2)^2 = 0$. A small perturbation of this problem has the form:

$$x^2 - 4x + 3.99 = 0 \tag{5.64}$$

Even though we've only perturbed one of the coefficients by 0.25%, the equation now has two *distinct* roots, at $x^* = 2.1$ and $x^* = 1.9$. Each of these has changed its value by 5% in comparison to the unperturbed problem. This effect is considerably larger than the change in the coefficient. Things can get even worse, though. Consider the perturbed equation:

$$x^2 - 4x + 4.01 = 0 \tag{5.65}$$

This one, too, has a single coefficient perturbed by 0.25%. Evaluating the roots in this case, however, leads to $x^* = 2 + 0.1i$ and $x^* = 2 - 0.1i$, where i is the *imaginary unit*! A small change in one coefficient of the polynomial changed the real double root into a complex conjugate pair, with a sizable imaginary part (i.e., there are no longer real roots).

Even so, we don't want you to get the impression that polynomials are ill-conditioned only for the case of multiple roots. To see that, we consider the infamous "Wilkinson polynomial":

$$\begin{aligned} W(x) &= \prod_{k=1}^{20}(x-k) = (x-1)(x-2)\dots(x-20) \\ &= x^{20} - 210x^{19} + 20\,615x^{18} + \cdots + 2\,432\,902\,008\,176\,640\,000 \end{aligned} \tag{5.66}$$

Clearly, this polynomial has the 20 roots $x^* = 1, 2, \dots, 20$. The roots certainly appear to be well-spaced, i.e., nowhere near being multiple. One would naively expect that a tiny change in the coefficients would not have a large effect on the locations of the roots. Wilkinson examined the effect of changing the coefficient of x^{19} from `-210` to:

```
-210 - 2**(-23) = -210.0000001192093
```

which is a single-coefficient modification of relative magnitude $\approx 6 \times 10^{-8}\%$. The result of this tiny perturbation ends up being dramatic, as shown in Fig. 5.12. Half the roots have now become complex, with large imaginary parts; the largest effect is on the roots that used to be $x^* = 15$ and 16. It's important to emphasize that these dramatic changes are not errors in our root-finding: we are calculating the true roots of the perturbed polynomial, these just happen to be very different from the true roots of the original polynomial. Whatever root-finding algorithm you employ, you should make sure the errors it introduces are small enough not to impact the interpretation of your actual polynomial equation.

You may recall that in problem 2.11 we studied the question of how to *evaluate* a polynomial when you have access to the coefficients c_i, using what is known as *Horner's rule*. In this section, we've been examining the problem of solving for the polynomial's zeros, when you have access to the coefficients c_i. Our conclusion has been that the problem of finding the roots starting from the coefficients can be very ill-conditioned, so this approach should be avoided if possible. While this is an extreme example, the general lesson is that you should be very careful in the earlier mathematical steps which lead to a polynomial problem. Given such idiosyncratic behavior, you will not be surprised to hear that there exist specialized algorithms, designed specifically for finding zeros of polynomials. Some of these techniques suppress an already-found root, as touched upon in problem 5.11. In what follows, we begin with a more pedestrian approach.

5.3.2 One Root at a Time: Newton's Method

The easiest way to solve for a polynomial's zeros is to treat the problem as if it was any other non-linear equation and employ one of the five methods we introduced in the previous section, or other ones like them. Specifically, you may recall that, if you have access to the function's derivative and you start close enough to the root, Newton's method is a very fast way of finding an isolated root. For a high-degree polynomial this will entail finding many roots one after the other. Thus, it becomes crucial that you have a good first estimate of where the roots lie. Otherwise, you will be forced to guess, for example picking a grid of starting values; many of these will likely end up finding the same roots anew, thereby wasting computational resources (cf. problem 5.11). Even worse, in order to find *all* the zeros of the polynomial, you may be forced to repeat the entire exercise with a finer grid.

We will address a specific problem, that of finding the roots of *Legendre polynomials*. As you may recall, we first encountered these in the context of the multipole expansion (section 2.5.2.1) and then saw them again in our discussion of Gram–Schmidt orthogonalization (problem 4.27). We will re-encounter them when we introduce the theory of

Gaussian quadrature (section 7.5.2.1); as it so happens, at that point we will need their roots, so here we are providing both scaffolding and a pedagogical example.[27]

Fortunately, Legendre polynomials have been extensively studied and tabulated, so we can use several of their properties. Take $P_n(x)$ to be the n-th degree Legendre polynomial[28] and denote its i-th root by x_i^*; as usual, we use 0-indexing, meaning that the n roots are labelled using $i = 0, 1, \ldots, n-1$. If we convert Eq. (22.16.6) from Ref. [1] to use our counting scheme, we learn that the true roots x_i^* obey the following inequality:

$$\cos\left(\frac{2i+1}{2n+1}\pi\right) \leq x_i^* \leq \cos\left(\frac{2i+2}{2n+1}\pi\right) \tag{5.67}$$

This allows us to start Newton's algorithm with the following initial guess:

$$x_i^{(0)} = \cos\left(\frac{4i+3}{4n+2}\pi\right) \tag{5.68}$$

where we are using a subscript to denote the cardinal number of the root[29] and a superscript in parentheses to denote the starting value in Newton's method. Note that we multiplied both numerator and denominator by 2 in order to avoid having a $3/2$ in the numerator.

We have now accomplished what we set out to do, namely to find a reasonably good starting estimate of the n roots of our polynomial. The other requirement that Newton's method has is access to the function derivative. Fortunately for us, we have already produced, in `legendre.py`, a function that returns both the Legendre polynomial and its first derivative at a given x. Intriguingly, we accomplished this without ever producing the coefficients of the Legendre polynomials, so we are not too worried about questions of conditioning. Thus, we are now ready to implement in Python a root-finder for Legendre polynomials; the result is code 5.4. We start out by importing our earlier `legendre()` and then define two new functions; let's discuss each of these in turn.

The function `legnewton()` is designed specifically to find a single root of a Legendre polynomial. In other words, it looks quite a bit like what a Newton's method function would have looked like, had we implemented it in section 5.2.5; however, it does *not* take in as parameters a general function `f` and its derivative `fprime`. Instead, it is designed to work only with `legendre()` which, in its turn, had been designed to return the values of a Legendre polynomial and its first derivative as a tuple.[30] The rest of the function, with its termination criterion and so on, looks very similar to what we saw in, say, `secant()`.

The next function, `legroots()`, takes care of some rather messy bookkeeping. In essence, all it does is to repeatedly call `legnewton()` with different initial guesses corresponding to each root, as per Eq. (5.68). The messiness appears because we have decided to halve our

[27] Note also how we are following the separation of concerns principle: Legendre-polynomial evaluation is different from Legendre-polynomial root-finding, and both of these are different from Gaussian quadrature, so there should be (at least) three Python functions involved here.

[28] Don't get confused: $P_n(x)$ involves x^n, just like our $(n-1)$-th degree polynomial in Eq. (5.62) involves x^{n-1}.

[29] Here x_i is referring to distinct roots, *not* to bracketing endpoints, like x_0 and x_1 above.

[30] Note that if you had tried to use a pre-existing general `newton()` function, you would have had to go through syntactic contortions in order to make the two interfaces match.

legroots.py Code 5.4

```
from legendre import legendre
import numpy as np

def legnewton(n,xold,kmax=200,tol=1.e-8):
    for k in range(1,kmax):
        val, dval = legendre(n,xold)
        xnew = xold - val/dval

        xdiff = xnew - xold
        if abs(xdiff/xnew) < tol:
            break

        xold = xnew
    else:
        xnew = None
    return xnew

def legroots(n):
    roots = np.zeros(n)
    npos = n//2
    for i in range(npos):
        xold = np.cos(np.pi*(4*i+3)/(4*n+2))
        root = legnewton(n,xold)
        roots[i] = -root
        roots[-1-i] = root
    return roots


if __name__ == '__main__':
    roots = legroots(9); print(roots)
```

workload: instead of calculating n roots for the n-th degree polynomial, we notice that these are symmetrically distributed around 0. Specifically, if n is even, then we can simply evaluate only $n/2$ roots (say, the positive ones) and we can rest assured that the other $n/2$ can be trivially arrived at. If n is odd, then 0 is also a root, so we need only evaluate $(n-1)/2$ positive roots. The code does just that, by starting out with an array full of zeros, `roots`, which all get overwritten in the case of even-n but provide us with the needed 0 for the case of odd-n. Notice that we employed Python's integer division to find the number of positive roots, `n//2`, which works correctly for both even-n and odd-n. We start `roots` out with the negative roots and then introduce the corresponding positive roots counting from the end: as you may recall, `roots[-1-i]` is the idiomatic way of accessing the last element, the second-to-last element, and so on.

Unsurprisingly, `numpy` has a function that carries out the same task as our `legroots()`; you may wish to use `numpy.polynomial.legendre.leggauss()` to compare with our results; this actually returns two arrays, but we are interested only in the first one. In the main program we try out the specific case of `n = 9`, in order to test what we said earlier about the case of odd-n. The output of running this program is:

```
[-0.96816024 -0.83603111 -0.61337143 -0.32425342 0. 0.32425342
  0.61337143  0.83603111  0.96816024]
```

As expected, the roots here are symmetrically distributed around 0; as a matter of fact, they also obey a "clustering" property, in that there are more points near the ends of the interval. You will see in problems 5.20 and 5.22 two other ways of arriving at the same roots.

5.3.3 All the Roots at Once: Eigenvalue Approach

The approach we covered in the previous subsection, of evaluating Legendre polynomial roots one at a time, seems perfect: we didn't have to go through the intermediate step of computing the polynomial coefficients, so we didn't get in trouble with the conditioning of the root-finding problem. However, there was one big assumption involved, namely that we had good initial guesses for each root available, see Eq. (5.68). But what if we didn't? More specifically, what if you are dealing with some new polynomials which have not been studied before? In that case, our earlier approach would be inapplicable. We now turn to a different method, which uses matrix techniques to evaluate all of a polynomial's roots at once. We state at the outset that this approach requires knowledge of the coefficients of the polynomial, so our earlier disclaimers regarding the conditioning of root-finding in that case also apply here. Another disclaimer: this technique is more costly than polynomial-specific approaches like Laguerre's method (which we won't discuss).

Qualitatively, the idea involved here is as follows: we said in section 4.4 that one shouldn't evaluate a matrix's eigenvalues by solving the characteristic equation; you now know that this was due to the conditioning issues around polynomial root-finding. However, there's nothing forbidding you from taking the reverse route: take a polynomial equation, map it onto an eigenvalue problem and then use your already-developed robust eigensolvers.

Motivated by Eq. (5.62), we define the following polynomial:

$$p(x) = c_0 + c_1 x + c_2 x^2 + c_3 x^3 + \cdots + c_{n-1} x^{n-1} + x^n \tag{5.69}$$

Note that $p(x)$ also includes an x^n term, with a coefficient of 1. This is known as a *monic polynomial*; you can think of it as a monomial plus a polynomial of one degree lower. Even though it may seem somewhat arbitrary, we now define the following *companion matrix* which has dimensions $n \times n$:[31]

$$\mathbf{C} = \begin{pmatrix} 0 & 1 & 0 & \dots & 0 \\ 0 & 0 & 1 & \dots & 0 \\ 0 & 0 & 0 & \ddots & \vdots \\ \vdots & \vdots & \ddots & 0 & 1 \\ -c_0 & -c_1 & \dots & -c_{n-2} & -c_{n-1} \end{pmatrix} \tag{5.70}$$

We picked this matrix because it has a very special *characteristic polynomial.* As you may recall from Eq. (4.13), the characteristic polynomial for our companion matrix $\mathbf{C}$ will be formally arrived at using a determinant, i.e., $\det(\mathbf{C} - \lambda\mathcal{I})$. Now here is the remarkable property of our companion matrix: if you evaluate this determinant, it will turn out to be (as a function of λ) equal to our starting polynomial (up to an overall sign), i.e., $p(\lambda)$! Therefore, if you are interested in the solutions of the equation $p(\lambda) = 0$, which is our starting problem in Eq. (5.69), you can simply calculate the eigenvalues using a method of your choosing. We spent quite a bit of time in section 4.4 developing robust eigenvalue-solvers, so you could use one of them; note that you have hereby mapped one non-linear equation with n roots to an $n \times n$ eigenvalue problem. Intriguingly, this provides you with *all* the roots at once, in contradistinction to what we saw in the previous subsection, where you were producing them one at a time.

In the previous paragraph we merely stated this remarkable property of companion matrices. Actually there were two interrelated properties: (a) the characteristic polynomial of $\mathbf{C}$ is (up to a sign) equal to $p(\lambda)$, and therefore (b) the roots of $p(\lambda)$ are the eigenvalues of $\mathbf{C}$. Here, we are really interested only in the latter property; let's prove it. Assume x^* is a root of our polynomial $p(x)$, i.e., $p(x^*) = 0$ holds. It turns out that the eigenvector of $\mathbf{C}$ is simply a tower of monomials in x^*, specifically $(1\ x^*\ (x^*)^2\ \dots\ (x^*)^{n-2}\ (x^*)^{n-1})^T$; observe that this is an $n \times 1$ column vector. Let's explicitly see that this is true, by acting with our companion matrix on this vector:

$$\mathbf{C}\begin{pmatrix} 1 \\ x^* \\ \vdots \\ (x^*)^{n-2} \\ (x^*)^{n-1} \end{pmatrix} = \begin{pmatrix} x^* \\ (x^*)^2 \\ \vdots \\ (x^*)^{n-1} \\ -c_0 - c_1 x^* - c_2 (x^*)^2 - \dots - c_{n-1}(x^*)^{n-1} \end{pmatrix} = \begin{pmatrix} x^* \\ (x^*)^2 \\ \vdots \\ (x^*)^{n-1} \\ (x^*)^n \end{pmatrix} = x^* \begin{pmatrix} 1 \\ x^* \\ \vdots \\ (x^*)^{n-2} \\ (x^*)^{n-1} \end{pmatrix} \tag{5.71}$$

In the first equality we simply carried out the matrix multiplication. In the second equality we used the fact that x^* is a root, so $p(x^*) = 0$, which can be re-expressed in the form $(x^*)^n = -c_0 - c_1 x^* - c_2 (x^*)^2 - \dots - c_{n-1}(x^*)^{n-1}$. In the third equality we pulled out a factor

[31] Amusingly enough, mathematicians have gone on to also define a "comrade matrix" and a "colleague matrix".

of x^*. Thus, we have managed to prove not only that the column vector we said was an eigenvector is, indeed, an eigenvector, but also that x^* is the corresponding eigenvalue![32]

To summarize, we can evaluate the roots of our polynomial $p(x)$ from Eq. (5.69) by computing the eigenvalues of the companion matrix from Eq. (5.70). Make sure you remember that Eq. (5.69) is a *monic* polynomial. If in your problem the coefficient of x^n is different from 1, simply divide all the coefficients by it; you're interested in the roots, and this operation doesn't impact them. From there onwards, everything we've said in this subsection applies. To be concrete, for a monic polynomial of degree 4, you will have a 4×4 companion matrix and a 4×1 eigenvector. Problem 5.19 asks you to implement this approach in Python, using the eigenvalue solvers we developed in the previous chapter, `qrmet.py` or `eig.py`.

5.4 Systems of Non-Linear Equations

We now turn to a more complicated problem, that of n simultaneous non-linear equations in n unknowns. Employing the notation of Eq. (5.10), this can be expressed simply as $\mathbf{f}(\mathbf{x}) = \mathbf{0}$. In section 5.2 on the one-variable problem, the approach that was the easiest to reason about and was also guaranteed to converge was the bisection method. Unfortunately, this doesn't trivially generalize to the many-variable problem: you would need to think about all the possible submanifolds and evaluate the function (which, remember, may be a costly operation) a very large number of times to check if it changed sign. On the other hand, the fixed-point iteration method generalizes to the n-variable problem straightforwardly (as you will discover in problem 5.21) but, as before, is not guaranteed to converge. What we would like is a fast method that can naturally handle an n-dimensional space. It turns out that Newton's method fits the bill, so in the present section we will discuss mainly variations of this general approach. It should come as no surprise that, in that process, we will use matrix-related functionality from chapter 4.

Writing down $\mathbf{f}(\mathbf{x}) = \mathbf{0}$ may appear misleadingly simple, so let's make things concrete by looking at a two-dimensional problem:

$$\begin{cases} f_0(x_0, x_1) = x_0^2 - 2x_0 + x_1^4 - 2x_1^2 + x_1 = 0 \\ f_1(x_0, x_1) = x_0^2 + x_0 + 2x_1^3 - 2x_1^2 - 1.5x_1 - 0.05 = 0 \end{cases} \tag{5.72}$$

We have two unknowns, x_0 and x_1, and two equations; the latter are given in the form $f_i = 0$, where each of the f_i's is a function of the two variables x_0 and x_1. Note that here we are truly faced with two variables, x_0 and x_1, in contradistinction to the previous section, where x_0, x_1, and so on were used to denote distinct roots of a single polynomial.[33] You should convince yourself that you cannot fully solve this pair of coupled non-linear equations by

[32] Note that we didn't have to evaluate a determinant anywhere.

[33] Also in contradistinction to the bracketing endpoints, x_0 and x_1, of earlier sections.

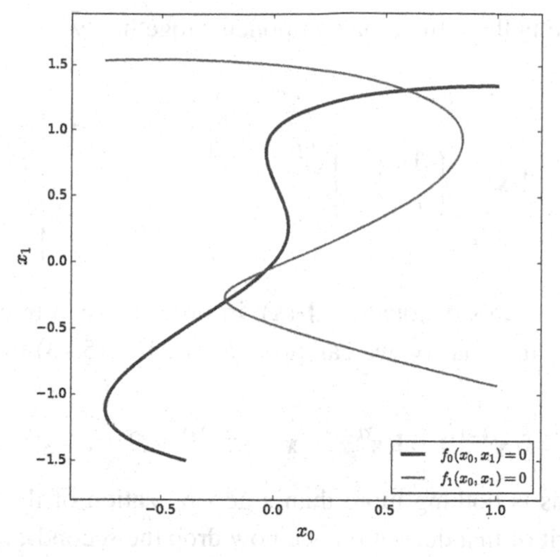

Fig. 5.13 Example of two coupled non-linear equations

hand; in this simple case, it's possible to solve, say, for x_0 in terms of x_1, and then get a single non-linear equation, but the question arises how you would be able to do the same for the case of a 10-dimensional system. To get some insight into the problem, Fig. 5.13 shows the curves described by our two equations. We find three intersection points between the two curves. (Actually, there is also a fourth one, if you look farther to the right.)

5.4.1 Newton's Method

The derivation of Newton's multidimensional method will basically be a straightforward generalization of what we saw in section 5.2.5, with a Taylor expansion at its core. We assume that **f** has bounded first and second derivatives; the actual solution of our problem is $\mathbf{x}^*$ and we will be trying to approximate it using iterates, which this time are themselves vectors, $\mathbf{x}^{(k)}$.[34] In order to make the transition to the general problem as simple as possible, let's start from a Taylor expansion of a single function component f_i around our latest iterate, $\mathbf{x}^{(k-1)}$:

$$f_i(\mathbf{x}) = f_i(\mathbf{x}^{(k-1)}) + \left(\nabla f_i(\mathbf{x}^{(k-1)})\right)^T \left(\mathbf{x} - \mathbf{x}^{(k-1)}\right) + O\left(\|\mathbf{x} - \mathbf{x}^{(k-1)}\|^2\right) \tag{5.73}$$

where, as usual, $i = 0, 1, \dots, n-1$. We can rewrite the second term on the right-hand side in terms of vector components:

$$\left(\nabla f_i(\mathbf{x}^{(k-1)})\right)^T \left(\mathbf{x} - \mathbf{x}^{(k-1)}\right) = \sum_{j=0}^{n-1} \left.\frac{\partial f_i}{\partial x_j}\right|_{x_j^{(k-1)}} \left(x_j - x_j^{(k-1)}\right) \tag{5.74}$$

[34] Just like in our discussion of the Jacobi iterative method for solving linear systems of equations in section 4.3.5.

With a view to collecting the n function components together, we now introduce the *Jacobian matrix*:

$$\mathbf{J}(\mathbf{x}) = \left\{\frac{\partial f_i}{\partial x_j}\right\} = \begin{pmatrix} \frac{\partial f_0}{\partial x_0} & \frac{\partial f_0}{\partial x_1} & \cdots & \frac{\partial f_0}{\partial x_{n-1}} \\ \frac{\partial f_1}{\partial x_0} & \frac{\partial f_1}{\partial x_1} & \cdots & \frac{\partial f_1}{\partial x_{n-1}} \\ \vdots & \vdots & \ddots & \vdots \\ \frac{\partial f_{n-1}}{\partial x_0} & \frac{\partial f_{n-1}}{\partial x_1} & \cdots & \frac{\partial f_{n-1}}{\partial x_{n-1}} \end{pmatrix} \quad (5.75)$$

You may sometimes see this denoted by $\mathbf{J_f}(\mathbf{x})$, in order to keep track of which function it's referring to. Using this matrix, we can now rewrite Eq. (5.73) to group together all n function components:

$$\mathbf{f}(\mathbf{x}) = \mathbf{f}(\mathbf{x}^{(k-1)}) + \mathbf{J}(\mathbf{x}^{(k-1)})\left(\mathbf{x} - \mathbf{x}^{(k-1)}\right) + O\left(\|\mathbf{x} - \mathbf{x}^{(k-1)}\|^2\right) \quad (5.76)$$

Keep in mind that this is nothing more than a generalization of the Taylor expansion in Eq. (5.37). In the spirit of that derivation, we now drop the second-order term and assume that we have found the solution, $\mathbf{f}(\mathbf{x}^*) = \mathbf{0}$:

$$\mathbf{0} = \mathbf{f}(\mathbf{x}^{(k-1)}) + \mathbf{J}(\mathbf{x}^{(k-1)})\left(\mathbf{x}^* - \mathbf{x}^{(k-1)}\right) \quad (5.77)$$

In practice, one iteration will not be enough to find the solution so, instead, we use our latest formula to introduce the *prescription* of Newton's method for the next iterate, $\mathbf{x}^{(k)}$:

$$\mathbf{J}(\mathbf{x}^{(k-1)})\left(\mathbf{x}^{(k)} - \mathbf{x}^{(k-1)}\right) = -\mathbf{f}(\mathbf{x}^{(k-1)}) \quad (5.78)$$

Since all quantities at the location of our previous iterate, $\mathbf{x}^{(k-1)}$, are known, this equation has the form of $\mathbf{A}\mathbf{x} = \mathbf{b}$, i.e., it is a *linear* system of n equations in n unknowns. Assuming $\mathbf{J}(\mathbf{x}^{(k-1)})$ is non-singular, we can solve this system and then we will be able to find all the $x_j^{(k)}$. This process is repeated, until we satisfy a termination criterion, which could be taken to be that in Eq. (4.122):

$$\sum_{j=0}^{n-1} \left| \frac{x_j^{(k)} - x_j^{(k-1)}}{x_j^{(k)}} \right| \le \epsilon \quad (5.79)$$

or something fancier.

In principle, you could further manipulate Eq. (5.78) so as to write it in the form:

$$\mathbf{x}^{(k)} = \mathbf{x}^{(k-1)} - \mathbf{J}^{-1}(\mathbf{x}^{(k-1)})\mathbf{f}(\mathbf{x}^{(k-1)}) \quad (5.80)$$

which looks very similar to Eq. (5.40) for the one-variable Newton's method. In practice, as you know from chapter 4, explicitly evaluating the inverse of a matrix is usually a bad idea.[35] Instead, you simply solve the linear system Eq. (5.78), say by Gaussian elimination or the Jacobi iterative method (if the Jacobian matrix is sparse). A straightforward generalization of the convergence study in section 5.2.5.2 will convince you that, even in many dimensions, Newton's method is *quadratically convergent*, assuming your initial guess (vector) $\mathbf{x}^{(0)}$ is close enough to the true solution $\mathbf{x}^*$.

[35] That said, the absolute condition number for the n-dimensional problem—recall Eq. (5.15)—is $\|\mathbf{J}^{-1}(\mathbf{x}^*)\|$.

5.4.2 Discretized Newton Method

In order to solve the linear system in Eq. (5.78), you need to have access to the Jacobian matrix, $\mathbf{J}(\mathbf{x})$. This is completely analogous to the one-variable version of Newton's method, where you needed to be able to evaluate $f'(x)$. For our two-dimensional problem in Eq. (5.72) this is very easy to do: you simply take the derivatives analytically. In general, however, this may not be possible, e.g., if your function components f_i are external complicated subroutines.

5.4.2.1 Algorithm

You may recall that in the case of the one-variable Newton's method we had advised against using a finite-difference approach to evaluate the first derivative $f'(x)$. This was both because we didn't know which spacing h to pick, and also because it implied two function evaluations at each iteration. In the present, multidimensional, case we are willing to deal with these problems, given the paucity of alternatives.

This gives rise to the *discretized Newton method*; this is simply a result of using the forward difference to approximate the derivatives that make up the Jacobian matrix:

$$\frac{\partial f_i}{\partial x_j} \approx \frac{f_i(\mathbf{x} + \mathbf{e}_j h) - f_i(\mathbf{x})}{h} = \frac{f_i(x_0, x_1, \ldots, x_j + h, \ldots, x_{n-1}) - f_i(x_0, x_1, \ldots, x_j, \ldots, x_{n-1})}{h} \tag{5.81}$$

As shown explicitly in the second equality, $\mathbf{e}_j$ is the j-th column of the identity matrix, as we had already seen back in Eq. (4.100). The algorithm of the discretized Newton method then consists of employing Eq. (5.78), where the Jacobian is approximated as per Eq. (5.81).

Since each of i and j takes up n possible values, you can see that in order to evaluate all the $f_i(\mathbf{x} + \mathbf{e}_j h)$ we need n^2 function-component evaluations. To that you must add another n function-component evaluations that lead to $f_i(\mathbf{x})$; you actually need those for the right-hand side of Eq. (5.78), so you compute them before you start forming the forward differences of Eq. (5.81). In addition to this cost, each iteration of the discretized Newton method requires the linear system of Eq. (5.78) to be solved, for which as you may recall the operation count is $\sim 2n^3/3$ for Gaussian elimination.[36]

5.4.2.2 Implementation

At this point, it should be relatively straightforward to implement the discretized Newton method in Python; the result is code 5.5. We start out by importing the Gaussian-elimination-with-pivoting function from the last chapter: since we have no guarantees about the structure of our matrices, we try to be as general as possible. We then introduce three new functions. Taking these in turn: first, we create a function that evaluates

[36] Which is the method you'd likely employ for a dense matrix.

Code 5.5 multi_newton.py

```
from gauelim_pivot import gauelim_pivot
from jacobi import termcrit
import numpy as np

def fs(xs):
    x0, x1 = xs
    f0 = x0**2 - 2*x0 + x1**4 - 2*x1**2 + x1
    f1 = x0**2 + x0 + 2*x1**3 - 2*x1**2 - 1.5*x1 - 0.05
    return np.array([f0,f1])

def jacobian(fs,xs,h=1.e-4):
    n = xs.size
    iden = np.identity(n)
    Jf = np.zeros((n,n))
    fs0 = fs(xs)
    for j in range(n):
        fs1 = fs(xs+iden[:,j]*h)
        Jf[:,j] = (fs1 - fs0)/h
    return Jf, fs0

def multi_newton(fs,jacobian,xolds,kmax=200,tol=1.e-8):
    for k in range(1,kmax):
        Jf, fs_xolds = jacobian(fs, xolds)
        xnews = xolds + gauelim_pivot(Jf, -fs_xolds)

        err = termcrit(xolds, xnews)
        print(k, xnews, err)
        if err < tol:
            break

        xolds = np.copy(xnews)
    else:
        xnews = None
    return xnews

if __name__ == '__main__':
    xolds = np.array([1.,1.])
    xnews = multi_newton(fs, jacobian, xolds)
    print(xnews); print(fs(xnews))
```

the n function components f_i given a position argument $\mathbf{x}$. We do this for our two-variable problem in Eq. (5.72), but the rest of the code is set up to be completely general. In other words, the dimensionality of the problem is located *only* in one part of the code, i.e., in the function `fs()`. Since the rest of the code will be manipulating matrices, we make sure to produce a one-dimensional `numpy` array to hold the values of $\mathbf{f}(\mathbf{x})$. Note that in `fs()` we start by unpacking the input argument `xs` into two local variables; this is not necessary, since we could have simply used `xs[0]` and `xs[1]` below. However, all those square brackets are error prone so we avoid them.

The function `jacobian()` builds up the Jacobian matrix `Jf` via the forward difference, as per Eq. (5.81); in (minor) violation of the separation of concerns principle, we are also returning `fs0` (computed only once, outside the loop, to avoid being wasteful). After this, we are reasonably idiomatic, producing one column of the Jacobian matrix, Eq. (5.75), at a time. To get all the columns, we are iterating through the j index in $f_i(\mathbf{x}+\mathbf{e}_j h)$, while using `numpy` functionality to step through the i's, without the need for a second explicit index. It's worth pausing to compare this to the way we evaluated the second derivative all the way back in section 3.5. At the time, we didn't have access to `numpy` arrays, so we would shift a given position by h, in order to avoid rounding error creep. Here, we have the luxury of using a new vector for each shifted component separately.

The function `multi_newton()` looks a lot like our earlier `legnewton()` but, of course, it applies to any multidimensional function. As a result, it uses `numpy` arrays throughout. After evaluating the Jacobian matrix at the latest iterate as per Eq. (5.81), we solve the linear system Eq. (5.78) and thereby produce a new iterate. As usual, we test against the termination criterion to see if we should keep going. The print-out statement could have included the value(s) of `fs(xnews)`, to highlight that each function component f_i gets closer to zero as the algorithm progresses. Before exiting the loop, we prepare for the next iteration by renaming our latest iterate; we could have also accomplished this by saying `xolds = xnews`, but it's good to get into the habit of using `numpy.copy()`.

The main part of the program simply makes up an initial guess and runs our multidimensional Newton function. We then print out the final solution vector, as well as the values of the function components f_i at the solution vector. Running this code, we succeed, after a small number of iterations, in finding a solution which leads to tiny function components. You should try different initial guesses to produce the other roots.

5.4.3 Broyden's Method

As it now stands, the discretized Newton method we just discussed works quite well for problems that are not too large; we will make use of it in the projects at the end of this and following chapters. However, let's back up and see which parts of the algorithm are wasteful. As you may recall, this method entailed two large costs: first, it involved n^2+n function-component evaluations, n^2 for the Jacobian matrix as per Eq. (5.81) and n for the right-hand side of the Newton update, as per Eq. (5.78). Second, the Newton update itself involved solving a linear system of equations, which for Gaussian elimination costs $\sim 2n^3/3$ operations.

A reasonable assumption to make at this point is that, of these two costs, the function (component) evaluations are the most time-consuming. You can imagine problems where n is, say, 10, so solving a 10×10 linear system is pretty straightforward, but each function-component evaluation could be a lengthy separate calculation involving many other mathematical operations; then, the n^2 evaluations required for the discretized Newton method will dominate the total runtime. To address this issue, we are inspired by the secant method for the one-variable case: what we did there was to use the function values at the two latest iterates in order to approximate the value of the derivative, see Eq. (5.50). The idea was that, since we're going to be evaluating the function value at each iterate, we might as well hijack it to also provide information on the derivative. As the iterates get closer to each other, this leads to a reasonably good approximation of the first derivative.

With the secant method as our motivation, we then come up with the idea of using the last two guess vectors, $\mathbf{x}^{(k)}$ and $\mathbf{x}^{(k-1)}$, and the corresponding function values, in order to approximate the Jacobian matrix. In equation form, this is:

$$\mathbf{J}(\mathbf{x}^{(k)})\left(\mathbf{x}^{(k)} - \mathbf{x}^{(k-1)}\right) \approx \mathbf{f}(\mathbf{x}^{(k)}) - \mathbf{f}(\mathbf{x}^{(k-1)}) \tag{5.82}$$

This *secant equation* has to be obeyed but, unfortunately, it cannot be solved on its own, since it is *underdetermined*. To see this, think of our problem as $\mathbf{Ax} = \mathbf{b}$, where we know $\mathbf{x}$ and $\mathbf{b}$ but not $\mathbf{A}$. In order to make progress, we need to impose some further condition.

This problem was tackled by Broyden, who had the idea of not evaluating $\mathbf{J}(\mathbf{x}^{(k)})$ from scratch at each iteration. Instead, he decided to take a previous estimate of the Jacobian, $\mathbf{J}(\mathbf{x}^{(k-1)})$, and update it. The notation can get messy, so let's define some auxiliary variables:

$$\mathbf{x}^{(k)} - \mathbf{x}^{(k-1)} \equiv \mathbf{q}^{(k)}, \qquad \mathbf{f}(\mathbf{x}^{(k)}) - \mathbf{f}(\mathbf{x}^{(k-1)}) \equiv \mathbf{y}^{(k)} \tag{5.83}$$

Our secant equation then becomes simply:

$$\mathbf{J}(\mathbf{x}^{(k)})\mathbf{q}^{(k)} \approx \mathbf{y}^{(k)} \tag{5.84}$$

Broyden chose to impose the further requirement that:

$$\mathbf{J}(\mathbf{x}^{(k)})\mathbf{p} = \mathbf{J}(\mathbf{x}^{(k-1)})\mathbf{p} \tag{5.85}$$

for any vector $\mathbf{p}$ that is orthogonal to $\mathbf{q}^{(k)}$, i.e., for which $\left(\mathbf{q}^{(k)}\right)^T \mathbf{p} = 0$ holds: if you change $\mathbf{x}$ in a direction perpendicular to $\mathbf{q}^{(k)}$ then you don't learn anything about the rate of change of $\mathbf{f}$. The last two equations are enough to uniquely determine the update from $\mathbf{J}(\mathbf{x}^{(k-1)})$ to $\mathbf{J}(\mathbf{x}^{(k)})$: this is a constrained-optimization problem, tackled as per Eq. (5.165). However, instead of constructing this update explicitly, we will follow an easier route: we'll state Broyden's prescription, and then see that it satisfies both the required conditions, Eq. (5.84) and Eq. (5.85). Broyden's update is:

$$\mathbf{J}(\mathbf{x}^{(k)}) = \mathbf{J}(\mathbf{x}^{(k-1)}) + \frac{\mathbf{y}^{(k)} - \mathbf{J}(\mathbf{x}^{(k-1)})\mathbf{q}^{(k)}}{\left(\mathbf{q}^{(k)}\right)^T \mathbf{q}^{(k)}} \left(\mathbf{q}^{(k)}\right)^T \tag{5.86}$$

(We state at the outset that we can get in trouble if $\mathbf{q}^{(k)} = \mathbf{x}^{(k)} - \mathbf{x}^{(k-1)} = 0$.) Let's check if our two requirements are satisfied.

First, we take Eq. (5.86) and multiply by $\mathbf{q}^{(k)}$ on the right. We have:

$$\begin{aligned}\mathbf{J}(\mathbf{x}^{(k)})\mathbf{q}^{(k)} &= \mathbf{J}(\mathbf{x}^{(k-1)})\mathbf{q}^{(k)} + \frac{\mathbf{y}^{(k)} - \mathbf{J}(\mathbf{x}^{(k-1)})\mathbf{q}^{(k)}}{(\mathbf{q}^{(k)})^T \mathbf{q}^{(k)}} \left(\mathbf{q}^{(k)}\right)^T \mathbf{q}^{(k)} \\ &= \mathbf{J}(\mathbf{x}^{(k-1)})\mathbf{q}^{(k)} + \mathbf{y}^{(k)} - \mathbf{J}(\mathbf{x}^{(k-1)})\mathbf{q}^{(k)} = \mathbf{y}^{(k)}\end{aligned} \tag{5.87}$$

In the second equality we cancelled the denominator and in the third equality we cancelled the other two terms. We find that our first requirement, Eq. (5.84), is satisfied.

Next, we take Eq. (5.86) again, this time multiplying by $\mathbf{p}$ on the right:

$$\mathbf{J}(\mathbf{x}^{(k)})\mathbf{p} = \mathbf{J}(\mathbf{x}^{(k-1)})\mathbf{p} + \frac{\mathbf{y}^{(k)} - \mathbf{J}(\mathbf{x}^{(k-1)})\mathbf{q}^{(k)}}{(\mathbf{q}^{(k)})^T \mathbf{q}^{(k)}} \left(\mathbf{q}^{(k)}\right)^T \mathbf{p} \tag{5.88}$$

Since we know that $\left(\mathbf{q}^{(k)}\right)^T \mathbf{p} = 0$ holds, we see that the second term on the right-hand side vanishes, so we are left with Eq. (5.85), as desired.

Thus, Broyden's update in Eq. (5.86) satisfies our two requirements in Eq. (5.84) and Eq. (5.85). In other words, we have been able to produce the next Jacobian matrix using only the previous Jacobian matrix and the last two iterates (and corresponding function values). Let us summarize the entire prescription using our original notation, which employs fewer extra variables:

$$\begin{aligned}&\mathbf{J}(\mathbf{x}^{(k-1)})\left(\mathbf{x}^{(k)} - \mathbf{x}^{(k-1)}\right) = -\mathbf{f}(\mathbf{x}^{(k-1)}) \\ &\mathbf{J}(\mathbf{x}^{(k)}) = \mathbf{J}(\mathbf{x}^{(k-1)}) + \frac{\mathbf{f}(\mathbf{x}^{(k)})\left(\mathbf{x}^{(k)} - \mathbf{x}^{(k-1)}\right)^T}{\|\mathbf{x}^{(k)} - \mathbf{x}^{(k-1)}\|_E^2}\end{aligned} \tag{5.89}$$

We start with a known $\mathbf{J}(\mathbf{x}^{(0)})$; this could be produced using a forward-difference scheme (just once at the start), or could even be taken to be the identity matrix. From then onwards, we simply apply Newton's step from Eq. (5.78), in the first line, and then use Broyden's formula, Eq. (5.86), in the second line to update the Jacobian matrix; we also took the opportunity to use the first line to cancel some terms in the numerator and to identify the Euclidean norm in the denominator. Remember, even though Broyden's update was designed to respect the secant equation, Eq. (5.82), this equation does not appear in the prescription itself. Observe that the first line requires n function-component evaluations for $\mathbf{f}(\mathbf{x}^{(k-1)})$ and the next line another n function-component evaluations, for $\mathbf{f}(\mathbf{x}^{(k)})$, but the latter will be re-used in the guise of $\mathbf{f}(\mathbf{x}^{(k-1)})$ the next time through the loop. Note, finally, that there is no derivative in sight. Keep in mind that we won't exhibit quadratic convergence (like Newton's method for analytical derivatives does) but superlinear convergence, similarly to what we saw in the one-variable case.

At the end of the day, our prescription above still requires us to solve a linear system of equations in the first line. Broyden actually did not carry out our cancellation in the numerator, proceeding, instead, to make the prescription even more efficient, using identities like Eq. (6.272); the main concept should be clear from our discussion. The crucial point is that we are avoiding the n^2 function-component evaluations; this approach is often much faster than the discretized Newton method we covered in the previous subsection. Problem 5.23 asks you to implement this version of Broyden's method in Python; you should keep in mind that the numerator in the second line of Eq. (5.89) involves the product of a column vector and a row vector, so it is an instance of the outer product in producing an $n \times n$ matrix.[37]

5.5 One-Dimensional Minimization

We turn to the last task of this chapter: instead of finding function zeros, we are now going to be locating function minima. Note that we will be studying the problem of *unconstrained minimization*, meaning that we will not be imposing any further constraints on our variables; we are simply looking for the variable values that minimize a scalar function. (The topic of *constrained minimization* is introduced in problems 5.38, 5.39, and 5.40.)

5.5.1 General Features

For simplicity, let's start from the case of a function of a single variable, $\phi(x)$.[38] As you may recall from elementary calculus, a *stationary point* (which, for a differentiable function, is also known as a *critical point*) is a point at which the derivative vanishes, namely:

$$\phi'(x^*) = 0 \tag{5.90}$$

where we are now using x^* to denote the stationary point. If $\phi''(x^*) > 0$ we are dealing with a *local minimum*, whereas if $\phi''(x^*) < 0$, a *local maximum*. Minima and maxima together are known as *extrema*.

A simple example is our function $\phi(x) = e^{x-\sqrt{x}} - x$ from Fig. 5.1; in an earlier section we saw that it has two zeros, at ≈ 1 and at ≈ 2.5, but we are now interested in the minimum, which is located at $x^* \approx 1.8$. It's easy to see that $\phi''(x^*) > 0$ (simply by looking) so we have a (single) minimum. From a practical perspective, to compute the location of the minimum we can use one of our five one-variable root-finding algorithms from section 5.2, this time applied not to $\phi(x)$ but to $\phi'(x)$. By simply applying root-finding algorithms to the $\phi'(x)$ function, you will find a stationary point, but you won't know if it is a minimum, maximum, or something else (see below).

As a second example, we look at Fig. 5.9, which was plotting the function:

$$\phi(x) = x^4 - 9x^3 + 25x^2 - 24x + 4 \tag{5.91}$$

As you may recall, that function had one root at ≈ 0.21, another one at ≈ 4.79, as well as a double root at 2. Of course, we are now not trying to evaluate roots but minima or

[37] So you should *not* use numpy's `@` here, since it would lead to a scalar; use `numpy.outer()` instead.
[38] You will soon see why we are switching our notation from $f(x)$ to $\phi(x)$.

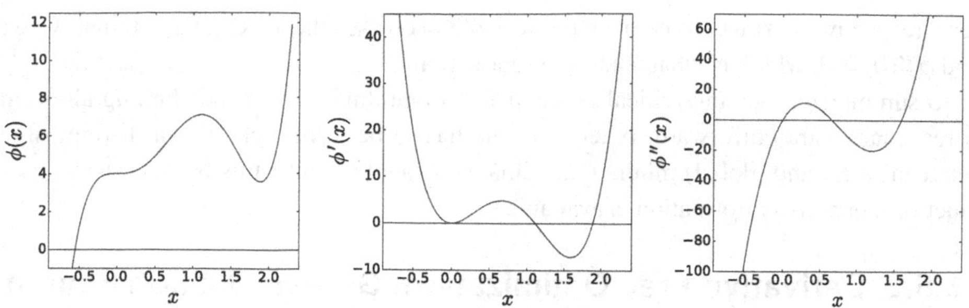

Fig. 5.14 Example illustrating the concept of a saddle point

maxima. Just by looking at the plot we can see that we have a local minimum at $x^* \approx 0.7$, a local maximum at $x^* = 2$ (which coincides with the location of the double root), as well as another minimum at $x^* \approx 4$. The second derivative of our function at each of these extrema, $\phi''(x^*)$, is positive, negative, and positive, respectively. This is a nice opportunity to point out that the minimum at $x^* \approx 4$ happens to be the *global minimum*, i.e., the point where the function reaches its smallest value not only locally but in general. It should be easy to see that our iterative root-finding algorithms, which start with an initial guess $x^{(0)}$ and proceed from there cannot guarantee that a global minimum (or, for other functions, a global maximum) will be reached. This applies even if you have somehow taken into account the value of the second derivative, i.e., even if you can guarantee that you are dealing with a local minimum, you cannot guarantee that you have found the global minimum. You can think of placing a marble (point) on this curve: it will roll to the minimum point that is accessible to it, but may not roll to the absolute minimum. We won't worry too much about this, but there are techniques that can move you out of a local minimum, in search of nearby (deeper) local minima.

At this point we realize that, when classifying critical points above, we forgot about one possibility, namely that $\phi''(x^*) = 0$. In this case, one must study the behavior of higher-order derivatives (this is called the *extremum test*):[39] the point may turn out to be a minimum, a maximum, or what is known as a *saddle point*.[40] In order to elucidate that last concept, let us look at a third example, namely the function:

$$\phi(x) = 4x^5 - 15x^4 + 14x^3 + 4 \tag{5.92}$$

In Fig. 5.14 we are plotting the function $\phi(x)$ itself in the left panel, the first derivative $\phi'(x)$ in the middle panel, and the second derivative $\phi''(x)$ in the right panel. We have a local minimum at $x^* \approx 1.89$, a local maximum at $x^* \approx 1.11$, and something new at $x^* = 0$: the function "flattens" but then keeps going in the same direction. Looking at the first derivative in the middle panel, we see that all three points are critical points (i.e., have vanishing first derivative). Then, turning to the second derivative in the right panel, examining these three points starting from the rightmost, $\phi''(x^*)$ is positive, negative, and

[39] This is actually more general, also covering the case where the first several derivatives are zero.

[40] There are situations where you can have an *inflection point* that is not a critical point (so it doesn't get called a saddle point). For example: $\phi(x) = \sin x$ leads to $\phi''(0) = 0$, but for $\phi'(0) \neq 0$.

zero, respectively. You can see from the way $\phi''(x)$ crosses the x axis at $x = 0$ that we will find $\phi'''(0) \neq 0$, which is what leads to a saddle point.

To summarize, locating critical points using a one-dimensional root-finding algorithm is reasonably straightforward. After this, one has to be a little careful in distinguishing between (local and global) minima, maxima, and saddle points; this is easier to do when function-derivative information is available.

5.5.2 Derivative-Free Optimization: Golden-Section Search

In the preceding subsection we didn't go over any new methods on one-dimensional minimization; the reason should be obvious: if you have information on the derivative(s) at your disposal, you may simply use one of the root-finding algorithms introduced earlier in this chapter (as problem 5.27 asks you to do). We now take a different tack: we will discuss how to minimize a function of one variable, $\phi(x)$, using only the function values themselves, i.e., no derivatives.

5.5.2.1 Algorithm and Interpretation

Roughly speaking, the technique we will introduce (known as *golden-section search*) will be analogous to the bisection method, in that we will arrive at a narrowing-down of the search interval that may not be lightning fast but, due to being pessimistic about where the solution may lie, is guaranteed to lead to a result without negatively impacting the rate of convergence. (In other words, it will be slow and steady.) This talk of a search interval, as well as the analogy drawn with the bisection method, should already be making you think of *bracketing* methods. This is appropriate, but a distinction has to be carefully made: for root-finding, two points bracketing a root means that these two points correspond to function values with opposite signs; for minimization, two points bracketing a minimum means that there exists a third point in between them where the function value is smaller. (This is different from the situation obtaining for a monotonic function, whose minimum would be located at one of the endpoints.)

Problem 5.24 addresses the question of the initial bracketing, i.e., here we will take it for granted that you have a starting interval (x_0, x_1) in which there exists a point x_2 for which $\phi(x_2)$ is smaller than both of $\phi(x_0)$ and $\phi(x_1)$. Thus, we are faced with the ordered triplet of points (x_0, x_2, x_1). As will soon become more clear, x_2 is *not* the midpoint; this means that it divides the interval (x_0, x_1) into a "small" subinterval and a "big" subinterval. To make things concrete, assume that (x_0, x_2) is the small subinterval (of width S) and (x_2, x_1) is the big subinterval (of width B). The guiding idea of the method we are discussing is that *the minimum is always to be found in the interval made up of the two points that are adjacent to the lowest-function-value point encountered so far*: at this early stage, $\phi(x_2)$ is the lowest function value so we know that the minimum will be located in (x_0, x_1).

The question then arises how to place a *new* intermediate point x_3.[41] This trial point x_3 has to be placed in the big subinterval (i.e., between x_2 and x_1), as you will soon come

[41] We will come back to the placement of x_2 itself later.

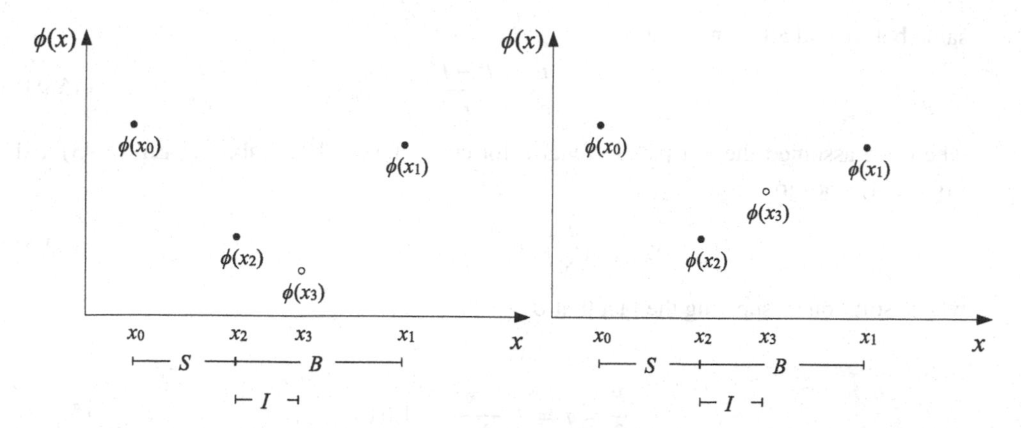

Golden-section search when the new point has a lower (left) or higher (right) value

Fig. 5.15

to appreciate. By now, things are starting to get hard to keep track of, so it may help you to look at Fig. 5.15. There are two scenarios at play. First, if $\phi(x_3)$ is the lowest function value encountered so far (left panel of Fig. 5.15), then we know that the minimum will be bracketed by the two adjacent points, i.e., the minimum lies in (x_2, x_1). We would then keep the triplet (x_2, x_3, x_1), renaming appropriately and re-applying the algorithm. Second, if $\phi(x_2)$ is the lowest function value encountered so far (right panel of Fig. 5.15), then we know that the minimum will be bracketed by the two adjacent points, i.e., the minimum lies in (x_0, x_3). We would then keep the triplet (x_0, x_2, x_3), renaming appropriately and re-applying the algorithm. For notational convenience, in the figure we are denoting the distance between the two intermediate points (x_2 and x_3) by I.

Now you can see why we placed x_3 in the "big" subinterval, i.e., in (x_2, x_1). Had we placed x_3 in the "small" subinterval, i.e., in (x_0, x_2), then the two possibilities would be imbalanced: if $\phi(x_3)$ was the lowest function value encountered then we would be faced with the small subinterval (x_0, x_2), but if $\phi(x_2)$ was the lowest function value encountered then the chosen subinterval (x_3, x_1) would be barely an improvement over our starting interval (x_0, x_1). The first possibility would be good, but the second one, bad; nothing could rule out a sequence of bad luck for several steps, needlessly slowing down our progress.[42]

While the above comments are intuitive, they don't actually tell us specifically where to place x_3 (or x_2, for that matter): we need to come up with some equations relating S, B, and I. We've already hinted at how to get the first condition: make sure that you don't leave things to chance, i.e., ensure that the two possibilities ($\phi(x_3)$ or $\phi(x_2)$ being the lowest function value to date) lead to a new interval of the same width:

$$B = S + I \tag{5.93}$$

The left-hand side corresponds to the left-panel scenario, whereas the right-hand side, to the right-panel one. Second, we will demand that the ratio of big subinterval to small is the

[42] This also helps to explain why we didn't place x_2 at the midpoint between x_0 and x_1: under this assumption, the next step would involve placing x_3 at the midpoint of one of the two halves and that would lead to the next interval being either one-half (good) or three-quarters (bad) of the original interval.

same before and after an update:

$$\frac{B}{S} = \frac{B - I}{I} \tag{5.94}$$

where we assumed the left-panel scenario for concreteness.[43] Combining Eq. (5.93) and Eq. (5.94) leads to:

$$\left(\frac{B}{S}\right)^2 - \frac{B}{S} - 1 = 0 \tag{5.95}$$

whose solution (respecting the fact that $B > S$) is:

$$\frac{B}{S} = \varphi = \frac{1 + \sqrt{5}}{2} \approx 1.618 \tag{5.96}$$

namely the *golden ratio* or *golden section*, whence the name of the minimization method.[44] We can now use this result to determine where to place x_2 and x_3, by solving Eq. (5.93) and Eq. (5.94) together with $S + B = x_1 - x_0$:

$$x_2 = x_0 + S = x_0 + \frac{x_1 - x_0}{\varphi + 1}, \qquad x_3 = x_0 + S + I = x_0 + \frac{(x_1 - x_0)\varphi}{\varphi + 1} \tag{5.97}$$

Observe that the values of x_2 and x_3 do not depend on which scalar function ϕ we're minimizing. All the elements are now in place for us to implement the golden-section search method.

5.5.2.2 Implementation

Before turning to the code, it may be helpful to summarize the above observations:

- Take the interval (x_0, x_1)—which we know brackets a minimum—and produce two intermediate points x_2 and x_3 as per Eq. (5.97).
- If $\phi(x_3) < \phi(x_2)$ (the left-panel scenario) then the minimum is in (x_2, x_1). We rename $x_0 \leftarrow x_2$ in preparation for repeating the entire process (which started with an x_0 and an x_1 that bracket the minimum).
- If $\phi(x_2) < \phi(x_3)$ (the right-panel scenario) then the minimum is in (x_0, x_3). We rename $x_1 \leftarrow x_3$ in preparation for repeating the entire process.
- This algorithm works by starting with an interval (x_0, x_1) and producing a narrower interval (re-using the same endpoint variables). To determine whether we've converged, we compare the width of $x_1 - x_0$ with a preset tolerance. In order to produce a single point as our estimate for the location of the minimum, we semi-arbitrarily produce the midpoint $(x_0 + x_1)/2$—even though this algorithm has nothing to do with bisection.

[43] The right-panel scenario would have led to $B/S = S/I$ and the rest of the derivation would have been identical.

[44] The golden ratio, widely held to lead to aesthetically pleasing art/architecture, makes another appearance in this chapter as the order of convergence of the secant method (see problem 5.12). Be sure to distinguish between the ϕ used for the scalar function and the φ used for the golden ratio.

golden.py

Code 5.6

```
from bisection import f as phi
from math import sqrt

def golden(phi, x0, x1, kmax=200, tol=1.e-8):
    varphi = 0.5*(1 + sqrt(5))
    for k in range(1,kmax):
        x2 = x0 + (x1-x0)/(varphi+1)
        x3 = x0 + (x1-x0)*varphi/(varphi+1)

        if phi(x3) < phi(x2):
            x0 = x2
        else:
            x1 = x3

        xnew = (x0+x1)/2
        xdiff = abs(x1-x0)
        rowf = "{0:2d} {1:1.16f} {2:1.16f} {3:1.16f}"
        print(rowf.format(k, xnew, xdiff, phi(xnew)))

        if abs(xdiff) < tol:
            break
    else:
        xnew = None
    return xnew

if __name__ == '__main__':
    val = golden(phi,0.,3.5); print(val)
```

Code 5.6, i.e., golden.py, shows a simple implementation of all the steps given in these bullet points. We introduce the golden ratio φ as a parameter that will be used to produce the values of x_2 and x_3 each time through the loop. Our printout is showing an iteration counter, the latest midpoint, the width of the latest interval, as well as the function value at the latest midpoint. Crucially, we are printing out the function value itself, i.e., *not* the absolute value; the latter was useful in the root-finding codes we studied earlier, but we are now interested in minimizing, so we wish to focus on the function value itself. This implementation is quite wasteful: we require two function evaluations each time through the loop. This is consistent with what the first bullet point recommended, but not actually necessary. One can, instead, store more information from the previous iteration, leading to only one new function evaluation per iteration (as you are asked to do in problem 5.25).

5.6 Multidimensional Minimization

Multidimensional minimization is, in general, much harder to handle; as the dimensionality grows one cannot even visualize what's going on very effectively. We start with some mathematical aspects of the problem and a two-dimensional example.

5.6.1 General Features

Consider a scalar function of many variables, i.e., $\phi(\mathbf{x})$, where $\mathbf{x}$ bundles together the variables $x_0, x_1, \ldots, x_{n-1}$ but ϕ produces scalar values. This is somewhat analogous to the single function components f_i that we encountered in section 5.4. As we did in Eq. (5.73), we will now employ a multidimensional Taylor expansion, this time going to one order higher. Also, in order to keep things general, we will not expand around our latest iterate, $\mathbf{x}^{(k-1)}$, since we are not introducing a specific method right now; we are simply trying to explore features of the problem of minimizing $\phi(\mathbf{x})$.[45]

We assume $\phi(\mathbf{x})$ has bounded first, second, and third derivatives. Then, employing notation inspired by Eq. (5.83), we Taylor expand in each of the components of the vector $\mathbf{x}$; these can be bundled together using vector notation, see Eq. (C.5):

$$\phi(\mathbf{x}+\mathbf{q}) = \phi(\mathbf{x}) + (\nabla\phi(\mathbf{x}))^T\,\mathbf{q} + \frac{1}{2}\mathbf{q}^T\mathbf{H}(\mathbf{x})\mathbf{q} + O\left(\|\mathbf{q}\|^3\right) \tag{5.98}$$

Here the first-order term involves $\nabla\phi(\mathbf{x})$, the *gradient* vector of ϕ at $\mathbf{x}$. This is:

$$\nabla\phi(\mathbf{x}) = \begin{pmatrix} \dfrac{\partial\phi}{\partial x_0} & \dfrac{\partial\phi}{\partial x_1} & \cdots & \dfrac{\partial\phi}{\partial x_{n-1}} \end{pmatrix}^T \tag{5.99}$$

Similarly to Eq. (5.74), we can re-express the first-order term:

$$(\nabla\phi(\mathbf{x}))^T\,\mathbf{q} = \sum_{j=0}^{n-1}\frac{\partial\phi}{\partial x_j}q_j \tag{5.100}$$

We will come back to this point below but, for now, note that $\nabla\phi(\mathbf{x})$ is the direction of *steepest ascent*. To see this, observe that, for small $\mathbf{q}$, the term linear in $\mathbf{q}$ is the dominant contribution, since $\|\mathbf{q}\|^2 \ll \|\mathbf{q}\|$. From elementary vector calculus we know that the dot product $(\nabla\phi(\mathbf{x}))^T\,\mathbf{q}$ will be maximized when $\mathbf{q}$ points in the direction of $\nabla\phi(\mathbf{x})$.

Assuming $\mathbf{x}^*$ is a local minimum of ϕ and ignoring higher-order terms in Eq. (5.98):

$$\phi(\mathbf{x}^*+\mathbf{q}) \approx \phi(\mathbf{x}^*) + (\nabla\phi(\mathbf{x}^*))^T\,\mathbf{q} \tag{5.101}$$

[45] Whatever we discover on the question of minimizing $\phi(\mathbf{x})$ will also apply to the problem of maximizing $-\phi(\mathbf{x})$. (Jeremy Bentham introduced the words "minimize" and "maximize" at the same time.)

Reversing our argument from the previous paragraph, the first-order term will lead to the largest possible decrease when $\mathbf{q}$ points in the direction of $-\nabla\phi(\mathbf{x}^*)$. If $\nabla\phi(\mathbf{x}^*) \neq \mathbf{0}$, we will have found a $\phi(\mathbf{x}^* + \mathbf{q})$ that is smaller than $\phi(\mathbf{x}^*)$; by definition, this is impossible, since we said that $\mathbf{x}^*$ is a local minimum of ϕ. This leads us to the following multidimensional generalization of our criterion for being a critical point in Eq. (5.90):

$$\nabla\phi(\mathbf{x}^*) = \mathbf{0} \tag{5.102}$$

where both the left-hand side and the right-hand side are vectors.

Having established that the gradient vector vanishes at a critical point, we now turn to the second-order term in Eq. (5.98), which involves the *Hessian matrix*, $\mathbf{H}(\mathbf{x})$. To see what this is, we expand the quadratic form as follows:

$$\frac{1}{2}\mathbf{q}^T\mathbf{H}(\mathbf{x})\mathbf{q} = \frac{1}{2}\sum_{i,j=0}^{n-1}\frac{\partial\phi}{\partial x_i\partial x_j}q_iq_j \tag{5.103}$$

In matrix form:

$$\mathbf{H}(\mathbf{x}) = \left\{\frac{\partial\phi}{\partial x_i\partial x_j}\right\} = \begin{pmatrix} \frac{\partial^2\phi}{\partial x_0^2} & \frac{\partial^2\phi}{\partial x_0\partial x_1} & \cdots & \frac{\partial^2\phi}{\partial x_0\partial x_{n-1}} \\ \frac{\partial^2\phi}{\partial x_1\partial x_0} & \frac{\partial^2\phi}{\partial x_1^2} & \cdots & \frac{\partial^2\phi}{\partial x_1\partial x_{n-1}} \\ \vdots & \vdots & \ddots & \vdots \\ \frac{\partial^2\phi}{\partial x_{n-1}\partial x_0} & \frac{\partial^2\phi}{\partial x_{n-1}\partial x_1} & \cdots & \frac{\partial^2\phi}{\partial x_{n-1}^2} \end{pmatrix} \tag{5.104}$$

which is sometimes denoted by $\mathbf{H}_\phi(\mathbf{x})$. Since for us the second partial derivatives will always be continuous, it is easy to see that our Hessian matrix will be *symmetric*. Let us now apply Eq. (5.98), for the case of $\mathbf{x}^*$, a local minimum of ϕ. We have:

$$\phi(\mathbf{x}^* + \mathbf{q}) = \phi(\mathbf{x}^*) + \frac{1}{2}\mathbf{q}^T\mathbf{H}(\mathbf{x}^*)\mathbf{q} + O\left(\|\mathbf{q}\|^3\right) \tag{5.105}$$

where we have not included the first-order term, since the gradient vector vanishes, as per Eq. (5.102). If we now further assume that $\mathbf{H}(\mathbf{x}^*)$ is *positive definite*,[46] then we can see that, indeed, $\phi(\mathbf{x}^* + \mathbf{q}) > \phi(\mathbf{x}^*)$, as it should, since $\mathbf{x}^*$ is a minimum.

To summarize: (a) a necessary condition for $\mathbf{x}^*$ being a local minimum is that it be a critical point, i.e., that its gradient vector vanish, and (b) a sufficient condition for the critical point $\mathbf{x}^*$ being a local minimum is that its Hessian matrix be positive definite.

5.6.2 A Two-Dimensional Example

As usual, it's easier to build your intuition via a two-variable problem, namely a (single) scalar function ϕ that takes in two variables, x_0 and x_1, and produces a single number, the function value. We decide to look at the f_0 from our earlier example, Eq. (5.72), this time not as a function component but as a single function; also, we are no longer looking at the zeros of this function, but its entire behavior, with a view to minimizing it. It is:

[46] Recall from page 163 that a matrix $\mathbf{A}$ is positive definite if $\mathbf{x}^T\mathbf{A}\mathbf{x} > 0$ for all $\mathbf{x} \neq \mathbf{0}$.

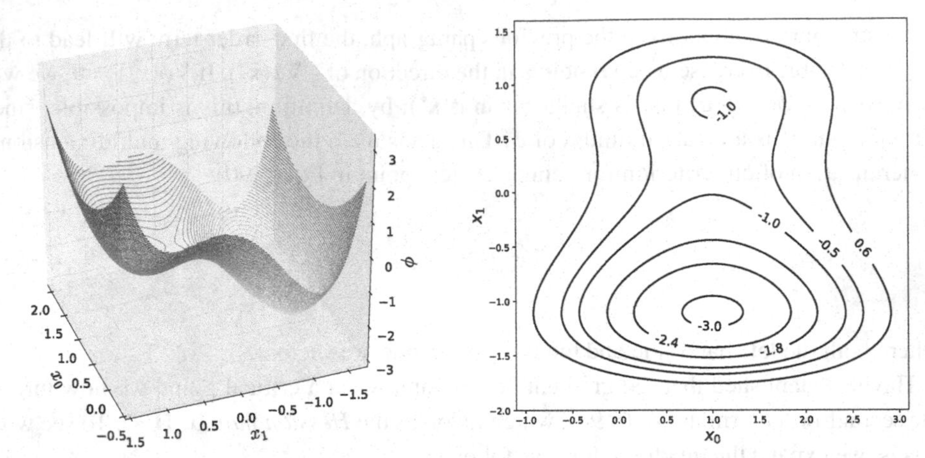

Fig. 5.16 Example of a scalar function of two variables in three (left) and two (right) dimensions

$$\phi(x_0, x_1) = x_0^2 - 2x_0 + x_1^4 - 2x_1^2 + x_1 \qquad (5.106)$$

To help you get more comfortable with higher dimensions, Fig. 5.16 (left) is attempting both to visualize the third dimension and to draw equipotential curves (also known as contour lines or level sets). The latter are directly visualized in the 2d plot in the right panel (e.g., if you take $\phi = 0$ then you will reproduce our earlier curve from Fig. 5.13).

We are dealing with two local minima. The one on the "right" in the 3d plot (and at the bottom of the 2d plot) leads to smaller/more negative function values, so it appears to be the global minimum. As a reminder of some of our earlier points: if you place a marble somewhere near these two wells, it will roll down to one of the minima; which of the two minima you end up in depends on where you start. Note, finally, that our discussion above on saddle points in one dimension needs to be generalized: you may have a saddle point when you are at a maximum along, say, the x_0 direction and at a minimum along the x_1 direction. Incidentally, this also explains why the word "saddle" is used in this context.

5.6.3 Gradient Descent

We turn to an intuitively clear approach to multidimensional minimization. This method, known as *gradient descent*, does not exhibit great convergence properties and can get in trouble for non-differentiable functions. Even so, it is a pedagogical, simple approach.

5.6.3.1 Algorithm and Interpretation

Recall from our discussion of general features that $\nabla\phi(\mathbf{x})$ is the direction of steepest ascent. This leads to the conclusion that $-\nabla\phi(\mathbf{x})$ is the direction of *steepest descent*: as in our discussion around Eq. (5.101), we know that choosing $\mathbf{q}$ to point along the negative gradient guarantees that the function value decrease will be the fastest. The method we are about to introduce, which employs $-\nabla\phi(\mathbf{x})$, is known as *gradient descent*.[47] Qualitatively, this approach makes use of *local* information: if you're exploring a mountainous region (with your eyes closed), you can take a small step downhill *at that point*; this doesn't mean that you're always actually moving in the direction that will most quickly bring you to a (possibly distant) local minimum, simply that you are moving in a downward direction.

Implicit in our discussion above is the fact that the steps we make will be *small*: while $-\nabla\phi(\mathbf{x})$ helps you pick the direction, it doesn't tell you how far in that direction you should go, i.e., how large a $||\mathbf{q}||$ you should employ. The simplest possible choice, analogous to our closed-eyes example, is to make small fixed steps, quantified by a parameter γ (called the *learning rate* in machine-learning circles). Using notation similar to that in Eq. (5.80), this leads to the following prescription:

$$\mathbf{x}^{(k)} = \mathbf{x}^{(k-1)} - \gamma\nabla\phi(\mathbf{x}^{(k-1)}) \tag{5.107}$$

The right-hand side involves an evaluation of the gradient, not of the function itself. At each step, this method picks the direction that is perpendicular to the contour line. Note also that there is no (costly) Jacobian/Hessian computation being carried out here; similarly, there is no matrix inversion or linear system solution: all that's being done is that the gradient is computed and a (small) step is taken along that direction.[48] Of course, the question arises what "small" means, i.e., of how to pick γ: we will take an empirical, i.e., trial-and-error, approach. In problem 5.29 you will explore a more systematic solution, where this magnitude parameter is allowed to change at each iteration, i.e., you will be dealing with $\gamma^{(k)}$ and will employ an extra criterion to help you pick this parameter at each step.

In practice, we may not know the gradient analytically. In complete analogy to Eq. (5.81), we typically approximate it using a forward-difference scheme. In equation form, this means that Eq. (5.99) becomes:

$$\nabla\phi(\mathbf{x}) = \begin{pmatrix} [\phi(\mathbf{x}+\mathbf{e}_0 h) - \phi(\mathbf{x})]/h \\ [\phi(\mathbf{x}+\mathbf{e}_1 h) - \phi(\mathbf{x})]/h \\ \dots \\ [\phi(\mathbf{x}+\mathbf{e}_{n-1} h) - \phi(\mathbf{x})]/h \end{pmatrix} \tag{5.108}$$

for a given spacing h. This involves "only" $n+1$ function evaluations, so it is not too costly (it doesn't involve any really demanding linear-algebra steps).

5.6.3.2 Implementation

Our prescription in Eq. (5.107) is quite simple, so you will not be surprised to hear that it is straightforward to implement. Code 5.7 is along the lines you'd expect. We first define our scalar function $\phi(\mathbf{x})$; again, notice that this is the only part of the code where the problem is

[47] Since it employs the direction of steepest descent, this method is also known by that name, though it should not be confused with the "method of steepest descent", which arises in the study of contour integration.

[48] The Hessian's condition number impacts the asymptotic error constant, but we don't *need* to compute that.

Code 5.7 descent.py

```
from jacobi import termcrit
import numpy as np

def phi(xs):
    x0, x1 = xs
    return x0**2 - 2*x0 + x1**4 - 2*x1**2 + x1

def gradient(phi,xs,h=1.e-6):
    n = xs.size
    phi0 = phi(xs)
    Xph = (xs*np.ones((n,n))).T + np.identity(n)*h
    grad = (phi(Xph) - phi0)/h
    return grad

def descent(phi,gradient,xolds,gamma=0.15,kmax=200,tol=1.e-8):
    for k in range(1,kmax):
        xnews = xolds - gamma*gradient(phi,xolds)

        err = termcrit(xolds,xnews)
        print(k, xnews, err, phi(xnews))
        if err < tol:
            break

        xolds = np.copy(xnews)
    else:
        xnews = None
    return xnews

if __name__ == '__main__':
    xolds = np.array([2.,0.25])
    xnews = descent(phi, gradient, xolds)
    print(xnews)
```

shown to be two-dimensional, implementing Eq. (5.106). In other words, changing `phi()` would be all you would need to do to solve a 10-dimensional problem.

We then introduce another function, `gradient()`, which computes the gradient using a forward-difference approximation, as per Eq. (5.108). It's important to keep in mind that

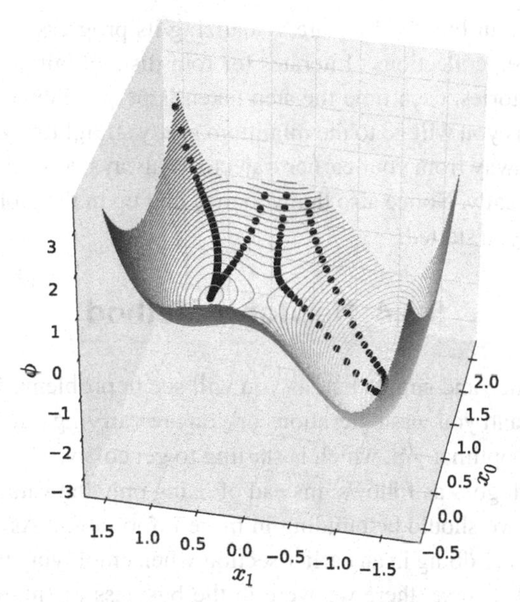

Fig. 5.17 Examples of gradient descent applied to a scalar function of two variables

ϕ is a scalar, i.e., `phi()` returns a single number; as a result, $\nabla\phi$ is a column vector, so `gradient()` returns a `numpy` array. We could have employed here an explicit loop, as you are asked to do in problem 5.28. Let's look at `Xph` a bit more closely: in essence, what we are doing is producing a matrix made up of n copies of the position $\mathbf{x}$ (or $\mathbf{x}^{(k-1)}$) and then adding in h to each of the position-components separately. The transposition is taking place because we intend to call `phi()` and wish to have each shifted position vector, $\mathbf{x}+\mathbf{e}_j h$, in its own *column*; then, each of `x0` and `x1` will end up being a `numpy` array instead of an individual number; this allows us to carry out all the needed calculations at once.

The function `descent()` takes in the ϕ to be minimized, as well as another function parameter, which will allow you to call a different gradient-computing prescription in the future, if you so desire. We then pass a γ with a reasonably small default value. Note that this is something you will need to tune by hand for each problem and possibly also for each initial-guess vector. The function body itself is quite similar to our earlier iterative codes. As a matter of fact, it's much less costly, since the main updating of Eq. (5.107) is so straightforward. At each iteration we are printing out the value of our latest iterate, $\mathbf{x}^{(k)}$, a measure of the change in the iterates, as well as the function value at our latest iterate, $\phi(\mathbf{x}^{(k)})$. Once again, note that we are *not* looking for function zeros, but for local minima. You can compare the final function value you find using different initial guesses, to see if the minimum you arrived at is lower than what you produced in previous runs.

The main part of the program simply makes an initial guess and runs our gradient-descent function. We then print out the final solution vector. Running this code, we succeed in finding the global minimum, but we need considerably more than a handful of iterations to get there. Of course, the number of iterations needed depends on the initial-guess vector and on the value of γ. In order to help you understand how the gradient-descent method reaches the minimum, in Fig. 5.17 we are visualizing its progress.[49] Shown are four distinct "trajectories", i.e., collections of iterates for four distinct initial guesses. As you can see from these trajectories, each time the step taken is perpendicular to the contour line. Sometimes that means you will go to the minimum pretty straightforwardly, whereas other times you will bend away from your earlier trajectory, always depending on what the contour line looks like locally. Notice also that you may end up in the global minimum or not, depending on where you started.

5.6.4 Newton's Method

Gradient descent is nice and simple but, as you will see in problems 5.28 and 5.29, either your γ is very small and you waste iterations or you are carrying out a line-search at each step to determine the optimal $\gamma^{(k)}$, which is starting to get costly.

A distinct approach goes as follows: instead of using only the value of the gradient at a given point, perhaps we should be building in more information. As you may recall, that is roughly what we were doing in an earlier section when employing the multidimensional Newton's method. Of course, there we were in the business of (multiple-equation) root-finding, whereas now we are trying to find (single) function minima. However, it turns out that these two problems can be mapped onto each other. Specifically, Eq. (5.102) told us that the gradient vanishes at a critical point. Combine that with the fact that the gradient of a scalar function is a column vector, and you can recast that equation as a set of n coupled non-linear equations:

$$\mathbf{f}(\mathbf{x}) = \nabla\phi(\mathbf{x}) = \mathbf{0} \tag{5.109}$$

In other words, finding a critical point is simply a special case of solving a non-linear system of equations. Thus, we can use approaches such as Newton's or Broyden's methods that we developed above. Importantly, by comparing the Jacobian matrix of Eq. (5.75), made up of first derivatives of function components, with the Hessian matrix of Eq. (5.104), made up of second derivatives of the scalar function, we realize that:

$$\mathbf{H}(\mathbf{x}) = \mathbf{J}_{\nabla\phi}(\mathbf{x}) \tag{5.110}$$

due to the fact that the Hessian is symmetric. Note that we have now seen the notation where the Jacobian gets a subscript pay off. In words, the Jacobian matrix of the gradient is the Hessian matrix of our scalar function. This enables us to apply Newton's method to the problem of scalar-function minimization and thereby produce the next iterate, $\mathbf{x}^{(k)}$. Thus, Eq. (5.78) now becomes:

$$\mathbf{J}_{\nabla\phi}(\mathbf{x}^{(k-1)})\left(\mathbf{x}^{(k)} - \mathbf{x}^{(k-1)}\right) = -\nabla\phi(\mathbf{x}^{(k-1)}) \tag{5.111}$$

Observe that this equation contains the gradient at the previous iterate, just like gradient

[49] We have changed the viewing angle, as `matplotlib` conveniently allows, but it's the same problem.

descent in Eq. (5.107), but now also builds in further information, in the form of the Hessian/Jacobian on the left-hand side. Of course, this is accomplished at the cost of having to solve a linear system of equations in order to produce the next iterate, $\mathbf{x}^{(k)}$.

In problem 5.31 you are asked to apply this method to our example scalar function of Eq. (5.106). The easiest way to do this is to analytically produce and code up the gradient. At that point, all the infrastructure we built in `multi_newton.py` above applies directly to our problem. Alternatively, you could envision evaluating both the gradient and the Hessian using finite-difference approximations. The qualitative point to realize is that we will have to carry out a larger number of function evaluations at each iteration, in addition to also having to solve a linear system each step of the way. On the other hand, Newton's method converges quadratically if you are close enough to the root/minimum, so the higher cost at each iteration is usually compensated by the smaller number of required iterations.

You might protest at this point: Newton's method as just described finds critical points, but how do you know that you reached a minimum? As it turns out, we (implicitly) addressed this question in our earlier discussion in section 5.6.1: once you've reached a critical point, you can test your Hessian matrix to see if it is *positive definite*;[50] as you may recall from page 163, since the Hessian is symmetric, this simply means that if you find all the eigenvalues of the Hessian to be *positive*, then you'll know you've found a minimum. Similarly, if they're all negative you are at a maximum. In problem 5.41 you will also explore the case of a multidimensional saddle point. Of course, there's another question that we haven't addressed: can you guarantee that you have found the global minimum? Generally, the answer is no. Many refinements of Newton's method exist that attempt to make it more safe, more efficient, as well as more globally oriented, but what we've covered should be enough for now.

5.6.5 Derivative-Free Optimization: Powell's Method

Recall that the gradient-descent method, see Eq. (5.107), starts at a position $\mathbf{x}^{(k-1)}$ and then takes a step along the direction of (the opposite of) the gradient (possibly also minimizing along that line, thereby determining the magnitude of the step, γ). Similarly, Newton's method for minimization, see Eq. (5.111), starts from $\mathbf{x}^{(k-1)}$ and then produces the next iterate by building in information on both the gradient and the Jacobian/Hessian.

5.6.5.1 Agenda and Concepts

We will now look at a somewhat different way of minimizing in many dimensions: we will manage to accomplish our goal *without* employing any information on the scalar-function's derivatives, i.e., we will use only the values of the function itself, $\phi(\mathbf{x})$. Qualitatively, our approach will be analogous to gradient descent with line-search minimization (problem 5.29), in that we will carry out a sequence of line minimizations (but, of course, this time around we won't be using the gradient to determine the direction each time). Basically, what we'll do is to start at a point $\mathbf{x}^{(k-1)}$, pick a direction $\mathbf{u}_{k-1}$, and minimize the

[50] Note, however, that the Hessian might *not* be positive definite away from the minimum.

single-variable function of γ, $\zeta(\gamma) = \phi(\mathbf{x}^{(k-1)} + \gamma\mathbf{u}_{k-1})$, along that direction. Thus, we take:

$$\mathbf{x}^{(k)} = \mathbf{x}^{(k-1)} + \gamma^{(k-1)}\mathbf{u}_{k-1}, \qquad k = 1, 2, \ldots \tag{5.112}$$

where the value of the coefficient $\gamma^{(k-1)}$ comes from the one-dimensional minimization. We will traverse the n-dimensional space through a sequence of such one-dimensional minimizations; the latter can be carried out via any one-dimensional technique. In the spirit of trying to avoid derivative evaluations, in this section we will use the golden-section search as a building block for our multidimensional-minimization approach.

Of course, a lot hinges upon how we pick the direction $\mathbf{u}_{k-1}$ each time. (Remember, we are not free to use derivative information to determine which way is down.) Also keep in mind that we are working in n-dimensional space: it would be nice if we could somehow guarantee that whatever progress (minimizing) we make along one direction, say $\mathbf{u}_0$, is not immediately undone when we turn to another direction, say $\mathbf{u}_1$. In other words, we are looking for a set of directions, each of which is somehow "aware" of the other ones. The simplest thing we could try is to use *mutually orthogonal* directions:

$$\mathbf{u}_j^T\mathbf{u}_k = 0, \qquad \text{for } j \neq k \tag{5.113}$$

in the spirit of what we saw for, say, the Gram–Schmidt orthogonalization prescription, Eq. (4.155). Instead of starting with n non-orthogonal vectors and orthogonalizing those, we could do something even simpler and start with the columns of the identity matrix, $\mathbf{e}_k$, see Eq. (4.100), which are already mutually orthogonal. The idea would then be to pick a γ that minimizes $\phi(\mathbf{x}^{(0)} + \gamma\mathbf{e}_0)$; this would give us $\gamma^{(0)}$ and therefore also $\mathbf{x}^{(1)}$. We would then pick the next orthogonal direction, $\mathbf{e}_1$, and minimize $\phi(\mathbf{x}^{(1)} + \gamma\mathbf{e}_1)$, and so on. At some point, we would encounter the last orthogonal vector, $\mathbf{e}_{n-1}$. Of course, in all likelihood these n one-dimensional minimizations wouldn't bring us to (anywhere near) the n-dimensional minimum, so we would have to "cycle" through our n directions again: $\mathbf{e}_0$, $\mathbf{e}_1$, and so on.

As you may have guessed, this approach (which you are asked to implement in problem 5.33) is not very efficient. To see why, picture a function that has a long narrow valley along a direction which doesn't quite match up with one of the $\mathbf{e}_k$'s. This means that for each cycle (i.e., each set of n one-dimensional minimizations) only one of the directions will help us make progress, proceeding down the valley, while the one-dimensional minimizations along the other directions are basically being wasted. We would then need a large number of cycles (each of which is made up of n one-dimensional minimizations) to get to the minimum. This is hardly surprising, since the $\mathbf{e}_k$'s are totally unrelated to the ϕ we're supposed to be minimizing. A better strategy is not to reach for mutually orthogonal directions, but *mutually conjugate* directions:

$$\mathbf{u}_j^T\mathbf{A}\mathbf{u}_k = 0, \qquad \text{for } j \neq k \tag{5.114}$$

where $\mathbf{A}$ is a symmetric $n \times n$ positive-definite matrix. Properly speaking, $\mathbf{u}_j$ and $\mathbf{u}_k$ are mutually conjugate with respect to $\mathbf{A}$, but more colloquially (especially when only a single symmetric positive-definite matrix is in play) one simply calls them mutually conjugate.

A couple of examples: first, recall that for the case of a symmetric matrix the eigenvalue problem of Eq. (4.124), $\mathbf{A}\mathbf{v}_i = \lambda_i \mathbf{v}_i$, leads to eigenvectors which are mutually orthogonal (whereas, as we noted on page 189, for non-symmetric matrices the eigenvectors are "only" linearly independent). This means that $\mathbf{v}_j^T \mathbf{v}_k = 0$ for $j \neq k$. However, we can do better:

$$\mathbf{v}_j^T \mathbf{A} \mathbf{v}_k = \mathbf{v}_j^T \lambda_k \mathbf{v}_k = \lambda_k \mathbf{v}_j^T \mathbf{v}_k = 0, \qquad \text{for } j \neq k \tag{5.115}$$

In the first equality we employed Eq. (4.124) and in the third equality the orthogonality of the eigenvectors. Thus, we have shown that the eigenvectors of a symmetric positive-definite matrix are mutually conjugate. We now turn to a second example: Eq. (5.114) refers to a general matrix $\mathbf{A}$; a special case is that of the identity matrix, $\mathcal{I}$, in which case $\mathbf{u}_j^T \mathbf{A} \mathbf{u}_k = 0$ becomes $\mathbf{u}_j^T \mathbf{u}_k = 0$, i.e., two vectors that are mutually conjugate with respect to the identity matrix are mutually orthogonal. Note that in what follows the mutually conjugate directions at work will be neither the eigenvectors (computing which would require explicit knowledge of $\mathbf{A}$) nor the columns of the identity matrix (which know nothing about a given function ϕ); these two examples were mentioned here simply so you can start to build some intuition around the idea of mutually conjugate directions.

5.6.5.2 Conjugate-Direction Minimization of a Quadratic

Let us turn to the interpretation of the matrix $\mathbf{A}$. Motivated by the multidimensional Taylor expansion of Eq. (5.98), we focus on a simple quadratic function $\phi(\mathbf{x})$ and derive exact results; these will later turn into an approximation/prescription for the general case where $\phi(\mathbf{x})$ is not quadratic. For now, the function to be minimized takes the form:

$$\phi(\mathbf{x}) = c + \mathbf{b}^T \mathbf{x} + \frac{1}{2} \mathbf{x}^T \mathbf{A} \mathbf{x} \tag{5.116}$$

where, as above, $\mathbf{A}$ is a symmetric $n \times n$ positive-definite matrix: our $\mathbf{A}$ is essentially the Hessian in disguise. Note that, even though our formalism below will make heavy use of $\mathbf{A}$, it is important that our final prescription be expressed in terms of ϕ (i.e., not $\mathbf{A}$) evaluations: the whole point is to minimize a function without access to its derivatives. We now go over how to minimize this way, *taking for granted* the existence of the mutually conjugate directions, i.e., the $\mathbf{u}_k$'s. Once we're done, we'll see how to produce this crucial ingredient (i.e., the $\mathbf{u}_k$'s themselves), at the same time also seeing how to generalize to a non-quadratic function ϕ.

Assume our search starts at the (arbitrary) point $\mathbf{x}^{(0)}$. We will expand the difference between the (unknown) true minimum $\mathbf{x}^*$ and our starting point in terms of the mutually conjugate $\mathbf{u}_k$'s (which, as you will show in problem 5.34, form a basis):

$$\mathbf{x}^* = \mathbf{x}^{(0)} + \sum_{k=0}^{n-1} \alpha_k \mathbf{u}_k \tag{5.117}$$

We will now determine the coefficient values, i.e., the α_k's. We do this by employing our criterion on what it means to be a critical point:

$$\nabla\phi(\mathbf{x}^*) = \mathbf{b} + \mathbf{A}\mathbf{x}^* = \mathbf{b} + \mathbf{A}\mathbf{x}^{(0)} + \sum_{k=0}^{n-1} \alpha_k \mathbf{A} \mathbf{u}_k = \mathbf{0} \tag{5.118}$$

In the first equality we took the derivative of the quadratic in Eq. (5.116).[51] In the second equality we plugged in our expansion from Eq. (5.117) and made a slight re-arrangement. The final equality is a result of our criterion in Eq. (5.102). We now take our last result and multiply it from the left by $\mathbf{u}_j^T$:

$$\mathbf{u}_j^T(\mathbf{b} + \mathbf{A}\mathbf{x}^{(0)}) + \sum_{k=0}^{n-1} \alpha_k \mathbf{u}_j^T \mathbf{A}\mathbf{u}_k = \mathbf{0} \tag{5.119}$$

Since the $\mathbf{u}_j$'s are mutually conjugate, see Eq. (5.114), only a single term in the sum survives, so we are now in a position to solve for the coefficients, finding:

$$\alpha_j = -\frac{\mathbf{u}_j^T(\mathbf{b} + \mathbf{A}\mathbf{x}^{(0)})}{\mathbf{u}_j^T \mathbf{A}\mathbf{u}_j} \tag{5.120}$$

So far, nothing too exciting is going on: we have computed the coefficients in the expansion relating our starting point to the true minimum, appearing in Eq. (5.117). We mentioned a starting point $\mathbf{x}^{(0)}$, but no actual iteration has taken place.

Here's where things get interesting: we recall that our overall agenda, see Eq. (5.112), is to set up a general iterative scheme as follows:

$$\begin{aligned} &\min \zeta(\gamma) = \phi(\mathbf{x}^{(k-1)} + \gamma\mathbf{u}_{k-1}) \quad \text{to find } \gamma^{(k-1)} \\ &\mathbf{x}^{(k)} = \mathbf{x}^{(k-1)} + \gamma^{(k-1)}\mathbf{u}_{k-1} \end{aligned} \tag{5.121}$$

This starts at the point $\mathbf{x}^{(k-1)}$ and determines the next point $\mathbf{x}^{(k)}$ by taking a step along the direction of $\mathbf{u}_{k-1}$; the step size $\gamma^{(k-1)}$ is determined by minimizing $\phi(\mathbf{x}^{(k-1)} + \gamma\mathbf{u}_{k-1})$, viewed as a function of a single variable (i.e., of γ). Let's formally carry out this minimization:

$$\begin{aligned} \frac{d}{d\gamma}\phi(\mathbf{x}^{(k-1)} + \gamma\mathbf{u}_{k-1})\bigg|_{\gamma^{(k-1)}} &= \mathbf{u}_{k-1}^T \nabla\phi(\mathbf{x}^{(k)}) = \mathbf{u}_{k-1}^T[\mathbf{b} + \mathbf{A}\mathbf{x}^{(k)}] \\ &= \mathbf{u}_{k-1}^T\left[\mathbf{b} + \mathbf{A}\left(\mathbf{x}^{(k-1)} + \gamma^{(k-1)}\mathbf{u}_{k-1}\right)\right] = 0 \end{aligned} \tag{5.122}$$

In the first equality we used the chain rule and employed the second line of Eq. (5.121); in the second equality we took the derivative of a quadratic, just like in the first step of Eq. (5.118); in the third equality we used the second line of Eq. (5.121) again. In the fourth equality we remembered that we are dealing with a (one-dimensional) critical point. Expanding the parentheses, we can solve for the $\gamma^{(k-1)}$ to find:

$$\gamma^{(k-1)} = -\frac{\mathbf{u}_{k-1}^T(\mathbf{b} + \mathbf{A}\mathbf{x}^{(k-1)})}{\mathbf{u}_{k-1}^T \mathbf{A}\mathbf{u}_{k-1}} \tag{5.123}$$

We now repeatedly step through Eq. (5.121) in order to write it out in terms of the starting point:

$$\mathbf{x}^{(k-1)} = \mathbf{x}^{(0)} + \sum_{j=0}^{k-2} \gamma^{(j)}\mathbf{u}_j \tag{5.124}$$

[51] If this is not immediately obvious, you should solve problem 6.38 on the relevant quadratic form.

Motivated by our term in the numerator of Eq. (5.123), we multiply Eq. (5.124) on the left by $\mathbf{u}_{k-1}^T\mathbf{A}$:

$$\mathbf{u}_{k-1}^T\mathbf{A}\mathbf{x}^{(k-1)} = \mathbf{u}_{k-1}^T\mathbf{A}\mathbf{x}^{(0)} + \sum_{j=0}^{k-2}\gamma^{(j)}\mathbf{u}_{k-1}^T\mathbf{A}\mathbf{u}_j = \mathbf{u}_{k-1}^T\mathbf{A}\mathbf{x}^{(0)} \tag{5.125}$$

where none of the terms in the sum survived, since the vectors are mutually conjugate. We can now take this result and plug it into Eq. (5.123) to find:

$$\gamma^{(k-1)} = -\frac{\mathbf{u}_{k-1}^T(\mathbf{b} + \mathbf{A}\mathbf{x}^{(0)})}{\mathbf{u}_{k-1}^T\mathbf{A}\mathbf{u}_{k-1}} = \alpha_{k-1} \tag{5.126}$$

Here we took the opportunity to emphasize that the γ's (namely the iterative coefficients determined by minimizing along a given direction each time) turned out to be identical to the α's of Eq. (5.120) (namely the expansion coefficients for the distance to the true minimum). This has the further consequence that, if we repeatedly step through Eq. (5.124), then we find that:

$$\mathbf{x}^{(n)} = \mathbf{x}^{(0)} + \sum_{j=0}^{n-1}\gamma^{(j)}\mathbf{u}_j = \mathbf{x}^{(0)} + \sum_{j=0}^{n-1}\alpha_j\mathbf{u}_j = \mathbf{x}^* \tag{5.127}$$

where in the last step we identified the true minimum from Eq. (5.117). Take a moment to marvel at what we've accomplished: we started at $\mathbf{x}^{(0)}$ and took n steps to get to the true minimum *exactly*! Crucially, we managed to do this via a step-by-step prescription, Eq. (5.121), which needs no $\mathbf{A}$ or $\mathbf{b}$ evaluations (*and no derivatives*); all we need to do is to carry out a sequence of one-dimensional minimizations of the ϕ function value, which could be done via the golden-section search method, thereby obviating any derivatives whatsoever. Each step in the multi-dimensional space is taken along a mutually conjugate direction, with the step size being determined by a one-dimensional algorithm (a line-search minimization).

5.6.5.3 Powell's Algorithm

Before turning to a full algorithm that can handle non-quadratic functions, let us explore some properties of conjugate directions (which also help explain why such approaches arose in the first place). A special case of the second step in Eq. (5.122) is that $\mathbf{u}_0^T\nabla\phi(\mathbf{x}^{(1)}) = 0$. In words, this is saying that the directional derivative at the (new) point $\mathbf{x}^{(1)}$ with respect to the conjugate direction that we used to get there (i.e., $\mathbf{u}_0$) is zero; this is essentially restating the fact that the new point (which was a minimum) is a critical point. Intriguingly, as you will show in problem 5.35, this continues to hold as you keep iterating, i.e., the latest gradient is perpendicular to all the previous conjugate directions (e.g., $\mathbf{u}_0^T\nabla\phi(\mathbf{x}^{(3)}) = 0$). Qualitatively, this means that our latest minimization doesn't spoil the earlier ones: that is one way of appreciating how useful mutually conjugate directions are.

We will now examine another important variation of essentially the same idea. Consider starting from the point $\mathbf{x}_I$ and doing a one-dimensional minimization along the direction $\mathbf{u}$; call $\mathbf{x}_I^*$ the point at which $\phi(\mathbf{x}_I + \gamma\mathbf{u})$ is minimized. Now imagine doing the same thing, this time starting from the point $\mathbf{x}_{II}$ (still minimizing along $\mathbf{u}$) with the minimum called $\mathbf{x}^*_{II}$. In equation form, these two minimizations are:

$$\begin{aligned}\frac{d}{d\gamma}\phi(\mathbf{x}_I + \gamma\mathbf{u}) = \mathbf{u}^T\nabla\phi(\mathbf{x}^*_I) = \mathbf{u}^T[\mathbf{b} + \mathbf{A}\mathbf{x}^*_I] = 0 \\ \frac{d}{d\gamma}\phi(\mathbf{x}_{II} + \gamma\mathbf{u}) = \mathbf{u}^T\nabla\phi(\mathbf{x}^*_{II}) = \mathbf{u}^T[\mathbf{b} + \mathbf{A}\mathbf{x}^*_{II}] = 0\end{aligned} \tag{5.128}$$

where the steps are identical to those in Eq. (5.122). Subtract these two equations to find:

$$\mathbf{u}^T\mathbf{A}\,(\mathbf{x}^*_{II} - \mathbf{x}^*_I) = 0 \tag{5.129}$$

In words, the vector $\mathbf{x}^*_{II} - \mathbf{x}^*_I$ is conjugate to the vector $\mathbf{u}$ (as always, with respect to the matrix $\mathbf{A}$); crucially, the latter (i.e., $\mathbf{u}$) was used to produce the two vectors making up the former (i.e., $\mathbf{x}^*_I$ and $\mathbf{x}^*_{II}$). This important result is known as the *parallel subspace property*.

We turn to *Powell's algorithm*, which minimizes in an n-dimensional space by carrying out a sequence of one-dimensional minimizations; this technique takes advantage of Eq. (5.129) to *generate* a set of mutually conjugate directions. Start at the point $\mathbf{x}^{(0)}$ and take $\mathbf{u}_0, \mathbf{u}_1, \ldots, \mathbf{u}_{n-1}$ to be the columns of the identity matrix. The following steps constitute one "cycle" of Powell's algorithm (in general, you would need several cycles):

(a) Apply Eq. (5.121) for $k = 1, 2, \ldots, n$, i.e., do n one-dimensional minimizations, each time using the new point as the starting point for the next minimization.
(b) Replace $\mathbf{u}_k$ by $\mathbf{u}_{k+1}$ for $k = 0, 1, \ldots, n-2$.
(c) Replace $\mathbf{u}_{n-1}$ by $\mathbf{x}^{(n)} - \mathbf{x}^{(0)}$.
(d) Carry out a one-dimensional minimization on $\phi(\mathbf{x}^{(0)} + \gamma\mathbf{u}_{n-1})$ (finding γ^*) and then replace $\mathbf{x}^{(0)}$ by $\mathbf{x}^{(0)} + \gamma^*\mathbf{u}_{n-1}$.

As you can see, the $\mathbf{u}_k$'s start out as the columns of the identity matrix (i.e., as mutually orthogonal). They slowly get overwritten: at the end of the first cycle, all but one of them are still columns of the identity matrix (but now renamed). One of them is totally new and constructed in a special way: each new cycle gives you a new conjugate vector.

Take a moment to understand this crucial fact, which is typically stated without proof (or passed over in silence) in the literature. The first new ϕ-aware vector (think of it as $\mathbf{u}^I_{n-1}$) emerges the first time (c) is encountered. The second new ϕ-aware vector, $\mathbf{u}^{II}_{n-1} = \mathbf{x}^{(n)}_{II} - \mathbf{x}^{(0)}_{II}$, is made up of two components: (i) $\mathbf{x}^{(0)}_{II}$, whose value is set up (along $\mathbf{u}^I_{n-1}$) the first time (d) is seen, and (ii) $\mathbf{x}^{(n)}_{II}$, which is created the second time (the last iteration of) (a) is encountered, also along $\mathbf{u}^I_{n-1}$. Thus, we can directly apply Eq. (5.129) to draw the conclusion that $\mathbf{u}^{II}_{n-1}$ is conjugate to $\mathbf{u}^I_{n-1}$. Obviously, the same argument can then be applied to the new conjugate vector arising in the third cycle, and so on.

Thus, n cycles later (each of which has carried out $n+1$ one-dimensional minimizations) you will have generated n mutually conjugate vectors and therefore you will be able to immediately reach the true minimum for a quadratic function. For a non-quadratic function you still follow the same prescription, but this time you will need more than n cycles to converge. Speaking of convergence, when implementing Powell's algorithm you would have to employ some criterion on when to call it a day; the most natural way to do this is to compare $\mathbf{x}^{(n)}$ and $\mathbf{x}^{(0)}$, right after you carry out the renaming mentioned in step (d).

A wrinkle: even for the simple case of a quadratic function, there is the danger that one of the γ's (produced in the one-dimensional minimizations) might become zero, in which case at least two of the directions will become linearly dependent. From that point onward the algorithm will only be exploring a proper subspace, not the whole space we are interested in. Even when a given γ is not exactly zero but tiny, the directions can become nearly linearly independent, thereby adversely affecting progress toward the minimum. We will employ the simplest remedy possible: after a fixed number of cycles, *reset* the search directions $\mathbf{u}_0, \mathbf{u}_1, \ldots, \mathbf{u}_{n-1}$ to (again) be the columns of the identity matrix. This is an effective way of avoiding the linear-dependency issue, though it's not very efficient (since it throws away all the ϕ-specific knowledge that had been built up). Problems 5.36 and 5.37 explore a couple of different strategies: first, instead of resetting $\mathbf{u}_0, \mathbf{u}_1, \ldots, \mathbf{u}_{n-1}$ to the identity matrix, you could reset to an orthogonal matrix that is ϕ-aware or, second, instead of arbitrarily discarding the old $\mathbf{u}_0$ in step (b), as we did above, you could be more judicious and pick the direction that is most likely to lead to issues with linear dependency.

5.6.5.4 Implementation

Our implementation of Powell's algorithm in `powell.py`, i.e., code 5.8 starts by importing our convergence criterion, our one-dimensional golden-section minimizer, as well as our two-dimensional scalar-function example from Eq. (5.106). The interface of the function `powell()` itself is mostly familiar-looking, with a couple of new features: `nreset` is a parameter that is specific to our strategy of handling the linear-dependency issue and `cycmax` controls the total number of *cycles* (i.e., it is analogous to the `kmax` of earlier codes, but not quite the same). Speaking of cycles, we immediately see that the body of the function contains one "big"/external loop, stepping through the cycles. Each given cycle first checks to see whether or not it should (re)set $\mathbf{u}_0, \mathbf{u}_1, \ldots, \mathbf{u}_{n-1}$ to the identity matrix and then proceeds with the steps of the algorithm itself.

Step (a) carries out n one-dimensional minimizations (i.e., it involves a "small"/internal loop) via golden-section search, namely `golden()`. Note that that function's signature was `golden(f,x0,x1,kmax=200,tol=1.e-8)`, i.e., it expected (at least) three parameters: the one-dimensional function to be minimized, followed by the endpoints of the interval. Since step (a) is carrying out the minimization of $\phi(\mathbf{x}^{(k-1)} + \gamma\mathbf{u}_{k-1})$, we need to find a way of appropriately mapping our n-dimensional ϕ argument, namely the $\mathbf{x}$'s and the $\mathbf{u}$'s. In other words, we need to somehow program the fact that, for a given $\mathbf{x}^{(k-1)}$ and a given $\mathbf{u}_{k-1}$, $\phi(\mathbf{x}^{(k-1)} + \gamma\mathbf{u}_{k-1})$ is a function of only γ. You should spend some time trying to tackle this programming challenge on your own. Our code opts for the use of an anonymous function via Python's `lambda` keyword (introduced in problem 1.4):[52] we quickly create a function right where we need it (i.e., as the first argument being passed into `golden()`). If it helps you, feel free to rewrite this using an intermediate variable:[53]

[52] The same problem also introduced nested functions, with which you may wish to experiment now.

[53] This is discouraged in PEP 8, but you should go easy on yourself while you're still mastering the concepts.

Code 5.8 powell.py

```
from jacobi import termcrit
from golden import golden
from descent import phi
import numpy as np

def powell(phi,xolds,nreset=4,cycmax=200,tol=1.e-8):
    n = xolds.size
    for icyc in range(cycmax):
        xnews = np.copy(xolds)
        if icyc%nreset==0:
            U = np.identity(n)

        for k in range(n):
            ga = golden(lambda g: phi(xnews+g*U[:,k]),-10.,+10.)
            xnews += ga*U[:,k]
        U[:,:-1] = U[:,1:]
        U[:,-1] = xnews - xolds
        ga = golden(lambda g: phi(xolds+g*U[:,-1]),-10.,+10.)
        xolds += ga*U[:,-1]

        err = termcrit(xolds,xnews)
        print(icyc, xnews, err, phi(xnews))
        if err < tol:
            break
    else:
        xnews = None
    return xnews

if __name__ == '__main__':
    xolds = np.array([2.,0.25])
    xnews = powell(phi, xolds); print(xnews)
```

```
func = lambda g: phi(xnews+g*U[:,k])
```

then passing func as the first argument to `golden()`. Note that the return value of `golden()` is $\gamma^{(k-1)}$, but we don't actually need to keep a "history" of all the γ's, i.e., we only keep the latest one; similarly, we don't need to keep a full trajectory of the positions $\mathbf{x}^{(k)}$, so we keep only the previous and the current ones. Finally, note that in our code the endpoints of the interval (passed in to `golden()`) are pulled out of a hat: if you've solved problem 5.24 on the bracketing of a one-dimensional minimum, you can do a better job here.

Steps (b) and (c) of Powell's algorithm are carried out idiomatically, with only a single line of code per step. (Note, incidentally, that we *do* need to keep track of all n of the $\mathbf{u}_k$'s.) As a result, even step (b), which contains an iteration in its statement above, doesn't involve a loop in our program. We then turn to step (d), still outside the inner loop (i.e., only once per cycle). We're a little careful with our variable names, but other than that, this $(n + 1)$-th one-dimensional minimization is very similar to the earlier n one-dimensional minimizations. The convergence test (as well as the use of for-else) is identical to what we've seen several times before.

Running this code, we see that it finds the minimum in a handful of iterations, i.e., it needs an order of magnitude fewer iterations to converge than our gradient-descent code did. Of course, each such "iteration" is actually a cycle, made up of $n + 1$ $(= 3)$ one-dimensional minimizations each of which, in its turn, will need a number of iterations to converge. Since we're not using a very efficient one-dimensional minimizer, our approach is not hyperefficient, but the essential ideas, and the power of Powell's algorithm, should be clear enough. Crucially, we did not compute any derivatives (via finite differences, automatic differentiation, etc.) anywhere.

Before concluding, we realize that our linear-algebra (and NumPy) machinery from chapter 4 has turned out to be pretty useful when dealing with the zeros of polynomials, with systems of non-linear equations, and with multidimensional minimization. These are quite different applications, some of which used Gaussian elimination and others an eigenvalue solver. What unites them is the power of linear algebra, even when you're solving problems that themselves are non-linear.

5.7 Project: Extremizing the Action in Classical Mechanics

As in previous chapters, we now turn to a physics application of the infrastructure we've developed in earlier sections. We will be studying a problem from classical mechanics, more specifically the dynamics of a single particle. Crucially, we will *not* be solving differential equations; that's the topic of chapter 8. Here, after some background material on Lagrangian mechanics, we will see how multidimensional minimization techniques, like those we just covered, can help you find the trajectory of a particle. This is a nice concretization of topics that students encounter in introductory courses; it's one thing to *say* that we minimize the action and quite another to see it happen.

5.7.1 Defining and Extremizing the Action

Working in a Cartesian system, let us study a single particle in one dimension. We can denote the particle's location by $x(t)$, where we're explicitly showing that the position is a function of time. For a more complicated problem, we would introduce generalized coordinates, but what we have here is enough for our purposes.

The kinetic energy of the particle will be a function of only $\dot{x}(t)$, i.e., of the time derivative of the position: $T = T(\dot{x}(t))$. Specifically, since we are dealing with a single particle, we know that:

$$T = \frac{1}{2}m\dot{x}^2 \tag{5.130}$$

where m is the mass of the particle. Similarly, in the absence of time-dependent external fields, the potential energy is a function of only $x(t)$: $V = V(x(t))$. The difference of these two quantities is defined as the *Lagrangian*:

$$L(x(t), \dot{x}(t)) \equiv T(\dot{x}(t)) - V(x(t)) \tag{5.131}$$

where, for our case, there is no explicit dependence of the Lagrangian on time.

We are interested in studying the particle from time $t = 0$ to time $t = \tau$. Then, one can define the *action functional* as the integral of the Lagrangian over time:

$$S[x(t)] \equiv \int_0^\tau dt\, L(x(t), \dot{x}(t)) = \int_0^\tau dt \left(\frac{1}{2}m\dot{x}^2 - V(x)\right) \tag{5.132}$$

where we applied Eq. (5.131) to our case. Notice that we called the action a *functional* and used square brackets on the left-hand side. Roughly speaking, a *functional* is a function of a function. A reminder: an ordinary function ϕ takes us from one number t to another number $\phi(t)$. A functional is an entity that takes in an entire function and gives back a number. In other words, a functional is a mapping from a space of functions onto the real (or complex) numbers.

In case you haven't encountered functionals before, let us start with a simple case: a functional F of $\phi(t)$ (where t is a regular variable – this is a one-dimensional problem): $F[\phi(t)]$. Being a functional, F depends simultaneously on the values of ϕ at all points t but *it does not depend on t itself*: we provide it with the entire function and it provides us with one number as a result. A trivial example: $F[\phi] = \int_0^1 dt\phi(t)$ gives us one number for $\phi(t) = t$ and a different number for $\phi(t) = t^2$. In both cases, however, the answer is an ordinary number that does not depend on t.

Applied to our mechanics problem, we see that the action S depends on the position of the particle $x(t)$ at *all* times from 0 to τ, but not on t directly, since t has been "integrated out". For a given trajectory $x(t)$ from $t = 0$ to $t = \tau$, the action produces a single number, S.[54] The question then arises: which trajectory $x(t)$ from $t = 0$ to time $t = \tau$ does the particle actually "choose"? The answer comes from *Hamilton's principle*: of all possible paths, the path that is actually followed is that which minimizes the action. As it so happens, the

[54] As an aside, observe that, since the Lagrangian has units of energy, the action has units of energy times time. Intriguingly, these are also the units of Planck's constant, $\hbar$, but here we're studying classical mechanics.

action only needs to be *stationary*, i.e., we are extremizing and not necessarily minimizing, but we will usually be dealing with a minimum. This extremization is taking place with the endpoints kept fixed, i.e., $x(0)$ and $x(\tau)$ are not free to vary.

5.7.2 Discretizing the Action

Our exposition will now diverge from that of a standard mechanics textbook: we will not proceed to write a possible trajectory as the physical trajectory plus a small perturbation. Instead, we will take the action from Eq. (5.132) and *discretize* it. This simply means that we will assume the positions $x(t)$ from $t = 0$ to $t = \tau$ can be accessed only at a discrete set of n points; applying our discussion of points on a grid from section 3.3.6, we have:

$$t_k = kh_t = k\frac{\tau}{n-1} \tag{5.133}$$

where, as usual, $k = 0, 1, \ldots, n-1$.[55] We will further employ the notation $x_k \equiv x(t_k)$ to denote the (possible) position of the particle at each of our time-grid points.

We promised to discretize the action from Eq. (5.132). This involves two separate steps. First, we have to discretize the integral: we haven't seen how to carry out numerical integration yet (the subject of chapter 7). Even so, you are already familiar with the simplest-possible way of carrying out an integral, namely the *rectangle rule*: assume that the area under L from t_k to t_{k+1} can be approximated by the area of a rectangle, with width h_t. Equivalently, the rectangle rule approximates L as a constant from t_k to t_{k+1}, namely a straight (horizontal) line. Then, the areas of all the individual rectangles have to be added up. Second, we have to discretize the time derivative $\dot{x}$. This is a process that you are even more familiar with, having studied chapter 3. The simplest thing to do is to employ a forward-difference scheme.[56] Putting these two discretization steps together allows us to introduce the finite-n analogue of Eq. (5.132):

$$S_n \equiv \sum_{k=0}^{n-2} h_t \left[\frac{1}{2}m\left(\frac{x_{k+1} - x_k}{h_t}\right)^2 - V(x_k)\right] \tag{5.134}$$

where the sum goes up to $n-2$: the last rectangle gets determined by the value of the function on the left. Since the continuum version of Hamilton's principle involved determining $x(t)$ with $x(0)$ and $x(\tau)$ kept fixed, we will now carry out the corresponding extremization of S_n, with x_0 and x_{n-1} kept fixed.[57] This means that we are dealing with n points in total, but only $n-2$ variable points.To keep thing simple, we define $n_{var} = n-2$. This means that

[55] This k is our discretization index, so it has nothing to do with the iteration counter (e.g., $\mathbf{x}^{(k)}$) we encountered earlier. The subscript t in h_t, on the other hand, is merely a reminder (i.e., not an index).

[56] Problem 5.44 invites you to do better.

[57] This is the simplest variety of *constrained minimization*, where we can use the *substitution method*.

our discrete action, S_n, is a function of these n_{var} variables:

$$S_n = S_n(x_1, x_2, \ldots, x_{n_{var}}) \tag{5.135}$$

There you have it: our problem has now become that of finding a minimum of S_n in this n_{var}-dimensional space. But this is precisely the problem we tackled in section 5.6.1, when we were faced with a scalar function of many variables, $\phi(\mathbf{x})$. Crucially, Eq. (5.135) shows us that S_n is a *function* of a discrete number of variables, x_1, x_2, and so on; there is no integral, no derivative, and no functional left anymore.

5.7.3 Newton's Method for the Discrete Action

We now decide to employ Newton's method from section 5.6.4 for this multidimensional minimization problem. While we haven't yet reached the stage of implementing things in Python, let us try to make sure we won't get confused by the notation. We are dealing with n positions x_k; of these, x_0 and x_{n-1} are kept fixed, so the only "true" variables are $x_1, x_2, \ldots, x_{n_{var}}$. These n_{var} variables will be determined by our Newton's method; the function we will employ to do that (`multi_newton()`) will expect our variables to employ the standard Python 0-indexing. Thus, we will employ distinct Python names for the first and last point: x_0 and x_{n-1} will be `xini` and `xfin`, respectively. The n_{var} actual variables, which are to be determined, will be stored in our usual `numpy` array by the name of `xs`, which will be indexed from `0` up to `nvar-1`. We attempt to summarize these facts in Fig. 5.18: notice how `xs[0]` corresponds to x_1, the first true variable; similarly, `xs[-1]` is another name for `xs[nvar-1]` and corresponds to x_{n-2}, the last true variable.

As mentioned, we plan to employ Newton's method for minimization. We recall from Eq. (5.111) that this will require the gradient vector $\nabla\phi(\mathbf{x})$ as well as the Hessian, i.e., the Jacobian of the gradient, $\mathbf{J}_{\nabla\phi}(\mathbf{x})$; of course, here we are faced with S_n instead of ϕ. Significantly, our approach in earlier sections was to evaluate any needed derivatives using a finite-difference scheme. In the present case, however, we have already employed a forward difference scheme once, in order to get to our discrete action, Eq. (5.134). More importantly, Eq. (5.134) contains the sum of a large number of contributions, but any derivative with respect to a given position will not involve so many contributions; in other words, if we try to evaluate the first and second derivatives of S_n numerically, we will be asking for trouble. Instead, we decide to take any required derivatives analytically; we will then code up our results in the following subsection.

To produce the components of the gradient vector, we need to take the derivative of S_n with respect to x_i, for $i = 1, 2, \ldots, n_{var}$. As repeatedly mentioned, x_0 and x_{n-1} are kept fixed so we will not be taking any derivatives with respect to them. Let's look at a given component:

$$\begin{aligned}\frac{\partial S_n}{\partial x_i} &= \sum_{k=0}^{n-2} h_t \left[\frac{m}{h_t^2}(x_{k+1} - x_k)(\delta_{i,k+1} - \delta_{i,k}) - \frac{\partial V(x_k)}{\partial x_i}\delta_{i,k}\right] \\ &= \frac{m}{h_t}(2x_i - x_{i-1} - x_{i+1}) - h_t\,\frac{\partial V(x_i)}{\partial x_i}\end{aligned} \tag{5.136}$$

In the first line, we saw the 1/2 get cancelled by taking the derivative of $(x_{k+1} - x_k)^2$;

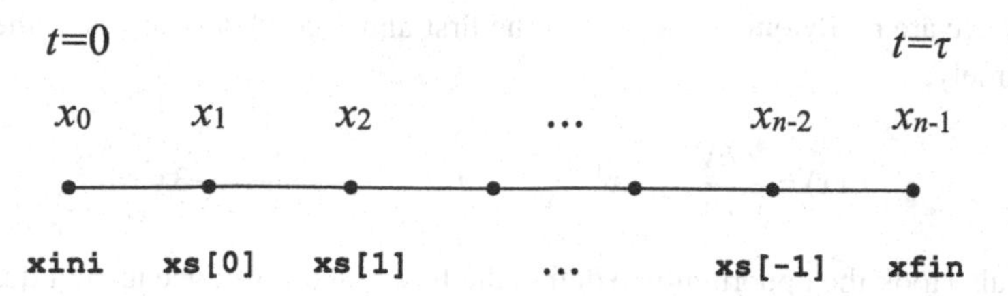

Fig. 5.18 Discretized coordinates, also showing the corresponding Python variables

observe that only specific values of k contribute to this sum. This fact is taken into account in the second line, where the Kronecker deltas are used to eliminate most of the terms in the sum, leaving us with a simple expression.[58] Note that, even though we're only taking the derivative with respect to x_i for $i = 1, 2, \ldots, n_{var}$, the result in Eq. (5.136) also involves x_0 and $x_{n_{var}+1} = x_{n-1}$, due to the presence of the x_{i-1} and x_{i+1} terms, respectively.

Similarly, we can find the Hessian by taking the derivative of the gradient vector components with respect to x_j, where $j = 1, 2, \ldots, n_{var}$. We have:

$$\frac{\partial S_n}{\partial x_i \partial x_j} = \frac{m}{h_t}(2\delta_{j,i} - \delta_{j,i-1} - \delta_{j,i+1}) - h_t \frac{\partial^2 V(x_i)}{\partial x_i^2} \delta_{j,i} \tag{5.137}$$

We see that the Hessian matrix will only have elements on the main diagonal and on the two diagonals next to it; in other words, it will be a tridiagonal matrix. Of course, this only holds for our specific prescription of how to discretize the action.

Armed with the gradient in Eq. (5.136) and the Hessian in Eq. (5.137), we are now ready to employ Newton's multidimensional minimization method from Eq. (5.111). Note that we haven't specified anything about $V(x_i)$ so far; this is a feature, not a bug. It means that our formalism of discretizing and then extremizing the action will apply to any physical problem where the potential energy at x_i depends only on x_i,[59] regardless of the specific form V takes.

5.7.4 Implementation

Before we implement our approach in Python, we have to make things concrete. Let's pick a specific form for the potential energy, that of the *quartic oscillator*:

$$V = \frac{1}{4}x^4 \tag{5.138}$$

where we didn't include an extra coefficient in front, for the sake of simplicity. Of course, the classical quartic oscillator is a problem that's been studied extensively; we are merely using it as a case where: (a) you won't be able to think of the analytical solution off the top off your head, and (b) the dependence on x is non-linear, so we'll get to test the robustness of our minimization methodology for a challenging case. You can generalize our approach to more complicated scenarios later. Given the forms of Eq. (5.136) and Eq. (5.137), we

[58] In the study of differential equations (the subject of chapter 8) our result is known as the *Verlet algorithm*.
[59] Of course, even this requirement can be loosened, by appropriately generalizing the above derivation.

realize that we are really only interested in the first and second derivatives of the potential energy, namely:

$$F(x) = -\frac{\partial V}{\partial x} = -x^3, \qquad F'(x) = -\frac{\partial^2 V}{\partial x^2} = -3x^2 \tag{5.139}$$

where we also took the opportunity to define the force (and introduce its first derivative).

Code 5.9 starts by importing our earlier function `multi_newton()`. We then define a function to hold several parameters; we do this here in order to help our future selves, who may be interested in trying out different endpoint values, total times, masses,[60] or numbers of variables, n_{var}. This way of localizing all the parameters in one function and then having other parts of the code call it to get the parameter values is somewhat crude,[61] but still better than using global variables (i.e., Python variables defined outside a function and never explicitly communicated to it). In our example we employ $n_{var} = 99$; this was chosen such that $n = 101$, which in its turn, see Eq. (5.133), leads to $h_t = 0.01$ in the appropriate units. Importantly, this means that we have to solve a minimization problem in $\sim$100 variables, which is a larger-scale problem than any we solved in earlier sections.

We then define a simple one-liner function (using Python's ternary operator) to evaluate the force-or-derivative. This is the only other place in our program where specifics about the physics problem are encoded. A point that will come in handy below: the parameter `x` of our function `fod()` can handle either single numbers or entire `numpy` arrays.

The bulk of our code is in the following two functions. We recall from code 5.5 that `multi_newton()` has a very specific interface: as its first argument it expects a function that returns a 1d `numpy` array and as its second argument a function that returns a 2d `numpy` array and a 1d array. These functions implement the gradient vector of Eq. (5.136) and the Jacobian/Hessian of Eq. (5.137). Crucially, both these functions do not employ *any* explicit loops or indices (like `i`, `j`, and so on). In most of our codes we have prioritized legibility over efficiency. Even here, despite the fact that our slicing is more efficient than the indexed alternative, our guiding principle was actually clarity: the grid in Fig. 5.18 can be confusing, so we'd rather avoid having to deal with indices that go up to `nvar-1` or perhaps `nvar-2` (or even worse: `n-3` or `n-4`) when we step through our arrays. As it so happens, problem 1.11 asked you to rewrite these two functions using explicit indices.

Since `actfs()` is following our prescription of fixed endpoints, x_0 and x_{n-1}, the lines setting up `arr[0]` and `arr[-1]` need to be different from what we do for the "middle" points.[62] Similarly, `actjac()` needs to treat the main and other diagonals differently. We accomplish the latter task by slicing the arrays we pass as the first argument to `numpy.fill_diagonal()`. For the main diagonal we pass the entire `xs` to `fod()`.

The main part of the program is incredibly simple. It sets up an initial guess vector, $\mathbf{x}^{(0)}$, designed to start near fixed endpoint x_0 and end near our other fixed endpoint x_{n-1}. It then calls our earlier `multi_newton()` function, passing in our tailored gradient and Hessian functions. Importantly, by using `params()` we have ensured that `actfs()` and `actjac()`

[60] For simplicity, we take $m = 1$.
[61] This is a case where object orientation would really help.
[62] This is analogous to our discussion of the central difference (for points on a grid) on page 96.

action.py Code 5.9

```
from multi_newton import multi_newton
import numpy as np

def params():
	nvar = 99; m = 1.
	xini, xfin = 2., 0.
	tt = 1.; dt = tt/(nvar+1)
	return nvar, m, xini, xfin, dt

def fod(der,x):
	return -x**3 if der==0 else -3*x**2

def actfs(xs):
	nvar, m, xini, xfin, dt = params()
	arr = np.zeros(nvar)
	arr[0] = (m/dt)*(2*xs[0]-xini-xs[1]) + dt*fod(0,xs[0])
	arr[1:-1] = (m/dt)*(2*xs[1:-1] - xs[:-2] - xs[2:])
	arr[1:-1] += dt*fod(0,xs[1:-1])
	arr[-1] = (m/dt)*(2*xs[-1]-xs[-2]-xfin) + dt*fod(0,xs[-1])
	return arr

def actjac(actfs,xs):
	nvar, m, xini, xfin, dt = params()
	Jf = np.zeros((nvar,nvar))
	np.fill_diagonal(Jf, 2*m/dt + fod(1,xs)*dt)
	np.fill_diagonal(Jf[1:,:], -m/dt)
	np.fill_diagonal(Jf[:,1:], -m/dt)
	actfs_xs = actfs(xs)
	return Jf, actfs_xs

if __name__ == '__main__':
	nvar, m, xini, xfin, dt = params()
	xolds = np.array([2-0.02*i for i in range(1,nvar+1)])
	xnews = multi_newton(actfs, actjac, xolds); print(xnews)
```

have the same interface as our earlier `fs()` and `jacobian()` functions, so everything works together smoothly. It may be surprising to see that we've been able to solve for the dynamical trajectory of a particle by minimizing in ~100 variables, using a code that takes up less

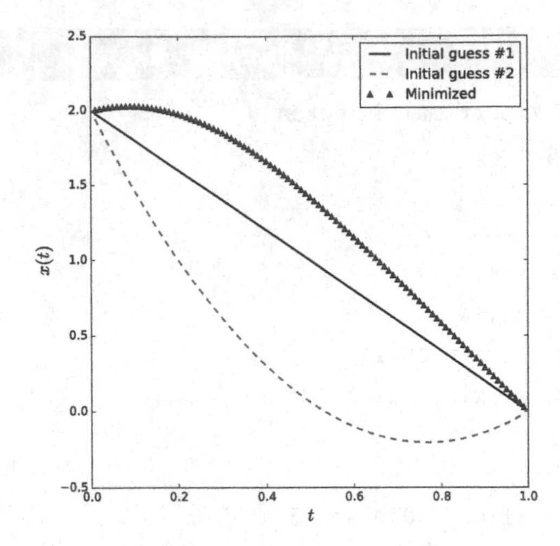

Fig. 5.19 Finding the trajectory of a particle via minimization of the action

than a page. Of course, this is slightly misleading, since we are also calling `multi_newton()`, which in its turn calls `gauelim_pivot()` and `termcrit()`. The former then calls `backsub()`. Thus, our latest program fits on one page, but only because we can so smoothly make use of several functions we developed earlier.

Instead of showing you the 99 numbers that get printed on the screen when you run this code, we have decided to plot them in Fig. 5.19. Our `xolds` is labelled with "Initial guess #1". The converged solution of `xnews` is labelled with "Minimized". You may be wondering how we know that our converged solution corresponds to a minimum; if so, compute the eigenvalues of the Hessian corresponding to `xnews`. Wanting to make sure that our result was not a fluke that was directly tied to a "magic" choice of our initial guess vector, we try out another `xolds`, labelled with "Initial guess #2": this leads to the exact same minimum as before, thereby increasing our confidence in our result.

It's worth pausing for a second to marvel at our accomplishment. In this Project, we went beyond the formal statement that the action "could" be minimized or "should" be minimized and actually minimized the action. We did this by using a (somewhat crude) discretization scheme and stepping through a many-variable minimization process which, incidentally, only required roughly half a dozen iterations to converge. You may become more convinced of the importance of this fact after you attempt problem 5.42 that asks you to carry out a minimization for the (easier) case of the *harmonic* oscillator using the gradient-descent method. You will discover that multidimensional minimization is a fickle procedure, with few guarantees. A combination of analytical work and earlier coding has allowed us to find a practical avenue toward making the action stationary and thereby determining the physical trajectory of the particle.

Now that we've marvelled at how great we did, let's tone things down: had we actually followed the standard-textbook route, we would have used Hamilton's principle to find the Euler–Lagrange equation(s), thereby getting an equation of motion, which would be a differential equation. For our case, this would have been simply $m\ddot{x} = -x^3$. This differential equation, in turn, can straightforwardly be solved using its own discretization scheme, as you will soon learn (in chapter 8). In contradistinction to this, in our present approach we had to repeatedly employ 99×99 matrices, even though the problem is reasonably simple. When all is said and done, it's nice to see that you can calculate dynamical trajectories without having to solve a differential equation.

Problems

5.1 Come up with an example where the relative termination criterion of Eq. (5.18) fails. Then, replace it with Eq. (5.19) and re-run your root-finding program.

5.2 This problem studies fixed-point iteration for the case of $g(x) = cx(1 - x)$. Examine:

(a) $c = 0.9$: play around with the number of iterations, tolerance, termination criterion, and initial guess to see where all your solutions tend to go to.
(b) $c = 1.5$: see how long it takes you to find a fixed point.
(c) $c = 2.8$: does this take longer than the $c = 1.5$ case?
(d) $c = 3.2$: start with $x^{(0)} = 0.2$ and observe that a new pattern emerges.
(e) $c = 3.5$: start with $x^{(0)} = 0.2$ and see another pattern emerge.
(f) $c = 3.6$: start with $x^{(0)} = 0.2$ and attempt to discern any pattern.

You may need to plot the $x^{(k)}$ as a function of k in order to (attempt to) discern the patterns. For future reference, this problem is known as a "logistic map".

5.3 Apply Aitken extrapolation, Eq. (2.107), to the sequence of iterates produced by the fixed-point method for $g(x) = \ln x + \sqrt{x}$ and $x^{(0)} = 2.499$. Compare the accuracy of the (untransformed) $x^{(60)}$ with the accelerated/transformed q_{30}.

5.4 Use bisection for $f(x) = 1/(x - 3)$ in the interval $(0, 5)$. Is your answer correct?

5.5 Code up Newton's method for our example function $f(x) = e^{x-\sqrt{x}} - x$ and reproduce Fig. 5.7. You should evaluate $f'(x)$ analytically.

5.6 [$\mathcal{A}$] Analytically use Newton's method to write $\sqrt{2}$ as a ratio of two *integers*. Provide a few (increasingly better) approximations.

5.7 Code up a *safe* version of Newton's method, by combining it with the bisection method. Specifically, start with a bracketing interval and carry out a Newton step: if the iterate "wants" to leave the interval, carry out a bisection step, instead. Repeat.

5.8 [$\mathcal{A}$] We will now recast Newton's method as a special case of the fixed-point iteration method. Specifically, if your problem is $f(x) = 0$, then one way of transforming it into the fixed-point form $x = g(x)$ is to take:

$$g(x) = x - \frac{f(x)}{f'(x)} \tag{5.140}$$

Assume x^* is a simple root (i.e., $f(x^*) = 0$ and $f'(x^*) \neq 0$). You should analytically evaluate $g'(x^*)$. You should then combine your result with our analysis of the convergence properties of the fixed-point iteration method to find the order of convergence for the method of Eq. (5.140).

5.9 [$\mathcal{A}$] We now build on problem 5.8, which recast Newton's method as a version of fixed-point iteration. Assume x^* is a root of multiplicity m, that is (for $q(x^*) \neq 0$):

$$f(x) = (x - x^*)^m q(x) \tag{5.141}$$

(a) Analytically derive a formula for the $g(x)$ of Eq. (5.140) where you've employed Eq. (5.141).

(b) Use that result to show that $g'(x^*) = (m-1)/m$. From there, since $m = 2$ or higher, draw the conclusion that Newton's method converges linearly for multiple roots.

(c) It should now be straightforward to see that if you had used:

$$g(x) = x - m\frac{f(x)}{f'(x)} \tag{5.142}$$

instead of Eq. (5.140), then you would have found $g'(x^*) = 0$, in which case Newton's method would converge quadratically (again).

5.10 In practice, we don't know the multiplicity of a root ahead of time, so the utility of Eq. (5.142) is limited. In the first part of problem 5.9, you incidentally also derived a formula for $w(x) = f(x)/f'(x)$ for the case of Eq. (5.141). Similarly, in the second part of that problem you found a formula for $w'(x)$, which leads to $w'(x^*) = 1/m$. But this means that $w'(x^*) \neq 0$, while $w(x^*) = 0$; in other words, $w(x)$ has a simple root at x^* regardless of what the multiplicity of $f(x)$ at x^* is. We can take advantage of this fact: whenever you think you may have a multiple root, you could apply Newton's method not to $f(x)$, but to $w(x)$. Code this approach up in Python for the case of:

$$f(x) = x^5 - 3x^4 + x^3 + 5x^2 - 6x + 2 \tag{5.143}$$

Compare your runs to how long it takes the (unmodified) Newton's method to converge and to the result of using Eq. (5.142)—you'll need to guess the value of m.

5.11 When using Newton's method, it can sometimes be frustrating to start from different initial guesses but end up producing the same root over and over. A trick that can be used to suppress an already-found root is to apply Newton's method not to $f(x)$ but to $u(x) = f(x)/(x - a)$, where a is the root you've already found and are no longer looking for. Implement this strategy for the case of our example function $f(x) = e^{x-\sqrt{x}} - x$, where you are suppressing the root $a = 1$. This means that you should always be converging to the other root, regardless of your initial guess.[63]

5.12 [$\mathcal{A}$] We now guide you toward deriving the secant method's order of convergence. First introduce notation similar to that in Eq. (2.5), i.e., $\Delta_k = x^{(k)} - x^*$. Now, Taylor expand $f(x^{(k-1)})$ and $f(x^{(k-2)})$ around x^*; note that this is different from what we did in Eq. (5.37) and elsewhere, where we Taylor expanded around $x^{(k-1)}$. Plug these relations into Eq. (5.51) to show that Δ_k is equal to the product of Δ_{k-1} and Δ_{k-2} (times another factor). Motivated by Eq. (5.16), take $\Delta_k = m\Delta_{k-1}^p$ and $\Delta_{k-1} = m\Delta_{k-2}^p$, and combine these with your earlier result. This should lead to a quadratic equation in p, which is solved by $p = (1 + \sqrt{5})/2 \approx 1.618$.

[63] Root suppression feels similar to (yet is distinct from) the trick of *deflation* applied to polynomials, which amounts to using *synthetic division* and then solving a smaller-sized problem (on the latter, see problem 3.11).

Hints: (a) it will probably make your life easier to also use a standard Taylor expansion for $1/(1+x)$, and (b) as usual in perturbation theory, higher-order terms can be dropped, but you have to be a little careful about when you drop them.

5.13 Implement Ridders' method from Eq. (5.59) for our example $f(x) = e^{x-\sqrt{x}} - x$.

5.14 There exists a method known as *regula falsi* or *false position* which is very similar to the secant method. The difference is that *regula falsi* is a bracketing method, so it starts with x_0 and x_1 such that $f(x_0)f(x_1) < 0$. Like the secant method, it then employs the line that goes through the two points (x_0, y_0) and (x_1, y_1), as per Eq. (5.52), and then finds the x axis intercept. The false-position method then ensures that it continues to bracket the root. Implement this approach for our example function $f(x) = e^{x-\sqrt{x}} - x$, starting with the interval $(0, 1.7)$. Do you see the method converging on both sides or not? Do you understand what that implies for the case where one of your initial points is "bad", i.e., very far from the root?

5.15 Another approach, known as *Steffensen's method*, iterates according to:

$$x^{(k)} = x^{(k-1)} - \frac{f(x^{(k-1)})}{g(x^{(k-1)})}, \quad g(x^{(k-1)}) = \frac{f\left(x^{(k-1)} + f(x^{(k-1)})\right) - f(x^{(k-1)})}{f(x^{(k-1)})} \tag{5.144}$$

This achieves quadratic convergence, at the cost of requiring two function evaluations per iteration. Implement it in Python for our example function $f(x) = e^{x-\sqrt{x}} - x$ and compare the required number of iterations to those of other methods.

5.16 Explore the roots of:

$$f(x) = -x^5 + 4x^4 - 4x^3 + x^2e^x - 2x^2 - 4xe^x + 8x + 4e^x - 8 \tag{5.145}$$

using a method of your choice.

5.17 Take the cubic equation:

$$x^3 - 21x^2 + 120x - 100 = 0 \tag{5.146}$$

and find its roots. Now perturb the coefficient of x^3 so that it becomes 0.99 (first) or 1.01 (next) and discuss how the roots are impacted.

5.18 In the Project of chapter 3 we computed the values of Hermite polynomials, see code 3.3, i.e., `psis.py`. You should now find the roots of Hermite polynomials, by writing a code similar to `legroots.py`. Digging up Eq. (6.31.19) in Ref. [140] and converting to our notation gives us:

$$\frac{\pi(4i + 3 - (-1)^n)}{4\sqrt{2n+1}} \le x_i^{(0)} \le \frac{4i + 6 - (-1)^n}{\sqrt{2n+1}} \tag{5.147}$$

for the positive Hermite-polynomial zeros. Figure out how to use this to produce a function that gives correct roots for n up to, say, 20. Compare with the output of the (alternatively produced) `numpy.polynomial.hermite.hermgauss()`.

5.19 Code up our companion matrix from Eq. (5.70) for a general polynomial. Then, use it to find all the roots of our polynomial from Eq. (5.143).

5.20 In `legroots.py` we computed the roots of a Legendre polynomial using Newton's method and calling `legendre.py` for the function values (and derivatives). We now see that it is possible to directly employ the recurrence relation from section 2.5.2.2

to evaluate the roots of a Legendre polynomial, by casting the problem into matrix form. This generalizes to other sets of polynomials, even if you don't have good initial guesses for the roots. It is part of the *Golub–Welsch algorithm*.

(a) Take Eq. (2.86) and write it in the form:

$$\frac{j+1}{2j+1}P_{j+1}(x) + \frac{j}{2j+1}P_{j-1}(x) = xP_j(x) \tag{5.148}$$

Evaluate this at x^*, one of the zeros of $P_n(x)$ (i.e., $P_n(x^*) = 0$) for $j = 1, \ldots, n-1$; together with $P_1(x^*) = x^*P_0(x^*)$, this becomes $\mathcal{J}\,\mathbf{p}(x^*) = x^*\mathbf{p}(x^*)$. Verify that $\mathcal{J}$, known as a *Jacobi matrix* (not to be confused with a *Jacobian matrix*), is tridiagonal. The matrix elements of the $\mathcal{J}$ you found can be immediately determined from the coefficients in the recurrence relation, Eq. (5.148).

(b) Observe that $\mathcal{J}\,\mathbf{p}(x^*) = x^*\mathbf{p}(x^*)$ is an eigenvalue problem for the zeros of $P_n(x)$. Using Python, compute the eigenvalues of $\mathcal{J}$ and verify that they coincide with the output of `legroots.py`.

5.21 We now generalize the fixed-point iteration method of Eq. (5.22) to the problem of n equations in n unknowns. In equation form, this is simply $\mathbf{x}^{(k)} = \mathbf{g}(\mathbf{x}^{(k-1)})$ where, crucially, all quantities involved are vectors. Implement this approach in Python and apply it to the following two-variable problem:

$$\begin{cases} f_0(x_0, x_1) = x_0^2 - 2x_0 - 2x_1^2 + x_1 = 0 \\ f_1(x_0, x_1) = x_0^2 + x_0 - 2x_1^2 - 1.5x_1 - 0.05 = 0 \end{cases} \tag{5.149}$$

This system has two solutions. Do you find them both using the fixed-point iteration method? Do you understand why (not)?

5.22 In our study of Gaussian quadrature in section 7.5.1, the brute-force approach will lead to (Vandermonde) systems of non-linear equations like the following:

$$2 = c_0 + c_1 + c_2, \qquad 0 = c_0x_0 + c_1x_1 + c_2x_2, \qquad \frac{2}{3} = c_0x_0^2 + c_1x_1^2 + c_2x_2^2,$$
$$0 = c_0x_0^3 + c_1x_1^3 + c_2x_2^3, \qquad \frac{2}{5} = c_0x_0^4 + c_1x_1^4 + +c_2x_2^4, \qquad 0 = c_0x_0^5 + c_1x_1^5 + +c_2x_2^5 \tag{5.150}$$

Apply `multi_newton.py` to solve these six equations for the six unknowns.

5.23 Implement our basic Broyden method from Eq. (5.89) for our two-variable problem in Eq. (5.72). Separately investigate the convergence for the two cases of $\mathbf{J}(\mathbf{x}^{(0)})$ being the identity matrix or a forward-difference approximation to the Jacobian.

5.24 Write a function that brackets a minimum, i.e., if you are provided with a function `phi`, a starting point `x0`, and an initial step size `h`, your function will return the endpoints `a` and `b` of an interval bracketing the minimum. Your strategy should be to take a sequence of downhill steps, until the function value starts increasing; the last two points you've encountered are the endpoints you are after. You should make sure that your steps are actually taking you downhill (you can flip the sign of the step if you find that the initial move takes you uphill). Also, it is more efficient to keep increasing the step size with each iteration, say by a constant factor (e.g., 1.5).

5.25 Make our implementation of the golden-section search in `golden.py` more efficient, i.e, have it carry out only a single function evaluation per iteration. Hint: inside the loop, you should not have any function evaluations outside a conditional statement.

5.26 One can employ an interpolatory scheme to go beyond the golden-section approach (minimizing faster, at the cost of potentially failing from time to time). Using a straight-line fit (as is done in the secant root-finding method) won't help here: we need a curve with a minimum between the endpoints. The simplest such scheme is *successive parabolic interpolation*: given three points x_0, x_1, and x_2 one fits a parabola and then finds the abscissa at which the parabola is minimized. You could use Lagrange interpolation for the parabolic fit, but it's easier to carry out both the fit and the computation of the minimum analytically, to find:

$$x_3 = x_2 - \frac{1}{2}\frac{(x_2 - x_0)^2[\phi(x_2) - \phi(x_1)] - (x_2 - x_1)^2[\phi(x_2) - \phi(x_0)]}{(x_2 - x_0)[\phi(x_2) - \phi(x_1)] - (x_2 - x_1)[\phi(x_2) - \phi(x_0)]} \tag{5.151}$$

where (x_0, x_1) initially brackets the minimum, and the starting x_2 is produced as per Eq. (5.97). Obviously, this is then repeatedly applied; after applying this formula, rename the variables, dropping the old value of x_0; be careful with our convention: the starting triplet is (x_0, x_2, x_1). Write a Python function that implements this minimization technique and test it on the same example as in `golden.py`.

5.27 Employ the secant method to minimize $\phi(x) = e^{x-\sqrt{x}} - x$, Eq. (5.46), and Eq. (5.92).

5.28 Rewrite the function `gradient()` from the code `descent.py` so that it uses an explicit loop. Compare with the original code. Then, study the dependence of the maximum required iteration number on the magnitude of γ; explore values from 0.02 to 0.30, in steps of 0.01.

5.29 In `descent.py` we studied the simplest case of constant γ; we now also include a "line-search minimization" to determine $\gamma^{(k)}$ at each iteration. If you're currently at position $\mathbf{x}$ and the gradient is $\nabla\phi(\mathbf{x})$, then you can define: $\zeta(\gamma) = \phi(\mathbf{x} - \gamma\nabla\phi(\mathbf{x}))$. Crucially, $\zeta(\gamma)$ is a function of one variable. What this approach does is to take the negative gradient direction, $-\nabla\phi(\mathbf{x})$, and then decide how large a step γ to make in that direction, by minimizing $\zeta(\gamma)$; since this minimization is taking place along the fixed line determined by the negative gradient, this approach is known as line-search minimization. Putting everything together, we have the following algorithm:

$$\begin{aligned} &\min \zeta(\gamma) = \phi\left(\mathbf{x}^{(k-1)} - \gamma\nabla\phi(\mathbf{x}^{(k-1)})\right) \quad \text{to find } \gamma^{(k-1)} \\ &\mathbf{x}^{(k)} = \mathbf{x}^{(k-1)} - \gamma^{(k-1)}\nabla\phi(\mathbf{x}^{(k-1)}) \end{aligned} \tag{5.152}$$

To implement the above prescription, combine `descent.py` with:

(a) a modified version of `secant.py` (applied to a derivative, which you can estimate using a finite-difference technique), and

(b) `golden.py`. It may help you to employ a nested function here.

Check both approaches on $\phi(\mathbf{x}) = x_0^2 + 5x_1^2$.

5.30 An interesting variation of what we saw in the main text is the approach known as *gradient descent with momentum*:

$$\begin{aligned}\mathbf{p}^{(k-1)} &= \eta\mathbf{p}^{(k-2)} + \gamma\nabla\phi(\mathbf{x}^{(k-1)})\\ \mathbf{x}^{(k)} &= \mathbf{x}^{(k-1)} - \mathbf{p}^{(k-1)}\end{aligned} \tag{5.153}$$

where η is a number (which you can take to be constant) from 0 to 1. Implement this for the same problem as in `descent.py`; feel free to start out with $\mathbf{p}^{(0)}$ being the zero vector. Interpret Eq. (5.153) and discuss why it can help accelerate convergence.

5.31 Carry out multidimensional minimization using Newton's method for our example in Eq. (5.106). After you locate both minima (how would you know they're minima?), try starting from `xolds = np.array([1.5,0.])`. Then, use that same initial guess vector in the gradient-descent method and compare the two results.

5.32 An intriguing generalization of Newton's method for multidimensional root-finding borrows ideas from multidimensional minimization. The goal is to ensure that Newton's method can do a good job of finding the solution to a system of equations, even if the initial guess was poor; in other words, we wish to make sure that each step we make decreases the function-component values.[64]

To do so, we now examine the quantity $(\mathbf{f}(\mathbf{x}^{(k-1)}))^T\mathbf{f}(\mathbf{x}^{(k-1)})$: it makes sense to require that after the step the magnitude of the functions is reduced, i.e., that $(\mathbf{f}(\mathbf{x}^{(k)}))^T\mathbf{f}(\mathbf{x}^{(k)})$ is smaller. It's easy to show that the Newton step is a descent direction; in other words, the step prescribed by Newton's method picks a direction that reduces the functions' values. Here's how to see this:

$$\begin{aligned}\frac{1}{2}\left\{\nabla\left[\left(\mathbf{f}(\mathbf{x}^{(k-1)})\right)^T\mathbf{f}(\mathbf{x}^{(k-1)})\right]\right\}^T\mathbf{q}^{(k)} &= \left(\mathbf{f}(\mathbf{x}^{(k-1)})\right)^T\mathbf{J}(\mathbf{x}^{(k-1)})\left[-\left(\mathbf{J}(\mathbf{x}^{(k-1)})\right)^{-1}\mathbf{f}(\mathbf{x}^{(k-1)})\right]\\ &= -\left(\mathbf{f}(\mathbf{x}^{(k-1)})\right)^T\mathbf{f}(\mathbf{x}^{(k-1)}) < 0\end{aligned} \tag{5.154}$$

On the first line, $\mathbf{q}^{(k)}$ is the Newton step as per Eq. (5.83). We also took the gradient of our $\mathbf{f}^T\mathbf{f}$ quantity using a standard vector identity, and also re-expressed the Newton step from Eq. (5.80). From there, we cancelled the Jacobians and were left with a negative value. Indeed, the Newton step is a descent direction.

Now the question arises how to pick the magnitude of our step in that direction (similarly to the gradient-descent method for minimization). We could, at this point, employ a line-search approach as we did in problem 5.29. Instead, we will try the empirical approach of multiplying the full Newton step by $\lambda = 0.25$, if the function value would have otherwise increased. Try out this *backtracking* strategy for our example system from Eq. (5.72): starting from `xolds = np.array([-100.,100.])`, compare the unmodified and modified Newton's runs. Make sure to print out a message each time the step is/would have been bad.

5.33 Implement the approach, described on page 290, employing the mutually orthogonal directions. This means that you should carry out multidimensional minimization (for multiple cycles) as per Eq. (5.112), where the $\mathbf{u}_k$'s are the columns of the identity matrix; use golden-section search to carry out the one-dimensional minimizations.

[64] In jargon, we are looking for a *globally* convergent method.

If you're feeling all-powerful after tackling our example from Eq. (5.106), try your hand at the *Rosenbrock function*:

$$\phi(x_0, x_1) = (1 - x_0)^2 + 100(x_1 - x_0^2)^2 \qquad (5.155)$$

5.34 [$\mathcal{A}$] Take the n vectors $\mathbf{u}_k$ which are mutually conjugate with respect to the symmetric positive definite matrix $\mathbf{A}$. First, consider a linear combination of them and set it to zero. Then, multiply from the left by $\mathbf{u}_j^T\mathbf{A}$. Use what you have to show that the n vectors $\mathbf{u}_k$ are linearly independent.

5.35 [$\mathcal{A}$] Use induction to show that $\mathbf{u}_j^T\nabla\phi(\mathbf{x}^{(k+1)}) = 0$ where $0 \le k \le n-1$ and $0 \le j \le k$. Hint: express the gradient in terms of $\mathbf{A}$ and $\mathbf{b}$, as in the third step in Eq. (5.122), and relate $\nabla\phi(\mathbf{x}^{(k+1)})$ to $\nabla\phi(\mathbf{x}^{(k)})$.

5.36 We will now improve Powell's algorithm such that it doesn't throw away the conjugate vectors $\mathbf{u}_0, \mathbf{u}_1, \ldots, \mathbf{u}_{n-1}$ it's built up every few cycles.

(a) Bundle together the vectors $\mathbf{u}_0, \mathbf{u}_1, \ldots, \mathbf{u}_{n-1}$ into a matrix $\mathbf{U}$. Show that:

$$\mathbf{U}^T\mathbf{A}\mathbf{U} = \mathbf{D} \qquad (5.156)$$

holds, where $\mathbf{D}$ is a diagonal matrix with positive elements.

(b) Define $\mathbf{V} = \mathbf{U}\mathbf{D}^{-1/2}$ and show that:

$$\mathbf{A}^{-1} = \mathbf{V}\mathbf{V}^T \qquad (5.157)$$

Thus, recalling the equation for the diagonalization of a matrix from Eq. (4.125), for the case of a symmetric $\mathbf{A}$:

$$\mathbf{Q}^T\mathbf{A}\mathbf{Q} = \mathbf{\Lambda} \qquad (5.158)$$

we can just as easily find the orthogonal matrix $\mathbf{Q}$ by diagonalizing $\mathbf{V}\mathbf{V}^T$:

$$\mathbf{Q}^T\left(\mathbf{V}\mathbf{V}^T\right)\mathbf{Q} = \mathbf{\Lambda}^{-1} \qquad (5.159)$$

This $\mathbf{Q}$ can be used to reset $\mathbf{U}$ (instead of using the identity matrix).

(c) As we saw in problem 4.42, the matrix $\mathbf{V}\mathbf{V}^T$ is much more ill-conditioned than the matrix $\mathbf{V}$. Explain the following steps (due to Brent) that show how to produce the desired orthogonal matrix $\mathbf{Q}$ without explicitly forming the product $\mathbf{V}\mathbf{V}^T$:

$$\mathbf{\Lambda}^{-1} = \mathbf{Q}^T\mathbf{V}\mathbf{V}^T\mathbf{Q} = \mathbf{Q}^T\mathbf{V}\mathbf{R}\mathbf{R}^T\mathbf{V}^T\mathbf{Q} = \left(\mathbf{Q}^T\mathbf{V}\mathbf{R}\right)\left(\mathbf{Q}^T\mathbf{V}\mathbf{R}\right)^T = \mathbf{\Sigma}^2 \qquad (5.160)$$

In words, the SVD of $\mathbf{V}$ immediately gives you the orthogonal matrix $\mathbf{Q}$.

(d) Implement the above approach programmatically. First hiccup: to compute the SVD of $\mathbf{V} = \mathbf{U}\mathbf{D}^{-1/2}$ you need to first find $\mathbf{D}$. Unfortunately, Eq. (5.156) doesn't help, since you don't actually know $\mathbf{A}$: remember that we prefer to express things in terms of ϕ evaluations (so they are also applicable to non-quadratic functions). To make further progress, set $\zeta(\gamma) = \phi(\mathbf{x} + \gamma\mathbf{u})$ and show the following:

$$\mathbf{u}^T\mathbf{A}\mathbf{u} = \frac{\zeta(+\gamma) + \zeta(-\gamma) - 2\zeta(0)}{\gamma^2} \qquad (5.161)$$

Derive this under the assumption that ϕ is quadratic, as per Eq. (5.116), but know that it can be applied more generally (by picking a specific value, e.g., $\gamma = 0.5$).[65] Second hiccup: if one or more of the elements of $\mathbf{D}$ are zero, this would wreak havoc when computing $\mathbf{V} = \mathbf{U}\mathbf{D}^{-1/2}$. Explicitly account for this possibility in your code by setting such $\mathbf{D}$ elements to a tiny value.

5.37 We now modify Powell's method so that it doesn't discard the old $\mathbf{u}_0$ in step (b), as we did in the main text. Instead, keep track of how large a decrease in the function ϕ's value each of the $\mathbf{u}_k$'s led to and get rid of the direction that produced the largest decrease. (The *largest*, not the smallest: the $\mathbf{u}_k$ that led to the largest decrease, i.e., the best so far, is likely to be similar to the new conjugate vector being produced, so it is smart from a linear-dependency perspective not to keep both around.) Apply your modifed Powell's method to the Rosenbrock function, Eq. (5.155).

5.38 This and the following two problems provide an elementary introduction to the topic of *constrained optimization*. The simplest non-trivial setting is that of minimizing a scalar function ϕ of n variables, subject to these variables satisfying a single equality constraint; thus, instead of Eq. (5.11) we are faced with:

$$\min \phi(\mathbf{x}), \quad \text{subject to} \quad g(\mathbf{x}) = 0 \tag{5.162}$$

where g is another (scalar) function. For concreteness, we will tackle the scalar function ϕ of Eq. (5.106), subject to the constraint $g(x_0, x_1) = 5x_0^2 - 3x_1^3 - 7 = 0$, but the formalism will be much more general.

(a) Visualize the level sets (i.e., make a contour plot) of ϕ. Then, overlay the equality constraint $g = 0$. You know what you'll find: the global minimum (at the bottom of the plot) is inaccessible, so the constrained minimum will be the point with the smallest-possible ϕ value that also lies on the curve corresponding to $g = 0$ (it has to, since we are dealing with an equality constraint).

(b) Geometrically argue why the constrained minimum satisfies:

$$\nabla\phi(\mathbf{x}^*) = \lambda^*\nabla g(\mathbf{x}^*) \tag{5.163}$$

where λ^* is a constant of proportionality. This equation, together with $g(\mathbf{x}^*) = 0$ (the equality constraint), generalize Eq. (5.102) for the case of constrained minimization. Explain why the above steps motivate the introduction of a *Lagrangian function* for this problem:

$$\mathcal{L}(\mathbf{x}, \lambda) = \phi(\mathbf{x}) - \lambda g(\mathbf{x}) \tag{5.164}$$

and discuss how this leads to $n + 1$ simultaneous non-linear equations. This λ is known as a *Lagrange multiplier*.

(c) Solve these $2 + 1 = 3$ simultaneous non-linear equations for our example. You have no guarantees that you are dealing with a minimum (or a maximum) in this three-variable space; luckily, the multidimensional Newton's method doesn't care. Try out a few initial guesses, to make sure you are locating the "lower" constrained minimum of ϕ.

[65] Observe that this is nothing other than the central-difference approximation to the second derivative, Eq. (3.32).

Hint (in the form of a generalization): later in life, you might wish to refer back to the general case where you have m distinct equality constraints; these can be bundled up into $\mathbf{g}(\mathbf{x})$ to produce the Lagrangian function:

$$\mathcal{L}(\mathbf{x}, \boldsymbol{\lambda}) = \phi(\mathbf{x}) - \mathbf{g}^T(\mathbf{x})\boldsymbol{\lambda} \tag{5.165}$$

The constrained minimum can be found by solving the simultaneous $n+m$ non-linear equations:

$$\nabla \mathcal{L}(\mathbf{x}, \boldsymbol{\lambda}) = \begin{pmatrix} \nabla_{\mathbf{x}} \mathcal{L}(\mathbf{x}, \boldsymbol{\lambda}) \\ \nabla_{\boldsymbol{\lambda}} \mathcal{L}(\mathbf{x}, \boldsymbol{\lambda}) \end{pmatrix} = \begin{pmatrix} \nabla \phi(\mathbf{x}) - \mathbf{J}_{\mathbf{g}}^T(\mathbf{x})\boldsymbol{\lambda} \\ \mathbf{g}(\mathbf{x}) \end{pmatrix} = \mathbf{0} \tag{5.166}$$

where the penultimate step involves the Jacobian.

5.39 We will now address the same setting as in problem 5.38; instead of employing the Lagrangian function of Eq. (5.164), we now introduce a *penalty function*:

$$\psi_\mu(\mathbf{x}) = \phi(\mathbf{x}) + \frac{\mu}{2} g^2(\mathbf{x}) \tag{5.167}$$

For a fixed value of μ, this is a weighted sum of our two objectives (minimize ϕ while satisfying $g = 0$); minimizing $\psi_\mu(\mathbf{x})$ with respect to $\mathbf{x}$ is a problem of *unconstrained* minimization. If μ is small, then the equality constraint is not respected all that much; crucially, we intend to take μ to infinity. For concreteness, we will tackle the scalar function ϕ of Eq. (5.106), subject to the constraint $g(x_0, x_1) = 5x_0^2 - 3x_1^3 - 7 = 0$.

(a) Visualize the level sets (i.e., make a contour plot) of $\psi_\mu(\mathbf{x})$ from Eq. (5.167) for $\mu = 0.5, 5, 50, 500$. Also overlay the equality constraint $g = 0$. Interpret the pattern you see as the value of μ is increased by separately visualizing the level sets of g^2.

(b) Use Powell's method to tackle the unconstrained-minimization problem for $\psi_\mu(\mathbf{x})$ when $\mu = 0.5, 5, 50, 500$. Take the initial guess to be the vector $(1\ 0.25)^T$; do yourself a favor and use the output from Powell's method at a given μ as the input guess vector at the next μ value. What do you find? In general, this approach leads to an increasingly ill-conditioned problem as the value of μ is increased; as a result, it becomes increasingly harder for Powell's method to find the minimum. (This is consistent with what you discovered in the previous part of this problem.) You can try other initial guesses (e.g., $(2\ 0.25)^T$), but even then you will not find a smooth progression.

5.40 We will address the same setting as in problems 5.38 and 5.39; we now introduce a combination of the Lagrangian function from Eq. (5.164) and of the penalty function from Eq. (5.167), known as an *augmented Lagrangian function*:

$$\mathcal{L}_\mu(\mathbf{x}, \lambda) = \phi(\mathbf{x}) - \lambda g(\mathbf{x}) + \frac{\mu}{2} g^2(\mathbf{x}) \tag{5.168}$$

Crucially, we will now set up a scheme in which, while we will tune μ, we do not intend to take the limit of μ going to infinity, i.e., we aim to satisfy the constraints rather well even for moderate μ values. For fixed values of λ and μ, in Eq. (5.168) we are faced with an *unconstrained* minimization problem in $\mathbf{x}$, similarly to Eq. (5.167).

(a) For given $\mu^{(k-1)}$ and $\lambda^{(k-1)}$ iterates, take the gradient of the augmented Lagrangian function $\mathcal{L}_{\mu^{(k-1)}}(\mathbf{x}, \lambda^{(k-1)})$ with respect to $\mathbf{x}$ and set it to zero. Compare with the constrained-minimum condition of Eq. (5.163) and thereby motivate the new Lagrange multiplier value:

$$\lambda^{(k)} = \lambda^{(k-1)} - \mu^{(k-1)} g(\mathbf{x}^{(k-1)}) \tag{5.169}$$

(b) We, again, turn to the scalar function ϕ of Eq. (5.106), subject to the constraint $g(x_0, x_1) = 5x_0^2 - 3x_1^3 - 7 = 0$. Implement an iterative scheme that starts with $\lambda^{(0)} = 1$, $\mu^{(0)} = 0.1$, and $\mathbf{x}^{(0)} = (2\ 0.25)^T$, using Powell's method to minimize Eq. (5.168) with respect to $\mathbf{x}$. Then, update the Lagrange multiplier as per Eq. (5.169), take $\mu^{(k)} = 5\mu^{(k-1)}$, and repeat the process. As usual, you should terminate when the $\mathbf{x}^{(k)}$ iterate stops changing. You might find it helpful to use a nested function here (as discussed in problem 1.4).

5.41 [$\mathcal{P}$] We will employ the same discretization of the action as in `action.py`, still for the case of the quartic oscillator. Imagine you made a mistake when trying to approach the fixed endpoints in your initial guess, leading you to start with $q^{(0)}(t) = 0.9375t^2 + 1.0625t - 2$. Plot the initial guess and converged solution together. Are you getting the same converged solution as in the main text? Why (or why not)? Compute the value of the action from Eq. (5.134) as well as the eigenvalues of the Jacobian for the two converged solutions and use them to justify your conclusions.

5.42 [$\mathcal{P}$] Modify `action.py` to address the *harmonic* oscillator. Take $x_0 = 0$, $x_{n-1} = 1$ and $\tau = 1$. Plot the coordinate of the particle as a function of time; also show the (analytically known) answer. Incidentally, what happens if you take τ to be larger? Then, try to solve the same problem using the gradient-descent method. You will have to tune `n`, `gamma`, `kmax`, and `tol` (as well as your patience levels) to do so.

5.43 [$\mathcal{P}$] Modify `action.py` to address the physical context of a vertically thrown ball; start by thinking what force this implies. Take $x_0 = x_{n-1} = 0$ and $\tau = 10$. Plot your solution for the height of the particle as a function of time.

5.44 [$\mathcal{P}$] Repeat the study of the quartic oscillator, but this time instead of employing Eq. (5.134) for the discretization, use a central-difference formula; you should analytically calculate the new gradient and Hessian, before turning to Python.

5.45 [$\mathcal{P}$] In problem 1.18 on the Kronig–Penney model, you were asked to plot E vs K by computing the right-hand side and then inverting the cosine.[66] You should now repeat the whole study from a root-finding perspective: pick a value of K and treat Eq. (1.7) as $f(E) = 0$; the zeros of this function will give you the energy solutions. Recall that you should be employing `cmath` functionality; as in problem 1.18, take $L = 1, M = 0.4, \hbar^2/m = 1, V_0 = 35$ and separately study several values of K from -3 to $+3$. Use the secant method to solve for the zeros of the above function, giving you a few E's per K value; plot E vs K.

5.46 [$\mathcal{P}$] We return to methane via the van der Waals equation of state, Eq. (5.2); take $T = 170$ K. First, plot the isotherm (i.e., P vs v). Then, for $P = 30$ atm find the three possible values of v. Finally, find the minimum P and the corresponding v.

[66] In problem 4.56 you approached the same setting via matrix techniques.

5.47 [$\mathcal{P}$] We turn to a one-dimensional electron–phonon system (mocking up a metal): the *Peierls instability* describes the situation where (typically for the half-filled case) a distortion/deformation of the lattice creates a band gap, lowering the energy of electrons near the Fermi surface. The total energy (per unit length) is:

$$E_L = \frac{a}{b^2}\Delta^2 + \frac{2E_F k_F}{\pi}\left[1 - \sqrt{1 + \frac{\Delta^2}{4E_F^2}} - \frac{\Delta^2}{4E_F^2}\operatorname{arcsinh}\left(\frac{2E_F}{\Delta}\right)\right] \tag{5.170}$$

namely the sum of the elastic energy and the electronic energy. Here Δ is (one-half of) the gap, E_F is the Fermi energy, and k_F is the Fermi wave number (a and b relate to the amplitude of the distortion and its effect on the electrons). Textbooks typically assume that the gap is small ($\Delta \ll 2E_F$) and simplify the above expression. Instead, you should take $E_F = 1/2$, $a = k_F = 1$, $b = 2$, and find the Δ that minimizes E_L.

5.48 [$\mathcal{P}$] We will now tackle the standard textbook problem mentioned in section 5.1.1, namely a single particle in a (one-dimensional) finite square well. For concreteness, take $V_0 = 20$, $a = 1$, and $\hbar^2/m = 1$. You will need to find all the (even) bound-state solutions supported by this well.

(a) Use the multidimensional Newton's method to explicitly solve the two coupled equations shown in Eq. (5.3) and Eq. (5.4). Once you've converged, you can find the energy via:

$$k = \sqrt{\frac{2mE}{\hbar^2}}, \qquad \kappa = \sqrt{\frac{2m(V_0 - E)}{\hbar^2}} \tag{5.171}$$

While you need only one of these to compute E, use both as a consistency check. Hint: given our setup, a bound-state energy may range from 0 to V_0.

(b) In the previous part, you probably had a hard time finding decent initial guess vectors. You can make your life easier by solving for κ in Eq. (5.4) and plugging the result into Eq. (5.3). You are now faced with only a single non-linear equation: in order to avoid convergence issues (as in the previous part of this problem), use the bisection method to tackle this equation. Hint: you will need to be a bit careful with the right endpoint of the bracket for the case of the most weakly bound state, in order to avoid issues with the argument of the square root.

5.49 [$\mathcal{P}$] In this problem we solve Eq. (5.5) together with Eq. (5.6), using Newton's method. For simplicity, work in units where $m = M_R = G = 1$ and $M_S = 2$. Assume we have made a measurement at $\mathbf{r}_0 = (2\ 1\ 3)^T$, finding:

$$\mathbf{F}_0 = (-0.016\,859\,96 \quad -0.040\,029\,72 \quad -0.014\,790\,52)^T \tag{5.172}$$

and another measurement at $\mathbf{r}_1 = (4\ 4\ 1)^T$, this time finding

$$\mathbf{F}_1 = (-0.143\,642\,92 \quad 0.120\,962\,46 \quad -0.264\,605\,37)^T \tag{5.173}$$

Set up the problem in a way that allows you to call `multi_newton.py`, with the understanding that the components of **R** are the first three elements of `xs` and the components of **S** are the next three elements of `xs`. If you're having trouble converging, try starting from `xolds = np.array([0.8,-1.1,3.1,3.5,4.5,-2.])`.

5.50 [$\mathcal{P}$] We now return to the Roche potential, Φ, of Eq. (1.6), which is of relevance to binary stars; we will still limit ourselves to the orbital plane ($z = 0$). If you did not solve problem 1.17, you should do so now. Our goal will be to determine the critical points of Φ (known as the *Lagrange points*); these are physically significant. For example, the L_1 point (located between m_1 and m_0) shows the easiest path for a star that has exceeded its Roche lobe to spill matter onto its companion star. Similarly, L_2 (located to the right of m_0) shows the easiest path for matter to escape from the entire binary system, and so on.
Use multidimensional extremization via Newton's method, as per Eq. (5.111), to determine the Lagrange points. You should analytically produce the gradient vector and employ finite differences for the Hessian, i.e., the Jacobian of the gradient. Employ appropriately placed initial guesses to find *five* critical points, for the case of $q = 0.4$. Overlay them on your earlier contour plot, using different symbols for extrema and for saddle points.

5.51 [$\mathcal{P}$] We now go over a physical application of derivative-free multidimensional minimization. Specifically, we will explore the geometric structure of *sodium chloride clusters* resulting from (a simplified version of) the *Tosi–Fumi potential*. For n_p ions experiencing pairwise (Coulomb plus exponential) interactions, the total potential energy is [121]:

$$E = \sum_{\substack{j,k=0 \\ j<k}}^{n_p-1} V(r_{jk}), \qquad V(r_{jk}) = a\frac{14.4}{r_{jk}} + b\,\exp(-r_{jk}/0.317) \tag{5.174}$$

where energy is measured in eV and length in Å; we are implicitly working in three spatial dimensions. Here a is $+1$ for like charges ($++$ or $--$) and -1 for opposite charges ($+-$ or $-+$). The b parameter takes on the value 423.63 for $++$, 1254.96 for $+-$ or $-+$, and 3485.35 for $--$.
You should employ Powell's method to find the energies of the lowest minima (you can find) for the clusters $(NaCl)Cl^-$ and $(NaCl)_2Cl^-$, i.e., for the cases of $n_p = 3$ and $n_p = 5$, respectively. Since Powell's method works with a (one-dimensional) vector/array, you should "unroll" the $3n_p$ components as follows:[67]

$$\mathbf{x} = \Big((\mathbf{r}_0)_x \quad (\mathbf{r}_0)_y \quad (\mathbf{r}_0)_z \quad (\mathbf{r}_1)_x \quad \dots \quad (\mathbf{r}_{n_p-1})_y \quad (\mathbf{r}_{n_p-1})_z \Big)^T \tag{5.175}$$

Crucially, your code should also follow some convention on how to order the positions of the particles according to charge (e.g., first the positively charged ions, then the negatively charged ones). You should carry out several runs for each of the two clusters, modifying the input-parameter values of `powell()` to find the lowest-possible minimum each time. Visualize the geometric structure of the clusters you find, using different symbols for positively and negatively charged ions. Hint: while your visualization machinery should be three-dimensional, in both cases the lowest-energy configuration will be a straight line.

[67] We will re-encounter this convention in Eq. (7.223), while studying many-particle quantum mechanics.

Approximation 6

The whole world believes in [the normal law of errors], the experimentalists imagining that it is a mathematical theorem and the mathematicians that it is an experimental fact.

Gabriel Lippmann

6.1 Motivation

6.1.1 Examples from Physics

Both in theoretical and in experimental physics we are often faced with a table of data points; we typically wish to manipulate physical quantities even for cases that lie "in between" the table rows. Here are a few examples:

1. **Velocity from unequally spaced points**
   When introducing numerical differentiation in section 3.1.1, our first motivational example came from classical mechanics. This, the most trivial example of differentiation in physics, takes the following form in one dimension:

$$v = \frac{dx}{dt} \tag{6.1}$$

   where v is the velocity and x is the position of the particle in a given reference frame. In chapter 3 we spent quite a bit of time discussing how to compute finite differences, and how to estimate the error budget.
   The question then arises how we should handle the case where we know the positions at *unequally spaced points*. To clarify what we mean by that, imagine we only have access to a set of n discrete data points (i.e., a table) of the form (t_j, x_j) for $j = 0, 1, \ldots, n-1$; the positions are known only at fixed times t_j not of our choosing, which are *not* equally spaced. One could simply revert to the general finite-difference formulas (i.e., those not on a grid); since you don't control the placement of the points, you'd likely have to use a forward or backward formula. However, simply using a non-central difference formula where the h is determined by whatever happened to be the distance between two t_j's is a recipe for disaster: what if there are many data points for early and late times, but few in between? Should we resign ourselves to not being able to extract the velocity at intermediate times?
   An alternative approach is to take the table of unequally spaced input data points, produce an *approximating function* $p(t)$ which captures the behavior of the data effectively,

and then evaluate the derivative of the approximating function *quasi-analytically* (see problem 6.42). As an aside, an analogous example arises when dealing with noisy data: one could first carry out, say, *Fourier smoothing* and then get a nicer looking derivative (see problem 6.31). This brings us to the next example, which also relates to Fourier analysis.

2. **Differentiating a wave function**

When taking an introductory quantum mechanics course, you may have wondered why we sometimes work in coordinate space and sometimes in momentum space. One of the reasons is that operations that may be difficult in one space are easy in another. A fascinating example of this is known as *Fourier differentiation* or *spectral differentiation*. To see what we mean, consider the following wave function expressed in terms of the inverse Fourier transform:

$$\Psi(x,t) = \frac{1}{2\pi}\int_{-\infty}^{\infty} dk\, \tilde{\Psi}(k,t)e^{ikx} \tag{6.2}$$

Taking the spatial derivative of $\Psi(x,t)$ becomes a simple arithmetic operation (multiplication by a scalar) in momentum space:

$$\frac{\partial}{\partial x}\Psi(x,t) = \frac{1}{2\pi}\int_{-\infty}^{\infty} dk\, \tilde{\Psi}(k,t)\frac{\partial}{\partial x}e^{ikx} = \frac{1}{2\pi}\int_{-\infty}^{\infty} dk\, \tilde{\Psi}(k,t)ike^{ikx} \tag{6.3}$$

In the first step we emphasized that only the complex exponential is position dependent and in the second step we took the derivative. As advertised, we replaced the derivative on the left-hand side with a simple multiplication on the right-hand side. We can take an extra step, expressing the $\tilde{\Psi}(k,t)$ itself in terms of a (direct) Fourier transform:

$$\frac{\partial}{\partial x}\Psi(x,t) = \frac{1}{2\pi}\int_{-\infty}^{\infty}\int_{-\infty}^{\infty} dk\, dy\, \Psi(y,t)ike^{ik(x-y)} \tag{6.4}$$

Take a moment to appreciate that this equation involves the coordinate-space wave function on both the left- and the right-hand sides.

Of course, these manipulations are here carried out for the continuous case, which involves integrals on the right-hand side; you may be wondering why we would bother with Eq. (6.4), since it's replaced one derivative by two integrals. In the present chapter we will learn about a related tool called the *discrete Fourier transform* (DFT), which involves sums. We will then see that it is possible to re-express the DFT so that it is carried out incredibly fast. Problem 6.30 asks you to develop and implement an analogous differentiation formula for the DFT. At that point you will appreciate the power of being able to produce a table of derivative values without having to worry about finite-difference errors. Even more crucially, in the n-dimensional case evaluating the gradient vector at one point via the forward difference requires n new function evaluations, see Eq. (5.108), but Fourier differentiation gives you the derivatives without *any* new function evaluations being necessary.

3. **Hubble's law**

According to observations, the universe displays remarkable large-scale uniformity. More specifically, at large scales the universe is *homogeneous*: this applies not only to the average density, but also to the types of galaxies, their chemical composition, and other features. Furthermore, the universe appears to be *isotropic* about every point, implying that there is no special direction. Another aspect of the uniformity of the universe is that even its expansion seems to be uniform: on average, galaxies are receding from us with a speed that is proportional to how far away from us they are. Of course, this doesn't apply only to us: any galaxy will see all other galaxies receding at a speed proportional to their distance from it.

The point we just made is significant enough to bear repeating: even though the universe is expanding, the homogeneity and isotropy are maintained, i.e., the universe does not exhibit a velocity anisotropy. This fact was first noticed by Edwin Hubble in 1929, in his study of 24 extra-galactic nebulae for which both distances and velocities were available. Plotting the velocities versus the distances, Hubble noticed a "roughly linear relation" and extracted the coefficient. Nowadays, we call this *Hubble's law*:

$$v = Hd \tag{6.5}$$

where v is the recessional velocity, d is the distance, and H is known as *Hubble's parameter*. The present value of this parameter is known as *Hubble's constant*, and has been measured to be $H_0 = (2.3 \pm 0.1) \times 10^{-18}\ \text{s}^{-1}$. Hubble's parameter has wider significance: if we denote by $R(t)$ the cosmological expansion parameter, then the following relation holds:

$$H = \frac{\dot{R}(t)}{R(t)} \tag{6.6}$$

i.e., at time t Hubble's parameter is the relative rate of expansion of the universe.

In modern terms, Hubble carried out a straight-line fit to the data. We encountered another instance of experimental data following a linear trend earlier on, when we discussed Millikan's 1916 experiment on the photoelectric effect (section 2.1—see also problem 6.65). In the Project at the end of the present chapter we will discuss yet another set of experimental data, this time *not* obeying a linear trend.

It is no coincidence that the first two items on our list both involve taking derivatives. The problem of *approximation*, the theme of this chapter, typically arises when one wishes to carry out operations that are otherwise hard or impossible.

6.1.2 The Problems to Be Solved

Basically, the goal of the present chapter is to learn how to approximate a function $f(x)$. There could be several reasons why this needs to happen: perhaps computing $f(x)$ is a costly procedure, which cannot be repeated at many points, or maybe $f(x)$ can only be experimentally measured. In any case, the point of the present chapter is that we wish to have access to $f(x)$ for general x but can only produce selected $f(x_j)$ values. What is commonly done is that a set of n *basis functions* $\phi_k(x)$ is chosen which, combined with a set of n undetermined parameters c_k $(k = 0, 1, \ldots, n-1)$, is used to produce the following *linear form*:

$$p(x) = \sum_{k=0}^{n-1} c_k \phi_k(x) \tag{6.7}$$

This is called a linear form because it constitutes a linear combination of basis functions; this does *not* mean that the basis functions themselves are linear. We assume that the $\phi_k(x)$'s are *linearly independent*. Note that we used $p(x)$ to denote our *approximating function*, since we don't know that it is the same as the underlying $f(x)$. Crucially, both $p(x)$ and the basis $\phi_k(x)$ apply to any x, not just points on a grid. This raises the question of how we would know that our approximating function, $p(x)$, is accurate, leading us to distinguish between the following two large classes of approach:[1]

1. **Interpolation**
Interpolation arises when we have as input a table of data points, (x_j, y_j) for $j = 0, 1, \ldots, n-1$, which we assume exactly represent the underlying $f(x)$.[2] We'll take the basis functions $\phi_k(x)$ as given and attempt to determine the n parameters c_k. Crucially, we have n unknowns and n data points, so we can determine all the unknown parameters by demanding that our approximating function $p(x)$ go *exactly* through the input data points (except for roundoff error), i.e.:

$$y_j = \sum_{k=0}^{n-1} c_k \phi_k(x_j) \tag{6.8}$$

where we used the fact that $p(x_j) = y_j$. This can be written in matrix form as follows:

$$\begin{pmatrix} \phi_0(x_0) & \phi_1(x_0) & \phi_2(x_0) & \cdots & \phi_{n-1}(x_0) \\ \phi_0(x_1) & \phi_1(x_1) & \phi_2(x_1) & \cdots & \phi_{n-1}(x_1) \\ \phi_0(x_2) & \phi_1(x_2) & \phi_2(x_2) & \cdots & \phi_{n-1}(x_2) \\ \vdots & \vdots & \vdots & \ddots & \vdots \\ \phi_0(x_{n-1}) & \phi_1(x_{n-1}) & \phi_2(x_{n-1}) & \cdots & \phi_{n-1}(x_{n-1}) \end{pmatrix} \begin{pmatrix} c_0 \\ c_1 \\ c_2 \\ \vdots \\ c_{n-1} \end{pmatrix} = \begin{pmatrix} y_0 \\ y_1 \\ y_2 \\ \vdots \\ y_{n-1} \end{pmatrix} \tag{6.9}$$

or, more compactly, $\mathbf{\Phi c} = \mathbf{y}$, where $\mathbf{\Phi}$ is an $n \times n$ matrix and we're solving for the $n \times 1$ column vector $\mathbf{c}$.[3] Thus, for a given choice of the $\phi_k(x)$, with the table (x_j, y_j) known, this is "simply" a linear system of equations for the c_k's; it can be solved using, say, Gaussian elimination after $O(n^3)$ operations.
An example of interpolation along these lines is shown in Fig. 6.1, where the solid curve smoothly connects the data points. Significantly, such an *interpolant* is not unique: even though every interpolant *has* to go through the data points, there are many shapes/forms

[1] A distinct strategy is to minimize the maximum error (*minimax approximation*). While this elegant approach has affinities with many of the tools we introduce, it is not as widely used so we do not discuss it further.
[2] Interpolation has been called an "exact approximation", but this is infelicitous (if not a contradiction in terms).
[3] We've built up $\mathbf{\Phi}$ taking $\Phi_{jk} = \phi_k(x_j)$, in that order. We'll do something analogous again below, in Eq. (6.140).

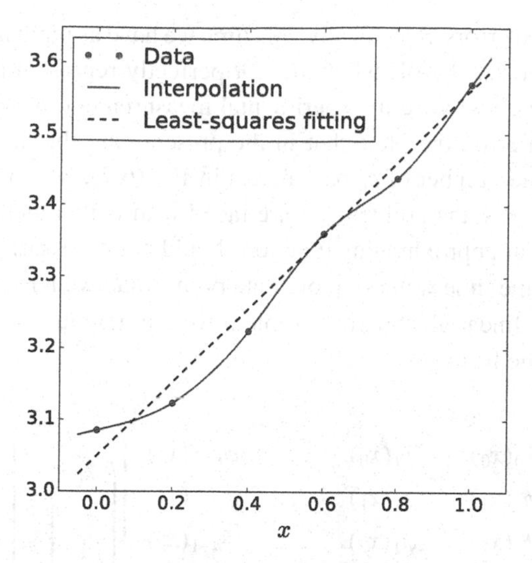

Illustration of the difference between interpolation and least-squares fitting

Fig. 6.1

it could have taken between them.[4] As a result, one needs to impose further criteria in order to choose an interpolant, such as smoothness, monotonicity, etc. The crudest way to put it is that the interpolant should behave "reasonably"; for example, wild fluctuations that are not supported by the data should generally be avoided.

In the present chapter, we'll examine three different approaches to the interpolants:

(a) **Polynomial interpolation**: this approach, covered in section 6.2, assumes that a single polynomial can efficiently and effectively capture the behavior of the underlying function. Despite popular myths to the contrary, this is a great assumption.
(b) **Piecewise polynomial interpolation**: here, as we'll see in section 6.3, one breaks up the data points into *subintervals* and employs a different low-degree polynomial in each subinterval.
(c) **Trigonometric interpolation**: in section 6.4 we'll find out how to carry out interpolation for the case of periodic data. This will also allow us to introduce one of the most famous algorithms in existence, the *fast Fourier transform*.

Regardless of how one picks the basis functions, having access to an interpolation scheme usually helps one before carrying out further manipulations; for example, one might wish to differentiate or integrate $p(x)$. Assuming the interpolation was successful, such tasks can be carried out much faster and with fewer headaches than before.

2. **Least-squares fitting**

We now turn to our next large class of approximation. There are two major differences from the case of interpolation: (a) we have more data than parameters, and (b) our input

[4] Actually, in Fig. 6.1 we are also showing our interpolant for x values slightly outside the interval where we have data; generally, this is a trickier problem, known as *extrapolation*. In cases where you have a theory about how the data should behave even in regions where you have no measurements, this is not too hard a problem; this is precisely the case of Richardson extrapolation, discussed in sections 3.3.7 and 7.4.1.

data have associated errors. Specifically, this time we have as input a table of data points, (x_j, y_j) for $j = 0, 1, \dots, N-1$, which do *not* perfectly represent the underlying $f(x)$; for example, these could arise in experimental measurement, in which case each y_j is associated with an error, σ_j. Note that in the present case the number of data points, N, is larger than the number of c_k parameters in Eq. (6.7), which was n, i.e., we have $N > n$. In other words, the problem we are faced with is that the input data cannot be fully trusted, i.e., our approximating function should not necessarily go exactly through them, but at the same time we have more data points than we have parameters.
In the language of linear algebra, the problem we are now faced with is an *overdetermined system*, of the form:

$$\begin{pmatrix} \phi_0(x_0) & \phi_1(x_0) & \dots & \phi_{n-1}(x_0) \\ \phi_0(x_1) & \phi_1(x_1) & \dots & \phi_{n-1}(x_1) \\ \phi_0(x_2) & \phi_1(x_2) & \dots & \phi_{n-1}(x_2) \\ \vdots & \vdots & \ddots & \vdots \\ \phi_0(x_{N-1}) & \phi_1(x_{N-1}) & \dots & \phi_{n-1}(x_{N-1}) \end{pmatrix} \begin{pmatrix} c_0 \\ c_1 \\ \vdots \\ c_{n-1} \end{pmatrix} \approx \begin{pmatrix} y_0 \\ y_1 \\ y_2 \\ \vdots \\ y_{N-1} \end{pmatrix} \tag{6.10}$$

Here our input y-values are grouped together in an $N \times 1$ column vector $\mathbf{y}$, $\mathbf{\Phi}$ is an $N \times n$ matrix, and we're solving for the $n \times 1$ column vector $\mathbf{c}$. As you can see, we are aiming for approximate equality, since this system cannot be generally solved (and even if it could, we know that our y_j values suffer from errors).
To tackle this overdetermined system, we'll do the next-best thing available; since we can't solve $\mathbf{\Phi c} = \mathbf{y}$, let's try to *minimize the norm of the residual vector*, i.e.,

$$\min \|\mathbf{y} - \mathbf{\Phi c}\| \tag{6.11}$$

where we're implicitly using the Euclidean norm. As we'll see in section 6.5 this approach is known as *least-squares fitting*, since it attempts to determine the unknown vector by minimizing the (sum of the) squares of the difference between the approximating function values and the input data.
An example of least-squares fitting is shown in Fig. 6.1: if we assume that the data points cannot be fully trusted, then it is more appropriate to capture their overall trend. As a result, the goal of our least-squares fitting procedure should not be to go through the points. We'll learn more about how to quantify the effect of the error in the input data later on but, for now, observe that the least-squares fit in Fig. 6.1 follows from the assumption that the approximating function $p(x)$ should be *linear in x*. Obviously, we could have made different assumptions, e.g., that $p(x)$ is quadratic in x, and so on. We'll need a goodness-of-fit measure to help us decide which form to employ.
In section 6.5 we'll first learn about straight-line fits, which involve only two parameters (as in Fig. 6.1); this is the simplest-possible choice of basis functions. We will then

turn to the more general problem of tackling the *normal equations* that arise in the case of polynomial least-squares fitting, employing machinery we developed in our study of linear algebra in chapter 4. It's important to realize that even polynomial least-squares fitting is a case of *linear least-squares fitting*: you can see from Eq. (6.7) that the dependence on the c_k parameters would be linear, even if the basis functions are, e.g., exponentials. In section 6.6 we employ the language of *statistical inference* to re-interpret and generalize our approach to fitting. In section 6.7 and in the Project of section 6.8 you will get a first exposure to the more general problem of *non-linear least-squares fitting*; in general this is a problem of multidimensional minimization, as in section 5.6.1, but one can also rewrite it in the form of a multidimensional root-finding problem as per section 5.4.[5]

6.2 Polynomial Interpolation

We now turn to our first major theme, namely *polynomial interpolation*; this refers to the case where the interpolating function $p(x)$ of Eq. (6.7) is a polynomial. Keep in mind that the underlying function $f(x)$ that you're trying to approximate does *not* have to be a polynomial, it can be anything whatsoever. That being said, we will study cases that are well-behaved, i.e., we won't delve too deeply into singularities, discontinuities, and so on; in other words, we focus on the behavior that appears most often in practice. As explained in the preceding overview, we start by using a single polynomial to describe an entire interval; in addition to its intrinsic significance, this is a topic that will also help us in chapter 7 when we study numerical integration.

At the outset, let us point out that polynomial interpolation is a subject that is misleadingly discussed in a large number of texts, including technical volumes on numerical analysis.[6] The reasons why this came to be are beyond the scope of our text; see N. Trefethen's book on approximation theory [147] for some insights on this and several related subjects. Instead, let us start with the final conclusion: *using a single polynomial to approximate a complicated function is an excellent approach, if you pick the interpolating points x_j appropriately*. This is not just an abstract question: knowing how easy, efficient, and accurate polynomial interpolation can be will benefit you in very practical ways in the future. Issues with single-polynomial interpolation arise when one employs equidistant points, an example which is all most textbooks touch upon.[7] We'll here follow a pedagogical approach, i.e., we will start with what works and only mention potential problems near the end of our discussion.

To be specific, let us immediately introduce our go-to solution for the placement of the

[5] This also helps explain why our discussion of approximating functions was not provided in chapter 2; our exposition will involve the linear algebra and minimization machinery that we developed in chapters 4 and 5.

[6] *Amicus Plato, sed magis amica veritas*, i.e., Plato is a friend, but truth is a better friend.

[7] The standard treatment, which employs equidistant points and gets into trouble, doesn't mesh well with the *Weierstrass approximation theorem*, which states that a polynomial *can* approximate a continuous function; the problem is that the tool used to provide a constructive proof of this theorem (*Bernstein polynomials*) converges very slowly, so it is not very useful from a practical perspective.

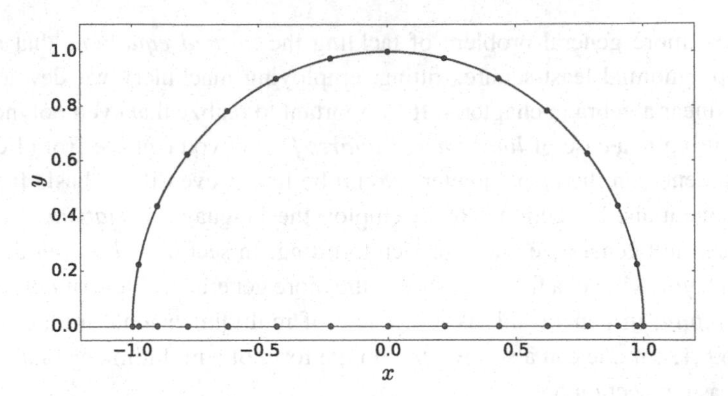

Fig. 6.2 Chebyshev nodes for $n = 15$, also showing the equidistant points on the unit circle

interpolation nodes. In much of what follows, we employ *Chebyshev points*, sometimes known as *Chebyshev nodes*. As you learned in problem 2.22, the extrema of *Chebyshev polynomials* take the simple form:

$$x_j = -\cos\left(\frac{j\pi}{n-1}\right), \qquad j = 0, 1, \ldots, n-1 \tag{6.12}$$

where we have included a minus sign in order to count them from left to right.[8] Given the importance of these points, we will mainly focus on the interval $[-1, 1]$; if you are faced with the interval $[a, b]$ you can scale appropriately:

$$t = \frac{b+a}{2} + \frac{b-a}{2}x \tag{6.13}$$

Spend some time to understand how this transformation takes us from x which is in the interval $[-1, 1]$ to t which is in the interval $[a, b]$: $x = -1$ corresponds to $t = a$, $x = 1$ corresponds to $t = b$ and, similarly, the midpoint $x = 0$ corresponds to $t = (b + a)/2$. Note that Eq. (6.13) is more general than Eq. (6.12), i.e., you can use it even if you are not employing Chebyshev points.

In order to help you build some intuition on the Chebyshev points, we are showing them in Fig. 6.2 on the x axis. Also shown are the equidistant points on the (upper half of the) unit circle; their projections onto the x axis are the Chebyshev points. The most conspicuous feature of Chebyshev points is that they cluster near -1 and 1 (i.e., near the ends of the interval). This will turn out to be crucial for the purposes of interpolation, as discussed in section 6.2.2.4. Despite not being equidistant on the x axis, Chebyshev points have the advantage of obeying *nesting*: when you double the number of points, n, you use, you can re-use the old points since you're simply adding new points between the old ones. This is

[8] Though the convention in the literature is not to include this minus sign.

generally desirable when you don't know ahead of time how many points you are going to need: start with some guess, keep doubling (while re-using the work you carried out for the old guesses), until you see that you can stop doubling (see, e.g., problem 7.24).

The attentive reader may recall encountering another set of points with a similar clustering property, namely the roots of *Legendre polynomials*, which we computed all the way back in section 5.3.2. For now, we will consider the placement of the nodes as settled, i.e., following Chebyshev points as per Eq. (6.12). This still leaves open the question of how to solve Eq. (6.9): we need to pick a specific set of basis functions $\phi_k(x)$ and then to compute the c_k parameters.

In order to avoid getting confused, one has to distinguish between the (possible) *ill-conditioning* of a *problem* and the *stability* of a given *algorithm* employed to solve that problem. For example, in section 4.2.3 we discussed how to quantify the conditioning of a linear system of equations; this had little to do with specific methods one could choose to apply to that problem. If a matrix is terribly ill-conditioned it will be so no matter which algorithm you use on it. Similarly, in section 4.3.4 we encountered a case (Gaussian elimination without pivoting) where the algorithm can be unstable even when the problem is perfectly well-conditioned. Turning to the present case of polynomial interpolation, one has to be careful in distinguishing between the conditioning of the problem itself (interpolating using a table (x_j, y_j)) and the method you employ to do that (which needs to evaluate the c_k parameters). We'll soon see a specific method that can be quite unstable, but this doesn't mean that the problem itself is pathological.

Another distinction one can make at a big-picture level relates to the practical implementation of polynomial interpolation. While we will always follow Eq. (6.7), we can distinguish between two stages in our process: (a) *constructing* the interpolating function is the first stage; after deciding on which basis functions $\phi_k(x)$ to employ, the main task here is to compute the c_k parameters, and (b) *evaluating* the interpolating function $p(x)$ at a given x, with the $\phi_k(x)$'s and c_k's in place. It stands to reason that the first stage, computing the c_k parameters, should only be carried out once, i.e., should be independent of the specific x where we may need to evaluate $p(x)$. As mentioned earlier, interpolation is supposed to make further manipulations easy, so the evaluation stage should be as efficient as possible; this would not be the case if the parameter values would need to be separately re-computed for each new x point.

6.2.1 Monomial Basis

The preceding discussion may be too abstract for your taste, so let us write down some equations. The most natural choice for the basis functions $\phi_k(x)$ is to use monomials:[9]

$$\phi_k(x) = x^k \tag{6.14}$$

It's not too early to highlight the fact that we use j for the spatial index and k for the index that keeps track of the basis functions. Plugging Eq. (6.14) into Eq. (6.7) allows you to write the interpolating polynomial $p(x)$ in the simplest way possible:

[9] In problem 4.27 you started with monomials before orthogonalizing to get to Legendre polynomials.

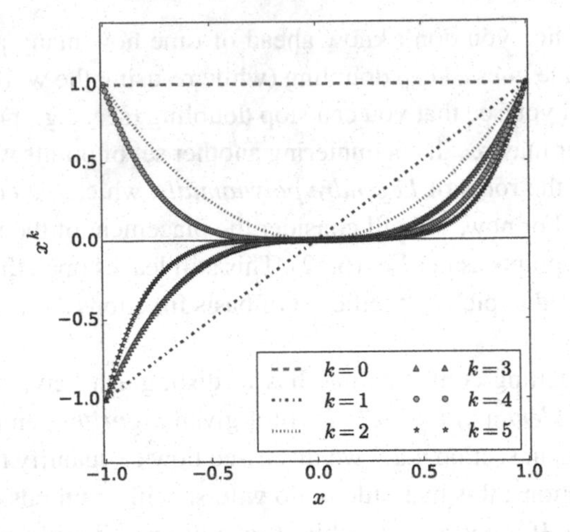

Fig. 6.3 Monomial basis functions for the first several degrees

$$p(x) = c_0 + c_1x + c_2x^2 + c_3x^3 + \cdots + c_{n-1}x^{n-1} \tag{6.15}$$

This employs notation similar to that we've used before, say in Eq. (5.62): this is a polynomial of degree $n-1$, in keeping with our 0-indexing convention. Remember, our input is a table of data points, (x_j, y_j) for $j = 0, 1, \ldots, n-1$; we have n unknown parameters and n input points.

The most naive way to carry out polynomial interpolation, which is the one you probably learned when you first saw how to find the line that goes through two specified points—see Eq. (5.52)—is to evaluate Eq. (6.15) at our grid points:

$$p(x_j) = y_j = c_0 + c_1x_j + c_2x_j^2 + c_3x_j^3 + \cdots + c_{n-1}x_j^{n-1} \tag{6.16}$$

and solve for the unknown parameters. Explicitly writing this out for all the values of j, i.e., $j = 0, 1, \ldots, n-1$, allows us to re-express our equations in matrix form:

$$\begin{pmatrix} 1 & x_0 & x_0^2 & \cdots & x_0^{n-1} \\ 1 & x_1 & x_1^2 & \cdots & x_1^{n-1} \\ 1 & x_2 & x_2^2 & \cdots & x_2^{n-1} \\ \vdots & \vdots & \vdots & \ddots & \vdots \\ 1 & x_{n-1} & x_{n-1}^2 & \cdots & x_{n-1}^{n-1} \end{pmatrix} \begin{pmatrix} c_0 \\ c_1 \\ c_2 \\ \vdots \\ c_{n-1} \end{pmatrix} = \begin{pmatrix} y_0 \\ y_1 \\ y_2 \\ \vdots \\ y_{n-1} \end{pmatrix} \tag{6.17}$$

This (non-symmetric) $n \times n$ coefficient matrix is known as a *Vandermonde matrix*.[10] As

[10] This is not the first Vandermonde matrix we've encountered: we mentioned one on page 101 in the context of finite-difference coefficient computation, while in problem 5.22 we solved a multidimensional system of non-linear equations, with the promise that it will re-emerge in our study of Gaussian quadrature in section 7.5.1.

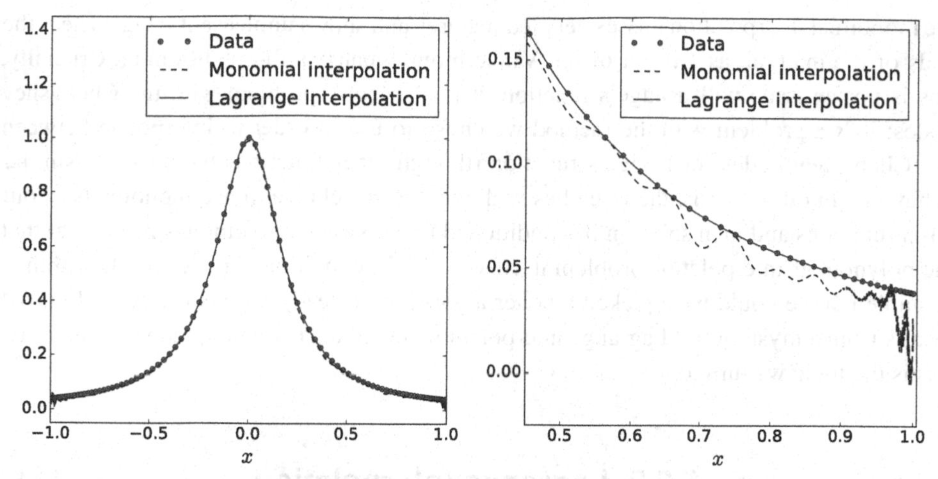

Fig. 6.4 Monomial interpolation for the case of $n = 101$ (left), also zoomed in (right)

you will show in problem 6.1, the determinant of a Vandermonde matrix is non-zero for distinct nodes x_j; thus, a Vandermonde matrix is *non-singular*. As a result, the columns of the Vandermonde matrix are linearly independent. This means that if you employ $O(n^3)$ operations (at most, since specialized algorithms also exist) you are guaranteed to be able to solve this linear system of equations for the c_k parameters; this is a *unique* solution. This is a general result: for real data points with distinct nodes, there exists a unique polynomial that goes through the points.

While the fact of the non-singularity of the Vandermonde matrix is good news, it's not great news. In Fig. 6.3 we show the monomial basis functions for the first several degrees. As the degree increases, different basis functions become closer in value to each other for positive-x; for negative-x odd degrees start to look alike, and similarly for even degree monomials. Even though in principle the columns of a Vandermonde matrix are linearly independent, in practice they are almost linearly dependent. Put another way, the Vandermonde matrix becomes increasingly ill-conditioned as n gets larger. In the past, textbooks would brush this off as a non-issue, since one didn't employ interpolating polynomials of a high degree anyway. However, as we're about to see, polynomials with $n = 100$, or even $n = 1000$, are routinely used in practice and therefore the Vandermonde ill-conditioning is a real problem.

Figure 6.4 shows data points drawn from a significant example, *Runge's function*:

$$f(x) = \frac{1}{1 + 25x^2} \tag{6.18}$$

In this figure, our data (x_j, y_j) are picked from the Chebyshev formula in Eq. (6.12) for this specific $f(x)$; we used $n = 101$ to stress the fact that n does not need to be small in practical applications. Note that our interpolation using the monomial basis/Vandermonde matrix does a reasonable job capturing the large-scale features of Runge's function (see left panel). This is starting to support our earlier claim that polynomial interpolation with Chebyshev nodes is an excellent tool. Unfortunately, once we zoom in (see right panel) we realize that the monomial interpolation misses detailed features and shows unpleasant wiggles near the ends of the interval, as a result of the Vandermonde matrix's ill-conditioning. Crucially, this is not an issue with Runge's function; it is also not a problem with the Chebyshev nodes; it *is* a problem with the method we chose to use in order to interpolate between our Chebyshev nodes for Runge's function (though other functions would show similar behavior). In other words, the issue lies with the "naive" choice to use monomials as our basis functions and then solve an ill-conditioned linear system. To emphasize the fact that the polynomial-interpolation problem itself was *not* ill-conditioned (i.e., our algorithm is unstable but we could have picked a better algorithm, instead), we also show in Fig. 6.4 results from a mysterious "Lagrange interpolation" method. Having piqued your curiosity, this is the topic we turn to next.

6.2.2 Lagrange Interpolation

Lagrange interpolation[11] is introduced in most textbooks on numerical methods. After some introductory comments, though, the approach is usually abandoned for other methods, e.g., Newton interpolation, typically after a claim that the Lagrange form is nice for proving theorems but not of practical value. As we'll see, Lagrange interpolation is helpful both formally and in practice. To emphasize this point, we will produce an implementation of Lagrange interpolation which leads to the (excellent) results in Fig. 6.4.[12]

6.2.2.1 Cardinal Polynomials

As before, we have as input a table of data points, (x_j, y_j) for $j = 0, 1, \ldots, n-1$. Before discussing how to interpolate through those points, let us introduce what are known as *cardinal* or *Lagrange* or *fundamental* polynomials:

$$L_k(x) = \frac{\prod_{j=0, j\neq k}^{n-1}(x - x_j)}{\prod_{j=0, j\neq k}^{n-1}(x_k - x_j)}, \qquad k = 0, 1, \ldots, n-1 \tag{6.19}$$

The denominator depends only on the x_j's, i.e., on the interpolation points, so it is clearly a constant (it doesn't depend on x). The numerator is a polynomial in x (of degree $n-1$), which for a given k goes to 0 at x_j when $j \neq k$. The Lagrange polynomial $L_k(x)$ goes to 1 at x_k, since the numerator and the denominator are equal to each other in that case.

[11] A typical example of the *Matthew effect*: Lagrange interpolation was actually discovered by Edward Waring.

[12] Incidentally, in section 5.3.1 we cautioned against finding the roots of a polynomial starting from its coefficients. Here, however, we're not finding roots but interpolatory behavior; also, we're starting from interpolation nodes, not from polynomial coefficients.

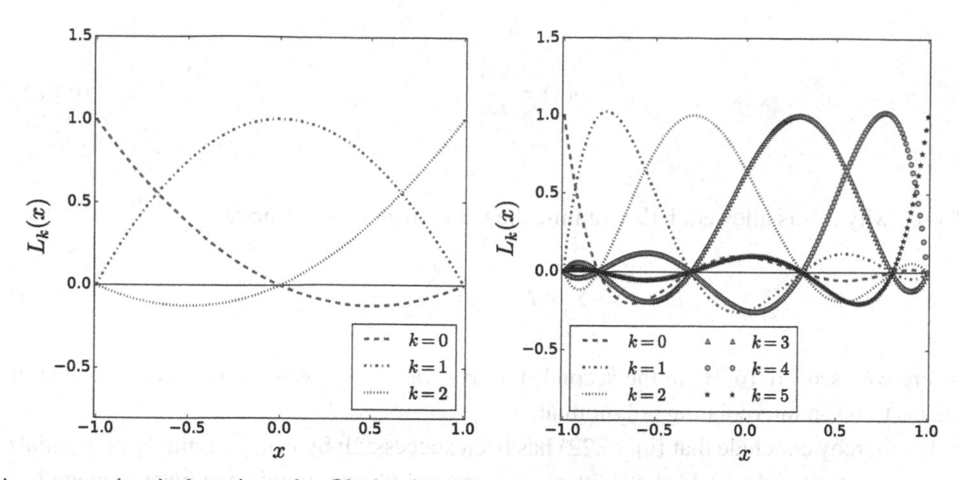

Fig. 6.5 Lagrange basis functions for Chebyshev 3-point (left) and 6-point (right) interpolation

If you've never encountered Lagrange polynomials before, you might benefit from seeing them explicitly written out for a simple case. Regardless of where the x_j's are placed, here's what the three Lagrange polynomials for the case of $n = 3$ look like:

$$L_0(x) = \frac{(x - x_1)(x - x_2)}{(x_0 - x_1)(x_0 - x_2)}, \quad L_1(x) = \frac{(x - x_0)(x - x_2)}{(x_1 - x_0)(x_1 - x_2)}, \quad L_2(x) = \frac{(x - x_0)(x - x_1)}{(x_2 - x_0)(x_2 - x_1)} \tag{6.20}$$

We could expand the parentheses here, but we won't, since that's less informative than the form the fractions are currently in. Clearly, for $n = 3$ the Lagrange polynomials are quadratic in x. Equally clearly, $L_0(x_1) = 0$, $L_0(x_2) = 0$, and $L_0(x_0) = 1$, and analogous relations hold for $L_1(x)$ and $L_2(x)$. These facts are illustrated in the left panel of Fig. 6.5: in the interval $[-1, 1]$ for $n = 3$ Chebyshev points and equidistant points are the same.

As mentioned before, these are general results: $L_k(x_j)$ is 0 if $j \neq k$, and 1 if $j = k$. This result is important enough that it deserves to be highlighted:

$$L_k(x_j) = \delta_{kj} \tag{6.21}$$

where δ_{kj} is the *Kronecker delta*. The right panel of Fig. 6.5 shows the example of $n = 6$, in which case the $L_k(x)$ are fifth degree polynomials (and there are six distinct Lagrange polynomials). No matter what they may do between one node and the next, these always go to 1 and to 0 in the appropriate places; unlike the monomials in Fig. 6.3, the L_k's are clearly linearly independent.

It is now time to use our cardinal polynomials as *basis functions*; this means that we take $\phi_k(x) = L_k(x)$ in Eq. (6.7). As will soon become clear, this also means that we don't need to use c_k in our definition of the interpolating polynomial, since we can simply write y_k (the y-values of our input data) in their place:

$$p(x) = \sum_{k=0}^{n-1} y_k L_k(x) \tag{6.22}$$

To see why this is allowed, let's examine the value of $p(x)$ at our nodes:

$$p(x_j) = \sum_{k=0}^{n-1} y_k L_k(x_j) = \sum_{k=0}^{n-1} y_k \delta_{kj} = y_j \tag{6.23}$$

where we used Eq. (6.21) in the second step. But $p(x_j) = y_j$ was the definition of what it means to be an interpolating polynomial.

We thereby conclude that Eq. (6.22) has been successful: by using cardinal polynomials (which go to 0 and to 1 in the right places) instead of monomials, we have managed to avoid encountering a (possibly ill-conditioned) Vandermonde matrix. As a matter of fact, the Lagrange form employs an ideally conditioned basis: in the language of Eq. (6.9), since $\phi_k(x_j) = L_k(x_j) = \delta_{kj}$, we see that $\mathbf{\Phi}$ is the identity matrix. Thus, we don't even have to solve $\mathbf{\Phi c} = \mathbf{y}$, since we'll always have $c_j = y_j$; this explains why we used y_k in Eq. (6.22). To summarize, in the Lagrange form we write the interpolating polynomial as a linear combination (not of monomials, but) of polynomials of degree $n-1$, see Eq. (6.22).

6.2.2.2 Barycentric Formula

At this point, you can understand why many books shun the Lagrange form: while it is true that it leads to a linear system in Eq. (6.9) which is perfectly conditioned, the approach still seems to be inefficient. In other words, while it is true that we don't have to solve a linear system, which in general costs $O(n^3)$ operations, our main formula in Eq. (6.22), combined with Eq. (6.19), appears to be highly inefficient: at a single x, we need $O(n)$ operations to compute $L_k(x)$ and must then carry out n such evaluations, leading to a total of $O(n^2)$ operations. When you change the value of x where you need to find the value of the interpolating polynomial, you'll need another $O(n^2)$ operations. To compare with the monomial basis: there it was costly to determine the c_k's as per Eq. (6.17) but, once you had done so, it took only $O(n)$ operations to evaluate $p(x)$ at a new x as per Eq. (6.15).

One can do much better, simply by re-arranging the relevant formulas. Observe that, as you change k, the numerator in Eq. (6.19) is always roughly the same, except for the missing $x - x_k$ factor each time. This motivates us to introduce the following *node polynomial*:

$$L(x) = \prod_{j=0}^{n-1} (x - x_j) \tag{6.24}$$

where there are no factors missing; this $L(x)$ is a (monic) polynomial of degree n. Clearly, $L(x_k) = 0$ for any k, i.e., the node polynomial vanishes at the nodes.

We now make a quick digression, which will not be used below but will help us in the following chapter, when we introduce Gauss–Legendre integration. The numerator in the equation for $L_k(x)$, Eq. (6.19), can be re-expressed as follows:

$$\prod_{j=0, j\neq k}^{n-1}(x - x_j) = \frac{\prod_{j=0}^{n-1}(x - x_j)}{x - x_k} = \frac{L(x)}{x - x_k} \tag{6.25}$$

where in the second equality we identified the node polynomial from Eq. (6.24). We now turn to the denominator in Eq. (6.19). This can be arrived at using Eq. (6.25), where we write the numerator as $L(x) - L(x_k)$ (since $L(x_k) = 0$) and then take the limit $x \to x_k$. This leads to:

$$\prod_{j=0, j\neq k}^{n-1}(x_k - x_j) = L'(x_k) \tag{6.26}$$

where the definition of the derivative appeared organically. Putting the last two equations together with Eq. (6.19) allows us to express a given cardinal polynomial $L_k(x)$ in terms of the node polynomial and its derivative:

$$L_k(x) = \frac{L(x)}{(x - x_k)L'(x_k)} \tag{6.27}$$

We will produce another (analogous) relation below.

Continuing with the main thread of our argument, we now introduce another quantity that captures the denominator in Eq. (6.19); we define the *barycentric weights* as follows:

$$w_k = \frac{1}{\prod_{j=0, j\neq k}^{n-1}(x_k - x_j)}, \qquad k = 0, 1, \ldots, n-1 \tag{6.28}$$

We are now in a position to re-express (once again) a given cardinal polynomial $L_k(x)$, see Eq. (6.19), in terms of the node polynomial and the weights:

$$L_k(x) = L(x)\frac{w_k}{x - x_k} \tag{6.29}$$

where we simply divided out the $x - x_k$ factor that is present in $L(x)$ but missing in $L_k(x)$.

A way to make further progress is to remember that, regardless of how you compute $L_k(x)$, the way to produce an interpolating polynomial $p(x)$ is via Eq. (6.22). Let us write out that relation, incorporating Eq. (6.29):

$$p(x) = L(x)\sum_{k=0}^{n-1} y_k \frac{w_k}{x - x_k} \tag{6.30}$$

where we pulled out $L(x)$ since it didn't depend on k. We now apply Eq. (6.22) again, this time for the special case where all the y_j's are 1: given the uniqueness of polynomial interpolants, the result must be the constant polynomial 1. In equation form, we have:

$$1 = \sum_{k=0}^{n-1} L_k(x) = L(x)\sum_{k=0}^{n-1}\frac{w_k}{x - x_k} \tag{6.31}$$

In the first step we show that adding up all the Lagrange polynomials gives us 1; in the second step we plugged in Eq. (6.29). If we now divide the last two equations, we can cancel the node polynomial $L(x)$, leading to:

$$p(x) = \sum_{k=0}^{n-1} \frac{w_k y_k}{x - x_k} \Bigg/ \sum_{k=0}^{n-1} \frac{w_k}{x - x_k} \tag{6.32}$$

where we get the w_k from Eq. (6.28). If you run into the special case $x = x_k$ for some k, you can simply take $p(x) = y_k$. Equation (6.32) is known as the *barycentric interpolation formula* and is both a beautiful and a practical result. Observe that the two sums are nearly identical: the first one involves the input data values y_k, whereas the second one doesn't. As x approaches one of the x_k's, a single term in the numerator and in the denominator become dominant; their ratio is y_k, so our interpolating polynomial has the right behavior. Of course, if x is actually equal to x_k this line of reasoning doesn't work, which is why we account for this scenario separately in our definition.

You may be wondering why we went through all the trouble. The form in Eq. (6.22) was clearly a polynomial, whereas Eq. (6.32), despite still being a polynomial, looks like it might be a rational function. The reason we bothered has to do with our earlier complaint that the Lagrange form of Eq. (6.22) necessitated $O(n^2)$ operations at each x value. In contradistinction to this, the barycentric formula splits the *construction stage* from the *evaluation stage*: we first compute the weights as per Eq. (6.28), using $O(n^2)$ operations; crucially, the weights don't rely on the y_j values but only on the placement of the nodes, x_j. When the weights have been computed (once and for all), we use Eq. (6.32) at each new x, requiring only $O(n)$ operations. As you may recall, this was the scaling exhibited by the monomial-basis evaluation stage, only this time we have no ill-conditioning issues.

Even though Eq. (6.32) looks like it might have issues with roundoff error, it is numerically stable. It actually has several more advantages, e.g., in how it handles the introduction of a new data pair (x_n, y_n). Another pleasant feature is that for Chebyshev nodes the weights w_k can be analytically derived, i.e., the $O(n^2)$ operations are not needed. Instead of getting into such details, however, we will now see the barycentric formula in action.

6.2.2.3 Implementation

We will study Runge's function from Eq. (6.18), namely the same example we plotted for the case of the monomial basis. In reality, interpolation deals with a table of data (x_j, y_j) as input, not with continuous functions. Even so, we want to test how well our interpolation is working, so it's good to start from a function and *produce* our data from it. Thus, code 6.1 starts by defining Runge's function and then introduces another function to generate the data pairs. We have coded up two distinct possibilities: Chebyshev points as per Eq. (6.12), or equidistant nodes. If you're still not very comfortable with `numpy.linspace()`, try to use `numpy.arange()` to accomplish the same task; for example, for Chebyshev nodes you would write down something like `np.arange(n)*np.pi/(n-1)`.

barycentric.py

Code 6.1

```
import numpy as np

def f(x):
    return 1/(1 + 25*x**2)

def generatedata(n,f,nodes="cheb"):
    if nodes=="cheb":
        dataxs = -np.cos(np.linspace(0,np.pi,n))
    else:
        dataxs = np.linspace(-1,1,n)
    datays = f(dataxs)
    return dataxs, datays

def weights(dataxs):
    n = dataxs.size
    ws = np.ones(n)
    for k in range(n):
        for j in range(n):
            if j == k:
                continue
            ws[k] *= (dataxs[k]-dataxs[j])
    return 1/ws

def bary(dataxs,datays,ws,x):
    k = np.where(x == dataxs)[0]
    if k.size == 0:
        nume = np.sum(ws*datays/(x-dataxs))
        denom = np.sum(ws/(x-dataxs))
        val = nume/denom
    else:
        val = datays[k[0]]
    return val


if __name__ == '__main__':
    dataxs, datays = generatedata(15, f)
    ws = weights(dataxs)
    x = 0.3; pofx = bary(dataxs, datays, ws, x)
    print(x, pofx, f(x))
```

Note that `dataxs` is a `numpy` array, so when we define `datays` we end up calling `f()` with an array as an argument; this works seamlessly for the operations involved in defining Runge's function (power, multiplication, addition, and division).

Having produced the table of data (x_j, y_j), we need to carry out the actual interpolation procedure. As hinted at above, we do this in two separate steps. We first define a function that computes the weights w_k; this knows nothing about the y_j values: if you need to carry out interpolation for a different problem later on, you can re-use the same weights. Similarly, our computation of the weights is not related to the specific x where we wish to interpolate: it only cares about where the nodes are located, as per Eq. (6.28). You can clearly see that this, construction, stage involves $O(n^2)$ operations, since it contains two nested loops. We eliminate the case $j = k$ from the product by employing `continue`, our first use of this keyword after our introduction to Python at the start of the book. Employing `numpy` functionality, we produce the denominator of w_k for each case, and then invert them all at once at the end.

With the table (x_j, y_j) and the weights w_k in place, we are in a position to turn to the second stage of the interpolation process, namely to use Eq. (6.32) in order to produce $p(x)$ at a given x. We start by considering the special case $x = x_k$ for some k, for which we should take $p(x) = y_k$ instead of using the full barycentric formula. We accomplish this by using `numpy.where()`; as you may recall from our introduction to `numpy` (or problem 1.10), `numpy.where()` allows us to find specific indices where a condition is met. Since it returns a *tuple* of arrays, we take its 0th element. If the element we're looking for is not there, the 0th element of the tuple returned by `numpy.where()` is an empty array; this means that its `size` is 0, so we should proceed with Eq. (6.32). That case is a straightforward application of `numpy` functionality: we carry out the two needed sums without employing any explicit loops. You might be wondering about the cost of traversing the array elements twice (once for the numerator and once for the denominator); the thing to note is that the evaluation of our interpolant is now a problem requiring $O(n)$ operations, so a factor of 2 doesn't make much of a difference. If `numpy.where()` did *not* return an empty array, we are dealing with the special case, so we index once again in order to return $p(x) = y_k$.

The main program produces the data, constructs the weights, and then evaluates the interpolating polynomial $p(x)$. In order to effectively distinguish the x_j (part of our input data) from the x (the point at which we wish to interpolate), we are calling the former `dataxs` and the latter `x`. The specific value of x we employ here is not important; the output of running this code is:

```
0.3 0.259275058184 0.3076923076923077
```

Since we have access to the underlying $f(x)$, we also compare its value to $p(x)$. The match is not great, so we then decide to use this code at thousands of points to produce a quasi-continuous curve. It's important to realize that we are not referring to the use of thousands of nodes (though we could have done that, as well): we pick, say, $n = 15$ Chebyshev nodes and then plot $p(x)$ for thousands of different x's. The result is shown in the left panel of Fig. 6.6 for both $n = 7$ and $n = 15$. In order to make this plot easier to understand, we are not showing the data points (x_j, y_j). We see that $n = 7$ captures only the peak height,

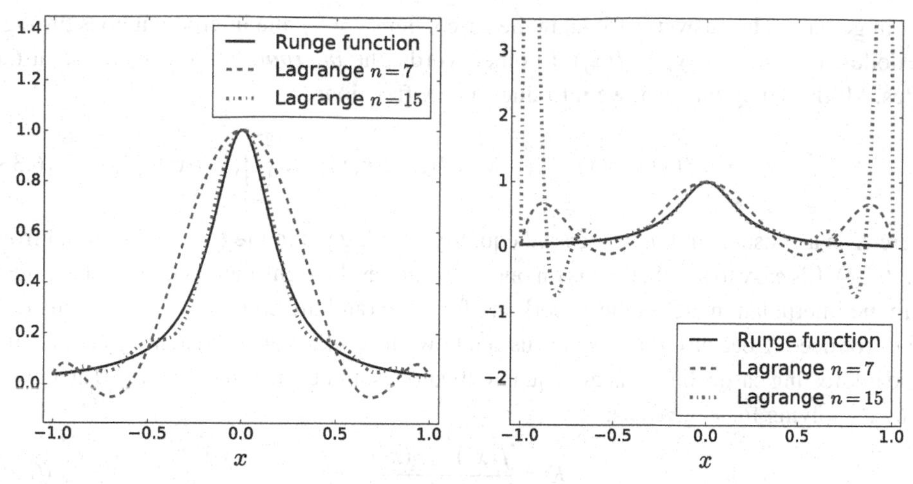

Lagrange interpolation for Chebyshev (left) and equidistant nodes (right)

Fig. 6.6

but misses the overall width of our curve, in addition to exhibiting strong oscillations. For $n = 15$ the width is better reproduced and the magnitude of the oscillations is considerably reduced. As you may recall from our comparison with the monomial basis, see Fig. 6.4, you can keep increasing n and you will keep doing a better job at approximating the underlying $f(x)$; those two panels showed that for $n = 101$ the match is nearly perfect. This cannot be over-emphasized: the canonical problem of misbehavior in polynomial interpolation (Runge's function) is easy to handle using Chebyshev points and the barycentric formula. As noted above, without the barycentric formula we would have had to keep re-evaluating the $L_k(x)$ from scratch at each of our thousands of x values, so this would have been prohibitively slow. To summarize, the accuracy resulted from the Chebyshev points and the efficiency from the barycentric formula.

We left the worst for last. If you use our code for equidistant nodes, you will find behavior like that shown in the right panel of Fig. 6.6. As n is increased, the width of our curve is captured better, but the magnitude of the oscillations at the edges gets *larger*! Even though the interpolating polynomial $p(x)$ goes through the input data points, the behavior near the ends of the interval is wholly unsatisfactory. It is this poor behavior that has often led to the generalization that polynomial interpolation should be avoided when n is large; you now know that this only applies to equidistant nodes. Of course, you may be in the unfortunate situation where the input data were externally provided at equidistant points; in that scenario, one typically resorts to a cubic-spline interpolation instead, as we will see in section 6.3. Before we get there, though, let's build some intuition into why different placements of the nodes lead to such dramatically different results.

6.2.2.4 Error Formula

As we just saw, polynomial interpolation behaves well for Chebyshev nodes and not so well for equidistant nodes. Instead of examining things on a case-by-case basis, it would be nice to see how well our interpolating polynomial $p(x)$ approximates the underlying

$f(x)$ in general. The answer is trivial in the case where $x = x_j$: the interpolant goes through the nodes, i.e., $p(x_j) = y_j = f(x_j)$. In other words, the *interpolation error* is zero at the nodes. Motivated by this fact, we introduce a new function:

$$F(x) = f(x) - p(x) - L(x)K = f(x) - p(x) - K\prod_{j=0}^{n-1}(x - x_j) \tag{6.33}$$

where K is a constant and in the second equality we plugged in the node polynomial from Eq. (6.24). It's easy to see that if x is on one of the nodes the right-hand-side vanishes, since there the interpolant matches the underlying function (and the node polynomial vanishes). We introduced K because it also helps us see how close $p(x)$ gets to matching $f(x)$ for the more interesting case where x is not equal to the nodes. Let us demand $F(x^*) = 0$ at a given point x^*; solving for K gives us:

$$K = \frac{f(x^*) - p(x^*)}{L(x^*)} \tag{6.34}$$

where we're allowed to divide by $L(x^*)$ since we know that it vanishes only at the nodes. In what follows we won't need this form, so we simply write K.

We have learned that $F(x)$ has n zeros at the x_j's and another zero at x^*. Thus, $F(x)$ has at least $n + 1$ zeros in the interval $[a, b]$. We now make use of *Rolle's theorem*, which you may recall from basic calculus as a special case of the *mean-value theorem*; Rolle's theorem tells us that between every two consecutive zeros of $F(x)$ there is a zero of $F'(x)$. Since $F(x)$ has at least $n+1$ zeros in our interval, we see that $F'(x)$ will have at least n zeros in the same interval. Similarly, $F''(x)$ will have at least $n - 1$ zeros, and so on. Repeatedly applying Rolle's theorem eventually leads to the result that $F^{(n)}(x)$ (the n-th derivative of $F(x)$) has at least one zero in the interval $[a, b]$; if we denote this zero by ξ, our finding is that $F^{(n)}(\xi) = 0$.

Let us examine the n-th derivative a bit more closely. Using Eq. (6.33), we have:

$$F^{(n)}(x) = f^{(n)}(x) - p^{(n)}(x) - L^{(n)}(x)K \tag{6.35}$$

Now, recall from our discussion of Eq. (6.22) that $p(x)$ is a linear combination of polynomials of degree $n - 1$, so it is itself a polynomial of degree at most $n - 1$. This implies that $p^{(n)}(x) = 0$. Similarly, we can see from the definition of the node polynomial in Eq. (6.24) that $L(x)$ is a polynomial of degree n. Actually, it happens to be a *monic polynomial*: as explained on page 266, this is what we call a polynomial whose leading power has a coefficient of 1. The fact that $L(x)$ is a monic polynomial makes it very easy to evaluate its n-th derivative: $L^{(n)}(x) = n!$ is the straightforward answer. Putting everything together:

$$F^{(n)}(\xi) = f^{(n)}(\xi) - Kn! = 0 \tag{6.36}$$

where $F^{(n)}(\xi) = 0$, since we're examining its zero. This equation can now be solved for K:

$$K = \frac{f^{(n)}(\xi)}{n!} \tag{6.37}$$

This is a more useful result than Eq. (6.34).

We are now in a position to take the expression for the constant from Eq. (6.37) and combine it with our earlier result $f(x^*) = p(x^*) + L(x^*)K$, to give:

$$f(x^*) = p(x^*) + \frac{f^{(n)}(\xi)}{n!} \prod_{j=0}^{n-1} (x^* - x_j) \tag{6.38}$$

where we've assumed all along that $f(x)$ has an n-th derivative. This is our desired general *error formula* for polynomial interpolation. It shows us that the error in approximating the underlying function is made up of two parts: the n-th derivative of the underlying function (at the unknown point ξ, divided by $n!$) and the node polynomial.

Of the two terms making up the error, the first one, $f^{(n)}(\xi)$, is complicated to handle, since ξ depends on the x_j's implicitly.[13] It's easier to focus on the second term, namely the node polynomial. In problems 6.7 and 6.8 you will see that Chebyshev points minimize the relevant product and also discover how differently equidistant nodes behave; this provides retroactive justification for our choice to employ Chebyshev nodes. Of course, in some cases the $f^{(n)}(\xi)$ will also play a role: in general, the smoother the function $f(x)$ is, the faster the interpolation will converge; the extreme case is Lagrange interpolation at Chebyshev points for *analytic* functions, in which case the interpolant converges geometrically, as we'll see in section 6.3.3. That being said, you don't necessarily have to limit yourself to Chebyshev nodes: problem 6.6 asks you to use the roots of *Legendre polynomials*, which as you'll find out also do a good job. What makes both sets of nodes good is the fact that they cluster near the ends of the interval; as a result, each node has roughly the same average distance from the others, something which clearly distinguishes them from equidistant points.

6.2.3 Hermite Interpolation

In the previous subsections we have been tackling the problem of interpolating through a set of data points, i.e., $p(x_j) = y_j$ for $j = 0, 1, \ldots, n-1$. A slightly different problem involves interpolating through both the points *and* the first derivatives at those points:[14]

$$p(x_j) = y_j, \qquad p'(x_j) = y'_j, \qquad j = 0, 1, \ldots, n-1 \tag{6.39}$$

The scenario involving these $2n$ conditions gives rise to what is known as *Hermite interpolation*. As you can guess, using a polynomial of degree $n-1$ won't work: this would have n undetermined parameters, but we need to satisfy $2n$ conditions. In other words, our Ansatz (a fancy way of saying "guess") for the interpolating polynomial should start from a form containing $2n$ undetermined parameters, i.e., a polynomial of degree $2n-1$.

We start from the following guess:

$$p(x) = \sum_{k=0}^{n-1} y_k \alpha_k(x) + \sum_{k=0}^{n-1} y'_k \beta_k(x) \tag{6.40}$$

[13] Even so, it's worth noting that $f^{(n)}(\xi)/n!$ is typically bounded.

[14] Observe that the prime in $p'(x_j)$ is a derivative, whereas that in y'_j just a notational choice—cf. Eq. (3.65).

where the $\alpha_k(x)$ are meant to capture the function values and the $\beta_k(x)$ the derivative values. In other words:

$$\begin{aligned} \alpha_k(x_j) &= \delta_{kj} & \beta_k(x_j) &= 0 \\ \alpha'_k(x_j) &= 0 & \beta'_k(x_j) &= \delta_{kj} \end{aligned} \tag{6.41}$$

for $k, j = 0, 1, \ldots, n-1$. This is merely re-stating our $2n$ conditions from Eq. (6.39).

We decide to write down $\alpha_k(x)$ and $\beta_k(x)$ in terms of $L_k(x)$, namely the cardinal polynomials of Eq. (6.19). As you may recall, $L_k(x)$ is a polynomial of degree $n-1$ which satisfies $L_k(x_j) = \delta_{kj}$, as per Eq. (6.21). We know that we need $\alpha_k(x)$ and $\beta_k(x)$ to be of degree $2n-1$. With that in mind, we realize that if we square $L_k(x)$ we get a polynomial of degree $2(n-1) = 2n-2$. Thus, multiplying $L_k^2(x)$ by a linear polynomial brings us up to degree $2n-1$, as desired:

$$\alpha_k(x) = u_k(x)L_k^2(x), \quad \beta_k(x) = v_k(x)L_k^2(x), \quad k = 0, 1, \ldots, n-1 \tag{6.42}$$

where both $u_k(x)$ and $v_k(x)$ are linear. To see how these are determined, we first write out two properties of the squared cardinal polynomials:

$$\begin{aligned} L_k^2(x_j) &= (\delta_{kj})^2 = \delta_{kj} \\ (L_k^2(x_j))' &= 2L_k(x_j)L'_k(x_j) = 2\delta_{kj}L'_k(x_j) \end{aligned} \tag{6.43}$$

We can now examine the first $\alpha_k(x)$-related condition in Eq. (6.41):

$$\delta_{kj} = \alpha_k(x_j) = u_k(x_j)L_k^2(x_j) = u_k(x_j)\delta_{kj} \tag{6.44}$$

where we used Eq. (6.43) in the last step. This implies that:

$$u_k(x_k) = 1 \tag{6.45}$$

Similarly, the next $\alpha_k(x)$-related condition in Eq. (6.41) gives us:

$$0 = \alpha'_k(x_j) = u'_k(x_j)L_k^2(x_j) + u_k(x_j)(L_k^2(x_j))' = u'_k(x_j)\delta_{kj} + 2u_k(x_j)\delta_{kj}L'_k(x_j) \tag{6.46}$$

where we, again, used Eq. (6.43) in the last step. Together with Eq. (6.45), this gives:

$$u'_k(x_k) + 2L'_k(x_k) = 0 \tag{6.47}$$

Since $u_k(x)$ is linear, we know that $u_k(x) = \gamma x + \delta$. Thus, we can use Eq. (6.45) and Eq. (6.47) to determine γ and δ, in which case we have fully determined $\alpha_k(x)$ to be:

$$\alpha_k(x) = \left[1 + 2L'_k(x_k)(x_k - x)\right] L_k^2(x) \tag{6.48}$$

A completely analogous derivation allows you to determine $\beta_k(x)$. Putting everything together, our interpolating polynomial[15] from Eq. (6.40) is then:

$$p(x) = \sum_{k=0}^{n-1} y_k \left[1 + 2L'_k(x_k)(x_k - x)\right] L_k^2(x) + \sum_{k=0}^{n-1} y'_k\, (x - x_k)L_k^2(x) \tag{6.49}$$

[15] More properly, this is called an *osculating polynomial* (from the Latin *osculum*, i.e., "kiss" or "small mouth").

Even if you didn't follow the derivation leading up to this equation, you should make sure to convince yourself that Eq. (6.49) satisfies our $2n$ conditions from Eq. (6.39).

Note that, using techniques similar to those in section 6.2.2.4, in problem 6.10 you will derive the following general *error formula* for Hermite interpolation:

$$f(x^*) = p(x^*) + \frac{f^{(2n)}(\xi)}{(2n)!}\prod_{j=0}^{n-1}(x^* - x_j)^2 \tag{6.50}$$

This looks very similar to Eq. (6.38) but, crucially, it contains a square on the right-hand side, as well as a $(2n)$-th derivative.

6.3 Cubic-Spline Interpolation

In the previous section, we covered interpolation using a single polynomial for an entire interval, staying away from the topic of how to deal with discontinuities. These are typically handled using *piecewise polynomial interpolation*, which is also useful when the placement of the nodes is irregular or, as mentioned earlier, when the nodes have been externally determined to be equidistant.

To be specific, our problem is the same as for Lagrange interpolation, namely we are faced with a table of input data points, (x_j, y_j) for $j = 0, 1, \ldots, n-1$. We wish to interpolate between these points, i.e., produce an easy-to-evaluate and well-behaved interpolant that can be computed at any x value. The approach known as *spline interpolation* (named after the thin strips used in building construction) cuts up the full interval into distinct panels, just like we did in section 3.3.6. A low-degree polynomial is used for each subinterval, i.e., $[x_0, x_1]$, $[x_1, x_2]$, and so on. To make things transparent, we will employ the notation:

$$p(x) = s_{k-1,k}(x), \qquad x_{k-1} \le x \le x_k, \qquad k = 1, 2 \ldots, n-1 \tag{6.51}$$

where $s_{k-1,k}(x)$ is the low-degree polynomial that is used (only) for the subinterval $[x_{k-1}, x_k]$; note that k starts at 1 and ends at $n-1$. These pieces are then stitched together to produce a continuous global interpolant; nodes are sometimes known as *knots* or *break points*.

Problem 6.13 sets up piecewise-linear interpolation, but in what follows we focus on the most popular case, namely piecewise-cubic interpolation, known as *cubic-spline interpolation*.[16] This leads to an interpolant that is smooth in the first derivative and continuous in

[16] Applying the Lagrange interpolation error formula from Eq. (6.38) to the function $e(x) = f(x) - p(x)$ with two points (taking $h = x_k - x_{k-1}$), we get: $|e(x)| \le |e''(\zeta)(x - x_{k-1})(x - x_k)/2 \le h^2|e''(\zeta)|/8$, where we first applied the triangle inequality and then found the maximum of the x-dependent term. An analogous argument leads to $|e''(\zeta)| \le h^2|f^{(4)}(\xi)|/2$—recall that $p''(x)$ is linear. Putting these two results together, we have arrived at an *error bound* for cubic-spline interpolation: $|f(x) - p(x)| \le h^4|f^{(4)}(\xi)|/16$. (To find $O(h^4)$ we implicitly assumed the case of clamped splines as per problem 6.15; natural splines—see Eq. (6.64)—are even worse.)

the second derivative. Before deriving the general case of n nodes, we go over an explicit example, that of $n = 3$, to help you build some intuition about how this all works.

6.3.1 Three Nodes

For $n = 3$ we have only three input-data pairs, (x_0, y_0), (x_1, y_1), and (x_2, y_2), so we are faced with only two panels, $[x_0, x_1]$ and $[x_1, x_2]$. In each of these subintervals, we will use a different cubic polynomial; to make things easy to grasp, we write these out in the monomial basis; that's not the best approach, but it's good enough to start with.

The first panel will be described by the cubic polynomial:

$$s_{0,1}(x) = c_0 + c_1 x + c_2 x^2 + c_3 x^3 \tag{6.52}$$

which is exactly of the form of Eq. (6.15). For the second panel we use a separate polynomial; this means that it will have distinct coefficients:

$$s_{1,2}(x) = d_0 + d_1 x + d_2 x^2 + d_3 x^3 \tag{6.53}$$

As was to be expected, each cubic polynomial comes with four unknown parameters, so we are here faced with eight parameters for our two cubic polynomials put together. We'll need eight equations to determine the unknown parameters.

The first polynomial should interpolate the data at the ends of its (sub)interval:

$$c_0 + c_1 x_0 + c_2 x_0^2 + c_3 x_0^3 = y_0, \qquad c_0 + c_1 x_1 + c_2 x_1^2 + c_3 x_1^3 = y_1 \tag{6.54}$$

Similarly, the second polynomial interpolates the data at the ends of its own subinterval:

$$d_0 + d_1 x_1 + d_2 x_1^2 + d_3 x_1^3 = y_1, \qquad d_0 + d_1 x_2 + d_2 x_2^2 + d_3 x_2^3 = y_2 \tag{6.55}$$

So far we have produced four equations, employing the definition of what it means to be an interpolating polynomial, $p(x_j) = y_j$. Obviously, that's not enough, so we'll have to use more properties. Specifically, we will impose the continuity of the first derivative at x_1:

$$c_1 + 2c_2 x_1 + 3c_3 x_1^2 = d_1 + 2d_2 x_1 + 3d_3 x_1^2 \tag{6.56}$$

and, similarly, the continuity of the second derivative at x_1:

$$2c_2 + 6c_3 x_1 = 2d_2 + 6d_3 x_1 \tag{6.57}$$

At this point, we have six equations for eight unknowns. This is a pattern that will re-emerge in the general case below. There are several ways via which we can produce two more constraints; here and in what follows, we will choose to make the second derivative go to 0 at the endpoints of the initial interval. This gives rise to what is known as a *natural spline*.[17] We get two more equations by demanding that the second derivative is 0 at x_0 and at x_2:

$$2c_2 + 6c_3 x_0 = 0, \qquad 2d_2 + 6d_3 x_2 = 0 \tag{6.58}$$

[17] Problem 6.15 investigates a different choice.

where, obviously, only the first polynomial can be used at x_0 and only the second one at x_2. The eight equations can be written in matrix form as follows:

$$\begin{pmatrix} 1 & x_0 & x_0^2 & x_0^3 & 0 & 0 & 0 & 0 \\ 1 & x_1 & x_1^2 & x_1^3 & 0 & 0 & 0 & 0 \\ 0 & 0 & 0 & 0 & 1 & x_1 & x_1^2 & x_1^3 \\ 0 & 0 & 0 & 0 & 1 & x_2 & x_2^2 & x_2^3 \\ 0 & 1 & 2x_1 & 3x_1^2 & 0 & -1 & -2x_1 & -3x_1^2 \\ 0 & 0 & 2 & 6x_1 & 0 & 0 & -2 & -6x_1 \\ 0 & 0 & 2 & 6x_0 & 0 & 0 & 0 & 0 \\ 0 & 0 & 0 & 0 & 0 & 0 & 2 & 6x_2 \end{pmatrix} \begin{pmatrix} c_0 \\ c_1 \\ c_2 \\ c_3 \\ d_0 \\ d_1 \\ d_2 \\ d_3 \end{pmatrix} = \begin{pmatrix} y_0 \\ y_1 \\ y_1 \\ y_2 \\ 0 \\ 0 \\ 0 \\ 0 \end{pmatrix} \tag{6.59}$$

where the only (minor) subtlety involved the fifth and sixth equations. This is analogous to, but quite distinct from, the matrix equation we were faced with in Eq. (6.17).[18]

To summarize, we employed three types of constraints: (a) interpolation conditions, (b) continuity conditions, and (c) natural-spline conditions. In problem 6.14 you are asked to implement Eq. (6.59) programmatically; instead of following that avenue, we now turn to a more systematic approach, which also has the (supervenient) benefit of involving a banded (tridiagonal) system of equations.

6.3.2 General Case

In the previous, explicit, example, we had three points and two panels. In the general case, we have n points (x_j for $j = 0, 1, \dots, n-1$) and $n-1$ panels. Similarly, the fact that there are n points in total means there are $n-2$ interior points (i.e., excluding x_0 and x_{n-1}). Each panel gets its own spline, i.e., corresponds to four unknown parameters; since there are $n-1$ panels in total, we are faced with $4n-4$ undetermined parameters. We will find these using the following constraints:

- A given spline should interpolate the data at its left endpoint:

$$s_{k-1,k}(x_{k-1}) = y_{k-1}, \qquad k = 1, 2 \dots, n-1 \tag{6.60}$$

  and at its right endpoint:

$$s_{k-1,k}(x_k) = y_k, \qquad k = 1, 2 \dots, n-1 \tag{6.61}$$

  Together, these are $2n-2$ conditions.
- For each of the $n-2$ interior points, the first derivative of the two splines on either side should match:

$$s'_{k-1,k}(x_k) = s'_{k,k+1}(x_k), \qquad k = 1, 2, \dots, n-2 \tag{6.62}$$

  where this k ends at $n-2$. We get another $n-2$ conditions this way.

[18] Incidentally, the coefficient matrix in Eq. (6.59) belongs to the class of non-symmetric matrices. As advertised on page 126, these show up quite often in practice.

- For each of the $n - 2$ interior points, the second derivative of the two splines on either side should match:

$$s''_{k-1,k}(x_k) = s''_{k,k+1}(x_k), \qquad k = 1, 2, \ldots, n-2 \tag{6.63}$$

  This provides us with another $n - 2$ conditions.
- So far, we have $4n - 6$ conditions. To find the missing two constraints, we decide to use a *natural spline*, i.e., make the second derivative go to 0 at x_0 and at x_{n-1}:

$$s''_{0,1}(x_0) = s''_{n-2,n-1}(x_{n-1}) = 0 \tag{6.64}$$

  This brings the total number of conditions up to $4n - 4$, which is equal to the number of undetermined parameters.

Let's start writing these conditions out. Our strategy will be to express our cubic spline $s_{k-1,k}(x)$ in terms of the input data (x_k and y_k) as well as the second-derivative values at the nodes. We call the latter c_k's; our relation giving the continuity of the second derivative, Eq. (6.63), is then simply:

$$s''_{k-1,k}(x_k) = s''_{k,k+1}(x_k) = c_k, \qquad k = 1, 2, \ldots, n-2 \tag{6.65}$$

We don't actually know these yet. For a natural spline, we do know that:

$$c_0 = c_{n-1} = 0 \tag{6.66}$$

as per Eq. (6.64). You may wish to think of the c_k's as y''_k's, though we denote them c_k's because we will end up solving a linear system of equations for them, as in earlier sections. This linear system, in its turn, will give us the c_k's in terms of the input data (x_k and y_k) directly. To belabor the obvious, in other sections of the present chapter the c_k's are coefficients of a polynomial, but here they are second-derivative values.

Remember that we are dealing with a $s_{k-1,k}(x)$ in Eq. (6.51) which is cubic, so $s''_{k-1,k}(x)$ is a straight line; thus, if we know the values of the second derivative at the left and right endpoint of a given subinterval, c_{k-1} and c_k at x_{k-1} and x_k, we can apply our standard Lagrange-interpolation formula from Eq. (6.22). This is:

$$s''_{k-1,k}(x) = c_{k-1}\frac{x - x_k}{x_{k-1} - x_k} + c_k\frac{x - x_{k-1}}{x_k - x_{k-1}} \tag{6.67}$$

As promised, we are expressing the second derivative of the spline in terms of the c_k's, namely the values of the second derivative at the nodes. The notation here can get a bit confusing, so make sure you remember the essential point: we are applying Lagrange interpolation for a straight line in $x_{k-1} \leq x \leq x_k$; just like in Eq. (6.51), we will eventually take $k = 1, 2 \ldots, n-1$.

If we now integrate Eq. (6.67) we can get an expression for the first derivative of the spline (at any x):

$$s'_{k-1,k}(x) = c_{k-1}\frac{x^2/2 - xx_k}{x_{k-1} - x_k} + c_k\frac{x^2/2 - xx_{k-1}}{x_k - x_{k-1}} + A \tag{6.68}$$

where we called the integration constant A. If we integrate another time, we can get the spline itself:

$$s_{k-1,k}(x) = c_{k-1}\frac{x^3/6 - x^2 x_k/2}{x_{k-1} - x_k} + c_k\frac{x^3/6 - x^2 x_{k-1}/2}{x_k - x_{k-1}} + Ax + B \tag{6.69}$$

where there is now another integration constant, B. If you use Eq. (6.69) twice to impose the conditions that the spline should interpolate the data at its left and right endpoints, Eq. (6.60) and Eq. (6.61), you can eliminate A and B (as you'll verify in problem 6.16). Thus, your spline is now written only in terms of the input data (x_k and y_k) and the second-derivative values at the nodes (c_k). In equation form, this is:

$$\begin{aligned} s_{k-1,k}(x) &= y_{k-1}\frac{x_k - x}{x_k - x_{k-1}} + y_k\frac{x - x_{k-1}}{x_k - x_{k-1}} \\ &\quad - \frac{c_{k-1}}{6}\left[(x_k - x)(x_k - x_{k-1}) - \frac{(x_k - x)^3}{x_k - x_{k-1}}\right] \\ &\quad - \frac{c_k}{6}\left[(x - x_{k-1})(x_k - x_{k-1}) - \frac{(x - x_{k-1})^3}{x_k - x_{k-1}}\right] \end{aligned} \tag{6.70}$$

where $k = 1, 2 \ldots, n-1$. This equation identically obeys the relation giving the continuity of the second derivative, Eq. (6.63) or Eq. (6.65).[19] It does *not* in general obey the relation giving the continuity of the first derivative, Eq. (6.62), so let's impose it explicitly and see what happens. First, differentiate Eq. (6.70) to get:

$$\begin{aligned} s'_{k-1,k}(x) = \frac{y_k - y_{k-1}}{x_k - x_{k-1}} &+ \frac{c_{k-1}}{6(x_k - x_{k-1})}\left(-3x^2 + 6xx_k - 2x_k^2 - 2x_k x_{k-1} + x_{k-1}^2\right) \\ &+ \frac{c_k}{6(x_k - x_{k-1})}\left(3x^2 - 6xx_{k-1} + 2x_{k-1}^2 + 2x_k x_{k-1} - x_k^2\right) \end{aligned} \tag{6.71}$$

This function can be immediately evaluated at x_k to give:

$$s'_{k-1,k}(x_k) = \frac{y_k - y_{k-1}}{x_k - x_{k-1}} + \frac{x_k - x_{k-1}}{6}(2c_k + c_{k-1}) \tag{6.72}$$

We can also take $k \to k+1$ in Eq. (6.71) and then evaluate at x_k to get:

$$s'_{k,k+1}(x_k) = \frac{y_{k+1} - y_k}{x_{k+1} - x_k} - \frac{x_{k+1} - x_k}{6}(c_{k+1} + 2c_k) \tag{6.73}$$

Equating the right-hand sides in the last two relations, as per Eq. (6.62), allows us to find an equation for the unknown parameters c_k:

$$(x_k - x_{k-1})c_{k-1} + 2(x_{k+1} - x_{k-1})c_k + (x_{k+1} - x_k)c_{k+1} = 6\left(\frac{y_{k+1} - y_k}{x_{k+1} - x_k} - \frac{y_k - y_{k-1}}{x_k - x_{k-1}}\right) \tag{6.74}$$

[19] If you're not seeing this, go back to Eq. (6.67) and check it there.

where $k = 1, 2 \dots, n-2$. Remember, as per Eq. (6.66), $c_0 = c_{n-1} = 0$ holds. When going over our strategy, we promised that we would produce a linear system that can be solved to find the c_k's in terms of the input data (x_k and y_k) directly; we have now set this up.

You might feel more convinced after seeing this in matrix form. Recall, the second derivative values go from c_0, c_1, all the way up to c_{n-2} and c_{n-1}, but we are only solving Eq. (6.74) for c_1 through c_{n-2}. The situation is similar to what we saw when discretizing the action around Fig. 5.18. We have n data points but only $n-2$ variables to solve for. The system of equations can be expressed as:

$$\begin{pmatrix} 2(x_2 - x_0) & x_2 - x_1 & 0 & \dots & 0 \\ x_2 - x_1 & 2(x_3 - x_1) & x_3 - x_2 & \dots & 0 \\ \vdots & \vdots & \ddots & \vdots & \vdots \\ 0 & \dots & x_{n-3} - x_{n-4} & 2(x_{n-2} - x_{n-4}) & x_{n-2} - x_{n-3} \\ 0 & \dots & 0 & x_{n-2} - x_{n-3} & 2(x_{n-1} - x_{n-3}) \end{pmatrix} \begin{pmatrix} c_1 \\ c_2 \\ \vdots \\ c_{n-3} \\ c_{n-2} \end{pmatrix} = \begin{pmatrix} b_1 \\ b_2 \\ \vdots \\ b_{n-3} \\ b_{n-2} \end{pmatrix} \tag{6.75}$$

where the b_k's are defined as per the right-hand side of Eq. (6.74): this makes the equation shorter, but also helps us think about how to implement things programatically below.

Our coefficient matrix here is symmetric tridiagonal (and also diagonally dominant); this means one could use very efficient solvers, but we'll go with our standard approach, Gaussian elimination with pivoting, despite it being overkill. Once we've solved the system in Eq. (6.75), we will have determined all the c_k's in terms of the input data. Then, we can use them in Eq. (6.70) to find the value of the interpolant at any x. There's actually a slightly subtle point involved here: Eq. (6.70) holds for a specific subinterval, namely $x_{k-1} \leq x \leq x_k$. When trying to find the value of the interpolant at x, we don't know ahead of time which subinterval x belongs to, so we'll also have to first determine that.

6.3.3 Implementation

Code 6.2 starts by importing the underlying $f(x)$, Runge's function from Eq. (6.18), as well as the function that produced a table of input data, (x_j, y_j), both from `barycentric.py`. We also import our Gaussian elimination function, which we'll need to determine the c_k's.

As before, the actual interpolation procedure will be carried out in two separate steps. First, we define a function that computes the values of the second derivative at the nodes, namely the c_k's, as per Eq. (6.74) or Eq. (6.75). This knows nothing about the specific x where we wish to interpolate: it only cares about the input data, (x_j, y_j). Since we have n data pairs, the cofficient matrix will have dimensions $(n-2)\times(n-2)$; similarly, the solution vector will be $(n-2) \times 1$, since we already know that $c_0 = c_{n-1} = 0$. We need to fill up a tridiagonal matrix, similarly to what we saw in section 5.7 when extremizing the action. We treat the main and other diagonals differently via slicing the arrays we pass as the first argument to `numpy.fill_diagonal()`, in addition to employing slicing to determine what makes up each diagonal. Setting up the right-hand side of the linear system is largely straightforward; once again `numpy`'s indexing makes things very convenient: `dataxs[1:-1]` is easier to read than `dataxs[1:n-1]`. Slicing is used yet again to ensure that all the c_j's

splines.py

Code 6.2

```
from barycentric import f, generatedata
from gauelim_pivot import gauelim_pivot
import numpy as np

def computecs(dataxs,datays):
    n = dataxs.size
    A = np.zeros((n-2,n-2))
    np.fill_diagonal(A, 2*(dataxs[2:]-dataxs[:-2]))
    np.fill_diagonal(A[1:,:], dataxs[2:-1]-dataxs[1:-2])
    np.fill_diagonal(A[:,1:], dataxs[2:-1]-dataxs[1:-2])
    b1 = (datays[2:]-datays[1:-1])/(dataxs[2:]-dataxs[1:-1])
    b2 = (datays[1:-1]-datays[:-2])/(dataxs[1:-1]-dataxs[:-2])
    bs = 6*(b1 - b2)

    cs = np.zeros(n)
    cs[1:-1] = gauelim_pivot(A, bs)
    return cs

def splineinterp(dataxs,datays,cs,x):
    k = np.argmax(dataxs>x)
    xk = dataxs[k]; xk1 = dataxs[k-1]
    yk = datays[k]; yk1 = datays[k-1]
    ck = cs[k]; ck1 = cs[k-1]

    val = yk1*(xk-x)/(xk-xk1) + yk*(x-xk1)/(xk-xk1)
    val -= ck1*((xk-x)*(xk-xk1) - (xk-x)**3/(xk-xk1))/6
    val -= ck*((x-xk1)*(xk-xk1) - (x-xk1)**3/(xk-xk1))/6
    return val

if __name__ == '__main__':
    dataxs, datays = generatedata(15, f, "equi")
    cs = computecs(dataxs, datays)
    x = 0.95; pofx = splineinterp(dataxs, datays, cs, x)
    print(x, pofx, f(x))
```

are stored in a single array: the solution of the linear system is an $(n-2) \times 1$ vector, so it's padded with $c_0 = c_{n-1} = 0$ to produce an $n \times 1$ vector.

With the table (x_j, y_j) and the c_j values in place, we turn to the second stage of the in-

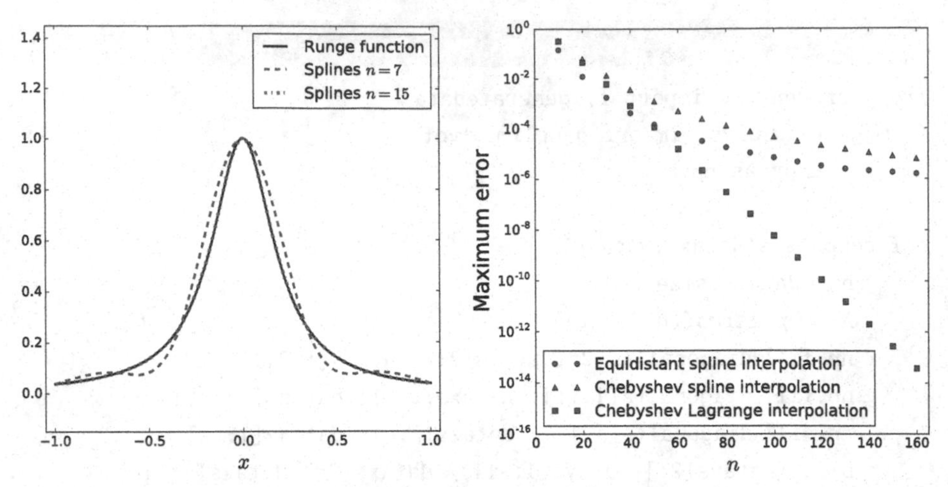

Fig. 6.7 Cubic-spline interpolation for equidistant nodes (left), vs other approaches (right)

terpolation process, namely the use of Eq. (6.70) in order to produce $p(x)$ at a given x. The main difficulty is determining which panel x falls in. We do so by using `numpy.argmax()`, which we also encountered in our implementation of Gaussian elimination with pivoting; as you may recall from our introduction to `numpy` (or problem 1.10), `numpy.argmax()` allows us to return the index of the maximum value. You should experiment to make sure you understand what's going on here: `dataxs>x` returns an array of boolean values (`True` or `False`). Since our x_j's are ordered, `dataxs>x` will give `False` for the x_j's that are smaller than x and `True` for the x_j's that are larger than x. Then, `numpy.argmax()` realizes that `True > False`, so it returns as `k` the index for the first x_j that is larger than x. We are fortunate that if the maximum value appears more than once, `numpy.argmax()` returns the first occurrence (all the other x_j's to the right of x_k are also larger than x). The rest of the function is near-trivial: Eq. (6.70) is algebra-heavy, so we first define some local variables to make our subsequent code match the mathematical expressions as closely as possible.

The main program produces the data, constructs the c_j's, and then evaluates the interpolating polynomial $p(x)$. In order to highlight that cubic-spline interpolation can handle equally spaced grids well, we produce equidistant grid points and also pass in $x = 0.95$, a value for which Lagrange interpolation (at equidistant points) was doing quite poorly in the right panel of Fig. 6.6. The output of running this code is:

```
0.95 0.0426343358892 0.042440318302387266
```

The match is very good, despite the fact that we used only 15 points. As before, we decide to be more thorough, using this code at thousands of points to produce a quasi-continuous curve. Once again, we'll use a reasonably small number of nodes ($n = 7$ or $n = 15$), but thousands of x's.

The result is shown in the left panel of Fig. 6.7; as we did earlier, we are not showing the data points (x_j, y_j) so you can focus on the curves. Note that we are only showing the case of equidistant points, where the splines are doing a better job (but see below for splines with Chebyshev nodes). Overall, both $n = 7$ and $n = 15$ do a better job than Lagrange interpolation did in Fig. 6.6. This bears repeating: cubic-spline interpolation does just fine when you use an equally spaced grid, in contradistinction to Lagrange interpolation. As a matter of fact, the $n = 15$ results are virtually indistinguishable from the underlying Runge function. While there were no singularities to be dealt with in this example, in practice it is very common to employ cubic-spline interpolation with an equidistant grid starting near the singularity, given little other knowledge about the behavior of the underlying function.

Our finding in the left panel of Fig. 6.7 may be a bit surprising given our earlier admiration for the barycentric formula at Chebyshev nodes. We decide to investigate things further, still for Runge's function using cubic-spline interpolation, only this time we will keep increasing n and also explore both equidistant and Chebyshev nodes. While we're at it, we will also repeat this exercise for Lagrange interpolation with Chebyshev nodes (we already know that equidistant nodes are bad for Lagrange interpolation, so we don't use them). Since we wish to plot the result for several n's, we must somehow quantify how well a given interpolant $p(x)$ matches $f(x)$; inspired by the *infinity norm*, see Eq. (4.23), we will calculate the maximum magnitude of the difference $f(x) - p(x)$ across all the x's we are interpolating at. The result for such a calculation with all three methods is shown in the right panel of Fig. 6.7. First, observe that spline interpolation with equidistant nodes consistently outperforms spline interpolation with Chebyshev nodes (by a little). Second, comparing spline interpolation with equidistant nodes to Lagrange interpolation with Chebyshev nodes we see that for small n the splines do a better job, as we also saw in the left panel. However, as n keeps increasing, the splines "level off", i.e., do as well as they're going to at a maximum error of $\approx 10^{-6}$. In contradistinction to this, the Lagrange-interpolation results keep getting better, reaching close to machine precision.

The results of the right panel of Fig. 6.7 are significant and bear emphasizing: a single polynomial of degree a couple of hundred can describe Runge's function to within machine precision, if you're picking the nodes at Chebyshev points. This is a linear trend on a semilog plot: as you add 10 more points, you improve by an order of magnitude! (This is sometimes called *spectral accuracy*, see also page 356.) Perhaps you can now appreciate the comments we made after deriving our general error formula for polynomial interpolation, Eq. (6.38): for analytic functions Lagrange interpolation at Chebyshev nodes converges geometrically, so it's likely the tool you are looking for.

6.4 Trigonometric Interpolation

What do you do if the data points you are faced with are *periodic*? In this case, obviously, your interpolatory function should also be periodic, so using a polynomial (as in previous sections) is not on the face of it appropriate.[20] In practice, one employs an expansion in

[20] Though, even here, the barycentric formula with Chebyshev nodes does a great job, as you'll see in problem 6.5. If you squint, you can see that polynomial interpolation at Chebyshev points is actually equivalent to trigonometric interpolation; one picks the x_j's trigonometrically and the other does the same for the ϕ_k's.

sines and cosines. The notation, as well as the concepts, involved in this context can get a bit confusing, so let's try to take things from the start. We will use a number of tools that have the name of Fourier attached to them but, since they serve different purposes, it's good to begin at a very basic level. This way (it is hoped) you will start getting comfortable with our symbols and their use, while things are still reasonably simple.

6.4.1 Fourier Series

Assume we are dealing with a function $f(x)$ that is periodic; for simplicity, let us focus on the interval $[0, 2\pi]$.[21] If $f(x)$ is reasonably continuous, then a standard result is that it can be decomposed in the following *Fourier series*:

$$f(x) = \frac{1}{2}a_0 + \sum_{k=1}^{\infty}(a_k \cos kx + b_k \sin kx) \tag{6.76}$$

Note that the sum here extends from 1 to ∞. We call the k's "wave numbers", or sometimes even "frequencies", a left-over from the use of Fourier analysis that relates time, t, to frequency, ω or f. Note also that the zero-frequency term a_0 has been singled out and treated differently (with a 2 in the denominator), for reasons that will soon become clear.[22]

Crucially, for well-behaved functions $f(x)$, this decomposition is *exact*, in the sense that the series on the right-hand side converges to $f(x)$; thus, a general periodic $f(x)$ is rewritten as a sum of sines and cosines. We can imagine a careful student, who solved problem 2.18, protesting at this point: what about the Gibbs phenomenon? As you may recall, this is a "ringing" effect which appears when you try to reconstruct a discontinuous function using a Fourier series; even as you take the number of terms in the series to infinity, this mismatch between the left-hand side and the right-hand side does not go away. In what follows, we will assume that our functions are sufficiently smooth so that such problems do not arise.

We can exploit the *orthogonality* of sines and cosines to extract the *Fourier coefficients* in Eq. (6.76), namely the a_k's and b_k's. As you will show in problem 6.18 multiplying Eq. (6.76) by $\cos jx$ (and, separately, with $\sin jx$) and integrating x from 0 to 2π we get:

$$a_k = \frac{1}{\pi}\int_0^{2\pi} dx f(x) \cos kx, \quad k = 0, 1, \ldots, \qquad b_k = \frac{1}{\pi}\int_0^{2\pi} dx f(x) \sin kx, \quad k = 1, 2, \ldots \tag{6.77}$$

In both of these equations we renamed $j \to k$ after we were done. You can now see why we included a 2 in the denominator for the case of a_0 in Eq. (6.76): it leads to the same expression, Eq. (6.77), for both $k = 0$ and $k > 0$.

In what follows, we will benefit from employing the Fourier series in a different form. First, recall *Euler's formula* (in its incarnation as *de Moivre's formula*):

$$e^{ikx} = \cos kx + i \sin kx \tag{6.78}$$

[21] You can trivially scale this to $[0, \tau]$ later on, by writing $x = 2\pi t/\tau$.

[22] We could have also included a b_0 term, but it wouldn't have made a difference.

where $i = \sqrt{-1}$ is the *imaginary unit*. Making (repeated) use of this formula, the Fourier series can be rewritten as:

$$f(x) = \sum_{k=-\infty}^{\infty} c_k e^{ikx} \tag{6.79}$$

as you will show in problem 6.19. It's important to notice that the k here goes from $-\infty$ to $+\infty$. In problem 6.18 you will also show how to solve for the coefficients c_k:

$$c_k = \frac{1}{2\pi}\int_0^{2\pi} dx f(x)e^{-ikx}, \qquad \text{integer } k \tag{6.80}$$

where this time we employed the orthogonality of the plane waves. Incidentally, this is one of the few points in the book where we are choosing to employ complex numbers.

Before we conclude this subsection, let us observe that there is an asymmetry at play here: we employ a sum to decompose our function $f(x)$, see Eq. (6.76) or Eq. (6.79), but we use integrals to get the coefficients a_k, b_k, or c_k in terms of the function $f(x)$, see Eq. (6.77), or Eq. (6.80).

6.4.2 Partial Sum: Trigonometric Interpolation

In the present subsection, we take the more practical route of assuming a set of n data pairs are known, and trying to reconstruct $f(x)$ using ideas borrowed from Fourier analysis. This approach is known as *trigonometric interpolation* for good reason. Since we're dealing with a finite number of data points, we will employ a partial sum (i.e., a finite number of terms) and make sure that no integrals appear at any point in our calculation.

6.4.2.1 Definition and Properties

We start from the same assumption as in earlier sections on interpolation: we are given n data points (also known as *support points*) where we know both the x and the y value: (x_j, y_j) for $j = 0, 1, \ldots, n-1$ using our standard Python-inspired 0-indexing.[23] As was also the case earlier in the chapter, we don't know if these points truly come from a given analytic function $f(x)$ (in which case $f(x_j) = y_j$), or were experimentally measured, and so on. All we know is that we need to produce an interpolating function, which *must* go through the points we are given, in addition to satisfying criteria relating to smoothness etc. Essentially, the only difference from earlier on is that our problem now involves x_j's which lie in the interval $[0, 2\pi]$ and y_j's which are periodic; thus, whatever interpolatory function we come up with had better be periodic as well.

Since we know only the value of y_j at n distinct values of x_j, it is unreasonable to expect that we will be able to determine infinitely many a_k and b_k coefficients from the integrals

[23] We will be interested only in the case where the y_j are real numbers.

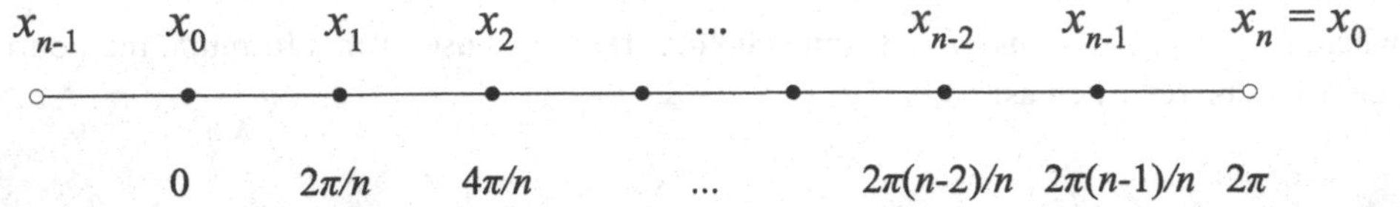

$$
\begin{array}{cccccccc}
x_{n-1} & x_0 & x_1 & x_2 & \cdots & x_{n-2} & x_{n-1} & x_n = x_0 \\
\circ & \bullet & \bullet & \bullet & \bullet\ \bullet & \bullet & \bullet & \circ \\
 & 0 & 2\pi/n & 4\pi/n & \cdots & 2\pi(n-2)/n & 2\pi(n-1)/n & 2\pi
\end{array}
$$

Fig. 6.8 Grid for the x_j's, with solid dots showing our actual points and open dots implied ones

of Eq. (6.77). Instead, what we do is to truncate the series in Eq. (6.76); since we have n data points, the corresponding interpolating polynomial will be:[24]

$$
p(x) = \frac{1}{2}a_0 + \sum_{k=1}^{m}(a_k \cos kx + b_k \sin kx) \tag{6.81}
$$

where we've implicitly assumed that we are dealing with the case of odd-n. To see why this is so, count the total number of undetermined parameters; these are $1+m+m = 2m+1$ which means that we'll be able to determine them by matching at our n data points, $p(x_j) = y_j$, if $n = 2m + 1$. Observe that so far we haven't assumed that the x_j are equally spaced.

For reasons that will later become clear, we prefer to focus on the case of n-even, so we cannot use Eq. (6.81). Instead, our main formula for trigonometric interpolation will be:

$$
p(x) = \frac{1}{2}a_0 + \sum_{k=1}^{m-1}(a_k \cos kx + b_k \sin kx) + \frac{1}{2}a_m \cos mx \tag{6.82}
$$

This time the total number of parameters is $1 + 2(m - 1) + 1 = 2m$; thus, we will be able to determine all of them by matching via $p(x_j) = y_j$, if $n = 2m$. We see that the first and last terms are treated separately. Since n (or m) is finite, this $p(x)$ will be an approximation to an (underlying) "true function" $f(x)$. A complementary viewpoint is that $f(x)$ is not part of our problem here: we are simply faced with a table of input data points, (x_j, y_j) for $j = 0, 1, \ldots, n-1$, and we are doing our best to produce a function, $p(x)$, which goes through those points and also is reasonably behaved everywhere else. Crucially, we will be interested in the case where both x_j and y_j are real; then, our $p(x)$ will also be real-valued.

Observe that applying Eq. (6.82) n times leads to an $n\times n$ linear system of equations, with $n = 2m$ unknowns. If there were no special tricks we could apply, solving this system would require $O(n^3)$ operations, just as we observed in our general comments in section 6.1.2.

While nothing so far assumes anything specific about the placement of the x_j values, it is very convenient to study the special case of equidistant values. Since we are dealing with the interval $[0, 2\pi]$ and n points in total, this naturally leads to a grid:

[24] It's not yet clear why this is called a *polynomial*, but bear with us.

$$x_j = \frac{2\pi j}{n}, \qquad j = 0, 1, \ldots, n-1 \tag{6.83}$$

As always, we are placing n points. While the value 0 is included in our x_j's (it's x_0), the value 2π isn't; this is because we know our signal is periodic, so if we had included an x_j at 2π then its corresponding y_j value would have been identical to y_0. Instead of storing needless information, we simply stop our points just before we get to 2π. This is illustrated in Fig. 6.8, where we are showing both the j index and the x_j value.

If we want to solve for the a_k and b_k parameters, the avenue that led to, say, Eq. (6.77) is now closed to us: we can't integrate Eq. (6.82) from 0 to 2π because we don't actually know $p(x)$; that is precisely what we are trying to compute. All we have at our disposal are the $p(x_j) = y_j$ values, available only at the grid points. With that in mind, we will, instead, employ the pleasant fact that sines and cosines are orthogonal to each other *even in the discrete case* for our equally spaced grid! In other words:

$$\begin{aligned}
&\sum_{j=0}^{n-1} \cos kx_j \cos lx_j = \begin{cases} 0, & k \neq l \\ m, & 0 < k = l < m \\ 2m, & k = l = 0 \text{ or } k = l = m \end{cases} \\
&\sum_{j=0}^{n-1} \cos kx_j \sin lx_j = 0 \\
&\sum_{j=0}^{n-1} \sin kx_j \sin lx_j = \begin{cases} 0, & k \neq l \\ m, & 0 < k = l < m \\ 0, & k = l = 0 \text{ or } k = l = m \end{cases}
\end{aligned} \tag{6.84}$$

You are guided toward these remarkable properties in problem 6.20. Armed with Eq. (6.84), we can immediately solve for our a_k and b_k parameters:

$$\begin{aligned}
a_k &= \frac{1}{m}\sum_{j=0}^{n-1} y_j \cos kx_j, \qquad k = 0, 1, \ldots, m \\
b_k &= \frac{1}{m}\sum_{j=0}^{n-1} y_j \sin kx_j, \qquad k = 1, 2, \ldots, m-1
\end{aligned} \tag{6.85}$$

Both of these are given in terms of y_j, the known input data. Pay close attention to the values k can take on in each case; these are fully consistent with what we need in Eq. (6.82). Speaking of which, we are now all set: we get our parameters from Eq. (6.85) and our interpolating polynomial $p(x)$ at any value x from Eq. (6.82); obviously, if we did things correctly, $p(x)$ should match the input data at the grid points, i.e., $p(x_j) = y_j$.[25]

[25] Incidentally, you should now be able to see why we did not include a $b_m \sin mx$ term in Eq. (6.82): such a term would vanish on the grid points, so we would have no way of determining the value of b_m.

Note that, having employed the discrete orthogonality, our problem of Eq. (6.85) requires only $O(n^2)$ operations to implement: we need $O(n)$ operations for a given k and are dealing with n values of k in total. This is already a major improvement (n^2 vs n^3); even so, we will soon find out that we can do even better than that.

6.4.2.2 Implementation

We remember that we dealt with a periodic function all the way back in chapter 3; this was $f(x) = e^{\sin(2x)}$, though the periodicity didn't play a role at the time. As in the case of polynomial interpolation earlier in this chapter, we first *produce* our data starting from our chosen function, followed by interpolating and comparing. Code 6.3 starts by defining a function to represent $f(x) = e^{\sin(2x)}$: observe that this employs `numpy` functionality; this shouldn't matter if we're passing in a single number but, as we'll soon discover, we can also use this function with a whole `numpy` array as an argument and it works equally seamlessly. We then define a function that picks the grid of x_j values as per Eq. (6.83) and produces the y_j values as $y_j = f(x_j)$; after this, $f(x)$ is no longer needed.

With the table of data (x_j, y_j) in place, we proceed to tackle the problem of trigonometric interpolation in two steps. We first create a function that evaluates our a_k and b_k parameters, as per Eq. (6.85). It is important to do this separately, since we don't want to re-evaluate these parameters each time we pick a different x at which to interpolate. The function `computeparams()` first sets up the dimensions of the two arrays, closely following Eq. (6.85); the values k can take for a_k and for b_k are different, so that is also reflected in the loops that evaluate our parameters. Note that, regardless of which of the two equations in Eq. (6.85) we are dealing with, the right-hand side always involves n terms in the sum; as a result, we have implemented both right-hand sides as `@` products of one-dimensional arrays, thereby obviating the need for a second loop and a `j` index. When storing the results in arrays, we have to be a bit careful in the case of `bparams`, since $k = 1, 2, \dots, m-1$ but `numpy` array indices start at 0; thus, all elements are shifted down by one (but not on the right-hand side, where Eq. (6.85) involves the actual k value).

The second step in trigonometric interpolation, now that the a_k and b_k parameters are in place, is to use Eq. (6.82) to compute the value of the interpolating polynomial $p(x)$ at a given x. In the function `triginterp()` we first take care of the a_0 and a_m terms, which need to be handled separately. We then step through the terms in the sum, handling $k = 1, 2, \dots, m-1$ one at a time. Note that we could have employed `@` multiplications here instead of a loop, as long as we were careful in treating the `bparams` appropriately. This would have involved a new array `ks = np.arange(1,m)` and slicing.

The main program simply calls three of the functions we defined. As before, we use `dataxs` for the input data x_j's and `x` for the x where we wish to interpolate; the value of the latter is picked at random. In Fig. 6.9 we are showing the $f(x)$ we started from. More importantly, we show the data (x_j, y_j) together with the result of our interpolating polynomial at many values of x. The left panel is for $n = 6$: the general trend is roughly captured, despite using so few points.[26] The $n = 8$ case (right panel) is even better. As you will find out

[26] You should try using $n = 4$ points. Did you see the result coming?

triginterp.py

Code 6.3

```
from math import pi
import numpy as np

def f(x):
    return np.exp(np.sin(2*x))

def generatedata(n,f):
    dataxs = 2*pi*np.arange(n)/n
    datays = f(dataxs)
    return dataxs, datays

def computeparams(dataxs,datays):
    n = dataxs.size
    m = n//2
    aparams = np.zeros(m+1)
    bparams = np.zeros(m-1)

    for k in range(m+1):
        aparams[k] = datays@np.cos(k*dataxs)/m
    for k in range(1,m):
        bparams[k-1] = datays@np.sin(k*dataxs)/m
    return aparams, bparams

def triginterp(aparams,bparams,x):
    n = aparams.size + bparams.size
    m = n//2
    val = 0.5*(aparams[0] + aparams[-1]*np.cos(m*x))
    for k in range(1,m):
        val += aparams[k]*np.cos(k*x)
        val += bparams[k-1]*np.sin(k*x)
    return val

if __name__ == '__main__':
    dataxs, datays = generatedata(6, f)
    aparams, bparams = computeparams(dataxs, datays)
    x = 0.3; pofx = triginterp(aparams, bparams, x)
    print(x,pofx)
```

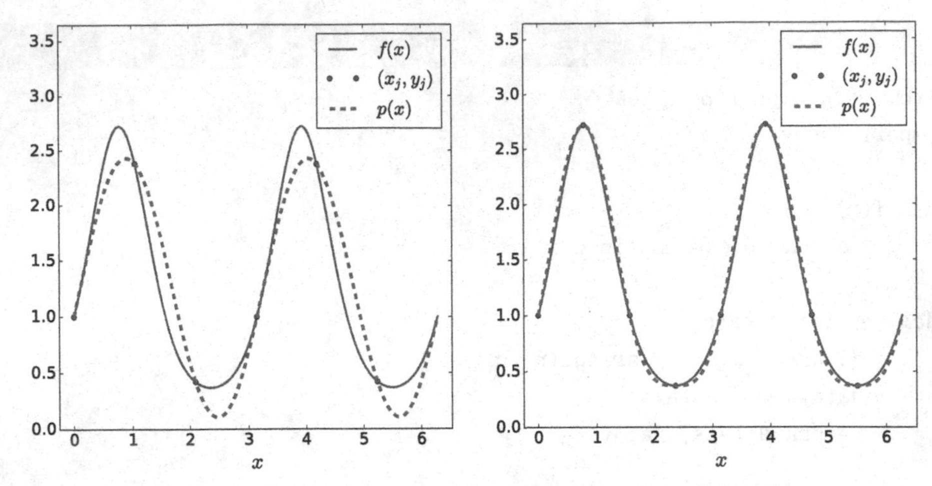

Fig. 6.9 Trigonometric interpolation for the cases of $n = 6$ (left) and $n = 8$ (right)

when you experiment for yourself, from 14 points and up you cannot tell the two curves apart with the naked eye.

Keep in mind that our equations and code above only work for the case of $n = 2m$, i.e., even-n. In problem 6.21 you are asked to extend the formalism to the case of odd-n but our main line of development will assume that n is even; as a matter of fact, below we will be yet more specific, requiring n to be a power of 2.

6.4.3 Discrete Fourier Transform

We could decide to stop here: combined with our earlier sections on Lagrange interpolation and cubic splines, we already have a toolbox that is sufficiently varied to handle a large number of approximation problems. However, the connections of trigonometric interpolation to the wider theme of Fourier transforms are too good to ignore; this will also allow us to introduce one of the most successful algorithms ever, the *fast Fourier transform* (FFT). Thus, we will now reformulate our earlier work with sines and cosines to use complex exponentials. However, we will not be doing this just for the sake of abstractly introducing the FFT, but will also see how to use this new algorithm to carry out interpolation for periodic problems. It's important to keep in mind that the literature on this subject is riddled with errors; we therefore choose to explicitly derive our interpolation formula step-by-step.

6.4.3.1 First Definition

When faced with the infinite Fourier series,[27] we saw that this could be given in either a real form (sines and cosines of Eq. (6.76)) or in a complex form (complex exponentials of Eq. (6.79)). In the former case our sum was over positive k's and in the latter over both positive and negative k's. When we moved to the real partial sum, Eq. (6.82), we summed

[27] Strictly speaking, an "infinite series" is a pleonasm, since a "finite series" would be a contradiction in terms.

over positive k's and were careful in handling the extra cosine that arose for even-n. We will now make an analogous transition from sines and cosines to complex exponentials for the partial sum, again for even-n (i.e., we are still taking $n = 2m$):

$$\begin{aligned}
p(x) &= \frac{1}{2}a_0 + \sum_{k=1}^{m-1}(a_k \cos kx + b_k \sin kx) + \frac{1}{2}a_m \cos mx \\
&= \frac{1}{2}a_0 + \sum_{k=1}^{m-1}\left(a_k \frac{e^{ikx}+e^{-ikx}}{2} + b_k \frac{e^{ikx}-e^{-ikx}}{2i}\right) + \frac{1}{2}a_m \cos mx \\
&= \frac{1}{2}a_0 + \sum_{k=1}^{m-1}\frac{1}{2}(a_k - ib_k)e^{ikx} + \sum_{k=1}^{m-1}\frac{1}{2}(a_k + ib_k)e^{-ikx} + \frac{1}{2}a_m \cos mx \\
&= \frac{1}{2}a_0 + \sum_{k=1}^{m-1}\frac{1}{2}(a_k - ib_k)e^{ikx} + \sum_{k=-m+1}^{-1}\frac{1}{2}(a_{-k} + ib_{-k})e^{ikx} + \frac{1}{2}a_m \cos mx \\
&= c_0 + \sum_{k=1}^{m-1} c_k e^{ikx} + \sum_{k=-m+1}^{-1} c_k e^{ikx} + c_{-m}\cos mx
\end{aligned} \tag{6.86}$$

In the first line we simply wrote down our interpolating polynomial from Eq. (6.82). In the second line we used de Moivre's formula (a couple of times) from Eq. (6.78). In the third line we grouped terms into two sums. In the fourth line we took $k \to -k$ in the second sum, also appropriately adjusting the values k can take on. In the fifth line we introduced a new set of c_k parameters, which are related to the a_k and b_k parameters in the following way:

$$\begin{aligned}
c_0 &= \frac{1}{2}a_0, \qquad & c_k &= \frac{1}{2}(a_k - ib_k), \qquad & k &= 1, 2, \ldots, m-1 \\
c_{-m} &= \frac{1}{2}a_m, \qquad & c_k &= \frac{1}{2}(a_{-k} + ib_{-k}), \qquad & k &= -m+1, -m+2, \ldots, -2, -1
\end{aligned} \tag{6.87}$$

The definition of c_{-m} was a bit arbitrary, since we could have just as well called this c_m (more on this below). We notice that the first three terms on the last line of Eq. (6.86) can be grouped into a single sum:

$$p(x) = \sum_{k=-m+1}^{m-1} c_k e^{ikx} + c_{-m}\cos mx \tag{6.88}$$

Crucially, this goes over both positive and negative k values, similarly to the infinite-series case in Eq. (6.79).[28] In contradistinction to Eq. (6.79), here we do not have only complex exponentials: the c_{-m} is treated separately and gets a cosine.

Recall that our problem started out with $n = 2m$ input data values y_j and Eq. (6.82) expressed $p(x)$ in terms of the a_k and b_k parameters: there were $n = 2m$ of these. In the present case of Eq. (6.88), counting up the c_k parameters, we find there are $2(m-1)+1+1 =$

[28] Incidentally, if you set $z = e^{ix}$, then the sum is over z^k: this is *polynomial interpolation on the unit circle*. Of course, k can also be negative, so this would be a *Laurent polynomial*.

$2m = n$ parameters, so everything is consistent. Note that we could have employed de Moivre's formula for the a_m term too in our derivation of Eq. (6.86); this would have led to our sum over k going from $-m$ to m, all for complex exponentials. Even so, the special treatment of this first/last term would not have been avoided: we would have been faced with a c_m and a c_{-m} that would have been equal. While it's not so pretty to single out c_{-m} as in Eq. (6.88), it would have been even stranger to use $2m + 1$ parameters c_k but have one of them be non-independent.[29]

Our main result is the new version of the interpolating polynomial $p(x)$, Eq. (6.88), that can be evaluated at any value of x we choose.[30] We now change gears, to address the more specific problem of using our grid of x_j values from Eq. (6.83). In other words, we will (once again) assume we are faced with a table of input data points, (x_j, y_j) for $j = 0, 1, \ldots, n-1$; we wish to compute the c_k's in terms of these known values. While we could produce an answer by using Eq. (6.85) to find the a_k and b_k parameters and then use those, in turn, via Eq. (6.87), to find the c_k parameters, we won't do that. Instead, we will try to work with Eq. (6.88) directly and attempt to find c_k in terms of x_j and y_j.

We start by evaluating the interpolating polynomial at the grid points:

$$p(x_j) = y_j = \sum_{k=-m+1}^{m-1} c_k e^{ikx_j} + c_{-m} \cos mx_j = \sum_{k=-m+1}^{m-1} c_k e^{ikx_j} + c_{-m} e^{-imx_j} = \sum_{k=-m}^{m-1} c_k e^{ikx_j} \tag{6.89}$$

The first equality assumed that the interpolation worked, i.e., we get the desired y_j values at x_j. The second equality plugged in x_j to Eq. (6.88). The third equality used the evenness of cosines, $\cos(-mx_j) = \cos mx_j$, and the vanishing of $\sin mx_j$ at our grid points:

$$\sin(-mx_j) = -\sin mx_j = -\sin\left(m\frac{2\pi j}{2m}\right) = -\sin \pi j = 0 \tag{6.90}$$

As a result, we are free to pretend that we were faced with a complex exponential, e^{-imx_j}, instead of only a cosine, $\cos mx_j$. This allows us to modify our sum so that it runs from $-m$ to $m-1$; this time, we are justified in doing so, because we are limited to points on the grid. Our main result is known as the *inverse discrete Fourier transform* (inverse DFT):

$$y_j = \sum_{k=-m}^{m-1} c_k e^{ikx_j}, \qquad j = 0, 1, \ldots, n-1 \tag{6.91}$$

It's important to realize that this equation relates y_j to c_k. In other words, it refers *only* to grid points as per Eq. (6.83); n numbers are transformed into n numbers. The problem, of course, is that we do not (yet) know the values of c_k; this is what we are trying to solve for.

[29] There is also the option of having the sum in Eq. (6.88) go from $-m$ to $m-1$ by fiat; this leads to an interpolating polynomial with an imaginary part and is therefore different from what we were doing in Eq. (6.82).

[30] We can produce an *error bound* for $|f(x) - p(x)|$: combine Eq. (6.79) with the odd-n version of Eq. (6.88), for which there is no cosine. The triangle inequality allows us to move the modulus inside the sum, getting rid of the complex exponentials, so we get $\sum_{|k|\geq m} |c_k|$. Now, integrate Eq. (6.80) by parts (p times) to show that $c_k = O(k^{-p})$. Together with the sum, our bound becomes $O(m^{-p})$ for any p: this is called *spectral accuracy*.

We now wish to have a formula that goes in the opposite direction: starting from the values of x_j and y_j, compute c_k. We will accomplish this by combining Eq. (6.91) with a(nother) remarkable property, namely the fact that complex exponentials are orthogonal to each other *even in the discrete case*:

$$\sum_{j=0}^{n-1} e^{ikx_j} e^{-ilx_j} = \begin{cases} n, & (k-l)/n \text{ is integer} \\ 0, & \text{otherwise} \end{cases} \tag{6.92}$$

Problem 6.20 guides you toward proving this. Knowing that sines and cosines are orthogonal to each other even for the discrete case, Eq. (6.84), this perhaps does not come as a total surprise. As advertised, we now multiply Eq. (6.91) by e^{-ilx_j} and sum over all the x_j's:

$$\sum_{j=0}^{n-1} e^{-ilx_j} y_j = \sum_{k=-m}^{m-1} c_k \sum_{j=0}^{n-1} e^{ikx_j} e^{-ilx_j} = \sum_{k=-m}^{m-1} c_k n \delta_{k,l} = nc_l \tag{6.93}$$

In the penultimate step we used the orthogonality property from Eq. (6.92), while acknowledging that our k and l values run only from $-m$ to $m-1$; thus, the only contribution comes from the case $k = l$. We now rewrite our result in Eq. (6.93) by renaming $l \to k$:

$$c_k = \frac{1}{n} \sum_{j=0}^{n-1} y_j e^{-ikx_j}, \qquad k = -m, -m+1, \ldots, m-2, m-1 \tag{6.94}$$

This, our desired result, is known as the *discrete Fourier transform* (DFT). It produces all $2m$ parameters c_k, starting from our input points y_j. These were precisely the c_k parameters we needed in order to evaluate the interpolating polynomial from Eq. (6.88). Once again, there is nothing continuous going on in Eq. (6.94): n numbers are transformed into n numbers. Our result also explains why Eq. (6.91) is known as the *inverse DFT*, since that equation recovers the data points y_j once you know the parameters c_k.

Let's quickly summarize a few points before things get out of hand; this paragraph will be somewhat dense, but may help you in the future. Keep in mind that in the case of an *infinite Fourier series* we are faced with a sum (expanding the function in terms of the Fourier coefficients) and an integral (computing the Fourier coefficients in terms of the function). On the other hand, in the case of *trigonometric interpolation* or the *discrete Fourier transform* we have a sum (direct) and a sum (inverse); interpolation deals with any x value whereas the DFT deals only with grid points x_j. Note that both the infinite Fourier series and the discrete Fourier transform apply to the case of a periodic function/signal. If you are faced with a non-periodic function, you can simply focus on a given interval (assuming the function *does* repeat outside) and try to decribe only that region; your predictions will only be applicable to the interval you focused on (since, in reality, the function *doesn't* repeat outside the region of interest).[31] Incidentally, non-periodic functions give rise to the *(continuous) Fourier transform*, which involves an integral (direct) and an integral (inverse). In this book we have little to say on continuous Fourier transforms, since we are always

[31] A more detailed study would also address topics like aliasing, windowing, and leakage.

discretizing. While we arrived at the discrete Fourier transform starting from trigonometric interpolation, you will not be surprised to hear that you can also view the DFT as a discretization of the continuous Fourier transform (see problem 6.25).

6.4.3.2 Shifted Version

We will now slightly tweak our results. This is done both in anticipation of introducing the fast Fourier transform in the following subsection, and also in order to make contact with notation that is standard on this subject. We will make three related changes.

First, we observe that our expressions in Eq. (6.94) and in Eq. (6.91) involve x_j, which are our grid points from Eq. (6.83). Let's use that equation to re-express the DFT:

$$c_k = \frac{1}{n}\sum_{j=0}^{n-1} y_j e^{-2\pi ikj/n}, \qquad k = -m, -m+1, \dots, m-2, m-1 \tag{6.95}$$

This ikj in the exponent may be confusing, so keep in mind that i is the imaginary unit, k is our wave number (or "frequency") index, and j is our spatial index. Second, we note that our expressions in Eq. (6.94) and in Eq. (6.91) exhibit an asymmetry: j is non-negative, going from 0 to $n-1$, whereas k can be either negative or positive, going from $-m$ to $m-1$. As it turns out, this asymmetry can be lifted; to do so, observe that Eq. (6.95) can help us show that the DFT is periodic in k, with period n:

$$c_{k+n} = \frac{1}{n}\sum_{j=0}^{n-1} y_j e^{-2\pi i(k+n)j/n} = \frac{1}{n}\sum_{j=0}^{n-1} y_j e^{-2\pi ikj/n} e^{-2\pi ij} = c_k \tag{6.96}$$

since $e^{-2\pi ij} = 1$. Given that up to this point our k could also take on negative values, we can use $c_{k+n} = c_k$ to re-express c_k for the negative frequencies $k = -m, -m+1, \dots, -1$ in terms of positive frequencies $k = m, m+1, \dots, n-1$.[32] Third, it is customary to have the $1/n$ in the definition of the *inverse* DFT, not in that of the DFT, cf. Eq. (6.91) and Eq. (6.94).

Putting these three modifications together, we arrive at a new definition of the DFT. In order to keep our notation straight, we will use a new symbol for the new version of our Fourier parameters, $\tilde{y}_k$, where k now goes from 0 to $n-1$:

$$\tilde{y}_k = \sum_{j=0}^{n-1} y_j e^{-2\pi ikj/n}, \qquad k = 0, 1 \dots, n-1 \tag{6.97}$$

As before, this is simply transforming n numbers into n numbers. Observe how there is no x_j left over here; the assumption that we have equally spaced points has already been used, so our expressions from now on will only involve y_j and $\tilde{y}_k$. Crucially, both our j and our k indices now run from 0 to $n-1$.

If you're having trouble seeing the correspondence between Eq. (6.94) and Eq. (6.97), keep in mind that (a) we've replaced x_j with j as per Eq. (6.83), (b) we've shifted the

[32] Observe that those slots were *not* previously taken: our positive frequencies were only going up to $m-1$.

negative frequencies so that they appear after the positive frequencies, and (c) there is no $1/n$ in the new definition of the DFT. The last two modifications can be summarized as follows:

$$\begin{aligned} \left(\tilde{y}_0 \quad \tilde{y}_1 \quad \ldots \quad \tilde{y}_{m-1}\right)^T &= n\left(c_0 \quad c_1 \quad \ldots \quad c_{m-1}\right)^T \\ \left(\tilde{y}_m \quad \tilde{y}_{m+1} \quad \ldots \quad \tilde{y}_{n-1}\right)^T &= n\left(c_{-m} \quad c_{-m+1} \quad \ldots \quad c_{-1}\right)^T \end{aligned} \tag{6.98}$$

In addition to the (arbitrary) choice to remove the $1/n$, this also shows that the (zero and) positive frequencies in c_k are still in the same place when using $\tilde{y}_k$, whereas the negative frequencies in c_k are now placed in the "second half" of the available frequency slots of $\tilde{y}_k$. Keep in mind that the c_{-m} can be viewed as corresponding to either the largest-magnitude negative frequency or the largest-magnitude positive frequency: $c_{-m} = c_m$, as a consequence of $c_{k+n} = c_k$. Note that the 0th-frequency component, $\tilde{y}_0$, was also the 0th-frequency component of c_k; it's sometimes called the DC component: you can see from Eq. (6.97) that it is simply the sum of all the y_j's. Overall, we have now switched to what is known as the "standard order" of the DFT; we will sometimes refer to it as the "shifted version", as in the current subsection's heading.

We can also use our new Fourier parameters $\tilde{y}_k$ to rewrite our definition of the *inverse DFT* from Eq. (6.91). This has now turned into:

$$y_j = \frac{1}{n}\sum_{k=0}^{n-1} \tilde{y}_k e^{2\pi i k j/n}, \qquad j = 0, 1 \ldots, n-1 \tag{6.99}$$

To see that this is true, you could start from Eq. (6.97) and employ the complex discrete orthogonality of Eq. (6.92). As in the case of the updated (direct) DFT, the x_j's are gone and the k's are now non-negative. Furthermore, there is now a $1/n$ term on the right-hand side: to see where this came from,[33] look at Eq. (6.91) and recall how we went from c_k to $\tilde{y}_k$ in Eq. (6.98).

A comment we will make for the fourth time in a row: Eq. (6.99) is simply transforming n numbers into n numbers. We keep repeating this fact in order to highlight that the direct and inverse DFT transforms are fundamentally discrete problems which refer to points on a grid. We will take this into consideration in what follows.

6.4.3.3 Fast Fourier Transform

As you will show in problem 6.31, the case of real input data, y_j, is special, in that you only need to evaluate $m + 1$ Fourier parameters; this is so because $\tilde{y}_{n-k} = \tilde{y}_k^*$ holds for this case, so the remaining coefficients are complex conjugates of already-evaluated ones. (Incidentally, the $\tilde{y}_k$ might be complex, even when the y_j are all real; this is OK, because the inverse DFT of Eq. (6.99) makes everything real again.) You could consider this to be a way of reducing the computational runtime of a DFT implementation. Instead of following

[33] If you didn't follow our admonition to employ the orthogonality explicitly.

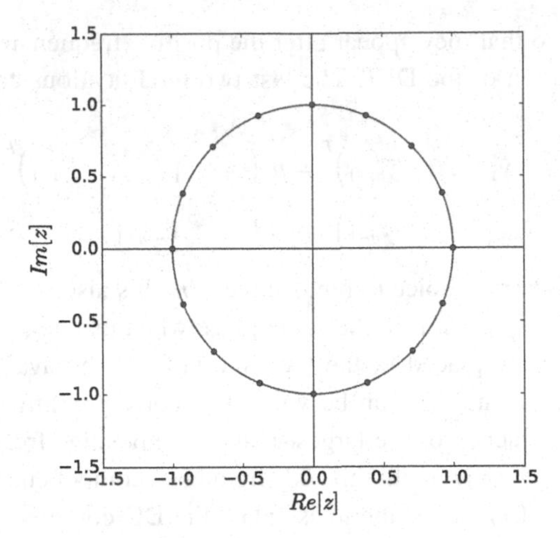

Fig. 6.10 Complex plane unit circle and 16th roots of unity, $e^{-2\pi ik/n}$ for $k = 0, 1, \ldots, 15$ and $n = 16$

this approach, however, we will now turn to a much more effective way of speeding the DFT up, regardless of whether or not the input data, y_j, are real. As is usually the case in computing, algorithmic breakthroughs matter much more than "small efficiencies".

Before we start accelerating things, let us observe that our latest definition of the DFT, Eq. (6.97), can be expressed in the form:

$$\tilde{y}_k = \sum_{j=0}^{n-1} (e^{-2\pi i/n})^{kj} y_j, \qquad k = 0, 1 \ldots, n-1 \tag{6.100}$$

But this, in its turn, can be written in matrix form:

$$\tilde{\mathbf{y}} = \mathbf{E}\mathbf{y} \tag{6.101}$$

where $\mathbf{E}$ is the $n \times n$ matrix[34] made up of $(\mathbf{E})_{kj}$, which are powers of $e^{-2\pi i/n}$. In short, we now see that the problem of computing the DFT is essentially *matrix–vector multiplication*! As we discovered in section 4.3.1.4,[35] this is a problem for which the operation count is $O(n^2)$. The main idea behind the *fast Fourier transform* (FFT) is that the matrix–vector multiplication of Eq. (6.100) involves specific symmetries (despite not being sparse) and therefore does not have to cost $O(n^2)$. In other words, we will now see how to exploit the specific properties of the $\mathbf{E}$ matrix in order to make the DFT dramatically faster. For the sake of simplicity, we assume that n is a power of 2.

Let us first turn to an elementary problem involving complex numbers. We are looking for the roots of the equation $z^n = 1$; the solutions are known as the *n-th roots of unity*. Here's one of them:

$$e^{-2\pi i/n} = \cos\left(\frac{2\pi}{n}\right) - i \sin\left(\frac{2\pi}{n}\right) \tag{6.102}$$

Obviously, $(e^{-2\pi i/n})^n = 1$, meaning that $e^{-2\pi i/n}$ is, indeed, an n-th root of unity. It is not the

[34] Intriguingly, this is a Vandermonde matrix giving rise to a unitary transformation (see problem 6.28).

[35] Of course, here we are dealing with complex numbers, but that doesn't change the overall scaling.

only one, however. It should be easy to see that $(e^{-2\pi i/n})^k = e^{-2\pi ik/n}$, for $k = 0, 1, \dots, n-1$, are all n-th roots of unity, since $(e^{-2\pi ik/n})^n = 1$ for all k's. For a given value of n, since k runs from 0 to $n-1$, there are n such distinct n-th roots of unity, $e^{-2\pi ik/n}$. They are illustrated in Fig. 6.10 for the case of $n = 16$. The crucial take-away here is that, for the case $n = 16$, there are 16 distinct roots *only*: even if you raise $e^{-2\pi ik/n}$ to another power, say j as in $(e^{-2\pi ik/n})^j$, you can only get back one of the 16 roots.[36]

Having grasped the meaning of the n-th roots of unity, we now return to our **E** matrix from Eq. (6.101). We realize that out of the n^2 matrix elements there are only n distinct ones (since many of them repeat); this **E** matrix is complex symmetric, but not Hermitian. To reiterate, the fast Fourier transform is essentially a clever way of exploiting the fact that our matrix–vector multiplication involves $(e^{-2\pi i/n})^{kj}$. It's easier to see this in a specific example, so let us write out Eq. (6.101) for the 4-element case:

$$\begin{aligned}
\tilde{y}_0 &= y_0(e^{-2\pi i/4})^0 + y_1(e^{-2\pi i/4})^0 + y_2(e^{-2\pi i/4})^0 + y_3(e^{-2\pi i/4})^0\\
\tilde{y}_1 &= y_0(e^{-2\pi i/4})^0 + y_1(e^{-2\pi i/4})^1 + y_2(e^{-2\pi i/4})^2 + y_3(e^{-2\pi i/4})^3\\
\tilde{y}_2 &= y_0(e^{-2\pi i/4})^0 + y_1(e^{-2\pi i/4})^2 + y_2(e^{-2\pi i/4})^4 + y_3(e^{-2\pi i/4})^6\\
\tilde{y}_3 &= y_0(e^{-2\pi i/4})^0 + y_1(e^{-2\pi i/4})^3 + y_2(e^{-2\pi i/4})^6 + y_3(e^{-2\pi i/4})^9
\end{aligned} \tag{6.103}$$

Were we to evaluate things at this stage, we would have to carry out $4^2 = 16$ multiplications. We notice that our exponents here go up to nine, even though we know that there are only four distinct roots of unity: $(e^{-2\pi i/4})^0$, $(e^{-2\pi i/4})^1$, $(e^{-2\pi i/4})^2$, and $(e^{-2\pi i/4})^3$. This means we can use elementary manipulations to re-express the 16 matrix elements of **E** in terms of only these four roots of unity. For example:

$$(e^{-2\pi i/4})^6 = (e^{-2\pi i/4})^4(e^{-2\pi i/4})^2 = (e^{-2\pi i/4})^2 \tag{6.104}$$

Using this fact, and regrouping, allows us to re-express our four equations in the form:

$$\begin{aligned}
\tilde{y}_0 &= [y_0(e^{-2\pi i/4})^0 + y_2(e^{-2\pi i/4})^0] + (e^{-2\pi i/4})^0[y_1(e^{-2\pi i/4})^0 + y_3(e^{-2\pi i/4})^0]\\
\tilde{y}_1 &= [y_0(e^{-2\pi i/4})^0 + y_2(e^{-2\pi i/4})^2] + (e^{-2\pi i/4})^1[y_1(e^{-2\pi i/4})^0 + y_3(e^{-2\pi i/4})^2]\\
\tilde{y}_2 &= [y_0(e^{-2\pi i/4})^0 + y_2(e^{-2\pi i/4})^0] + (e^{-2\pi i/4})^2[y_1(e^{-2\pi i/4})^0 + y_3(e^{-2\pi i/4})^0]\\
\tilde{y}_3 &= [y_0(e^{-2\pi i/4})^0 + y_2(e^{-2\pi i/4})^2] + (e^{-2\pi i/4})^3[y_1(e^{-2\pi i/4})^0 + y_3(e^{-2\pi i/4})^2]
\end{aligned} \tag{6.105}$$

In addition to factoring out and simplifying some of the exponents, observe that we first write the even y_j elements (y_0 and y_2) and then the odd ones (y_1 and y_3).

Introducing new notation, which we hope is self-explanatory, this takes the form:

$$\begin{aligned}
\tilde{y}_0 &= \tilde{y}_0^{\text{even}} + (e^{-2\pi i/4})^0\tilde{y}_0^{\text{odd}}\\
\tilde{y}_1 &= \tilde{y}_1^{\text{even}} + (e^{-2\pi i/4})^1\tilde{y}_1^{\text{odd}}\\
\tilde{y}_2 &= \tilde{y}_0^{\text{even}} + (e^{-2\pi i/4})^2\tilde{y}_0^{\text{odd}}\\
\tilde{y}_3 &= \tilde{y}_1^{\text{even}} + (e^{-2\pi i/4})^3\tilde{y}_1^{\text{odd}}
\end{aligned} \tag{6.106}$$

where we noticed that the y_j's and the powers of $e^{-2\pi i/4}$ inside the square brackets appear in only four distinct combinations, which we denoted by $\tilde{y}_0^{\text{even}}$, $\tilde{y}_0^{\text{odd}}$, $\tilde{y}_1^{\text{even}}$, and $\tilde{y}_1^{\text{odd}}$. To

[36] As hinted at in footnote 20, the similarity between Fig. 6.10 and Fig. 6.2 is hard to miss.

compute these, we need to carry out only $2 \times (4/2)^2 = 8$ multiplications. Make sure you understand what these entities are; for example:

$$\tilde{y}_1^{\text{even}} = y_0(e^{-2\pi i/4})^0 + y_2(e^{-2\pi i/4})^2 = y_0(e^{-2\pi i/2})^0 + y_2(e^{-2\pi i/2})^1 \tag{6.107}$$

The $e^{-2\pi i/4}$ always appears raised either to the 0th or the 2nd power, so we can rewrite the exponential to have 2 in the denominator instead of 4. (This also applies to the $\tilde{y}^{\text{odd}}$'s.)

As per Eq. (6.106), once you have the two $\tilde{y}^{\text{even}}$'s and the two $\tilde{y}^{\text{odd}}$'s, you do four more multiplications in order to combine them together and produce the $\tilde{y}_k$. Even in this small problem, we have managed to reduce the total number of multiplications required.

We now take yet another step, rewriting Eq. (6.106) as follows:

$$\begin{aligned}
\tilde{y}_0 &= \tilde{y}_0^{\text{even}} + (e^{-2\pi i/4})^0 \tilde{y}_0^{\text{odd}} \\
\tilde{y}_1 &= \tilde{y}_1^{\text{even}} + (e^{-2\pi i/4})^1 \tilde{y}_1^{\text{odd}} \\
\tilde{y}_{2+0} &= \tilde{y}_0^{\text{even}} - (e^{-2\pi i/4})^0 \tilde{y}_0^{\text{odd}} \\
\tilde{y}_{2+1} &= \tilde{y}_1^{\text{even}} - (e^{-2\pi i/4})^1 \tilde{y}_1^{\text{odd}}
\end{aligned} \tag{6.108}$$

The first two relations are unchanged; the last two were rewritten to highlight the fact that all the relevant indices go from 0 to 1. On the right-hand sides of the last two equations we wrote $2 = 2 + 0$ and $3 = 2 + 1$ in the exponents and then made use of the fact that $e^{-\pi i} = -1$. In short, we have rewritten our 4-element DFT problem from Eq. (6.103) in terms of two smaller (2-element) DFT problems, one for the even-j indices and one for the odd-j indices (recall Eq. (6.107)), which are combined with the appropriate prefactors as per Eq. (6.108).

Having worked out the 4×4 example in gory detail, we hope that you will now be able to understand without much effort the general case, which goes by the name of the *Danielson–Lanczos lemma*. Let's start from Eq. (6.97), splitting the sum over j into a sum over the even components, $j = 2l$, and the odd components, $j = 2l + 1$:

$$\begin{aligned}
\tilde{y}_k &= \sum_{j=0}^{n-1} y_j e^{-2\pi i k j/n} = \sum_{l=0}^{m-1} y_{2l} e^{-2\pi i k(2l)/n} + \sum_{l=0}^{m-1} y_{2l+1} e^{-2\pi i k(2l+1)/n} \\
&= \sum_{l=0}^{m-1} y_{2l} e^{-2\pi i k l/m} + e^{-2\pi i k/n} \sum_{l=0}^{m-1} y_{2l+1} e^{-2\pi i k l/m} = \tilde{y}_k^{\text{even}} + e^{-2\pi i k/n} \tilde{y}_k^{\text{odd}}
\end{aligned} \tag{6.109}$$

Note that in the second equality our sums switched to using l and therefore the maximum value became $m - 1$. In the exponents of the third equality we moved the 2 from the numerators to the denominators, thereby getting m; in the second term, we also factored out the extra term (which has an n in the denominator of the exponent). In the fourth equality we identified our new m-element DFTs, the ones for the even and odd components.

As we saw above, this is already a gain; however, let's take one more step, to make our general result match what we did for the 4×4 case. As you may recall, we managed to express the "second half" of our equations in terms of quantities that had already appeared in the first half. Right now, in Eq. (6.109) the k goes from 0 to $n - 1$, so we seem to be using more $\tilde{y}^{\text{even}}$'s and $\tilde{y}^{\text{odd}}$'s than we actually need. To remedy the situation, introduce a new wave number index q which runs from 0 to $m - 1$, i.e., it appears in the combination

$k = m + q$. We have:

$$\tilde{y}_{m+q}^{\text{even}} = \sum_{l=0}^{m-1} y_{2l} e^{-2\pi iml/m} e^{-2\pi iql/m} = \sum_{l=0}^{m-1} y_{2l} e^{-2\pi iql/m} = \tilde{y}_q^{\text{even}} \tag{6.110}$$

since $e^{-2\pi il} = 1$. Now both the sum over l and the index q only cover half the range from 0 to $n-1$ (i.e., go from 0 to $m-1$). A fully analogous derivation can be carried out for the odd case, leading to:

$$\tilde{y}_{m+q}^{\text{odd}} = \tilde{y}_q^{\text{odd}} \tag{6.111}$$

There is only one other ingredient missing: the factored-out coefficients in Eq. (6.109) still employ k which goes up to $n-1$. As in Eq. (6.108), we can also halve those:

$$e^{-2\pi ik/n} = e^{-2\pi i(m+q)/n} = e^{-2\pi i/2} e^{-2\pi iq/n} = -e^{-2\pi iq/n} \tag{6.112}$$

Putting everything together, we have accomplished what we set out to do:

$$\begin{aligned} \tilde{y}_q &= \tilde{y}_q^{\text{even}} + e^{-2\pi iq/n} \tilde{y}_q^{\text{odd}} \\ \tilde{y}_{m+q} &= \tilde{y}_q^{\text{even}} - e^{-2\pi iq/n} \tilde{y}_q^{\text{odd}} \end{aligned} \tag{6.113}$$

where we left the first-half of the factored-out coefficients untouched. For both equations, we have $q = 0, 1, \ldots, m-1$. This closely matches our earlier result, Eq. (6.108).

As you may have suspected, the fast Fourier transform doesn't carry out such a halving process only once: it employs a *divide-and-conquer* approach, continually halving, until the problem becomes sufficiently small that it cannot be cut in half. At that point, the problem will be simple enough that the DFT can be trivially arrived at: you can see from Eq. (6.97) that when $n = 1$ we have $\tilde{y}_0 = y_0$. You may now realize why we said that our n has to be a power of 2: this was in order to enable us to keep halving until we get down to $n = 1$ without a problem. Each stage of this algorithm involves n multiplications and there are $\log_2 n$ such stages in total. In all, the fast Fourier transform operation count is $O(n \log_2 n)$, a dramatic improvement over the "naive" implementation of the DFT, which involved $O(n^2)$ operations. You may also recall, from problem 1.1, that divide-and-conquer algorithms are most naturally implemented using recursion. This is fortunate, because it means that we do not have to produce separate notation for each stage of the algorithm: it is enough to show one halving step and state that we repeat the process recursively. We will return to these considerations soon.

As usual, things can be made even more efficient than that. You could think about where the different quantities are stored, how many of them can be re-used from one stage to the next, how you would go about reformulating the algorithm iteratively (problem 6.29), and so on. However, the crucial point is that the FFT introduces a different *scaling*: the matrix multiplication of Eq. (6.101) is no longer quadratic! When n is large, this can have a dramatic effect on how long things take; problem 6.26 asks you to compare the runtime of DFT and FFT for the case of $n = 8192$. Viewed from another perspective, the FFT enables calculations that would have been impossible using only the old-fashioned DFT.

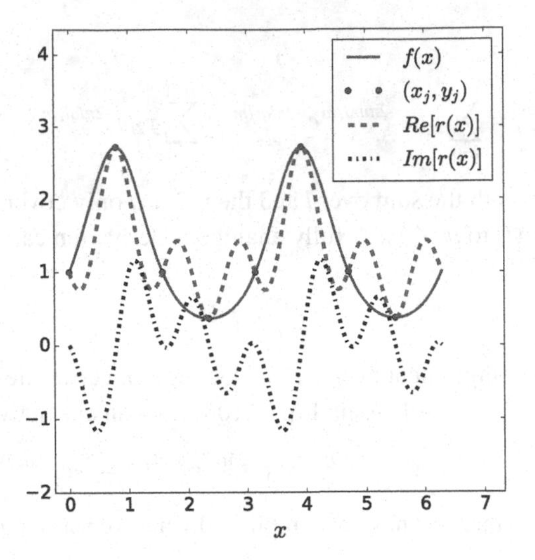

Fig. 6.11 Naive attempt to carry out shifted-DFT interpolation for the case of $n = 8$

This is one of the reasons why Fourier transforms are so prevalent in a very large number of applications. For example, in the study of differential equations, FFTs are routinely used to go to wave-number space and (trivially) take derivatives there (see section 8.6). Instead of following that route, however, we now recall that our goals in this chapter were more humble: we return to the problem of trigonometric interpolation.

6.4.3.4 Interpolation Using the FFT

Backing up for a second, we realize that the fast Fourier transform, Eq. (6.113), is just a fast way of producing the discrete Fourier transform, Eq. (6.97). In other words, both approaches simply evaluate the $\tilde{y}_k$'s for $k = 0, 1, \ldots, n-1$. A few subsections ago, once we calculated the a_k and b_k parameters we were ready to produce the interpolating polynomial $p(x)$ given in Eq. (6.82); similarly, in Eq. (6.88) we re-expressed $p(x)$ in terms of the c_k parameters. Wishing to produce the shifted-DFT analogue, once you've computed the $\tilde{y}_k$'s, you may be tempted to write down an interpolating polynomial of the form:

$$r(x) = \frac{1}{n}\sum_{k=0}^{n-1} \tilde{y}_k e^{ikx} \tag{6.114}$$

This bears a passing resemblance to our infinite Fourier series, Eq. (6.79), and looks pretty similar to our shifted inverse DFT, Eq. (6.99). As a matter of fact, it basically *is* Eq. (6.99) but this time evaluated not at $x_j = (2\pi j)/n$ but at any x. It is certainly plausible that $r(x)$ would do a good job interpolating, since we already know (by construction) that $r(x_j) = y_j$: evaluating Eq. (6.114) at the grid points turns $r(x)$ into the y values of our input data.

As you will discover when you solve problem 6.24, this polynomial doesn't actually do a very good job of interpolating between the (x_j, y_j) points; see Fig. 6.11 for the $n = 8$ case. Our $r(x)$ does go through the input data (x_j, y_j), as it should. However, comparing our new polynomial to our earlier trigonometric interpolation, see the right panel of Fig. 6.9, the general behavior is very different. Not only is the real part of $r(x)$ exhibiting wild fluctuations between the grid points, but $r(x)$ also comes with a sizable imaginary part, which also fluctuates. As it so happens, we already knew that the "interpolating" polynomial of Eq. (6.114) would give these different results, which is why we made sure to call it $r(x)$, i.e., not $p(x)$.

The reason $r(x)$ is so different from $p(x)$ has nothing to do with the $1/n$ that appears in Eq. (6.114). As you may recall, that was simply a standard choice in how to define the Fourier coefficients $\tilde{y}_k$ in comparison to our earlier c_k parameters. The $1/n$ should be there if you're using $\tilde{y}_k$ as input. The actual problems with using $r(x)$ as per Eq. (6.114) are two: (a) we assumed that a sum over non-negative k's, as per the shifted version of the DFT, would work, and (b) we didn't treat the largest/smallest frequency term, $\tilde{y}_m$, specially.

Both of these issues can be addressed in one go. Instead of producing an *ad hoc* guess, let us start with our interpolating polynomial $p(x)$ from Eq. (6.88) and update its definition to use our shifted Fourier coefficients, $\tilde{y}_k$. To refresh your memory, what we had there was:

$$p(x) = \sum_{k=-m+1}^{m-1} c_k e^{ikx} + c_{-m} \cos mx \tag{6.115}$$

We now recall that the correspondence between shifted and unshifted parameters was, as per Eq. (6.98):

$$\begin{aligned} \begin{pmatrix} \tilde{y}_0 & \tilde{y}_1 & \dots & \tilde{y}_{m-1} \end{pmatrix}^T &= n \begin{pmatrix} c_0 & c_1 & \dots & c_{m-1} \end{pmatrix}^T \\ \begin{pmatrix} \tilde{y}_m & \tilde{y}_{m+1} & \dots & \tilde{y}_{n-1} \end{pmatrix}^T &= n \begin{pmatrix} c_{-m} & c_{-m+1} & \dots & c_{-1} \end{pmatrix}^T \end{aligned} \tag{6.116}$$

This immediately leads us to the following formulation of our interpolating polynomial:

$$p(x) = \frac{1}{n} \sum_{k=0}^{m-1} \tilde{y}_k e^{ikx} + \frac{1}{n} \tilde{y}_m \cos mx + \frac{1}{n} \sum_{k=m+1}^{n-1} \tilde{y}_k e^{i(k-n)x} \tag{6.117}$$

where you will notice that we are using the same symbol, $p(x)$, since it *is* the same polynomial, this time expressed in terms of our shifted Fourier coefficients, $\tilde{y}_k$. The positive terms in the sum remain unchanged, the negative ones are handled separately, as is true of the midpoint (which corresponds to the largest/smallest wave number). The only subtle point here is the exponent in the second sum: to match the exponents for the negative terms of the sum in Eq. (6.115), we need to shift; make sure to check that $k - n$ goes from $-m + 1$ to -1, when k goes from $m + 1$ to $n - 1$. Finally, we are also including a $1/n$ term to ensure the two definitions match.

6.4.3.5 Implementation

We are now ready to implement trigonometric interpolation using the FFT. If you've been following along, you will realize that we could have done something analogous at several points along the way: we could have implemented the unshifted DFT, Eq. (6.94), to find the

Code 6.4 fft.py

```
from triginterp import f, generatedata
from math import pi
import numpy as np

def fft(ys):
    n = ys.size
    m = n//2
    if n==1:
        ytils = ys
    else:
        evens = fft(ys[::2])
        odds = fft(ys[1::2])
        coeffs = np.exp(-2*pi*np.arange(m)*1j/n)
        first = evens + coeffs*odds
        second = evens - coeffs*odds
        ytils = np.concatenate((first, second))
    return ytils

def fftinterp(ytils,x):
    n = ytils.size
    m = n//2
    val = ytils[:m]@np.exp(np.arange(m)*x*1j)
    val += ytils[m]*np.cos(m*x)
    val += ytils[m+1:]@np.exp(np.arange(-m+1,0)*x*1j)
    return val/n

if __name__ == '__main__':
    n = 8
    dataxs, datays = generatedata(n, f)
    ytils = fft(datays)
    x = 0.3; pofx = fftinterp(ytils, x)
    print(x,pofx.real)
```

c_k's and then used that to produce $p(x)$ as per Eq. (6.88). We could have also implemented the shifted DFT, Eq. (6.97), to find the $\tilde{y}_k$'s and then computed $p(x)$ from Eq. (6.117). Both of these are easier programming tasks, which you are asked to carry out in the problem set. Instead, we will now capitalize on all our progress so far and implement the FFT,

Eq. (6.113), to find the $\tilde{y}_k$'s and then plug those in to Eq. (6.117) to compute $p(x)$. Since we are faced with two separate tasks, computing the $\tilde{y}_k$'s and $p(x)$, we define two functions in code 6.4. Our first function implements the fast Fourier transform. As discussed in problem 1.1, and around Eq. (6.113), it is natural to code up divide-and-conquer algorithms using recursion. We start from our *base case*, $n = 1$, for which the answer can be directly given: we know from Eq. (6.97) that in this case we have $\tilde{y}_0 = y_0$. For any other value of n (always a power of 2), we cut the problem up into two halves using `numpy`'s slicing: `ys[::2]` and `ys[1::2]` select the even and odd components, respectively. Then, `fft()` calls itself to solve that simpler problem. Once an answer for `evens` and `odds` is known, our program simply multiplies by $+e^{-2\pi iq/n}$ or $-e^{-2\pi iq/n}$, exactly as in Eq. (6.113). It's interesting to note that we are not employing any loops, so the q index doesn't explicitly appear. We combine our two halves ($\tilde{y}_q$ and $\tilde{y}_{m+q}$) together by using `numpy.concatenate()` to make a larger array. In all, even though we are implementing a state-of-the-art algorithm, this function is very short and reasonably straightforward to understand. You may wish to print out its output or intermediate values, in order to be certain that everything is clear to you.

Specifically, you should make sure to understand that, even though Eq. (6.113) is written in terms of q and $m+q$, the end result is that we've managed to compute the $\tilde{y}_k$ coefficients, with k going from 0 to $n-1$. This may be even easier to grasp when you solve problem 6.26 that asks you to evaluate the $\tilde{y}_k$'s using the (slow) DFT approach, Eq. (6.97). Having computed the $\tilde{y}_k$'s, our second function uses them to implement Eq. (6.117). Nothing exciting is going on here: once again, we are employing `numpy` functionality, i.e., the dot product of one-dimensional arrays using `@`, to avoid having explicit loops and indices. Our three lines of code directly correspond to the three terms on the right-hand side of Eq. (6.117).

The main program first creates some data points using the functions we defined in `triginterp.py`. It then calls our two new functions to compute the $\tilde{y}_k$'s and then the interpolating polynomial $p(x)$. Once again, we pick a value of x at random, just to show that we can evaluate $p(x)$ at any point we choose. We end by printing out the real part of the value of $p(x)$; crucially, the imaginary part that we are dropping is *always* zero. Our $\tilde{y}_k$ and the exponentials in Eq. (6.117) are complex, but they combine to produce a real interpolating polynomial at any value of x. This is so by construction, since we've been trying to match our trigonometric-interpolation polynomial every step of the way. As a result, we can use code 6.4 to reproduce the right panel of Fig. 6.9 exactly.

6.5 Linear Least-Squares Fitting

As we saw in our introduction in section 6.1.2, *least-squares fitting* is different from what we've spent the last several pages doing: we will now no longer trust our input table (x_j, y_j) of N data pairs blindly. This time around, each data point will have an associated input uncertainty, σ_j; you may wish to think of the y-values of your data as $y_j \pm \sigma_j$. In other words, we will *not* demand that our approximating function, $p(x)$, go through the data points; instead, we will try to come up with a $p(x)$ that "roughly" captures the behavior of the data. Of course, there are infinitely many choices we could make; in practice, one is guided by other knowledge of the system, say from the physical theory that describes that observable. If no extra knowledge is available, one typically picks a $p(x)$ that is a polynomial in x, but this time a low-degree polynomial should be enough, since the higher the degree the larger the number of undetermined parameters: since we don't fully trust the input data, we don't want to capture the "scatter", but only the underlying trend. An example of this was shown in Fig. 6.1, where we had assumed that the data points could not be fully trusted, and taken the overall trend to be linear.

6.5.1 Chi-Squared Statistic

Let's study this problem a bit more systematically. Recall that we are still expanding our $p(x)$ in terms of n parameters c_k, as per Eq. (6.7):

$$p(x) = \sum_{k=0}^{n-1} c_k \phi_k(x) \tag{6.118}$$

where the basis functions $\phi_k(x)$ may be polynomials (or not). This is a *linear* problem, in the sense that $p(x)$ is a linear combination of the c_k's. In the context of least-squares fitting, our approximating function $p(x)$ is sometimes known as a *theory* or a *model.*

As we mentioned around Eq. (6.10), we have N data points and n undetermined parameters (where $N > n$), so this could be thought of as an *overdetermined system.* As we saw there, since $\mathbf{\Phi c} = \mathbf{y}$ cannot be solved exactly, what we'll do is the closest thing available, i.e., we'll try to minimize the norm of the residual vector, as per Eq. (6.11):

$$\min \|\mathbf{y} - \mathbf{\Phi c}\| \tag{6.119}$$

Of course, this task seems to only depend on the basis functions and the input data: where did the input uncertainties σ_j go? Qualitatively, what this is asking is: suppose some of your input data have large uncertainties, and perhaps also appear to show different behavior than the rest. Obviously, the points with larger input uncertainty should somehow be *weighted* differently, i.e., they shouldn't impact our final determination as much as points that were measured much more accurately.

Since we wish to take the input uncertainties into account, we choose to minimize:

$$\chi^2 = \sum_{j=0}^{N-1} \left(\frac{y_j - p(x_j)}{\sigma_j} \right)^2 \tag{6.120}$$

which is known as the *chi-squared statistic* ("statistic" means a function of the data). It goes without saying (but we'll say it anyway) that (x_j, y_j) are variables (independent and dependent, respectively) whereas the c_k's are parameters; in other words, we'll never *measure* the c_k's directly. Note that this sum extends over all the N data points; the approximating function Eq. (6.118) is (implicitly) dependent on the values of the parameters c_k. When a given input uncertainty is large, the contribution of that term to the sum is small, as it should be. Presumably you can now see where the name *least-squares fitting* comes from:

as shown by Eq. (6.120), minimizing χ^2 minimizes the distance between the theory and the data ($p(x_j)$ and y_j, respectively), weighted by the size of the error bar in the data.

Since we will choose the c_k parameters that *minimize* χ^2, we are free to take the derivative of Eq. (6.120) with respect to a given parameter c_k and set the result to zero:

$$\frac{\partial \chi^2}{\partial c_k} = -2 \sum_{j=0}^{N-1} \left(\frac{y_j - p(x_j)}{\sigma_j^2} \right) \frac{\partial p(x_j)}{\partial c_k} = 0 \tag{6.121}$$

where $k = 0, 1, \ldots, n-1$. In $\partial p(x_j)/\partial c_k$ we are taking the derivative with respect to c_k, since x_j and y_j are considered externally given (and "frozen"). Assuming that the approximating function $p(x)$ is linear in the c_k's, as per Eq. (6.118), means that these derivatives can be trivially taken, leading to n linear equations in n unknowns (the c_k's).

If we now assume that the measurement errors are normally distributed, then our prescription provides *maximum likelihood parameter estimation* for the c_k's; for (much) more on the subject, as well as an introduction to the Bayesian approach, have a look at section 6.6 on *statistical inference*. Our goals here are more limited: we try to find the c_k's that mimize the χ^2, without worrying too much about questions regarding goodness-of-fit.[37] Even so, we would like to have some guidance on the question of how to compare different theories. To reiterate, minimizing the χ^2 of Eq. (6.120) tells you how to find the best c_k's you can, *for a given theory*. But what if your theory is wrong? For example, imagine that you should have picked $p(x) = c_0 + c_1 x + c_2 x^2$ but instead you picked $p(x) = c_0 + c_1 x$. The χ^2-minimization process still tries to do its best, but the results may end up being poor, as you could see simply by inspection (i.e., by comparing $p(x)$ to the input data: it is rare that a parabola can be approximated by a straight line).

Here, we give some practical advice that will help you decide which theory to pick, always under the assumption that the measurement errors are normally distributed. If the (minimum) χ^2 is very large, you can see from its definition in Eq. (6.120) that you have not been able to produce a $p(x)$ that goes "near" the data points; this could simply be due to the fact that your model is wrong. Another explanation could be that whoever produced the σ_j's underestimated them, i.e., they are actually much larger, in which case your model would do a better job since the χ^2 would be smaller. On the other hand, if $\chi^2 \approx 0$, then from the same definition, Eq. (6.120), you see that your theory curve essentially goes through the data points. This is not always a good outcome: you may be using too many parameters, thereby "overfitting", i.e., trying to capture the random scatter in the data. Another explanation could be, as before, related to whoever produced the σ_j's: perhaps these were overestimated, i.e., in reality the σ_j's should be much smaller. As always, the question arises how to distinguish between "large" and "small" chi-squared values. As a rule of thumb, typically $\chi^2 \approx N - n$ implies a reasonably good fit.[38] The difference between the

[37] Which means that we also don't go into important alternative scenarios like Poisson-distributed data.

[38] If you were not given the σ_j's as input, you could assume that they are all equal, $\sigma_j = \sigma$, and also that you did a good job fitting, in which case $\chi^2 \approx N - n$. Then, you could find σ^2 from Eq. (6.120) but, of course, this does not help you check how well you did in your fit, since you started by assuming you did well.

number of data points and the number of parameters, $N - n$, is known as the number of *degrees of freedom*. Thus, the rule of thumb can be restated as saying that for a fit that is neither too good nor too bad you should expect the *chi-squared per degree of freedom to be roughly one*. (This is justified in problems 6.45 and 6.46; see, however, page 390.[39])

It turns out that the quantity to be minimized in Eq. (6.120) is equivalent to what we were faced with in Eq. (6.119), only this time we're also appropriately dividing by the input uncertainties. We will return to the matrix notation below but, for now, we will help you build some intuition on least-squares fitting, by showing Eq. (6.121) explicitly applied to a simple case.

6.5.2 Straight-Line Fit

You have probably encountered the task of fitting data to a straight line in an introductory lab course. Here, we will essentially repeat what you saw there, employing the notation that we introduced above; in the following section, we go over a more general form, which will map the least-squares problem to a linear-algebra one.

Our task is to fit a straight-line model to N data points (x_j, y_j):

$$p(x) = c_0 + c_1 x \tag{6.122}$$

We need to determine the two parameters c_0 and c_1. For this theory, the chi-squared definition of Eq. (6.120) takes the form:

$$\chi^2 = \sum_{j=0}^{N-1} \left(\frac{y_j - c_0 - c_1 x_j}{\sigma_j} \right)^2 \tag{6.123}$$

where we are explicitly showing the dependence on the c_k parameters.

6.5.2.1 Parameter Estimates and Variances

The vanishing-derivative condition of Eq. (6.121) for our theory, Eq. (6.122), gives:

$$\frac{\partial \chi^2}{\partial c_0} = -2 \sum_{j=0}^{N-1} \left(\frac{y_j - c_0 - c_1 x_j}{\sigma_j^2} \right) = 0, \qquad \frac{\partial \chi^2}{\partial c_1} = -2 \sum_{j=0}^{N-1} \left(\frac{y_j - c_0 - c_1 x_j}{\sigma_j^2} \right) x_j = 0 \tag{6.124}$$

where the derivatives of $p(x_j)$ with respect to each c_k were easy to produce. Each numerator can be split into three terms; to do so, let us define the following six helper variables:

$$S \equiv \sum_{j=0}^{N-1} \frac{1}{\sigma_j^2}, \quad S_x \equiv \sum_{j=0}^{N-1} \frac{x_j}{\sigma_j^2}, \quad S_y \equiv \sum_{j=0}^{N-1} \frac{y_j}{\sigma_j^2},$$
$$S_{xx} \equiv \sum_{j=0}^{N-1} \frac{x_j^2}{\sigma_j^2}, \quad S_{xy} \equiv \sum_{j=0}^{N-1} \frac{x_j y_j}{\sigma_j^2}, \quad \Delta \equiv S S_{xx} - S_x^2 \tag{6.125}$$

[39] "Such is the influence of custom, that, where it is strongest, it not only covers our natural ignorance, but even conceals itself" (David Hume, *An Enquiry concerning Human Understanding*, paragraph 4.8).

Note that all the right-hand sides here are expressed in terms of known quantities, which means that they can be computed from the data. We can now re-express our vanishing-derivative results from Eq. (6.124) in terms of these helper variables:

$$\begin{aligned} S c_0 + S_x c_1 &= S_y \\ S_x c_0 + S_{xx} c_1 &= S_{xy} \end{aligned} \tag{6.126}$$

This is a simple 2×2 linear system of equations:

$$\begin{pmatrix} S & S_x \\ S_x & S_{xx} \end{pmatrix} \begin{pmatrix} c_0 \\ c_1 \end{pmatrix} = \begin{pmatrix} S_y \\ S_{xy} \end{pmatrix} \tag{6.127}$$

If you're wondering why we went through the trouble of defining the Δ, the answer is that it helps us compactly write down the solution to our 2×2 system:

$$c_0 = \frac{S_{xx}S_y - S_x S_{xy}}{\Delta}, \qquad c_1 = \frac{S S_{xy} - S_x S_y}{\Delta} \tag{6.128}$$

Since the helper variables of Eq. (6.125) were computed in terms of the data, everything here is known: we have managed to find the c_0 and c_1 values that minimize the χ^2.

At this point, we could immediately proceed to a Python implementation of Eq. (6.128) for a given dataset. Instead, let us take a minute to think about the values of c_0 and c_1 we just determined. We recall that these are *estimates* of the parameter values: since our input data suffered from uncertainties, σ_j, a corresponding uncertainty has been introduced in the determination of the c_k parameters. In short, we now need to make a quick detour through the subject of *propagation of error* which, as you may recall, is a topic we spent quite a bit of time discussing all the way back in section 2.2.2; however, back then we were referrring to "maximal error" propagation, whereas we are now interested in employing "standard error" propagation. In short, this time around we will add the errors in quadrature, in contradistinction to what we did, say, in Eq. (2.41). Assuming the data are independent, propagation of error for any function g takes the form (see problem 6.32):

$$\sigma_g^2 = \sum_{j=0}^{N-1} \left(\frac{\partial g}{\partial y_j}\right)^2 \sigma_j^2 \tag{6.129}$$

where in our specific case g will be either c_0 or c_1. We can evaluate the needed derivatives from our answer in Eq. (6.128):

$$\frac{\partial c_0}{\partial y_j} = \frac{S_{xx} - S_x x_j}{\sigma_j^2 \Delta}, \qquad \frac{\partial c_1}{\partial y_j} = \frac{S x_j - S_x}{\sigma_j^2 \Delta} \tag{6.130}$$

It may help you to use different dummy summation variables in the definitions of our helpers, Eq. (6.125), to see where these results came from. The first equation helps us

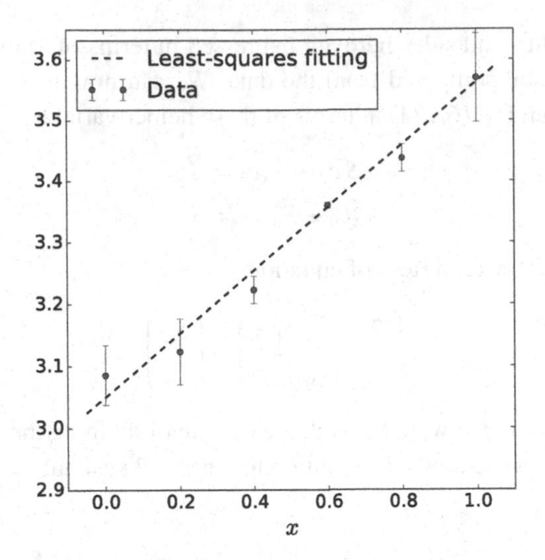

Fig. 6.12 Least-squares fitting for a straight-line theory

determine $\sigma_{c_0}^2$ and the second one $\sigma_{c_1}^2$. Plugging these two equations into Eq. (6.129), expanding the parentheses, and collecting terms leads to:

$$\sigma_{c_0}^2 = \frac{S_{xx}}{\Delta}, \qquad \sigma_{c_1}^2 = \frac{S}{\Delta} \tag{6.131}$$

These are the *variances* in the estimates for our two parameters. If you want the *standard deviations*, you'll have to take the square root(s). As you will see in problem 6.33 one can also compute the *covariance* of c_0 and c_1, but what we have here is enough for our purposes.

6.5.2.2 Implementation

To see our least-squares straight-line fitting algorithm at work, we'll need a set of input data. We decide to re-use the data shown in Fig. 6.1: the input uncertainties, σ_j, were suppressed there (since interpolation doesn't need them), but can be taken into account now. Code 6.5 first computes the helper variables of Eq. (6.125). We then define another function to evaluate the c_k's from Eq. (6.128) and the σ_{c_k}'s from Eq. (6.131). We store these in two `numpy` arrays: this is overkill for our 2×2 problem, but it sets the stage for what is to follow.[40] Having determined the c_k's, we then define another function to compute the (minimum) χ^2 from Eq. (6.123).

The main program sets up the data corresponding to Fig. 6.1 and then proceeds to call the last two functions, first computing the c_k's and then using them to compute the χ^2.

[40] Note that we could have opted in favor of *not* solving the 2×2 system of Eq. (6.127) by hand: instead, we could have simply set it up and then used Gaussian elimination to solve it programmatically.

linefit.py

Code 6.5

```
import numpy as np

def helpers(dataxs,datays,datasigs):
    S = np.sum(1/datasigs**2)
    Sx = np.sum(dataxs/datasigs**2)
    Sy = np.sum(datays/datasigs**2)
    Sxx = np.sum(dataxs**2/datasigs**2)
    Sxy = np.sum(dataxs*datays/datasigs**2)
    Del = S*Sxx - Sx**2
    return S, Sx, Sy, Sxx, Sxy, Del

def computecs(dataxs,datays,datasigs):
    S,Sx,Sy,Sxx,Sxy,Del = helpers(dataxs,datays,datasigs)
    cs = np.zeros(2); dcs = np.zeros(2)
    cs[0] = (Sxx*Sy - Sx*Sxy)/Del
    cs[1] = (S*Sxy - Sx*Sy)/Del
    dcs[0] = np.sqrt(Sxx/Del)
    dcs[1] = np.sqrt(S/Del)
    return cs, dcs

def computechisq(dataxs,datays,datasigs,cs):
    chisq = np.sum((datays-cs[0]-cs[1]*dataxs)**2/datasigs**2)
    return chisq

dataxs = np.linspace(0,1,6)
datays = np.array([3.085, 3.123, 3.224, 3.360, 3.438, 3.569])
datasigs = np.array([0.048, 0.053, 0.02, 0.005, 0.023, 0.07])
cs, dcs = computecs(dataxs, datays, datasigs)
print(cs); print(dcs)
chisq = computechisq(dataxs, datays, datasigs, cs)
print(chisq/(dataxs.size - cs.size))
```

Printed out are the parameters and their standard deviations, as well as the χ^2 per degree of freedom, i.e., $\chi^2/(N-n)$. The output of running this program is:

```
[ 3.04593186 0.5189044 ]
[ 0.02927752 0.04896135]
```

```
1.09916819554
```

We see that the χ^2 per degree of freedom is pretty close to 1, signifying a decent fit. This is also reflected in the error bars we have placed on the parameters. Speaking of which, we observe that the error in the offset parameter, c_0, is much smaller than that in the slope, c_1.

In Fig. 6.12 we are showing the data along with the input uncertainties, σ_j. We're also plotting the straight-line model, Eq. (6.122), that results from our best-fit parameters c_0 and c_1.[41] Even if you take the error bars into consideration, the straight-line model does not need to go through all the points. We are minimizing the χ^2 from Eq. (6.123), trying to take into account all the data points and associated errors; our procedure does the best job it can, always under the assumption that our model is a straight line.

6.5.3 General Linear Fit: Normal Equations

The (standard) material that we developed in the previous section is good as far as it goes. However, it's easy to see that solving things "by hand" doesn't take you very far. For example, imagine you wanted to check whether you picked the right model. One way to do so would be to try a quadratic theory:

$$p(x) = c_0 + c_1 x + c_2 x^2 \tag{6.132}$$

and repeat the entire least-squares fitting procedure.[42] This involves three unknown parameters; in this case Eq. (6.127) generalizes to:

$$\begin{pmatrix} S & S_x & S_{xx} \\ S_x & S_{xx} & S_{xxx} \\ S_{xx} & S_{xxx} & S_{xxxx} \end{pmatrix} \begin{pmatrix} c_0 \\ c_1 \\ c_2 \end{pmatrix} = \begin{pmatrix} S_y \\ S_{xy} \\ S_{xxy} \end{pmatrix} \tag{6.133}$$

where you should be able to guess how the new terms like S_{xxxx} are defined. Problem 6.35 asks you to derive this, starting from the zero-derivative conditions in Eq. (6.121).

A moment's thought will convince you that the coefficient matrix here is an extension of the Vandermonde matrix we encountered in Eq. (6.17). As you may recall from that discussion, such matrices can be very ill-conditioned, especially for large dimensions. Since Lagrange interpolation for $n = 100$ or even $n = 1000$ is routinely used in practice, this is a real issue. However, when carrying out *least-squares fitting*, you would generally not like to use a polynomial of very high degree: as mentioned in section 6.5.1, this mostly leads to overfitting, namely to capturing the random scatter in the data, which is an unwelcome outcome. In other words, you will typically not get into too much trouble if you attack the 3×3 problem of Eq. (6.132) via the naive approach of Eq. (6.133).[43] Of course, what happens if you then decide to try out the 4×4 or the 5×5 case? It's easy to see that you will

[41] We could have also used the σ_{c_k}'s and thereby produced a *band* instead of a straight line (see problem 6.53).

[42] As you will see in problem 6.35, in this case you find that the coefficient of the quadratic term is very small, thereby increasing your confidence in your decision to carry out a straight-line fit for this set of data.

[43] Even so, problem 6.40 introduces an alternative approach, involving a generalization of the QR decomposition we encountered in section 4.4.3.1 to rectangular matrices; this leads to a more robust prescription, which is routinely used in libraries. We will also encounter theories with many parameters in section 6.7.2.

have to give up on solving the linear system analytically, opting for Gaussian elimination or another method, instead.

So far in this subsection we have been considering the possibility of employing a model that is a polynomial, raising the question of what degree we should go up to. A related question arises when your basis functions are not monomials/polynomials. For example, your model could be:

$$p(x) = c_0 + c_1 e^{-x^2} \tag{6.134}$$

Note that this is still linear (in the c_k's), as per Eq. (6.118); it's only one of the basis functions that is not linear (in x). As you will discover in problem 6.34, trying to carry out a least-squares fit for this model will lead you to introduce helper variables like:

$$S_x \equiv \sum_{j=0}^{N-1} \frac{e^{-x_j^2}}{\sigma_j^2} \tag{6.135}$$

in direct generalization of what we saw in the straight-line case. Obviously, if you had three (or, say, seven) parameters, this way of doing things "by hand" would no longer be very practical.

6.5.3.1 Normal Equations

Instead of studying models on a case-by-case basis, we will now tackle the general (linear) problem of Eq. (6.118):

$$p(x) = \sum_{k=0}^{n-1} c_k \phi_k(x) \tag{6.136}$$

It should be easy to see that polynomial fitting is merely a special case of this model, that in which the $\phi_k(x)$'s are monomials. Thus, we will develop and implement the general machinery for any set of basis functions; then, in problem 6.36 you will specialize our code to the case of polynomial fitting.

For this model, the χ^2 of Eq. (6.120) takes the form:

$$\chi^2 = \sum_{j=0}^{N-1} \frac{1}{\sigma_j^2} \left(y_j - \sum_{i=0}^{n-1} c_i \phi_i(x_j) \right)^2 \tag{6.137}$$

Similarly, the zero-derivative conditions of Eq. (6.121) now become:

$$\frac{\partial \chi^2}{\partial c_k} = -2 \sum_{j=0}^{N-1} \frac{1}{\sigma_j^2} \left[y_j - \sum_{i=0}^{n-1} c_i \phi_i(x_j) \right] \phi_k(x_j) = 0 \tag{6.138}$$

where $k = 0, 1, \ldots, n-1$, as usual. If we now drop the -2 coefficient, re-arrange terms, and interchange the order of summation, we find:

$$\sum_{i=0}^{n-1} \sum_{j=0}^{N-1} \frac{\phi_k(x_j)}{\sigma_j} \frac{\phi_i(x_j)}{\sigma_j} c_i = \sum_{j=0}^{N-1} \frac{\phi_k(x_j)}{\sigma_j} \frac{y_j}{\sigma_j} \tag{6.139}$$

Keep in mind that typically N is large and n is small. On both sides, we are summing over the j index, which corresponds to the (large) number of data points, N. The way we have chosen to write Eq. (6.139) clearly motivates the introduction of two new entities:

$$A_{jk} = \frac{\phi_k(x_j)}{\sigma_j}, \qquad b_j = \frac{y_j}{\sigma_j} \tag{6.140}$$

You can see that $\mathbf{A} = \{A_{jk}\}$ is a rectangular matrix made up of $N \times n$ elements; the row index goes from 0 to $N-1$ and the column index from 0 to $n-1$. This is usually called the *design matrix* of the problem. Similarly, $\mathbf{b}$ is an $N \times 1$ column vector, made up of the data scaled by the input uncertainties. Crucially, all the elements of $\mathbf{A}$ and $\mathbf{b}$ can be immediately computed from the input data.

In terms of our two new entities, Eq. (6.139) becomes:

$$\sum_{i=0}^{n-1}\sum_{j=0}^{N-1} A_{jk}A_{ji}c_i = \sum_{j=0}^{N-1} A_{jk}b_j \tag{6.141}$$

We now recognize that the sums over j can be replaced by matrix–matrix multiplication (on the left-hand side) and matrix–vector multiplication (on the right-hand side):

$$\sum_{i=0}^{n-1}(\mathbf{A}^T\mathbf{A})_{ki}c_i = (\mathbf{A}^T\mathbf{b})_k \tag{6.142}$$

where we appropriately introduced the transpose of $\mathbf{A}$ whenever we needed two indices to appear in different order.[44] We now realize that even the summation over i on the left-hand side can be expressed in terms of a matrix–vector multiplication, leading us to write:

$$(\mathbf{A}^T\mathbf{A}\mathbf{c})_k = (\mathbf{A}^T\mathbf{b})_k \tag{6.143}$$

Since this holds for any component k, we have thereby derived the *normal equations*:

$$\mathbf{A}^T\mathbf{A}\mathbf{c} = \mathbf{A}^T\mathbf{b} \tag{6.144}$$

Given that $\mathbf{A}$ and $\mathbf{b}$ are already known (computed in terms of the input data), this equation is a simple linear system of equations: we can solve it using, say, Gaussian elimination to find $\mathbf{c}$. Make sure you understand the dimensions of each entity: $\mathbf{A}$ is $N \times n$, which means $\mathbf{A}^T$ is $n \times N$ and therefore $\mathbf{A}^T\mathbf{A}$ is $n \times n$. This is appropriate since $\mathbf{c}$ is an $n \times 1$ vector containing all the c_k's. Similarly, $\mathbf{A}^T$ is $n \times N$ and $\mathbf{b}$ is $N \times 1$, so $\mathbf{A}^T\mathbf{b}$ is $n \times 1$. In other words, we have managed to transform our problem into square form, with a coefficient matrix of dimension $n \times n$: this is in general a much easier problem than dealing with $\mathbf{A}$ directly (which has dimensions $N \times n$), since typically n will be much smaller than N.

[44] If you are uncomfortable with this, you should introduce an intermediate variable $D_{kj} = A_{jk}$, follow the rest of the derivation, and realize that $\mathbf{D} = \mathbf{A}^T$ at the end.

6.5.3.2 Interpreting the Normal Equations

Let us now build some intuition on the $\mathbf{A}^T\mathbf{A}$ matrix. Our main goal is to show that the $\mathbf{c}$ that we get from Eq. (6.144) is not only a critical point, but a *minimum*. You might want to brush up on the multidimensional minimization material in section 5.6.1 at this point.

First, it's easy to see that $\mathbf{A}^T\mathbf{A}$ is symmetric:

$$(\mathbf{A}^T\mathbf{A})^T = \mathbf{A}^T(\mathbf{A}^T)^T = \mathbf{A}^T\mathbf{A} \tag{6.145}$$

Since $\mathbf{A}^T\mathbf{A}$ is equal to its transpose, it is symmetric. Second, $\mathbf{A}^T\mathbf{A}$ is positive semidefinite, because for any non-trivial vector $\mathbf{x}$ we have:

$$\mathbf{x}^T\mathbf{A}^T\mathbf{A}\mathbf{x} = (\mathbf{A}\mathbf{x})^T\mathbf{A}\mathbf{x} = \|\mathbf{A}\mathbf{x}\|^2 \geq 0 \tag{6.146}$$

where, as usual, we're implicitly using the Euclidean norm. We can do even better: write $\mathbf{A}$ in terms of its columns:

$$\mathbf{A} = \begin{pmatrix} \mathbf{a}_0 & \mathbf{a}_1 & \dots & \mathbf{a}_{n-1} \end{pmatrix} \tag{6.147}$$

If we assume that the columns of $\mathbf{A}$ are linearly independent, we can re-express $\mathbf{A}\mathbf{x}$, whose norm we were taking in Eq. (6.146), as:

$$\mathbf{A}\mathbf{x} = \sum_{i=0}^{n-1} \mathbf{a}_i x_i \neq 0 \tag{6.148}$$

where the last step follows from the definition of being linearly independent. That means that we have shown $\|\mathbf{A}\mathbf{x}\|^2 \neq 0$, which implies $\|\mathbf{A}\mathbf{x}\|^2 > 0$, i.e., $\mathbf{A}^T\mathbf{A}$ is positive definite if the columns of $\mathbf{A}$ are linearly independent.

We now realize that our new $\mathbf{A}$ and $\mathbf{b}$ entities from Eq. (6.140) can help us rewrite the χ^2 from Eq. (6.137) in matrix form:

$$\chi^2(\mathbf{c}) = (\mathbf{b} - \mathbf{A}\mathbf{c})^T(\mathbf{b} - \mathbf{A}\mathbf{c}) \tag{6.149}$$

where we also took the opportunity to explicitly note that χ^2 depends on the values of the c_k parameters. To see that $\chi^2(\mathbf{c})$ involves a quadratic form in $\mathbf{c}$, expand things out:

$$\chi^2(\mathbf{c}) = \mathbf{b}^T\mathbf{b} - \mathbf{b}^T\mathbf{A}\mathbf{c} - \mathbf{c}^T\mathbf{A}^T\mathbf{b} + \mathbf{c}^T\mathbf{A}^T\mathbf{A}\mathbf{c} \tag{6.150}$$

In problem 6.38 you will show that the Hessian of a quadratic form $\mathbf{x}^T\mathbf{B}\mathbf{x}$ is $\mathbf{H} = 2\mathbf{B}$ when $\mathbf{B}$ is symmetric. Recall that the Hessian for our case would be defined by analogy to Eq. (5.103):

$$H_{ij} = \frac{\partial \chi^2(\mathbf{c})}{\partial c_i \partial c_j} \tag{6.151}$$

From Eq. (6.150) you can see that only the quadratic form $\mathbf{c}^T\mathbf{A}^T\mathbf{A}\mathbf{c}$ contributes to the second derivative of $\chi^2(\mathbf{c})$; thus, $\mathbf{H} = 2\mathbf{A}^T\mathbf{A}$ for our case. The reason we care about the Hessian is that it makes an appearance in the multidimensional Taylor expansion (see Eq. (5.98)):

$$\chi^2(\mathbf{c} + \mathbf{q}) = \chi^2(\mathbf{c}) + \left(\nabla\chi^2(\mathbf{c})\right)^T \mathbf{q} + \frac{1}{2}\mathbf{q}^T\mathbf{H}\mathbf{q} \tag{6.152}$$

where now there is no term $O(||\mathbf{q}||^3)$ since the third derivative of a quadratic form vanishes. Recall that our main result in Eq. (6.144) gives us the $\mathbf{c}$ value that corresponds to the zero-derivative conditions, Eq. (6.138). In other words, it is the analogue of Eq. (5.102):

$$\nabla\chi^2(\mathbf{c}^*) = \mathbf{0} \tag{6.153}$$

This $\mathbf{c}^*$ is the set of c_k's that satisfies Eq. (6.144).[45] But this means that the vanishing gradient leads to:

$$\chi^2(\mathbf{c}^* + \mathbf{q}) = \chi^2(\mathbf{c}^*) + \frac{1}{2}\mathbf{q}^T\mathbf{H}\mathbf{q} > \chi^2(\mathbf{c}^*) \tag{6.154}$$

which is essentially a repeat of our argument in Eq. (5.105). The last step, showing that $\mathbf{c}^*$ is, indeed, a minimum as desired, follows from the fact that our $\mathbf{H}$ is positive definite. This is all consistent with what we said in section 5.6.1: a necessary condition for $\mathbf{c}^*$ being a local minimum is that it be a critical point and a sufficient condition for the critical point $\mathbf{c}^*$ being a local minimum is that its Hessian matrix be positive definite.

Before concluding, let us return to the cases where the normal-equations approach can get in trouble (and therefore another algorithm should be preferred). The main takeaway from the last few pages has been that Eq. (6.144) has transformed the problem such that we are no longer faced with the rectangular $\mathbf{A}$ matrix but with the square $\mathbf{A}^T\mathbf{A}$ matrix (incidentally, recall page 185 on normal matrices). As you will explicitly show in problem 6.39, the condition numbers of the two matrices are straightforwardly related:[46]

$$\kappa(\mathbf{A}^T\mathbf{A}) \approx \kappa(\mathbf{A})^2 \tag{6.155}$$

Thus, if $\mathbf{A}$ is an ill-conditioned matrix, then $\mathbf{A}^T\mathbf{A}$ will be a terribly ill-conditioned matrix: the issue has gotten exacerbated by the method we chose to employ. This is a nice example of the interplay between the ill-conditioning of a problem and the (possible) instability of an algorithm (cf. page 325). There exist other, more robust algorithms, typically involving orthogonalization, as you will find out in problem 6.40.

6.5.3.3 Variances

For the case of the straight-line fit of section 6.5.2, after determining the estimated parameters c_0 and c_1, we proceeded to find the uncertainties in these estimates, σ_{c_0} and σ_{c_1}. You will not be surprised to hear that something analogous can be done in the present case, also. The general relation in Eq. (6.129) has the form:

$$\sigma_{c_i}^2 = \sum_{j=0}^{N-1}\left(\frac{\partial c_i}{\partial y_j}\right)^2\sigma_j^2 \tag{6.156}$$

[45] If you're careful, you could simply *derive* Eq. (6.144) by taking the gradient of Eq. (6.150).

[46] This is an exact equality for the case of the 2-norm, see problem 4.42.

where we have, implicitly, assumed that the uncertainties in any two data points are uncorrelated, i.e., that the measurements are independent. To compute $\sigma_{c_i}^2$ we see that we'll need the derivative of the parameter c_i with respect to the datum y_j, i.e., $\partial c_i/\partial y_j$. To get that, in its turn, we first need to write out c_i as a function of y_j.

Multiplying Eq. (6.144) by $(\mathbf{A}^T\mathbf{A})^{-1}$ on the left, we find:

$$\mathbf{c} = (\mathbf{A}^T\mathbf{A})^{-1}\mathbf{A}^T\mathbf{b} \tag{6.157}$$

The inverse matrix is important enough that it gets its own symbol:

$$\mathbf{V} = (\mathbf{A}^T\mathbf{A})^{-1}, \qquad \mathbf{U} = \mathbf{A}^T\mathbf{A} \tag{6.158}$$

where we also went ahead and introduced a new symbol for $\mathbf{A}^T\mathbf{A}$ itself, for later use. We can write down the relation in Eq. (6.157) for the i-th component:

$$c_i = \sum_{k=0}^{n-1} V_{ik}(\mathbf{A}^T\mathbf{b})_k = \sum_{k=0}^{n-1} V_{ik} \sum_{l=0}^{N-1} A_{lk} b_l = \sum_{k=0}^{n-1} V_{ik} \sum_{l=0}^{N-1} \frac{\phi_k(x_l)}{\sigma_l}\frac{y_l}{\sigma_l} \tag{6.159}$$

In the second equality we expressed $(\mathbf{A}^T\mathbf{b})_k$ as in Eq. (6.141), and in the third equality we wrote out A_{lk} and b_l as per their definitions in Eq. (6.140). The final expression is the result we were after, relating c_i to y_l.

Crucially, V_{ik} does not depend on y_j. This means that we can straightforwardly evaluate the needed derivative:

$$\frac{\partial c_i}{\partial y_j} = \sum_{k=0}^{n-1} V_{ik} \frac{\phi_k(x_j)}{\sigma_j^2} \tag{6.160}$$

where you'll notice that the sum over l is now gone. We now plug this result into Eq. (6.156):

$$\sigma_{c_i}^2 = \sum_{k=0}^{n-1}\sum_{l=0}^{n-1} V_{ik}V_{il}\left(\sum_{j=0}^{N-1} \frac{\phi_k(x_j)\phi_l(x_j)}{\sigma_j^2}\right) \tag{6.161}$$

where we interchanged the order of summation and cancelled one of the σ_j^2's. Now, the term inside the parentheses can also be re-expressed:

$$\sum_{j=0}^{N-1} \frac{\phi_k(x_j)\phi_l(x_j)}{\sigma_j^2} = \sum_{j=0}^{N-1} A_{jk}A_{jl} = (\mathbf{A}^T\mathbf{A})_{kl} = U_{kl} \tag{6.162}$$

In the first equality we used the definitions in Eq. (6.140). In the second equality we replaced the sum by matrix–matrix multiplication. In the third equality we identified our new matrix from Eq. (6.158). Now we can take this result and re-introduce it into Eq. (6.161):

$$\sigma_{c_i}^2 = \sum_{k=0}^{n-1}\sum_{l=0}^{n-1} V_{ik}V_{il}U_{kl} = \sum_{l=0}^{n-1} V_{il} \sum_{k=0}^{n-1} V_{ik}U_{kl} = \sum_{l=0}^{n-1} V_{il}\delta_{il} \tag{6.163}$$

In the second equality we re-ordered the sums and terms, and in the third equality we realized that $\mathbf{V}$ is the inverse of $\mathbf{U}$, so their product is the identity. This leads us to the significant result that:

$$\sigma_{c_i}^2 = V_{ii} = \left[(\mathbf{A}^T\mathbf{A})^{-1}\right]_{ii} \tag{6.164}$$

namely, the diagonal elements of the $\mathbf{V}$ matrix are the variances of the estimates of the $\mathbf{c}$ parameters. In problem 6.41 you will show that the off-diagonal matrix elements of $\mathbf{V}$ are the covariances between different parameters, i.e., V_{ik} is the covariance between parameters c_i and c_k. As a result, the symmetric, positive definite $\mathbf{V}$ is known as the *variance-covariance matrix* (or simply the *covariance matrix*); crucially, it does *not* depend on the y_j's.

6.5.3.4 Implementation

We are now ready to implement our general linear least-squares fitting procedure in Python, see code 6.6. We will only determine the c_k's, leaving the determination of the variances, which requires a matrix inversion, as problem 6.37.

Our code will store the x_j's, the y_j's, and the σ_j's in a 2d NumPy array, `data`. To highlight the generality of our machinery, we decide to generate the data from a sinusoidal function. We intend to randomly perturb the y_j's, i.e., add in some noise; to ensure that we'll get the same answer every time we call our `generatedata()` function, we first seed the random-number generator, via `numpy.random.seed()`. (We will re-encounter random numbers in the following chapter, in our discussion of stochastic integration, see section 7.7.1.) We pick some standard deviations σ_j out of a hat (i.e., we sample them from a uniform distribution via `numpy.random.random()`); then, for each data point we produce a normally distributed error term, by employing `numpy.random.normal()` with zero mean and standard deviation σ_j. As usual, we bundle all the data-related code lines in the same function; if you're dealing with a different dataset, simply don't call this one function.

We then have to think about how to program our basis functions, namely the $\phi_k(x)$. In the spirit of showing that our formalism is truly general, we implement two distinct sets of basis functions: (a) monomials, and (b) sinusoidal behavior. Of course, for both cases the dependence on the c_k's is linear, as per Eq. (6.136). Since the data were drawn from a sinusoidal function, it should come as no surprise that using an analogous function will do a good job capturing what's going on (modulo the "noise" that we added in by hand). In the function `phi()` we are choosing to differentiate between the two different types of theory based on which value of the size of the problem, n, was passed in. We are hard-coding only two possibilities: (a) $n = 5$ is interpreted as monomials up to fourth degree, and (b) $n = 2$ is interpreted as either 1 or $\sin(x)$—on one line of code via Python's ternary operator.

While the two functions we just discussed were special, the next function is completely general. We define a function called `normalfit()` which computes the c_k's and calculates the minimum χ^2, in (minor) contravention of the separation of concerns principle. Comparing to our earlier function `computecs()` from `linefit.py`, we note that this time we are storing the entire dataset in one variable, we are passing in the $\phi_k(x)$ function, and we are also explicitly passing in the number of parameters, n. Note that we don't need to pass in N, since we're passing in the dataset. We immediately create the rectangular $N \times n$ matrix `A`. This is (almost) the first rectangular matrix we've seen in this book, but observe

newnormal.py

Code 6.6

```
from gauelim_pivot import gauelim_pivot
import numpy as np

def generatedata(N,a=0.,b=9,sts=(2,5,0.5,1)):
	sa, sb, sc, sd = sts
	np.random.seed(7921)
	data = np.zeros((3,N))
	data[0,:] = np.linspace(a,b,N)
	data[1,:] = sa + sb*np.sin(data[0,:])
	data[2,:] = sc + sd*np.random.random(N)
	data[1,:] += np.random.normal(0,data[2,:])
	return data

def phi(n,k,x):
	if n==5:
		val = x**k
	elif n==2:
		val = 1. if k==0 else np.sin(x)
	return val

def normalfit(data,phi,n):
	N = data.shape[1]
	A = np.zeros((N,n))
	for k in range(n):
		A[:,k] = phi(n,k,data[0,:])/data[2,:]
	bs = data[1,:]/data[2,:]

	cs = gauelim_pivot(A.T@A, A.T@bs)
	chisq = np.sum((bs - A@cs)**2)
	return cs, chisq

if __name__ == '__main__':
	data = generatedata(8)
	for n in (5, 2):
		cs, chisq = normalfit(data, phi, n)
		print(cs)
		print(chisq/(data.shape[1]-cs.size))
		print("")
```

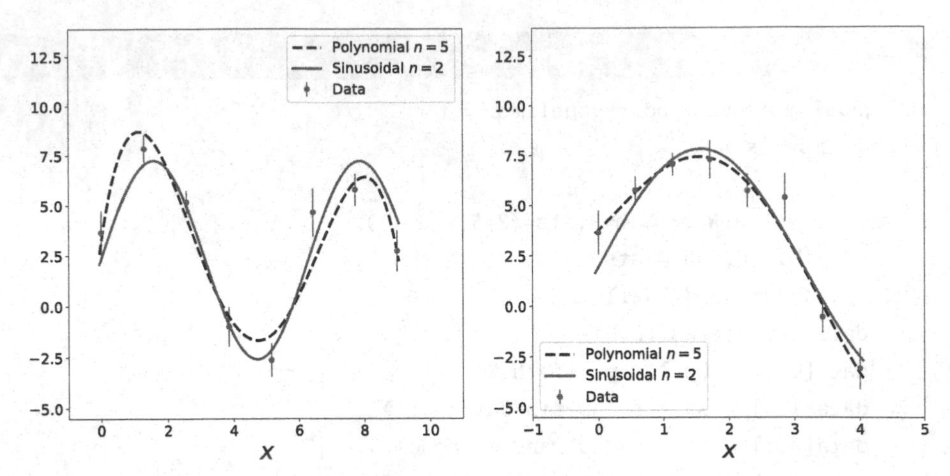

Fig. 6.13 Least-squares fitting for a polynomial and a sinusoidal theory

that numpy handles everything as smoothly as before. We iterate through the columns (of which there are n), directly implementing the definition of A_{jk} from Eq. (6.140); we employ numpy functionality and therefore only need a column index: the rows are iterated over implicitly, by manipulating the input data arrays. We similarly define the $\mathbf{b}$ column vector as per Eq. (6.140). Having computed $\mathbf{A}$ and $\mathbf{b}$, we both set up and solve the linear system $\mathbf{A}^T\mathbf{A}\mathbf{c} = \mathbf{A}^T\mathbf{b}$ of Eq. (6.144) in a single line. This would have been unpleasant without the convenience of using `@` to carry out matrix–matrix or matrix–vector multiplication. We call our usual routine that does Gaussian elimination with pivoting and then returns the c_k's. Having produced the c_k's, we can then compute the minimum χ^2, again in one line. This line implements Eq. (6.149): in problem 6.37 you are asked to calculate χ^2 starting from Eq. (6.137) and calling `phi()`; as you'll discover there, that is a messier task. Instead, here we take advantage of the fact that we've already computed $\mathbf{A}$ and $\mathbf{b}$.

The main program picks $N = 8$ and then tries out the two theories, printing out the c_k's and minimum χ^2. Running this code, we see that the sinusoidal theory did a better job capturing the behavior of the data, as reflected in the smaller minimum χ^2: despite involving five parameters, the polynomial was not able to describe the data as well. (This is not immediately obvious in the left panel of Fig. 6.13.) You may be thinking this was inevitable, given that we already knew which function the data were drawn from (roughly). That being said, you shouldn't be too quick to dismiss the accomplishments of our polynomial theory. The right panel of Fig. 6.13 shows the result of choosing the data from 0 to 4 (instead of from 0 to 9), with every other aspect of the code left unchanged. The polynomial seems to have done a good job; if you compare the minimum χ^2 (per degree of freedom) for the two models, the polynomial seems to actually be doing *better*. This raises the question: should you trust a smaller minimum χ^2 blindly? Generally speaking, you should try to balance the magnitude of the minimum χ^2 with the complexity of your model: if you use many parameters you may end up overfitting, i.e., capturing some of the noise in the data. Backing up for a second, we realize that we were fitting to an $N = 8$ dataset using an $n = 5$ model. Problem 6.36 asks you to study our earlier straight-line data using polynomial models of increasing degree, where this issue is thrown into sharp relief. Problem 6.54 discusses the balance between underfitting and overfitting in terms of the *bias–variance decomposition*.

6.6 Linear Statistical Inference

In section 6.5 we treated linear regression largely as an optimization problem: we qualitatively motivated the χ^2 of Eq. (6.120) and then proceeded to minimize it, thereby arriving at the normal equations of Eq. (6.144). (Etymologically speaking, the term *fitting* means that there is an exact relationship between two measured quantities, whereas *regression* refers to extracting some sort of correlation between two quantities, even in the absence of a model that explains this dependence; in practice, the two terms—fitting or regression—are often used interchangeably, despite their distinct origins.) We briefly mentioned our assumption that the input data uncertainties are normally distributed and its relation to maximum-likelihood parameter estimation (saying even less on the question of model selection). While statistics is not our main focus in this book, it is worthwhile to see how to recover the same results from a new perspective (and therefore grasp how to go beyond them).[47] It may help you to keep the traditional summarization in mind: *probability* takes us from the data-generating process to the data, while *statistics* goes the other way around. Our discussion will help show why normal distributions (or their absence) are so important. It will also allow the reader, in conjunction with the problem set, to see the provenance of the aforementioned rule of thumb (using χ^2 for model selection). As part of this probabilistic outlook, we will also provide an introduction to the fascinating approach of *Bayesian inference* as it relates to parameter estimation and model selection.

6.6.1 Maximum-Likelihood Approach

To get things going, let's recall Eq. (6.8), which reflected the fact that for the case of interpolation the approximating function goes exactly through the input data points:

$$y_j = \sum_{k=0}^{n-1} c_k \phi_k(x_j) \tag{6.165}$$

When we turned to fitting, we didn't provide a corresponding equation, resorting instead to a more hand-waving expression, Eq. (6.11):

$$\min \|\mathbf{y} - \mathbf{\Phi}\mathbf{c}\| \tag{6.166}$$

After realizing that the input dataset was imperfect, we stated our practical task as the minimization of the χ^2 statistic, Eq. (6.120): this involved the distance between the dependent variable of the data, y_j, and the prediction of the model at the independent variable of the data, $p(x_j)$, divided by the input data uncertainty, σ_j. Qualitatively, this prescription was plausible, but at the time we didn't really justify it.

[47] At this point, you should read over appendix C.3 to refresh your memory; the same ideas will also be important in section 7.7. Be forewarned that section 6.6 includes (preternaturally) many footnotes on conceptual or notational subtleties; while you may safely skip these in a first reading, you should eventually return to them.

6.6.1.1 Data-Generating (Parent) Distribution

We now decide to explicitly model the discrepancy between theory and experiment:

$$\mathscr{Y}_j = \sum_{k=0}^{n-1} c_{\star,k}\phi_k(x_j) + \mathscr{E}_j \tag{6.167}$$

For the time being, we take it for granted that we have the correct theory (i.e., that we've picked the ϕ_k's correctly). We are also assuming that there exists a way to pick the true c_k values (even though we don't yet know how to do that from our new perspective); we are using new notation for the data value $\mathscr{Y}_j$ and for the true parameters $c_{\star,k}$, which will pay off later on. Note that, like Eq. (6.165) but unlike Eq. (6.166), in Eq. (6.167) we are faced with an equality. The error term $\mathscr{E}_j$ (also known as the *noise*) is new here and to interpret it we will need to adjust our perspective somewhat. Before doing so, we realize that the sum in Eq. (6.167) can be re-expressed in terms of a dot product, so we have:

$$\mathscr{Y}_j = \boldsymbol{\phi}_j \mathbf{c}_\star + \mathscr{E}_j \tag{6.168}$$

where we made use of the j-th *row*[48] of the full $(N \times n)$ matrix $\boldsymbol{\Phi}$. This more compact notation will help us when we turn to all possible j values below. As above, we take as given the true theory and true parameter values.

We now have to think a bit more carefully about the meaning of the input data. The data values $\mathscr{Y}_j$ don't have to be viewed as single numbers that are set in stone; instead, you can think of each individual measurement $\mathscr{Y}_j$ as being itself drawn from a specific *parent distribution*[49] with mean $\boldsymbol{\phi}_j \mathbf{c}_\star$ (for the true theory) and standard deviation σ_j; you could imagine a different $\mathscr{Y}_j$ value having been drawn from the same distribution at the same x_j. In other words, the difference between $\mathscr{Y}_j$ and $\boldsymbol{\phi}_j \mathbf{c}_\star$ is now taken to be *a continuous random variable*: this is precisely the role played by $\mathscr{E}_j$ in Eq. (6.168). Take some time to adjust your perspective: since $\mathscr{E}_j$ is a random variable (and the x_j, the ϕ_k's, and the $c_{\star,k}$'s are frozen) the output, i.e., the $\mathscr{Y}_j$ given on the left-hand side of Eq. (6.168), is itself a random variable. For a given x_j, there exists only one true theory value $\boldsymbol{\phi}_j \mathbf{c}_\star$, but several (to put it mildly) $\mathscr{Y}_j$ data values that could be drawn from it.[50]

We now make the further *assumption* that the parent distribution of $\mathscr{Y}_j$ is *Gaussian*. We emphasize that there are important situations, most notably when the data are generated via counting/discrete processes, where the normality assumption is not warranted. However, as we'll learn in problem 6.59 when we introduce *asymptotic normality*, if you are dealing with sufficiently many counts, the normal distribution often does emerge even in such scenarios. In any case, we are explicit here about the fact that the Gaussianity of the parent distribution is an *assumption*. Incidentally, we treat the terms *distribution* and *density function* as synonyms; to establish the notation, we write down the form of a *Gaussian distribution* (also known as a *normal distribution*):

[48] Since $\Phi_{jk} = \phi_k(x_j)$, this is one of the rare cases where we produce the dot product without a transposition.
[49] In the idiom of section 7.7.2, the underlying entity is the parent *population*.
[50] We could denote some specific realizations by, e.g., $y_j^{(0)}, y_j^{(1)}, y_j^{(2)}$ etc.; in what follows we use simply y_j.

$$N(\mu, \sigma) = \frac{1}{\sqrt{2\pi}\sigma} \exp\left[-\frac{1}{2}\left(\frac{x-\mu}{\sigma}\right)^2\right] \tag{6.169}$$

where μ (the mean) controls the location of the peak and σ (the standard deviation) its width. Note that the (popular) notation employed here,[51] $N(\mu, \sigma)$, lists only these two parameters, i.e., it does not mention the name of the independent variable (for the case of Eq. (6.169), x on the right-hand side) which has to be determined based on the context. The normal distribution has a number of appealing properties, which you encountered in an introductory course on statistics. For example, for a normal distribution 99.7% of the values lie within three standard deviations (3σ) of the mean; this is sometimes referred to as the *68–95–99.7 rule* or the *empirical rule* and is explored in problem 6.50.

Having introduced the notation of Eq. (6.169), we can recast our earlier comments about the parent distribution of $\mathcal{Y}_j$ in probabilistic language. Recalling the distinction (from appendix C.3) between the random variable $\mathcal{Y}_j$ and a specific realization y_j, we are now ready to write down the following *probability density function*:[52]

$$P(y_j; \mathbf{c}_\star, \boldsymbol{\phi}_j, \sigma_j) = N\left(\boldsymbol{\phi}_j \mathbf{c}_\star, \sigma_j\right) = \frac{1}{\sqrt{2\pi}\sigma_j} \exp\left[-\frac{1}{2}\left(\frac{y_j - \boldsymbol{\phi}_j \mathbf{c}_\star}{\sigma_j}\right)^2\right] \tag{6.170}$$

This makes it explicit that we are taking the x_j, the σ_j, the ϕ_k's, and the $c_{\star,k}$'s as given (fixed). In the second step we employed Eq. (6.169) in order to emphasize that we are dealing with the probability density for a given measurement y_j to be near the "actual" value $\boldsymbol{\phi}_j \mathbf{c}_\star$. We are taking each $\mathcal{Y}_j$ as being associated with its own standard deviation σ_j: this is typically the case in physics, where a separate experiment is carried out at each x_j.[53]

As a consequence of what we've developed so far, we can also determine the distribution of the error term in Eq. (6.168), $\mathcal{E}_j$. Since $\mathcal{Y}_j$ is a normally distributed random variable and $\boldsymbol{\phi}_j \mathbf{c}_\star$ is a constant (everything in it is fixed—it is also known as the *deterministic component*), we see that $\mathcal{E}_j$, too, is a normally distributed random variable: its mean is shifted (downward) by the same constant, i.e., $\mathcal{E}_j$ obeys the distribution $N(0, \sigma_j)$;[54] this is sometimes summarized by saying that one assumes that σ_j is normally distributed. Of course, you should be able to see by this point that this is, strictly speaking, wrong, since the σ_j is constant: it's the error term $\mathcal{E}_j$ (which is a random variable) that one assumes to be normally distributed. (While the argument doesn't always work as smoothly—from normal $\mathcal{E}_j$ to normal $\mathcal{Y}_j$ or vice versa—this won't be an issue here.)

While Eq. (6.170) corresponds to the j-th data point, we are more interested in the entity $P(y_0, y_1, \ldots, y_{N-1}; \mathbf{c}_\star, \boldsymbol{\phi}_0, \boldsymbol{\phi}_1, \ldots, \boldsymbol{\phi}_{N-1}, \sigma_0, \sigma_1, \ldots, \sigma_{N-1})$, i.e., the joint probability

[51] An alternative convention shows the variance instead of the standard deviation, $N(\mu, \sigma^2)$.

[52] This is *not* a conditional probability density, since $\mathbf{c}_\star$, $\boldsymbol{\phi}_j$, and σ_j are constants (i.e., not random variables). This explains why we are using $P(y_j; \mathbf{c}_\star, \boldsymbol{\phi}_j, \sigma_j)$ instead of the popular abuse of notation $P(y_j|\mathbf{c}_\star, \boldsymbol{\phi}_j, \sigma_j)$: combining the latter with the definition of a conditional density, Eq. (6.181), is likely to confuse you.

[53] The literature on machine learning typically assumes a fixed σ across all j's. Thus, physicists should seek out works on statistics (or econometrics!) where *weighted least squares* are applied to *heteroskedastic* errors.

[54] We could have also taken x_j to be the realization of a random variable, in which case $\mathbb{E}(\mathcal{E}_j|\mathcal{X}_j = x_j) = 0$ and $\mathbb{V}(\mathcal{E}_j|\mathcal{X}_j = x_j) = \sigma_j^2$. Instead, here we consider x_j to be frozen, i.e., *not* arising from a random variable.

density corresponding to the entire dataset. Qualitatively, this tells us how probable it is for the observed dataset to have occurred, *fixing* the true theory and the true parameters, $\mathbf{c}_\star$. The notation for the data-generating parent distribution is made somewhat less clunky if we bundle together the y_j data points and (separately) the $\boldsymbol{\phi}_j$ rows and σ_j values using matrix/vector notation:

$$\begin{aligned} P(\mathbf{y}; \mathbf{c}_\star, \boldsymbol{\Phi}, \boldsymbol{\Sigma}_d) &= \prod_{j=0}^{N-1} P(y_j; \mathbf{c}_\star, \boldsymbol{\phi}_j, \sigma_j) = \prod_{j=0}^{N-1} \mathcal{N}\left(\boldsymbol{\phi}_j \mathbf{c}_\star, \sigma_j\right) \\ &= \left(\prod_{j=0}^{N-1} \frac{1}{\sqrt{2\pi}\sigma_j}\right) \exp\left[-\frac{1}{2}\sum_{j=0}^{N-1}\left(\frac{y_j - \boldsymbol{\phi}_j \mathbf{c}_\star}{\sigma_j}\right)^2\right] \end{aligned} \tag{6.171}$$

On the left-hand side we used our new vector notation for the joint density; this is also the first occurrence of the diagonal matrix $\boldsymbol{\Sigma}_d$, which serves to bundle together the input data variances, σ_j^2. In the first equality we factorized the joint density as a product of each individual density; this is justified because it is reasonable to assume that the error term at each data point is independent from other error terms.[55] In the second equality we simply applied the first step from Eq. (6.170), telling us that each of the densities is normal. In the third equality we applied the second step from Eq. (6.170) and multiplied the exponentials through, collecting the exponents together. For future reference, observe that if we express things in terms of *random vectors*, our starting equation in Eq. (6.168) becomes:

$$\mathscr{Y} = \boldsymbol{\Phi}\mathbf{c}_\star + \mathscr{E} \tag{6.172}$$

which bundles together N random variables for the data points and for the error terms.[56]

6.6.1.2 The Likelihood and Its Maximization

Take a moment to appreciate that while the $P(\mathbf{y}; \mathbf{c}_\star, \boldsymbol{\Phi}, \boldsymbol{\Sigma}_d)$ appearing in Eq. (6.171) is useful formally, things are more complicated from a practical perspective: even if we know that the data have been drawn from a Gaussian parent distribution, this doesn't help much, since we don't actually know the true parameters $\mathbf{c}_\star$. With that in mind, we decide to write out the analogous probability density of the entire dataset arising from a general set of parameters c_k:

$$L(\mathbf{c}) = P(\mathbf{y}; \mathbf{c}, \boldsymbol{\Phi}, \boldsymbol{\Sigma}_d) = \left(\prod_{j=0}^{N-1} \frac{1}{\sqrt{2\pi}\sigma_j}\right) \exp\left[-\frac{1}{2}\sum_{j=0}^{N-1}\left(\frac{y_j - \boldsymbol{\phi}_j \mathbf{c}}{\sigma_j}\right)^2\right] \tag{6.173}$$

Two intermediate steps are not shown here, since they are identical to those in Eq. (6.171),

[55] Elsewhere, we employ the term *independent and identically distributed* (IID). Crucially, here the σ_j's are different, so the different data points $\mathscr{Y}_j$ are independent but not identically distributed (I but not ID). Incidentally, this use of "independent" is unrelated to the term used to describe x ("independent variable") in Eq. (6.169).

[56] As per appendix C.3, we use $\mathscr{Y}$ instead of the standard notation for a random vector, $\mathbf{Y}$; a bold uppercase symbol like $\mathbf{Y}$ here would have been misidentified as a matrix (see appendix C.2). Be sure to distinguish between the single dataset $\mathbf{y}$ at our disposal (also used in earlier sections) and the random vector $\mathscr{Y}$.

the only change being that we are now dealing with $\mathbf{c}$ instead of $\mathbf{c}_\star$. Since the ϕ_k's, x_j's, and σ_j's are all fixed, if you also think of the $\mathscr{Y}_j$'s as fixed at the values y_j, then $P(\mathbf{y}; \mathbf{c}, \mathbf{\Phi}, \mathbf{\Sigma}_d)$ is a function of the c_k parameters alone, $L(\mathbf{c})$. We now set a practical goal: *maximize* $L(\mathbf{c})$, which is called the *likelihood of the parameters* $\mathbf{c}$ (given the dataset $\mathbf{y}$): since we are still dealing with the correct theory (i.e., the ϕ_k's have been chosen wisely), we hope that by focusing on the (argument leading to the) *maximum likelihood* we will be approximating the true parameters $\mathbf{c}_\star$ well. Thus, we have managed to express our task in terms of maximizing a single scalar function, $L(\mathbf{c})$, which encapsulates how probable it is for the entire dataset to occur; at this stage, we take the set of parameters which makes the observed data most probable to be the best set of parameters available to us. As discussed below, this line of reasoning is complicated by (a) the finiteness of the dataset, i.e., the fact that N might be small, and (b) the fact that we have access to only a single dataset $\mathbf{y}$.

A subtle point: with a view to estimating parameter values, we produced $P(\mathbf{y}; \mathbf{c}, \mathbf{\Phi}, \mathbf{\Sigma}_d)$; this is the probability density of the entire dataset $\mathbf{y}$ occurring given the parameters $\mathbf{c}$, but we interpret it as the likelihood *of the parameters*; this brings to mind what philosophers call *abductive reasoning* (as opposed to deductive or inductive reasoning). This quantity is often[57] denoted by $L(\mathbf{c}|\mathbf{y})$—instead of our $L(\mathbf{c})$—but that notation may trick you into thinking that you are computing the probability (density) that the parameters take the value $\mathbf{c}$, which you most certainly are not: an entity like $P(\mathbf{c}|\mathbf{y}; \mathbf{\Phi}, \mathbf{\Sigma}_d)$ does not even make sense in the present context, since the parameters are not considered to be random variables, in contradistinction to the data. (Rest assured that we will attack this question head-on in the following subsections.) If you're getting confused, just remember that Eq. (6.173) allows us to compute $L(\mathbf{c}) = P(\mathbf{y}; \mathbf{c}, \mathbf{\Phi}, \mathbf{\Sigma}_d)$ for different values of the parameters $\mathbf{c}$ and thereby attain the maximum; whether that leads to $\mathbf{c}_\star$ is still to be determined.[58]

Instead of directly maximizing the likelihood $L(\mathbf{c})$ from Eq. (6.173), it is customary to maximize the natural logarithm of the likelihood, $\ln[L(\mathbf{c})]$. That avenue is followed in problem 6.51, where you are asked to recover the normal equations of Eq. (6.144). Here, we (mundanely) realize that all the prefactors in Eq. (6.173) are constant, being experimentally provided; since the exponential is monotonic, to maximize $L(\mathbf{c})$ we can minimize what's in the exponent. Observe that we've seen this before (modulo a factor of 2):

$$\chi^2 = \sum_{j=0}^{N-1} \left(\frac{\mathscr{Y}_j - \boldsymbol{\phi}_j \mathbf{c}}{\sigma_j} \right)^2 \tag{6.174}$$

where we used $\mathscr{Y}_j$ to emphasize that the experiment/dataset could be regenerated. This is nothing other than the χ^2 statistic of Eq. (6.120), expressed using our updated notation! You can now see the truth of our claim on page 369 that chi-squared minimization is equivalent to maximum-likelihood parameter estimation for the case where the errors are normally distributed. While Eq. (6.174) is not really new, our perspective on it has changed:[59] in an experiment, you draw a normally distributed $\mathscr{Y}_j$ about the *actual* value $\boldsymbol{\phi}_j \mathbf{c}_\star$—which is given by the true theory and true parameters, recall Eq. (6.168)—with standard deviation

[57] Advanced works (e.g., Ref. [83]) typically employ something like $L(\mathbf{y}|\mathbf{c})$, *contra* footnote 52.

[58] Be sure to realize that the maximum of $L(\mathbf{c})$ could be *greater* than the $P(\mathbf{y}; \mathbf{c}_\star, \mathbf{\Phi}, \mathbf{\Sigma}_d)$ of Eq. (6.171).

[59] "The road up and the road down is one and the same" (Heraclitus, fragment 60).

σ_j, for each of the N values of j. When you minimize the χ^2, you maximize the likelihood $L(\mathbf{c})$, and thereby produce an estimate of the true parameters $\mathbf{c}_\star$.

Since there exists a lot of confusion on the subject in the literature, let us summarize things here. Following the above maximization/minimization procedure, we arrive at the normal-equation solution of Eq. (6.157) or, equivalently, the *maximum-likelihood (ML) estimator*:

$$\hat{\mathbf{c}} = (\mathbf{\Phi}^T \mathbf{\Sigma}_d^{-1} \mathbf{\Phi})^{-1} \mathbf{\Phi}^T \mathbf{\Sigma}_d^{-1} \mathscr{Y} \tag{6.175}$$

where we updated the notation (see Eq. (6.269) for the correspondence);[60] note that the only inversion one actually needs to carry out here is that for $\mathbf{\Sigma}_d^{-1}$, which is trivial since that is a diagonal matrix. If we limit ourselves to the single dataset $\mathbf{y}$ our experimentalist colleagues have provided us with, then Eq. (6.175) gives rise to a *point estimate*, i.e., a single number for each of the parameters in our model. Of course, since here we're taking $\mathscr{Y}$ to be a random vector as per Eq. (6.172), it clearly follows that $\hat{\mathbf{c}}$ is also a random vector. (There's no contradiction with our earlier claim on the parameters $\mathbf{c}$ not being a random vector in the present approach: what is random is the result of positing a general likelihood $L(\mathbf{c})$, maximizing it to get $\hat{\mathbf{c}}$, and then considering alternative datasets. As per our definition on page 368, since $\hat{\mathbf{c}}$ is a function of the data $\mathscr{Y}$, it is a *statistic*.) Given that $\hat{\mathbf{c}}$ inherits the randomness of the data, we are faced with the corresponding variances, see Eq. (6.164) or:

$$\hat{\Sigma}_{kk} = \left(\mathbf{\Phi}^T \mathbf{\Sigma}_d^{-1} \mathbf{\Phi}\right)^{-1}_{kk} \tag{6.176}$$

where we are using shorthand to denote the diagonal elements, see also Eq. (6.271); it's harder to avoid actually having to carry out the inversion needed for the variance-covariance matrix. The variance in each (ML estimate of each) parameter clearly depends on the basis functions ϕ_k, the input-data uncertainties σ_j, the input-data abscissas x_j (but equally clearly does *not* depend on the data values $\mathscr{Y}_j$ or on the estimator $\hat{\mathbf{c}}$). As you will show in problem 6.52, if you include more x_j's, the variances will be smaller (they asymptotically tend to zero).[61] Since the only probability distributions we've encountered in this section have been densities in $\mathscr{Y}$, you won't be surprised to discover in the same problem that the variances arise when you integrate across *possible datasets*.[62]

Above we cast our problem in the form of multidimensional minimization, taking it for granted that each density was independently of Gaussian form, see Eq. (6.173). For the sake of completeness we now note that maximum-likelihood estimation is a much more general technique. (For example, there are exotic likelihood functions for which the maximum is not unique, or doesn't even exist.) Imposing very few regularity conditions, ML obeys several pleasing properties, some of which you will explore in problem 6.52: $\hat{\mathbf{c}}$ converges "in probability" (i.e., as N goes to infinity) to the true value $\mathbf{c}_\star$: we say that *the ML estimate is consistent.* Another related property: $\hat{\mathbf{c}}$ minimizes the statistical distance between the general likelihood $P(\mathbf{y}; \mathbf{c}, \mathbf{\Phi}, \mathbf{\Sigma}_d)$ and the parent distribution $P(\mathbf{y}; \mathbf{c}_\star, \mathbf{\Phi}, \mathbf{\Sigma}_d)$.

[60] We are calling $\hat{\mathbf{c}}$ what was denoted by $\mathbf{c}^*$ in Eq. (6.153); this is *different* from the $\mathbf{c}_\star$ used to denote the true parameters. While $\hat{\mathbf{c}}$ is reminiscent of a quantum-mechanical operator, this is standard statistical notation.

[61] We will re-encounter the same idea when we discuss the variance of the sample mean in Eq. (7.180).

[62] Here is an intuitive summary: the *magnitude* of the variances can be reduced by increasing the number of points N in the single dataset $\mathbf{y}$ at your disposal, while the *interpretation* of the variances is across datasets $\mathscr{Y}$.

Finally, for us (and—asymptotically—for nearly everyone else, too) the expectation of $\hat{\mathbf{c}}$ is equal to $\mathbf{c}_\star$ (where such statements are interpreted in terms of integrations across other datasets): we say that *the ML estimator is unbiased.*[63]

6.6.1.3 Parameter Estimation and Model Selection

With all the material on random variables under our belt, let us now try to interpret more deeply the χ^2 quantity appearing in Eq. (6.174). We start by observing that (for $\mathbf{c} = \mathbf{c}_\star$) each numerator is nothing other than the error term, $\mathscr{E}_j$, as you can see from Eq. (6.168). Next, we realize that each term in the sum is associated with the corresponding σ_j. To see the effect this has, we investigate the variance of the random variable $\mathscr{E}_j/\sigma_j$:

$$\mathbb{V}\left(\frac{\mathscr{E}_j}{\sigma_j}\right) = \frac{1}{\sigma_j^2}\mathbb{V}(\mathscr{E}_j) = \frac{\sigma_j^2}{\sigma_j^2} = 1 \tag{6.177}$$

In the first equality we used a property of the variance from appendix C.3. In the second equality we recalled from page 385 that $\mathscr{E}_j$ obeys the distribution $\mathcal{N}(0, \sigma_j)$, so its variance is simply σ_j^2. In the third equality we reached the conclusion that each term appearing in the sum for χ^2 in Eq. (6.174) has unit variance! But since $\mathscr{E}_j$ had zero mean, so will $\mathscr{E}_j/\sigma_j$: we have shown that $\mathscr{E}_j/\sigma_j$ obeys the distribution $\mathcal{N}(0, 1)$, i.e., it is a *standard normal* random variable; to produce the χ^2 statistic we simply square and sum. In problem 6.45 you will show that a random variable that is produced by squaring ν independent standard normal random variables (and then adding the squares together) obeys the *chi-squared distribution*:

$$f_\nu(x) = \frac{1}{2^{\nu/2}\Gamma(\nu/2)}e^{-x/2}x^{(\nu/2)-1} \tag{6.178}$$

The χ^2 of Eq. (6.174) obeys the distribution of Eq. (6.178); this is why the statistic and the distribution have the same name. In Eq. (6.174) we are summing N terms and adjusting n parameters so (when $\mathbf{\Sigma}_d$ is diagonal and the basis functions are independent) $\nu = N - n$.

Still on the subject of further interpreting the χ^2 statistic: up to this point in the present subsection, we were always assuming that we had access to the true theory, i.e., that the ϕ_k's (which give rise to $\mathbf{\Phi}$) were chosen wisely. Of course, in practice you typically don't know the true theory, so you are reduced to postulating an Ansatz (i.e., expanding in a set of basis functions) and hoping for the best; as was pithily stated by George Box [24] (of Box–Muller fame, see problem 7.37): *all models are wrong, but some models are useful.* Even if you don't have the true theory, you can still try to do your best at maximizing the likelihood $L(\mathbf{c})$ of Eq. (6.173), which is equivalent to minimizing the χ^2 of Eq. (6.174) for the dataset $\mathbf{y}$, within a given parameter space. We can now make contact with the rule of thumb from page 370, which informed us which values of χ^2 we should expect if we have carried out a good fit. Problem 6.46 asks you to compute integrals related to the chi-squared distribution, Eq. (6.178), one of which is the mean of this probability density. There you will learn that its mean is $\nu = N - n$ (the number of degrees of freedom): this is the source of the standard advice that a good fit is synonymous with $\chi^2 \approx N - n$. As a result, one

[63] The estimator $\hat{\mathbf{c}}(\mathscr{Y})$ is a random vector (*alias* $\hat{\mathscr{C}}$), whereas the estimate $\hat{\mathbf{c}}(\mathbf{y})$ corresponds to a given dataset.

option that is very often pursued in practice is to use the value of chi-squared per degree of freedom for the purposes of model selection; for example, one can separately carry out fits with two different theories (as we did in `newnormal.py`) and compare the values of $\chi^2/(N-n)$, picking as superior that with value closest to 1 (from above).

We conclude this subsection with a plot twist. The further points we are about to make are not so dramatic in connection with assessing the goodness of fit after carrying out maximum-likelihood estimation (namely, what we've been doing for most of this subsection). We simply admonish you to keep in mind that there is an ambiguity at play here, since you could get a large value for $\chi^2/(N-n)$ even if you have the correct model: first, the input-data uncertainties σ_j may have been underestimated or, second, you may be dealing with an improbable statistical fluctuation. (Perhaps even more disquieting is the fact that you could get $\chi^2 \approx N-n$ even though your model is bad.)

We now turn to much more serious issues regarding the earlier advice on model selection. First, imagine using the true theory with the true parameters twice to generate two datasets which differ only in the error terms, i.e., in the $\mathscr{E}_j$'s: for these two sets of data you would produce two distinct values of χ^2 but, of course, you would have trouble trying to use those values to determine which theory is correct (since you had only one theory to begin with!). Second, let us understand a bit more carefully the statistical fluctuations mentioned in the previous paragraph. The aforementioned problem 6.46, in which you will study integrals related to the chi-squared distribution, also explores the variance in the χ^2—remember: this is a random variable itself, obeying the chi-squared distribution of Eq. (6.178). This variance, which we repeat applies to the χ^2 for the true theory, is $2N$;[64] thus, χ^2/N has a mean of 1 and a variance of $2/N$ (again due to the property of the variance of random variables). For large N the chi-squared distribution is approximately normal,[65] so we can estimate the width of the Gaussian peak by taking the square root, i.e., $\sqrt{2/N}$. For the case of $N = 50$ data points, this leads to a 3σ interval[66] of $0.4 \leq \chi^2/N \leq 1.6$, always for the true theory with the true parameters! Obviously, you cannot even think of comparing alternative models within this interval. Even worse, the chi-squared distribution is a probability distribution supported on a semi-infinite interval so, while improbable, it is still possible to find a χ^2 value even outside the 3σ interval, i.e., in the tail(s). Third, and perhaps most important, our argument from Eq. (6.177) onward, allowing us to go from the sum of squares of standard normal variables to a chi-squared-distributed random variable, completely breaks down if we do not have the true theory with the true parameters. For other theories, the $\mathscr{E}_j/\sigma_j$'s do *not* obey a standard normal distribution, so the $\chi^2/(N-n)$ does not obey a chi-squared distribution, therefore we cannot quantify its spread via $\sqrt{2/(N-n)}$.

6.6.2 Bayes' Rule

As part of the intellectual approach that we were following in the previous subsection (known as *frequentist statistics* or *classical statistics*) there exist improved techniques (such as cross-validation and bootstrapping) that cure the illnesses we encountered when

[64] For the true theory there is no fit involved, so the number of degrees of freedom is $N - 0 = N$.
[65] As a result of the central limit theorem, which is derived in problem 6.47 and explored in problem 6.48.
[66] As mentioned above, 99.7% of the values lie within three standard deviations of the mean; see problem 6.50.

discussing model selection. Instead of delving into them, however, we will now take the opportunity to introduce a markedly different approach, known as *Bayesian statistics*. The differences between frequentist and Bayesian statistics are fascinating and largely reflect distinct philosophical positions. Under the frequentist interpretation, probability represents the long-run relative frequency of occurrences (e.g., after many coin flips, roughly half will be heads and half tails). Under the Bayesian interpretation, probability represents a reasonable expectation about something given current knowledge (e.g., our belief that the next coin flip is equally likely to be heads or tails). Once you master the basics, it's OK if you remain "of twosome twiminds" and practise methodological polytheism.

In the spirit of this book, we will try to steer clear of most subtleties that still divide statisticians (*facilis descensus Averno*), but some conceptual questions have to be addressed when juxtaposing Bayesian with frequentist attitudes to regression. First, let us disabuse you of the notion that the crucial distinction between the two approaches is whether or not the true parameters $\mathbf{c}_\star$ are deterministic quantities: it is defensible from a Bayesian perspective[67] (just like from a frequentist one) to view the data as being drawn, as per Eq. (6.172), from the true theory with fixed (unknown) true parameters $\mathbf{c}_\star$; to put it bluntly, would you even be estimating a given quantity in the first place if it wasn't fixed? (A stranger misconception has it that in Bayesian regression the data are not random; it is hard to reconcile that with the fact that such an approach employs a...probability density for $\mathbf{y}$ to arise from a given parameter set—roughly what we called the likelihood above). The two approaches do differ on how they treat parameter estimates, their uncertainty, and other quantities that follow from them. In frequentist regression, the parameters are typically approximated via the ML estimator and the uncertainty in the associated estimate is interpreted across alternative datasets (i.e., in N dimensions). In Bayesian regression, in addition to the likelihood one also factors in any knowledge one has about the parameters *before* encountering the data; this gives rise to an overall probability distribution encapsulating all that has been gleaned on the parameter values (i.e., in n dimensions).[68]

To understand what all this actually means, we will start by deriving what is known as *Bayes' rule* (also known as *Bayes' theorem*) for a simple case, namely for two continuous random variables, the unobserved variable $\mathscr{C}$ and the observed variable $\mathscr{Y}$. While the following derivation will use only elementary properties of probability theory (i.e., will be general), the symbols $\mathscr{C}$ and $\mathscr{Y}$ are here chosen in order to evoke the parameters and data (respectively) that appear in the problem of linear regression. After you grasp the essential features, we will extend the present (toy version) case to our real problem, which involves many data points, many parameters, input-data uncertainties, and so on.

For the continuous random variables $\mathscr{C}$ and $\mathscr{Y}$ (with realizations c and y, respectively), the definition of the *marginal probability density function* $P(y)$ takes the *joint probability density function* $P(c, y)$ and integrates out (or "marginalizes over") the values of c:

$$P(y) = \int P(c, y)dc \tag{6.179}$$

[67] Specifically, both from an *objective* and from a *tempered/weak subjective* Bayesian viewpoint. While those of an *operational/strong subjective* bent [17] typically repudiate an absolute conception of reality [155], even an extreme anti-realist guards against falling into a ditch (Aristotle, *Metaphysics*, 1008b15).

[68] This is the distinction between frequentist *confidence* regions and Bayesian *credible* regions (keep reading).

We can now write down the definition of the *conditional probability density function* $P(c|y)$ in terms of the joint density $P(c, y)$ and the marginal density $P(y)$:

$$P(c, y) = P(c|y)P(y) \tag{6.180}$$

Here we "condition" on y, which is another way of saying that the right-hand side takes y as given. Of course, there is nothing special about y, as we could have just as well written out the definition of the conditional probability density function $P(y|c)$ in the form of:

$$P(c, y) = P(y|c)P(c) \tag{6.181}$$

conditioning on c. We can now combine the definition of the marginal density $P(y)$ from Eq. (6.179) with the definition of the conditional density $P(y|c)$ from Eq. (6.181) to find:

$$P(y) = \int P(y|c)P(c)dc \tag{6.182}$$

This is known as the *law of total probability*. We can now equate the right-hand sides of Eq. (6.180) and Eq. (6.181) and solve for $P(c|y)$, where the denominator is given by Eq. (6.182). Putting it all together, we have arrived at *Bayes' rule*:[69]

$$\underbrace{P(c|y)}_{\text{posterior}} = \frac{\overbrace{P(y|c)}^{\text{likelihood}}\ \overbrace{P(c)}^{\text{prior}}}{\underbrace{P(y)}_{\text{marginal likelihood}}}, \quad \text{where } P(y) = \int P(y|c)P(c)dc \tag{6.183}$$

This involves some further terminology, which we now elaborate on:

- The *likelihood* $P(y|c)$ is analogous to what we were dealing with above. In the Bayesian case, it is the conditional probability density to observe the data given the parameters. Crucially, while the likelihood may be viewed as a function of c, it is a probability distribution in y (but *not* in c, i.e., you won't get 1 if you integrate over all c's).
- The *prior* $P(c)$ encapsulates our knowledge about the unobserved variable $\mathscr{C}$ *before* we encounter the data. In the regression problem, the prior could, e.g., represent the fact that we expect $\mathscr{C}$ values to be positive. If you don't know that much about the parameter values ahead of time, you could employ what is called a *non-informative prior*.
- The *marginal likelihood* $P(y)$ is clearly independent of the unobserved variable $\mathscr{C}$: as you can see in Eq. (6.182), c is integrated out.[70] In Bayes' rule it's basically a normalization factor, though, as you'll discover in problem 6.57, the marginal likelihood plays an important role in model selection.

[69] We are committing a near-universal abuse of notation: we are dropping the subscript in what is properly $P_{\mathscr{C}}(c)$, and so on. It is important to remember that $P(c)$ and, say, $P(y)$ will usually have different functional forms; it may help you to think in terms of different symbols for each entity, e.g., $\mathcal{P}(c|y) = \mathcal{L}(y|c)P(c)/M(y)$. As with all notational conventions, one must balance clarity against the risk of being too distracting.

[70] We repeat: even though both entities have "likelihood" in the name, the marginal likelihood, in contradistinction to the likelihood, is not a function of c. You can see where the modifier "marginal" came from, however, since the parameter values are marginalized over. Instead of using the term marginal likelihood, some texts speak of the *model evidence*, others simply of the *evidence*, yet others of the *prior predictive distribution*.

- The *posterior* $P(c|y)$ is the central entity in Bayesian statistics, summarizing all we know about the unobserved variable $\mathscr{C}$ *after* we've encountered the data instance y. Crucially, the posterior is a *distribution*, i.e., after you apply Bayes' rule you do not arrive at a *point estimate* for $\mathscr{C}$ but at a probability density function, $P(c|y)$—compare the numerator and denominator in Eq. (6.183) to see that if you integrate over all c values you get 1.

A lot of the time, the marginal likelihood is either hard to compute or uninteresting, so Bayes' rule for the problem of regression can be summarized in the following form:

$$P\,(\text{parameters}|\text{data}) \propto P\,(\text{data}|\text{parameters})\,P\,(\text{parameters}) \qquad (6.184)$$

This is simply saying that the posterior is proportional to the product of the likelihood and the prior, where all three quantities are viewed as functions of the parameters. Bayes' rule has enabled us to *invert* the conditioning: we produced the posterior (in which the parameters are conditioned on the data) starting from the likelihood (in which the data were conditioned on the parameters). When tackling complicated problems involving many different variables, it may help you to keep in mind the big picture as summarized in Eq. (6.184).

6.6.3 Bayesian Approach to Linear Regression

We now briefly summarize what a frequentist approach to linear regression amounts to and thereby motivate the alternative Bayesian outlook. In the frequentist way of doing things, expectations are taken with respect to the probability density $P(\mathbf{y}; \mathbf{c}_\star, \boldsymbol{\Phi}, \boldsymbol{\Sigma}_d)$ of the parent distribution given in Eq. (6.171)—see, e.g., Eq. (6.270)—carrying out N-dimensional integrations over the $\mathbf{y}$'s (i.e., in terms of imagining further experiments). This allows one to, e.g., prove that $\hat{\mathbf{c}}$ (produced by positing a general likelihood function $P(\mathbf{y}; \mathbf{c}, \boldsymbol{\Phi}, \boldsymbol{\Sigma}_d)$ as per Eq. (6.173) and maximizing it) is an unbiased estimator and to quantify the uncertainty in the corresponding estimate. These are intriguing formal results, but when you need to carry out some practical computations you always fall back on using $\hat{\mathbf{c}}$ (evaluated at a given dataset $\mathbf{y}$) instead of $\mathbf{c}_\star$ (which you still don't know)—see, e.g., Eq. (6.274). As a matter of fact, the simulation technique known as the *parametric bootstrap* (see problem 6.68) generates new datasets after plugging in $\hat{\mathbf{c}}$ instead of $\mathbf{c}_\star$ in Eq. (6.172). In other words, it is a bit awkward to be proving results using integrals over all $\mathbf{y}$'s, when all you actually have at your disposal is the single $\mathbf{y}$ that was experimentally produced (this being the realization of the random vector $\boldsymbol{\mathscr{Y}}$, no doubt, but a single dataset—made up of N distinct y_j's—nonetheless). Put yet another way, the $\hat{\boldsymbol{\Sigma}}$ of Eq. (6.176) cannot give rise to a neat probabilistic interpretation in connection with $\mathbf{c}_\star$ for your single dataset $\mathbf{y}$: frequentist approaches, like ML, always refer to long-run performance. For the one dataset at your disposal, $\mathbf{y}$, you may produce, e.g., a 3σ interval (known as a frequentist *confidence region*)[71]

[71] While we often use the one-dimensional term *interval*, in many dimensions one should speak of a *region*. The modifier *confidence* interval is clearly a misnomer (as one's confidence is typically *low* when the interval is big). Alternative proposals (e.g., uncertainty interval or compatibility interval) have not been widely adopted.

by combining $\hat{\mathbf{c}}$ and the uncertainties in $\hat{\boldsymbol{\Sigma}}$, but you cannot say there is an 89% probability[72] that your interval contains nature's true parameters $\mathbf{c}_\star$; all you can say is that $\mathbf{c}_\star$ either is or isn't in the given interval, i.e., the probability is either unity or zero (a rather unsatisfying assertion). Instead, what *does* hold is that (in the absence of systematic errors) as you keep producing new datasets (drawn from the same parent distribution) and the corresponding 3σ intervals, 89% of your intervals will contain $\mathbf{c}_\star$, a considerably different statement.

A related, yet distinct, issue has to do with what happens when you increase the number of data points, N, for your single dataset $\mathbf{y}$ (recall footnote 62): this makes unpleasant effects like overfitting less of an issue (another cure goes by the name of *regularization*). Speaking of overfitting, you may recall that our advice on page 382 was to adjust the number of your model's parameters to the size of the dataset. This is somewhat disconcerting: shouldn't we rather pick the complexity of our model based on the complexity of the *problem* we are trying to solve (and not on how many data points we happen to currently have)?

6.6.3.1 The Prior and Its Role

This is a natural segue into the Bayesian approach to regression, according to which *we can have more undetermined parameters than data points*! This little bit of magic is accomplished via the use of a *prior distribution*. The prior encapsulates the probability (density) that our model parameters $\mathscr{C}$ (the only thing at our disposal) take on specific values $\mathbf{c}$, before we've seen any data. Note that nature's true parameters $\mathbf{c}_\star$ are still fixed, but we are acknowledging that we have no direct access to them. If you know roughly where $\mathbf{c}_\star$ lies, you could pick your prior for $\mathscr{C}$ to be, e.g., a Gaussian centered at your best estimate for $\mathbf{c}_\star$. If you are fully certain that you are at the "physical point", then your probability distribution would be of Dirac-delta form, $\delta^{(n)}(\mathbf{c} - \mathbf{c}_\star)$,[73] but in that case (where you already know the true parameters) you don't have a problem to solve in the first place (the problem would have been how to estimate the *unknown* true parameters). If you know almost nothing about the parameters before you encounter any data, then you should use a less informative prior. (Unfortunately, a uniform prior is not: (a) invariant under transformations, and (b) a probability density function, as it integrates to infinity over the reals.) A common approach in this context is to employ a *Jeffreys prior* (introduced in problem 6.58).

For now, let's keep things general by writing the prior as $P(\mathbf{c}; \boldsymbol{\mu}_0, \boldsymbol{\Sigma}_0)$, i.e., in terms of its mean vector and covariance matrix but without specifying a specific distribution. Note that $P(\mathbf{c}; \boldsymbol{\mu}_0, \boldsymbol{\Sigma}_0)$ doesn't depend on $\mathbf{y}$, $\boldsymbol{\Phi}$, or $\boldsymbol{\Sigma}_d$: how could it, since it is *prior* to encountering the data? The prior is combined with a statistical model reflecting what we know about how a given $\mathbf{c}$ gives rise to $\mathbf{y}$'s: this is precisely what we called the *likelihood* $P(\mathbf{y}|\mathbf{c}; \boldsymbol{\Phi}, \boldsymbol{\Sigma}_d)$ in Eq. (6.173), where we tweaked the notation (keep reading). This (time around, this) is the conditional probability density of the entire dataset $\mathbf{y}$ occurring given the parameters $\mathbf{c}$. The Bayesian approach to linear regression amounts to combining the prior and the likelihood together (after observing the single dataset $\mathbf{y}$) to update our knowledge via the application

[72] The empirical rule takes the form 68–95–99.7 only for a univariate Gaussian. Here we are working with an $n = 5$ multivariate Gaussian for which, as you'll show in problem 6.50, the 3σ interval is less constraining. When the data are non-normally distributed, these are *asymptotic* confidence intervals (see problem 6.59).

[73] Violating Cromwell's rule, according to which you should *think it possible that you may be mistaken*.

of Bayes' rule; the result of this is the *posterior distribution*:

$$P(\mathbf{c}|\mathbf{y};\boldsymbol{\Phi},\boldsymbol{\Sigma}_d,\boldsymbol{\mu}_0,\boldsymbol{\Sigma}_0) = \frac{P(\mathbf{y}|\mathbf{c};\boldsymbol{\Phi},\boldsymbol{\Sigma}_d)P(\mathbf{c};\boldsymbol{\mu}_0,\boldsymbol{\Sigma}_0)}{P(\mathbf{y};\boldsymbol{\Phi},\boldsymbol{\Sigma}_d,\boldsymbol{\mu}_0,\boldsymbol{\Sigma}_0)} = \frac{P(\mathbf{y}|\mathbf{c};\boldsymbol{\Phi},\boldsymbol{\Sigma}_d)P(\mathbf{c};\boldsymbol{\mu}_0,\boldsymbol{\Sigma}_0)}{\int d^n c P(\mathbf{y}|\mathbf{c};\boldsymbol{\Phi},\boldsymbol{\Sigma}_d)P(\mathbf{c};\boldsymbol{\mu}_0,\boldsymbol{\Sigma}_0)} \tag{6.185}$$

This is the n-dimensional generalization of Eq. (6.183), also taking into account that we are faced with more entities than in our earlier toy problem.[74] Note that the denominator in the first equality, $P(\mathbf{y};\boldsymbol{\Phi},\boldsymbol{\Sigma}_d,\boldsymbol{\mu}_0,\boldsymbol{\Sigma}_0)$, does *not* condition on $\mathbf{c}$; this is what we called above the *marginal likelihood*; we write this out in the second equality, where you can see that $\mathbf{c}$ is integrated out. This is only our first example of integrating over the parameter values (instead of the datasets, as in the earlier maximum-likelihood approach): we will re-encounter such integrations over $\mathbf{c}$ both in the main text and in the problem set. For now, the qualitative takeaway is that the posterior $P(\mathbf{c}|\mathbf{y};\boldsymbol{\Phi},\boldsymbol{\Sigma}_d,\boldsymbol{\mu}_0,\boldsymbol{\Sigma}_0)$ is a probability distribution in $\mathbf{c}$: whenever we wish to compute an actual probability or make a prediction we will *not* be simply maximizing the posterior but taking the entire distribution into consideration.

An excursus on the conceptual underpinnings of our notation: one could choose to call the posterior in Eq. (6.185) $P(\mathbf{c}_\star|\mathbf{y};\boldsymbol{\Phi},\boldsymbol{\Sigma}_d,\boldsymbol{\mu}_0,\boldsymbol{\Sigma}_0)$. In that case, as stressed by Edwin Jaynes [78], one should be careful to always speak of the "probability distribution for $\mathbf{c}_\star$", as opposed to the "posterior distribution of $\mathbf{c}_\star$" (since what is distributed is the probability, not the true parameters). That alternative notation would have had a jarring effect, however, since it would have involved an integral over $\mathbf{c}_\star$ in the denominator; recall from Eq. (6.172) that nature's $\mathbf{c}_\star$ are fixed (though unknown), i.e., the data-generating distribution is still the (unknown) $P(\mathbf{y};\mathbf{c}_\star,\boldsymbol{\Phi},\boldsymbol{\Sigma}_d)$ from Eq. (6.171). Here we are denoting our model parameters by $\mathscr{C}$ (with realization $\mathbf{c}$): after the Bayesian update, our knowledge of the probability (density) for them to take on specific values is encapsulated in $P(\mathbf{c}|\mathbf{y};\boldsymbol{\Phi},\boldsymbol{\Sigma}_d,\boldsymbol{\mu}_0,\boldsymbol{\Sigma}_0)$. This allows us to straightforwardly integrate over $\mathbf{c}$ values in Eq. (6.185)—or, say, in Eq. (6.195)—thereby avoiding the specious implication that nature's $\mathbf{c}_\star$ might somehow not be fixed.

It's time to address the elephant on the page: if you've never encountered the concept of a prior before, you may be feeling uncomfortable about it, thinking that it is arbitrary or just a way of biasing the answer. In other words, you may be worried that the entire Bayesian approach is built on quicksand; we will now discuss how you can learn to stop worrying about the prior.[75] First, in the study of physics we often truly do have prior knowledge about a given parameter, based on decades of earlier work on the subject. For example, when extracting a neutron star's mass from observational data you can safely assume that it won't be 100 solar masses. Another example at the opposite extreme in scale: if you're extracting the gyromagnetic ratio of the muon from experimental data, it's OK to assume that it won't be 50 (the magnetic moment is not *that* anomalous!). The Bayesian attitude is that we shouldn't have to pretend we know nothing about the parameter in question, even if we haven't started digging into our latest set of data. (To be fair, this is skirting the subtle question of how to turn a *constraint* into a prior *distribution*.) Second, the likelihood $P(\mathbf{y};\mathbf{c},\boldsymbol{\Phi},\boldsymbol{\Sigma}_d)$ of Eq. (6.173) clearly scales with the value of N: we alluded to this fact above, when we noted that overfitting is assuaged in the ML approach as you add in more

[74] Crucially, we are using | for random-variable conditioning and ; to denote the fixing of constants (as earlier).

[75] And, in due course, love how naturally systematic errors are handled (marginalizing over nuisance parameters).

data points. Now, a Bayesian approach still makes use of the likelihood $P(\mathbf{y}|\mathbf{c}; \mathbf{\Phi}, \mathbf{\Sigma}_d)$, see Eq. (6.185); the new contribution it involves, the prior, clearly does *not* scale with the dataset size (it doesn't even depend on the data). Thus, as you introduce more data points you will rely less and less on the prior you started with; if you chose a poor (but not pathological) prior, this will get washed away after the Bayesian update, i.e., after you apply Bayes' rule (see problem 6.59). To reverse this argument, for a small dataset size the prior may actually be important: but in that case, where you don't have much data, wouldn't you want to factor in as much knowledge as possible in carrying out the regression? Third, you certainly have the right (some would say the duty) to try out different parameters $\boldsymbol{\mu}_0$ and $\mathbf{\Sigma}_0$ for your prior $P(\mathbf{c}; \boldsymbol{\mu}_0, \mathbf{\Sigma}_0)$, to see if the posterior distribution (or another result you produce based on it) is impacted or not (or by how much). Better yet, you could try out a totally different form of prior, examining how different your final answers look when, e.g., you first ignore your knowledge of the parameters and use a non-informative prior, turning to a more realistic prior next. In other words, you can carry out a prior *sensitivity analysis*; of course, you should also check the choices you made regarding your model's likelihood. Such investigations form part of what is known as *robust Bayesian analysis*.

In the previous paragraph we spoke of "the prior you started with" because you may very well not wish to apply Bayes' rule from Eq. (6.185) only once. The $\mathbf{y}$ appearing in this equation corresponds to N data points, but there's nothing keeping you from dividing up your full dataset into smaller batches. For example, you could process a dataset of 100 points in batches of 10: the first time you apply Bayes' rule your prior will be based on whatever *a priori* knowledge you have on the parameter values (if any). After you've processed the first batch (of $N = 10$ data points) and produced the posterior, you could turn around and repeat this process, interpreting that posterior as the prior for the second batch of points. (There's no double-counting involved here: your new prior truly knows nothing about the second batch of data, which has not been processed yet.) You could even set up the problem as processing data points one by one (i.e., in batches of $N = 1$) as they arrive. Such a *sequential* approach (also referred to as *online learning*) is an appealing feature of Bayesian computation. This is to be compared with the normal-equations/ML approach described above, which does not trivially lend itself to a sequential treatment: you can't immediately use your older $\hat{\mathbf{c}}$ if a new data point shows up. (Again, to be fair, in section 6.7.2 we will describe a method that is implicitly based on the maximum-likelihood approach, in which the parameters are updated after processing each new data point, but that is certainly not how introductory presentations cast the problem of likelihood maximization.)

6.6.3.2 Determining the Posterior

The above discussion might be too abstract for you, so let's make it concrete by addressing a specific likelihood, a specific prior, and the resulting posterior. To continue the earlier exposition, we will take our statistical model for the likelihood to be the same as in Eq. (6.173), using our updated notation:

$$P(\mathbf{y}|\mathbf{c}; \mathbf{\Phi}, \mathbf{\Sigma}_d) = \mathcal{N}(\mathbf{\Phi}\mathbf{c}, \mathbf{\Sigma}_d) = \frac{1}{(2\pi)^{N/2}|\mathbf{\Sigma}_d|^{1/2}} \exp\left[-\frac{1}{2}(\mathbf{y} - \mathbf{\Phi}\mathbf{c})^T \mathbf{\Sigma}_d^{-1} (\mathbf{y} - \mathbf{\Phi}\mathbf{c})\right] \quad (6.186)$$

We took advantage of the fact that a product of univariate Gaussian densities is a multivariate Gaussian density.[76] You will show this in problem 6.49, together with the normalization (which holds even when $\boldsymbol{\Sigma}_d$ is not diagonal) involving the determinant, $|\boldsymbol{\Sigma}_d|$.

The question now arises regarding the form of the prior. It should be easy to see from Eq. (6.185) that if the prior and the likelihood are of different form (which they generally will be), then the posterior will (customarily) take a third form. As mentioned after Eq. (6.185), once we have the posterior we typically make predictions by taking expectations of some quantity with respect to the posterior: this translates into an integral over $\mathbf{c}$ values. If the posterior is complicated, then we would be in a quandary, as we don't generally know how to compute n-dimensional integrals (though we will learn how to do so in section 7.7.5.2, when we introduce the Metropolis algorithm). Thus, for now we are limited to simple scenarios. Specifically, we'll pick a *conjugate prior*, which is so called because it gives rise to a posterior of the same form as the prior; this is quite convenient because it means that the Bayesian update simply modifies the parameter values of the prior. There exist several well-studied cases of conjugate priors, e.g., if you have a likelihood obeying the binomial distribution (applicable to discrete random variables) and you choose a prior obeying the beta distribution (applicable to continuous random variables), then your posterior will also be of beta form.

For us, choosing a conjugate prior means that we will take the prior to be a Gaussian:

$$P(\mathbf{c};\boldsymbol{\mu}_0,\boldsymbol{\Sigma}_0) = \mathcal{N}(\boldsymbol{\mu}_0,\boldsymbol{\Sigma}_0) = \frac{1}{(2\pi)^{N/2}|\boldsymbol{\Sigma}_0|^{1/2}}\exp\left[-\frac{1}{2}(\mathbf{c}-\boldsymbol{\mu}_0)^T\boldsymbol{\Sigma}_0^{-1}(\mathbf{c}-\boldsymbol{\mu}_0)\right] \tag{6.187}$$

This is an n-dimensional Gaussian density (a function of $\mathbf{c}$), as advertised. *Prima facie* the likelihood of Eq. (6.186) is an N-dimensional Gaussian function of $\mathbf{y}$, so the fact that both the likelihood and the posterior can be written as n-dimensional Gaussian functions of $\mathbf{c}$ is not totally trivial. This is often implicitly assumed in the literature; let's derive it explicitly, to increase your confidence in handling multivariate Gaussians:[77]

$$\begin{aligned}
P(\mathbf{c}|\mathbf{y};\boldsymbol{\Phi},\boldsymbol{\Sigma}_d,\boldsymbol{\mu}_0,\boldsymbol{\Sigma}_0) &\propto P(\mathbf{y}|\mathbf{c};\boldsymbol{\Phi},\boldsymbol{\Sigma}_d)P(\mathbf{c};\boldsymbol{\mu}_0,\boldsymbol{\Sigma}_0)\\
&\propto \exp\left[-\frac{1}{2}(\mathbf{y}-\boldsymbol{\Phi}\mathbf{c})^T\boldsymbol{\Sigma}_d^{-1}(\mathbf{y}-\boldsymbol{\Phi}\mathbf{c})-\frac{1}{2}(\mathbf{c}-\boldsymbol{\mu}_0)^T\boldsymbol{\Sigma}_0^{-1}(\mathbf{c}-\boldsymbol{\mu}_0)\right]\\
&\propto \exp\left[-\frac{1}{2}\mathbf{c}^T\left(\boldsymbol{\Phi}^T\boldsymbol{\Sigma}_d^{-1}\boldsymbol{\Phi}+\boldsymbol{\Sigma}_0^{-1}\right)\mathbf{c}+\mathbf{c}^T\left(\boldsymbol{\Phi}^T\boldsymbol{\Sigma}_d^{-1}\mathbf{y}+\boldsymbol{\Sigma}_0^{-1}\boldsymbol{\mu}_0\right)\right]\\
&\equiv \exp\left[-\frac{1}{2}\mathbf{c}^T\boldsymbol{\Lambda}_\mathbf{c}\mathbf{c}+\mathbf{c}^T\mathbf{q}_\mathbf{c}\right]\\
&= \exp\left[-\frac{1}{2}\mathbf{c}^T\boldsymbol{\Lambda}_\mathbf{c}\mathbf{c}+\mathbf{c}^T\boldsymbol{\Lambda}_\mathbf{c}\boldsymbol{\Lambda}_\mathbf{c}^{-1}\mathbf{q}_\mathbf{c}\right]\\
&= \exp\left[-\frac{1}{2}\left(\mathbf{c}-\boldsymbol{\Lambda}_\mathbf{c}^{-1}\mathbf{q}_\mathbf{c}\right)^T\boldsymbol{\Lambda}_\mathbf{c}\left(\mathbf{c}-\boldsymbol{\Lambda}_\mathbf{c}^{-1}\mathbf{q}_\mathbf{c}\right)\right]\exp\left[\frac{1}{2}\mathbf{q}_\mathbf{c}^T\boldsymbol{\Lambda}_\mathbf{c}^{-1}\mathbf{q}_\mathbf{c}\right]\\
&\propto \exp\left[-\frac{1}{2}(\mathbf{c}-\boldsymbol{\mu}_\mathbf{c})^T\boldsymbol{\Sigma}_\mathbf{c}^{-1}(\mathbf{c}-\boldsymbol{\mu}_\mathbf{c})\right]
\end{aligned} \tag{6.188}$$

[76] We write $\mathcal{N}(\boldsymbol{\phi}_j\mathbf{c},\sigma_j)$ but also $\mathcal{N}(\boldsymbol{\Phi}\mathbf{c},\boldsymbol{\Sigma}_d)$, where $\boldsymbol{\Sigma}_d$ contains the σ_j^2's (with emphasis on the square).

[77] We are taking the product of two normal *densities*; we are *not* saying that the product of two normal random variables is another normal variable. That would actually be a (linear combination of) chi-squared variable(s).

In the first step we applied Bayes' rule from Eq. (6.185), dropping the marginal likelihood (since we're trying to determine the **c**-dependence and the denominator doesn't depend on **c**). In the second step we plugged in Eq. (6.186) and Eq. (6.187), dropping the prefactors for the same reason; we also took the opportunity to collect the two contributions onto the same exponent. In the third step we expanded the parentheses and regrouped terms, once again dropping whatever did not depend on **c**; in the second term we used (in two variations) the fact that $\mathbf{c}^T\boldsymbol{\Sigma}_0^{-1}\boldsymbol{\mu}_0$ is a number (so taking the transpose doesn't change anything) and also that $\boldsymbol{\Sigma}_0^{-1} = (\boldsymbol{\Sigma}_0^{-1})^T$ (since a covariance matrix is always symmetric—recall page 380). In the fourth step we introduced the definitions:

$$\boldsymbol{\Lambda}_\mathbf{c} = \boldsymbol{\Phi}^T\boldsymbol{\Sigma}_d^{-1}\boldsymbol{\Phi} + \boldsymbol{\Sigma}_0^{-1}, \qquad \mathbf{q}_\mathbf{c} = \boldsymbol{\Phi}^T\boldsymbol{\Sigma}_d^{-1}\mathbf{y} + \boldsymbol{\Sigma}_0^{-1}\boldsymbol{\mu}_0 \tag{6.189}$$

In the fifth step we observed that $\boldsymbol{\Lambda}_\mathbf{c}$ is symmetric and invertible, so we introduced an identity matrix in the second term. In the sixth step we completed the square (in matrix form): this is a prototypical operation when handling multivariate Gaussians and will repeatedly re-appear in the problem set. We also separated out the term that didn't depend on **c** so you could focus on the term that did. Appealingly enough, even the normalization term (the one that doesn't depend on **c**) *is of Gaussian form* (this will come in handy in problem 6.57). In the seventh step we remembered our definition of the multivariate Gaussian in Eq. (6.260), or the analogous Eq. (6.187), to identify the posterior's mean vector and covariance matrix; we also dropped the term that didn't depend on **c**. Take a moment to appreciate the fact that our result is (proportional to) an n-dimensional Gaussian density, as advertised. This last step assumes that:

$$\boldsymbol{\Sigma}_\mathbf{c} = \boldsymbol{\Lambda}_\mathbf{c}^{-1}, \qquad \boldsymbol{\mu}_\mathbf{c} = \boldsymbol{\Sigma}_\mathbf{c}\mathbf{q}_\mathbf{c} \tag{6.190}$$

This is our first encounter with a *precision matrix*, namely the inverse of the corresponding covariance matrix, $\boldsymbol{\Lambda}_\mathbf{c} = \boldsymbol{\Sigma}_\mathbf{c}^{-1}$. Cleaning things up, Eq. (6.190) together with Eq. (6.189) has led to the following results for these two most crucial characteristics of our Gaussian posterior density, the covariance matrix and the mean vector:

$$\boldsymbol{\Sigma}_\mathbf{c} = \left(\boldsymbol{\Phi}^T\boldsymbol{\Sigma}_d^{-1}\boldsymbol{\Phi} + \boldsymbol{\Sigma}_0^{-1}\right)^{-1}, \qquad \boldsymbol{\mu}_\mathbf{c} = \boldsymbol{\Sigma}_\mathbf{c}\left(\boldsymbol{\Phi}^T\boldsymbol{\Sigma}_d^{-1}\mathbf{y} + \boldsymbol{\Sigma}_0^{-1}\boldsymbol{\mu}_0\right) \tag{6.191}$$

The thing that immediately jumps out is that, as a result of the conjugate prior $\mathcal{N}(\boldsymbol{\mu}_0, \boldsymbol{\Sigma}_0)$, carrying out the Bayesian update has modified the prior's parameters $\boldsymbol{\mu}_0$ and $\boldsymbol{\Sigma}_0$: the new values are $\boldsymbol{\mu}_\mathbf{c}$ and $\boldsymbol{\Sigma}_\mathbf{c}$, but the type of distribution (n-dimensional Gaussian) is still the same for $\mathcal{N}(\boldsymbol{\mu}_\mathbf{c}, \boldsymbol{\Sigma}_\mathbf{c})$. As noted earlier, as you increase the number of data points N, the prior's contribution starts to matter less. At the other extreme, if you take $N = 0$ (no data)[78] the $\boldsymbol{\Phi}$ drops out of the equations above and the posterior is identical to the prior.

Looking at the $(n \times n)$ posterior covariance matrix $\boldsymbol{\Sigma}_\mathbf{c}$ in more detail, we realize that it is somewhat similar to the ML covariance matrix $\hat{\boldsymbol{\Sigma}}$ from Eq. (6.176), but tweaked to also take into account the prior's covariance matrix $\boldsymbol{\Sigma}_0$. Just like its predecessor $\hat{\boldsymbol{\Sigma}}$, our new $\boldsymbol{\Sigma}_\mathbf{c}$ does *not* depend depend on the data values $\mathscr{Y}$ or $\mathbf{y}$ (which is a bit more surprising this

[78] As Lucien Le Cam put it [95], it's hard to design experiments with a strictly negative number of observations.

time around, since we most certainly did not carry out any integrations over $\mathbf{y}$) or on the new mean vector $\boldsymbol{\mu}_\mathbf{c}$. Observe that the interpretation of our new covariance matrix $\boldsymbol{\Sigma}_\mathbf{c}$ is quite different from what we had before: the posterior $P(\mathbf{c}|\mathbf{y}; \boldsymbol{\Phi}, \boldsymbol{\Sigma}_d, \boldsymbol{\mu}_0, \boldsymbol{\Sigma}_0)$ is a probability distribution for the parameters $\mathscr{C}$ (though we didn't bother normalizing it), so $\boldsymbol{\Sigma}_\mathbf{c}$ has the corresponding meaning as the spread around $\boldsymbol{\mu}_\mathbf{c}$. In other words, given the single dataset $\mathbf{y}$ (which is the only one we've been given), we can make specific claims about the probability of bracketing the true parameters $\mathbf{c}_\star$ (we return to this point below).

Something analogous holds regarding the $(n \times 1)$ posterior mean vector $\boldsymbol{\mu}_\mathbf{c}$, which is modified, compared to $\hat{\mathbf{c}}$ from Eq. (6.175), to take into account the new covariance matrix $\boldsymbol{\Sigma}_\mathbf{c}$ and (inside the parentheses) the prior's properties. Intriguingly, if you take an infinitely broad prior ($\boldsymbol{\Sigma}_0^{-1} = \alpha \boldsymbol{I}$ with $\alpha \to 0$) then the prior drops out of the problem and the mean of the posterior becomes the ML estimate, $\boldsymbol{\mu}_\mathbf{c} = \hat{\mathbf{c}}$. In the general case of a reasonable prior, since the posterior is Gaussian, its mode (i.e., the single most probable value) coincides with its mean; put another way, if you maximized the posterior you would find $\boldsymbol{\mu}_\mathbf{c}$. This gives rise to the *maximum a posteriori* (MAP) estimate; we just explained that for us $\mathbf{c}_{\text{MAP}} = \boldsymbol{\mu}_\mathbf{c}$. Carrying out MAP estimation is one way to help you avoid overfitting for a small dataset. Of course, MAP is still just a point estimate, which seems wasteful, since we have an entire posterior distribution at our disposal. Thus, crucially, when you want to, say, compute the posterior predictive distribution or carry out model selection (studied in section 6.6.3.4 and in problem 6.57, respectively) you employ the full posterior distribution of the parameters $\mathscr{C}$.

Returning to the interpretation of parameter intervals: now that we have a posterior *distribution* $P(\mathbf{c}|\mathbf{y}; \boldsymbol{\Phi}, \boldsymbol{\Sigma}_d, \boldsymbol{\mu}_0, \boldsymbol{\Sigma}_0)$ for our parameters $\mathscr{C}$, which we know is Gaussian from Eq. (6.188), we can use the mean vector $\boldsymbol{\mu}_\mathbf{c}$ and covariance matrix $\boldsymbol{\Sigma}_\mathbf{c}$ from Eq. (6.191) to construct a 3σ interval for the uncertainty in our model's parameters $\mathscr{C}$. This, called a Bayesian *credible region*, has a much more satisfying interpretation than what we encountered for the ML case on page 394: given our single dataset $\mathbf{y}$, we can state with 89% probability[79] (for a non-pathological prior) that nature's true parameters $\mathbf{c}_\star$ lie within our parameter interval. Observe how the meaning of "probability" has changed: we are no longer talking about what would happen if we produced many datasets via $\mathscr{Y}$, but about our rational degree of belief given the single dataset $\mathbf{y}$ at our disposal.[80] Of course, there's no free lunch: if we generate many datasets from the same parent distribution, it is generally not true that 89% of the corresponding Bayesian credible regions will contain $\mathbf{c}_\star$ (recall that the guaranteed *coverage* was the selling point of frequentist confidence regions).[81]

6.6.3.3 Implementation

Code 6.7, i.e., `bayes.py`, first does what we had warned against all the way back in chapter 4: it sets up a function that inverts a general (square) matrix. This is done by applying

[79] As stated in footnote 72, this is the empirical rule for a 3σ interval and an $n = 5$ multivariate Gaussian.

[80] As in the frequentist case, increasing N decreases the variances (see problem 6.59); this time around, both the magnitude and the interpretation of the variances refer to our single dataset $\mathbf{y}$ (cf. footnote 62). If we are given *more* datasets, we can either completely pool them into a larger one, or employ a Bayesian *hierarchical model*.

[81] A sardonic way of putting things: Bayesians address the question everyone is interested in with assumptions no one believes, while frequentists use impeccable logic to deal with an issue of no interest to anyone [101].

Eq. (4.99), i.e., $\mathbf{Ax}_i = \mathbf{e}_i$ column by column; the transposition of the identity matrix is not necessary here, but a good habit. Actually, we are not even opting for the more efficient option of first LU-decomposing the matrix before solving the n linear systems, because we want to guard against pivoting-related issues (and we cannot assume you solved problem 4.16). As touched upon earlier in this section, it is typically necessary to compute a matrix inverse when you need the covariance matrix, as is clearly the case for us in Eq. (6.191). The fact that we are using a slow method is not really a problem, since we will be dealing with only a handful of parameters, i.e., we'll be inverting very small matrices.

The core of our new code is the function `bayes()` which takes in the full dataset, a tuple with the number of data points in each batch, as well as the prior's mean vector $\boldsymbol{\mu}_0$ and covariance matrix $\boldsymbol{\Sigma}_0$ (more on these below). Comparing with code 6.6, one big difference is already apparent: `bayes()` is set up to process the dataset in batches, in contradistinction to `normalfit()` which tackles the entire dataset at one go. We set up a loop over the number of data points in a given batch, using the variable name `N`, in order to keep things similar to the normal-equations code: if you pass in a single batch containing the total number of data points, the final answer is still going to be the same. Of course, the beauty of the present approach is that we can process the dataset step by step. Inside the loop over values of N, we have the same loop over n columns as in `normalfit()`. There are two further changes to note: first, as was just explained, we slice the dataset (via `data[0,i:i+N]` and its relatives) to process one batch at a time. Second, we are no longer solving the normal equations of Eq. (6.144), but are carrying out explicitly the inversions in Eq. (6.191) using our newly developed `inv()`. Incidentally, if you're having trouble seeing the correspondence between our earlier design matrix $\mathbf{A}$ and its right-hand side $\mathbf{b}$, on the one hand, and the new entities $\boldsymbol{\Phi}$, $\boldsymbol{\Sigma}_d$, and $\mathbf{y}$ employed in the present section, on the other hand, you should (again) have a look at Eq. (6.269); while the latter are easier to reason about, the former are more compact and therefore easier to code up. The loop body then re-assigns new variables to old, and increments a counter, in preparation for the next iteration.

The main program employs the same dataset and two models used in `newnormal.py`. It splits up the eight data points into three batches, simply in order to illustrate our new iterative powers. Since we don't really know anything much about what the parameters for these models should be in the absence of data, we take the mean vectors for both models to be zero. We also take the covariance matrices to be diagonal, corresponding to the assumption that the parameters are independent in the prior (of course, as we'll soon see, they can be dependent in the posterior). Regarding the elements on the diagonal (the variance of each individual parameter), since we don't know much, we basically pick these out of a hat. In lieu of a justification, we note that in the $n = 5$ (polynomial) model the last coefficient is expected to be somewhat small (otherwise we are likely overfitting), so we allow that variance to be the smallest, but still not tiny since we don't really know what to expect; in any case, you should feel free to play around with these prior parameters and see what happens. The code output (printouts of the posterior mean and covariance) may be a bit underwhelming, but in what follows we will examine plots that were produced with this very program (and which you are asked to reproduce in problem 6.55). Let us first discuss the $n = 2$ model which, as you may recall, is a sinusoidal theory (just like the one that generated the dataset). Since this has only two parameters, visualizing the

bayes.py Code 6.7

```
from gauelim_pivot import gauelim_pivot
from newnormal import generatedata, phi
import numpy as np

def inv(A):
    n = A.shape[0]
    invA = np.zeros((n,n))
    for i,bs in enumerate(np.identity(n).T):
        invA[:,i] = gauelim_pivot(A,bs)
    return invA

def bayes(data, batches, primus, priS):
    n = primus.size
    i = 0
    for N in batches:
        A = np.zeros((N,n))
        for k in range(n):
            A[:,k] = phi(n,k,data[0,i:i+N])/data[2,i:i+N]
        bs = data[1,i:i+N]/data[2,i:i+N]

        priSinv = inv(priS)
        postS = inv(A.T@A + priSinv)
        postmus = postS@(A.T@bs + priSinv@primus)

        primus, priS = postmus, postS
        i += N
        print(n, i, postmus, postS)
    return postmus, postS

if __name__ == '__main__':
    batches = 1, 4, 3
    data = generatedata(np.sum(batches))
    for n in (5, 2):
        primus = np.zeros(n)
        priS = np.zeros((n,n))
        np.fill_diagonal(priS, np.linspace(10,2,n))
        postmus, postS = bayes(data, batches, primus, priS)
        print(postmus)
```

posterior distribution is straightforward. This is shown in Fig. 6.14: there are three rows corresponding to the three batches and two columns corresponding to a contour plot of the posterior (left) and approximation results (right). Note that, in order to keep things interesting, we have randomized the order in which points are read in (i.e., `data[0,:]` values are not being processed from left to right). Recall that our code in `bayes()` is sequential, in the sense that it treats a given posterior (produced after a number of batches has been processed) as the prior for the next batch. As mentioned on page 396, this does *not* constitute double-counting: our prior encapsulates all earlier information, including that coming from previous experiments.

Starting from the left panel of the first row (when only a single point has been processed), we see that this is quite spread out, i.e., we haven't learned that much (after encountering a single point); even so, it's worth highlighting a few features in the posterior: first, the values around which $\mathscr{C}_0$ and $\mathscr{C}_1$ are centered are roughly 4 and 0.1, already different from the values on the prior's mean vector (0 and 0). Second, the same thing holds for the first element on the diagonal of the posterior covariance matrix: while the variance in $\mathscr{C}_1$ hasn't really changed after the Bayesian update, the variance in $\mathscr{C}_0$ (which was quite large in our prior) has been reduced drastically, as reflected in the horizontal extent/width. Third, the prior covariance matrix was diagonal, but the posterior covariance matrix is not: this is reflected in the characteristic tilting of the contour plot.

The right panel is a bit busy, so ignore the gray band(s) for now. We are showing the data points processed so far (a single one in the first row), the MAP predictions (which result from using $\mathbf{c}_{\text{MAP}} = \boldsymbol{\mu}_\mathbf{c}$, as per page 399), as well as random samples drawn from the posterior ("samples" in the sense of page 514, i.e., specific parameter realizations). Thus, we use the values of $\boldsymbol{\mu}_\mathbf{c}$ and $\boldsymbol{\Sigma}_\mathbf{c}$ produced by the code and draw 15 samples from the multivariate Gaussian $\mathcal{N}(\boldsymbol{\mu}_\mathbf{c}, \boldsymbol{\Sigma}_\mathbf{c})$ shown in the last line of Eq. (6.188). (For now, we can do this by calling `np.random.multivariate_normal()`; we will learn how to handle a general multivariate distribution in section 7.7.5.) In either case (i.e., MAP or random sample), we end up with a given vector $\mathbf{c}$, which we then use to plot our approximating function, Eq. (6.7), as a function of x:

$$p(x) = \sum_{k=0}^{n-1} c_k \phi_k(x) \tag{6.192}$$

for a given model. Since we've only processed a single data point, there is a very large spread in the possible predictions that are all consistent with our posterior distribution; note, however, that nearly all of the samples go through that one data point. Observe, also, that these samples help us understand the covariance between predictions at different x values (a single sample corresponds to fixed c_0 and c_1, so the shape of the curve is dictated by the theory/model). Perhaps it's worth noting that the left and right panels are always consistent: the contour plot told us that c_1 was expected to be small and the MAP prediction on the right panel shows a nearly flat curve.

In the next row, after we've processed four more data points, the contour plot continues to tilt and tighten; similarly, the MAP prediction is no longer nearly flat; finally, the 15 random samples are still spread out but are now starting to exhibit a pattern: the scatter is largest in the regions where we haven't processed any data points (yet). It just so happened

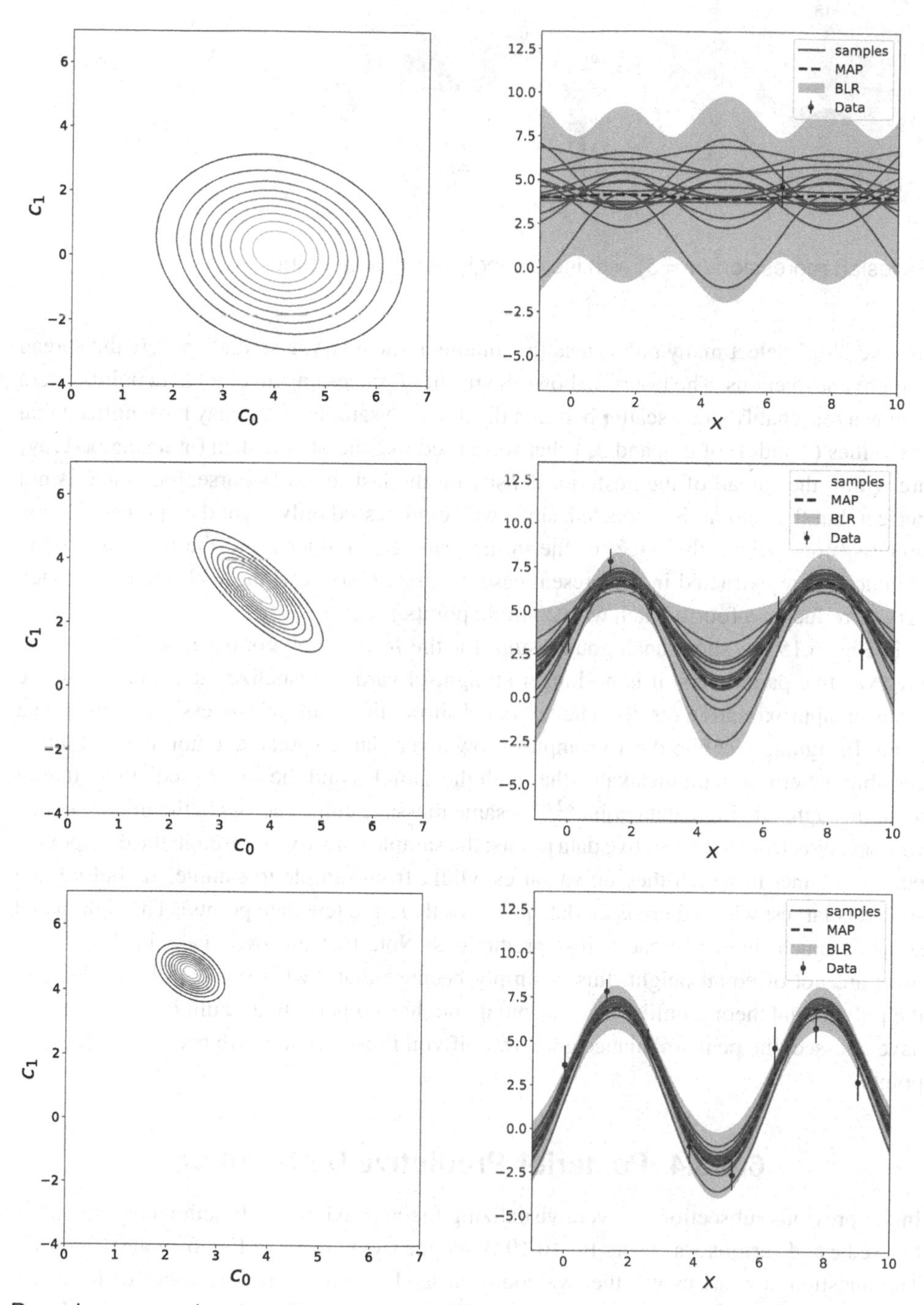

Bayesian regression ($n = 2$) as you add data: posterior (left) and fitting (right)

Fig. 6.14

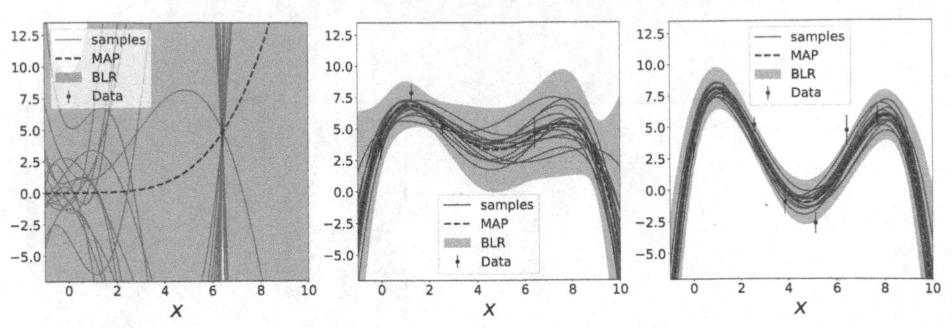

Fig. 6.15 Bayesian regression ($n = 5$) with increasingly many data points

that we didn't select many points near the middle of the interval, so that's where the spread is more conspicuous. The last row shows the result of processing all eight data points: there is now a reasonably small scatter between the different samples. You may have noticed that the values (2 and 5) of $c_{\star,0}$ and $c_{\star,1}$ that were used to generate the data (in `newnormal.py`) are within the spread of the posterior density on the last row; of course, the match is not perfect, but that should be expected, since we've processed only eight data points. As you process more points, the effect of the initial prior gets reduced and the true underlying parameters are extracted in the present case, where we are working with the true model. (Try introducing a fourth batch with 20 more points.)

In Fig. 6.15 we show analogous results for the $n = 5$ polynomial model. Since this involves five parameters, it is no longer straightforward to visualize the posterior, so we focus on approximation results. The left panel shows the result of processing a single data point. Intriguingly, while the 15 samples show a very large spread and not much of a discernible pattern, it is unmistakable that both the samples and the MAP prediction attempt to go through our single data point.[82] The same message comes across in the middle panel, which results from a total of five data points: the samples mostly go through the data points, but the manner in which they do so varies wildly from sample to sample. As before, the spread is largest where there is no data point (or there are few data points). The right panel shows a much tighter spread across predictions. Note that the two peaks in the predictions are not of equal height: this is simply because that's what our data look like, and the polynomial theory, unlike the sinusoidal one, has no periodicity built in. (As you may have guessed, the peak imbalance goes away if you introduce a fourth batch with 20 more points.)

6.6.3.4 Posterior Predictive Distribution

In the previous subsection we were visualizing the approximating function corresponding to fixed $\mathbf{c}$ and various x's, as per Eq. (6.192), see the right panels of Fig. 6.14 and Fig. 6.15. The question now arises whether we could, instead, use the entire posterior distribution, i.e., the full multivariate Gaussian $P(\mathbf{c}|\mathbf{y}; \boldsymbol{\Phi}, \boldsymbol{\Sigma}_d, \boldsymbol{\mu}_0, \boldsymbol{\Sigma}_0)$ instead of isolated samples drawn

[82] Incidentally, this plot and the first row of Fig. 6.14 clearly show that our Bayesian approach is able to handle more parameters than data points (as mentioned on page 394).

from it; even the MAP prediction is an isolated sample: it corresponds to the most likely set of parameters, but that's only a single value of $\mathbf{c}$, nonetheless. The notation can get pretty confusing, since $\mathbf{y}$ refers to the data values (the y_j's) and, similarly, $\boldsymbol{\Phi}$ has the x_j's buried inside it; on the contrary, in Eq. (6.192) we are interested in prediction, i.e., we want to study a general x value. With that in mind, let us introduce some new notation: we wish to find out which $\tilde{y}$ value corresponds to a given $\tilde{x}$ value, with the latter being previously unseen, i.e., not part of the input data x_j's. Thus, we are led to introduce a new random variable $\tilde{\mathcal{Y}}$ whose realization is $\tilde{y}$, rewriting Eq. (6.192) for random parameters:

$$\tilde{\mathcal{Y}} = \sum_{k=0}^{n-1} \mathcal{C}_k \phi_k(\tilde{x}) = \tilde{\boldsymbol{\phi}} \boldsymbol{\mathcal{C}} \tag{6.193}$$

Here $\tilde{\boldsymbol{\phi}}$ is the row vector of all the ϕ_k's evaluated at $\tilde{x}$, i.e., $\phi_k(\tilde{x})$ for all k's.

At this point, textbooks typically introduce a probabilistic model for $P(\tilde{y}|\mathbf{c}; \tilde{\boldsymbol{\phi}})$—typically with an *ad hoc* Gaussian noise term—and take the expectation of this Gaussian with respect to the posterior (i.e., an integral over the $\mathbf{c}$'s of a product of two Gaussians). This has the advantage that it uses nice properties of multivariate Gaussians, but the disadvantage that we already have a well-defined model for $P(\tilde{y}|\mathbf{c}; \tilde{\boldsymbol{\phi}})$, i.e., the conditional probability density that we would produce a given $\tilde{y}$ taking for granted the parameters $\mathbf{c}$ and the row vector $\tilde{\boldsymbol{\phi}}$: our Eq. (6.193) tells us exactly what value $\tilde{y}$ should take (with no uncertainty whatsoever).[83] In equation form, this is saying:[84]

$$P(\tilde{y}|\mathbf{c}; \tilde{\boldsymbol{\phi}}) = \delta(\tilde{y} - \tilde{\boldsymbol{\phi}}\mathbf{c}) \tag{6.194}$$

Since these are continuous quantities, we are dealing with a *Dirac* delta here (i.e., not a Kronecker delta); this is a *one-dimensional* Dirac delta, since we are producing a single number, $\tilde{y}$. While it's true that you will get a different value of $\tilde{y}$ for different parameters $\mathbf{c}$, here we are dealing with a *conditional* probability (density), taking $\mathbf{c}$ as given.

What we would like to do is to start from the density $P(\tilde{y}|\mathbf{c}; \tilde{\boldsymbol{\phi}})$ and produce another probability density which is not conditioned on $\mathbf{c}$. We do this as follows:

$$\begin{aligned} P(\tilde{y}|\mathbf{y}; \tilde{\boldsymbol{\phi}}, \boldsymbol{\Phi}, \boldsymbol{\Sigma}_d, \boldsymbol{\mu}_0, \boldsymbol{\Sigma}_0) &= \mathbb{E}[P(\tilde{y}|\mathbf{c}; \tilde{\boldsymbol{\phi}})] = \int d^n c P(\tilde{y}|\mathbf{c}; \tilde{\boldsymbol{\phi}}) P(\mathbf{c}|\mathbf{y}; \boldsymbol{\Phi}, \boldsymbol{\Sigma}_d, \boldsymbol{\mu}_0, \boldsymbol{\Sigma}_0) \\ &= \int d^n c\, \delta(\tilde{y} - \tilde{\boldsymbol{\phi}}\mathbf{c})\, \mathcal{N}(\boldsymbol{\mu}_\mathbf{c}, \boldsymbol{\Sigma}_\mathbf{c}) \\ &\propto \int d^n c\, \delta(\tilde{y} - \tilde{\boldsymbol{\phi}}\mathbf{c}) \exp\left[-\frac{1}{2}(\mathbf{c} - \boldsymbol{\mu}_\mathbf{c})^T \boldsymbol{\Sigma}_\mathbf{c}^{-1} (\mathbf{c} - \boldsymbol{\mu}_\mathbf{c})\right] \end{aligned} \tag{6.195}$$

In the first step we took the expectation of our model with respect to the posterior density (using the notation of appendix C.3); note that the left-hand side is, indeed, giving us a predictive distribution for $\tilde{y}$ which is *not* conditioned on $\mathbf{c}$, since we are integrating over the $\mathbf{c}$'s. In the second step we explicitly wrote out the expectation as an integral. In the third step we plugged in Eq. (6.194) for the conditional predictive model and our main

[83] While some authors use *prediction* for a noisy quantity and *regression* for a noise-free quantity, other authors do the exact opposite; quite often, a given author does both (!). For us, prediction will always be noise-free.

[84] Cromwell's rule, mentioned in footnote 73, does not apply here, since Eq. (6.194) is not a prior distribution but a logically true statement of what it means to make a prediction (which is noise-free, as per footnote 83).

result from Eq. (6.188) for the posterior. In the fourth step we spelled out the last result from Eq. (6.188), to emphasize that the posterior is a multivariate Gaussian in $\mathbf{c}$. We are integrating the product of a one-dimensional Dirac delta function with a multivariate Gaussian; this integral is precisely of the form studied in problem 6.49, see Eq. (6.267). Without getting into the full argument here (involving rotation matrices and the fact that Gaussians are closed under marginalization), we can cite the conclusion that Eq. (6.195) leads to a *univariate Gaussian in* $\tilde{y}$.

Let us investigate the properties of this new distribution. First, we examine $\tilde{\mathcal{Y}}$'s mean:

$$\mathbb{E}(\tilde{\mathcal{Y}}) = \int d\tilde{y}\, \tilde{y}\, P(\tilde{y}|\mathbf{y}; \tilde{\boldsymbol{\phi}}, \boldsymbol{\Phi}, \boldsymbol{\Sigma}_d, \boldsymbol{\mu}_0, \boldsymbol{\Sigma}_0) = \int d\tilde{y}\, \tilde{y} \int d^n c\, \delta(\tilde{y} - \tilde{\boldsymbol{\phi}}\mathbf{c})\, \mathcal{N}(\boldsymbol{\mu}_\mathbf{c}, \boldsymbol{\Sigma}_\mathbf{c})$$
$$= \int d^n c\, \tilde{\boldsymbol{\phi}}\mathbf{c}\, \mathcal{N}(\boldsymbol{\mu}_\mathbf{c}, \boldsymbol{\Sigma}_\mathbf{c}) = \tilde{\boldsymbol{\phi}}\mathbb{E}(\mathscr{C}) = \tilde{\boldsymbol{\phi}}\boldsymbol{\mu}_\mathbf{c} \tag{6.196}$$

In the first step we took the expectation with respect to our new univariate Gaussian. In the second step we plugged in the penultimate step from Eq. (6.195). In the third step we interchanged the order of the integrations and carried out the integral over $\tilde{y}$. In the fourth step we pulled the $\tilde{\boldsymbol{\phi}}$ outside the integral and identified an expectation, this time with respect to the posterior (since we were integrating over $\mathbf{c}$). In the fifth step we replaced the expectation of $\mathscr{C}$ by the mean of the Gaussian posterior. Observe that $\tilde{\boldsymbol{\phi}}\boldsymbol{\mu}_\mathbf{c}$ is a scalar, as expected based on the fact that we were computing $\mathbb{E}(\tilde{y})$.

We now turn to $\tilde{\mathcal{Y}}$'s variance:

$$\mathbb{V}(\tilde{\mathcal{Y}}) = \mathbb{E}\left[\tilde{\mathcal{Y}}^2 - \mathbb{E}(\tilde{\mathcal{Y}})^2\right] = \int d\tilde{y}\left[\tilde{y}^2 - \mathbb{E}(\tilde{\mathcal{Y}})^2\right] P(\tilde{y}|\mathbf{y}; \tilde{\boldsymbol{\phi}}, \boldsymbol{\Phi}, \boldsymbol{\Sigma}_d, \boldsymbol{\mu}_0, \boldsymbol{\Sigma}_0)$$
$$= \int d\tilde{y}\left[\tilde{y}^2 - \left(\tilde{\boldsymbol{\phi}}\boldsymbol{\mu}_\mathbf{c}\right)^2\right] \int d^n c\, \delta(\tilde{y} - \tilde{\boldsymbol{\phi}}\mathbf{c})\, \mathcal{N}(\boldsymbol{\mu}_\mathbf{c}, \boldsymbol{\Sigma}_\mathbf{c})$$
$$= \int d^n c\, \left[\left(\tilde{\boldsymbol{\phi}}\mathbf{c}\right)^2 - \left(\tilde{\boldsymbol{\phi}}\boldsymbol{\mu}_\mathbf{c}\right)^2\right] \mathcal{N}(\boldsymbol{\mu}_\mathbf{c}, \boldsymbol{\Sigma}_\mathbf{c})$$
$$= \int d^n c\, \left[\tilde{\boldsymbol{\phi}}\mathbf{c}\mathbf{c}^T\tilde{\boldsymbol{\phi}}^T - \tilde{\boldsymbol{\phi}}\boldsymbol{\mu}_\mathbf{c}\boldsymbol{\mu}_\mathbf{c}^T\tilde{\boldsymbol{\phi}}^T\right] \mathcal{N}(\boldsymbol{\mu}_\mathbf{c}, \boldsymbol{\Sigma}_\mathbf{c}) = \tilde{\boldsymbol{\phi}}\, \mathbb{E}(\mathscr{C}\mathscr{C}^T - \boldsymbol{\mu}_\mathbf{c}\boldsymbol{\mu}_\mathbf{c}^T)\, \tilde{\boldsymbol{\phi}}^T = \tilde{\boldsymbol{\phi}}\boldsymbol{\Sigma}_\mathbf{c}\tilde{\boldsymbol{\phi}}^T \tag{6.197}$$

In the first step we used Eq. (C.30) together with $\mathbb{E}[\mathbb{E}(\tilde{\mathcal{Y}})^2] = \mathbb{E}(\tilde{\mathcal{Y}})^2$. In the second step we explicitly showed what this means with respect to our new univariate Gaussian. In the third step we plugged in our result for the mean from Eq. (6.196) and the penultimate step from Eq. (6.195). In the fourth step we interchanged the order of the integrations and carried out the integral over $\tilde{y}$. In the fifth step we re-expressed the squares for later use. In the sixth step we pulled the $\tilde{\boldsymbol{\phi}}$ and $\tilde{\boldsymbol{\phi}}^T$ out of the integral and identified an expectation, this time with respect to the posterior (since we were integrating over $\mathbf{c}$). In the seventh step we applied the definition of the covariance matrix from Eq. (6.271). As above, observe that $\tilde{\boldsymbol{\phi}}\boldsymbol{\Sigma}_\mathbf{c}\tilde{\boldsymbol{\phi}}^T$ is a scalar, as expected based on the fact that we were computing $\mathbb{V}(\tilde{\mathcal{Y}})$.

To summarize, we know from Eq. (6.195) that we have a univariate Gaussian distribution in $\tilde{y}$; we determined its mean in Eq. (6.196) and its variance in Eq. (6.197). Putting it all together, we have arrived at the *posterior predictive distribution* (sometimes also known as the *posterior over functions*):

$$P(\tilde{y}|\mathbf{y};\tilde{\boldsymbol{\phi}},\boldsymbol{\Phi},\boldsymbol{\Sigma}_d,\boldsymbol{\mu}_0,\boldsymbol{\Sigma}_0) = \mathcal{N}\left(\tilde{\boldsymbol{\phi}}\boldsymbol{\mu}_c, \tilde{\boldsymbol{\phi}}\,\boldsymbol{\Sigma}_c\,\tilde{\boldsymbol{\phi}}^T\right) \quad (6.198)$$

It is so called because it tells us how to produce a $\tilde{y}$ corresponding to an $\tilde{x}$, after integrating over *all* **c** parameter values using the posterior distribution. We reiterate that the mean and the variance (just like $\tilde{y}$) are *numbers* (i.e., not vectors/matrices).

A comment that may appear irrelevant to you now, but will probably help you later in life: you can also write down the last steps of Eq. (6.196) and Eq. (6.197) by treating $\tilde{\mathcal{Y}}$ as a random variable starting with Eq. (6.193), taking expectations with respect to the posterior, and therefore without introducing $P(\tilde{y}|\mathbf{c};\tilde{\boldsymbol{\phi}})$ at any point. This means that you can get the same answers for $\mathbb{E}(\tilde{\mathcal{Y}})$ and $\mathbb{V}(\tilde{\mathcal{Y}})$ without having to bring up any Dirac delta functions. The only subtlety involved in this derivation is in showing that $P(\tilde{y}|\mathbf{y};\tilde{\boldsymbol{\phi}},\boldsymbol{\Phi},\boldsymbol{\Sigma}_d,\boldsymbol{\mu}_0,\boldsymbol{\Sigma}_0)$ is a univariate Gaussian; that is precisely why we invoked the corresponding argument after Eq. (6.195).[85]

We will now see how to visualize the posterior predictive distribution of Eq. (6.198). The crucial point is that it takes in a dataset (i.e., many x_j's and y_j's, with the associated σ_j's), a set of basis functions (i.e., many ϕ_k's), as well as a prior distribution and a given $\tilde{x}$; it folds in the posterior's mean vector and covariance matrix—$\boldsymbol{\mu}_c$ and $\boldsymbol{\Sigma}_c$ from Eq. (6.191)—on the right-hand side to tell us which $\tilde{y}$ values we should expect. Inspired by our earlier drawing of samples from the multivariate Gaussian of Eq. (6.188), you might choose to do the same for the posterior predictive distribution of Eq. (6.198). You should keep in mind that for each $\tilde{x}$ a given sample drawn from $P(\tilde{y}|\mathbf{y};\tilde{\boldsymbol{\phi}},\boldsymbol{\Phi},\boldsymbol{\Sigma}_d,\boldsymbol{\mu}_0,\boldsymbol{\Sigma}_0)$ gives rise to a single $\tilde{y}$, i.e., a single number (i.e., *not* a vector/set of numbers). This is quite different from our earlier samples from the posterior of Eq. (6.188): these were drawing a vector **c** which was then used in Eq. (6.192): a single set of c_k's was employed for all $\tilde{x}$'s. This time around we have integrated out the **c**'s, so we are certainly not drawing specific parameter values. Drawing samples from Eq. (6.198) is an option that is explored in problem 6.56; as one should expect given its provenance, it gives rise to jagged results.

Instead, here we opt for a different approach to visualizing Eq. (6.198), motivated by the empirical rule of problem 6.50: for several $\tilde{x}$'s, we generate three curves, corresponding to $\tilde{\boldsymbol{\phi}}\boldsymbol{\mu}_c$ and to $\tilde{\boldsymbol{\phi}}\boldsymbol{\mu}_c \pm 3\tilde{\boldsymbol{\phi}}\boldsymbol{\Sigma}_c\tilde{\boldsymbol{\phi}}^T$; note that all three of these curves are $\tilde{x}$-dependent. We've already discussed the curve corresponding to $\tilde{\boldsymbol{\phi}}\boldsymbol{\mu}_c$: this is the MAP prediction shown in the right panels of Fig. 6.14 and in Fig. 6.15; plug $\mathbf{c} = \boldsymbol{\mu}_c$ into Eq. (6.192) to convince yourself of this. As you may have realized, the other two curves give rise to a 3σ uncertainty band, which is precisely what is shown, in the same figures, with the gray bands labelled by BLR (standing for Bayesian linear regression); this is known as a *prediction interval.* (This is related to, yet distinct from, the 3σ interval in the parameters $\mathscr{C}$, mentioned on page 399: most obviously, here we are dealing with a univariate Gaussian, so we can employ the 68–95–99.7 rule directly.) As advertised, nearly all of the samples drawn (from either the posterior distribution or directly from the posterior predictive distribution) will fall within this interval; the figures show samples from the posterior distribution, as discussed earlier.

[85] In problem 6.53 we employ a different argument, a version of which also applies here.

As promised, the width of these bands is $\tilde{x}$-dependent: they are wide when unconstrained by data and narrow when there's plenty of data going into their construction. To drive the point home, problem 6.56 employs such bands for the problem of *extrapolation*, i.e., outside the interval where the data were produced.

6.6.3.5 Summary

Before concluding this section, it may be worthwhile to summarize what is special about the Bayesian approach to linear regression. You should avoid the (widespread) misapprehension that plots like Fig. 6.14 and Fig. 6.15 have something inherently Bayesian about them. You can employ a frequentist approach to draw samples from the distribution of the ML estimator $\hat{\mathbf{c}}$ and to plot the ML point estimate itself (see problem 6.52), as well as to generate a likelihood-based predictive distribution and 3σ prediction intervals (see problem 6.53); as a matter of fact, Eq. (6.274) bears a striking formal similarity to the posterior predictive distribution of Eq. (6.198). Similarly, while the sequential learning shown in Fig. 6.14 and in Fig. 6.15 is certainly appealing, one can generate analogous approaches in an ML setting. Another common claim is that a Bayesian approach doesn't have to employ a normal likelihood; that is a distinction without a difference, as ML can also handle non-normal distributions. Finally, while it is true that Bayesian regression helps to avoid overfitting, the same effect can also be accomplished in a frequentist context.[86]

The real difference between the two outlooks is that the frequentist approach focuses on the likelihood[87] and interprets results in terms of integrations across other datasets (i.e., in N dimensions), whereas the Bayesian approach combines the likelihood with a prior distribution for the parameters and interprets results in terms of integrations across possible parameter values (i.e., in n dimensions)—always limiting itself to the single dataset input. If you *do* have constraints on (i.e., prior knowledge of) the parameters, frequentist confidence intervals can get you in trouble, so a Bayesian approach then becomes necessary (see problem 6.67).

6.7 Non-Linear Least-Squares Fitting

Our general least-squares fitting formalism, whether in maximum-likelihood (equivalently: normal-equations) guise or following from a Bayesian approach, was not truly general: everything we have said so far applies to the case of linear dependence on the parameters (see Eq. (6.136) and Eq. (6.186), respectively). However, in practice you may be faced with more complicated scenarios, e.g.:

$$p(x) = c_0 e^{-c_1 x} \tag{6.199}$$

The dependence on c_1 is non-linear, as you will discover immediately when you try to take the derivative of $p(x)$ with respect to c_1. There is a time-honored trick you could employ

[86] Regularization (recall page 394) introduces extra conditions/terms to assuage ill-conditioning/overfitting.
[87] We are passing over in silence the (indubitable) distinction between frequentist and likelihoodist statistics.

here: taking the natural logarithm of both sides, you are led to:

$$q(x) = d_0 + d_1 x \tag{6.200}$$

where $q(x) = \ln[p(x)]$, and the d_k's are straightforwardly related to the c_k's. Then, you could simply carry out a straight-line fit for $q(x)$ and later translate what you've learned back to the $p(x)$ model. That being said, you should keep in mind that in both the frequentist and the Bayesian setting we had assumed that the errors were normally distributed; manipulating your theory so that it becomes linear in the parameters will quite likely remove the Gaussianity of your measurement errors. Even so, the result of a χ^2 minimization might still be helpful to you, practically speaking.

There are endless cases of non-linearities that are not easy to transform away. For example, here is the *Breit–Wigner form*:

$$p(x) = \frac{c_0}{(x - c_1)^2 + c_2} \tag{6.201}$$

which you will explore in problem 6.74. For this case, if you have a set of input data $(x_j, y_j \pm \sigma_j)$ and wish to determine c_0, c_1, and c_2 our earlier discussion cannot help you very much. The task at hand is to minimize the χ^2 in the general case, namely Eq. (6.120):

$$\chi^2 = \sum_{j=0}^{N-1} \left(\frac{y_j - p(x_j)}{\sigma_j} \right)^2 \tag{6.202}$$

As you may recall, for the linear case this problem was reformulated into a simple expression, $\chi^2(\mathbf{c}) = (\mathbf{b} - \mathbf{Ac})^T(\mathbf{b} - \mathbf{Ac})$—where $\mathbf{Ac}$ is clearly linear in $\mathbf{c}$—minimizing which led to a linear system of equations, $\mathbf{A}^T\mathbf{Ac} = \mathbf{A}^T\mathbf{b}$. For a general $p(x)$ this is no longer true; put another way, the derivative $\partial p(x_j)/\partial c_k$ of Eq. (6.121) no longer leads to a clean contribution of a basis function alone, $\phi_k(x_j)$. Even so, there's nothing keeping you from analytically evaluating the derivative and then treating the zero-derivative conditions of Eq. (6.121) as n non-linear equations in n unknowns, to which you can apply, say, the multidimensional Newton's method for root-finding (problem 6.73 invites you to do this). Of course, such an approach is quite inefficient: we would be applying a very general technique (multidimensional Newton's method for root-finding) to a very specific problem (χ^2 minimization for the case where the approximating function $p(x)$ has a non-linear dependence on the parameters $\mathbf{c}$). Instead, below we will turn to a more tailored approach, borrowing ideas from multidimensional minimization, as discussed in section 5.6, and specializing them for the specific problem at hand.

6.7.1 Gauss–Newton Method

As mentioned, we are given N data points $(x_j, y_j \pm \sigma_j)$, we choose an approximating function $p(x)$ which depends on the n parameters c_k non-linearly, and we wish to minimize the χ^2 of Eq. (6.202). Let's introduce some new notation to make the following manipulations easier to grasp:

$$\rho_j = \frac{p(x_j)}{\sigma_j}, \qquad b_j = \frac{y_j}{\sigma_j} \tag{6.203}$$

The definition of b_j is the same as in Eq. (6.140), but ρ_j is new here: as noted above, we cannot write the approximating function $p(x)$ simply as a linear combination of the parameters and basis functions. In terms of these quantities, the χ^2 of Eq. (6.202) becomes:

$$\chi^2(\mathbf{c}) = (\mathbf{b} - \boldsymbol{\rho})^T(\mathbf{b} - \boldsymbol{\rho}) \tag{6.204}$$

where the column vector $\boldsymbol{\rho}$ collects all of the ρ_j's.

Next, we turn to Newton's method *for minimization*, from section 5.6.4. The role of the scalar function $\phi(\mathbf{x})$ is played by $\chi^2(\mathbf{c})$: we are explicitly marking the dependence of χ^2 on the parameters (everything else—x_j's, y_j's, σ_j's, and the choice of $p(x)$—being frozen). It's important to realize that in Eq. (6.204) this dependence on $\mathbf{c}$ is buried inside $\boldsymbol{\rho}$, which contains $p(x_j)$ which, in turn, involves a dependence on the parameters (see, e.g., Eq. (6.201)). Thus, Newton's method for minimization from Eq. (5.111) now takes the form:

$$\mathbf{J}_{\nabla\chi^2}(\mathbf{c}^{(k-1)})\left[\mathbf{c}^{(k)} - \mathbf{c}^{(k-1)}\right] = -\nabla\chi^2(\mathbf{c}^{(k-1)}) \tag{6.205}$$

Here $\nabla\chi^2(\mathbf{c})$ is the gradient of our scalar function and the left-hand side involves the corresponding Jacobian $\mathbf{J}_{\nabla\chi^2}(\mathbf{c})$; as is standard in Newton's method, both of these are evaluated at the previous iterate $\mathbf{c}^{(k-1)}$. The notation is still somewhat unwieldy, so let's set up some more scaffolding and we'll return to Eq. (6.205) later.

First, we introduce the Jacobian matrix corresponding to $\boldsymbol{\rho}$ (not to the gradient, as above):

$$\mathbf{K}_\rho(\mathbf{c}) = \left\{\frac{\partial\rho_j}{\partial c_k}\right\} = \begin{pmatrix} \frac{\partial\rho_0}{\partial c_0} & \frac{\partial\rho_0}{\partial c_1} & \cdots & \frac{\partial\rho_0}{\partial c_{n-1}} \\ \frac{\partial\rho_1}{\partial c_0} & \frac{\partial\rho_1}{\partial c_1} & \cdots & \frac{\partial\rho_1}{\partial c_{n-1}} \\ \vdots & \vdots & \ddots & \vdots \\ \frac{\partial\rho_{N-1}}{\partial c_0} & \frac{\partial\rho_{N-1}}{\partial c_1} & \cdots & \frac{\partial\rho_{N-1}}{\partial c_{n-1}} \end{pmatrix} \tag{6.206}$$

which you should compare with Eq. (5.75); crucially, this $\mathbf{K}_\rho$ is an $N \times n$ matrix. In terms of this new matrix, the gradient of χ^2 takes the form:

$$\nabla\chi^2(\mathbf{c}) = -2\mathbf{K}_\rho^T[\mathbf{b} - \boldsymbol{\rho}] \tag{6.207}$$

as you can verify starting from Eq. (6.204) and writing out the components of the gradient vector. Next, we introduce the Hessian matrix corresponding to the scalar function ρ_j:

$$\mathbf{L}_{\rho_j}(\mathbf{c}) = \left\{\frac{\partial\rho_j}{\partial c_k \partial c_l}\right\} = \begin{pmatrix} \frac{\partial^2\rho_j}{\partial c_0^2} & \frac{\partial^2\rho_j}{\partial c_0\partial c_1} & \cdots & \frac{\partial^2\rho_j}{\partial c_0\partial c_{n-1}} \\ \frac{\partial^2\rho_j}{\partial c_1\partial c_0} & \frac{\partial^2\rho_j}{\partial c_1^2} & \cdots & \frac{\partial^2\rho_j}{\partial c_1\partial c_{n-1}} \\ \vdots & \vdots & \ddots & \vdots \\ \frac{\partial^2\rho_j}{\partial c_{n-1}\partial c_0} & \frac{\partial^2\rho_j}{\partial c_{n-1}\partial c_1} & \cdots & \frac{\partial^2\rho_j}{\partial c_{n-1}^2} \end{pmatrix} \tag{6.208}$$

which you should compare with Eq. (5.104); note that such an $\mathbf{L}_{\rho_j}$ matrix has dimensions $n \times n$ (and, since you can write one down for each ρ_j, there are N such matrices). In terms of these two new entities, the Jacobian corresponding to $\nabla\chi^2(\mathbf{c})$ takes the form:

$$\mathbf{J}_{\nabla\chi^2}(\mathbf{c}) = 2\mathbf{K}_\rho^T\mathbf{K}_\rho - 2\sum_{j=0}^{N-1}(b_j - \rho_j)\mathbf{L}_{\rho_j} \tag{6.209}$$

as you can verify by producing, as per Eq. (5.75), the Jacobian of the $\nabla\chi^2(\mathbf{c})$ of Eq. (6.207)—which is also the Hessian of $\chi^2(\mathbf{c})$. You should take a moment to write out both the right-hand side and the left-hand side here.

We are now ready to update the main Newton's method formula, Eq. (6.205), using the gradient from Eq. (6.207) and the corresponding Jacobian from Eq. (6.209):

$$\left[\mathbf{K}_\rho^T\mathbf{K}_\rho - \sum_{j=0}^{N-1}(b_j - \rho_j)\mathbf{L}_{\rho_j}\right]\left[\mathbf{c}^{(k)} - \mathbf{c}^{(k-1)}\right] = \mathbf{K}_\rho^T[\mathbf{b} - \boldsymbol{\rho}] \tag{6.210}$$

where we cancelled the factor of 2. Crucially, the left-hand side here involves both first derivatives (in $\mathbf{K}_\rho$) and second derivatives (in $\mathbf{L}_{\rho_j}$). Note also that both $\mathbf{K}_\rho$ and $\mathbf{L}_{\rho_j}$ are (implicitly) evaluated at $\mathbf{c}^{(k-1)}$, as you can see from Eq. (6.205). So far, all we've been doing is beautifying, i.e., we have applied Newton's method to our problem, but we haven't yet assumed anything specific about the different terms involved. We decide to make an approximation: since each second term on the left-hand side of Eq. (6.210) involves the residual $b_j - \rho_j$, it should make a small contribution if $p(x_j)$ is close to y_j, i.e., if the approximating function $p(x)$ does a good job reproducing the data; this motivates us to drop the second term, leading to the *Gauss–Newton method*:

$$\mathbf{K}_\rho^T\mathbf{K}_\rho\left[\mathbf{c}^{(k)} - \mathbf{c}^{(k-1)}\right] = \mathbf{K}_\rho^T[\mathbf{b} - \boldsymbol{\rho}] \tag{6.211}$$

Observe that this is a linear system of equations with a particularly simple ($n\times n$) coefficient matrix on the left-hand side; thus, our resulting equation is similar in form to the normal equations *for each iteration*. Of course, the normal equations of Eq. (6.144) could be solved directly in one step, i.e., one could tell ahead of time the necessary operation count, whereas the Gauss–Newton method is inherently *iterative*.

Take a moment to appreciate the fact that, since we dropped the second derivatives in Eq. (6.211), this approach is considerably less computationally expensive than Newton's method. (As usual, there's no free lunch: sufficiently close to the solution, Newton's method exhibits quadratic convergence, but that will not generally hold for the Gauss–Newton method.) At this point, we could show a code implementing the Gauss–Newton method, but we postpone that until the Project appearing in the following section; there, the first derivatives that are needed to set up $\mathbf{K}_\rho$ will be taken analytically, i.e., we won't need to make any use of a finite-difference approximation.

Another intriguing property of our new method is that it gives rise to a *descent direction*:

$$\left[\nabla\chi^2(\mathbf{c}^{(k-1)})\right]^T\left[\mathbf{c}^{(k)} - \mathbf{c}^{(k-1)}\right] = -2\mathbf{K}_\rho^T[\mathbf{b} - \boldsymbol{\rho}]\left(\mathbf{K}_\rho^T\mathbf{K}_\rho\right)^{-1}\mathbf{K}_\rho^T[\mathbf{b} - \boldsymbol{\rho}] \equiv -2\mathbf{v}\left(\mathbf{K}_\rho^T\mathbf{K}_\rho\right)^{-1}\mathbf{v} < 0 \tag{6.212}$$

Our starting expression is for the dot product of the gradient (of our scalar function χ^2) with the step in parameter space; if this is negative, then we are dealing with a descent direction. In the first step we plugged in our result for the gradient from Eq. (6.207) and also solved Eq. (6.211) for the parameter step. In the second step we realized that the same vector $\mathbf{K}_\rho^T[\mathbf{b} - \boldsymbol{\rho}]$ was appearing on the left and on the right, so we gave it a new name.

In the final step, we realized that, since $\mathbf{K}_\rho^T\mathbf{K}_\rho$ is positive definite,[88] from the argument of Eq. (6.146), then its inverse will also be positive definite and therefore our final expression will be negative. Thus, the Gauss–Newton step is guaranteed to be a descent direction, i.e., to point downhill (see, however, problem 6.74). From this perspective, the Gauss–Newton approach is superior to Newton's method: $\mathbf{J}_{\nabla\chi^2}(\mathbf{c})$ doesn't have to be positive definite, so Newton's method might fail to converge.[89]

Speaking of the gradient and descent, you may recall from section 5.6.3 that the gradient-descent method is a simpler technique, employing only the gradient—i.e., not employing the $\mathbf{K}_\rho^T\mathbf{K}_\rho$ of Eq. (6.211). Instead of starting from Newton's method and dropping the second derivatives, we could have started from the gradient-descent method; we do this in problem 6.75, where we also take the opportunity to discuss an important combination of the Gauss–Newton and gradient-descent methods, known as the Levenberg–Marquardt method. We now turn to a somewhat different way of approaching the problem of non-linear least-squares fitting which also employs the gradient-descent technique.

6.7.2 Artificial Neural Networks

A situation that arises quite often in practice is when you have access to a large dataset but, even though your points do suffer from "noise", no error bars are provided; a natural approach, in that scenario, is to trust each data point equally. In that case, the χ^2 statistic of Eq. (6.202) takes the form:

$$\chi^2 = \frac{1}{N}\sum_{j=0}^{N-1}\left[y_j - p(x_j)\right]^2 \equiv \frac{1}{N}\sum_{j=0}^{N-1} s_j \tag{6.213}$$

that is, the σ_j's have dropped out of the problem; we introduced a $1/N$ term for later convenience. As usual, the x_j's and y_j's are provided as input and $p(x)$ is your approximating function; we also took the opportunity to define a new variable s_j denoting the squared distance between the true and approximate values, for later reference.[90] We will now see an interesting approach to "learning" from the data what the true underlying function is; we will also look at a specific technique for doing so, which is then implemented in Python.

6.7.2.1 Feedforward Neural Network

In the previous subsection, we saw how to carry out non-linear least-squares fitting via the Gauss–Newton method; implicit in that approach was the fact that we would start from a given non-linear Ansatz, e.g., the Breit–Wigner form of Eq. (6.201). Obviously, for the same dataset one could make several very different choices for the functional form. Then, if you were given a new dataset, you might have to consider completely new analytical

[88] Actually, positive *semi*definite, but we won't have to deal with the case where it is non-invertible.

[89] A careful reader might protest: didn't we show in Eq. (5.154) that Newton's method led to a descent direction? Answer: these are different Newton's method steps. In problem 5.32 the step was $-(\mathbf{J}(\mathbf{x}^{(k-1)}))^{-1}\mathbf{f}(\mathbf{x}^{(k-1)})$, with $-\mathbf{f}(\mathbf{x}^{(k-1)})$ on the right-hand side, corresponding to Eq. (6.205) with $-(\mathbf{b}-\boldsymbol{\rho})$ on the right-hand side.

[90] Each s_j is a *residual* squared, whereas a non-linear generalization of Eq. (6.168)—see problem 6.76—would involve the *error terms* (for the parent distribution). If you take $\mathbf{c} = \mathbf{c}_\star$, then you can relate the two: $s_j = \varepsilon_j^2$.

structures, some linear and some non-linear. Instead of pulling such Ansätze out of a hat, it would be psychologically pleasing if one could extract in a "universal" manner the underlying function that gave rise to the data, i.e., without having to keep adjusting the functional form every time. If you're thinking that this sounds too good to be true, you are right; as you may have noticed while reading earlier sections of this chapter, universal approximation theorems usually cannot help you solve a given problem in practice.

Even so, what we said above serves to motivate a general approach to function approximation, that of *artificial neural networks*:[91] this basically amounts to combining linearity and non-linearity in a specific way, while also introducing what is typically a large number of parameters. The (also typically large) dataset is used to find "good" values of the parameters, i.e., values that lead to a small mean-squared of the residuals as per Eq. (6.213). In the field of machine learning, the process of determining good parameter values is known as *training* or as *learning*; as you know from earlier sections, this is followed by the *predictive* stage, in which the already-obtained values of the parameters are used to predict $p(x)$ at x's which have not been encountered before.[92] As you should expect by now, training will be time consuming, whereas prediction (with the parameters in place) will be a very straightforward (fast) operation.

We will employ an Ansatz for the approximating function $p(x)$. We will *not* be writing $p(x)$ down as a linear combination of parameters and basis functions as per Eq. (6.118):

$$p(x) = \sum_{k=0}^{n-1} c_k \phi_k(x) \tag{6.214}$$

but we *will* be writing down a *single* functional form, regardless of which dataset we are faced with. Thus, this approach is quite different from what we were doing both in the linear combination of Eq. (6.214) and in the *ad hoc* non-linear functions like the Breit–Wigner form of Eq. (6.201), which would not be applicable to a different dataset. We will employ a single non-linear "basis" function ϕ (known as the *activation function*) and many parameters in the following combination:

$$p(x) = \phi\left(\sum_{k=0}^{M-1} a_k\, \phi(b_k x + c_k) + d\right) \tag{6.215}$$

This may be hard to take in at one go, so let's split it up into two expressions:

$$p(x) = \phi\left(\sum_{k=0}^{M-1} a_k z_k + d\right), \qquad z_k = \phi(b_k x + c_k) \tag{6.216}$$

Whether you look at Eq. (6.215) or Eq. (6.216), the interpretation of our new approach is the same: we have a single input x and a single output $p(x)$, the latter trying to approximate the true underlying function at that x. It's important to keep in mind that the sum above

[91] The name reflects the initial aim of such approaches, i.e., modelling (biological) neural circuits.

[92] In actuality we should be dividing our data into a *training set* (to learn the parameters), a *validation set* (to tune the hyperparameters), and a *test set* (to check generalization performance against previously unseen data).

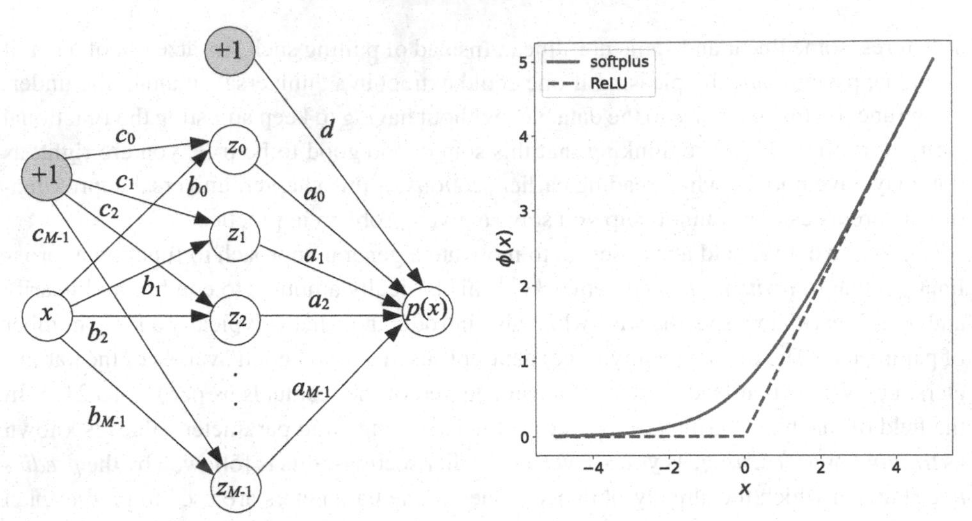

Fig. 6.16 Neural-network architecture (left) and activation function (right)

has nothing to do with the number of data points, which is why we are using k (from 0 to $M - 1$) here; note also that we are faced with $3M + 1$ parameters in total (**a**, **b**, **c**, and d), which play an analogous role to the n parameters in Eq. (6.214). The parameters a_k and b_k are known as *weights*, while the parameters c_k and d are known as *biases*. Note, finally, that in Eq. (6.214) we were expanding in terms of the basis functions $\phi_k(x)$ which were presumed to be simple and externally given. Here we are dealing with a single such activation function, ϕ, which is being used twice in our expression(s); crucially, in neural networks aimed at non-linear regression the ϕ is taken to be a *non-linear function*.

In Eq. (6.215) we wrote things down in keeping with the present chapter's overall theme, namely function approximation. It may benefit you to think in terms of our neural network's architecture, i.e., in terms of the (standard) concepts of the input, hidden, and output layers. The diagram shown in the left panel of Fig. 6.16 visualizes the operations involved in Eq. (6.216): the circles are known as *neurons* or *nodes* or *units*. We are dealing with a single input variable x (shown in the *input layer* on the left) which is combined with the b_k weights to give rise to the intermediate variables z_k, which appear in the *hidden layer* (in the middle). The input layer also involves a filled circle corresponding to the bias: as per Eq. (6.216), you can think of the input going into a given z_k as being $b_k \times x + c_k \times 1$. Similarly, each of the intermediate units z_k is then combined with a_k and used as input to produce the single output variable $p(x)$, appearing in the *output layer* (on the right). We also employ a bias unit in the hidden layer which, again, you may think of as being involved in $a_k \times z_k + d \times 1$. Our diagram doesn't explicitly show the application(s) of the activation function, i.e., it/they are implied. Since we are employing M hidden neurons (not counting the bias), it may help you to think of our network's architecture as 1–M–1. (We repeat that, in total, we are dealing with $n = 3M + 1$ parameters.) This is an example of a *feedforward neural network*,[93] in that connections between nodes do not form a cycle; information al-

[93] Also known as a *multilayer perceptron*; this is probably the most jargon-heavy subsection in the book.

ways moves from left to right. Since we are employing a single hidden layer, applying our neural network will involve what we may call "shallow learning". We could have, instead, employed two (or three, and so on) hidden layers: this would have involved more intermediate variables/hidden neurons and calls to the activation function ϕ (and would give rise to "deep learning"); as an example, for three hidden layers the architecture could be summarized as 1–M–M–M–1. Similarly, it should be straightforward to imagine how to tackle a scalar function of two variables: for the case of one hidden layer, this would lead to an architecture of the type 2–M–1. What we have here should be enough to give you a taste of the salient concepts.

At this point, you may be (understandably) wondering what the activation function ϕ looks like. For example, why did we need to include the filled circles/biases in the diagram? Qualitatively, as you could guess based on its name, an activation function serves to activate a given neuron, i.e., turn it on or off. In other words, you toggle a neuron (continuously) between two extremes; in older applications, this meant going from -1 to $+1$ (or from 0 to $+1$). You can probably now see why a bias term would have to be employed, if your target (or intermediate) values need to vary outside these two extremes. More recently, activation functions are taken to have values that go from 0 to infinity. Regardless of which specific activation function we end up employing, for us ϕ will be a non-linear function of one variable; as a matter of fact, in the machine-learning literature activation functions are also called *non-linearities*. In Eq. (6.216) we employ terms like $\phi(b_k x + c_k)$, in which a single number (each time) is passed into ϕ.

This discussion may be too abstract for you, so let's look at a specific example. The following activation function is known as *softplus*:

$$\phi(x) = \ln(1 + e^x) \tag{6.217}$$

This is visualized in the right panel of Fig. 6.16: as advertised, its values range from 0 to infinity. It maps positive arguments onto positive values and negative arguments onto increasingly tiny values. Crucially, softplus is a smooth function. To drive this point home, in Fig. 6.16 we are also showing another popular activation function, known as a *rectified linear unit* or *ReLU*, which is defined as:

$$\phi(x) = \max(0, x) \tag{6.218}$$

As you can see in the figure, ReLU maps its input onto 0 or a positive number, i.e., it is more sudden than softplus. (As a matter of fact, softplus can be viewed as a smooth version of ReLU.) You may be wondering how it is that ReLU can be considered to be a non-linear function, given that its plot clearly shows it to be piecewise linear. The explanation is that "non-linear function" is here used in contradistinction to the linear-algebra concept of a linear function, i.e., a *linear map*: such an entity satisfies additivity ($f(\mathbf{x}+\mathbf{y}) = f(\mathbf{x}) + f(\mathbf{y})$) and homogeneity ($f(a\mathbf{x}) = af(\mathbf{x})$). The same terminology is used when the values being taken in are in the scalar field; clearly, ReLU does not satisfy the first of these properties (take x and y to be -1 and 1 to see this), so it is not a linear function in this sense.[94]

Before proceeding further, we emphasize that the form we picked in Eq. (6.215), involving two calls to the activation function ϕ, is more commonly used when studying the problem of *classification*; of course, here we are actually interested in *regression*. To understand why Eq. (6.215) could get you in trouble, imagine a dataset that includes some (or all) data points with negative y_j. Since softplus in Eq. (6.217) can give rise only to positive values, it will be impossible for you to describe that dataset effectively. A more "traditional" approach to neural-network regression is tackled in problem 6.62: for regression, it is common to use the identity as an "activation function" in the output layer. Here we are employing the form of Eq. (6.215) to keep things interesting, with the disclaimer that we only expect it to work (well) for datasets that do not involve negative y_j values.

Given the feedforward neural-network Ansatz of Eq. (6.215) and an activation function of the form of, say, softplus in Eq. (6.217), everything is in place: we can now proceed to determine the values of the $3M + 1$ parameters ($\mathbf{a}, \mathbf{b}, \mathbf{c}$, and d) by minimizing the χ^2 of Eq. (6.213). You may be thinking that the specific technique employed to carry out this optimization is simply an implementation aspect and you would, in a sense, be right. There's nothing keeping you from employing, say, Powell's derivative-free method from section 5.6.5 to find the best minimum you can. As a matter of fact, you are asked to do precisely that in problem 6.63; as you will discover there, however, this approach becomes increasingly inefficient as the number of parameters is increased. State-of-the-art techniques employ hundreds of billions of parameters, but a derivative-free approach to optimization typically starts to be forbiddingly slow as you employ more than a few dozen parameters. This motivates the use of other optimization techniques for neural-network training, which *do* employ derivatives, a topic we now turn to.

6.7.2.2 Stochastic Gradient Descent and Backpropagation

As explained above, we will now see how to learn from our dataset what the parameter values should be. The simplest derivative-based approach to our problem is the gradient-descent method of Eq. (5.107). Bundling the $3M + 1$ parameters ($\mathbf{a}, \mathbf{b}, \mathbf{c}$, and d) together into a vector $\mathbf{w}$, gradient descent takes the form:

$$\mathbf{w}^{(i)} = \mathbf{w}^{(i-1)} - \gamma \nabla \chi^2(\mathbf{w}^{(i-1)}) = \mathbf{w}^{(i-1)} - \frac{\gamma}{N} \sum_{j=0}^{N-1} \nabla s_j(\mathbf{w}^{(i-1)}) \tag{6.219}$$

Here we are employing i for the iteration counter, so as not to get confused with the neuron index k or the data-point index j; for simplicity, we are using a constant learning rate γ. The initial values of the parameters $\mathbf{w}^{(0)}$ can be taken to be random. In the first equality we are explicitly showing that the scalar function we'll be minimizing is the χ^2 of Eq. (6.213); here ∇ is taking the gradient with respect to $\mathbf{w}$. In the second equality we re-expressed things in terms of each individual residual term. The approach of Eq. (6.219), which uses the full $\nabla \chi^2$ arising from the entire dataset is known as *batch gradient descent*.

[94] Of course, $f(x) = c_0 + c_1 x$ doesn't satisfy this criterion, either, if $c_0 \neq 0$.

Observe that to implement Eq. (6.219), as you are asked to do in problem 6.61, you will need to process the entire dataset *for each iteration* (labelled by i); if the dataset size N is large, this can be a very costly operation. As we learned when discussing the Bayesian approach to linear regression in section 6.6.3, this is also highly inefficient in a different way: if new data points come in, you will need to reprocess the augmented dataset (again, in order to take a single step from $i-1$ to i). An appealing approach that is used instead of Eq. (6.219) is to employ ∇s_j instead of $\nabla\chi^2$ in order to take a step in parameter space:

$$\mathbf{w}^{(i)} = \mathbf{w}^{(i-1)} - \gamma\nabla s_j(\mathbf{w}^{(i-1)}) \tag{6.220}$$

After you reach convergence (stepping through the i's) for a given j, you can turn to a different data point. You can view this approach as sampling the full gradient of Eq. (6.219); let's take a moment to justify this fact, which is more often stated than shown:[95]

$$\mathbb{E}(\nabla s_j) = \sum_{j=0}^{N-1}\frac{1}{N}\nabla s_j = \nabla\left(\frac{1}{N}\sum_{j=0}^{N-1} s_j\right) = \nabla\chi^2 \tag{6.221}$$

The left-hand side takes the expectation of an individual gradient. This is then expanded out in the first equality, with each term getting a weight of $1/N$. In the second equality the order of the summation and differentiation is interchanged. The final result is the gradient of our full χ^2, so ∇s_j is an *unbiased estimator* of $\nabla\chi^2$, i.e., the individual gradients will be correct on average (see problem 6.52 and section 7.7.2.2). Given that it randomly samples the gradient, our new prescription goes by the (rather grandiose) name of *stochastic gradient descent* (SGD). Regardless of the interpretation, it should be clear that Eq. (6.220) is a *sequential* or *online* approach: a new set of parameters can be produced using only a single new data point.

Qualitatively, the reason SGD works well in practice is that when you are starting out it is reasonable to expect that many/all the data-point residuals squared (s_j) will lead you to move in roughly the same direction (away from the random initial point and toward a minimum), so it is wasteful to wait until you've processed them all before you start producing new parameter values. Of course, as a result of processing only a single data-point term at a time, the convergence toward the minimum can be somewhat noisy; this fact can be exploited, in that SGD can help you escape from local minima: in general, it is certainly plausible that a stationary point of χ^2 will not coincide with a stationary point of s_j. Of course, this cuts both ways: the jerkiness of SGD implies that the iterates might try to leave even the global minimum of χ^2 if it happens to be reached, something that would, obviously, be an undesirable outcome (we discuss a possible remedy below). A compromise between the two extremes (of processing the entire dataset or only a single data point at a time) that is often used in practice is known as *mini-batch gradient descent*,

[95] This is analogous to Eq. (6.270) or Eq. (7.179), but here (in addition to being faced with gradients) we are taking the expectation with respect to a discrete set of outcomes, using Eq. (C.18) instead of Eq. (C.29). While this argument focuses on *empirical risk minimization*, intriguingly, the stochastic gradient descent approach also optimizes the data-generating/population/expected risk (if there are no repeated samples—see page 421).

in which the dataset is sliced up and correspondingly the gradient approximated using a collection of residual terms (s_j's) each time.

Regardless of whether one wishes to employ Eq. (6.219) or Eq. (6.220), it is clear that it will be necessary to compute ∇s_j. To spell things out, Eq. (6.220) is shorthand for the following four equations:

$$a_m^{(i)} = a_m^{(i-1)} - \gamma \frac{\partial s_j}{\partial a_m}\bigg|^{(i-1)}, \qquad b_m^{(i)} = b_m^{(i-1)} - \gamma \frac{\partial s_j}{\partial b_m}\bigg|^{(i-1)},$$
$$c_m^{(i)} = c_m^{(i-1)} - \gamma \frac{\partial s_j}{\partial c_m}\bigg|^{(i-1)}, \qquad d^{(i)} = d^{(i-1)} - \gamma \frac{\partial s_j}{\partial d}\bigg|^{(i-1)} \tag{6.222}$$

where we are using a new counter m to denote a specific component of each parameter set (in addition to using i for the iterations, j for the data point/residual term, and k for the hidden neurons). If you are using a large number of neurons M, as is typically the case in practice, approximating the gradient using a finite-difference approach would be quite inefficient (not to mention that it can also be inaccurate, as you learned in chapter 2). For a given j in the SGD prescription of Eq. (6.220), you would need to evaluate s_j at shifted values of the parameters (for each parameter), leading to (at least) an additional $n = 3M + 1$ evaluations of s_j.[96] While this is not the end of the world for our toy example, it is essentially forbidding for neural networks made up of many hidden layers (or a single hidden layer with many neurons) when tackling large datasets.

We will now see how to compute these derivatives using the chain rule for differentiation without necessitating *any* additional evaluations of (the quantities entering the definition of) s_j; this approach is known as *backpropagation*.[97] Let's examine in detail how this works for the case of the derivative with respect to a_m; we start from the definition of the residual squared term s_j in Eq. (6.213):

$$\frac{\partial s_j}{\partial a_m} = 2[p(x_j) - y_j]\frac{\partial p(x_j)}{\partial a_m} \tag{6.223}$$

We then employ the definition of our approximating function $p(x)$ in Eq. (6.216) to find:

$$\frac{\partial p(x_j)}{\partial a_m} = \phi'\left(\sum_{k=0}^{M-1} a_k z_k + d\right) z_m \tag{6.224}$$

where we suppressed the iteration counter $(i-1)$ for ease of reading (since we were already dealing with three other indices) and it is implied that we are evaluating the z's at x_j. Note that here we need to have (analytical) access to the derivative of the activation function, ϕ': below we'll be using the softplus function of Eq. (6.217), so we can trivially write down/code up its derivative. Plugging Eq. (6.224) into Eq. (6.223) gives:

$$\frac{\partial s_j}{\partial a_m} = 2[p(x_j) - y_j]\,\phi'\left(\sum_{k=0}^{M-1} a_k z_k + d\right) z_m \tag{6.225}$$

[96] Recall the similar comments we made in connection with the discretized Newton's method after Eq. (5.81).

[97] Backpropagation is a special case of (reverse-mode) automatic differentiation; in section 3.4 we provided an introduction to the approach of *forward*-mode automatic differentiation (highlighting its less involute nature).

Similarly, applying the chain rule for the other needed derivatives gives:

$$\frac{\partial s_j}{\partial b_m} = 2[p(x_j) - y_j]\, \phi'\left(\sum_{k=0}^{M-1} a_k z_k + d\right) a_m\, x_j\, \phi'(b_m x_j + c_m)$$

$$\frac{\partial s_j}{\partial c_m} = 2[p(x_j) - y_j]\, \phi'\left(\sum_{k=0}^{M-1} a_k z_k + d\right) a_m\, \phi'(b_m x_j + c_m), \tag{6.226}$$

$$\frac{\partial s_j}{\partial d} = 2[p(x_j) - y_j]\, \phi'\left(\sum_{k=0}^{M-1} a_k z_k + d\right)$$

We can now introduce Eq. (6.225) and Eq. (6.226) into Eq. (6.222) to get a complete prescription for how to carry out each step in our SGD approach; recall that all this has been referring to a single data pair, x_j and y_j. If you're working with stochastic gradient descent, as per Eq. (6.220), then you have all you need; if you are employing batch gradient descent, as per Eq. (6.219), then you will need to add together the gradient contributions coming from each of the s_j's.

6.7.2.3 Implementation

We are now in a position to implement in Python our neural-network approach to non-linear least-squares fitting, see `neural.py`, i.e., code 6.8. We start by defining a function, `soft()`, corresponding to the activation function ϕ from Eq. (6.217), also including the function-derivative ϕ' as a possible output; this uses the ternary operator for compactness. The next function computes the intermediate variables z_k and the approximating function $p(x)$ as per Eq. (6.216). We pass in the activation function as an argument, in case we change our mind in the future. The only intricacy involved here is that the $n = 3M + 1$ parameters **a**, **b**, **c**, and d are stored together in a 2d NumPy array `W` with dimensions $4 \times M$:[98] this is in the spirit of the vector **w** used in Eq. (6.220), but it is much more convenient to employ one index to keep track of which parameter set we are dealing with (**a**, **b**, **c**, or d) and another index to keep track of which hidden neuron we are referring to, k. By now, you may be used to the fact that the neuron index k itself does not need to explicitly appear in our code. The next function, `backprop()`, takes in the parameter array `W`, the data pair x_j and y_j, and the activation function. It straightforwardly implements Eq. (6.225) and Eq. (6.226); as expected, we call `soft()` with `der=1` when we need to evaluate ϕ'. The only subtlety is that we noticed the right-hand side in the equation giving $\partial s_j/\partial d$ is a factor that appears in the other derivatives, so we compute the various terms in the appropriate order.

The core of this program is in the function `sgd()` which, if you squint, is similar to the function `descent()` from code 5.7, e.g., the last three parameters are the learning rate γ, the maximum number of gradient-descent iterations, and the desired tolerance. The first parameter, `W`, plays the role of `xolds` in the earlier code; we also pass in the data pair x_j and y_j—since we'll need it for the backpropagation/computation of the gradient, i.e., we'll be passing it in to `backprop()`—as well as the activation function we wish to use. The gradient-descent step of Eq. (6.222) is carried out on a single line: a list

[98] Admittedly, it is a bit wasteful to use an entire row for the single value d.

Code 6.8 neural.py

```
from jacobi import termcrit
from newnormal import generatedata
from numpy import exp
import numpy as np

def soft(der, x):
    return np.log(1+exp(x)) if der==0 else exp(x)/(1+exp(x))

def getzspofx(W, x, act):
    zs = act(0, W[1,:]*x + W[2,:])
    pofx = act(0, W[0,:]@zs + W[3,0])
    return zs, pofx

def backprop(W, x, y, act):
    zs, pofx = getzspofx(W, x, act)
    ders = np.zeros((4,zs.size))
    ders[3,0] = 2*(pofx - y)*act(1, W[0,:]@zs + W[3,0])
    ders[0,:] = ders[3,0]*zs
    ders[1,:] = ders[3,0]*W[0,:]*x*act(1,W[1,:]*x+W[2,:])
    ders[2,:] = ders[3,0]*W[0,:]*act(1,W[1,:]*x+W[2,:])
    return ders

def sgd(W, x, y, act=soft, g=0.006, kmax=300, tol=1.e-4):
    for k in range(1,kmax):
        ders = backprop(W, x, y, act)
        Wn = np.array([W[i,:]-g*ders[i,:] for i in range(4)])
        _, pofxnew = getzspofx(Wn, x, act)
        conds = [termcrit(W[i,:],Wn[i,:]) for i in range(3)]
        if np.all([val<tol for val in conds[:3]]):
            break
        W = np.copy(Wn)
    return Wn, (pofxnew - y)**2

if __name__ == '__main__':
    sts = (1.5,1.2,0.15,0.08)
    data = generatedata(400,0.2,4.2,sts)
    np.random.seed(5179)
    W = np.random.uniform(-1,1,(4,10))
    W, s = sgd(W, data[0,155], data[1,155]); print(s)
```

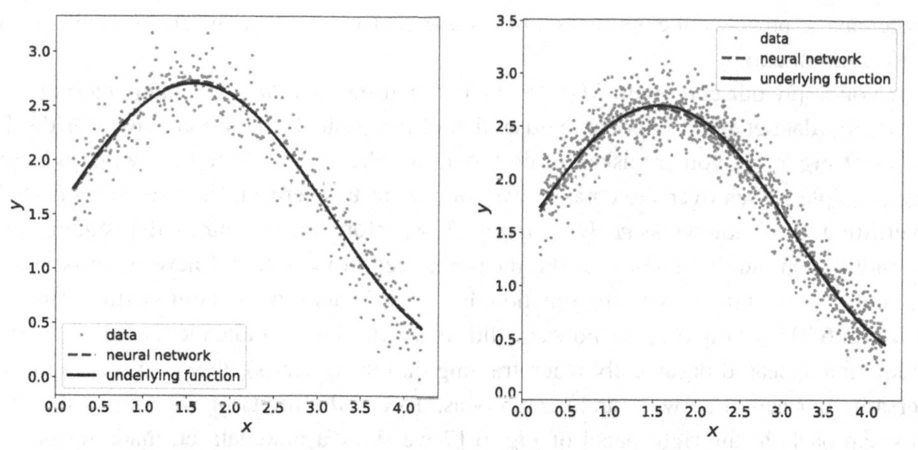

Fig. 6.17 Neural-network predictions for 400 (left) and 1200 (right) data points

comprehension sets up the four sets of updated parameters and the result is converted into a new NumPy array, `Wn` (i.e., the updated **a**, **b**, **c**, or d are stored as rows of `Wn`, as you would expect). We then compute the $p(x_j)$ corresponding to the updated parameters; note that, this time around, we won't be using the intermediate variables z_k, so we employ the Python convention of using an underscore to denote a placeholder variable (recall page 171). The function `descent()` from code 5.7 checked for convergence using `termcrit()`; here we do the same thing (applied separately to **a**, **b**, and **c**), once more employing a list comprehension to store the convergence-test results. We use a third list comprehension as input to NumPy's `all()`, which was introduced in problem 1.10. As usual, the new parameters are copied over, in preparation for the next iteration of the loop. Note that this time we do not employ the for-else idiom returning `None` to indicate failure to reach convergence; we fail "gracefully", hoping that the next SGD run (for a new data point) will fix things up. We also return the latest value of s_j, which the main program may use to keep track of the progress of our iterative approach.

Speaking of the main program, this starts out by generating a quite noisy dataset of $N =$ 400 points, using the same functionality we've employed elsewhere (from `newnormal.py`); the data, along with the underlying function, are visualized in the left panel of Fig. 6.17. The main program then randomly initializes the parameters and calls `sgd()` for an arbitrarily selected data point x_j and y_j. A more systematic approach would be to step through the dataset using something like `numpy.random.choice()` to select new indices. At this point, we need to introduce some more jargon: a single pass over the entire (training) dataset is called an *epoch*.[99] In practical applications, one typically revisits the full dataset several times, i.e., employs multiple epochs. Yet more terminology: the SGD approach typically works best if you employ a *learning-rate schedule*: this means changing the value of γ from epoch to epoch, e.g., $0.001/l^{1.3}$ where l is the epoch counter; this allows you to take large(-ish) steps when you're starting out, but then decrease their size as you have learned

[99] While epochs are typically used to sample *without* replacement, you should try out both options.

more/have approached the optimal set of parameters (in order to avoid accidentally leaving a good minimum).

If you apply our code with $M = 10$ hidden neurons (i.e., $n = 31$ parameters) to this $N = 400$ dataset for 40 epochs, you will find the dashed prediction shown in the left panel of Fig. 6.17; you are asked to do this in problem 6.60. (Note that, while we carry out multiple passes over the dataset, we don't want to overdo it, i.e., we wish to avoid overfitting; this is known as *early stopping*.) The performance of our model is quite good: despite the considerable noise in the data, our feedforward neural network was able to accurately extract the underlying function. It's worth reiterating that our starting Ansatz is still Eq. (6.215) or Eq. (6.216): nowhere did we assume a sinusoidal theory, but the correct prediction appeared organically when training on our dataset(s); this is evidence of how versatile our neural-network architecture was. To test the limits of what's possible with this approach, in the right panel of Fig. 6.17 we show a new dataset, made up of $N = 1200$ points (but with the same data-generating philosophy) together with the underlying function. If you plot the prediction corresponding to $M = 10$ hidden neurons (not shown here), you'll find that this time it won't do as good a job: it's more difficult to learn from the larger dataset. The right panel of Fig. 6.17 shows the result of employing $M = 80$ hidden neurons (i.e., $n = 241$ parameters): once again, we can nicely capture the underlying trend, despite being faced with a messier dataset.

Changing the number of neurons is a natural segue into some parting thoughts: when training on a specific dataset, one simply has to make a number of choices. How many hidden layers should you employ? How many neurons in each hidden layer? Which activation function should you use? If you chose to use gradient descent, as we did here, then you would also need to specify the learning rate (or the associated schedule), as well as the specific algorithm flavor employed: are you using batch gradient descent or stochastic gradient descent? Better yet, should you employ a different approach, e.g., gradient descent with momentum (see problem 5.30)? Finally, given that you process the data in randomized form, how many epochs/passes should you employ? Such aspects of your fitting, attention to which is necessary for you to get decent results, involve what are known as *hyperparameters* (recall footnote 92). Observe that regardless of which hyperparameters you employ, the task at hand is still the same—go from x to $p(x)$. As usual, here we make some choices regarding all these hyperparameters in order to streamline the presentation, but you will have the opportunity to explore alternative scenarios when you tackle the problem set.

6.8 Project: Testing the Stefan–Boltzmann Law

As you may have expected, this Project will discuss an application of the concepts described earlier in this chapter, for a case where a computer is necessary. Specifically, we will apply some of our material on non-linear least-squares fitting. The physics problem we have chosen to study is the extraction of the exponent in the *Stefan–Boltzmann law* for the power radiated by a black body. This is a result that preceded early quantum theory and in which the fitting to experimental data played an important role historically.

6.8.1 Total Power Radiated by a Black Body

We now turn to our physics theme, the total amount of radiation emitted by a black body. As you may recall, a *black body* is a perfect absorber of radiation. For reasons that will soon become clear, we will discuss in more detail than usual the history of this subject [77, 97], whose major events all took place in the second half of the nineteenth century.

6.8.1.1 Stefan's Stab in the Dark

In the early 1860s John Tyndall reported on his experiments heating a platinum wire and measuring the deflection of a galvanometer. Adolph Wüllner, in the 1875 edition of his textbook on experimental physics, summarized Tyndall's findings, providing estimates for the temperatures at which the platinum wire was observed. Thus, Tyndall's "faint red" became 525 °C and his "full white" turned into 1200 °C. The "intensity of the radiation" was observed to increase from 10.4 to 122, i.e., by a factor of 11.7. The qualitative point is that, even though the temperature increased by a factor of roughly 2, the radiation emitted increased considerably more.

In 1879 Josef Stefan published an article in which he converted (Wüllner's estimates of) Tyndall's temperatures into absolute temperatures. Stefan also noticed that by raising the ratio of the two absolute temperatures to the fourth power, one gets a value of 11.6, which is very similar to the 11.7 of the previous paragraph. Explicitly:

$$\frac{(273+1200)^4}{(273+525)^4} \approx 11.6 \tag{6.227}$$

From this, Stefan drew the general conclusion that the "heat radiation" is proportional to the fourth power of the absolute temperature.

You may be thinking that drawing a general conclusion about a power-law dependence based on only two points is stretching the limits of plausibility. You may also be thinking that Wüllner's temperatures were merely estimates. It turns out that Stefan was even luckier than that: Tyndall's measurements did *not* probe black-body radiation; repeating Tyndall's experiment today would lead to a ratio of 18.6, instead of the 11.7 following from Tyndall's/Wüllner's numbers. This is a pretty striking example of serendipity: Stefan made the right inference, using very little (and incorrect) experimental data.

6.8.1.2 Boltzmann's Thermodynamic Derivation

In 1884 Ludwig Boltzmann, who had earlier been Stefan's doctoral student, published a paper in which he derived the Stefan law from purely thermodynamic arguments. Nowadays, we call it the *Stefan–Boltzmann law* and write it as follows:

$$I = \sigma T^4 \tag{6.228}$$

where I is the radiated energy per second per surface area, at absolute temperature T.

The proportionality factor, σ, is now known as the Stefan–Boltzmann constant. This law describes the energy emitted *at all wavelengths*. It is also significant to note that, as shown by Boltzmann's derivation, this law is a *classical result*. In problem 6.71 you are asked to recover this result starting from Planck's formula for the black-body spectrum, thereby deriving the Stefan–Boltzmann constant, σ, in terms of fundamental constants.

Given the beauty and simplicity of Boltzmann's derivation, we now briefly go over its main steps. Suppose we have a closed system, made up of a "gas" of electromagnetic radiation. The fundamental thermodynamic relation is:

$$dU = TdS - PdV \tag{6.229}$$

where U is the internal energy, T the temperature, S the entropy, P the pressure, and V the volume. Dividing by dV at fixed T gives us:[100]

$$\left(\frac{\partial U}{\partial V}\right)_T = T\left(\frac{\partial S}{\partial V}\right)_T - P = T\left(\frac{\partial P}{\partial T}\right)_V - P \tag{6.230}$$

where the second step followed from one of the most common *Maxwell relations*.

Maxwell's name is also relevant to the next step, since he came up with the formula for *radiation pressure* (which you will explicitly show in problem 6.69), $P = \mathcal{E}/3$. Here $\mathcal{E} = U/V$ is the energy density of radiation (assumed to depend only on T). Plugging Maxwell's result and $U = \mathcal{E}V$ into Eq. (6.230), we get:

$$\mathcal{E} = \frac{1}{3}T\frac{d\mathcal{E}}{dT} - \frac{1}{3}\mathcal{E} \tag{6.231}$$

We wrote $d\mathcal{E}$ since $\mathcal{E}$ is a function of the temperature solely. This can be integrated to give:

$$\mathcal{E} = bT^4 \tag{6.232}$$

where b is a constant. You can immediately see the formal similarity with the Stefan–Boltzmann law of Eq. (6.228). Problem 6.70 guides you toward relating this b constant with the Stefan–Boltzmann constant, σ.

6.8.1.3 Lummer and Pringsheim's Experiments

In 1897, well-controlled experiments were carried out by Otto Lummer and Ernst Pringsheim [99]. In their experimental setup, Lummer and Pringsheim employed vessels coated with platinum black and surrounded by baths of water and niter; they used a bolometer to measure the radiant energy. They checked that the deflection in the galvanometer was proportional to the energy of the radiation.

Their results are shown in Fig. 6.18. Notice immediately that the y axis is labelled "reduced deflection": we don't get into details on the manipulations of the data the authors carried out. Instead, we will treat the numbers in this figure as the raw data, (x_j, y_j) for $j = 0, 1, \ldots, 12$. Another thing that immediately stands out is that the y_j's are not accompanied by input uncertainties, σ_j's. This is standard in early experimental papers, which generally only made a few qualitative statements about the errors in the main text, instead.

[100] It would be more accurate to say that we are using a standard property of exact differentials (cf. page 520).

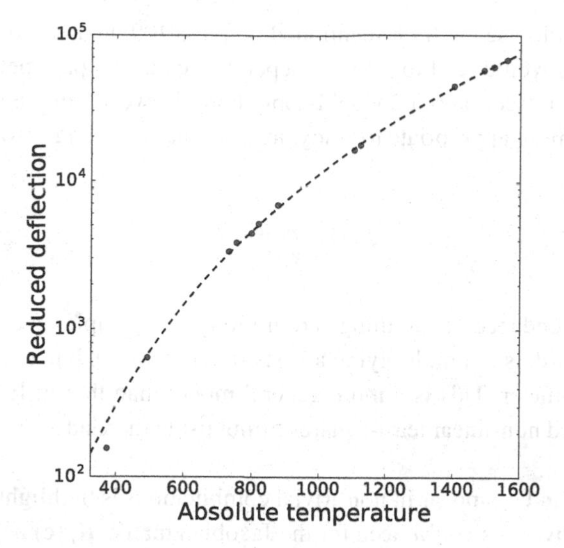

Fig. 6.18 Black-body radiation data from the 1897 paper by Lummer and Pringsheim

For us, this will be an opportunity to learn what to do (and how to interpret our results) when there are no input uncertainties available. We are also showing in Fig. 6.18 our least-squares fitting results pre-emptively, which we will produce in the following subsection. Note that we are using a logarithmic scale for our y axis and a linear scale for our x axis. You can probably already guess that on a log-log plot the points would largely end up on a straight line; this follows immediately from the Stefan–Boltzmann law, Eq. (6.228), which does look like a straight line in a log-log plot.

6.8.2 Fitting to the Lummer and Pringsheim Data

Since, as we already saw, Stefan's conjecture was based on little data which did not apply to the case in hand, the Lummer and Pringsheim work constitutes the first piece of experimental evidence confirming Stefan's intuition and Boltzmann's derivation. This motivates our choice to study this set of data in some detail, trying to extract the physics ourselves.

For most of their paper, Lummer and Pringsheim *assumed* the Stefan–Boltzmann law of Eq. (6.228) was true. In other words, they assumed that the data followed a power-law with temperature raised to the fourth power. Then, they tested to see how universal the b (or σ) constant is. In what follows, we will implement a more general strategy: don't assume that T^4 gives you the energy; instead, keep the exponent general and see if the experimental measurements lead you to the Stefan–Boltzmann law.

6.8.2.1 Setting up the Equations

As noted, we will take the (x_j, y_j) points of Fig. 6.18 as our input data. It would be nice if we could assume that the data could be modelled by the theory:

$$p(x) = c_0 x^{c_1} \tag{6.233}$$

In that case, we could use the trick mentioned on page 409, namely to take the logarithm on both sides. This would lead to a linear dependence on two parameters, in which case everything we said in section 6.5.2 would apply here. However, in the spirit of letting the data guide us it is more appropriate to study, instead, the following theory:

$$p(x) = c_0 + c_1 x^{c_2} \tag{6.234}$$

Except for the dependence on c_0, things are no longer so simple: even the c_1, which appears straightforward, is not multiplying a basis function $\phi_k(x)$, but x^{c_2}, which involves an undetermined parameter. This is a more general model than that in Eq. (6.233). We will carry out the needed non-linear least-squares fitting using the Gauss–Newton method from section 6.7.1.

The fact that c_1 and c_2 appear in non-trivial combinations is highlighted when you evaluate the partial derivatives you'll need for the Jacobian matrix $\mathbf{K}_\rho(\mathbf{c})$ of Eq. (6.206):

$$\frac{\partial p(x_j)}{\partial c_0} = 1, \qquad \frac{\partial p(x_j)}{\partial c_1} = x_j^{c_2}, \qquad \frac{\partial p(x_j)}{\partial c_2} = c_1 x_j^{c_2} \ln x_j \tag{6.235}$$

You should understand that when we refer to non-linearities, we are talking about the c_k's; keep in mind that x_j here refers to input data: the y_j's and x_j's are known, but the c_k's are unknown. You may be thinking: couldn't one simply define variables like S_x, S_y, and so on, as in Eq. (6.125), to get a *linear* problem? The problem is that here we cannot introduce such sums that depend *only* on the input data: the c_2-dependent exponent and the c_1 in $\partial p(x_j)/\partial c_2$ get in the way.

One last point: the Gauss–Newton method also involves the input uncertainties, σ_j's. As you can see from Fig. 6.18, though, this time we were provided only with a set of (x_j, y_j) points. What we'll do is to take $\sigma_j = 1$ for each data pair: this actually simplifies our equations. This assumption is equivalent to trusting each data pair the same; as you can see from Eq. (6.202), it is the *uniformity* of the variance that matters here: even if we took $\sigma_j = A$ for some fixed value A, the minimization process is not impacted. Just to warn you ahead of time, since we are dealing with a non-linear model and no input errors, the minimum χ^2 we find from Eq. (6.202) will be nowhere near $N - n$ (which in our case is $13 - 3 = 10$).[101] That's not an issue, though, since right now we are not interested in *model selection* but in *parameter estimation*, taking the model in Eq. (6.234) as given. As usual, while there is much more to be said on the statistical concepts involved in non-linear fitting, here we take the practical route of minimizing the χ^2, which for our case is simply minimizing the sum of the squares of the residuals.

6.8.2.2 Implementation

Since the main part of our calculation for the Gauss–Newton method, see Eq. (6.211), will involve a linear system of equations, we start code 6.9 by importing the Gaussian

[101] Of course, for a non-linear model it's not even obvious that we are dealing with $N - n$ degrees of freedom.

elimination function. In order to avoid the use of global variables,[102] we define a function, `generatedata()`, which sets up the `numpy` arrays for the x_j's, y_j's, and σ_j's.[103] We just mentioned that for us all the σ_j's are equal to 1, but we're still keeping things general: this might help you if you were to estimate them in some other way later on: you would need to change only a single function, `generatedata()`. Similarly, if you were to study another set of data, you would have to change only this one location in the code.

In the same spirit of producing a separate function for each task, we define a function, `model()`, to evaluate our theory, $p(x)$ from Eq. (6.234). This is called in our next function, `getKrs()`, which evaluates the $\mathbf{K}_\rho(\mathbf{c})$ and the residuals $\mathbf{b} - \boldsymbol{\rho}$ we'll need for Eq. (6.211). Have a look at Eq. (6.203) and Eq. (6.206) if you don't quite recall what these entities are. One should be careful not to overinterpret the modularity of our code: while it's nice that we have `model()` at our disposal and don't need to rewrite it out each time we need it, the lines of code computing the columns of `K` are using Eq. (6.235), which *knows* that our model is that of Eq. (6.234). In other words, if you wish to try out a different model, you will need to modify both `model()` and `getKrs()`.

That being said, the next function, `gaussnewton()`, *is* a modular implementation of the Gauss–Newton method of Eq. (6.211). It calls `getKrs()` to get the needed variables and `gauelim_pivot()` to solve the linear system. This is done iteratively, until convergence is reached. Notice how the line calling `gauelim_pivot()` is formally similar to the corresponding line in `normalfit()` from the normal-equations code 6.6. Of course, the Gauss–Newton method is an iterative approach, so we have to repeatedly solve Eq. (6.211) until we reach convergence. As advertised, the problem-specific aspects have been hidden in `model()` and `getKrs()`, leaving `gaussnewton()` a fully general function.

The main program writes down an initial guess for the c_k's, as needed by the Gauss–Newton method to get going. How does one pick this initial guess? At this point, textbooks usually state general facts, e.g., that non-linear fitting (or multidimensional minimization or root-finding) is tricky business, so you may not be able to find a solution unless you start sufficiently close to it. That is certainly true, but it doesn't help us much in our practical task of deciding what to pass to `gaussnewton()` as an initial guess. Experiment with crude guesses like `colds = np.array([1., 2, 3])` and see what happens.

A slightly more systematic way of going about things is to assume the Stefan–Boltzmann exponent is 4, *only* to get things going. In other words, you could take the last two data points and assume c_0 and c_1 can be chosen to make the curve go through the points:

$$c_0 + c_1 1522^4 = 60600, \qquad c_0 + c_1 1561^4 = 67800 \tag{6.236}$$

This gives $c_0 \approx -700$ and $c_1 \approx 1.26 \times 10^{-8}$. These rough estimates should be good enough to start with. We could also take the initial c_2 to be 4, but we don't want to do all the work of the Gauss–Newton method for it, so we start with a guess of 4.5 to see what happens. It's already looking like the three parameters will have very different magnitudes. As an aside, if you are worried that c_0 is non-zero (whereas the Stefan–Boltzmann law has no offset), don't be: we are describing the reduced data, which have been manipulated such that the energy would not go to zero if you extrapolated down to zero temperature.

[102] This is another case where object orientation would be useful.

[103] The larger y_j values seem awfully round, making us wonder if we should have trusted all data points equally.

Code 6.9 newblack.py

```
from gauelim_pivot import gauelim_pivot
from jacobi import termcrit
import numpy as np

def generatedata():
    data = np.zeros((3,13))
    data[0,:] = np.array([373.1, 492.5, 733, 755, 799, 820,
                877, 1106, 1125, 1403, 1492, 1522, 1561])
    data[1,:] = np.array([156., 638, 3320, 3810, 4440, 5150,
                6910, 16400, 17700, 44700, 57400, 60600, 67800])
    data[2,:] = np.ones(data.shape[1])
    return data

def model(cs,x):
    return cs[0] + cs[1]*x**cs[2]

def getKrs(data, cs):
    K = np.zeros((data.shape[1], cs.size))
    K[:,0] = 1/data[2,:]
    K[:,1] = data[0,:]**cs[2]/data[2,:]
    K[:,2] = cs[1]*data[0,:]**cs[2]*np.log(data[0,:])/data[2,:]
    rs = (data[1,:] - model(cs, data[0,:]))/data[2,:]
    return K, rs

def gaussnewton(data, colds, kmax=50, tol=1.e-8):
    for k in range(1,kmax):
        K, rs = getKrs(data, colds)
        cnews = colds + gauelim_pivot(K.T@K, K.T@rs)
        err = termcrit(colds,cnews)
        if err < tol:
            break
        colds = np.copy(cnews)
    return cnews

if __name__ == '__main__':
    data = generatedata()
    colds = np.array([-700, 1.26e-8, 4.5])
    cs = gaussnewton(data,colds); print(cs)
```

After roughly a dozen iterations, our code outputs:

```
[2.36554562e+01 8.33558626e-09 4.04306527e+00]
```

The converged parameters indeed exhibit very different orders of magnitude, as was also true of the initial-guess parameters. The converged values of c_0 and c_1 are quite different from our guesses, but that's not an issue for us. The main result for us is the value of c_2: minimizing the χ^2 for the Lummer–Pringsheim data led to an exponent of 4.04. Together with the associated standard error (recall page 37), namely the standard deviation in our estimate for this parameter (which is several orders of magnitude smaller, see problem 6.72), this is clear experimental evidence for the Stefan–Boltzmann exponent being (roughly) 4, just as intuited by Stefan and derived by Boltzmann.

Problems

6.1 [$\mathcal{A}$] Show that the determinant of the Vandermonde matrix from Eq. (6.17) obeys:

$$\det(\mathbf{X}) = \prod_{i>j}(x_i - x_j) \qquad (6.237)$$

The easiest way to show this is via induction. Our result shows that if $x_i \neq x_j$ for all pairs then the determinant is non-zero, and therefore the matrix is *non-singular*.

6.2 Reproduce Fig. 6.4. Investigate the ill-conditioning by: (a) evaluating the condition number of the Vandermonde matrix, and (b) defining the Chebyshev points without the minus in front (i.e., in opposite order) and comparing the zoomed-in plots.

6.3 Implement Lagrange interpolation without the barycentric formula, i.e., the slow way, as per Eq. (6.22) and Eq. (6.19); use the same data as in `barycentric.py`.

6.4 Build intuition on equispaced vs Chebyshev nodes by plotting the cardinal polynomials for $n = 10$. What do you observe regarding their magnitude?

6.5 Use Chebyshev points and Lagrange interpolation for (a) $f(x) = e^{\sin(20x)}$ (for $n = 15$ and $n = 60$), and (b) $f(x) = 100(x-1)^3\cos(4(x-1))e^{5(x-1)}$ (for $n = 7$ and $n = 15$).

6.6 Implement Lagrange interpolation with the nodes being the roots of Legendre polynomials, given by `legroots.py`. Plot $n = 7$ and $n = 15$ for Runge's function.

6.7 We know from Eq. (6.38) that the error of the interpolant at the point x is proportional to the node-polynomial value—Eq. (6.24)—at that point. In problem 6.8 we formally show that Chebyshev nodes minimize the maximum value of the node polynomial. Here you should plot $L(x)$ for $n = 7$ and $n = 15$ for the two choices of equidistant and Chebyshev nodes and draw the corresponding conclusions.

6.8 [$\mathcal{A}$] In problem 6.7 we explored "experimentally" the difference between the use of equidistant and Chebyshev nodes in the node polynomial, namely what appears in the expression for the error in Lagrange interpolation, Eq. (6.38). We will now help you show that Chebyshev nodes actually minimize the maximum value of the node polynomial; this is known as the *minimax property of Chebyshev polynomials*.

(a) Use the recurrence relation in Eq. (2.122) to show that the leading coefficient of the Chebyshev polynomial $T_n(x)$ is 2^{n-1}.
(b) Define the *monic Chebyshev polynomial* $\tilde{T}_n(x) = 2^{1-n}T_n(x)$ and compute the values of $\tilde{T}_n(x)$ at the $n+1$ extrema of $T_n(x)$—modify Eq. (2.124).
(c) Prove by contradiction that a monic polynomial $p_n(x)$ of degree n obeys:

$$\max_{-1\leq x\leq +1} |p_n(x)| \geq 2^{1-n} \tag{6.238}$$

To do so, examine how many roots the polynomial $\tilde{T}_n(x) - p_n(x)$ has in $[-1, +1]$. Hint: distinguish between minima and maxima (i.e., values of j).
(d) Use Eq. (2.123) to show that $|\tilde{T}_n(x)| \leq 2^{1-n}$ and thereby conclude that the minimum of the maximum shown in Eq. (6.238) is attained by $\tilde{T}_n(x)$. We have accomplished what we were after, i.e., we showed that using Chebyshev nodes leads to the minimization of the maximum value of the node polynomial.[104]

6.9 [$\mathcal{A}$] We will now see how Lagrange interpolation can be used to recover finite-difference formulas like those we encountered all the way back in chapter 3. Apply the error formula of Eq. (6.38) for the case of $n = 3$, i.e., for the Lagrange polynomials of Eq. (6.20). Differentiate this equation with respect to x^* and then take $x^* = x_0$, $x_1 = x_0 + h$, and $x_2 = x_0 + 2h$. Compare your final answer with Eq. (3.27).

6.10 [$\mathcal{A}$] Following techniques similar to those used in section 6.2.2.4, you should derive the general error formula for Hermite interpolation, Eq. (6.50). Specifically, you should start in the spirit of Eq. (6.33) and define:

$$F(x) = f(x) - p(x) - K\prod_{j=0}^{n-1}(x - x_j)^2 \tag{6.239}$$

and then follow the rest of the derivation.

6.11 [$\mathcal{A}$] Imagine we hadn't introduced Hermite interpolation in the main text. Try to determine an interpolating polynomial that satisfies the following conditions:

$$p(0) = 2 \qquad p(1) = -5 \qquad p'(0) = -5 \qquad p'(1) = -8 \tag{6.240}$$

Then, repeat the calculation using Hermite interpolation, Eq. (6.49).

6.12 You were asked to (and could) solve problem 6.11 analytically. The present problem will address the more general case where an automated approach pays off.

(a) Some scaffolding: take the logarithm of Eq. (6.19) and differentiate, to show:

$$L_k'(x) = L_k(x)\sum_{j=0, j\neq k}^{n-1}\frac{1}{x - x_j} \tag{6.241}$$

(b) Taking $x = x_k$ in Eq. (6.241) you now have an expression for $L_k'(x_k)$. Combined with Eq. (6.19), which gives you $L_k(x)$, you now have everything you need to implement Hermite interpolation in general, see Eq. (6.49). Test your function

[104] The monic polynomial used in this problem is anchored at the Chebyshev-polynomial *zeros*, not at the extrema which we employ elsewhere. For the sake of completeness, we note that the monic Chebyshev polynomial which is anchored at our usual Chebyshev nodes, Eq. (2.124), is $\tilde{T}_n(x) = 2^{1-n}[T_n(x) - T_{n-2}(x)]$.

for the x_k's $-1, -0.5, 0, 0.5, 1$, the y_k's $1, 1.72265625, 2, 1.78515625, 3$, and the y'_k's $-1, 1.25, 0, -0.625, 11$ for a large number of x's (between -1 and $+1$), comparing with the function $f(x) = x^8 + x^5 - x^2 + 2$.

6.13 The method of *piecewise-linear interpolation* is essentially an application of Lagrange interpolation in each panel, this time not to the second derivative, as in Eq. (6.67), but to the function itself. Specifically, we have:

$$s_{k-1,k}(x) = y_{k-1}\frac{x_k - x}{x_k - x_{k-1}} + y_k\frac{x - x_{k-1}}{x_k - x_{k-1}} \tag{6.242}$$

where this is to be interpreted as in Eq. (6.51). A moment's attention will highlight that this expression is nothing but Eq. (6.70) where we have dropped all the c terms Implement this in Python for Runge's function, for $n = 15$ and $n = 150$.

6.14 Implement the monomial three-node spline approach, as per Eq. (6.59), and compare with the output of `splines.py` for $n = 3$. Do this for many x values, leading to an $n = 3$ version of the left panel in Fig. 6.7.

6.15 Instead of using *natural splines* in Eq. (6.59), you could have implemented *clamped cubic splines*, in which the value of the first derivative is externally provided at the ends of the interval. Implement this and, as in problem 6.14, do so for many x values, leading to the $n = 3$ version of Fig. 6.7. Take $f'(x_0) = 0.1$ and $f'(x_2) = -0.1$ as your clamped boundary conditions and discuss the effect this choice has.

6.16 [$\mathscr{A}$] Derive Eq. (6.70); you may choose to employ a symbolic algebra package to help with the manipulations. While you're at it, also check for the continuity of the second derivative, Eq. (6.65), for Eq. (6.70).

6.17 Produce the right panel in Fig. 6.7. Repeat this exercise for $f(x) = |x| - x/2 - x^2$; do you understand why Lagrange interpolation is not doing as well?

6.18 [$\mathscr{A}$] Use the orthogonality of sines and cosines in the continuous case to prove Eq. (6.77). Then, use the orthogonality of complex exponentials to show Eq. (6.80).

6.19 [$\mathscr{A}$] Go from the real Fourier series, Eq. (6.76), to the complex series, Eq. (6.79).

6.20 [$\mathscr{A}$] Show the orthogonality of the complex exponentials in the discrete case, as per Eq. (6.92); the trick here is to identify a geometric series. Then, use Eq. (6.92) to show the orthogonality of sines and cosines in the discrete case, Eq. (6.84); simply employ de Moivre's formula and write out the real and imaginary parts separately.[105]

6.21 Carry out trigonometric interpolation, similarly to `triginterp.py`, but this time for odd-n, as per Eq. (6.81). The formulas in Eq. (6.82) and Eq. (6.85) now become:

$$\begin{aligned}
a_k &= \frac{2}{n}\sum_{j=0}^{n-1} y_j \cos kx_j, \qquad k = 0, 1, \ldots, m \\
b_k &= \frac{2}{n}\sum_{j=0}^{n-1} y_j \sin kx_j, \qquad k = 1, 2, \ldots, m \\
p(x) &= \frac{1}{2}a_0 + \sum_{k=1}^{m}(a_k \cos kx + b_k \sin kx)
\end{aligned} \tag{6.243}$$

[105] As the equation numbers cited here show, in the main text we went in the opposite direction i.e., started with sines and cosines and then went to complex exponentials.

where $n = 2m + 1$. Produce one plot for $n = 7$ and another one for $n = 9$.

6.22 Produce plots like in Fig. 6.9, this time for $f(x) = e^{\sin x + \cos x}$ using $n = 6$ and $n = 8$.

6.23 Reproduce our plots in Fig. 6.9, this time using the "unshifted" discrete Fourier transform, Eq. (6.88) and Eq. (6.94). Also show the imaginary part.

6.24 Reproduce our plot in Fig. 6.11 by implementing $r(x)$ from Eq. (6.114).

6.25 We turn to the *connection between the DFT and the continuous Fourier transform.*

(a) Justify each step, where DFT[...] refers to the operation contained in Eq. (6.97):

$$\tilde{f}(k_p) = \int_{-\infty}^{+\infty} f(x)e^{-ik_p x}dx = \int_{-L/2}^{+L/2} f(x)e^{-ik_p x}dx \approx \frac{L}{n}\sum_{q=0}^{n-1} f(x_q)e^{-ik_p x_q}$$
$$= (-1)^p \frac{L}{n} \text{DFT}\left[(-1)^q f(x_q)\right] \tag{6.244}$$

The variables used above are as follows:

$$x_q = \frac{L}{n}\left(q - \frac{n}{2}\right), \qquad k_p = \frac{2\pi}{L}\left(p - \frac{n}{2}\right), \qquad p, q = 0, 1, \ldots, n-1 \tag{6.245}$$

(b) Apply our final result from Eq. (6.244) to the integral given on the left-hand side of Eq. (4.287), for $L = 5$ and a few values of $n - n'$. Compare to the analytically derived answer. Hint: n will need to be a large power of 2; be sure to distinguish between the n used there and the DFT-n employed here.

6.26 Compare the slow DFT with the FFT: (a) implement the (shifted) slow DFT of Eq. (6.97), and (b) carry out timing runs for the slow DFT of part (a) and the `fft()` function from `fft.py`. To make things more systematic, print out runtimes for 2^9 up to 2^{13} points, using the `default_timer()` function from the `timeit` module.

6.27 [$\mathcal{A}$] In our derivation from Eq. (6.103) to Eq. (6.108) we showed how the FFT approach reduces the 4-element problem to two 2-element ones. Write out the analogous derivation for breaking down the 8-element problem into two 4-element ones.

6.28 [$\mathcal{A}$] We now view the DFT as a unitary transformation: (a) Write out the DFT matrix $\mathbf{E}$ of Eq. (6.101) and verify that it is a Vandermonde matrix. (b) Define $\mathbf{U} = \mathbf{E}/\sqrt{n}$ and show that $\mathbf{U}$ is a unitary matrix (see appendix C.2) after writing Eq. (6.99) in matrix/vector notation. (c) Part (b) implies that we can cast the DFT as a *unitary transformation.*[106] Prove the *Plancherel theorem*, i.e., that the unitary DFT preserves the length of a vector (employing $\tilde{\mathbf{v}} = \mathbf{U}\mathbf{v}$):

$$\sum_{k=0}^{n-1} |\tilde{v}_k|^2 = \sum_{j=0}^{n-1} |v_j|^2 \tag{6.246}$$

6.29 In the main text we split the computation of an n-element DFT into that of two $(n/2)$-element DFTs (see, e.g., problem 6.27); the next step would be to produce four $(n/4)$-element DFTs, and so on. Such a depth-first approach naturally gives rise to a recursive implementation, employed in `fft()`. Here we tackle the problem from the opposite end, via a breadth-first approach. Start by writing out the 2-element

[106] This fortuitous fact is why the DFT is perfectly well-conditioned despite involving a Vandermonde matrix.

DFTs ($n/2$ of them in total) and placing these in a $2 \times (n/2)$ matrix, which we call $\mathbf{Y}_2$. The idea is to combine these 2-element DFTs with the appropriate prefactors in such a way as to produce $(n/4)$ 4-element DFTs (i.e., a $4 \times (n/4)$ matrix): this is accomplished by taking the left half of $\mathbf{Y}_2$ and adding/subtracting the right half of $\mathbf{Y}_2$, remembering to include prefactors (analogously to Eq. (6.109)) and stacking the results. Continue in this way until you get to an $n \times 1$ matrix, at which point you're done. This is naturally implemented in an *iterative* (i.e., not recursive) manner. Implement this iterative technique in Python for $n = 128$. It may help you to know that the following three lines from `fft()` carry over completely unchanged:

```
first = evens + coeffs*odds
second = evens - coeffs*odds
ytils = np.concatenate((first, second))
```

but, of course, these entities are now 2d arrays. It will probably be easiest for you to first tackle the $n = 8$ case, for which your starting point is the following 2×4 matrix:

$$\mathbf{Y}_2 = \begin{pmatrix} y_0 + y_4 & y_1 + y_5 & y_2 + y_6 & y_3 + y_7 \\ y_0 - y_4 & y_1 - y_5 & y_2 - y_6 & y_3 - y_7 \end{pmatrix} \tag{6.247}$$

There are only two intermediate stages, involving 2×4 and 4×2 matrices.

6.30 This problem builds up to spectral differentiation using the FFT.

(a) We'll need a routine implementing the inverse FFT. Thankfully, we can simply re-use the direct FFT. To see this, take the complex conjugate of Eq. (6.99): this shows that the direct DFT of $\tilde{y}_k^*$ divided by n gives you y_j^*; taking the complex conjugate of *that* then leads to y_j. Write a Python function that does this.

(b) To evaluate derivatives, we'll start with $p(x)$ from Eq. (6.117). If you take the first derivative of $p(x)$ and then evaluate at the grid points, the cosine term (which has turned into minus sine) vanishes,[107] leading to:

$$y'_j = \frac{1}{n}\sum_{k=0}^{m-1} ik\tilde{y}_k e^{ikx_j} + \frac{1}{n}\sum_{k=m+1}^{n-1} i(k-n)\tilde{y}_k e^{i(k-n)x_j} = \frac{1}{n}\sum_{k=0}^{n-1} \tilde{y}'_k e^{ikx_j} \tag{6.248}$$

To get to the second equality, we employed Eq. (6.83) for the x_j (in the second sum) and also introduced a new entity, $\tilde{y}'_k$, as follows:

$$\tilde{y}'_k = \begin{cases} ik\tilde{y}_k, & 0 \le k \le m-1 \\ 0, & k = m \\ i(k-n)\tilde{y}_k, & m+1 \le k \le n-1 \end{cases} \tag{6.249}$$

Thus, to implement Eq. (6.248) you simply need the inverse DFT (or inverse FFT) of $\tilde{y}'_k$, which is defined as per Eq. (6.249). Plot the derivative for $n = 8$ and $n = 128$ and draw a conclusion about what's going on.

6.31 This problem addresses the question of real input data in the DFT and examines as an application the concept of *Fourier smoothing*.

[107] Analogously, when taking the second derivative the cosine term does *not* vanish at the grid points. Thus, counterintuitively, the second spectral derivative is *not* equivalent to taking the first spectral derivative twice.

(a) For real input data, we can cut the number of required computations in half, because $\tilde{y}_{n-k} = \tilde{y}_k^*$ holds. Show this relationship by writing out $\tilde{y}_{n-k}$ explicitly.

(b) We will apply a filter that captures the essential features of the noisy function of Eq. (3.96). Take as input the same set of points as in problem 3.25 (function values for 128 points from 0 to 2π). Use `fft()` to transform your input data. Keep the values $\tilde{y}_0$ to $\tilde{y}_m$; for the rest, explicitly impose $\tilde{y}_{n-k} = \tilde{y}_k^*$. Set all but the first five independent amplitudes to zero;[108] you have to be careful when handling the corresponding amplitudes which appear in the second half of your array. Finally, transform back using the inverse FFT; if you didn't solve problem 6.30, you may use `np.fft.ifft()`. Plot the result, comparing with the original set of data, and marvel at the smoothness of the filtered set of points.

6.32 [$\mathcal{A}$] Show that, for the propagation of error in a statistical context, the formula corresponding to Eq. (2.41) is Eq. (6.129). Hint: start from the case of a single data point y_j with input uncertainty σ_j; use the definition of the variance from appendix C.3.

6.33 [$\mathcal{A}$] Start from the covariance between the two parameters of the straight-line theory:

$$\sigma^2_{c_0,c_1} = \sum_{j=0}^{N-1} \frac{\partial c_0}{\partial y_j}\frac{\partial c_1}{\partial y_j}\sigma_j^2 \tag{6.250}$$

and show that $\sigma^2_{c_0,c_1} = -S_x/\Delta$ holds.

6.34 [$\mathcal{A}$] As sketched on page 375, set up the linear fit for $p(x) = c_0 + c_1 e^{-x^2}$.

6.35 This problem uses the straight-line data of `linefit.py`. We will employ a quadratic model, as per Eq. (6.132). You should analytically set up the problem, making it take the form of Eq. (6.133). Then, code it up in Python and print out $\chi^2/(N-n)$.

6.36 We now wish to generalize the approach of problem 6.35 to handle cubic, and so on, models. Obviously, doing everything by hand is not the way to go.

(a) You should convert `newnormal.py` to do general *polynomial fitting*.

(b) Apply your new code to the straight-line data of `linefit.py`. Show a plot with the data, as well as the least-squares fits using polynomials with two, three, and four parameters. Observe that the three-parameter case gives essentially a straight line, increasing our confidence in the straight-line fit (whereas the four-parameter case appears to be overfitting).

(c) Plot the residuals for the three cases (of polynomials with two, three, and four parameters) and interpret the results.

6.37 For the data in `newnormal.py`, compare the two models already contained there, with the model $p(x) = c_0+c_1\sin(x)+c_2\sin(2x)$. Implement Eq. (6.164) to also compute the standard deviations. Also, write a new function `computechisq()`, using Eq. (6.137) instead of Eq. (6.149), and calling `phi()` as the need arises.

6.38 [$\mathcal{A}$] This problem studies the quadratic form $\beta = \mathbf{x}^T\mathbf{B}\mathbf{x}$.

(a) Take the gradient of β, i.e., $\nabla\beta$. It is probably a good idea to break everything down in terms of components.

[108] Given the simplicity of our example you need even fewer amplitudes, but the lesson holds more generally.

(b) Now specialize to the case where $\mathbf{B}$ is a symmetric matrix.

(c) Your previous result should be of the form $\nabla\beta = \mathbf{Cx}$. Now take the derivative of *this*, thereby getting the Hessian of β.

6.39 This problem studies the condition number for the normal-equations approach:

(a) Derive Eq. (6.155). To do this, start from:

$$(\mathbf{A} + \Delta\mathbf{A})^T(\mathbf{A} + \Delta\mathbf{A})(\mathbf{c} + \Delta\mathbf{c}) = (\mathbf{A} + \Delta\mathbf{A})^T\mathbf{b} \tag{6.251}$$

and then follow an approach similar to that of section 4.2.3 to show that:

$$\frac{||\Delta\mathbf{c}||}{||\mathbf{c}||} \lessapprox \kappa(\mathbf{A})^2 \frac{||\Delta\mathbf{A}||}{||\mathbf{A}||} \tag{6.252}$$

(b) Compare $\kappa(\mathbf{A}^T\mathbf{A})$ and $\kappa(\mathbf{A})^2$ for a specific ill-conditioned rectangular $\mathbf{A}$.

6.40 [$\mathscr{A}$] We will now see how QR decomposition can help with the least-squares problem. Assume $\mathbf{A}$ is a rectangular matrix, of dimension $N \times n$. There exists a generalization of what we saw in section 4.4.3.1, known as the *reduced QR decomposition* (or sometimes as the *thin QR decomposition*): $\mathbf{A} = \mathbf{QR}$ where $\mathbf{Q}$ is an $N \times n$ matrix with orthogonal columns and $\mathbf{R}$ is an $n \times n$ upper triangular matrix.

(a) Start from the normal equations, Eq. (6.144). Introduce the reduced QR decomposition to derive an $n \times n$ system where the left-hand side will be $\mathbf{Rc}$. Take a moment to observe that this is an upper-triangular system, which can therefore be trivially solved via back substitution.

(b) Multiply by $\mathbf{R}^T$ on the left and manipulate the equation such that $\mathbf{Q}$ no longer appears. You have now derived what are known as the *seminormal equations*. These are helpful for the case where $\mathbf{A}$ is large and sparse and you need to repeat the calculation for several different $\mathbf{b}$'s; in this scenario you'd rather avoid producing and using $\mathbf{Q}$ (which can be dense) each time.

6.41 [$\mathscr{A}$] This problem deals with the covariance between two parameters in the normal-equations (linear) least-squares fitting problem. Starting from:

$$\sigma^2_{c_i,c_l} = \sum_{j=0}^{N-1} \frac{\partial c_i}{\partial y_j}\frac{\partial c_l}{\partial y_j}\sigma_j^2 \tag{6.253}$$

show that $\sigma^2_{c_i,c_l} = V_{il}$ holds.

6.42 We revisit the task of taking the derivative of a noisy function, as per problem 3.25. Armed with the machinery of `newnormal.py` (with the $n = 2$ model) carry out a least-squares fit to the $(x_i, f(x_i))$ data. Then, compute the forward-difference approximation (with $h = 10^{-3}$) of $f'(x)$ and compare to the analytically computed $g'(x)$ (which had completely dropped the third, highly oscillatory term). Once again, we emphasize that this problem, too, employs only the crudest finite-difference approximation (the forward difference). This has not kept us from doing an excellent job in reproducing the underlying pattern (that was buried in our data).

6.43 In Eq. (6.157) we got $\mathbf{c} = (\mathbf{A}^T\mathbf{A})^{-1}\mathbf{A}^T\mathbf{b}$. Introducing the *pseudoinverse*:

$$\mathbf{A}^+ = (\mathbf{A}^T\mathbf{A})^{-1}\mathbf{A}^T \tag{6.254}$$

the solution to the normal equations can be simply expressed as $\mathbf{c} = \mathbf{A}^+\mathbf{b}$. We will now see how one can use the SVD, $\mathbf{A} = \mathbf{U\Sigma V}^T$, to carry out least-squares fitting. First, show that the pseudoinverse can be expressed as $\mathbf{A}^+ = \mathbf{V\Sigma}^+\mathbf{U}^T$.[109] We produce $\mathbf{\Sigma}^+$ by taking the reciprocal of each *non-zero* singular value. Then, combine this result with $\mathbf{c} = \mathbf{A}^+\mathbf{b}$ to update `newnormal.py` such that no linear system of equations needs to be solved. You will be using the SVD of a rectangular matrix, so feel free to employ `np.linalg.svd()`.

6.44 [$\mathscr{A}$] In the spirit of page 409, come up with analytical manipulations that will allow you to determine the parameters of the models $p(x) = 1/(c_0x + c_1)$ and $p(x) = c_0x/(x + c_1)$ using only straight-line fitting.

6.45 [$\mathscr{A}$] In section 6.5.1 we introduced the chi-squared statistic. This is so named because it obeys the *chi-squared distribution* with probability density function:

$$f_\nu(x) = \frac{1}{2^{\nu/2}\Gamma(\nu/2)}e^{-x/2}x^{(\nu/2)-1} \tag{6.255}$$

where Γ is the *gamma function* (unsurprisingly) and we are using a new symbol for the number of degrees of freedom, ν. The distribution in Eq. (6.255) arises when you square independent random variables that obey a standard normal distribution (and then add the squares together). We will now help you derive it.

(a) Start from the case of a single degree of freedom. In other words, let $\mathscr{Z}$ be a standard normal random variable and investigate its square, $\mathscr{X} = \mathscr{Z}^2$. The probability density function of $\mathscr{X}$ is, by definition, equal to the derivative of its cumulative distribution function, so for $x \geq 0$:

$$f_1(x) = \frac{d}{dx}\left[\int_{-\sqrt{x}}^{\sqrt{x}} \frac{1}{\sqrt{2\pi}}e^{-z^2/2}dz\right] \tag{6.256}$$

Analytically evaluate this to find an answer of the form of Eq. (6.255).

(b) For the *moment-generating function* of a random variable $\mathscr{X}$ obeying a chi-squared distribution, show the last step in:

$$M_{\mathscr{X}}(t) = \int_0^\infty e^{tx}f_\nu(x)dx = \frac{1}{2^{\nu/2}\Gamma(\nu/2)}\int_0^\infty e^{tx}e^{-x/2}x^{(\nu/2)-1}dx = (1-2t)^{-\nu/2} \tag{6.257}$$

Feel free to employ the definition of the gamma function:

$$\Gamma(z) = \int_0^\infty x^{z-1}e^{-x}dx \tag{6.258}$$

(c) Examine two independent random variables, $\mathscr{X}_A$ and $\mathscr{X}_B$, which obey chi-squared distributions with (possibly) different numbers of degrees of freedom (ν_A and ν_B, respectively). Take the sum $\mathscr{X} = \mathscr{X}_A + \mathscr{X}_B$ and show that its moment-generating

[109] Note that for the case of a square, full-rank matrix this is the same as the matrix inverse, as per Eq. (4.274).

function $M_{\mathscr{X}}(t) = M_{\mathscr{X}_A}(t)M_{\mathscr{X}_B}(t)$ is also of the chi-squared form, Eq. (6.257). How many degrees of freedom does $\mathscr{X}$ have? The argument here was for two variables, but it should be easy to see that it generalizes: the sum of many (mutually independent) chi-squared random variables is also a chi-squared random variable. It is now trivial to combine the results from parts (a) and (c) and thereby draw the conclusion that adding together the squares of independent standard normal random variables gives you a chi-squared random variable. Combined with Eq. (6.177), this shows why the χ^2 statistic of Eq. (6.120) obeys the chi-squared distribution of Eq. (6.255).

6.46 We address the chi-squared distribution, Eq. (6.255), for $\nu = 4, 8, 12$—for this case, $\Gamma(\nu/2) = ((\nu/2) - 1)!$. You may use `numpy.polynomial.legendre.leggauss()` here (Gaussian quadrature is introduced in section 7.5).

(a) Compute the integral $\int_0^{+\infty} x f_\nu(x)dx$ and compare with the analytical expectation for the mean, ν.
(b) Compute the integral $\int_0^{+\infty} (x-\nu)^2 f_\nu(x)dx$ and compare with the analytical expectation for the variance, 2ν.
(c) Use Gauss–Legendre quadrature and a root-finder to solve $\int_m^{+\infty} f_\nu(x)dx = 0.5$ for m; compare with the approximate formula for the median, $\nu(1 - [2/(9\nu)])^3$, and with the output of `scipy.stats.chi2.median()`.
(d) Compute the integral $\int_0^M f_\nu(x)dx$ for $M = 1, 4, 9$ and compare with the output of `scipy.stats.chi2.cdf()` for the cumulative distribution function.

The rule of thumb of page 370 says that you should be pleased when a given χ^2 is near the mean value, i.e., ν (which, as per part (c), is slightly larger than the median).

6.47 [$\mathscr{A}$] Assume that the random variables $\mathscr{X}_0, \mathscr{X}_1, \mathscr{X}_2, \ldots, \mathscr{X}_{N-1}$ are independent and identically distributed (IID) according to $P(x)$, which could be any distribution (i.e., *not* necessarily Gaussian) with mean μ and finite variance[110] σ^2. The *central limit theorem* tells us that the sample mean $\overline{\mathscr{X}} = \sum_{i=0}^{N-1} \mathscr{X}_i/N$ (heavily used in section 7.7) obeys a Gaussian distribution asymptotically. To see this, first define:

$$\mathscr{Y}_i = \frac{\mathscr{X}_i - \mu}{\sigma}, \qquad \mathscr{Z} = \frac{1}{\sqrt{N}}\sum_{i=0}^{N-1} \mathscr{Y}_i \tag{6.259}$$

Then, take the *moment-generating function*[111] (MGF) of the random variable $\mathscr{Y}_i$, $\varphi(t) = \mathbb{E}(e^{t\mathscr{Y}_i})$, and show that the MGF of $\mathscr{Z}$ satisfies: $M_{\mathscr{Z}}(t) = \mathbb{E}(e^{t\mathscr{Z}}) = [\varphi(t/\sqrt{N})]^N$. Then, use the Maclaurin expansion of $\varphi(t)$ (and one of the definitions of the exponential) to show that $M_{\mathscr{Z}}(t) \to e^{t^2/2}$ as $N \to \infty$: this is the MGF of a standard normal distribution. Finally, use this fact to show that $\overline{\mathscr{X}}$ obeys a normal distribution with mean μ and variance σ^2/N.

6.48 Problem 6.47 may have been too abstract for you, so let's see the central limit theorem in action: generate $p = 10^5$ numbers, each of which is a sample mean of N uniformly distributed random numbers from 0 to 1, i.e., $\overline{\mathscr{X}} = \sum_{i=0}^{N-1} \mathscr{X}_i/N$. Be sure

[110] Thus, the Breit–Wigner/Cauchy distribution, of Eq. (6.201), would not work here; see also problem 7.35.
[111] Used in problem 6.45 to investigate the sums of squares of standard normal variables.

to differentiate between the N buried inside $\overline{\mathscr{X}}$ (how many uniform numbers we are averaging) and the overall p (which controls how many such sample means we are producing). Take N from 1 to 15 and plot: (a) normalized histogram counts (with 100 bins), and (b) the corresponding normal distribution with mean μ and variance σ^2/N (with $\mu = 1/2$ and $\sigma^2 = 1/12$ since we started with uniform numbers from 0 to 1). Use the appropriate NumPy/Matplotlib functionality for the visualization.

6.49 [$\mathscr{A}$] This problem studies the multivariate Gaussian (in N dimensions):

$$\mathcal{N}(\boldsymbol{\mu}, \boldsymbol{\Sigma}) = \frac{1}{(2\pi)^{N/2}|\boldsymbol{\Sigma}|^{1/2}} \exp\left[-\frac{1}{2}(\mathbf{x}-\boldsymbol{\mu})^T \boldsymbol{\Sigma}^{-1}(\mathbf{x}-\boldsymbol{\mu})\right] \tag{6.260}$$

where $|\boldsymbol{\Sigma}|$ is the determinant.

(a) Consider a diagonal $\boldsymbol{\Sigma}$ and show the equivalence of Eq. (6.260) and of:

$$\prod_{j=0}^{N-1} \mathcal{N}\left(\mu_j, \sigma_j\right) = \prod_{j=0}^{N-1} \left(\frac{1}{\sqrt{2\pi}\sigma_j} \exp\left[-\frac{1}{2}\left(\frac{x_j - \mu_j}{\sigma_j}\right)^2\right]\right) \tag{6.261}$$

by starting from the former and expanding. In words, this means that the product of univariate Gaussians is a multivariate Gaussian.

(b) The previous part was made (much) easier by the fact that $\boldsymbol{\Sigma}$ was diagonal. We will now help you show that the prefactor in Eq. (6.260) is correct even when $\boldsymbol{\Sigma}$ is not diagonal (but is still symmetric). To see this, integrate Eq. (6.260) over all $\mathbf{x}$'s, diagonalize $\boldsymbol{\Sigma}$ as per Eq. (4.125), and introduce a new variable $\mathbf{z} = \mathbf{V}(\mathbf{x}-\boldsymbol{\mu})$ into the exponent (where $\mathbf{V}$ is the eigenvector matrix). Argue why the determinant of the Jacobian is 1 and then integrate each of the z_j's. Finally, use problem 4.1: the product of the eigenvalues is equal to the determinant.

(c) We now turn to the important fact that *Gaussians are closed under marginalization*: when you integrate out some of the variables in a multivariate Gaussian, you get another Gaussian (for the variables that you didn't integrate out). For simplicity, start from the two-dimensional (non-diagonal, but zero-mean) case:

$$\begin{aligned} &\int_{-\infty}^{+\infty} dx_1 \frac{1}{2\pi\sigma_0\sigma_1\sqrt{1-\rho^2}} \exp\left(-\frac{1}{2(1-\rho^2)}\left[\left(\frac{x_0}{\sigma_0}\right)^2 - 2\rho\frac{x_0x_1}{\sigma_0\sigma_1} + \left(\frac{x_1}{\sigma_1}\right)^2\right]\right) \\ &= \frac{1}{\sqrt{2\pi}\sigma_0} \exp\left[-\frac{1}{2}\left(\frac{x_0}{\sigma_0}\right)^2\right] \end{aligned} \tag{6.262}$$

To show this, focus on the terms depending on x_1, complete the square, and then carry out the one-dimensional Gaussian integral (over x_1).

(d) We will now help you show that this property (closedness under marginalization) holds more generally; this will require some more scaffolding. Group together the first n elements of $\mathbf{x}$ into $\mathbf{x}_A$ and then the next m elements of $\mathbf{x}$ into $\mathbf{x}_B$; similarly, $\boldsymbol{\mu}$ is split into $\boldsymbol{\mu}_A$ and $\boldsymbol{\mu}_B$. The analogous partitioning of the covariance matrix $\boldsymbol{\Sigma}$ is into blocks:

$$\boldsymbol{\Sigma} = \begin{pmatrix} \boldsymbol{\Sigma}_{AA} & \boldsymbol{\Sigma}_{AB} \\ \boldsymbol{\Sigma}_{BA} & \boldsymbol{\Sigma}_{BB} \end{pmatrix} = \begin{pmatrix} \boldsymbol{\Lambda}_{AA} & \boldsymbol{\Lambda}_{AB} \\ \boldsymbol{\Lambda}_{BA} & \boldsymbol{\Lambda}_{BB} \end{pmatrix}^{-1} = \boldsymbol{\Lambda}^{-1} \tag{6.263}$$

where we also took the opportunity to introduce (and partition) the *precision matrix* $\mathbf{\Lambda} = \mathbf{\Sigma}^{-1}$. Note that the relationship between, say, $\mathbf{\Sigma}_{AA}$ and the blocks of $\mathbf{\Lambda}$ is not totally trivial (more on this below). Use Eq. (6.263) to re-express the N-dimensional Gaussian of Eq. (6.260) in the form:

$$\frac{1}{(2\pi)^{N/2}|\mathbf{\Sigma}|^{1/2}} \exp\Big[-\frac{1}{2}(\mathbf{x}_A - \boldsymbol{\mu}_A)^T \mathbf{\Lambda}_{AA} (\mathbf{x}_A - \boldsymbol{\mu}_A) - \frac{1}{2}(\mathbf{x}_A - \boldsymbol{\mu}_A)^T \mathbf{\Lambda}_{AB} (\mathbf{x}_B - \boldsymbol{\mu}_B) \\ - \frac{1}{2}(\mathbf{x}_B - \boldsymbol{\mu}_B)^T \mathbf{\Lambda}_{BA} (\mathbf{x}_A - \boldsymbol{\mu}_A) - \frac{1}{2}(\mathbf{x}_B - \boldsymbol{\mu}_B)^T \mathbf{\Lambda}_{BB} (\mathbf{x}_B - \boldsymbol{\mu}_B)\Big] \tag{6.264}$$

Integrate over $\mathbf{x}_B$ by focusing on those terms and completing the square (note the formal similarity with what you did in Eq. (6.262)). You should find:

$$\frac{(2\pi)^{m/2}|\mathbf{\Lambda}_{BB}|^{1/2}}{(2\pi)^{N/2}|\mathbf{\Sigma}|^{1/2}} \exp\left[-\frac{1}{2}(\mathbf{x}_A - \boldsymbol{\mu}_A)^T \left(\mathbf{\Lambda}_{AA} - \mathbf{\Lambda}_{AB}\mathbf{\Lambda}_{BB}^{-1}\mathbf{\Lambda}_{BA}\right)(\mathbf{x}_A - \boldsymbol{\mu}_A)\right] \tag{6.265}$$

Note that this is of Gaussian form with mean $\boldsymbol{\mu}_A$. Even more intriguingly, if you employ the *Schur complement* (which arises in block Gaussian elimination) you can re-express the term in the middle:

$$\mathbf{\Lambda}_{AA} - \mathbf{\Lambda}_{AB}\mathbf{\Lambda}_{BB}^{-1}\mathbf{\Lambda}_{BA} = \mathbf{\Sigma}_{AA}^{-1} \tag{6.266}$$

Our result is that after integrating Eq. (6.264) over $\mathbf{x}_B$, you get a Gaussian for $\mathbf{x}_A$ with mean $\boldsymbol{\mu}_A$ and covariance $\mathbf{\Sigma}_{AA}$, as advertised. (If you're feeling courageous you can have a look at the prefactors.)

(e) We will now see how to tackle integrals of the form:

$$f(a) = \int d^N x\, \delta(a - \mathbf{v}^T\mathbf{x}) \exp\left[-\frac{1}{2}(\mathbf{x} - \boldsymbol{\mu})^T \mathbf{\Sigma}^{-1} (\mathbf{x} - \boldsymbol{\mu})\right] \tag{6.267}$$

where δ is the Dirac delta function and $\mathbf{v}$ is a column vector. We are integrating over $\mathbf{x}$ a multivariate Gaussian while imposing a linear constraint: the delta function helps us integrate out all the $\mathbf{x}$ directions that are orthogonal to $\mathbf{v}$. In other words, we are carrying out an N-dimensional integration but the delta function eliminates one of those dimensions. More concretely, similarly to part (b), carry out a change of variables to $\mathbf{x}' = \mathbf{R}(\mathbf{x} - \boldsymbol{\mu})$; here $\mathbf{R}$ is a rotation matrix (therefore orthogonal, with unit determinant), chosen such that $\mathbf{R}\mathbf{v}$ is non-zero only along the first component. Use part (d) to justify why the integrations over $x'_1, x'_2, \ldots x'_{N-1}$ lead to a *univariate* Gaussian in x'_0; then, use the delta function to carry out the integration over x'_0 and thereby find a univariate Gaussian in a.

6.50 We now discuss the empirical rule for a univariate or multivariate Gaussian.

(a) Start from the univariate normal distribution of Eq. (6.169). You should compute:

$$\int_{\mu - j\sigma}^{\mu + j\sigma} dx \frac{1}{\sqrt{2\pi}\sigma} \exp\left[-\frac{1}{2}\left(\frac{x - \mu}{\sigma}\right)^2\right] \tag{6.268}$$

for $j = 1, 2, 3$. (This is similar to the integral we will need for the Maxwell–Boltzmann distribution for the velocities, Eq. (7.5).) For concreteness, take $\mu =$

12 and $\sigma = 4$; we'll employ Gaussian quadrature[112] (introduced in section 7.5); you may use `numpy.polynomial.legendre.leggauss()`.

(b) In the previous part you should have arrived at the 68–95–99.7 rule directly. Now turn to the multivariate Gaussian of Eq. (6.260), for which the problem is more complicated: use SciPy (or the approach of problem 6.46) to determine the cumulative distribution function of the chi-squared distribution in different dimensions/for different numbers of degrees of freedom n (call it F_n). For $j\sigma$ intervals (where $j = 1, 2, 3$), plot $F_n(j^2)$ vs n (for $n = 1$ to 20). Hint: the $n = 1$ results should agree with what you found in part (a).

6.51 [$\mathscr{A}$] In section 6.5 we set up the normal equations in terms of the entities **A** and **b** of Eq. (6.140). In section 6.6.1 we recast the same approach using $\boldsymbol{\Phi}$, $\boldsymbol{\Sigma}_d$, and **y** in, e.g., Eq. (6.173). To see that these two formulations are equivalent, start by showing that the relevant expressions appearing in Eq. (6.144) can be written as follows:

$$\mathbf{A}^T\mathbf{A} = \boldsymbol{\Phi}^T\boldsymbol{\Sigma}_d^{-1}\boldsymbol{\Phi}, \qquad \mathbf{A}^T\mathbf{b} = \boldsymbol{\Phi}^T\boldsymbol{\Sigma}_d^{-1}\mathbf{y} \tag{6.269}$$

by expanding things out in terms of components. Then, write out the natural logarithm of the multivariate Gaussian of Eq. (6.186). Now (analytically) maximize it by setting the gradient (with respect to **c**) to zero. (You may use problem 6.38 on the quadratic form.) You should arrive at the normal equations of Eq. (6.144).

6.52 We now discuss properties of the maximum-likelihood estimator and estimate.

(a) Use the $\hat{\mathbf{c}}$ of Eq. (6.175) to show that (using the notation of appendix C.3):

$$\mathbb{E}(\hat{\mathbf{c}}) = \int P(\mathbf{y}; \mathbf{c}_\star, \boldsymbol{\Phi}, \boldsymbol{\Sigma}_d)\, \hat{\mathbf{c}}\, d^N y = \mathbf{c}_\star \tag{6.270}$$

For us the parent distribution is Gaussian, $P(\mathbf{y}; \mathbf{c}_\star, \boldsymbol{\Phi}, \boldsymbol{\Sigma}_d) = \mathcal{N}(\boldsymbol{\Phi}\mathbf{c}_\star, \boldsymbol{\Sigma}_d)$, as per Eq. (6.171); this is consistent with Eq. (6.172), which implies that $\mathbb{E}(\mathscr{Y}) = \boldsymbol{\Phi}\mathbf{c}_\star$, i.e., these datasets are drawn from the same parent distribution with that mean. Here expressions like $\mathbb{E}(\hat{\mathbf{c}})$ are interpreted as integrations *across all datasets* (i.e., integrating $y_0, y_1, \ldots, y_{N-1}$ over all possible values): in a frequentist setting only the $\mathscr{Y}$ is a random vector, (i.e., we are not integrating across possible **c** values). Our estimator $\hat{\mathbf{c}}$ is associated with $\mathscr{Y}$, so since $\mathscr{Y}$ is random then so is $\hat{\mathbf{c}}$. In words, Eq. (6.270) is saying that $\hat{\mathbf{c}}$ is an *unbiased estimator*, i.e., it is correct on average.

(b) Appendix C.3 introduces the covariance of two random variables $\mathscr{X}$ and $\mathscr{Y}$. Generalize that approach to the case of the components of a random vector $\mathscr{U}$ to show the third step in the following:

$$\boldsymbol{\Sigma}_{\mathscr{U}} \equiv \text{cov}(\mathscr{U}) = \mathbb{E}\left[(\mathscr{U} - \mathbb{E}(\mathscr{U}))(\mathscr{U} - \mathbb{E}(\mathscr{U}))^T\right] = \mathbb{E}(\mathscr{U}\mathscr{U}^T) - \mathbb{E}(\mathscr{U})\mathbb{E}(\mathscr{U})^T \tag{6.271}$$

In $\mathscr{U}\mathscr{U}^T$ we are multiplying a column vector by a row vector, producing a matrix. To interpret Eq. (6.271), place indices in the first equality (i in the first term, j in the second, and no need for a transposition) to produce $\text{cov}(\mathscr{U})_{ij}$.

(c) Apply Eq. (6.271) to $\mathscr{Y}$ and solve for $\mathbb{E}(\mathscr{Y}\mathscr{Y}^T)$.

[112] Observe that there is terminology overloading at play here: we are using *Gaussian* quadrature for the *Gaussian* distribution. Things would have been even worse if we were also employing *Gaussian* elimination.

(d) Apply Eq. (6.271) to the $\hat{\mathbf{c}}$ appearing in Eq. (6.175) and use the result from the previous part; as above, feel free to also employ Eq. (6.269). Observe that your final answer for $\hat{\boldsymbol{\Sigma}}$ is equivalent to our earlier result for the covariance matrix, $(\mathbf{A}^T\mathbf{A})^{-1}$—i.e., the matrix version of $\sigma^2_{c_i,c_l} = V_{il}$—but this time we did not start from Eq. (6.253). It's worth repeating that, as you can see from Eq. (6.269), $\hat{\boldsymbol{\Sigma}}$ does *not* depend on the data $\mathscr{Y}$ (how could it, since we've integrated out all the $y_0, y_1, \ldots, y_{N-1}$'s?) or on the parameter estimate $\hat{\mathbf{c}}$.

(e) It is frequently stated without proof that $\hat{\boldsymbol{\Sigma}}$ goes to zero asymptotically (i.e., as N is taken to infinity). Less frequently, $\hat{\boldsymbol{\Sigma}}$ is explicitly shown to obey a characteristic $1/N$ trend—just like Eq. (7.180)—by invoking the central limit theorem and either introducing the *Fisher information matrix* or assuming that $\boldsymbol{\Phi}$ itself is stochastically produced. Since all of those concepts are, likely, unfamiliar to you at this point, let's take a direct path: augment the design matrix $\mathbf{A}$ with a new row $\mathbf{a}_N$ (corresponding to a new measurement at x_N) such that it becomes $\mathbf{A}_{\text{aug}}$. Then, show that the *Sherman–Morrison formula*—a special case of the Woodbury matrix identity of Eq. (6.279)—can be used to write:

$$(\mathbf{A}^T_{\text{aug}}\mathbf{A}_{\text{aug}})^{-1} = (\mathbf{A}^T\mathbf{A} + \mathbf{a}^T_N\mathbf{a}_N)^{-1} = (\mathbf{A}^T\mathbf{A})^{-1} - \frac{(\mathbf{A}^T\mathbf{A})^{-1}\mathbf{a}^T_N\mathbf{a}_N(\mathbf{A}^T\mathbf{A})^{-1}}{1 + \mathbf{a}_N(\mathbf{A}^T\mathbf{A})^{-1}\mathbf{a}^T_N} \qquad (6.272)$$

Finally, argue why this implies that the diagonal elements of the augmented covariance matrix are smaller than those of the original covariance matrix, i.e., as you make new measurements the variances are decreased. As you take N to infinity, this is saying that you will get zero variance, i.e., *your estimate is consistent.*

(f) The *Kullback–Leibler distance*[113] or *relative entropy* between the parent distribution $P(\mathbf{y}; \mathbf{c}_\star, \boldsymbol{\Phi}, \boldsymbol{\Sigma}_d)$ of Eq. (6.171) and the likelihood $P(\mathbf{y}; \mathbf{c}, \boldsymbol{\Phi}, \boldsymbol{\Sigma}_d)$ of Eq. (6.173), for a general set of parameters $\mathbf{c}$, can be defined as follows:

$$D_{KL}(\mathbf{c}) = \int P(\mathbf{y}; \mathbf{c}_\star, \boldsymbol{\Phi}, \boldsymbol{\Sigma}_d) \log \frac{P(\mathbf{y}; \mathbf{c}_\star, \boldsymbol{\Phi}, \boldsymbol{\Sigma}_d)}{P(\mathbf{y}; \mathbf{c}, \boldsymbol{\Phi}, \boldsymbol{\Sigma}_d)} d^N y \qquad (6.273)$$

You are asked to show another pleasing property of maximum-likelihood estimation, namely that the point estimate $\hat{\mathbf{c}}$ of Eq. (6.175) *minimizes the Kullback–Leibler distance to the parent distribution.* Feel free to employ the asymptotic limit, wherein the expectation (here an integral) can be approximated by a finite sum—as for the sample mean of Eq. (7.211); see also problem 7.33.

(g) In summary, $\hat{\mathbf{c}}$ obeys the n-dimensional sampling distribution $\mathcal{N}\left(\mathbf{c}_\star, \hat{\boldsymbol{\Sigma}}\right)$.[114] Note that this is not a distribution of the parameter $\mathbf{c}$, but of our estimator $\hat{\mathbf{c}}$, which is a statistic;[115] since we don't generally know $\mathbf{c}_\star$, approximate it by $\hat{\mathbf{c}}$ to get the distribution $\mathcal{N}\left(\hat{\mathbf{c}}, \hat{\boldsymbol{\Sigma}}\right)$.[116] For the data (and each of the models) of `newnormal.py`,

113 More commonly known as the Kullback–Leibler *divergence* since, while it is a statistical distance, it is not a metric—it is not symmetric with respect to $P(\mathbf{y}; \mathbf{c}_\star, \boldsymbol{\Phi}, \boldsymbol{\Sigma}_d)$ and $P(\mathbf{y}; \mathbf{c}, \boldsymbol{\Phi}, \boldsymbol{\Sigma}_d)$.

114 Since the $\hat{\mathbf{c}}$ of Eq. (6.175) is a linear transformation of the normal variable $\mathscr{Y}$ then it, too, is normal.

115 This is analogous to how the χ^2 statistic of Eq. (6.174) obeyed the chi-squared distribution of Eq. (6.178).

116 We are saying that the random vector $\hat{\mathbf{c}}$ roughly follows the distribution $\mathcal{N}\left(\hat{\mathbf{c}}, \hat{\boldsymbol{\Sigma}}\right)$. But how can this be? The confusion arises because, as mentioned in footnote 63, one often uses the same symbol $\hat{\mathbf{c}}$ to denote both the estimator (which is a random vector involving $\mathscr{Y}$) and the estimate (which corresponds to a given dataset $\mathbf{y}$).

draw 15 samples from this distribution and plot the corresponding approximating functions together with the ML estimate and the data.

6.53 We now turn to individual *predictions* based on the maximum-likelihood approach: having established our estimates for $\hat{\mathbf{c}}$ and $\hat{\boldsymbol{\Sigma}}$ (as per problem 6.52), we would like to find out which $\tilde{\mathscr{Y}}$ value corresponds to a (previously unseen, i.e., not part of the input data) $\tilde{x}$ value. Similarly to Eq. (6.193), we write: $\tilde{\mathscr{Y}} = \tilde{\boldsymbol{\phi}}\hat{\mathbf{c}}$.

(a) Examine the expectation and variance of $\tilde{\mathscr{Y}}$. First, show that $\mathbb{E}(\tilde{\mathscr{Y}}) = \tilde{\boldsymbol{\phi}}\mathbf{c}_\star$, using Eq. (6.270). (We are working in a frequentist context, so expressions like $\mathbb{E}(\tilde{\mathscr{Y}})$ are interpreted as integrations across all datasets, given the Gaussian parent distribution $P(\mathbf{y};\mathbf{c}_\star,\boldsymbol{\Phi},\boldsymbol{\Sigma}_d) = \mathcal{N}(\boldsymbol{\Phi}\mathbf{c}_\star,\boldsymbol{\Sigma}_d)$ from Eq. (6.171).) Then, show that $\mathbb{V}(\tilde{\mathscr{Y}}) = \tilde{\boldsymbol{\phi}}\,\hat{\boldsymbol{\Sigma}}\,\tilde{\boldsymbol{\phi}}^T$. You will need to use the definition of the variance from appendix C.3, the ML estimator of Eq. (6.175), as well as cov($\mathscr{Y}$) from problem 6.52. Putting everything together, the predictive distribution is:

$$P(\tilde{y};\mathbf{c}_\star,\tilde{\boldsymbol{\phi}},\boldsymbol{\Phi},\boldsymbol{\Sigma}_d) = \mathcal{N}\left(\tilde{\boldsymbol{\phi}}\mathbf{c}_\star,\tilde{\boldsymbol{\phi}}\,\hat{\boldsymbol{\Sigma}}\,\tilde{\boldsymbol{\phi}}^T\right) \tag{6.274}$$

To see why the relevant distribution is normal, look at $\tilde{\mathscr{Y}} = \tilde{\boldsymbol{\phi}}\hat{\mathbf{c}}$ and note that the only random component is $\hat{\mathbf{c}}$; this, in turn, is a linear transformation of $\mathscr{Y}$, as mentioned in footnote 114. The mean and the variance (just like $\tilde{y}$) are *numbers* (i.e., not vectors/matrices): this is a *univariate* Gaussian.

(b) Visualize Eq. (6.274) for the $n = 2$ and $n = 5$ models (and the dataset) of `newnormal.py` by picking 200 $\tilde{x}$'s from -1 to $+10$. Pretend you don't know the parent distribution, i.e., determine both the mean and the variance of the Gaussian in Eq. (6.274) for each model separately, using the corresponding $\hat{\mathbf{c}}$ and $\hat{\boldsymbol{\Sigma}}$; here we are making use of the plug-in principle. For each model, plot: (i) the data, (ii) the ML mean (i.e., the same curve as shown in the left panel of Fig. 6.13), and (iii) a band showing $\pm$ three standard deviations around the ML mean (at each $\tilde{x}$). Observe that the width of the prediction interval depends on $\tilde{x}$.

(c) A desideratum for a predictive theory is that it be able to tell you when its predictions start to become untrustworthy. Focus on our polynomial theory and check to see if the uncertainty band width starts to increase as you move away from the data. Compare the models with $n = 4, 5, 6, 7$, both in terms of how well they capture the data and in terms of the error bands, for $\tilde{x}$ going from -5 to $+15$.

6.54 [$\mathscr{A}$] We will now examine a general estimator $\mathscr{C}_\diamond$ (which could be the ML $\hat{\mathbf{c}}$ or something else), for which $\mathbb{E}(\mathscr{C}_\diamond) = \mathbf{c}_\blacklozenge$. Note that $\mathbf{c}_\blacklozenge$ might not equal $\mathbf{c}_\star$, so this may be a biased estimator; appropriately, we now define the *bias* in its prediction at the point x_j: $\mathcal{B} = \boldsymbol{\phi}_j\mathbf{c}_\blacklozenge - \boldsymbol{\phi}_j\mathbf{c}_\star$. Start from the expectation of the numerator in a single term in the sum of Eq. (6.174), evaluated for the estimator $\mathscr{C}_\diamond$, i.e., $\mathbb{E}[(\mathscr{Y}_j - \boldsymbol{\phi}_j\mathscr{C}_\diamond)^2]$. Your task is to show the *bias–variance decomposition*:

$$\mathbb{E}[(\mathscr{Y}_j - \boldsymbol{\phi}_j\mathscr{C}_\diamond)^2] = \mathbb{V}(\mathscr{E}_j) + \mathcal{B}^2 + \mathbb{V}(\boldsymbol{\phi}_j\mathscr{C}_\diamond) \tag{6.275}$$

Start on the left-hand side, adding and subtracting $\boldsymbol{\phi}_j\mathbf{c}_\star$ inside the parentheses; then, use the addition rule from appendix C.3, arguing why one term cancels. One of the surviving terms should be $\mathbb{E}[(\boldsymbol{\phi}_j\mathbf{c}_\star - \boldsymbol{\phi}_j\mathscr{C}_\diamond)^2]$; use the same trick again, this time adding and subtracting $\boldsymbol{\phi}_j\mathbf{c}_\blacklozenge$. Our final result in Eq. (6.275) contains: the variance of the noise (σ_j^2—not at our disposal), the square of the bias, and the variance in our model's prediction (at x_j). One needs to balance a model's bias (the relation between $\boldsymbol{\phi}_j\mathbf{c}_\blacklozenge$ and $\boldsymbol{\phi}_j\mathbf{c}_\star$) against its variance (the relation between $\boldsymbol{\phi}_j\mathscr{C}_\diamond$ and $\boldsymbol{\phi}_j\mathbf{c}_\blacklozenge$). You could minimize the $\mathbb{V}(\boldsymbol{\phi}_j\mathscr{C}_\diamond)$ by using a crude prediction, but then you'd get a large $\mathcal{B}$ (underfitting); conversely, you could minimize the $\mathcal{B}$ by using a model that captures all the scatter in the data, but that would lead to a large $\mathbb{V}(\boldsymbol{\phi}_j\mathscr{C}_\diamond)$ (overfitting).

6.55 Reproduce Fig. 6.14 and Fig. 6.15 for the dataset and two models of `newnormal.py`. (Note the order in which points are read in.) Plot the data and use the Gaussian posterior of Eq. (6.188) with parameters given in Eq. (6.191) to also plot the MAP prediction and the curves corresponding to 15 samples randomly drawn from the posterior distribution. You should also visualize the posterior as in the left panels of Fig. 6.14, but don't bother with the BLR uncertainty bands (which are tackled in problem 6.56); for the $n = 2$ model you should also visualize the initial prior. Finally, you should generate plots for a fourth batch made up of 20 data points.

6.56 This problem visualizes the posterior predictive distribution, Eq. (6.198), for the dataset and two models of `newnormal.py`. You should also generate plots for a fourth batch made up of 20 data points.

(a) Reproduce the BLR prediction intervals in Fig. 6.14 and Fig. 6.15 as discussed in the main text. (Note the order in which points are read in.)

(b) Draw 100 samples directly from the posterior predictive distribution, as discussed in the main text. Hint: only a tiny fraction of these (no more than 0.3%) should fall outside the 3σ bands you produced in part (a).

(c) As mentioned in problem 6.53, a desideratum for a predictive theory is that it be able to tell you when its predictions start to become untrustworthy. Compare the polynomial models with $n = 4, 5, 6, 7$, both in terms of how well they capture the data and in terms of error-band width, for $\tilde{x}$ going from -5 to $+15$; use both prediction intervals and samples from the posterior predictive distribution.

6.57 We will now discuss the marginal likelihood for our (Gaussian) case.

(a) Recall from Eq. (6.185) that the marginal likelihood is:

$$P(\mathbf{y}; \boldsymbol{\Phi}, \boldsymbol{\Sigma}_d, \boldsymbol{\mu}_0, \boldsymbol{\Sigma}_0) = \int d^n c P(\mathbf{y}|\mathbf{c}; \boldsymbol{\Phi}, \boldsymbol{\Sigma}_d) P(\mathbf{c}; \boldsymbol{\mu}_0, \boldsymbol{\Sigma}_0) \tag{6.276}$$

The integrand is the same as what we explored in Eq. (6.188) though, crucially, we are now integrating over $\mathbf{c}$ and the result will be a function of $\mathbf{y}$. This means that we can focus only on the $\mathbf{y}$-dependent terms of our earlier derivation, appearing in the second and sixth lines of Eq. (6.188):

$$P(\mathbf{y}|\mathbf{c}; \boldsymbol{\Phi}, \boldsymbol{\Sigma}_d) P(\mathbf{c}; \boldsymbol{\mu}_0, \boldsymbol{\Sigma}_0) \propto \exp\left[-\frac{1}{2}\mathbf{y}^T\boldsymbol{\Sigma}_d^{-1}\mathbf{y} + \frac{1}{2}\mathbf{q}_\mathbf{c}^T\boldsymbol{\Sigma}_\mathbf{c}\mathbf{q}_\mathbf{c}\right] \tag{6.277}$$

where $\mathbf{q}_\mathbf{c}$ and $\boldsymbol{\Sigma}_\mathbf{c}$ are given by Eq. (6.189) and Eq. (6.191). This is already starting to look like a multivariate Gaussian in $\mathbf{y}$. At this point we could ask you to complete the square (in $\mathbf{y}$ this time) and thereby produce $\boldsymbol{\mu}_\mathbf{y}$ and $\boldsymbol{\Sigma}_\mathbf{y}$. Instead, you should carry out the (easier) task of verifying that the following expression for the marginal likelihood is correct:

$$P(\mathbf{y}; \boldsymbol{\Phi}, \boldsymbol{\Sigma}_d, \boldsymbol{\mu}_0, \boldsymbol{\Sigma}_0) = \mathcal{N}\left(\boldsymbol{\Phi}\boldsymbol{\mu}_0, \boldsymbol{\Sigma}_d + \boldsymbol{\Phi}\boldsymbol{\Sigma}_0\boldsymbol{\Phi}^T\right) \tag{6.278}$$

by comparing the terms linear and quadratic in $\mathbf{y}$ with what you had in Eq. (6.277). You will need to employ the *Woodbury matrix identity*:

$$\left(\mathbf{A} + \mathbf{B}\mathbf{D}^{-1}\mathbf{C}\right)^{-1} = \mathbf{A}^{-1} - \mathbf{A}^{-1}\mathbf{B}\left(\mathbf{D} + \mathbf{C}\mathbf{A}^{-1}\mathbf{B}\right)^{-1}\mathbf{C}\mathbf{A}^{-1} \tag{6.279}$$

(b) Evaluate Eq. (6.278), an N-dimensional Gaussian density function, for the dataset $\mathbf{y}$ of `newnormal.py` (with $N = 8$). As models, use our sinusoidal $n = 2$ theory and the polynomial $n = 4, 5, 6, 7$ ones. Note that our earlier codes were expressed in terms of $\mathbf{A}$ and $\mathbf{b}$, but this time you will need to produce $\boldsymbol{\Phi}$. Compare the answers and draw a conclusion about how you should rank these models (for this problem). Does the answer change if you use $N = 28$ data points?
This is an example of how the marginal likelihood can be used for model selection: once you've integrated out the parameters $\mathbf{c}$, the only factor making the marginal likelihood large or small is the model you chose.

(c) Actually, the prior covariance matrix in part (b) was pulled out of a hat. One could, instead, try to *maximize* the marginal likelihood, for each model separately, by varying the elements on the diagonal of the prior covariance matrix. Carry out such a study for the same models as above, assuming a diagonal prior covariance matrix of the form $\boldsymbol{\Sigma}_0 = \sigma_0^2 \mathcal{I}$. Take σ_0^2 to range from 0.1 to 50 and apply (a modified version of) `golden()` to minimize the opposite of the marginal likelihood. Does the answer change if you use $N = 28$ data points?

6.58 [$\mathcal{A}$] The *Fisher information* is important in both frequentist and Bayesian inference. To help you grasp the idea, we address the case of a single parameter $c_\star$. Qualitatively, the Fisher information measures how much information on $c_\star$ can be gleaned from the dataset(s), $\mathbf{y}$ or $\mathscr{Y}$. For one parameter $c_\star$ and one basis function ϕ, the data-generating distribution of Eq. (6.171) becomes $P(\mathbf{y}; c_\star, \boldsymbol{\phi}, \boldsymbol{\Sigma}_d)$—where $\boldsymbol{\phi}$ bundles together all the $\phi(x_j)$'s. We define the Fisher information as follows:

$$I(c_\star) = \mathbb{V}\left(\frac{\partial \ln P(\mathbf{y}; c_\star, \boldsymbol{\phi}, \boldsymbol{\Sigma}_d)}{\partial c_\star}\right) \tag{6.280}$$

(a) Use the definition of the variance from appendix C.3 and elementary calculus to show that the following two formulas are equivalent to Eq. (6.280):

$$\begin{aligned} I(c_\star) &= \int \left(\frac{\partial \ln P(\mathbf{y}; c_\star, \boldsymbol{\phi}, \boldsymbol{\Sigma}_d)}{\partial c_\star}\right)^2 P(\mathbf{y}; c_\star, \boldsymbol{\phi}, \boldsymbol{\Sigma}_d) d^N y \\ &= -\int \frac{\partial^2 \ln P(\mathbf{y}; c_\star, \boldsymbol{\phi}, \boldsymbol{\Sigma}_d)}{\partial c_\star^2} P(\mathbf{y}; c_\star, \boldsymbol{\phi}, \boldsymbol{\Sigma}_d) d^N y \end{aligned} \tag{6.281}$$

(b) We will now examine a general unbiased estimator $\mathscr{C}_\diamond$ of $c_\star$; this $\mathscr{C}_\diamond$ may be the $\hat{c}$

of Eq. (6.175), but it might also take a different form. In the spirit of Eq. (6.270), our unbiased estimator $\mathscr{C}_\diamond$ obeys:

$$\int P(\mathbf{y}; c_\star, \boldsymbol{\phi}, \boldsymbol{\Sigma}_d)\, \mathscr{C}_\diamond\, d^N y = c_\star \tag{6.282}$$

If you take the derivative with respect to $c_\star$ and then multiply and divide by $P(\mathbf{y}; c_\star, \boldsymbol{\phi}, \boldsymbol{\Sigma}_d)$, you can make the derivative of the natural logarithm appear, as in Eq. (6.280) and Eq. (6.281). Square the resulting integral and then use the Cauchy–Schwarz inequality to show that:

$$\mathbb{V}(\mathscr{C}_\diamond) \geq \frac{1}{I(c_\star)} \tag{6.283}$$

This is known as the *Cramér–Rao bound.*

(c) In Eq. (6.281) we were taking the expectation of the *observed information*:

$$J(c_\star) = -\frac{\partial^2 \ln P(\mathbf{y}; c_\star, \boldsymbol{\phi}, \boldsymbol{\Sigma}_d)}{\partial c_\star^2} \tag{6.284}$$

This corresponds to a single dataset, $\mathbf{y}$. Take a Gaussian parent distribution, as per Eq. (6.171), and show that, for this case, $J(c_\star)$ is independent of $\mathbf{y}$, i.e., it is equal to the Fisher information (which is the expectation of the observed information). Next, observe that for this case both J and I are actually independent of $c_\star$. Finally, specialize the result of Eq. (6.176) to the case of a single parameter and thereby show that the ML variance (i.e., the only element contained in the 1×1 matrix $\hat{\boldsymbol{\Sigma}}$) saturates the Cramér–Rao bound of Eq. (6.283).

(d) We now change gears, slightly, turning to the question of Bayesian priors. On page 394 we stated that a uniform prior over an infinite interval can get you into trouble; this, however, is not its biggest problem. Imagine taking a parameter $\mathscr{C}$ and the associated prior $P(c)$ and carrying out a transformation to another parameter, $\mathscr{V}$. If $P(c)$ is uniform (and in most other cases) the new prior, $P(v)$, will in general look very different from $P(c)$ and will therefore lead to a correspondingly different posterior (hence to different predictions). Surely a simple reparametrization should not have such a dramatic impact on our conclusions; this is one of the reasons Bayesian statistics did not become more popular in its early days. One way out is to employ a *Jeffreys prior*:

$$P(c) \propto \sqrt{I(c)} \tag{6.285}$$

Use Eq. (6.281) to show that this leads to transformation-invariant results:[117]

$$P(v) = P(c)\left|\frac{dc}{dv}\right| \tag{6.286}$$

We'll revisit this when we introduce the inverse transform sampling, Eq. (7.195).

[117] The multiparameter generalization of Eq. (6.285) is $P(\mathbf{c}) \propto \sqrt{\det[\mathbf{I}(\mathbf{c})]}$, where $\mathbf{I}(\mathbf{c})$ is the *Fisher information matrix*; this is designed to take care of the determinant of the Jacobian matrix. (In general relativity, the same idea—the square root of the metric determinant—is employed to produce an integration measure that is invariant under coordinate transformations.) The generalization of the Cramér–Rao bound of Eq. (6.283) is that the variances are greater than or equal to the elements on the diagonal of the *matrix inverse* of $\mathbf{I}(\mathbf{c})$.

6.59 [$\mathcal{A}$] We turn to *asymptotic normality*, whereby the Gaussian distribution emerges (in both the ML and Bayesian settings) even if the data are not normally distributed.

(a) The asymptotic distribution of $\hat{c}$ is $\mathcal{N}[c_\star, 1/\sqrt{I(c_\star)}]$; this involves the Fisher information of Eq. (6.280). To see this, call the log-likelihood $l(c)$, Taylor expand $l'(\hat{c})$ around $c_\star$ and solve for $\hat{c} - c_\star$. First, show that the numerator $(l'(c_\star)/N)$ has an expectation of 0 and a variance of $I(c_\star)/N^2$ so (by the central limit theorem)[118] its asymptotic distribution is $\mathcal{N}[0, \sqrt{I(c_\star)}/N]$. Next, show that the denominator $(-l''(c_\star)/N)$ is a sample mean, so by the weak law of large numbers (see problem 7.33) converges to an expectation—see Eq. (7.192)—which we know from Eq. (6.281) is equal to $I(c_\star)/N$. You're done.

(b) Asymptotically, the posterior distribution becomes $\mathcal{N}[\hat{c}, 1/\sqrt{I(\hat{c})}]$. To see this, call the log-posterior $q(c)$ and the log-prior $p(c)$. Taylor expand $q(c)$ around $\hat{c}$ and then exponentiate. As a function of c, the zeroth-order term turns into a prefactor and the quadratic term into $\exp\{-[-p''(\hat{c}) - l''(\hat{c})](c - \hat{c})^2/2\}$. In the exponent $l''(\hat{c})$—a sum over N terms—dominates over $p''(\hat{c})$ as N grows larger. You can write $-l''(\hat{c}) = 1/[-1/l''(\hat{c})]$ and use similar steps as in part (a) to show that the variance turns into $1/I(\hat{c})$.

6.60 Reproduce Fig. 6.17 using the hyperparameters mentioned in the main text. For each epoch, print out the cumulative sum of the s_j's encountered.

6.61 Implement batch gradient descent, Eq. (6.219), for the dataset of `neural.py`. Play around with the tolerance, the learning-rate schedule, and the number of epochs.

6.62 Implement stochastic gradient descent for a feedforward neural network of the form:

$$p(x) = \sum_{k=0}^{M-1} a_k \, \phi(b_k x + c_k) + d \tag{6.287}$$

for a similar dataset as in `neural.py`, but passing into `generatedata()` a first strength of 0.25. You will have to first work out the corresponding expressions for backpropagation. Compare the predictions with those resulting from the neural network of Eq. (6.215). Do you understand why the two models behave differently?

6.63 For the neural network of Eq. (6.215) and the dataset from `neural.py`, minimize the χ^2 using Powell's method (i.e., without using SGD and backpropagation) as described at the end of section 6.7.2.1. Recall that `powell()` takes in a 1d NumPy array, whereas our functions in `neural.py` work with 2d arrays. Separately explore both the $N = 400, M = 10$ case and the $N = 1200, M = 80$ case. (For the latter case, which is especially time-consuming, you may need to tweak `powell()`.)

6.64 [$\mathcal{P}$] We return to the van der Waals equation of state, Eq. (5.2). This time we are told that $b = 0.03985$, i.e., we are not dealing with methane. You can find the dataset (i.e., the v_j's, the P_j's, and the σ_j's) at `www.numphyspy.org`; this is in the format of `data` from `newnormal.py`. Crucially, since we know the value of b, this problem is one of

[118] Here we need to invoke the Lindeberg–Feller central limit theorem (a generalization of what we encountered in problem 6.47) since our variables are independent but not identically distributed (I but not ID).

linear least-squares fitting. Use `normalfit()` to extract T and a; then plot the data together with your best-fit curve.

6.65 [$\mathcal{P}$] We turn to the data on the photoelectric effect, shown in Fig. 2.1. Millikan's measurements correspond to wavelengths of 2535, 3126, 3650, 4047, 4339, 5461 Å. Convert these to frequencies using $c = +2.997\,924\,58 \times 10^8$ m/s. The y-axis values are $V_s = 0.52, -0.38, -0.91, -1.29, -1.49, -2.04$ V, respectively. In keeping with Millikan's estimates, take the uncertainty in the stopping potential to be 0.02 V for each point except the 2535 Å one, for which you can take the uncertainty to be 0.05 V.

(a) Carry out a linear least-squares fit to Millikan's data as per Eq. (2.2); you should extract both the coefficient h/e and the standard deviation of this coefficient. Use the elementary charge value $e = 1.602\,176\,634 \times 10^{19}$ C to extract Planck's constant h and the associated uncertainty.

(b) Repeat the fit, this time employing a more general dependence of the stopping potential V_s on the frequency ν, namely allowing for the possibility of a quadratic term. (Again, extract both the coefficient of the linear term h/e and the standard deviation of this coefficient.) Check to see that: (i) the coefficient of the quadratic term is many orders of magnitude smaller than the linear term, (ii) the standard deviation in the quadratic-term coefficient is larger than the coefficient itself, and (iii) the chi-squared per degree of freedom for the quadratic theory is larger than for the linear theory. All three of these results confirm Einstein's intuition that the dependence is actually linear.

(c) Above, we anachronistically used modern values for the speed of light and charge of the electron. You should repeat the extractions, employing $+3 \times 10^8$ m/s and $1.592\,434\,536 \times 10^{19}$ C, respectively.

6.66 [$\mathcal{P}$] A phenomenological approach to nuclear binding energies considers the nucleus as being an incompressible, charged *liquid drop*. Generalized, this gives rise to the *semi-empirical mass formula* for the binding energy per nucleon:

$$\frac{E_B}{N+Z} = a_V - \frac{a_S}{(N+Z)^{1/3}} - a_C\frac{Z(Z-1)}{(N+Z)^{4/3}} - a_{sym}\frac{(N-Z)^2}{(N+Z)^2} - \lambda\frac{a_p}{(N+Z)^{3/2}} \tag{6.288}$$

Here N is the number of neutrons and Z is the number of protons in a given nucleus whose binding energy is E_B. The structure of each term is motivated by physical considerations; the coefficients are, accordingly, labelled to correspond to the volume, surface, Coulomb, symmetry, and pairing terms. The latter is not always present, since the λ parameter takes the values:

$$\lambda = \begin{cases} +1, & \text{odd-}N\text{–odd-}Z \\ 0, & \text{odd-}N\text{–even-}Z\text{, or even-}N\text{–odd-}Z \\ -1, & \text{even-}N\text{–even-}Z \end{cases} \tag{6.289}$$

At `www.numphyspy.org` you will find an appropriately pruned dataset from an atomic mass evaluation [10]; the three columns are the N, Z, and $E_B/(N+Z)$ values, respectively. Your task is to fit the model in Eq. (6.288) to the data, thereby determining the a_V, a_S, a_C, a_{sym}, and a_p coefficients.

(a) Observe that you will need to carry out *linear* least-squares fitting; the dependence on N and Z is clearly non-linear, but the dependence on the a's is just as clearly linear. A couple of new features jump out when you think of how to generalize `newnormal.py` so it can handle Eq. (6.288): first, note that $E_B = E_B(N, Z)$, i.e., we are faced with *two* independent variables. It may help you to think of the approximating function in Eq. (6.118) as being a function of a vector, i.e., $p(\mathbf{x})$; in the present case this $\mathbf{x}$ would be made up of two numbers (N and Z). Second, you will have to figure out how to programmatically encode λ.

(b) Note that the input data table that we provided does not come with associated error bars. Your first instinct may be to take all the σ_j's to be equal to 1. However, it should be easy to see that this would treat $\sigma_j = 1$ as an absolute error (but what if the quantities you were trying to describe had a magnitude of, say, 10^{-3}?). This is where footnote 38 comes in, telling you to assume that your error bars are all equal (so $\sigma_j = \sigma$) and that you did a good job fitting (so $\chi^2 \approx N - n$).[119] Practically speaking, this means that you can set $\sigma_j = 1$ in your code, produce a value of $\chi^2/(N - n)$ as per Eq. (6.120) and of the standard deviation in each parameter, σ_{c_i}, as per Eq. (6.156); finally, rescale σ_{c_i} by multiplying it by $\sqrt{\chi^2/(N - n)}$.

(c) In the same figure, plot the input binding energies per nucleon ($E_B^{\text{data}}/(N + Z)$) and the binding energies per nucleon corresponding to the a_V, a_S, a_C, a_{sym}, and a_p values you determined earlier in this problem ($E_B^{\text{model}}/(N + Z)$). In both cases, your x axis should be $N + Z$ and your y axis should be $E_B/(N + Z)$.

(d) Plot the residuals in your predictions for the binding energies. That means that your x axis should be $N + Z$ and your y axis should be $E_B^{\text{data}} - E_B^{\text{model}}$. Observe that the residuals exhibit systematic trends at given $N + Z$ values, reflecting the absence of shell corrections in Eq. (6.288).

6.67 [$\mathcal{P}$] This problem addresses data for the relativistic momentum p vs velocity v of a single particle, in order to extract the particle's mass $m_\star$ as tightly as possible. You can find the dataset (i.e., the v_j's, the p_j's, and the σ_j's) in the online supplement at `www.numphyspy.org`; this is in the same format as `data` from `newnormal.py`. From earlier experiments we know that $0.1 \le m_\star \le 0.3$ holds. Our goal is to extract a 99.7% confidence or credible interval for the mass parameter $m_\star$.

(a) Start from a normal-equations/maximum-likelihood approach. Similarly to problem 6.58, for a one-parameter problem the likelihood of Eq. (6.186) turns into:

$$P(\mathbf{p}; m, \boldsymbol{\phi}, \boldsymbol{\Sigma}_d) = \frac{1}{(2\pi)^{N/2}|\boldsymbol{\Sigma}_d|^{1/2}} \exp\left[-\frac{1}{2}(\mathbf{p} - \boldsymbol{\phi}m)^T \boldsymbol{\Sigma}_d^{-1} (\mathbf{p} - \boldsymbol{\phi}m)\right] \quad (6.290)$$

where, for us, the single basis function takes the form $\phi(v) = v/\sqrt{1 - v^2}$ and is evaluated at all the v_j's to make up $\boldsymbol{\phi}$ (incidentally, for a problem like this one, $m\phi(v)$, we need to carry out *regression through the origin*). Naively construct a 3σ confidence interval around $\hat{m}$ for this univariate problem: you will arrive at

[119] Inevitably, there is a degeneracy in N here, between the number of data points (employed in the rest of the chapter) and the number of neutrons (employed in Eq. (6.288)).

an interval that is *wider* than $0.1 \le m_\star \le 0.3$ and even has a negative (mass!) value at the left endpoint. You could, simply, conclude that $0.1 \le m_\star \le 0.3$, but (i) that doesn't tell you anything you didn't already know, and (ii) the most you could say is that *at least* 99.7% of your intervals will bracket $m_\star$.

(b) While the confidence-interval approach struggled, a Bayesian treatment is quite straightforward. First, combine the likelihood from Eq. (6.290) with your prior knowledge that $0.1 \le m \le 0.3$ to show that for this problem the posterior is:

$$P(m|\mathbf{p};\boldsymbol{\phi},\boldsymbol{\Sigma}_d) = \frac{\exp\left[-\frac{1}{2}(\mathbf{p}-\boldsymbol{\phi}m)^T\boldsymbol{\Sigma}_d^{-1}(\mathbf{p}-\boldsymbol{\phi}m)\right]}{\int_{0.1}^{0.3} dm \exp\left[-\frac{1}{2}(\mathbf{p}-\boldsymbol{\phi}m)^T\boldsymbol{\Sigma}_d^{-1}(\mathbf{p}-\boldsymbol{\phi}m)\right]}, \qquad 0.1 \le m \le 0.3 \tag{6.291}$$

Then, to find a 99.7% credible interval you have to solve the two equations:

$$\int_{0.1}^{a} dmP(m|\mathbf{p};\boldsymbol{\phi},\boldsymbol{\Sigma}_d) = \int_{b}^{0.3} dmP(m|\mathbf{p};\boldsymbol{\phi},\boldsymbol{\Sigma}_d) = \frac{1-0.997}{2} \tag{6.292}$$

for a and b. Here the MAP solution is identical to the ML one and does *not* lie at the midpoint of the interval from 0.1 to 0.3 (or from a to b). In Eq. (6.292) we opt for the *central interval*, in which the probabilities in the two tails are equal to each other. Use Gauss–Legendre quadrature (introduced in section 7.5; you may use `numpy.polynomial.legendre.leggauss()` here) and the bisection method. The final result is not much tighter than the starting interval of $0.1 \le m \le 0.3$ but, crucially, it *is* tighter: this is progress.

(c) Assume that, instead of $0.1 \le m \le 0.3$, your prior knowledge is:

$$P(m) = \begin{cases} 0, & m < 0 \\ e^{-5m}, & m \ge 0 \end{cases} \tag{6.293}$$

Use Gauss–Legendre quadrature and the bisection method to produce a 99.7% credible interval. This time employ a *one-sided interval*, i.e., the tail (on the right) should contain 0.03% of the total contribution.

6.68 [$\mathcal{P}$] For the physical setting and dataset of problem 6.67 (but this time without any prior knowledge), examine the relativistic dispersion relation in natural units ($c = 1$):

$$E_\star(\tilde{p}) = \sqrt{\tilde{p}^2 + m_\star^2} \tag{6.294}$$

We are after $E_\star(0.6)$; since the ML estimator is *equivariant*, we may conclude that $\hat{E}(\tilde{p}) = \sqrt{\tilde{p}^2 + \hat{m}^2}$. To produce confidence intervals for $E_\star(0.6)$ we would have to know the distribution of $\hat{E}(0.6)$. As per footnote 114, $\hat{m}$ is a linear transformation of the normally distributed variable $\mathcal{P}$—see Eq. (6.290) and Eq. (6.175)—and therefore $\hat{m}$, too, is normally distributed. But what about $\hat{E}(0.6)$?

We turn to the *parametric bootstrap*, which employs the plug-in principle: generate a fictitious dataset $\mathbf{p}_b^*$ from the distribution $P(\mathbf{p};\hat{m},\boldsymbol{\phi},\boldsymbol{\Sigma}_d)$; do this B times in total (i.e., produce $N = 20$ samples making up one dataset, and do that $B = 500$ times overall). For each fictitious dataset $b = 0, 1, \ldots, B-1$ produce an ML estimate $\hat{m}_b^*$

and compute $\hat{E}_b^*(0.6) = \sqrt{0.6^2 + (\hat{m}_b^*)^2}$. Estimate the standard error via:

$$\sigma_{\text{BOOT}} = \sqrt{\frac{\sum_{b=0}^{B-1}\left[\hat{E}_b^*(0.6) - \hat{E}(0.6)\right]^2}{B-1}} \tag{6.295}$$

Finally, produce a 99.7% confidence interval via $\hat{E}(0.6) \pm 3\sigma_{\text{BOOT}}$.

6.69 [$\mathcal{A}, \mathcal{P}$] Maxwell derived his result for the radiation pressure, $P = \mathcal{E}/3$, using only electromagnetic fields, but it is easier to use the kinetic theory of a gas of photons. For N non-interacting particles in a volume V and in thermodynamic equilibrium at temperature T, the pressure P obeys: $PV = \sum_j \mathbf{p}_{x,j}\mathbf{v}_{x,j}$, where $\mathbf{p}$ is the momentum, $\mathbf{v}$ is the velocity, and x, j refers to the x component of the j-th particle. Apply this expression to photons to derive Maxwell's result.[120]

6.70 [$\mathcal{A}, \mathcal{P}$] We will relate the energy density of black-body radiation, $\mathcal{E}$, to the energy emitted per unit area per second, I. The energy in the volume element d^3r is $\mathcal{E}d^3r$. The fraction of this energy that passes through an elementary area at angle θ is $\cos\theta/(4\pi r^2)$. Thus, by integrating $\cos\theta\ \mathcal{E}\ d^3r/(4\pi r^2)$ in the upper hemisphere (ϕ from 0 to 2π, θ from 0 to $\pi/2$, and r from 0 to c), you should show that $I = c\mathcal{E}/4$.

6.71 [$\mathcal{A}, \mathcal{P}$] Re-derive Boltzmann's Eq. (6.232) starting from the Planck distribution:

$$\mathcal{E}_\nu = \frac{8\pi h\nu^3}{c^3}\frac{1}{e^{h\nu/k_BT} - 1} \tag{6.296}$$

integrating over all (positive) frequencies ν to get an expression for b. Use the result of problem 6.70, $I = c\mathcal{E}/4$, to find an expression for σ from Eq. (6.228).

6.72 [$\mathcal{P}$] We now re-do the fits to the Lummer–Pringsheim data of `newblack.py`.

(a) Combine Eq. (6.156) and Eq. (6.211) to generalize Eq. (6.164) for the variances of a non-linear problem; implement this in Python.
(b) Carry out the χ^2 minimization using Powell's method from section 5.6.5.
(c) Use the theory of Eq. (6.233), i.e., the one with no offset. Take the logarithm on both sides, do a straight-line fit, and then translate back to the original quantities.
(d) Use the theory of Eq. (6.233) in a non-linear fit. If the value of the c_1 doesn't agree with that from the previous part, check the variance (in the previous part).
(e) Use the theory $p(x) = c_0 x^{c_1} - c_0 290^{c_1}$, which is inspired by Lummer and Pringsheim's manipulations, in a non-linear fit.

6.73 [$\mathcal{P}$] Approach the theory of Eq. (6.234) and the data of `newblack.py` from a different perspective than in the main text: tackle the vanishing-derivative conditions of Eq. (6.121) directly, via the multidimensional Newton's method. You should find that, due to the different orders of magnitude of the parameters, c_1 is essentially left untouched. Vary the converged-c_1 by hand and show what minimum χ^2 you get for each value of c_1. On a separate plot, track the converged value of the exponent, c_2, as you vary the c_1. Use a vertical line to mark the c_1 that minimized the χ^2 in the previous plot. (Your final answer should be the same as that found using `newblack.py`.)

[120] If you're familiar with the covariant formulation of Maxwell electrodynamics, note that $P = \mathcal{E}/3$ can be straightforwardly derived from the vanishing of the trace of the electromagnetic stress-energy tensor.

6.74 [$\mathcal{P}$] In the study of nuclear reactions, the energy dependence of the cross section near a single, isolated *resonance* is described using the *Breit–Wigner formula*:

$$f(E) = \frac{c}{(E - E_0)^2 + \Gamma^2/4} \tag{6.297}$$

where E_0 tells us where the resonance is centered, Γ how wide it is (in turn related to its lifetime), and c is an overall strength parameter. As part of the online supplement at `www.numphyspy.org`, you will find a dataset containing energies, cross-sections, and the standard deviations in the latter (i.e., the E_j's, the f_j's, and the σ_j's); this is in the format of `data` from `newnormal.py`. Observe that this is a *non-linear* least-squares fitting problem. Use the Gauss–Newton method to extract c, E_0, and Γ. Plot the data together with your best-fit curve.

Then, carry out a separate fit using a Gaussian form:

$$f(E) = ce^{-(x-E_0)^2/(2\Gamma^2)} \tag{6.298}$$

It should be pretty easy to see (both visually and based on the χ^2 values) which theory does a better job here. Note that, even though the Gauss–Newton method employs a descent direction, you will still need to play with the initial guesses; a descent direction is not guaranteed to go towards the *global* minimum.

6.75 [$\mathcal{P}$] We now examine different methods that tackle non-linear least-squares fitting, starting from the data of problem 6.74 and the Breit–Wigner model of Eq. (6.297).

(a) For our problem, the simple gradient-descent method of Eq. (5.107) turns into:

$$\mathbf{c}^{(k)} = \mathbf{c}^{(k-1)} + \gamma \mathbf{K}_\rho^T[\mathbf{b} - \boldsymbol{\rho}] \tag{6.299}$$

This is analogous to the Gauss–Newton method of Eq. (6.211) but does not involve $\mathbf{K}_\rho^T\mathbf{K}_\rho$ (and does involve a parameter, γ, controlling the step size). Implement this approach; you will need to carry out a quite large number of iterations.

(b) In its simplest version, the Levenberg–Marquardt technique combines the Gauss–Newton method of Eq. (6.211) and the gradient-descent method of Eq. (6.299):

$$\left[\mathbf{K}_\rho^T\mathbf{K}_\rho + \lambda \mathcal{I}\right]\left[\mathbf{c}^{(k)} - \mathbf{c}^{(k-1)}\right] = \mathbf{K}_\rho^T[\mathbf{b} - \boldsymbol{\rho}] \tag{6.300}$$

Compare Eq. (6.211), Eq. (6.299), and Eq. (6.300) to see that $\lambda = 1/\gamma$. Instead of picking that value, since this technique is more robust, in your implementation you should use a value of λ that leads to convergence in (dramatically) fewer iterations than were necessary for the gradient-descent method.

(c) An improved version of the Levenberg–Marquardt method takes the form:

$$\left[\mathbf{K}_\rho^T\mathbf{K}_\rho + \lambda \mathrm{diag}(\mathbf{K}_\rho^T\mathbf{K}_\rho)\right]\left[\mathbf{c}^{(k)} - \mathbf{c}^{(k-1)}\right] = \mathbf{K}_\rho^T[\mathbf{b} - \boldsymbol{\rho}] \tag{6.301}$$

This uses the diagonal elements of $\mathbf{K}_\rho^T\mathbf{K}_\rho$ instead of the identity matrix, allowing you to differently impact the step size for each parameter. In addition to this, you should modify your iterative scheme such that you multiply the value of λ by 3 if the χ^2 is increased (i.e., you made a bad step) and divide the value of λ by 3 otherwise. (You should also investigate the dependence on the initial λ value.)

(d) To confirm that the improved technique from the previous part is, indeed, robust, apply it to the data (and model) of `newblack.py`.

6.76 [$\mathcal{P}$] We will now address *non-linear statistical inference*; much of what we learned regarding linear statistical inference carries over. We model the discrepancy between theory and experiment by analogy to Eq. (6.168): $\mathscr{Y}_j = p(x_j; \mathbf{c}_\star) + \mathscr{E}_j$, where $\mathscr{E}_j$ obeys the distribution $\mathcal{N}(0, \sigma_j)$—and σ_j is frozen, being experimentally provided. The new feature here is that the theory term is not the—linear—$\boldsymbol{\phi}_j \mathbf{c}_\star$ but the—possibly non-linear—$p(x_j; \mathbf{c}_\star)$ which also appeared in, say, Eq. (6.202): we have updated our notation to also show that the (true) theory is evaluated at the data point x_j with the true parameters $\mathbf{c}_\star$. Be sure to appreciate that here $p(x_j; \mathbf{c}_\star)$ will be neither linear in $\mathbf{c}_\star$ nor Gaussian in x_j (and neither Gaussian in $\mathbf{c}_\star$ nor linear in x_j), yet the assumption that the data-generating distribution is normal persists. In other words, a general likelihood for $\mathbf{c}$, written for the entire dataset, will be similar to Eq. (6.173) in that the dependence on $\mathbf{y}$ is Gaussian:

$$L(\mathbf{c}) = P(\mathbf{y}; \mathbf{c}, \boldsymbol{\Phi}, \boldsymbol{\Sigma}_d) = \left(\prod_{j=0}^{N-1} \frac{1}{\sqrt{2\pi}\sigma_j} \right) \exp\left[-\frac{1}{2} \sum_{j=0}^{N-1} \left(\frac{y_j - p(x_j; \mathbf{c})}{\sigma_j} \right)^2 \right] \tag{6.302}$$

yet different from Eq. (6.173) in that the numerator in the exponent is no longer linear in $\mathbf{c}$. Tackle the dataset (in the format of `data` from `newnormal.py`) in the online supplement at `www.numphyspy.org`, using the (non-linear) Ansatz:

$$p(x_j; \mathbf{c}) = c_0 + (c_1 x_j^2 + c_2 x_j + c_3) e^{-c_4 x_j^2} \tag{6.303}$$

The functional form of a polynomial times a Gaussian appears in, e.g., the eigenfunctions of the quantum harmonic oscillator, Eq. (3.72)—and therefore in the functional integrals appearing in quantum field theory—in continuous wavelet transforms, and in the steerable filters used as part of edge detection in image processing [50].

(a) Carry out a χ^2 minimization using the Gauss–Newton method and (equivalently, yet separately) a likelihood maximization using Powell's derivative-free method. You may need to be a little careful with the initial guesses.

(b) Turn to a Bayesian study, focusing on the posterior distribution of Eq. (6.185), dropping the normalization (for now). Think about whether or not the derivation of Eq. (6.188) carries over to the present case. Your task is to employ Powell's method to produce maximum *a posteriori* estimates using: (i) a Gaussian prior with a zero mean vector and a diagonal covariance matrix with 30's on the diagonal, and (ii) a prior like that in (i) but with the modification that the mean and variance corresponding to the parameter $\mathscr{C}_3$ are 1.9 and 0.5, respectively.

(c) Turn to the denominator in Eq. (6.185), i.e., use the marginal likelihood to decide among the two priors of part (b). To compute the marginal likelihood for the model of Eq. (6.303) you will need to evaluate five-dimensional integrals. Study section 7.7.5.2 and then implement the Metropolis algorithm for this problem. You have to be a bit careful with the initialization of the parameters; also, do yourself a favor and make sure that you only read in the dataset once.

7 Integrals

The arrow would be indeed valuable if it picked out the good from the rest.

Thucydides

7.1 Motivation

We now turn to *numerical integration*, also known as *quadrature*. At a big-picture level, it's good to keep in mind that in numerical integration "throwing more points at the problem" typically pays off; it is common that one can provide the answer for a definite integral (approximately) to within machine precision.

7.1.1 Examples from Physics

In the spirit of starting each chapter with appropriately motivated problems, we now discuss a few integrals that pop up in different areas of physics.

1. **Electrostatic potential away from a uniformly charged rod**
   Our first (quasi-random) example comes from classical electromagnetism. When calculating the electrostatic potential at a distance d away from (the start of) a finite uniformly charged rod, we are faced with the calculation:

$$V = \int \frac{kdq}{r} = \int_0^a \frac{k\lambda dx}{\sqrt{x^2+d^2}} = \int_0^a \frac{k\lambda dx}{d\sqrt{\left(\frac{x}{d}\right)^2+1}} \tag{7.1}$$

   where λ is the linear density of electric charge and a is the length of the rod. Changing variables to $y = x/d$ and assuming $a/d = 1$, leads to the integral:

$$\Phi = \int_0^1 dx \frac{1}{\sqrt{x^2+1}} \tag{7.2}$$

   where we renamed the dummy integration variable y back to x. This integral can be evaluated analytically: substitute $x = \tan\theta$ and then take $w = \tan\theta + \sec\theta$. The answer turns out to be:

$$\Phi = \ln(x + \sqrt{x^2+1})\Big|_0^1 = \ln(1+\sqrt{2}) \approx 0.881\,373\,587\,019\,542\ldots \tag{7.3}$$

   Here, the indefinite integral happens to be an inverse hyperbolic sine. (This has the fascinating implication that $\int_0^{1/2} dx/\sqrt{x^2+1} = \ln\varphi$, where φ is the golden ratio of Eq. (5.96).) We will employ this example in what follows, for the sake of concreteness.

2. **Maxwell–Boltzmann distribution for the velocities**
While it's not totally trivial to come up with such substitutions, you could always use a symbolic math package or even a table of integrals. However, in some cases there may be no simple analytical answer. A well-known example appears in the theory of statistical mechanics. Take the Maxwell–Boltzmann distribution for the velocities of an ideal gas in one dimension (see problem 7.44 for the three-dimensional case):

$$P(v_x) = \sqrt{\frac{\beta m}{2\pi}} e^{-\beta m v_x^2/2} \tag{7.4}$$

where v_x is the velocity and the temperature is hidden inside $\beta = 1/k_B T$. If we're interested in the probability of the particles having velocities from, say, $-A$ to A, then the integral we need to evaluate can be recast as:

$$I = \int_{-A}^{A} P(v_x) dv_x = \frac{1}{\sqrt{2\pi}} \int_{-A\sqrt{\beta m}}^{A\sqrt{\beta m}} e^{-x^2/2} dx \tag{7.5}$$

The answer here cannot be expressed in terms of elementary functions; instead, this is precisely the definition of the error function, $\text{erf}(A\sqrt{\beta m})$. This is important enough that it bears repeating: the indefinite integral of a simple Gaussian cannot be expressed in terms of elementary functions.[1]

3. **Statistical mechanics starting from elementary degrees of freedom**
There is a case where simple analytical evaluation or standard quadrature methods both fail: multidimensional integrals. For example, the (interaction part of the) classical partition function of a gas of n atoms is:

$$Z = \int d^3 r_0 d^3 r_1 \ldots d^3 r_{n-1} e^{-\beta V(\mathbf{r}_0, \mathbf{r}_1, \ldots, \mathbf{r}_{n-1})} \tag{7.6}$$

Here, $\beta = 1/k_B T$ is the inverse temperature and $V(\mathbf{r}_0, \mathbf{r}_1, \ldots, \mathbf{r}_{n-1})$ contains the interactions between particles (or with external fields). This is clearly a $3n$-dimensional integral, evaluating which analytically (or via quadrature) is hopeless for all but the smallest values of n.

7.1.2 The Problem to Be Solved

More generally, the task of quadrature is to evaluate $\int_a^b f(x)dx$, for a function $f(x)$ which may look like that in Fig. 7.1. Numerical integration at its most fundamental consists of approximating a definite integral by a sum, as follows:

$$\int_a^b f(x)dx \approx \sum_{i=0}^{n-1} c_i f(x_i) \tag{7.7}$$

[1] If we take the limit $A \to \infty$, then we *can* carry out the integral analytically: this gives $\sqrt{2\pi}$ and therefore $I = 1$. This is a very important integral, which you will re-encounter in problem 7.1.

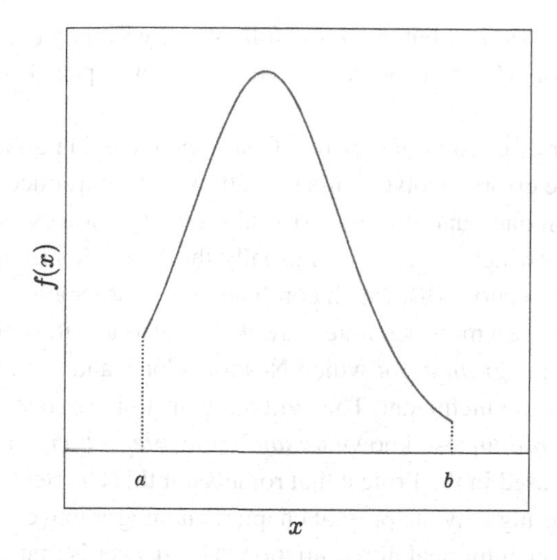

Fig. 7.1 The definite integral is the area under the curve, when x goes from a to b

Here, the x_i are known as the *nodal abscissas* and the c_i are known as the *weights*. Note that the i runs from 0 to $n-1$, so we are dealing with n abscissas (and n weights) in total.

Such quadrature methods (and much of this chapter) can be divided into two large categories. First, we have *closed methods*, where the endpoints of our interval (a and b) are included as abscissas (x_i). Of the approaches we discuss below, the trapezoid and Simpson's methods are closed. Second, we have *open methods*, where the endpoints of our interval (a and b) are *not* included as abscissas (x_i). Of the approaches we discuss below, the midpoint method and Gaussian quadrature are open. Most categorizations are too neat and this one is no exception: there exists a third category, namely the case of *half-open* methods, where one of the endpoints of our interval (a or b) is not included in the abscissas (x_i). Of the approaches we discuss below, the rectangle method is half-open. It's easy to see that open or half-open methods are to be preferred in the case where the integrand has a singularity at an endpoint.

Above, we mentioned several integration methods by name, grouping them according to whether or not they are closed, open, or half-open. In practice, it is much more common to group them, instead, according to a different criterion, namely whether they are:

- **Newton–Cotes methods**: these make the assumption that the integral can be approximated by summing up the areas of elementary shapes (e.g., rectangles); such methods typically involve *equally spaced abscissas*: these are useful (if not necessary) when $f(x)$ has already been evaluated at specific points on a grid.
- **Gaussian quadrature**: these make use of *unequally spaced abscissas*: these methods choose the x_i in such a way as to provide better accuracy. As a result, they typically require fewer abscissas and therefore fewer function evaluations, making them an attractive option when an $f(x)$ evaluation is costly.

Keep in mind that some of the Newton–Cotes methods we discuss below (e.g., the rectangle rule) don't, strictly speaking, require equally spaced abscissas (e.g., we could have adjacent rectangles of different widths), but in all that follows we will only use points on an equally spaced grid for Newton–Cotes methods, reserving unequally spaced abscissas for Gaussian quadrature methods.

A large part of our discussion of Newton–Cotes and Gaussian quadrature methods will involve a study of the errors involved; this will allow us to introduce *adaptive integration* (section 7.3), wherein one quantifies the error in the integration even when the analytical answer is not known (which in practice is usually the case). A similar idea is involved in *Romberg integration* (section 7.4), which combines a low-order approach with Richardson extrapolation, to produce a more accurate answer. The second half of the chapter will focus on *multidimensional integration*, for which Newton–Cotes and even Gaussian quadrature methods are typically too inefficient. This will allow us to introduce the important subject of *Monte Carlo integration*, also known as *stochastic integration*, which employs random numbers; this is also used in the Project that rounds out this chapter.

You may be wondering why the present chapter, on integration, comes much later in the book than chapter 3 on numerical differentiation. The answer is that we will be using a lot of the machinery developed in earlier chapters, most notably our infrastructure on interpolation from chapter 6 and on root-finding from chapter 5. Since this is the penultimate chapter of the book, we assume the reader has developed some maturity from working through the earlier material. Similarly, our codes will use `numpy` functionality repeatedly, which we had opted against in chapter 3.

7.2 Newton–Cotes Methods

As already noted, Newton–Cotes methods evaluate the integral as a sum of elementary areas (rectangles, trapezoids, etc.). In this book, we use Newton–Cotes methods that employ an equally spaced grid: this is similar (nay, identical) to what we saw in chapter 3, when we examined derivatives for points on a grid in section 3.3.6. As a reminder, the assumption is that we have access to a set of n discrete data points (i.e., a table) of the form $(x_i, f(x_i))$ for $i = 0, 1, \dots, n-1$. The points x_i are on an equally spaced grid, from a to b. The n points then are given by the following relation:

$$x_i = a + ih \tag{7.8}$$

where, as usual, $i = 0, 1, \dots, n-1$. The h is clearly given by:

$$h = \frac{b-a}{n-1} \tag{7.9}$$

Recall that we are dealing with n points in total, so we are faced with $n-1$ subintervals from a to b. As always, we are using the terms *subinterval* and *panel* interchangeably. (A final reminder, which will come in handy in sections 7.3 and 7.4: you may want to introduce a new variable containing the number of panels, $N = n - 1$, in which case $h = (b - a)/N$; then, the i in x_i would go from 0 to N, since we have $N + 1 = n$ points in total.)

For each of the several methods covered here, we will first start out with a version of the answer for a small problem (e.g., for one panel, an approximation to the integral $\int_{x_i}^{x_{i+1}} f(x)dx$). This includes a graphic interpretation as well as an appropriately motivated formula. In each case, we then turn to what is known as the *composite* formula, which is nothing other than a sum of all the "small-problem" answers like $\int_{x_i}^{x_{i+1}} f(x)dx$ and is thereby an approximation to the integral $\int_a^b f(x)dx$. That is then followed by an expression for the approximation error made by employing that specific composite formula. In three cases (rectangle, trapezoid, Simpson's methods) we also provide a derivation of the expression for the error as well as a full Python implementation.

7.2.1 Rectangle Rule

We start with the one-panel version of the rectangle rule; it may help to periodically look at Fig. 7.2 while reading. For the one-panel version, we are interested in approximating $\int_{x_i}^{x_{i+1}} f(x)dx$. The *rectangle rule* makes the simplest assumption possible, namely that the area under $f(x)$ from x_i to x_{i+1} can be approximated by the area of a rectangle, with width h (the distance from x_i to x_{i+1}) and height given by the value of $f(x)$ either at x_i or at x_{i+1}. An equivalent way of seeing this is that the rectangle rule approximates $f(x)$ as a *constant* from x_i to x_{i+1}, namely a straight (horizontal) line: this means that instead of evaluating the area under the curve it evaluates the area under that straight line.

Let's provide a formula for this before we elaborate further. The analytical expression for the *one-panel version of the rectangle rule* is simply:

$$\int_{x_i}^{x_{i+1}} f(x)dx \approx hf(x_i) \tag{7.10}$$

As noted, we've taken the distance from x_i to x_{i+1} to be fixed, given by Eq. (7.9).

Implicit both in the figure and in the formula above is an assumption, which is easier to elucidate in a specific case. Take the figure, which illustrates the case of five points (and therefore four panels), namely $n = 5$, and focus on any one of the rectangles. Observe we have determined the height of the rectangle as the value of $f(x)$ at the left abscissa, namely $f(x_i)$. This can be referred to as the *left-hand rectangle rule*. We could just as easily have taken the height of the rectangle as the value of $f(x)$ at the right abscissa, namely $f(x_{i+1})$, giving rise to the right-hand rectangle rule.

We now turn to the approximation for the total integral from a to b, namely $\int_a^b f(x)dx$. This is nothing other than the sum of all the one-panel rectangle areas, giving rise to the *composite version of the rectangle rule*:

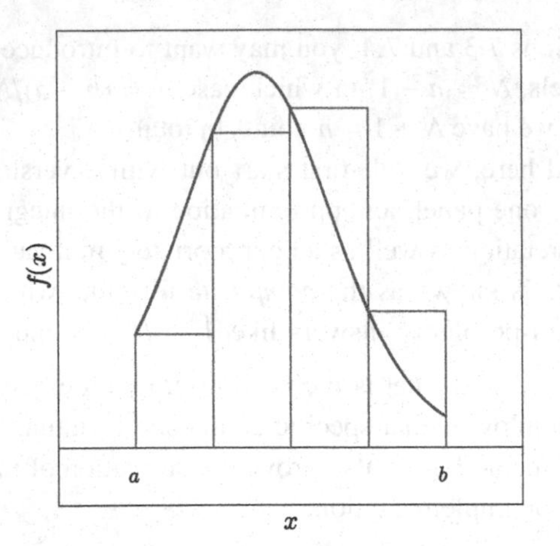

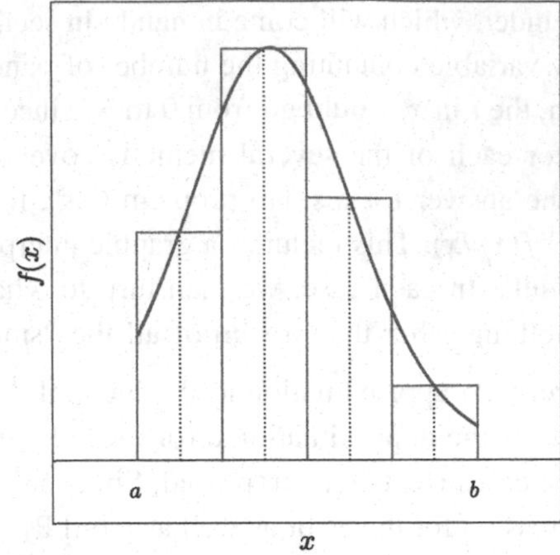

Fig. 7.2 Composite version of rectangle rule (left) and midpoint rule (right)

$$\int_a^b f(x)dx = \sum_{i=0}^{n-2} \int_{x_i}^{x_{i+1}} f(x)dx$$
$$\approx hf(x_0) + hf(x_1) + hf(x_2) + \cdots + hf(x_{n-2}) \quad (7.11)$$

The first line is simply taking the sum and the second line has plugged in the one-panel expression from Eq. (7.10). We can translate this result into the language of weights c_i from Eq. (7.7). The answer is $c_i = h\{1, 1, \ldots, 1, 0\}$, namely, the weights are all equal to h, except for the endpoint at b, where the weight is zero: the rectangle formula is half-open, as it doesn't include one of the two endpoints.[2]

The result in the second line of Eq. (7.11) brings to mind the definition of the Riemann integral: the difference is that here we are taking h as fixed by n, i.e., we are not taking the limit of $h \to 0$ (even if we later try out a larger n, we will never take the actual limit). This is similar to what we saw in section 3.3 on derivatives: the simplest finite-difference formula is just the definition of the derivative, without the limit. One main difference, however, is that in that case making the h smaller led to problems (due to catastrophic cancellation), whereas this won't be an issue for integration.

7.2.1.1 Error Analysis

We start from a discussion of the approximation error in the one-panel version of the rectangle formula, Eq. (7.10). Take the Taylor expansion of $f(x)$ around x_i:

$$f(x) = f(x_i) + (x - x_i)f'(x_i) + \cdots \quad (7.12)$$

[2] Note that the sum in Eq. (7.7) goes from 0 to $n - 1$ whereas the sum in Eq. (7.11) goes from 0 to $n - 2$.

If we stop at first order, we can rewrite this expression as:

$$f(x) = f(x_i) + (x - x_i)f'(\xi_i) \tag{7.13}$$

where, as usual, ξ_i is a point between x_i and x.

To belabor the obvious: in the previous subsection, when introducing the rectangle rule, we were dealing with a constant approximation to $f(x)$, *not* with a Taylor series. Thus, we then took it for granted that $f(x)$ could be approximated as being a constant from x_i to x_{i+1}, whereas now we are Taylor expanding the actual function $f(x)$, which in general is much more complicated than a simple constant (this is why in what follows we carry around a first derivative, which in general is non-zero).

We now integrate this Taylor series from x_i to x_{i+1} to find:

$$\int_{x_i}^{x_{i+1}} f(x)dx = \int_{x_i}^{x_{i+1}} dx\,[f(x_i) + (x - x_i)f'(\xi_i)] \tag{7.14}$$

where the last term will turn out to be the error term for us. This integral is easy enough that it can be evaluated by hand, but in the interest of establishing the notation for coming sections, we define a helper variable:

$$u = \frac{x - x_i}{h} \tag{7.15}$$

Expressed in terms of u, the integral from x_i to x_{i+1} becomes:

$$\int_{x_i}^{x_{i+1}} f(x)dx = h\int_0^1 du\,[f(x_i) + huf'(\xi_i)] = hf(x_i) + \frac{1}{2}h^2 f'(\xi_i) \tag{7.16}$$

In the second equality we evaluated the integral over u. Comparing this result with the one-panel version of the rectangle formula, Eq. (7.10), we find that the *absolute error in the one-panel rectangle formula* is:

$$\mathcal{E}_i = \frac{1}{2}h^2 f'(\xi_i) \tag{7.17}$$

where we introduced the notation $\mathcal{E}_i$ for the error incurred by approximating one panel.[3]

We now turn to a discussion of the approximation error in the composite version of the rectangle formula, Eq. (7.11). Just like in the first line of Eq. (7.11), we will evaluate the absolute error for the full interval by summing up all the subinterval contributions:

$$\begin{aligned}\mathcal{E} &= \sum_{i=0}^{n-2} \mathcal{E}_i = \frac{1}{2}h^2\left(f'(\xi_0) + f'(\xi_1) + f'(\xi_2) + \cdots + f'(\xi_{n-2})\right) \\ &= \frac{n-1}{2}h^2\left(\frac{f'(\xi_0) + f'(\xi_1) + f'(\xi_2) + \cdots + f'(\xi_{n-2})}{n-1}\right) = \frac{n-1}{2}h^2 f'(\xi)\end{aligned} \tag{7.18}$$

If you're uncomfortable with the first equality, you could instead take Eq. (7.16) and sum all

[3] In chapter 2 we defined the absolute error as "approximate minus exact", $\Delta x = \tilde{x} - x$, see Eq. (2.5). Our definition of the absolute error in the present chapter differs by an overall minus sign, which is why we're using a different symbol, $\mathcal{E}$, to denote it (section 6.6's $\mathscr{E}_j$ constitutes a third case, closer to chapter 2's convention).

the panels separately: this will clearly show that the subinterval error contributions simply add up. In the second equality we plugged in the result for $\mathcal{E}_i$ from Eq. (7.17). In the third equality we multiplied and divided by $n-1$, which is the number of terms in the sum. In the fourth equality we identified the term in the parentheses as the arithmetic mean and remembered from elementary calculus that there exists a ξ (from a to b) for which $f'(\xi)$ is equal to the arithmetic mean of f'. Combining the last result with our definition of h in Eq. (7.9), recast as $(n-1)h = b-a$, we find our final result for the *absolute error in the composite rectangle formula*:

$$\mathcal{E} = \frac{b-a}{2} h f'(\xi) \tag{7.19}$$

A point that will re-emerge below: it is wrong to infer from Eq. (7.19) that $\mathcal{E} = ch$ where c is a constant, since $f'(\xi)$ is actually not independent of h (this will become crystal clear in section 7.2.4). If we *do* assume that c is a constant, then we will draw the conclusion that the error in the composite rule is $O(h)$: this is actually correct, if you interpret it as giving the *leading error*. Another pattern that will re-emerge below: the leading error in the composite rule ($O(h)$ here) is worse by one degree compared to the elementary-interval error ($O(h^2)$ here); we say "worse" because for small h we have $h^2 < h$.

7.2.2 Midpoint Rule

We now briefly turn to an improved rule, which is surprisingly similar to the rectangle rule. As expected, we start with the one-panel version; it may help to keep the right panel of Fig. 7.2 in mind. The *midpoint rule* makes the same assumption as the rectangle rule, namely that the area under $f(x)$ from x_i to x_{i+1} can be approximated by the area of a rectangle, with width h (the distance from x_i to x_{i+1}) and height given by the value of $f(x)$. The only difference is that the midpoint rule uses the value of $f(x)$ not at the left or at the right, but at the *midpoint* of the panel. In other words, just like for the rectangle rule, the midpoint rule approximates $f(x)$ as a *constant* from x_i to x_{i+1}, namely a straight (horizontal) line: instead of evaluating the area under the curve it evaluates the area under that straight line. Again, not a great assumption but, as we'll soon see, one that works much better than the rectangle rule. The analytical expression for the *one-panel version of the midpoint rule* is:

$$\int_{x_i}^{x_{i+1}} f(x)dx \approx hf\left(x_i + \frac{h}{2}\right) \tag{7.20}$$

Thinking about the composite version of the midpoint rule, we realize there's a problem here: we would need $f(x)$ evaluated at $x_i + h/2$; if all we have is a table of the form $(x_i, f(x_i))$ for $i = 0, 1, \ldots, n-1$, then we simply don't have access to those function values. This is another way of saying that we cannot really translate our result into the language of weights c_i from Eq. (7.7), since here we are not evaluating the function at the grid points. Incidentally, you should have realized by now that the midpoint formula is open, since it doesn't include either of the two endpoints. Problem 7.2 guides you toward the following approximation error in the one-panel version of the midpoint formula:

$$\mathcal{E}_i = \frac{1}{24}h^3 f''(\xi_i) \quad (7.21)$$

Similarly, adding up all the one-panel errors gives us the final result for the *absolute error in the composite midpoint formula*:

$$\mathcal{E} = \frac{b-a}{24}h^2 f''(\xi) \quad (7.22)$$

7.2.3 Integration from Interpolation

Both the rectangle rule and the midpoint rule approximated the area under $f(x)$ from x_i to x_{i+1} in the simplest way possible, namely by the area of a rectangle. The natural next step is to assume that the function is *not* approximated by a constant, i.e., a horizontal line, from x_i to x_{i+1}, but by a straight line, a quadratic, a cubic, and so on. In other words, we see the problem of *interpolation*, studied extensively in chapter 6, emerge here.

It should be straightforward to see that one panel, made up of two consecutive abscissas, is enough to define a general straight line (i.e., not necessarily a flat, horizontal line). Similarly, two panels, made up of three consecutive abscissas, can "anchor" a quadratic, and so on. This is precisely the problem Lagrange interpolation solved, see section 6.2.2. In the language of the problem we are now faced with, we have as input a table of q data points $(x_{i+j}, f(x_{i+j}))$ for $j = 0, 1, \dots, q-1$ and wish to find the interpolating polynomial that goes through them. For $q = 2$ we get a straight line, for $q = 3$ a quadratic, and so on. Make sure you keep the notation straight: $q = 3$ leads to the three abscissas x_i, x_{i+1}, and x_{i+2}.

Thus, for a given approach the *elementary interval* will depend on the value of q: for $q = 2$ the elementary interval has a width of one panel, for $q = 3$ of two panels, for $q = 4$ of three panels, and so on.[4] More generally, for the case of q points in the elementary interval we wish to approximate the integral:

$$\int_{x_i}^{x_{i+q-1}} f(x)dx \quad (7.23)$$

The way we do this in general is to employ an interpolating polynomial as per Eq. (6.22):

$$p(x) = \sum_{j=0}^{q-1} f(x_{i+j})L_{i+j}(x) \quad (7.24)$$

[4] We use q for an elementary interval and n, as before, for the composite case, i.e., for the full interval.

Since our nodes go from x_i to x_{i+q-1}, the cardinal polynomials $L_{i+j}(x)$ take the form:

$$L_{i+j}(x) = \frac{\prod_{k=0,k\neq j}^{q-1}(x - x_{i+k})}{\prod_{k=0,k\neq j}^{q-1}(x_{i+j} - x_{i+k})}, \qquad j = 0, 1, \ldots, q-1 \tag{7.25}$$

which is simply a translation of Eq. (6.19) into our present notation. Once again, make sure to disentangle the notation: for $q = 4$ we have $L_i(x)$, $L_{i+1}(x)$, $L_{i+2}(x)$, $L_{i+3}(x)$, each of which is a cubic polynomial. The points they are interpolating over are x_i, x_{i+1}, x_{i+2}, and x_{i+3}.

Using $p(x)$, Newton–Cotes methods in an elementary interval are cast as:

$$\int_{x_i}^{x_{i+q-1}} f(x)dx \approx \int_{x_i}^{x_{i+q-1}} p(x)dx = \sum_{j=0}^{q-1}\left(f(x_{i+j})\int_{x_i}^{x_{i+q-1}} L_{i+j}(x)dx\right) = \sum_{j=0}^{q-1} w_{i+j}f(x_{i+j}) \tag{7.26}$$

We plugged in Eq. (7.24) and then defined the *weights for the elementary interval*:

$$w_{i+j} = \int_{x_i}^{x_{i+q-1}} L_{i+j}(x)dx \tag{7.27}$$

The crucial point is that these weights depend *only* on the cardinal polynomials, *not* on the function $f(x)$ that is being integrated. Thus, for a given q, implying an elementary interval with a width of $q-1$ panels, these weights can be evaluated once and for all, and employed to integrate any function you wish, after the fact.

Before seeing the details worked out, let us make a general comment: by focusing on an elementary interval and employing a low-degree polynomial in it, in essence what we're doing in Eq. (7.26) is *piecewise polynomial interpolation and then integration of each interpolant*. This is completely analogous to what we did in section 6.3. Interpolation on an equally spaced grid suffers from serious issues, but focusing on a few points at a time and integrating in elementary intervals mostly gets rid of these problems.

7.2.4 Trapezoid Rule

It may help to look at the left panel of Fig. 7.3 in what follows, but keep in mind that it is showing the composite version of the trapezoid rule, whereas we will, as before, first discuss our new rule in its elementary-interval version.

The previous discussion may be too abstract, so let us make it tangible. We start from the case of $q = 2$, giving rise to what is known as the *trapezoid rule*.[5] We have the two

[5] Strangely enough, this is typically called the *trapezoidal* rule. But we don't say *rectangular* rule, do we? More importantly, the rule itself is most certainly *not* shaped like a trapezoid. How could it? It's a rule. Note that we're not being pedantic here. A purist would point out that Proclus introduced the term *trapezoid* to denote a quadrilateral figure no two of whose sides are parallel. The same purist would insist that this be called the *trapezium* rule and would cringe were anyone to refer to it as the *trapeziform* rule.

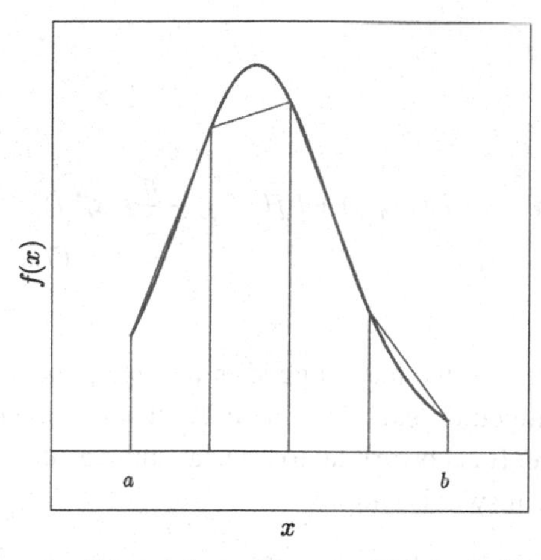

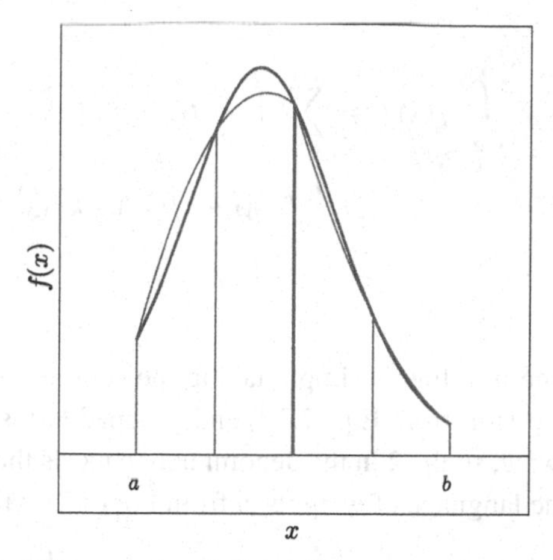

Fig. 7.3 Composite version of trapezoid rule (left) and Simpson's rule (right)

points $(x_i, f(x_i))$ and $(x_{i+1}, f(x_{i+1}))$ and the two cardinal polynomials from Eq. (7.25):

$$L_i(x) = \frac{x - x_{i+1}}{x_i - x_{i+1}} = -\frac{x - x_{i+1}}{h}, \qquad L_{i+1}(x) = \frac{x - x_i}{x_{i+1} - x_i} = \frac{x - x_i}{h} \tag{7.28}$$

We can immediately evaluate the two elementary weights from Eq. (7.27):

$$\begin{aligned} w_i &= \int_{x_i}^{x_{i+1}} L_i(x)dx = -\frac{1}{h}\int_{x_i}^{x_{i+1}} (x - x_{i+1})\, dx = \frac{h}{2} \\ w_{i+1} &= \int_{x_i}^{x_{i+1}} L_{i+1}(x)dx = \frac{1}{h}\int_{x_i}^{x_{i+1}} (x - x_i)\, dx = \frac{h}{2} = w_i \end{aligned} \tag{7.29}$$

where we used the fact that $x_{i+1} = x_i + h$. It's worth noting at this point that the sum of weights in the one-panel version of the trapezoid rule is h ($h/2$ for the point at x_i and $h/2$ for the point at x_{i+1}). This is actually a general result, which you will show in problem 7.3: *for any Newton–Cotes integration rule, the sum of all the weights is equal to the width of the elementary interval.* Here the elementary interval is simply one panel, so the sum of the weights is h. We will see more examples of this below.

Taking these elementary weights and plugging them into Eq. (7.26) leads to the *one-panel version of the trapezoid rule*:

$$\int_{x_i}^{x_{i+1}} f(x)dx \approx \frac{h}{2}[f(x_i) + f(x_{i+1})] \tag{7.30}$$

which merely reiterates the fact that each of the two points making up our elementary interval gets the same weight, $h/2$.

We now turn to the approximation for the total integral from a to b, namely $\int_a^b f(x)dx$. This is nothing other than the sum of all the one-panel trapezoid areas, giving rise to the *composite version of the trapezoid rule*:

$$\int_a^b f(x)dx = \sum_{i=0}^{n-2} \int_{x_i}^{x_{i+1}} f(x)dx$$
$$\approx \frac{h}{2}f(x_0) + hf(x_1) + hf(x_2) + \cdots + hf(x_{n-3}) + hf(x_{n-2}) + \frac{h}{2}f(x_{n-1}) \tag{7.31}$$

The first line is simply taking the sum; the second line has plugged in the one-panel expression from Eq. (7.30) and grouped terms together: each intermediate term is counted twice, so the 2 in the denominator cancels there. It is now trivial to translate this result into the language of weights c_i from Eq. (7.7). The answer is simply:

$$c_i = h\left\{\frac{1}{2}, 1, \ldots, 1, \frac{1}{2}\right\} \tag{7.32}$$

namely, the weights are all equal to h, except for the endpoints, where the weight is $h/2$; the trapezoid formula is closed, meaning that it includes both endpoints.[6] Note, finally, that we have been careful to use different symbols for the weights in an elementary interval and for the overall Newton–Cotes weights: the former are denoted by w_{i+j} and the latter by c_i.

Just to avoid confusion, we observe that the trapezoid rule may approximate a function by a sequence of straight lines, but that's not something you need to worry about when using the rule: Eq. (7.31) *is* the composite trapezoid rule so you could, if you want, forget about where it came from. (Remember: the composite rule is made up of many trapezoids, whereas the one-panel version, of a single trapezoid.)

7.2.4.1 Error Analysis

In problem 7.5 you are asked to carry out an error analysis for the trapezoid rule by generalizing what we did for the rectangle rule in an earlier section. However, since we just introduced the trapezoid rule using the Lagrange-interpolation machinery, it makes sense to turn to the latter for guidance on the error behavior.

As you may recall, in section 6.2.2.4 we derived a general error formula for polynomial interpolation, Eq. (6.38). For the present case of $q = 2$ this takes the form:

$$f(x) - p(x) = \frac{1}{2!}f''(\zeta_i)(x - x_i)(x - x_{i+1}) \tag{7.33}$$

after slightly updating the notation. Integrating this difference from x_i to x_{i+1} will then give us the error in the one-panel version of the trapezoid rule. We have:

$$\mathcal{E}_i = \int_{x_i}^{x_{i+1}} \left(\frac{1}{2}f''(\zeta_i)(x - x_i)(x - x_{i+1})\right)dx = \frac{1}{2}f''(\xi_i)\int_{x_i}^{x_{i+1}} (x - x_i)(x - x_{i+1})dx \tag{7.34}$$

This might seem like an improper thing to do: after all, the ξ in Eq. (6.38) implicitly

[6] Note that the sum in Eq. (7.7) goes from 0 to $n - 1$ whereas the sum in the first line of Eq. (7.31) goes from 0 to $n - 2$, but each term in that sum includes both $f(x_i)$ and $f(x_{i+1})$, as per Eq. (7.30).

depended on the placement of the nodes. However, you can use the extended version of the integral mean-value theorem: this is justified because $(x - x_i)(x - x_{i+1})$ has a constant sign in our interval. If you then do the integral in the last step and, once again, employ the fact that $x_{i+1} = x_i + h$, you find the following *absolute error in the one-panel trapezoid formula*:

$$\mathcal{E}_i = -\frac{1}{12}h^3 f''(\xi_i) \tag{7.35}$$

Thus, the leading term in the error for the one-panel case is $O(h^3)$.

We now turn to a discussion of the approximation error in the composite version of the trapezoid formula, Eq. (7.31). Just like in the first line of Eq. (7.31), we will evaluate the absolute error for the full interval by summing up all the subinterval contributions:

$$\begin{aligned}\mathcal{E} = \sum_{i=0}^{n-2} \mathcal{E}_i &= -\frac{1}{12}h^3 \left(f''(\xi_0) + f''(\xi_1) + f''(\xi_2) + \cdots + f''(\xi_{n-2})\right) \\ &= -\frac{n-1}{12}h^3 \left(\frac{f''(\xi_0) + f''(\xi_1) + f''(\xi_2) + \cdots + f''(\xi_{n-2})}{n-1}\right) = -\frac{n-1}{12}h^3 f''(\xi)\end{aligned} \tag{7.36}$$

In the second equality we simply plugged in the result for $\mathcal{E}_i$ from Eq. (7.35). In the third equality we multiplied and divided by $n - 1$, which is the number of terms in the sum. In the fourth equality we identified a ξ (from a to b) for which $f''(\xi)$ is equal to the arithmetic mean of f''. Combining the last line with our definition of h in Eq. (7.9), we find our final result for the *absolute error in the composite trapezoid formula*:

$$\mathcal{E} = -\frac{b-a}{12}h^2 f''(\xi) \tag{7.37}$$

Observe that this error contains a second derivative, so *for polynomials of up to first degree, the composite trapezoid rule is exact.* Since we derived the trapezoid rule by approximating $f(x)$ by straight lines, this isn't really surprising. This approximation error, $O(h^2)$, is of the same order as the approximation error of the midpoint rule, Eq. (7.22). As a matter of fact, in absolute value, the error in the midpoint rule, Eq. (7.22), is smaller by a factor of 2 than the error in the trapezoid rule, Eq. (7.37). This is quite interesting: the midpoint rule uses a cruder shape (a rectangle) than the trapezoid one, but manages to be more accurate, by using the midpoint instead of the endpoints of each panel. Of course, that is at the price of not employing the function values at the grid points, $f(x_i)$.

7.2.4.2 Beyond the Leading Error

It is wrong to infer from Eq. (7.37) that $\mathcal{E} = ch^2$ where c is a constant, since $f''(\xi)$ is actually not independent of h. The question then arises if we can do better, i.e., if we can quantify the error to higher orders. This will turn out to be a very fruitful avenue: section 7.4

will essentially rely on the behavior of the trapezoid error beyond the leading term. Thus, it is important to investigate the trapezoid-rule error in more detail.

Help comes from a (somewhat) unexpected place: in problem 2.21 we introduced Bernoulli numbers and the *Euler–Maclaurin summation formula*, Eq. (2.121):[7]

$$\int_0^1 g(t)dt = \frac{g(0)}{2} + \frac{g(1)}{2} - \sum_{j=1}^{m} \frac{1}{(2j)!} B_{2j} \left[g^{(2j-1)}(1) - g^{(2j-1)}(0) \right] + R \tag{7.38}$$

where the B_{2j}'s are Bernoulli numbers and the R is the higher-order error term, given in terms of an integral over a Bernoulli polynomial. We wish to use this equation to get guidance on the behavior of the trapezoid rule. With that in mind, we make the transformation $x = x_i + ht$ and take $g(t) = f(x) = f(x_i + ht)$. From this definition, together with the chain rule, we find for the first derivative $g'(t) = f'(x)h$, and similarly $g^{(k)}(t) = f^{(k)}(x)h^k$ for higher derivatives. Putting everything together, the one-panel integral of $f(x)$ becomes:

$$\int_{x_i}^{x_{i+1}} f(x)dx = h \int_0^1 g(t)dt = \frac{h}{2}(f(x_i)+f(x_{i+1})) - \sum_{j=1}^{m} \frac{1}{(2j)!} B_{2j} [f^{(2j-1)}(x_{i+1}) - f^{(2j-1)}(x_i)] h^{2j} \tag{7.39}$$

where we dropped the R term, since we're only interested in establishing the order pattern beyond the leading term. The first term on the right-hand side already matches the one-panel version of the trapezoid rule, Eq. (7.30). This means that we've already computed the one-panel error, $\mathcal{E}_i$, beyond leading order.

Instead of focusing on the one-panel result, we will press on: adding together all the one-panel expressions, as in Eq. (7.31) above, we find:

$$\begin{aligned}\int_a^b f(x)dx &= \frac{h}{2} f(x_0) + hf(x_1) + hf(x_2) + \cdots + hf(x_{n-3}) + hf(x_{n-2}) + \frac{h}{2} f(x_{n-1}) \\ &\quad - \sum_{j=1}^{m} \frac{1}{(2j)!} B_{2j} \left[f^{(2j-1)}(b) - f^{(2j-1)}(a) \right] h^{2j} \end{aligned} \tag{7.40}$$

where on the first line we identify the trapezoid-rule approximation to the integral; on the second line we are happy to see the "intermediate" series all cancel; this is a version of a telescoping series,[8] only this time each term that cancelled was a sum over j. In the end, we're left with a single sum over derivative values at the *endpoints* of our integral, only.

Plugging in the values of the first few Bernoulli numbers, which you computed in problem 2.21, the *absolute error in the trapezoid rule to all orders* takes the form:

$$\begin{aligned}\mathcal{E} &= -\frac{h^2}{12} [f'(b) - f'(a)] + \frac{h^4}{720} [f'''(b) - f'''(a)] \\ &\quad - \frac{h^6}{30\,240} \left[f^{(5)}(b) - f^{(5)}(a) \right] + \frac{h^8}{1\,209\,600} \left[f^{(7)}(b) - f^{(7)}(a) \right] + \cdots \end{aligned} \tag{7.41}$$

[7] If you didn't solve that problem, and therefore would prefer not to rely on it, you can solve problem 7.6, which guides you toward proving our main result in Eq. (7.41) "by hand".

[8] We encountered another telescoping series in problem 4.9; there's one more waiting for us in problem 7.19.

This is a significant formula, which tells us several things. First, the error of the trapezoid rule contains only *even* powers in h and, second, that the coefficient of each h^{2j} term depends only on a *difference* of (odd-order) derivative values at the *endpoints*. The latter point means that the error is not impacted by the behavior of the derivatives at intermediate points. We will return to the first point when we introduce Romberg integration in section 7.4 and to the second point in our discussion of analytical features in section 7.6.

7.2.5 Simpson's Rule

As you may have guessed, *Simpson's rule* is the natural continuation of the Lagrange-interpolation process we saw above; it uses $q = 3$ in an elementary interval, namely the three abscissas x_i, x_{i+1}, and x_{i+2}. The trapezoid rule fit a straight line through two points, so Simpson's rule fits a quadratic through three points (making up two panels); in problem 7.4 you are asked to employ "naive", i.e., monomial, interpolation, but here we will capitalize on the infrastructure developed in chapter 6. It may help to look at the right panel of Fig. 7.3 from here onward; this shows the case of four panels, i.e., two elementary intervals each of which has width $2h$.

We are dealing with three points $(x_i, f(x_i))$, $(x_{i+1}, f(x_{i+1}))$, and $(x_{i+2}, f(x_{i+2}))$, and three cardinal polynomials from Eq. (7.25):

$$\begin{aligned}
L_i(x) &= \frac{(x - x_{i+1})(x - x_{i+2})}{(x_i - x_{i+1})(x_i - x_{i+2})} \\
L_{i+1}(x) &= \frac{(x - x_i)(x - x_{i+2})}{(x_{i+1} - x_i)(x_{i+1} - x_{i+2})} \\
L_{i+2}(x) &= \frac{(x - x_i)(x - x_{i+1})}{(x_{i+2} - x_i)(x_{i+2} - x_{i+1})}
\end{aligned} \tag{7.42}$$

Each of these is a quadratic polynomial. We would now like to compute the three elementary weights; as you can see from Eq. (7.27), this will require integrating from x_i to x_{i+2}, i.e., in an elementary interval. In order to streamline the evaluation of these integrals, we define a helper variable:[9]

$$u = \frac{x - x_{i+1}}{h} \tag{7.43}$$

and notice that the terms in the denominators in Eq. (7.42) are simply $-h$, $-2h$, h, and so on. Then, the elementary weights from Eq. (7.27) are:

$$\begin{aligned}
w_i &= \int_{x_i}^{x_{i+2}} L_i(x)dx = \frac{h}{2}\int_{-1}^{1} du\, u(u-1) = \frac{h}{3} \\
w_{i+1} &= \int_{x_i}^{x_{i+2}} L_{i+1}(x)dx = -h\int_{-1}^{1} du\, (u+1)(u-1) = \frac{4h}{3} \\
w_{i+2} &= \int_{x_i}^{x_{i+2}} L_{i+2}(x)dx = \frac{h}{2}\int_{-1}^{1} du\, (u+1)u = \frac{h}{3}
\end{aligned} \tag{7.44}$$

[9] This is similar to what we did when deriving the leading error in the rectangle rule, Eq. (7.15), only this time we employ the midpoint of our elementary interval in the numerator, x_{i+1}.

Taking these elementary weights and plugging them into Eq. (7.26) leads to the *two-panel version of Simpson's rule*:

$$\int_{x_i}^{x_{i+2}} f(x)dx \approx \frac{h}{3}\left[f(x_i) + 4f(x_{i+1}) + f(x_{i+2})\right] \tag{7.45}$$

As we observed when introducing the one-panel version of the trapezoid rule in Eq. (7.30), we notice that the sum of all the weights in this elementary interval is $(1 + 4 + 1)h/3 = 2h$, consistent with the fact that the elementary interval in this case has width $2h$.

We now turn to the approximation for the total integral from a to b, namely $\int_a^b f(x)dx$. This is nothing other than the sum of all the two-panel elementary areas, giving rise to the *composite version of Simpson's rule*:

$$\begin{aligned}
\int_a^b f(x)dx &= \sum_{i=0,2,4,\ldots}^{n-3} \int_{x_i}^{x_{i+2}} f(x)dx \\
&\approx \frac{h}{3}f(x_0) + \frac{4h}{3}f(x_1) + \frac{2h}{3}f(x_2) + \cdots + \frac{2h}{3}f(x_{n-3}) + \frac{4h}{3}f(x_{n-2}) + \frac{h}{3}f(x_{n-1})
\end{aligned} \tag{7.46}$$

The first equality is simply taking the sum, but this time we're careful since we're moving in steps of $2h$: this is why the sum over i goes in steps of 2 and ends at $n - 3$ (so the last x_{i+2} can be x_{n-1}, which is our last point). The second equality has plugged in the two-panel expression from Eq. (7.45) and grouped terms together: each intermediate term is counted twice, so we go from $h/3$ to $2h/3$ for those points. It is now trivial to translate this result into the language of weights c_i from Eq. (7.7). The answer is simply:

$$c_i = \frac{h}{3}\{1, 4, 2, 4\ldots, 2, 4, 1\} \tag{7.47}$$

namely, the weights for the two endpoints are $h/3$, for each "middle" point the weight is $4h/3$, whereas the intermediate/matching points get a weight of $2h/3$ as already mentioned. Obviously, Simpson's formula is closed, as it includes both endpoints.

Here's a very important point that was implicit up to here: to use Simpson's rule, you have to use an even number of panels. This is because the "elementary interval" version gives us the two-panel answer: $\int_{x_i}^{x_{i+2}} f(x)dx$. We actually took this fact for granted in the right panel of Fig. 7.3, which shows the case of $n = 5$ points and therefore four panels. We used $n = 5$ in the earlier methods, too, but there this was by choice, since those methods could use an even or an odd number of panels and work just fine. To summarize this significant point, *Simpson's rule requires an even number of panels and therefore an odd number of points n*.

7.2.5.1 Error Analysis

The natural next step would be to generalize our error analysis for the trapezoid rule, leading up to Eq. (7.35), for the case of Simpson's rule. You are guided toward such an analysis using the general interpolation error formula, Eq. (6.38), in problem 7.7. Here we will follow a different route: we will carry out the error analysis for Simpson's rule by employing a Taylor-series approach, similarly to what we did for the rectangle rule starting in Eq. (7.12). Since Simpson's rule is a more accurate method, we'll have to keep more terms in the Taylor series expansion this time.

As usual, we start from the elementary interval, corresponding to the two-panel version of Simpson's rule in Eq. (7.45). Taylor expanding $f(x)$ around x_{i+1} gives us:

$$f(x) = f(x_{i+1}) + (x - x_{i+1})f'(x_{i+1}) + \frac{1}{2}(x - x_{i+1})^2 f''(x_{i+1}) + \frac{1}{6}(x - x_{i+1})^3 f'''(x_{i+1}) + \frac{1}{24}(x - x_{i+1})^4 f^{(4)}(\xi_{i+1}) \tag{7.48}$$

where we have stopped at fourth order; as usual, ξ_{i+1} is a point between x_{i+1} and x (we're calling it ξ_{i+1}, instead of ξ_i, for purely aesthetic reasons). We wish to integrate this Taylor series from x_i to x_{i+2}: to make this calculation easier, we employ u, the helper variable from Eq. (7.43). Expressed in terms of u, the integral from x_i to x_{i+2} becomes:

$$\begin{aligned}\int_{x_i}^{x_{i+2}} f(x)dx &= h\int_{-1}^{1} du\Big[f(x_{i+1}) + huf'(x_{i+1}) + \frac{1}{2}h^2u^2 f''(x_{i+1}) \\ &\quad + \frac{1}{6}h^3u^3 f'''(x_{i+1}) + \frac{1}{24}h^4u^4 f^{(4)}(\xi_{i+1})\Big] \\ &= 2hf(x_{i+1}) + \frac{1}{3}h^3 f''(x_{i+1}) + \frac{1}{60}h^5 f^{(4)}(\xi_{i+1})\end{aligned} \tag{7.49}$$

where, crucially, the odd-derivative terms cancelled. We're not done yet. The first term on the last line of our result contains $2hf(x_{i+1})$, which looks different from the two-panel version of Simpson's formula, Eq. (7.45). In order to make further progress, we will plug in an approximation for the second derivative term $f''(x_{i+1})$. We get this from Eq. (3.39):

$$f''(x_{i+1}) = \frac{f(x_i) + f(x_{i+2}) - 2f(x_{i+1})}{h^2} - \frac{h^2}{12}f^{(4)}(\xi_{i+1}) \tag{7.50}$$

where we assumed $x = x_{i+1}$ and took the liberty of writing $f(x_{i+1})$ instead of $f(x_i+h)$—and $f(x_{i+2})$ instead of $f(x_i + 2h)$. We also dropped all higher-order terms (at the cost of using ξ_{i+1} in the last term): the actual error term came from Eq. (3.31) with $h \to 2h$.[10]

Now, plugging our result for $f''(x_{i+1})$ from Eq. (7.50) into our expression for the two-panel integral, Eq. (7.49), we find:

$$\int_{x_i}^{x_{i+2}} f(x)dx = 2hf(x_{i+1}) + \frac{1}{3}h^3\left[\frac{f(x_i) + f(x_{i+2}) - 2f(x_{i+1})}{h^2} - \frac{h^2}{12}f^{(4)}(\xi_{i+1})\right] + \frac{1}{60}h^5 f^{(4)}(\xi_{i+1})$$

[10] Strictly speaking, this ξ_{i+1} is different from the one in the Taylor expansion Eq. (7.49). However, even if we used a different symbol for it now, the end result would be the same, so we don't bother.

$$= \frac{h}{3}[f(x_i) + 4f(x_{i+1}) + f(x_{i+2})] - \frac{1}{90}h^5 f^{(4)}(\xi_{i+1}) \tag{7.51}$$

If you're troubled by this, you should observe that the next term in the finite-difference expansion would be proportional to h^4 which, together with the h^3 outside the brackets would give a h^7; we drop this, since it's of higher order than our leading error.

We're now happy to see that this result loooks exactly like the two-panel version of Simpson's formula, Eq. (7.45), with the addition of the *absolute error in the two-panel Simpson's formula*:

$$\mathcal{E}_i = -\frac{1}{90}h^5 f^{(4)}(\xi_{i+1}) \tag{7.52}$$

Thus, the leading term in the two-panel error is $O(h^5)$.

We now turn to a discussion of the approximation error in the composite version of Simpson's formula, Eq. (7.46). Just like in the first line of Eq. (7.46), we will evaluate the absolute error for the full interval by summing up all the elementary contributions:

$$\mathcal{E} = \sum_{i=0,2,4,\ldots}^{n-3} \mathcal{E}_i = -\frac{1}{90}h^5 \left(f^{(4)}(\xi_1) + f^{(4)}(\xi_3) + f^{(4)}(\xi_5) + \cdots + f^{(4)}(\xi_{n-2})\right)$$
$$= -\frac{(n-1)/2}{90}h^5 \left(\frac{f^{(4)}(\xi_1) + f^{(4)}(\xi_3) + f^{(4)}(\xi_5) + \cdots + f^{(4)}(\xi_{n-2})}{(n-1)/2}\right) = -\frac{n-1}{180}h^5 f^{(4)}(\xi) \tag{7.53}$$

In the second equality we plugged in the result for $\mathcal{E}_i$ from Eq. (7.52). In the third equality we multiplied and divided by $(n-1)/2$, which is the number of terms in the sum. In the fourth equality we employed a ξ (from a to b) for which $f^{(4)}(\xi)$ is equal to the arithmetic mean of $f^{(4)}$. Combining the last expression with our definition of h, we find our final result for the *absolute error in the composite Simpson's formula*:

$$\mathcal{E} = -\frac{b-a}{180}h^4 f^{(4)}(\xi) \tag{7.54}$$

Observe that this error contains a fourth derivative, so *for polynomials of up to third degree, the composite Simpson's rule is exact.* Given what we know about Simpson's rule, this is surprising: we found that this method is exact for all cubic polynomials, even though we derived it using a quadratic polynomial (remember when the third-derivative term in Eq. (7.49) vanished?). This approximation error, $O(h^4)$, is much better than any of the other quadrature errors we've encountered up to this point (e.g., midpoint or trapezoid). We will offer some advice in section 7.6.5 on which integration method you should choose in general; for now, we note that if you're not dealing with poorly behaved functions (which may have Taylor series that behave improperly, in the sense of having successive terms grow in magnitude) Simpson's is a good method.

Table 7.1 Features of selected Newton–Cotes methods in an elementary interval

Method	Panels	Polynomial	Weights	Absolute error
Rectangle	1	Constant	h	$\frac{1}{2}h^2 f'(\xi_i)$
Trapezoid	1	Straight line	$\frac{h}{2}(1,1)$	$-\frac{1}{12}h^3 f''(\xi_i)$
Simpson's 1/3	2	Quadratic	$\frac{h}{3}(1,4,1)$	$-\frac{1}{90}h^5 f^{(4)}(\xi_i)$
Simpson's 3/8	3	Cubic	$\frac{3h}{8}(1,3,3,1)$	$-\frac{3}{80}h^5 f^{(4)}(\xi_i)$
Boole	4	Quartic	$\frac{2h}{45}(7,32,12,32,7)$	$-\frac{8}{945}h^7 f^{(6)}(\xi_i)$

7.2.6 Summary of Results

It shouldn't be hard to see how this process continues: we keep raising the power of the polynomial used to approximate $f(x)$ in an elementary interval and hope for the best. Using a cubic polynomial to approximate the function within the elementary area gives rise to *Simpson's 3/8 rule*[11] (which needs three panels) and, similarly, using a quartic polynomial in the four-panel problem leads to *Boole's rule*. These are totally straightforward, so you are asked to implement them in problem 7.12. As you learned in the previous chapter, this process cannot go on *ad infinitum*: interpolation using high-degree polynomials on an equally spaced grid will eventually misbehave.

What we have so far, however, already allows us to notice a pattern: *a Newton–Cotes method that requires an even number of panels as an elementary interval gives the exact answer for the full interval a to b for polynomials of one degree higher than the number of panels in that elementary interval* (e.g., Simpson's rule needs two panels and is exact for polynomials up to third degree, whereas Boole's rule needs four panels and is exact for polynomials up to fifth degree). On the other hand, Newton–Cotes methods that contain an odd number of panels in an elementary interval are exact for the full interval a to b for polynomials of the same degree as the number of panels in that elementary interval (e.g., the trapezoid rule uses one panel and is exact for polynomials up to first degree, whereas Simpson's 3/8 rule needs three panels and is exact for polynomials up to third degree).[12]

Since we've introduced a fair number of Newton–Cotes formulas, exploring the underlying approximation (e.g., quadratic polynomial), the weights involved, as well as the error behavior, it might be helpful to produce a compendium of the most important results, see table 7.1. In this table "panels" refers to the number of panels included in an elementary interval; "polynomial" is referring to the highest degree of the polynomial used to approximate the function in each elementary interval; the "weights" here are the coefficients of the function values in an elementary interval; the "absolute error" is $\mathcal{E}_i$, i.e., the error only

[11] Simpson's rule from Eq. (7.46) is sometimes called "Simpson's 1/3 rule" because the first weight is $1/3$ (times h). As you can imagine, "Simpson's 3/8 rule" employs different weights. Given the popularity of "Simpson's 1/3 rule" it is often called simply "Simpson's rule".

[12] We further observe that all the closed formulas we've seen (trapezoid, the two Simpson's rules, and Boole's rule) have an error coefficient that is negative, whereas the half-open or open methods (rectangle and midpoint rules) have an error coefficient that is positive.

in one elementary interval; as a result, the powers of h shown are one degree higher than those of the composite rules—e.g., the composite trapezoid error is $O(h^2)$ whereas the elementary trapezoid error listed here is $O(h^3)$.

7.2.7 Implementation

We are now in a position to implement the composite rectangle rule, Eq. (7.11), the composite trapezoid rule, Eq. (7.31), as well as the composite Simpson's rule, Eq. (7.46). To make things tangible, we will evaluate the integral giving the electrostatic potential in Eq. (7.2), where we know the answer analytically. An implementation is given in code 7.1.

Our grid of abscissas `xs` is the one from Eq. (7.8). The only exception is in the function `rectangle()`, where we don't include x_{n-1} (also known as b) in the sum. Note that if we were really efficiency-oriented, we would have stored neither the x_i nor the $f(x_i)$: we would have simply summed up the latter. Instead, we are here opting for a more reader-friendly solution, as usual. We don't separately multiply each $f(x_i)$ by h, as per Eq. (7.11): instead of having several multiplications, we simply factor the h out and have only one multiplication at the end. We didn't bother storing the weights c_i in a separate array, since they're so straightforward. Turning to the `trapezoid()` function: the main difference here is that we are now explicitly storing the weights c_i in a separate array, in order to avoid any bookkeeping errors (and for ease of reading). The first and last elements of the array have weight $h/2$ and all other ones are h (where we again multiply by h only at the end, to avoid roundoff error creep). You should observe that we are using a `numpy` array for the weights `cs` that contains `n` elements and similarly an array for the grid points `xs` that also contains `n` elements. The implementation of Simpson's rule involves slightly more subtle bookkeeping: we distinguish between odd and even slots (and use different values for the two endpoints at a and at b). Our function does *not* check to see if n is odd, so would give wrong answers if employed for an even value of n.

The output of running this code is:

```
0.88137358702
0.884290734036   0.881361801848   0.881373587255
```

Thus, for $n = 51$ points, the absolute error for: (a) the rectangle rule is in the third digit after the decimal point, (b) the trapezoid rule is in the fifth digit after the decimal point, and (c) Simpson's rule is in the tenth digit after the decimal point. Remember: for 51 points, we have $h = 0.02$—the error in the rectangle rule being $O(h)$—while $h^2 = 0.0004$: since the error in the trapezoid rule is $O(h^2)$, we have an improvement of two orders of magnitude. Similarly, the error in Simpson's rule is $O(h^4)$; $h^4 = 1.6 \times 10^{-7}$, but there's also a 180 in the denominator of Eq. (7.54).

We now decide to employ this code for many different values of h (or, equivalently, of n). The result is shown in Fig. 7.4. Note that the x axis is showing n on a logarithmic scale, whereas the corresponding figure in the chapter on derivatives, Fig. 3.2, had h on a logarithmic scale. (In other words, here as the grid gets finer we are moving to the right, whereas in the derivatives figure a finer grid meant moving to the left.)

newtoncotes.py

Code 7.1

```
import numpy as np

def f(x):
    return 1/np.sqrt(x**2 + 1)

def rectangle(f,a,b,n):
    h = (b-a)/(n-1)
    xs = a + np.arange(n-1)*h
    fs = f(xs)
    return h*np.sum(fs)

def trapezoid(f,a,b,n):
    h = (b-a)/(n-1)
    xs = a + np.arange(n)*h
    cs = np.ones(n); cs[0] = 0.5; cs[-1] = 0.5
    contribs = cs*f(xs)
    return h*np.sum(contribs)

def simpson(f,a,b,n):
    h = (b-a)/(n-1)
    xs = a + np.arange(n)*h
    cs = 2*np.ones(n)
    cs[1::2] = 4; cs[0] = 1; cs[-1] = 1
    contribs = cs*f(xs)
    return (h/3)*np.sum(contribs)

if __name__ == '__main__':
    ans = np.log(1 + np.sqrt(2))
    print(ans)

    for integrator in (rectangle, trapezoid, simpson):
        print(integrator(f, 0., 1., 51), end=" ")
```

You should interpret this figure as follows: a decrease in the absolute error (moving down the y axis) means we're doing a good job; moving to the right on the x axis means we're adding more points. Let us start from discussing the rectangle-rule results: we see that this is a pretty slow method, as the absolute error for this formula decreases only very slowly. Even when using a million points, the absolute error is $\approx 10^{-7}$, which is quite large. Even so,

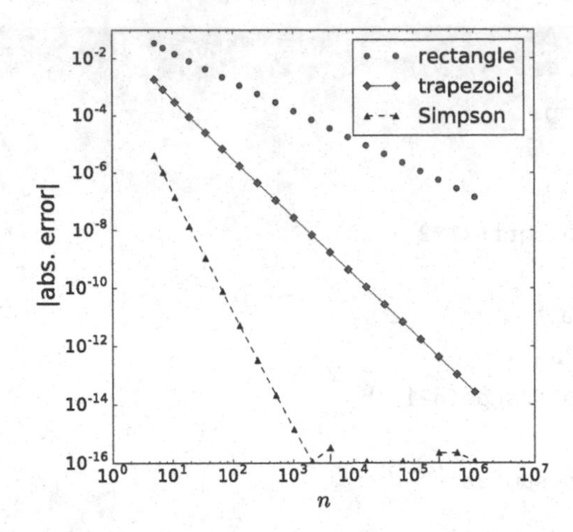

Fig. 7.4 Log-log plot resulting from the rectangle, trapezoid, and Simpson's rules

it's worth observing that the rectangle-rule results fall on a straight line, implying that we are only witnessing the approximation error here, not any roundoff error. In other words, as n gets larger by an order of magnitude, the absolute error gets smaller by an order of magnitude: recall that for this approach the approximation error is $O(h)$. You should keep in mind Eq. (7.9), which tells us that $h = (b - a)/(n - 1)$.

The results for the trapezoid rule are analogous: as n gets larger by an order of magnitude, the absolute error gets smaller by two orders of magnitude: for this approach the approximation error is $O(h^2)$. This process doesn't seem to have reached a minimum: we simply stopped after roughly a million points, where the absolute error is $\approx 10^{-14}$.

The results for Simpson's (1/3) rule are not that strange, either: as n gets larger by an order of magnitude, the absolute error gets smaller by roughly four orders of magnitude: recall that for this approach the approximation error is $O(h^4)$. This process is so fast that after about 1000 points or so we have reached a minimum absolute error of $\approx 10^{-16}$, which is as good as we're going to get. After that, we run up against numerical roundoff issues, so as n is increased beyond that point the absolute error wobbles.[13]

7.3 Adaptive Integration

In the previous subsection we examined the absolute error by comparing the output of each Python function (corresponding to a quadrature rule) with the analytically known answer for the integral. However, there are going to be (several) situations where you cannot evaluate an integral analytically, so it will be impossible to quantify the absolute error in this manner.

[13] It's worth observing that in Fig. 3.2 the central-difference formula (crudely speaking the analogue to the trapezoid rule) reached a minimum absolute error of $\approx 10^{-10}$, which is four orders of magnitude worse than what we found for the trapezoid rule in Fig. 7.4 (which hadn't even reached a minimum).

In practice we are often faced with a very different problem: we know that we can only tolerate an answer that is accurate to at least, say, 10 decimal places, and are now wondering which n to pick and whether or not we should trust our answer. One solution to this problem is to keep increasing the number of points n by hand and seeing if the answer appears to be converging toward a given value. This is fine for simple "back of the envelope" calculations, but for dependable answers there should be a way of automating this entire process. This can be done as part of what is called *adaptive integration*: you provide your function, the region where you want it integrated, and your error tolerance; your method then determines the optimal n (without needing to inform you of this) and gives you a value for the integral along with an estimate of how accurate it is.

7.3.1 Doubling the Number of Panels

As usual, our task is to evaluate $\int_a^b f(x)dx$, which we now call I for ease of reference. In the present section, it will be helpful to keep in mind both the number of points n and the number of panels $N = n - 1$ separately. We will now examine elementary consequences of the error analyses for the rectangle, trapezoid, and Simpson's rules that we discussed above. Regardless of which quadrature method we employ, we denote the result for the value of the integral I_N when using N panels. We also modify our notation on the panel width: as per Eq. (7.9), we define:

$$h_N = \frac{b-a}{N} \tag{7.55}$$

The only differences are that we now use the number of panels N in the denominator (instead of using $n - 1$) and that we called the panel width h_N, to ensure that we keep track of which width we're talking about. For example, the estimate for our integral will be called I_{16} for the case of $N = 16$ panels (for which we have $n = N + 1 = 17$ points), making use of panel width h_{16}; similarly, I_{128} for the case of $N = 128$ panels (for which we have $n = N + 1 = 129$ points), making use of panel width h_{128}.

We will now focus on what happens when we start from a given panel number N and then change the panel number to $N' = 2N$. We're simply doubling the panel number, but it will turn out to be convenient for us to use a new variable (N') to denote the new panel number. This is almost too trivial to write out explicitly but, just in case, here's how we can relate the new panel width $h_{N'}$ to the old panel width h_N:

$$h_{N'} = \frac{b-a}{N'} = \frac{b-a}{2N} = \frac{1}{2}\frac{b-a}{N} = \frac{h_N}{2} \tag{7.56}$$

This isn't rocket surgery: doubling the number of panels, you halve the panel width.

We will now see what happens to the error as we go from N panels to N' panels. In essence, we are doing (for a given integration rule) two calculations of the integral, one giving I_N and one giving $I_{N'}$. To keep things manageable, we will study only Simpson's rule, leaving the analogous derivation for the rectangle and trapezoid rules as problem 7.11.

Using our new notation, we know that to leading order in the error, the total absolute error for a Simpson's-rule calculation with N panels is $\mathcal{E}_N = ch_N^4$, as per Eq. (7.54); as above, c actually depends on h_N, but we here take it to be constant. Thus, our estimate for the value of I when employing N panels can be related to its total absolute error:

$$I = I_N + ch_N^4 \tag{7.57}$$

This is nothing other than Eq. (7.54), but this time using our new notation. Similarly, we can apply the same equation to the case of $N' = 2N$ panels:

$$I = I_{N'} + ch_{N'}^4 \tag{7.58}$$

Equating the right-hand sides of the last two equations gives us:

$$I_N + ch_N^4 = I_{N'} + ch_{N'}^4 \tag{7.59}$$

which can be manipulated to give:

$$I_{N'} - I_N = ch_N^4 - ch_{N'}^4 \tag{7.60}$$

If we now plug in our result from Eq. (7.56), recast as $h_N = 2h_{N'}$, this gives us:

$$I_{N'} - I_N = 15ch_{N'}^4 \tag{7.61}$$

But we already know that $\mathcal{E}_{N'} = ch_{N'}^4$ for Simpson's rule. Thus, we have been able to derive an expression for $\mathcal{E}_{N'}$ in terms of the two results I_N and $I_{N'}$:

$$\mathcal{E}_{N'} = \frac{1}{15}(I_{N'} - I_N) \tag{7.62}$$

This is good news: it means that we've been able to provide an estimate of the accuracy of our result $I_{N'}$ by using $I_{N'}$ itself (a calculation for N' panels) and I_N (a calculation for N panels). We've managed to provide a quantitative estimate of our ignorance, without knowing the correct answer ahead of time.

Something analogous ends up holding for the trapezoid rule:

$$\mathcal{E}_{N'} = \frac{1}{3}(I_{N'} - I_N) \tag{7.63}$$

and for the rectangle rule:

$$\mathcal{E}_{N'} = I_{N'} - I_N \tag{7.64}$$

as you will explicitly show. The only difference is in the denominator. Incidentally, a larger number in the denominator means that our absolute error is smaller: this is as it should be, since Simpson's rule is the most accurate method of the three.

7.3.2 Thoughts before Implementing

First, what do we do if going from N to $N' = 2N$ panels doesn't give us an absolute error that is small enough for our purposes? The answer is simple: we double again, and keep doing so until we have reached the accuracy we want. Notation-wise, you can envision schemes where you label the different iterations as N_i or something like that, but from a programming perspective it's probably just easier to use N and $N' = 2N$ and then keep renaming your variables.

Second, how do we pick the starting value of N? This isn't too big a deal, either: simply start with any value you want (if you're using Simpson's rule, make sure that's an even number) and double it to $N' = 2N$ as described above. If you start with N too small, you might take too long (for your purposes) to meet your error tolerance. If you start with an N that is too large, you might be wasting your time if $\mathcal{E}_{N'}$ is already much smaller than you need.

Third, you should keep in mind that our integration routines were written in terms of n (the number of points), not N (the number of panels). Thus, if you want to calculate I_N and $I_{N'}$ using those routines you should make sure to pass in n and n' instead of N and N'. This is quite straightforward to do, too: for our starting value, we simply have $N = n - 1$, so we pass in $n = N + 1$. When we double the number of panels, $N' = 2N$, we don't quite double the number of points.[14] We simply carry out this elementary bookkeeping exercise: $N' = 2N$ combined with $N = n - 1$ and $N' = n' - 1$ leads to:

$$n' = 2n - 1 \tag{7.65}$$

As a trivial check, note that if n is odd, this expression gives an n' that is also odd.

7.3.3 Implementation

Code 7.2 is a Python implementation of an adaptive integrator. There are quite a few things to unpack here. First of all, note that the main program produces `ans`, the analytically known answer, but *only* so that we may print it out and compare at the end, i.e., this is not used anywhere inside the function we are defining.

Observe that we are employing a dictionary, using function names as the *keys*. Since functions are first-class objects in Python, we can use them as we would any other entity. This allows us to select a denominator value according to which integrating function is passed in as an argument *when we call* `adaptive()`! This is an incredibly powerful idiom: notice how the rest of the code simply uses `integrator()`, i.e., the code looks the same whether we want to create an adaptive integrator for the rectangle, trapezoid, or Simpson's rules (except for the value of the denominator).[15] To appreciate how nice this is, you should try to implement your own version without looking at this solution: if you're still a beginner, you will likely end up having a cascade of `if` and `elif`, the bodies of which are carrying out the same operations by a different name. This is an invitation for trouble if, e.g., you later want to modify the algorithm, include an extra integrator, etc.

This code is making use of the `for-else` idiom, as usual; we set the return value to `None` if we are unsuccessful in determining the answer. Speaking of the error tolerance, our program sets that (arbitrarily) to 10^{-12}. We use a relative error ($\epsilon_{N'}$), not an absolute one ($\mathcal{E}_{N'}$). This is in order to make sure that we always conserve roughly 11 or 12 significant figures, but don't try to accomplish the impossible.

[14] This would be an especially egregious mistake for Simpson's rule, since we need to start with an odd number of points n and also *end up* with an odd number of points n'.

[15] Taking separation of concerns seriously, it would be best to pass in the dictionary as a parameter.

Code 7.2 adaptive.py

```
from newtoncotes import f, rectangle, trapezoid, simpson
import numpy as np

def adaptive(f,a,b,integrator,kmax = 20,tol = 1.e-12):
    functodenom = {rectangle:1, trapezoid:3, simpson:15}
    denom = functodenom[integrator]

    n = 2
    val = integrator(f,a,b,n)
    for k in range(kmax):
        nprime = 2*n-1
        valprime = integrator(f,a,b,nprime)
        err = abs(valprime-val)/denom
        err /= abs(valprime)
        print(nprime, valprime, err)
        if err<tol:
            break

        n, val = nprime, valprime
    else:
        valprime = None
    return valprime

if __name__ == '__main__':
    ans = np.log(1 + np.sqrt(2))
    print(ans); print("")
    for integrator in (rectangle, trapezoid, simpson):
        print(adaptive(f, 0., 1., integrator)); print("")
```

As you will see when you run this code, the rectangle rule fails to reach an absolute error below our tolerance (thus returning `None`), even after doubling the number of panels 20 times. This is consistent with what we saw in Fig. 7.4: for a million points, the rectangle rule has an absolute error of order $\approx 10^{-7}$. The output for the trapezoid rule is analogous: we reach our tolerance levels with a bit over a quarter of a million points, just like in Fig. 7.4. Things are also fully analogous for Simpson's rule: 257 points (i.e., 256 panels) are enough to get the tiny absolute error we desire, highlighting the superiority of Simpson's rule.[16]

[16] You shouldn't put too much faith in the Simpson's rule error bar for $n = 3$, as it follows from the Simpson's $n = 2$ answer and we already know that we should never apply Simpson's rule for an even number of points.

This output also serves to provide justification for our earlier claims, around Fig. 7.4, about the order-of-magnitude behavior of the error as h is decreased. For example, Simpson's rule (which never gets up to 1000 points) jumps by more than four orders magnitude when the number of panels goes from roughly 10 to roughly 100.

Before concluding, we note that here we have presented only a simple choice for an adaptive integration scheme: keep doubling the number of panels, until the answer for the integral has an acceptable error. In the implementation above, we throw away our earlier calculations, but in problem 7.13 you will see how to be less wasteful. In any case, we have to keep redoing the integral from a to b again and again. Another scheme might be to divide the interval a to b in two: we then have two separate integrals, $\int_a^{(a+b)/2} f(x)dx$ and $\int_{(a+b)/2}^b f(x)dx$, to carry out; however, we can take the tolerance in each subinterval to be one-half of the total desired tolerance. If that's not good enough, we further sub-divide, and continue doing this until we reach our desired sub-tolerance in each subinterval. This would be an especially good idea for a function that changes a lot in one region and is "boring" in another region; there's no point in doubling the number of your panels everywhere if you've already done a good enough job in a region where it's easy to integrate the function.[17]

7.4 Romberg Integration

We now turn to Romberg integration, which combines the composite trapezoid rule together with Richardson extrapolation. The latter was first introduced in section 3.3.7, where we proceeded to apply it to the problem of estimating derivatives via finite differences. Here we do much the same, this time applied to numerical integration. Some of the equations that follow will be very similar to what we saw in section 7.3.1, when we were doubling the number of panels in the context of adaptive integration.

7.4.1 Richardson Extrapolation

We start from summarizing our earlier results from section 3.3.7 on Richardson extrapolation. The task there was to evaluate the quantity G. We made the assumption that the error term could be written as a sum of powers of h:

$$G = g(h) + Ah^p + Bh^{p+q} + Ch^{p+2q} + \cdots \tag{7.66}$$

where A, B, C are constants, p denotes the order of the leading error term, and q is the increment in the order for the error terms after that. The main idea behind Richardson extrapolation was to apply Eq. (7.66) twice, once for a step size h and once for a step size $h/2$. This gave us:

$$G = \frac{2^p g\left(\frac{h}{2}\right) - g(h)}{2^p - 1} + O(h^{p+q}) \tag{7.67}$$

Thus, starting from a formula with a leading error of $O(h^p)$, we used two calculations (one

[17] See also section 8.2.3.5, where we distinguish between a "global adaptive" and a "local adaptive" approach.

for a step size h and once for a step size $h/2$) and managed to eliminate the error $O(h^p)$. This meant that we were left with a formula that has error $O(h^{p+q})$. At the time, we observed that this process could be repeated: by starting with two calculations each of which has error $O(h^{p+q})$, one can arrive at an answer with error $O(h^{p+2q})$, and so on. We will soon carry out precisely such a repetition.

Before we do that, we first re-express the main result of Richardson extrapolation in the language of integrals, using our new notation. Our task is to evaluate $I = \int_a^b f(x)dx$. The result for the value of the integral is I_N when using N panels. Remember that the panel width for this case is simply:

$$h_N = \frac{b-a}{N} \tag{7.68}$$

As above, we are interested in what happens when we start from a given panel number N and then change the panel number to $N' = 2N$. We saw in Eq. (7.56) that the new panel width $h_{N'}$ can be trivially related to the old panel width h_N, via $h_{N'} = h_N/2$. Thus, the Richardson-extrapolated estimate, as per Eq. (7.67), can be rewritten as:

$$I = \frac{2^p I_{N'} - I_N}{2^p - 1} + O(h_N^{p+q}) \tag{7.69}$$

where, as always, $N' = 2N$. This applies to a method for which the leading error goes as h_N^p: the extrapolation cancels that leading term, leaving us with an error of h_N^{p+q}.

In the spirit of being fully explicit, let us apply Eq. (7.69) for a few specific cases, which will come in handy in the next subsection. First, we examine the case of the trapezoid rule, where the absolute error behaves as:

$$\mathcal{E} = c_2 h_N^2 + c_4 h_N^4 + c_6 h_N^6 + \cdots \tag{7.70}$$

as per Eq. (7.41). In this case, Eq. (7.69) gives us:

$$I = \frac{2^2 I_{N'} - I_N}{2^2 - 1} + O(h_N^4) = \frac{4}{3} I_{N'} - \frac{1}{3} I_N + O(h_N^4) \tag{7.71}$$

where $N' = 2N$. Since the leading error goes as h_N^p, we have $p = 2$. Using two estimates, each of which has an error of order $O(h_N^2)$, we end up with an estimate with error $O(h_N^4)$.

Second, imagine we were dealing with a situation where the absolute error behaves as:

$$\mathcal{E} = c_4 h_N^4 + c_6 h_N^6 + c_8 h_N^8 + \cdots \tag{7.72}$$

In this case, Eq. (7.69) gives us:

$$I = \frac{2^4 I_{N'} - I_N}{2^4 - 1} + O(h_N^6) = \frac{16}{15} I_{N'} - \frac{1}{15} I_N + O(h_N^6) \tag{7.73}$$

where $N' = 2N$. Since the leading error goes as h_N^p, we have $p = 4$; employing two estimates with error of order $O(h_N^4)$, we get an estimate with error $O(h_N^6)$.

Third, imagine we were dealing with a situation where the absolute error behaves as:

$$\mathcal{E} = c_6 h_N^6 + c_8 h_N^8 + c_{10} h_N^{10} + \cdots \tag{7.74}$$

In this case, Eq. (7.69) gives us:

$$I = \frac{2^6 I_{N'} - I_N}{2^6 - 1} + O(h_N^8) = \frac{64}{63} I_{N'} - \frac{1}{63} I_N + O(h_N^8) \tag{7.75}$$

where $N' = 2N$. Since the leading error goes as h_N^p, here we have $p = 6$. Thus, using two estimates with error $O(h_N^6)$, we end up with an estimate with error $O(h_N^8)$.

7.4.2 Romberg Recipe

Romberg integration amounts to using the trapezoid rule along with Richardson extrapolation. As discussed in the previous subsection, Richardson extrapolation involves doubling the number of panels, just like we did when discussing adaptive integration. In short, we will combine (a small number of) actual trapezoid-rule integrations with repeated Richardson-extrapolation steps. We note at this early stage that Romberg integration will give disappointing results when our function (or its derivatives) exhibits pathological behavior. Instead, here we will study only our standard example, which does not suffer from such problems.

Romberg integration uses some new notation, which can be confusing if you're not paying close attention. The confusion may arise from the fact that the Romberg approach involves some trapezoid rule results and some extrapolated results. In order to keep these distinct, we will be very explicit in our derivation in what follows. Start from a trapezoid-rule calculation that uses one panel. This was denoted by I_1 in the previous section. Introduce the new notation $R_{0,0} = I_1$: as a mnemonic, think of this as your starting point in two axes (which will be explained soon) the numbering for which starts at 0, as usual. This $R_{0,0}$ is the result of a real calculation, not an extrapolation (it's simply the trapezoid-rule result for one panel I_1). In order to carry out a Richardson extrapolation, we will need another "raw" calculation. Take the case of the trapezoid-rule result for two panels, denoted by I_2. Introduce the new notation $R_{1,0} = I_2$. Importantly, our new $R_{i,j}$ entity has two indices, and we've only increased the value of the first one. This will continue to be the case below every time we carry out a new trapezoid-rule computation: *whenever we carry out a new trapezoid rule step the first index in $R_{i,0}$ will go up by one and the second index will be 0.*

Now that we have access to the two trapezoid-rule results $R_{0,0}$ and $R_{1,0}$ (which we used to call I_1 and I_2) we can carry out a Richardson-extrapolation step. The situation we're faced with is identical to what we were facing in the "first case" in the previous subsection. We have at our disposal two results with errors that go as $c_2h^2 + c_4h^4 + c_6h^6 + \cdots$ so we can apply Richardson extrapolation in the form of Eq. (7.71). Since we're already introducing new notation, let's keep at it. We call the result of this first extrapolation step $R_{1,1}$:

$$R_{1,1} = \frac{2^2 R_{1,0} - R_{0,0}}{2^2 - 1} = \frac{4}{3} R_{1,0} - \frac{1}{3} R_{0,0} \tag{7.76}$$

In this definition, we don't include the error term. Note how the second index on the left-hand side increased by one. This will continue to happen below: *whenever we carry out a Richardson-extrapolation step the second index in $R_{i,j}$ will go up by one.*

Our situation can be conveniently recast using matrices. Our two trapezoid-rule results belong in the same column (the 0th one, since they both have the second index set to 0):

$$\begin{pmatrix} R_{0,0} \\ R_{1,0} \end{pmatrix} \tag{7.77}$$

As we saw in Eq. (7.76), carrying out a Richardson extrapolation allows us to move into a new column:

$$\begin{pmatrix} R_{0,0} & \\ R_{1,0} & R_{1,1} \end{pmatrix} \tag{7.78}$$

Importantly, both elements in the 0th column had a leading error term of the form $O(h^2)$, so they allowed us to produce a new estimate with a leading error term of the form $O(h^4)$, placed in the 1st column.

At this point, we decide to produce yet another "raw" calculation, namely the trapezoid-rule result for four panels, denoted by I_4 in the previous section. Since this is a new trapezoid result, we increase the first index in $R_{i,0}$ and store $R_{2,0} = I_4$, obtaining:

$$\begin{pmatrix} R_{0,0} & \\ R_{1,0} & R_{1,1} \\ R_{2,0} & \end{pmatrix} \tag{7.79}$$

It is still the case that all the results in the 0th column have an error of $O(h^2)$. However, that means that we are now free to perform yet another Richardson extrapolation: instead of combining $R_{0,0}$ and $R_{1,0}$ (which gave us $R_{1,1}$ above), we combine $R_{1,0}$ and $R_{2,0}$, to produce an answer which we store in $R_{2,1}$:

$$R_{2,1} = \frac{2^2 R_{2,0} - R_{1,0}}{2^2 - 1} = \frac{4}{3}R_{2,0} - \frac{1}{3}R_{1,0} \tag{7.80}$$

As our notation clearly shows, this new value belongs in the 1st column. It goes under $R_{1,1}$ and has an error $O(h^4)$ just like that value. But we now realize that we have access to two values in the 1st column, $R_{1,1}$ and $R_{2,1}$, each of which has a leading error of $O(h^4)$. The situation we're faced with is identical to the "second case" in the previous subsection. We have at our disposal two results with errors that go as $c_4 h^4 + c_6 h^6 + \cdots$ so we can apply Richardson extrapolation in the form of Eq. (7.73):

$$R_{2,2} = \frac{2^4 R_{2,1} - R_{1,1}}{2^4 - 1} = \frac{16}{15}R_{2,1} - \frac{1}{15}R_{1,1} \tag{7.81}$$

where you should observe that by combining two estimates with error $O(h^4)$ we have produced a new estimate with error $O(h^6)$. Thus, we have:

$$\begin{pmatrix} R_{0,0} & & \\ R_{1,0} & R_{1,1} & \\ R_{2,0} & R_{2,1} & R_{2,2} \end{pmatrix} \tag{7.82}$$

in matrix form. For reasons that will soon become clear, we now observe that we went from the matrix in Eq. (7.78) to that in Eq. (7.82) by carrying out three steps: (a) a new trapezoid rule calculation, giving us $R_{2,0}$, (b) a Richardson extrapolation as per Eq. (7.80) giving us $R_{2,1}$, and (c) a Richardson extrapolation as per Eq. (7.81) giving us $R_{2,2}$. The most accurate answer at our disposal is currently in $R_{2,2}$, which is the bottom element on the diagonal.

You are probably starting to see how this works, but let's evaluate another row explicitly, to make things crystal clear. The 0th element in each new row is simply another "raw" calculation. In this case we will have the trapezoid-rule result for eight panels, denoted by I_8 in the previous section. Since this is a new trapezoid result, we increase the first index in $R_{i,0}$ and store $R_{3,0} = I_8$, obtaining:

$$\begin{pmatrix} R_{0,0} & & \\ R_{1,0} & R_{1,1} & \\ R_{2,0} & R_{2,1} & R_{2,2} \\ R_{3,0} & & \end{pmatrix} \tag{7.83}$$

As always, the results in the 0th column have an error of $O(h^2)$. Just like we did before, we will now employ the previous row and the 0th element of the current row to carry out several Richardson-extrapolation steps. First, we combine $R_{2,0}$ and $R_{3,0}$, to produce an answer which we store in $R_{3,1}$:

$$R_{3,1} = \frac{2^2 R_{3,0} - R_{2,0}}{2^2 - 1} = \frac{4}{3}R_{3,0} - \frac{1}{3}R_{2,0} \tag{7.84}$$

This new value belongs in the 1st column because it has an error $O(h^4)$. Second, we now realize that we have access to two values in the 1st column, $R_{2,1}$ and $R_{3,1}$, each of which has a leading error of $O(h^4)$. This means that we can carry out yet another Richardson-extrapolation step:

$$R_{3,2} = \frac{2^4 R_{3,1} - R_{2,1}}{2^4 - 1} = \frac{16}{15}R_{3,1} - \frac{1}{15}R_{2,1} \tag{7.85}$$

Thus, by combining two estimates with error $O(h^4)$ we have produced a new estimate with error $O(h^6)$, which belongs in the 2nd column. Third, we are now able to carry out yet another step because the situation we're faced with is identical to the "third case" in the previous subsection. We have at our disposal two results with errors that go as $c_6h^6 + c_8h^8 + \cdots$ so we can apply Richardson extrapolation in the form of Eq. (7.75):

$$R_{3,3} = \frac{2^6 R_{3,2} - R_{2,2}}{2^6 - 1} = \frac{64}{63}R_{3,2} - \frac{1}{63}R_{2,2} \tag{7.86}$$

This new estimate belongs in the 3rd column because it has an error $O(h^8)$. Thus, we have:

$$\begin{pmatrix} R_{0,0} & & & \\ R_{1,0} & R_{1,1} & & \\ R_{2,0} & R_{2,1} & R_{2,2} & \\ R_{3,0} & R_{3,1} & R_{3,2} & R_{3,3} \end{pmatrix} \tag{7.87}$$

in matrix form. We now reiterate our main point about how to produce a new row in this matrix, by observing how we went from the matrix in Eq. (7.82) to that in Eq. (7.87).

First, we doubled the number of panels and carried out a new trapezoid-rule calculation and, then, we carried out three Richardson-extrapolation steps in Eq. (7.84), Eq. (7.85), and Eq. (7.86). The most accurate answer at our disposal is currently in $R_{3,3}$, which is the bottom element on the diagonal.

In summary, to produce a new row we first carry out a new trapezoid-rule calculation for twice as many panels as before (placing that in the 0th column) and then carry out Richardon extrapolation steps *which require the previous row and the newly produced 0th-column trapezoid result*. This leads to a lower-triangular matrix. The 0th column contains trapezoid-rule results, which therefore have error $O(h^2)$. The 1st column contains results for one Richardson-extrapolation step, which therefore have error $O(h^4)$. The 2nd column contains results of two Richardson-extrapolation steps, which therefore have error $O(h^6)$. The pattern should be clear by now. As you go down the 0th column the number of panels is doubled so the h becomes smaller. As you go to the right on any row you carry out increasingly more Richardson-extrapolation steps. As a result, the bottom element on the diagonal is the best estimate each time, so we compare our best value on this row with our best value on the previous row and terminate this process when the difference between these two "best estimates" is smaller than some set tolerance level. In equation form:

$$
\begin{aligned}
R_{i,0} &= I_N & &(N = 2^i, i = 0, 1, 2, \ldots) \\
R_{i,j} &= \frac{4^j R_{i,j-1} - R_{i-1,j-1}}{4^j - 1} & &(i = 1, 2, \ldots \text{ and } j = 1, 2, \ldots, i)
\end{aligned}
\tag{7.88}
$$

Let's try to unpack this. The first line simply says that the elements of the 0th column are trapezoid-rule results, with the number of panels being doubled as we move down the column: $R_{0,0}$ is I_1, $R_{1,0}$ is I_2, $R_{2,0}$ is I_4, $R_{3,0}$ is I_8, and so on.[18] The second line simply encapsulates how we carry out the Richardson extrapolation: we always use (a) the element that's in the previous column on the same row, and (b) the element that's in the previous column on the previous row. This equation on the second line gives us all the elements of the lower-triangular matrix except for the 0th column. It's also worth observing that the second line in this pair of equations uses 4^j for the coefficient in the numerator and the denominator. This is simply a result of noticing that the powers that show up are 2^2, 2^4, 2^6, and so on, which are conveniently re-expressed by remembering that $(2^2)^j = 4^j$.

As you can imagine from the expected scaling with h as you move to the right of the matrix, Romberg integration is a nice way of starting from a crude integration method (the trapezoid rule) and producing very accurate results with very few function evaluations (as usual, assuming the integrand is not pathological). As a result, the very simple trapezoid rule ends up being the workhorse of some very accurate calculations.

Before we turn to a Python implementation, applied to a specific problem, we note a general feature of Romberg integration. This is that the 1st column (i.e., the one containing results of one Richardson-extrapolation step) of the Romberg matrix turns out to be iden-

[18] Note that we've avoided the use of the confusing-looking I_{2^i}.

tical to Simpson's rule results. To be more specific, as you will show in problem 7.15, $R_{i,1}$ contains the Simpson's 1/3 answer for 2^i panels, for $i = 1, 2, \ldots$.

7.4.3 Implementation

Thinking about the simplest way of implementing Romberg integration, a few things immediately come to mind. First, since the 0th column elements will be trapezoid-rule results while doubling the number of panels, it seems reasonable to use the adaptive-integration code `adaptive.py` as a starting point. Second, in addition to carrying out these trapezoid-rule calculations, we'll also need to carry out the Richardson-extrapolation steps shown in the second line of Eq. (7.88). Since these implement a specific equation, it stands to reason that we should define a separate function encapsulating that process. What should that use as parameters and as a return value? As already emphasized, to carry out the sequence of Richardson-extrapolation steps that fill out an entire row, what we need (as parameters) are the previous row and the newly produced 0th-column trapezoid estimate. The question naturally arises as to how to store the $R_{i,j}$. Since these make up a matrix, it seems natural to reach for a `numpy` array. However, we don't really need to store the entire matrix: at any point in time, all we need is the current row and the previous one, so we can simply use two (one-dimensional) lists.

The above considerations lead to the implementation in code 7.3. We skip (for now) the `prettyprint()` function. As advertised, this code includes a function `richardson()` which implements the second line of Eq. (7.88). It takes in a list containing the previous row (`Rprev`) and a float containing the 0th element on the current row (`Rincurr0`).

As promised, the major function here, `romberg()`, merely keeps track of the current row and the previous one (`Rcurr` and `Rprev`, respectively), not the entire matrix. The doubling of the panels (implemented via `n` and `nprime`) is identical to what we had in the earlier code. This code calls `trapezoid()` whenever a new 0th-column element needs to be produced and `richardson()` for everything else. Note that the loop starting value is different than in code 7.2, since `i` actually matters here: it is passed in as an argument to `richardson()`.

The error is estimated by using the last element on the current and the previous row. Note that there's no denominator here, since we are not employing an analytical expectation of how these two elements are related: we simply expect that when the last diagonal elements stop changing our work is done.

The only other point is the line saying `Rprev = Rcurr[:]` whose logic should be familiar enough by now: in the next iteration our current row will be treated as the previous row. Copying the entire list (as opposed to renaming it) is not really necessary here (since a new list is created inside `richardson()` each time), but it's a good habit to maintain.

Running this code, we notice that the (absolute) error of our estimate is 10^{-11}, even though we only hard-coded a requirement for a (relative) error of 10^{-8} inside `romberg()`. This is good but, without knowing how many function evaluations were necessary, not necessarily impressive. To find out how good our approach is, we could print out the value of `nprime` as we did before. Instead, we could just uncomment the two lines containing

Code 7.3 romberg.py

```
from newtoncotes import f, trapezoid
import numpy as np

def prettyprint(row):
    for elem in row:
        print("{0:1.11f} ".format(elem),end="")
    print("")

def richardson(Rprev, Rincurr0, i):
    Rcurr = [Rincurr0]
    for j in range(1, i+1):
        val = (4**j*Rcurr[j-1] - Rprev[j-1])/(4**j - 1)
        Rcurr.append(val)
    return Rcurr

def romberg(f,a,b,imax = 20,tol = 1.e-8):
    n = 2
    val = trapezoid(f,a,b,n)
    Rprev = [val]
    #prettyprint(Rprev)
    for i in range(1,imax):
        nprime = 2*n-1
        Rincurr0 = trapezoid(f,a,b,nprime)
        Rcurr = richardson(Rprev, Rincurr0, i)
        #prettyprint(Rcurr)
        err = abs(Rprev[-1] - Rcurr[-1])/abs(Rcurr[-1])
        valprime = Rcurr[-1]
        if err < tol:
            break
        n = nprime
        Rprev = Rcurr[:]
    else:
        valprime = None
    return valprime

if __name__ == '__main__':
    ans = np.log(1 + np.sqrt(2))
    print(ans)
    print(romberg(f,0.,1.))
```

calls to our beautifying `prettyprint()` function in the present code to get all the $R_{i,j}$'s that came into the picture:

```
0.85355339059
0.87399029080 0.88080259086
0.87953077043 0.88137759698 0.88141593072
0.88091314184 0.88137393231 0.88137368800 0.88137301748
0.88125849242 0.88137360928 0.88137358774 0.88137358615 0.88137358838
0.88134481442 0.88137358842 0.88137358703 0.88137358702 0.88137358702 0.88137358702
```

Here, the first row corresponds to a trapezoid-rule calculation with one panel, I_1. Going down the 0th column we then encounter I_2, I_4, I_8, I_{16}, and finally I_{32}. This means that in order to arrive at our very accurate final result (the rightmost one in the last row) we needed to carry out only 33 function evaluations (32 panels means 33 points). This is truly spectacular, if you compare to the output of `adaptive.py`, where an error of 10^{-11} for the trapezoid rule needed 65 537 points!

Well, not quite. As it so happens, our implementation of the Romberg recipe is wasteful: it throws away the older trapezoid calculations when producing the new ones (e.g., the function evaluations used to produce I_8 are discarded when evaluating I_{16}). As a result, we actually carried out $2 + 3 + 5 + 9 + 17 + 33 = 69$ function evaluations. Still, that's roughly a factor of a thousand fewer function evaluations than the non-Romberg trapezoid approach. As a matter of fact, even Simpson's rule needed roughly 100 points to produce an estimate with an absolute error of 10^{-11}. In problem 7.16 you will write and test your own non-wasteful Romberg integration code.

7.5 Gaussian Quadrature

The Newton–Cotes formulas we encountered have been cast in the same form, that of Eq. (7.7), where we were dealing with n nodal abscissas x_i and n weights c_i. All of these Newton–Cotes methods employ an evenly spaced grid, as per Eq. (7.8). In other words, such methods all use predetermined nodal abscissas x_i and differ only in which weights c_i they employ; the latter are chosen in order to exactly integrate a low-degree polynomial. In contradistinction to this, what the method(s) of Gaussian quadrature accomplish is both simple and impressive: instead of limiting yourself to equally spaced nodal abscissas, *expand your freedom so that both the n abscissas x_i and the n weights c_i are at your disposal.* Then, you will be able to *integrate all polynomials up to degree $2n - 1$ exactly*! It should be straightforward to see why this is impressive: it implies that even using a very small number of points (say $n = 5$) allows us to integrate all polynomials up to quite high order exactly (up to ninth order for $n = 5$). As you may have already guessed, the reason this is possible is that we have doubled the number of parameters at our disposal: we can use the $2n$ parameters (the x_i's and the c_i's) to handle polynomials up to degree $2n - 1$.

In our discussion of Newton–Cotes methods, we always started from an elementary interval and later generalized to a composite integration rule, which could handle the full integral from a to b. Given the accuracy that Gaussian quadrature is capable of, we will employ it *directly* in the full interval from a to b.[19] Just like for the problem of interpolation (chapter 6), using unequally spaced abscissas allows you to eliminate the issues that arise with an equally spaced grid; as a result, you can use a single polynomial interpolant for the entire interval instead of employing piecewise polynomial interpolation as we did for Newton–Cotes integration above.

It is standard to take $a = -1$ and $b = 1$ at this stage: this is completely analogous to what we did in our discussion of polynomial interpolation in section 6.2. Thus, we start by focusing on the standard interval $[-1, 1]$; as we saw in Eq. (6.13) and will discuss again below, it is straightforward to scale to the interval $[a, b]$ after the fact. It is important to understand that our use of the standard interval $[-1, 1]$ is unrelated to an elementary interval: as already mentioned, we will set up Gauss–Legendre quadrature directly on the full interval (which happens to be $[-1, 1]$), i.e., we will not first study an elementary rule and later a composite rule. Thus, our defining equation will be:

$$\int_{-1}^{1} f(x)dx \approx \sum_{k=0}^{n-1} c_k f(x_k) \tag{7.89}$$

Our earlier claim was that by an intelligent choice of *both* the x_k's and the c_k's, we will be able to integrate polynomials up to degree $2n - 1$ exactly. It's probably not too early to point out that *all Gaussian quadrature methods are open*, so the x_k's are *not* to be identified with -1 and $+1$. Let's see how this all works.

7.5.1 Gauss–Legendre: *n*=2 Case

We start from explicitly addressing the exact integrability of polynomials up to degree $2n - 1$ for the simplest non-trivial case, namely that of two points.[20] We take Eq. (7.89) and apply it to the case of $n = 2$:

$$\int_{-1}^{1} f(x)dx = c_0 f(x_0) + c_1 f(x_1) \tag{7.90}$$

The abscissas x_0 and x_1 along with the weights c_0 and c_1 are the four quantities we need to now determine. Since Gaussian quadrature methods are open, x_0 and x_1 will lie in (a, b). As advertised, we will determine the four unknown parameters by demanding that all polynomials up to degree $2n - 1 = 3$ can be integrated exactly.

It is simplest to use monomials, i.e., single powers of x, instead of polynomials (which

[19] In practice, one sometimes slices up the integration interval "by hand", using more points in regions that exhibit more structure, and fewer points when the behavior of the integrand is pretty simple.

[20] Problem 7.17 invites you to study the even more basic case of $n = 1$, which is equivalent to the midpoint rule!

can later be arrived at as linear combinations of monomials). Thus, we will take:

$$f(x) = \begin{cases} x^0 \\ x^1 \\ x^2 \\ x^3 \end{cases} \tag{7.91}$$

and assume that for all these cases Eq. (7.90) holds. This leads to the following four equations in four unknowns:

$$\begin{aligned} &\int_{-1}^{1} 1dx = 2 = c_0 + c_1, &&\int_{-1}^{1} xdx = 0 = c_0x_0 + c_1x_1, \\ &\int_{-1}^{1} x^2dx = \frac{2}{3} = c_0x_0^2 + c_1x_1^2, &&\int_{-1}^{1} x^3dx = 0 = c_0x_0^3 + c_1x_1^3 \end{aligned} \tag{7.92}$$

In each of these cases the left-hand side is that of Eq. (7.90), the second step gives the answer for that integral (calculated analytically), and the third step gives the right-hand side of Eq. (7.90), applied to that specific monomial.

We are now faced with a set of non-linear equations. The second of these gives us $c_0x_0 = -c_1x_1$; this result, when combined with the fourth relation, leads to $x_0^2 = x_1^2$. Thus, since x_0 and x_1 are different from each other, we find $x_0 = -x_1$. Together with the second relation, this implies $c_0 = c_1$. This result, together with the first relation in Eq. (7.92) gives $c_0 = c_1 = 1$. The third relation in Eq. (7.92) now gives us $x_0^2 + x_0^2 = 2/3$, which can be solved by taking $x_0 = -1/\sqrt{3}$. Thus, we have arrived at the solution:

$$c_0 = c_1 = 1, \qquad x_0 = -\frac{1}{\sqrt{3}}, \qquad x_1 = \frac{1}{\sqrt{3}} \tag{7.93}$$

This can now be used to recast Eq. (7.90) as:

$$\int_{-1}^{1} f(x)dx \approx f\left(-\frac{1}{\sqrt{3}}\right) + f\left(\frac{1}{\sqrt{3}}\right) \tag{7.94}$$

This employs two points and is exact for polynomials up to third degree; it is approximate for other functions. It's worth pausing for a second to realize that our accomplishment is already noteworthy: Simpson's rule could integrate polynomials up to third degree exactly, but it needed at least three points in order to do that.

7.5.2 Gauss–Legendre: General Case

There are several ways one could go about introducing Gaussian quadrature. First, you could try to generalize the approach of the previous subsection to the case of larger n. As you can imagine, this will lead to a system of increasingly non-linear equations; this is precisely the task we tackled in problem 5.22. This approach to Gauss–Legendre integration involves a Vandermonde matrix and is therefore increasingly unstable as n gets larger. Second, one could employ the standard-textbook approach, which introduces the theory of orthogonal polynomials by fiat right about now, typically leaving students scratching their heads. This involves a number of auxiliary entities and is almost never carried through to the end, resorting instead to the dreaded phrase "it can be shown"; for the sake of completeness, we briefly go over that alternative argument in problem 7.22. Intriguingly, there exists a third approach to implementing Gauss–Legendre quadrature, known as the *Golub–Welsch algorithm*; this involves Jacobi matrices, which were touched upon in problem 5.20 and are further elaborated on in problem 7.23. This happens to be a robust approach which is employed in state-of-the-art libraries, but it assumes one is already convinced of the importance of orthogonal polynomials to this problem.

In contradistinction to these approaches, in the present section we will follow a fourth option, tackling Gaussian quadrature using *Hermite interpolation*, a subject we introduced in section 6.2.3. As you may recall, Hermite interpolation arises when you wish to interpolate through both function and derivative values. At the simplest level, Hermite interpolation is the only tool we've encountered that can handle polynomials of degree $2n-1$. While the question of exactly how Hermite interpolation accomplishes the task at hand may initially be a bit nebulous, everything will become clear soon enough and will then follow straightforwardly from this one choice. Crucially, our approach will allow us to explain in a unified manner: (a) how to pick the nodes x_k, thereby justifying the presence of the name "Legendre" in Gauss–Legendre quadrature, (b) how to compute the weights c_k, and (c) the overall error scaling of Gauss–Legendre quadrature. We discuss these topics in turn, in the following subsections.

7.5.2.1 Node Placement

In section 7.2.3 we saw how to produce Newton–Cotes quadrature rules starting from Lagrange interpolation, writing the interpolating polynomial $p(x)$ of degree $n-1$ in terms of the function values at the grid points, as well as the cardinal polynomials, see Eq. (7.24). We now write down the analogous formula for the case of Hermite interpolation, since we are interested in studying polynomials of degree up to $2n-1$. This is Eq. (6.40):

$$p(x) = \sum_{k=0}^{n-1} f(x_k)\alpha_k(x) + \sum_{k=0}^{n-1} f'(x_k)\beta_k(x) \tag{7.95}$$

where we have updated the notation to employ $f(x_k)$ and $f'(x_k)$ for our integrand and derivative values at the nodes; the $\alpha_k(x)$ and $\beta_k(x)$ are determined in terms of the cardinal polynomials as per Eq. (6.49). You may be worried about the fact that we are trying to interpolate through the $f'(x_k)$ values, which we don't actually know; our goal is to express an integral in terms of function values as per Eq. (7.89), not in terms of function-derivative values. For now, you should be patient.

We are actually interested in integrating $f(x)$: if we pick $f(x)$ to be a polynomial of degree $2n-1$ or less, then Eq. (7.95) will *exactly* match $f(x)$. To see this, remember that our error formula for Hermite interpolation from Eq. (6.50) involved a $2n$-th derivative, $f^{(2n)}$. For a polynomial of degree $2n-1$ or less, that derivative is always zero, implying that $p(x)$ perfectly matches $f(x)$ for this case. We are therefore free to integrate the $p(x)$ of Eq. (7.95) from -1 to $+1$:

$$\int_{-1}^{1} f(x)dx = \int_{-1}^{1} p(x)dx = \sum_{k=0}^{n-1} f(x_k) \int_{-1}^{1} \alpha_k(x)dx + \sum_{k=0}^{n-1} f'(x_k) \int_{-1}^{1} \beta_k(x)dx$$
$$= \sum_{k=0}^{n-1} c_k f(x_k) + \sum_{k=0}^{n-1} d_k f'(x_k) \tag{7.96}$$

In the last step we defined:

$$c_k = \int_{-1}^{1} \alpha_k(x)dx, \qquad d_k = \int_{-1}^{1} \beta_k(x)dx \tag{7.97}$$

If we can now somehow make the d_k's vanish, we will have accomplished what we set out to: the left-hand side of Eq. (7.96) is the integral we wish to evaluate and the last step in Eq. (7.96) would express it (*exactly*) as a sum of weights times function values, precisely as in Eq. (7.89). Also, if the d_k's dropped out of the problem then we would no longer have to worry about the fact that we don't know the $f'(x_k)$ values.

To see how we could possibly make the d_k's vanish, we express the $\beta_k(x)$ in terms of the cardinal polynomials as per Eq. (6.49):[21]

$$d_k = \int_{-1}^{1} \beta_k(x)dx = \int_{-1}^{1} (x - x_k)L_k^2(x)dx = \frac{1}{L'(x_k)} \int_{-1}^{1} L(x)L_k(x)dx \tag{7.98}$$

The x_k in the second step are the interpolation nodes and similarly $L_k(x)$ are the cardinal polynomials defined for those nodes. In the last step, we expressed (one of) the $L_k(x)$ in terms of the node polynomial and its derivative, as per Eq. (6.27); we pulled the $L'(x_k)$ outside the integral since it doesn't depend on x.

We now recall the definition of the node polynomial from Eq. (6.24):

$$L(x) = \prod_{j=0}^{n-1} (x - x_j) \tag{7.99}$$

This is a monic polynomial in x of degree n, that vanishes at the interpolation nodes x_j. Similarly, we remember the definition of a cardinal polynomial from Eq. (6.19):

$$L_k(x) = \frac{\prod_{j=0, j\neq k}^{n-1} (x - x_j)}{\prod_{j=0, j\neq k}^{n-1} (x_k - x_j)} \tag{7.100}$$

This is a polynomial in x of degree $n - 1$. The way to make progress is to demand that the integral in Eq. (7.98) vanishes: we recall from problem 4.27 that we had derived:

$$(r_{n-1}, q_n) \equiv \int_{-1}^{1} r_{n-1} q_n dx = 0 \tag{7.101}$$

in Eq. (4.258). In this equation, $r_{n-1}(x)$ was a general polynomial of degree $n - 1$ and q_n was an orthogonal polynomial of degree n. This would be precisely the situation we are faced with in Eq. (7.98) *if* $L(x)$ was an orthogonal polynomial. However, we recall that the

[21] Since we're integrating over the entire interval, we're using entities like $L_k(x)$, rather than the more awkward notation of $L_{i+j}(x)$ in Eq. (7.25).

interpolation nodes, i.e., the x_k's, are still at our disposal: *if we take them to be the zeros of an orthogonal polynomial of degree n*, then $L(x)$ will be the unique[22] monic polynomial of degree n that is orthogonal (in the interval $[-1, +1]$) to all polynomials of degree $n-1$ or less. In that case all the d_k's will vanish and Eq. (7.96) will take the form:

$$\int_{-1}^{1} f(x)dx = \sum_{k=0}^{n-1} c_k f(x_k) \tag{7.102}$$

where the c_k can be determined from Eq. (7.97). As the preceding derivation shows, this relation is *exact* when $f(x)$ is a polynomial of degree up to $2n-1$ and the x_k's are taken to be the roots of the orthogonal polynomial of degree n. As noted, there is no error term here, since the $2n$-th derivative, $f^{(2n)}$, vanishes. Similarly, there is no term involving $f'(x_k)$ since the d_k's vanish. In short, we have managed to employ Hermite interpolation to integrate polynomials of degree $2n-1$ or less using n nodal abscissas, with the derivatives dropping out of the problem.[23] Crucially, this was a result following from *orthogonality*, i.e., *integration*: the d_k's (which involve an integral) vanish, whereas the $\beta_k(x)$'s (which appear in Hermite interpolation) do not vanish (outside the interpolation nodes).

At this point, we remember that in problem 4.27 we had observed that the orthonormal q_j we had been finding there were multiples of the orthogonal *Legendre polynomials*, $P_n(x)$. The same holds for our monic orthogonal polynomial $L(x)$: with the exception of an overall multiplicative constant, $P_n(x)$ is the same polynomial as $L(x)$. We actually computed this multiplicative factor in Eq. (3.88), using Rodrigues' formula:

$$P_n(x) = \frac{(2n)!}{2^n(n!)^2}L(x) \tag{7.103}$$

Most importantly for our purposes, the *zeros of $L(x)$ are the zeros of the Legendre polynomial $P_n(x)$* since, as we mentioned on page 268, you will still find the same roots if you multiply a polynomial by a number. As it so happens, we've already written a Python code that computes the zeros of Legendre polynomials, `legroots.py`. You can now see why we have been speaking of Gauss–*Legendre* quadrature: this is Gaussian quadrature when the nodal abscissas are taken to be the roots of Legendre polynomials. This also explains why we decided to focus on the interval $[-1, +1]$ in the first place.

The main advantage of our derivation is that the theory of orthogonal polynomials (and the placement of the nodes at their roots) occurs organically: we didn't start the section using Legendre polynomials, but these emerged naturally when we needed to make the d_k's vanish, thereby managing to integrate exactly all polynomials of degree up to $2n-1$. It's worth observing that in this one subsection we have employed Legendre polynomials from chapter 2, their leading coefficient from chapter 3, their orthogonality from chapter 4, their zeros from chapter 5, as well as Hermite interpolation and cardinal polynomials from chapter 6. This is a prominent example of the fact that the presentation in this book is

[22] Problem 7.18 guides you toward showing this uniqueness property.
[23] Thus, the input is not x_k, $f(x_k)$, and $f'(x_k)$, but solely x_k and $f(x_k)$.

cumulative: we didn't need to state anything without proof, but you will need to master the earlier material in order to grasp how everything fits together.

7.5.2.2 Weight Computation

After deriving Eq. (7.102), we observed that the c_k's can be determined from Eq. (7.97), which in its turn can make use of Eq. (6.48); this is true, and is an avenue you will explore in problem 7.20; here, we realize that we can easily arrive at a simpler expression. Our Eq. (7.102) holds for any polynomial of degree up to $2n-1$, which means it will also hold for $f(x) = L_j(x)$: since $L_j(x)$ is a polynomial of degree $n-1$, it can be exactly integrated via Gauss–Legendre quadrature. Thus:

$$\int_{-1}^{1} L_j(x)dx = \sum_{k=0}^{n-1} c_k L_j(x_k) = \sum_{k=0}^{n-1} c_k \delta_{jk} \tag{7.104}$$

where the last step followed from our general result on cardinal polynomials, Eq. (6.21). Thus, we have been able to show that:

$$c_j = \int_{-1}^{1} L_j(x)dx \tag{7.105}$$

Intriguingly, this is formally identical to what we had found for Newton–Cotes integration, see Eq. (7.27). The only difference is that here we are not integrating over an elementary interval, but directly over the full interval.

For the Newton–Cotes case, we studied only small n values, given that we were only interested in an elementary interval, so we were able to explicitly carry out the integrals giving the weights for the trapezoid and Simpson's rules. In the Gauss–Legendre case, we're keeping n general, so we should try to see if we can re-express the integral involved. It is straightforward to start doing so:

$$c_j = \int_{-1}^{1} L_j(x)dx = \int_{-1}^{1} \frac{L(x)}{(x-x_j)L'(x_j)}dx = \frac{1}{P_n'(x_j)} \int_{-1}^{1} \frac{P_n(x)}{x-x_j}dx \tag{7.106}$$

In the second step we expressed the cardinal polynomial in terms of the node polynomial and its derivative, as per Eq. (6.27), as in Eq. (7.98). In the third step we remembered that $L(x)$ is simply a multiple of the Legendre polynomial $P_n(x)$, as per Eq. (7.103); since the multiplicative factor appears on both the numerator and the denominator, it cancels.

This is progress, but we're not quite there yet: Eq. (7.106) involves an integral, which we cannot do numerically, since that's why we're computing the c_j's in the first place. To proceed, we invoke the *Christoffel–Darboux identity* for Legendre polynomials:

$$\sum_{k=0}^{n} (2k+1)P_k(x)P_k(y) = (n+1)\,\frac{P_{n+1}(x)P_n(y) - P_n(x)P_{n+1}(y)}{x-y} \tag{7.107}$$

which you are guided toward deriving in problem 7.19. We now pick $y = x_j$, where x_j is a root of $P_n(x)$, i.e., $P_n(x_j) = 0$. That makes the n-th term in the sum on the left-hand side as well as the first term in the numerator on the right-hand side vanish, leaving us with:

$$\sum_{k=0}^{n-1}(2k+1)P_k(x_j)P_k(x) = -(n+1)P_{n+1}(x_j)\,\frac{P_n(x)}{x-x_j} \tag{7.108}$$

We now recall from Eq. (2.79) that $P_0(x) = 1$, so we can think of the left-hand side of Eq. (7.108) as already containing $P_0(x)$. We can then integrate from -1 to $+1$ to find:

$$\sum_{k=0}^{n-1}(2k+1)P_k(x_j)\int_{-1}^{1} P_0(x)P_k(x)dx = -(n+1)P_{n+1}(x_j)\int_{-1}^{1}\frac{P_n(x)}{x-x_j}dx \tag{7.109}$$

At this point, we employ the orthogonality of Legendre polynomials, which has the form:

$$(P_n(x), P_m(x)) \equiv \int_{-1}^{1} P_n(x)P_m(x)dx = \frac{2}{2n+1}\delta_{nm} \tag{7.110}$$

as we saw in Eq. (4.259). Using this on the left-hand side of Eq. (7.109) we get:

$$2 = -(n+1)P_{n+1}(x_j)\int_{-1}^{1}\frac{P_n(x)}{x-x_j}dx \tag{7.111}$$

where we dropped the $P_0(x_j)$ on the left-hand side, since it's equal to 1.

We now realize that the integral in Eq. (7.111) is the same as that appearing in Eq. (7.106); putting these two equations together leads to:

$$c_j = -\frac{2}{(n+1)P_{n+1}(x_j)P'_n(x_j)} \tag{7.112}$$

This is, in principle, the end of our journey: as you may recall, `legendre.py` computed not only the value of the Legendre polynomial, but also the value of its first derivative. Thus, we could now proceed to a Python implementation of Eq. (7.112). However, this formula involves a Legendre polynomial and its derivative for two different degrees; let us further simplify our expression for the weights. From Eq. (2.89) we get:

$$(1-x^2)P'_n(x) = -nxP_n(x) + nP_{n-1}(x) \tag{7.113}$$

We can also write down Eq. (2.86) for the case of $j = n$:

$$(n+1)P_{n+1}(x) = (2n+1)xP_n(x) - nP_{n-1}(x) \tag{7.114}$$

Adding the last two equations together, we eliminate $nP_{n-1}(x)$ to find:

$$(1-x^2)P'_n(x) = (n+1)xP_n(x) - (n+1)P_{n+1}(x) \tag{7.115}$$

If we now apply this relation at $x = x_j$, where as usual $P_n(x_j) = 0$, we get:

$$(1-x_j^2)P'_n(x_j) = -(n+1)P_{n+1}(x_j) \tag{7.116}$$

which, combined with Eq. (7.112), gives for the *weights in Gauss–Legendre quadrature*:

$$c_j = \frac{2}{(1 - x_j^2)[P'_n(x_j)]^2} \tag{7.117}$$

As desired, this requires the value of the Legendre polynomial (derivative) of a single degree. The plan of action, were one to implement things programmatically at this point, should be clear: Gauss–Legendre quadrature amounts to employing Eq. (7.102), where the x_k's are produced by finding the roots of $P_n(x)$ and the c_k's are computed using Eq. (7.117).

We now produce an identity that must be obeyed by the Gauss–Legendre weights. We have showed that Eq. (7.102) holds for all polynomials of degree $2n - 1$ or less, so it also holds for the case of a constant polynomial, $f(x) = 1$. This leads to:

$$\sum_{k=0}^{n-1} c_k = 2 \tag{7.118}$$

The 2 on the right-hand side is the result of working in the standard interval; it would have been $b - a$ in the general case. As mentioned after Eq. (7.29), in problem 7.3 you will show something analogous for Newton–Cotes methods. Speaking of the weights and Newton–Cotes, there is another difference between those methods and Gaussian quadrature ones: as you will show in problem 7.20, in Gauss–Legendre integration *all the weights are positive*. This turns out to be different from the behavior of Newton–Cotes weights, where for higher degrees we tend to see both positive and negative weights (bringing to mind the finite-difference formulas from chapter 3 and the accompanying error issues).

7.5.2.3 Error Analysis

Similarly to what we did in the case of the Newton–Cotes rules, we will now discuss the error scaling in Gauss–Legendre quadrature. More specifically, just as we did for the trapezoid rule in section 7.2.4.1 we will start from an interpolation error formula and then will integrate that. Of course, this time we are dealing with Hermite (not Lagrange) interpolation, and we are also dealing with the full (not elementary) interval.

In our derivation of Gauss–Legendre quadrature above, we were dealing with polynomials of degree up to $2n - 1$, which is why there was no error term in Eq. (7.96). We will now address the general case, where our interpolating polynomial $p(x)$ of degree up to $2n - 1$ may not fully capture the function $f(x)$, which could be non-polynomial (or of higher degree). Let us start from the Hermite-interpolation error formula, Eq. (6.50):

$$f(x) = p(x) + \frac{f^{(2n)}(\zeta)}{(2n)!} \prod_{j=0}^{n-1} (x - x_j)^2 \tag{7.119}$$

where we slightly updated the notation. We move the $p(x)$ to the left-hand side and identify the product over $(x - x_j)^2$ with the square of the node polynomial from Eq. (6.24):

$$f(x) - p(x) = \frac{f^{(2n)}(\zeta)}{(2n)!} L^2(x) \tag{7.120}$$

Integrating this difference from −1 to +1 gives us the error in Gauss–Legendre quadrature:

$$\mathcal{E} = \int_{-1}^{1} \frac{f^{(2n)}(\zeta)}{(2n)!} L^2(x)dx = \frac{f^{(2n)}(\xi)}{(2n)!} \int_{-1}^{1} L^2(x)dx \tag{7.121}$$

In the second step we used the extended version of the integral mean-value theorem to effectively pull out the derivative term. Similarly to the situation in Eq. (7.34), this is here justified because $L^2(x)$ is non-negative.

If we now express the node polynomial $L(x)$ in terms of the corresponding Legendre polynomial $P_n(x)$ as per Eq. (7.103), we will be left with an integral over $P_n^2(x)$. But that, in its turn, can be straightforwardly carried out using the orthogonalization/normalization relation for Legendre polynomials from Eq. (7.110). Putting it all together, we find the following error formula for Gauss–Legendre quadrature:

$$\mathcal{E} = \frac{2^{2n+1}(n!)^4}{(2n+1)[(2n)!]^3} f^{(2n)}(\xi) \tag{7.122}$$

This is more inscrutable than the corresponding formulas for Newton–Cotes methods, for two reasons: first, the $2n$-th derivative is typically hard to estimate and, second, all the powers and factorials make it hard to intuitively grasp how the error scales with n.

To elucidate things, we take one of the $(2n)!$ terms from the denominator and group it with the derivative term. The remaining prefactor can be approximated as follows:[24]

$$\frac{2^{2n+1}(n!)^4}{(2n+1)[(2n)!]^2} \sim \frac{\pi}{4^n} \tag{7.123}$$

Using this asymptotic relation, we can now give our final expression for the (approximation) *error scaling for Gauss–Legendre quadrature*:

$$\mathcal{E} \sim \frac{\pi}{4^n} \frac{f^{(2n)}(\xi)}{(2n)!} \tag{7.124}$$

The derivative term has been grouped in this way motivated by a standard Taylor series: as you will recall, the n-th term is divided by $n!$; we have therefore divided our $2n$-th derivative by $(2n)!$; this ratio is quite often bounded (see also page 337). If that is indeed the case, then the simple prefactor, $\pi/4^n$, gives us a practical criterion for the scaling of the error with n: as we increase n to $n+1$, we expect the error to decrease by a factor of roughly 4. This is a dramatically better scaling than what we had for Newton–Cotes methods: for example, the absolute error in the composite Simpson's rule was proportional to h^4, see Eq. (7.54), i.e., asymptotically $1/n^4$; compare that to the $1/4^n$ that we found in Eq. (7.124) for Gauss–Legendre quadrature. This is the difference between a power law and an exponential scaling. Thus, it should come as no surprise that Gaussian quadrature does a great job for many well-behaved functions.

[24] To prove this, use *Stirling's formula*, $n! \sim \sqrt{2\pi} e^{-n} n^{n+1/2}$.

7.5.2.4 Integrating from *a* to *b*

Up to this point we've only discussed Gauss–Legendre integration from −1 to 1, as per Eq. (7.89). We generally need to integrate from a to b. The solution is simply to carry out a change of variables:

$$t = \frac{b+a}{2} + \frac{b-a}{2}x \tag{7.125}$$

just like we did in our discussion of polynomial interpolation, see Eq. (6.13). We now use Eq. (7.125) to get the integration measure:[25]

$$dt = \frac{b-a}{2}dx \tag{7.126}$$

At this point, we are able to recast our general task of computing $\int_a^b f(t)dt$ into a form that makes use of our earlier results on $\int_{-1}^1 f(x)dx$:

$$\int_a^b f(t)dt = \frac{b-a}{2}\int_{-1}^1 f\left(\frac{b+a}{2} + \frac{b-a}{2}x\right)dx \tag{7.127}$$

If we now employ Eq. (7.89) to go from integral to sum, this becomes:

$$\int_a^b f(t)dt \approx \frac{b-a}{2}\sum_{k=0}^{n-1} c_k f\left(\frac{b+a}{2} + \frac{b-a}{2}x_k\right) \tag{7.128}$$

Here the x_k and c_k are the standard quantities we derived above. You may wish to view Eq. (7.128) as summing up rectangle areas by analogy to the rectangle or midpoint methods, Fig. 7.2 (cf. Fig. 7.6 below); here each height is determined by the value of the function at the unequally spaced (clustered) t_k's and, crucially, the widths are variable, $(b-a)c_k/2$.

7.5.2.5 Implementation

A Python implementation of Gauss–Legendre quadrature is given in code 7.4. As before, we will compute the integral giving the electrostatic potential in Eq. (7.2); we import the integrand from `newtoncotes.py`. Similarly, we bring in the Legendre polynomial roots from `legroots.py` and derivatives from `legendre.py`. Observe that these two functions do all the heavy lifting; as a result, our code is very short.

The first function in this code produces the x_k's by finding the roots of $P_n(x)$ and the c_k's by using Eq. (7.117). Observe that `legroots()` returns a `numpy` array, which we now call `xs`. This is very convenient in the following line where, as usual, we don't need to employ any indices. Note also that we are calling `legendre()` passing in an array as the second argument: even though this function was written in chapter 2, before we had started using `numpy` arrays, it works seamlessly.

The next function, `gauleg()`, implements Eq. (7.128), taking us from x to t. This is

[25] As usual, no differential forms were harmed in the making of this book—see also Eq. (7.195).

Code 7.4 gauleg.py

```
from legendre import legendre
from legroots import legroots
from newtoncotes import f
import numpy as np

def gauleg_params(n):
    xs = legroots(n)
    cs = 2/((1-xs**2)*legendre(n,xs)[1]**2)
    return xs, cs

def gauleg(f,a,b,n):
    xs, cs = gauleg_params(n)
    coeffp = 0.5*(b+a)
    coeffm = 0.5*(b-a)
    ts = coeffp + coeffm*xs
    contribs = cs*f(ts)
    return coeffm*np.sum(contribs)

if __name__ == '__main__':
    ans = np.log(1 + np.sqrt(2))
    print(ans)
    for n in range(2,10):
        print(n, gauleg(f,0.,1,n))
```

intentionally written to have the same interface as our Newton–Cotes functions: we pass in the integrand, the integration limits, and n. As always, we carry out operations over entire arrays, so no indices are necessary here, either. The output of running this code is:

```
0.88137358702
2 0.881789806445
3 0.881331201938
4 0.881375223073
5 0.881373570699
6 0.881373584915
7 0.881373587172
8 0.881373587015
9 0.88137358702
```

The function we're integrating is non-polynomial, so we don't expect Gauss–Legendre to give the exact answer for the first few values of n. Overall, the results are quite impressive. Gauss–Legendre gives us the correct answer to three decimal digits using only $n = 2$ points and to four decimal digits with only $n = 3$ points. By the time we get up to $n = 9$ we have already determined the right answer for all printed digits: this is pretty spectacular. To compare, recall that Simpson's rule needed over 100 points to get to the same accuracy and even (non-wasteful) Romberg integration required 33 points. Of course, as our discussion of the Gauss–Legendre error formula showed, the convergence to the right answer is always impacted by specific properties of the function we choose to integrate.

On the other hand, it is worth pointing out that each value of n necessitates completely different values of the abscissas x_k. This is not a problem if your $f(x)$ is a simple function, as in the present example. However, you can imagine that in practical applications computing $f(x)$ may be rather costly. In problem 7.13 you are guided toward implementing a non-wasteful adaptive trapezoid integrator; something analogous *cannot* be straightforwardly done for Gauss–Legendre integration. There are schemes, most notably *Gauss–Kronrod quadrature*, where the Gauss–Legendre points are kept and accompanied by new ones. Another option is *Clenshaw–Curtis quadrature*, which expands the integrand using Chebyshev polynomials and can therefore compute the integration weights very efficiently, via the FFT; this method's abscissas obey *nesting*, so they can be re-used as you increase n. Even so, the plain Gauss–Legendre approach will be enough for us (for more, see problem 7.24). From a practical perspective, you can start with, say, $n = 10$ and then increase n by steps of your choosing, to empirically test the convergence for your problem.

7.5.3 Other Gaussian Quadratures

In the previous subsections, we focused on Gauss–*Legendre* quadrature only. This was partly because that's the more general and widely employed technique and partly for pedagogical reasons: instead of giving a laundry list of formulas for many different cases, we studied one case carefully, deriving everything by ourselves. In the present short subsection, we will provide a qualitative argument of how one goes about generalizing things; problem 7.25 investigates some of these new formulas further.[26]

The general form of the problem at hand is:

$$\int_a^b w(x)f(x)dx \approx \sum_{i=0}^{n-1} c_k f(x_k) \tag{7.129}$$

where $w(x)$ is a non-negative *weight function*. As you can see, the presence of $w(x)$ on the left-hand side (only) is new. The weight function can be singular, but it needs to be integrable; crucially, it can be non-polynomial. As before, the Gaussian quadrature approach allows you to find the exact answer when $f(x)$ is a polynomial of degree up to $2n-1$: in that case, the full integrand does *not* need to be a polynomial. Table 7.2 shows a few standard

[26] As usual, our discussion in the main text is carried out *pros ten chreian ikanos*, i.e., sufficiently as to need, to quote Aristotle (*Nicomachean Ethics*, 1133b); the alternative would be for us to continue *ad taedium*.

Table 7.2 Summary of selected Gaussian quadrature methods

Method	Weight function $w(x)$	Standard interval $[a, b]$
Gauss–Legendre	1	$[-1, 1]$
Gauss–Chebyshev	$\frac{1}{\sqrt{1-x^2}}$	$[-1, 1]$
Gauss–Laguerre	$x^\alpha e^{-x}$	$[0, \infty)$
Gauss–Hermite	e^{-x^2}	$(-\infty, \infty)$

choices for the weight function, which have been worked out in the literature.[27] The line of reasoning goes as follows: start from the Hermite interpolation formula of Eq. (7.95), this time multiplying by $w(x)$ before integrating over the interval $[a, b]$, to find:

$$\begin{aligned}\int_a^b w(x)f(x)dx &= \int_a^b w(x)p(x)dx \\ &= \sum_{k=0}^{n-1} f(x_k)\int_a^b w(x)\alpha_k(x)dx + \sum_{k=0}^{n-1} f'(x_k)\int_a^b w(x)\beta_k(x)dx \\ &= \sum_{k=0}^{n-1} c_k f(x_k) + \sum_{k=0}^{n-1} d_k f'(x_k)\end{aligned} \tag{7.130}$$

In the last step we defined:

$$c_k = \int_a^b w(x)\alpha_k(x)dx, \qquad d_k = \int_a^b w(x)\beta_k(x)dx \tag{7.131}$$

As before, we will demand that the d_k's vanish. These are:

$$d_k = \int_a^b w(x)\beta_k(x)dx = \int_a^b w(x)(x - x_k)L_k^2(x)dx = \frac{1}{L'(x_k)}\int_a^b w(x)L(x)L_k(x)dx \tag{7.132}$$

Once again, $L(x)$ is a monic polynomial of degree n and $L_k(x)$ is a polynomial of degree $n-1$. In Eq. (7.98) we ensured that this integral vanished by taking the nodes, x_k, to be the roots of a Legendre polynomial. We did this because, for $w(x) = 1$ and in the interval $[-1, +1]$, it is precisely the Legendre polynomials that obey such an orthogonality criterion.

Our new approach leads us to the roots of polynomials that are *w-orthogonal in the interval from a to b*. These turn out to be Chebyshev polynomials, Laguerre polynomials, and Hermite polynomials for the cases listed in table 7.2 (which also shows the corresponding standard interval). Incidentally, you should make sure not to get confused by the presence of the name *Hermite* in Gauss–Hermite: we have derived *all* Gaussian quadratures using Hermite *interpolation*. What's special about Gauss–Hermite quadrature is that it involves Hermite *polynomials*.[28] It should be relatively obvious that the weights can now be com-

[27] Since Gauss–Legendre is equivalent to taking $w(x) = 1$, a crude strategy for avoiding other Gaussian quadratures is to treat the entire integrand $w(x)f(x)$ as what used to be called your $f(x)$ and then apply Eq. (7.128).

[28] We met these in section 3.5 and in problem 5.18; you may produce the Hermite polynomials by orthogonalizing the monomials with $w(x) = e^{-x^2}$ (cf. $w(x) = 1$ giving the Legendre polynomials in problem 4.27).

puted by a generalization of Eq. (7.105), possibly augmented by a Christoffel–Darboux identity for the specific set of polynomials, as well as the corresponding normalization and recurrence relation. Similarly, the error scaling will follow from the generalization of Eq. (7.121) to include a $w(x)$ in the integrand. In broad strokes, that's all there is to it.

7.6 Complicating the Narrative

Most statements about errors that we've made in this chapter so far assumed we were dealing with well-behaved Taylor series, with each term being smaller in magnitude than the previous one. In practice, this is equivalent to saying that our function can be reasonably well approximated locally by a polynomial. However, there are many cases, even non-pathological ones, which cannot be effectively captured by polynomial behavior. For example, as Forman Acton liked to point out, polynomials don't have asymptotes [2]. In what follows, we go over some cases where the error budget behaves differently than what we saw above, we discuss analytical manipulations one can carry out *before* computing an integral, discuss the problem of integrating over more than one variable, and close with some advice on which integration approach(es) you should prefer.

7.6.1 Periodic Functions

Given the ability to use Gaussian quadrature or even Simpson's rule, you may be tempted to say that a method as simple as the trapezoid rule, with its $O(h^2)$ approximation error as per Eq. (7.37), should *never* be employed. As you'll see when you solve problem 7.27, this conclusion is wrong; the problem studies the function $f(x) = e^{\sin(2x)}$, which we've repeatedly encountered in earlier chapters, but the argument explaining what's going on is much more general.

As we saw when introducing the Romberg recipe, our conclusion for the trapezoid rule was merely that the *leading* error is $O(h^2)$; in Eq. (7.70) we saw that the full expression contains many more (even) degrees:

$$\mathcal{E} = c_2 h^2 + c_4 h^4 + c_6 h^6 + \cdots \tag{7.133}$$

This fact, namely that only even degrees survive, was all we needed to employ Richardson extrapolation repeatedly. This time around, we are interested in investigating the error budget in more detail: Eq. (7.70) was actually merely a way of condensing our result employing the Euler–Maclaurin summation formula; this was Eq. (7.41), which explicitly listed the coefficients of the powers of h:

$$\begin{aligned}\mathcal{E} = &-\frac{h^2}{12}\left[f'(b) - f'(a)\right] + \frac{h^4}{720}\left[f'''(b) - f'''(a)\right] \\ &- \frac{h^6}{30\,240}\left[f^{(5)}(b) - f^{(5)}(a)\right] + \frac{h^8}{1\,209\,600}\left[f^{(7)}(b) - f^{(7)}(a)\right] + \cdots\end{aligned} \tag{7.134}$$

These coefficients involve Bernoulli numbers, factorials and, crucially, *the differences between odd-order derivatives at the endpoints.*

If you take a periodic function and integrate it over one full period (or more periods), you will have $f'(a) = f'(b)$, $f'''(a) = f'''(b)$, and so on; thus, the error estimate from Eq. (7.134) will have a zero coefficient for each power.[29] This means that in such a case the composite trapezoid rule will have spectral accuracy, i.e., *dramatically* better error behavior than usual; as you will discover in problem 7.27, in some cases it even outperforms Gauss–Legendre quadrature, whose praises we were singing in the previous section. Even if your setting is not periodic, you can transform it into one that is: this is precisely what lies behind the power of the Clenshaw–Curtis technique (see page 499 and problem 7.24).

7.6.2 Singularities

A major complication arises when you are faced with an integrand that has a singularity, or a singular derivative at one of the endpoints. If the singularity is not at an endpoint, you can slice up your interval so that the singularity is at an endpoint. We will discuss five examples here to give you a flavor of the techniques involved; the problems invite you to apply analogous tricks to other integrals.

Singularity at the Left Endpoint

Examine the integral:

$$I_A = \int_0^2 \frac{\sin x}{\sqrt{x}} dx \tag{7.135}$$

This has a singularity[30] at the left endpoint, $x = 0$, so closed integration methods (e.g., the trapezoid rule) naively applied to this problem will give "not a number" (along with a `RuntimeWarning`) or a `ZeroDivisionError`; furthermore, if you plot the integrand you will see that it is vertical near the origin.

The solution is to make a change of variables; taking $u = \sqrt{x}$ leads to:

$$I_A = \int_0^{\sqrt{2}} 2\sin(u^2) du \tag{7.136}$$

This has no singularities and no vertical slopes, so closed methods will work just fine.

Discontinuous Derivative

Now take the following integral:

$$I_B = \int_0^2 \sqrt{x} \sin x dx \tag{7.137}$$

This time, the integrand is not singular, but its derivative(s) will be singular. Problem 7.28 asks you to apply the trapezoid rule to a similar integral; there you will see in action that the error does not behave according to Eq. (7.133), i.e., it is much worse. To see why, recall that Eq. (7.37) told us that the leading error for the trapezoid rule is not simply $O(h^2)$, but actually $-h^2 f''(\xi)(b-a)/12$. This means that you get in trouble if you are dealing with non-smooth functions. We will now see three different ways of addressing this situation.

29 The only possible exception being the remainder term that was dropped from Eq. (7.39).
30 The integrand has a *numerical* singularity, since floats don't know about L'Hôpital's rule.

First, we can make the same change of variables as above, namely $u = \sqrt{x}$. This gives:

$$I_B = \int_0^{\sqrt{2}} 2u^2 \sin(u^2) du \tag{7.138}$$

Our new integrand is arbitrarily often differentiable in our interval.

Second, you can choose to slice up the integral:

$$I_B = \int_0^{0.01} \sqrt{x} \sin x dx + \int_{0.01}^{2} \sqrt{x} \sin x dx \tag{7.139}$$

where the 0.01 is, obviously, not set in stone but merely indicative of a "small" value. Notice that for the second term, whose endpoints do not include the origin, the integrand is arbitrarily often differentiable. For the first integral, since it only involves small values of x, you are allowed to Taylor expand $\sin x$ in it:

$$\int_0^{0.01} \sqrt{x} \sin x dx = \int_0^{0.01} \sqrt{x} \left(x - \frac{x^3}{6} + \frac{x^5}{120} - \cdots \right) dx \tag{7.140}$$

Each of these integrals can now be carried out analytically.

Third, we often try to *subtract off the singularity*, which in our case translates to subtracting off the singularity in the derivative. Practically speaking, what this means is we add in and subtract out another function that can be analytically integrated *and* exhibits the same misbehavior as our integrand. In our case, this can be done as follows:

$$I_B = \int_0^2 \sqrt{x} \sin x dx = \int_0^2 \sqrt{x}(\sin x - x) dx + \int_0^2 \sqrt{x} x dx = \int_0^2 \sqrt{x}(\sin x - x) dx + \frac{8\sqrt{2}}{5} \tag{7.141}$$

The integrand we are left with in the last step has a continuous third derivative, so an approach like the trapezoid rule won't have as hard a time tackling it.[31]

Singularity at the Right Endpoint

Examine the following integral:

$$I_C = \int_0^1 \frac{\sin x}{\sqrt{1 - x}} dx \tag{7.142}$$

This one has a singularity at the right endpoint. The trick is to move the singularity to the origin, by using the change of variables $u = 1 - x$, which leads to:

$$I_C = \int_0^1 \frac{\sin(1 - u)}{\sqrt{u}} du \tag{7.143}$$

where the minus signs from the differential and the new limits of integration have balanced each other out. We now have the same problem with a square root at the origin as in I_A, so we will make a further change of variables, namely $v = \sqrt{u}$, leading to:

$$I_C = \int_0^1 2 \sin(1 - v^2) dv \tag{7.144}$$

which has no issues with singularities or smoothness.

[31] Of course, as you discovered in chapter 2, $\sin x - x$ suffers from subtractive cancellation for small x, so you would benefit from employing a Taylor expansion here.

Singularities at Both Endpoints

We now turn to a slightly different integral:

$$I_D = \int_0^1 \frac{\sin x}{\sqrt{x(1-x)}} dx \tag{7.145}$$

This looks similar to I_C, but now has problems at both endpoints. The trick is to split it into two integrals by hand:

$$I_D = \int_0^{1/2} \frac{\sin x}{\sqrt{x(1-x)}} dx + \int_{1/2}^1 \frac{\sin x}{\sqrt{x(1-x)}} dx \tag{7.146}$$

where the matching point was arbitrarily chosen to be the midpoint. Now the first integrand has a singularity at the left endpoint and the second integrand a singularity at the right endpoint. The first integral should be tackled by saying $u = \sqrt{x}$ and the second integral by taking $u = 1 - x$ followed by $v = \sqrt{u}$.

Again, Singularities at Both Endpoints

As our final example we look at another integral that raises concerns at both endpoints:

$$I_E = \int_{-1}^1 \frac{\sin x}{\sqrt{1-x^2}} dx \tag{7.147}$$

This time taking $u = \sqrt{x}$ complicates things at the left endpoint. What we'll do, instead, is to make the change of variables $u = \sqrt{1-x}$, which leads to:

$$I_E = \int_0^{\sqrt{2}} 2\frac{\sin(1-u^2)}{\sqrt{2-u^2}} du \tag{7.148}$$

The denominator may look similar to the one we started from, but now it only has an issue at the right endpoint.

You could now imagine employing yet another trick, most likely chosen from the list of those we've already encountered. Instead, we recognize that I_E is an example of a standard Gaussian quadrature, namely *Gauss–Chebyshev*, as per table 7.2; the weights have already been studied and tabulated, so this integral can be straightforwardly evaluated. The lesson of this story is to always examine your integrals to see if they are of standard form *before* you start using the tricks of the trade.

7.6.3 Infinite Intervals

We now turn to another scenario, one where an infinity appears not in the integrand but in the endpoints of integration; as you may recall, these are called *improper integrals*.

One Infinity

Here's a straightforward example:

$$I_F = \int_1^\infty \frac{e^{-x}}{x+1} dx \tag{7.149}$$

If you wanted to use a standard (Newton–Cotes or Gaussian) quadrature approach, you would be faced with the question of how to handle the right endpoint. One possible solution is to take it to be large but finite, say, 5 or 10; but is that enough? You could try increasing it further to, say, 20 or 50 and see if the result of numerical integration keeps changing. The problem is that once you do that, you need to also make sure you employ more integration abscissas. But a simple plot of this integrand shows you that its values are tiny when x is above 10 or so. This means that you would be employing the majority of the abscissas to integrate a tiny tail of the function.

In the spirit of the previous subsection, what we do, instead, is a change of variables; taking $u = 1/x$ our integral becomes:

$$I_F = \int_0^1 \frac{e^{-1/u}}{u(u+1)} du \tag{7.150}$$

Note how the new integration interval is *not* infinite. Of course, now there is a new singularity at the origin, but that can be tackled using the techniques of the previous subsection.

Another Infinity

Of course, the trick we just introduced worked only because one of the endpoints was infinite *and* the other endpoint was non-zero; you can immediately see that if we had been faced with, say:

$$I_G = \int_0^\infty \frac{1}{\sqrt{x^8 + x}} dx \tag{7.151}$$

then the change of variables $u = 1/x$ would still lead to an infinite endpoint. One approach is to manually slice this integral up:

$$I_G = \int_0^1 \frac{1}{\sqrt{x^8 + x}} dx + \int_1^\infty \frac{1}{\sqrt{x^8 + x}} dx \tag{7.152}$$

The first integral is over a finite interval and the second one is amenable to the change of variables $u = 1/x$ and will, crucially, have no singularities. The former integral doesn't have an infinite endpoint, but it does have a singularity at the origin; as in the previous subsection, we can take $u = \sqrt{x}$ to eliminate that.

Another approach involves a single transformation for the entire interval. This one is sufficiently important that it deserves to be emphasized:[32]

$$u = \frac{1}{1+x} \tag{7.153}$$

It is straightforward to see that this leads to a finite interval (0 gets mapped onto 1 and ∞ onto 0). Similarly, it can be solved to give $x = (1-u)/u$ and $dx = -du/u^2$ (recall footnote 25). Using this transformation, I_G turns into:

$$I_G = \int_0^1 \frac{u^2}{\sqrt{(1-u)^8 + u^7(1-u)}} du \tag{7.154}$$

[32] Of course, this is not unique; for example, $u = x/(1+x)$ is another choice.

This has taken care of the infinite interval, but is suffering from a singularity at the right endpoint. However, the sequence of steps discussed earlier can cure all issues here, as you will discover in problem 7.29.

Two Infinities

Sometimes we are faced with an integration from minus infinity to plus infinity:

$$I_H = \int_{-\infty}^{\infty} \frac{e^{-x^2}}{\sqrt{x^2+1}} dx \tag{7.155}$$

What's typically done in these cases is to slice up the integral into two pieces, one going from $-\infty$ to 0 and the other one from 0 to $+\infty$. In the present case, this allows us to cut the work in half, since the integrand is even, but the technique is of wider applicability.

Yet another approach is to recognize that I_H is an example of a standard Gaussian quadrature, namely *Gauss–Hermite*, as per table 7.2; that means that the weights have already been studied and tabulated, so this integral can be straightforwardly evaluated. This is a general feature: when faced with integrals from $-\infty$ to $+\infty$ you should always consider Gauss–Hermite and, similarly, for integrals from 0 to ∞ (such as I_G) you should think of Gauss–Laguerre.

7.6.4 Multidimensional Integrals

It is, in principle, straightforward to generalize our earlier discussion to the case of multidimensional problems. The easiest way to do so is to use *Fubini's theorem*, whereby a multidimensional integral is written as a sequence of iterated one-dimensional integrals.

To be specific, let's look at the following two-dimensional integral:

$$I = \int_a^b \int_a^b f(x,y)dxdy \tag{7.156}$$

We can re-express this by first defining:

$$F(y) = \int_a^b f(x,y)dx \tag{7.157}$$

This keeps y fixed and carries out the integration over x. Then, we can use this new entity to rewrite our original integral as follows:

$$I = \int_a^b F(y)dy \tag{7.158}$$

which is also a one-dimensional integral.

Numerically, we can use either Newton–Cotes or Gauss–Legendre quadrature, both of which take the form Eq. (7.7), to express the above steps as follows:

$$F(y) \approx \sum_{i=0}^{n-1} c_i f(x_i, y) \tag{7.159}$$

and then:

$$I \approx \sum_{j=0}^{n-1} c_j F(y_j) \tag{7.160}$$

Putting the last two equations together leads to:

$$I \approx \sum_{i=0}^{n-1} \sum_{j=0}^{n-1} c_i c_j f(x_i, y_j) \tag{7.161}$$

Implementing this should be straightforward. Problem 7.38 asks you to employ Gaussian quadrature for a five-dimensional integral: you simply generalize Eq. (7.161) to the five-variable case, which also means you will employ five sets of abscissas and weights.

As you will learn when you attempt that problem, as you increase the number of dimensions, you are faced with the following issue: should you keep the number of points in one dimension fixed, or the total number of points fixed? Let's be specific: assume d is the number of dimensions, n is the number of points in each dimension (just like in our one-dimensional studies earlier in this chapter), and $\mathfrak{y} = n^d$ is the total number of points across all dimensions. As is shown in Eq. (7.161), this $\mathfrak{y}$ gives the total number of function evaluations you will need to perform.[33] As we've noted repeatedly throughout this volume, in the real world it is precisely a function evaluation that is the slowest part of a computation: this is typically the result of separately running another lengthy calculation.

Since the practical limitation is "total wall-clock time", we now investigate the scaling with respect to the total number of function evaluations, $\mathfrak{y}$. For all the (one-dimensional) quadrature methods we've encountered, we can describe the leading error as $O(h^p)$, see also section 7.4.1. Using the definition of h from Eq. (7.9), we can express this leading error as $O(n^{-p})$. The first use of O applies to small h, whereas when we switch to n we are interested in large n. This is the error we make in one dimension, so in d dimensions we should probably multiply it by d; however, if we are interested in examining the scaling with n (or $\mathfrak{y}$), then d can be considered a constant prefactor which is therefore omitted. If we now take our leading error, $O(n^{-p})$, and express it in terms of the total number of function evaluations, $\mathfrak{y} = n^d$, we arrive at the *total error in d-dimensional integration*:

$$\mathcal{E} = O\left(\mathfrak{y}^{-p/d}\right) \tag{7.162}$$

To reiterate, p is the power of the leading error in the quadrature rule we're employing ($O(h^p)$), d is the number of dimensions, and $\mathfrak{y}$ is the number of function evaluations. When d is large, this scaling is very poor; for example, take Simpson's rule, $p = 4$, applied to a 100-dimensional integral: the scaling is $O(\mathfrak{y}^{-1/25})$, so as we double the total work, i.e.,

[33] In the present chapter, n denotes the number of abscissas and N the number of panels in the one-dimensional case; asymptotically, these two are equal: $O(n) = O(N)$. We also need to denote the normal distribution of Eq. (6.169) by $\mathcal{N}$. This is why we are using a new symbol, $\mathfrak{y}$, for the total number of function evaluations.

double η, the error is divided by $2^{1/25} \approx 1.03$. This is known as the *curse of dimensionality*: when d is large, doubling our effort is basically a waste of time.

This curse of dimensionality provides a natural entry point to section 7.7, which studies an approach to integration known as *Monte Carlo*; as you will see there, Monte Carlo has an error that scales as $O(\eta^{-1/2})$. Crucially, Monte Carlo exhibits this scaling *regardless of the dimensionality!* (See, however, footnote 57.) To return to our example of a 100-dimensional integral: in Monte Carlo when you double the total work the error is divided by $\sqrt{2} \approx 1.41$. This means that the effort you put in toward doubling η actually pays off. In practice, for one (or a few) dimension(s) it is standard to employ conventional quadrature methods like those discussed above, but for anything more complicated one has to choose a Monte Carlo algorithm. Before we see how that all works, let's do a brief recap.

7.6.5 Evaluating Different Integration Methods

We have encountered several quadrature techniques and seen their strengths and weaknesses. Here we summarize these comments:

- There is no substitute for plotting your integrand and thinking about it before you numerically integrate.
- If you can use an analytical trick to eliminate a singularity or a singularity in the derivative or an infinite interval, you should do so ahead of time.
- If your integrand is periodic and you are integrating over a full period (or several periods), the composite trapezoid rule can do a great job.
- If your (proper or improper) integral is of a standard Gaussian quadrature form, you should use the tabulated nodes and weights.
- For most smooth integrands where you need high accuracy, Romberg integration or Gauss–Legendre integration are the go-to solutions.
- If you need your abscissas to nest, employ a Newton–Cotes rule (or Clenshaw–Curtis quadrature).
- If you're not too worried about the accuracy but need a dependable error[34] (or if you are externally constrained to use equally spaced nodes), you should employ an adaptive Simpson's rule; if you have an even number of points, remove one point and use the trapezoid rule on the last panel.
- If you're dealing with a multidimensional integral, use Gaussian quadrature in a few dimensions and Monte Carlo in many dimensions.

7.7 Monte Carlo

The section's title comes from the casino in Monaco and reflects the "chance" (or random) aspect of such approaches. Monte Carlo techniques employ random numbers to

[34] Of course, the dependability of the error rests on a well-behaved Taylor expansion, i.e., a smooth integrand.

tackle either naturally stochastic (from the Greek for "aim" or "guess") processes or non-probabilistic problems. In keeping with this chapter's theme, we will focus on the latter, namely numerical integration. We start by discussing what "random" numbers are and how to produce them; we then turn to a detailed discussion of one-dimensional Monte Carlo quadrature, before addressing the real-world problem of multidimensional integration.

7.7.1 Random Numbers

Determining the exact meaning of "random" can become complicated. A nice example of patterns that we notice which may (or may not) really be there has to do with the decimal digits of the number π. For most practical purposes, the digits of π are random. However, when looking at two-digit repeats, we notice that the first such pair to re-appear is 26:

$$3.14159265358979323846264338327950 \tag{7.163}$$

We also notice that the second occurrence of 26 shows up in the middle of a strange repetition pattern: 79, 32, and 38 are repeated in reverse order contiguously, as 38, 32, and 79. The moral of this story is that checking for patterns/correlations can easily devolve into numerology. This is why one needs quantitative tests of the randomness of a given sequence. We will only scratch the surface of this topic here.

Computers produce *pseudorandom numbers*: the use of the modifier "pseudo" is due to the fact that computers are (supposed to be) deterministic systems. Thus, they produce sequences where each number is completely determined by its predecessor(s). However, if someone who does not have access to the random-number generation algorithm is led to believe that the sequence is unpredictable (truly random), then we have a "good" random-number generator.[35] Thus, when dealing with good pseudorandom number sequences we tend to simply drop the "pseudo" and speak simply of random-number sequences.

7.7.1.1 Linear Congruential Generator

We start with a simple approach that can produce reasonably good sequences of uniformly distributed random numbers, namely a *linear congruential generator*. While high-quality libraries typically employ more complicated algorithms, what we discuss here should be enough to give you a flavor of what's involved.

Integers

We first see how to produce randomly distributed integers and turn to floating-point numbers next. A linear congruential generator (LCG) produces integers from 0 to $m-1$, where m is an integer appearing in the formula:

$$u_i = (pu_{i-1} + c) \bmod m \tag{7.164}$$

[35] We do not broach the subject of algorithmic randomness, Turing machines, martingales, oracles, etc.

Here p and c are also integers. The way this works is that (for a fixed choice of p, c, and m) we start with an integer u_{i-1} and this equation instructs us how to construct the next integer in the sequence, u_i. This sequence has to start somewhere: we call the first/0th number (the one that jumpstarts the generation of numbers) the *seed* and denote it by u_0.

As an example, assume $p = 4$, $c = 1$, and $m = 15$ and pick the seed to be $u_0 = 5$:

$$
\begin{aligned}
&u_0 = 5, \quad u_1 = (4 \times 5 + 1) \bmod 15 = 6, \quad u_2 = (4 \times 6 + 1) \bmod 15 = 10,\\
&u_3 = (4 \times 10 + 1) \bmod 15 = 11, \quad u_4 = (4 \times 11 + 1) \bmod 15 = 0,\\
&u_5 = (4 \times 0 + 1) \bmod 15 = 1, \quad u_6 = (4 \times 1 + 1) \bmod 15 = 5
\end{aligned}
\tag{7.165}
$$

We have stopped here, because we know that as soon as we get a repeated number (5 in this case) all the numbers to follow will be obeying the same trend: this is a deterministic algorithm that will keep repeating the same sequence over and over again. The distinct numbers are those from u_0 to u_5, so we say we have a *period* of 6. To be explicit, the parameters we used here lead to the sequence:

$$
5, 6, 10, 11, 0, 1, 5, \ldots \tag{7.166}
$$

which is infinitely repeating. Obviously, this is not good enough for practical purposes.

Note that, in our example above, we divided by $m = 15$ but got a period of 6. It's fairly easy to see that an LCG sequence cannot have a period that is larger than m: since we're taking mod m, we're producing numbers from 0 to $m - 1$. But there only exist m distinct numbers from 0 to $m - 1$: the minute we get a repeat, we know the entire sequence will repeat after that.[36] Given that a fixed m leads to numbers from 0 to $m - 1$ and can therefore have a maximum period of m, it stands to reason that we would like to have m be large; it is standard to pick m to be 2 raised to a large power. However, simply increasing m is not enough: it's generally a good idea to also pick p to be large. Of course, using a large m and a large p only addresses the question of the large period, not that of the "quality" of our random number sequence; we return to this question below.

So far, we've been discussing Eq. (7.164): once the parameters p, c, and m are set, this formula requires knowledge of only the current number in order to produce the next one. However, one could concoct more complicated generators, which depend on a larger part of the sequence's history, for example:

$$
u_i = \left(p u_{i-j} + c u_{i-k}\right) \bmod m \tag{7.167}
$$

for fixed values of j and k. In this case, to jumpstart the generation of numbers one would have to start not with one number (the seed), but with a sequence of such numbers.

Floats

In applications, we often need not integer random numbers, but floats/real numbers. The most common variety in which these are needed is (uniformly distributed) random numbers from 0 to 1. It's easy to see how to produce these numbers starting from a sequence of integers generated using Eq. (7.164); simply divide by m:

36 Note that this holds for the LCG algorithm, not for random sequences in general.

$$r_i = \frac{u_i}{m} \tag{7.168}$$

This automatically ensures that the output floats will be from 0 (inclusive) to 1 (exclusive). Since our integers go from 0 to $m-1$, we'll never get an r_i that is all the way up to 1 itself, but that's generally not a problem, for very large values of m.

To be explicit, the sequence of integers in Eq. (7.166) will (upon division by $m = 15$) lead to the sequence of roughly:

$$0.3333, 0.4, 0.6667, 0.7333, 0.0, 0.0667, 0.3333, \ldots \tag{7.169}$$

This is still a random-number sequence (this time, of floats) of period 6.

In some applications we need random floats not in [0, 1) but in $[a, b)$. Again, it's easy to do this. Starting from r_i in [0, 1), we transform as follows:

$$x_i = a + (b-a)r_i \tag{7.170}$$

where x_i is now in $[a, b)$. We can use this transformation when we need floats from -1 to 1: the relation above gives us $x_i = -1 + 2r_i$, which can be re-expressed as $x_i = 2(r_i - 0.5)$.

7.7.1.2 Tests of Random Number Generators

In the preceding discussion, we saw that pseudorandom number generators eventually repeat, so one's goal should be to have as large a period as possible. However, one should also consider if the random-number generator has good statistical quality; this basically means that there should not be glaring *correlations* along the sequence of random numbers. For example, the sequence:

$$1, 2, 3, 4, 5, 6, \ldots \tag{7.171}$$

is clearly not random, since you can easily predict the next number in the sequence (i.e., you know that if your current number is 37 then your next number will be 38). Here's another example:

$$u_0, 101 - u_0, u_1, 101 - u_1, u_2, 101 - u_2, \ldots \tag{7.172}$$

This is also not a very random sequence: very often, if you know one number you can predict the next one. For example, if you're at the start of a new pair and the number 33 appears, then you know that the next number will be 68. This sequence contains strong *pairwise* correlations. Both these examples are hinting at what a random-number sequence is by exhibiting its opposite: non-random number sequences, which are often predictable (at least in part).

One way of testing for correlations is by plotting pairs of numbers, $(x_i, y_i) = (u_{i-1}, u_i)$.

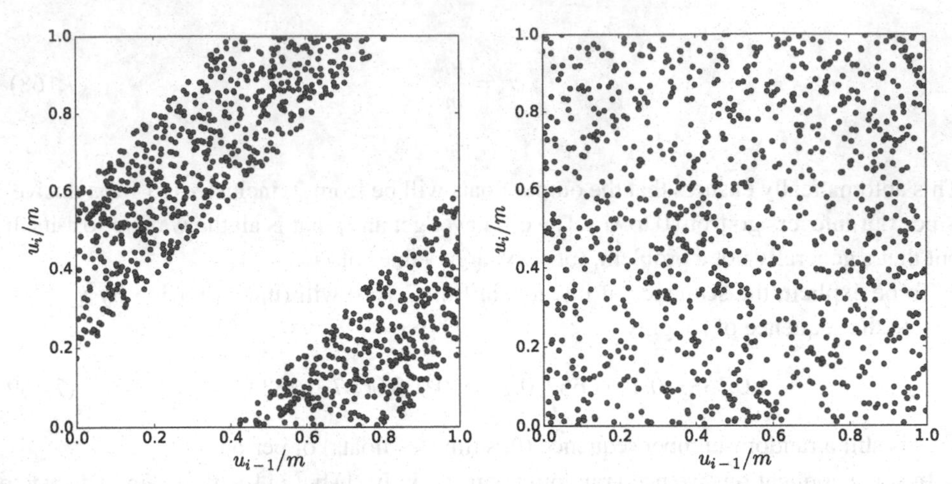

Fig. 7.5 Random numbers plotted pairwise for a "bad" (left) and a "good" (right) generator

If such a plot exhibits regularity, something is wrong. The left panel of Fig. 7.5 shows the result of generating 1000 uniformly distributed random numbers from the generator:

$$u_i = \left[(2^{18} + 1)u_{i-1} + 7\right] \bmod 2^{32} \tag{7.173}$$

and plotting them pairwise; the right panel employs a generator that is almost identical: the only change is in the coefficient, $p = 1\,812\,433\,253$. In both cases we started from the seed $u_0 = 314\,159$ and have divided the u's by 2^{32}. The left panel exhibits a clear pattern: all the pairs end up in one of two bands; in problem 7.31 you are asked to reproduce these plots using Python. Of course, this test involves "visual inspection", so it is still prone to human error (e.g., is it obvious that the right panel is exhibiting no patterns whatsoever?); in reality, this is merely the visual component of a systematic approach known as the *serial test*, which checks tuples of successive numbers to see if they are independent of each other. Problem 7.32 explores the *equidistribution test*: you are asked to divide up the region from 0 to $m - 1$ into "bins" and then count how many numbers fall into each bin.

Even though up to this point we've been producing (or trying to produce) uniformly distributed random numbers, it's important to realize that *random* is not the same thing as *uniform*. Random in the sense that we're using it here means that there are few/no correlations and that there is a very large period. If you're dealing with numbers from, say, 1 to 100, "random" does not necessarily mean that all numbers from 1 to 100 are equally likely to occur. For example, we can have a random number sequence that follows a normal/Gaussian distribution centered around 50: that means that numbers around 50 are much more likely to occur than, say, numbers around 10 or 90, which are close to the tails of the distribution. Such numbers are sometimes called *normal deviates*, in contradistinction to random numbers that are uniformly distributed, which are called *uniform deviates*. We discuss how to produce non-uniformly distributed random numbers in later subsections.

7.7.1.3 Random Numbers in Python

For industrial-scale applications, we typically do not rely on our own random-number generators. Python itself includes a high-quality generator in the random module:

```
>>> import random
>>> random.seed(314159)
>>> random.random()
0.19236379321481523
>>> random.random()
0.2868424512347926
```

We first call the random.seed() function to provide the seed. We then see that repeated invocations to the random.random() function lead to new random numbers, uniformly distributed in $[0, 1)$. Python's random.random() uses a Mersenne Twister algorithm with very good statistical properties and a period of $2^{19\,937} - 1$, which is large enough for most practical applications.[37] You will compare its properties to LCG ones in problem 7.31.

If you need several random numbers stored in an array, you could hand-roll a solution using random.random(). However, it's probably best to directly employ the functionality contained in numpy.random.uniform(); this function takes in three parameters: the first two determine the interval $[a, b)$ while the third one contains the shape of the output array:

```
>>> import numpy as np
>>> np.random.seed(314159)
>>> np.random.uniform(-1,1,4)
array([ 0.63584662, 0.10209259, -0.16044929, -0.80261629])
>>> np.random.uniform(-1,1,(2,3))
array([[ 0.6220415 , 0.93471281, -0.80358661],
       [ 0.60372074, 0.20980423,  0.16953762]])
```

where we also showed the seeding taking place through a call to np.random.seed().[38]

7.7.2 Monte Carlo Quadrature

We now turn to the question of how random numbers can be used to compute integrals, starting from the one-dimensional case for simplicity. At this point, you would be well-advised to have a look at our brief recap on probability theory in appendix C.3. If you wish to arrive at a deeper understanding of the conceptual foundations, you should also study the statistical material in section 6.6.

[37] Such a number is known as a Mersenne prime; this explains the presence of Descartes' chief correspondent (Marin Mersenne) in the name of an algorithm developed in 1997.

[38] While np.random.seed() is fine for our limited needs here, if you think you may end up needing independent streams in the future you should use np.random.RandomState() instead.

7.7.2.1 Population Mean and Population Variance

Our starting point will be the law of the unconscious statistician, Eq. (C.31), giving us the *expectation* (which is an integral) of a function f of a continuous random variable $\mathscr{X}$:

$$\mathbb{E}[f(\mathscr{X})] = \int_{-\infty}^{+\infty} P(x)f(x)dx \tag{7.174}$$

For now, we take $P(x)$, the probability density function, to be uniform from a to b, with the value $P(x) = 1/(b-a)$, and zero elsewhere[39] (we'll generalize this in section 7.7.3 below):

$$\mu_f \equiv \mathbb{E}[f(\mathscr{X})] = \frac{1}{b-a}\int_a^b f(x)dx \tag{7.175}$$

Except for the $1/(b-a)$ prefactor, the right-hand side is exactly of the form of Eq. (7.7), which is the problem we've been solving this entire chapter. To keep the terminology straight, we will call this μ_f the *population mean*. Similarly, we can take Eq. (C.32), giving us the *variance* of a function of a random variable in terms of an integral:

$$\mathbb{V}[f(\mathscr{X})] = \mathbb{E}\left\{(f(\mathscr{X}) - \mathbb{E}[f(\mathscr{X})])^2\right\} = \mathbb{E}[f^2(\mathscr{X})] - \mathbb{E}[f(\mathscr{X})]^2 \tag{7.176}$$

and specialize to the case where $P(x) = 1/(b-a)$, to find:

$$\sigma_f^2 \equiv \mathbb{V}[f(\mathscr{X})] = \frac{1}{b-a}\int_a^b f^2(x)dx - \left(\frac{1}{b-a}\int_a^b f(x)dx\right)^2 \tag{7.177}$$

To keep the terminology consistent, we call this σ_f^2 the *population variance*. Obviously, its square root gives us the *population standard deviation*, σ_f.

Crucially, both the population mean and the population variance, μ_f and σ_f^2, are unknown,[40] i.e., we don't know the value of either $\int_a^b f(x)dx$ or $\int_a^b f^2(x)dx$, which appear in Eq. (7.175) and Eq. (7.177); this is precisely why we wish to employ Monte Carlo integration. In what follows, we will learn how to *estimate* both μ_f and σ_f^2.

7.7.2.2 Sample Mean and Its Variance

Assume that the random variables $\mathscr{X}_0, \mathscr{X}_1, \mathscr{X}_2, \ldots, \mathscr{X}_{n-1}$ are drawn from $P(x)$, which for now is uniform from a to b. We say that the $\mathscr{X}_i$'s are *independent and identically distributed (IID)*. Note that many statisticians use the term "sample" to refer to all the random variables $\mathscr{X}_0, \mathscr{X}_1, \mathscr{X}_2, \ldots, \mathscr{X}_{n-1}$ put together (i.e., these would make up a single sample of size n); this nomenclature would get very confusing below, so we prefer to call a given $\mathscr{X}_i$ (e.g., $\mathscr{X}_5$) a sample, i.e., we are dealing with n samples. For each of these random variables $\mathscr{X}_i$, we act with the function f, leading to n new random variables, $f(\mathscr{X}_0), f(\mathscr{X}_1), f(\mathscr{X}_2), \ldots, f(\mathscr{X}_{n-1})$.

[39] We do this to ensure that the probability density is normalized: $\int_{-\infty}^{+\infty} P(x)dx = \int_a^b 1/(b-a)dx = 1$.

[40] In the language of section 6.6, they refer to the *parent* population.

(Remember, a function of a random variable is another random variable.) Motivated by the law of the unconscious statistician, this time for the discrete-variable equiprobable case as per Eq. (C.22), we define:[41]

$$\overline{f} \equiv \frac{1}{n}\sum_{i=0}^{n-1} f(\mathscr{X}_i) \tag{7.178}$$

We call this the *sample mean*: it is the arithmetic average of the function value over n random samples. Crucially, the random variable $\overline{f}$ is given in terms of a finite number of samples (the n random variables $\mathscr{X}_i$), so it is quite different from the population mean, μ_f, which as we see in Eq. (7.175) is an integral over continuous values of x (and is a *number*). In a similar vein: Eq. (7.178) refers to random variables, $\mathscr{X}_i$, whereas Eq. (C.22) to specific outcomes, x_i—a distinction that, alas, is often obscured in the literature.[42]

We will now see that the sample mean, $\overline{f}$, can be used to *estimate* the population mean, μ_f. Let us examine the expectation of the sample mean:[43]

$$\mathbb{E}(\overline{f}) = \mathbb{E}\left[\frac{1}{n}\sum_{i=0}^{n-1} f(\mathscr{X}_i)\right] = \frac{1}{n}\sum_{i=0}^{n-1}\mathbb{E}[f(\mathscr{X}_i)] = \mathbb{E}[f(\mathscr{X})] = \mu_f \tag{7.179}$$

In the first equality we substituted the definition of the sample mean from Eq. (7.178). In the second equality we used (repeatedly) the addition rule for expectations of random variables, from appendix C.3. In the third equality we noticed that all n terms in the sum are identical (as per Eq. (7.174), you can call the dummy integration variable anything you'd like) and therefore cancelled the denominator. In the fourth equality we used our definition of the population mean from Eq. (7.175). Our result is that *the expectation of the sample mean is equal to the population mean*. This motivates our choice to use $\overline{f}$ as an *estimator* of μ_f. As discussed in problem 6.52, an unbiased estimator $\mathscr{J}$ of a quantity J is one for which the expectation is equal to the quantity you're trying to estimate, $\mathbb{E}(\mathscr{J}) = J$; thus, $\overline{f}$ is an *unbiased estimator* of μ_f. In problem 7.33 you will prove the *weak law of large numbers*, which tells us that as n gets larger there is a high probability that the sample mean will be close to the population mean. This is good news, but we would like to know how *fast* the sample mean approaches the population mean, so we can decide when to stop.

In order to quantify how well we're doing, we turn to the *variance of the sample mean*:

$$\sigma_{\overline{f}}^2 \equiv \mathbb{V}(\overline{f}) = \mathbb{V}\left[\frac{1}{n}\sum_{i=0}^{n-1} f(\mathscr{X}_i)\right] = \frac{1}{n^2}\sum_{i=0}^{n-1}\mathbb{V}[f(\mathscr{X}_i)] = \frac{1}{n}\mathbb{V}[f(\mathscr{X})] = \frac{1}{n}\sigma_f^2 \tag{7.180}$$

In the first equality we substituted the definition of the sample mean from Eq. (7.178). In the second equality we used (repeatedly) the addition rule for variances of random variables, from appendix C.3. In the third equality we noticed that all n terms in the sum are

[41] In actuality, here and in Eq. (7.182) we are inspired by the *plug-in method* of estimating functionals in a non-parametric manner. A discussion of *empirical distribution functions* is beyond the ambit of this work.

[42] Recall from page 386 that we use $\mathscr{X}$ to denote a random variable, instead of (the near ubiquitous) X. Far too many introductory Monte Carlo treatments fail to discriminate between X (i.e., our $\mathscr{X}$) and a realization x.

[43] Since the sample mean is a random variable, we take its expectation to see if it gives the population mean.

identical and therefore cancelled one of the n's in the denominator. The last step used Eq. (7.177). Since σ_f^2 is fixed and given by Eq. (7.177), our result is telling us that *the variance of the sample mean decreases as* $1/n$. If we take the square root, we find:

$$\sigma_{\bar{f}} = \frac{1}{\sqrt{n}}\sigma_f \tag{7.181}$$

that is, the standard deviation of the sample mean goes as $1/\sqrt{n}$: as you quadruple n, the standard deviation is halved. As we saw in our discussion of Eq. (7.162), this scaling persists in the multidimensional version of Monte Carlo, which is why Monte Carlo vastly outperforms conventional quadrature methods when integrating over many dimensions.

While this, too, is good news, it's not great news: Eq. (7.180) relates $\sigma_{\bar{f}}^2$ (the variance of the sample mean) to σ_f^2 (the population variance) which, as you may recall from our discussion around Eq. (7.177), we don't actually know. This means that we will need to come up with an estimator for σ_f^2. Once we have a usable approximation for it, we can employ Eq. (7.180) to produce an actual number that approximates the variance of the sample mean. Motivated by the definition of the variance in the discrete-variable equiprobable case, as per Eq. (C.23), we propose an estimator for the population variance of the following form:

$$\mathscr{V} \equiv \overline{f^2} - \bar{f}^2 = \frac{1}{n}\sum_{i=0}^{n-1} f^2(\mathscr{X}_i) - \left[\frac{1}{n}\sum_{i=0}^{n-1} f(\mathscr{X}_i)\right]^2 \tag{7.182}$$

which introduces the new notation $\overline{f^2}$, whose meaning should be intuitive. It's important to note that the final expression makes use of finitely many instances of $f(\mathscr{X}_i)$, so it can readily be evaluated. As you will show in problem 7.34, the expectation of this new estimator $\mathscr{V}$ is close to, but not quite the same as, the population variance:[44]

$$\mathbb{E}(\mathscr{V}) = \frac{n-1}{n}\sigma_f^2 \tag{7.183}$$

This implies that it is *not* an *unbiased estimator* (i.e., it is a biased estimator).[45] It's trivial to see how to produce an unbiased estimator of the variance: simply multiply both sides by n and divide by $n-1$ (this is known as *Bessel's correction*), i.e., $n\mathscr{V}/(n-1)$ is an unbiased estimator of the population variance (known as the *sample variance*). We can now use this unbiased estimator to get:

$$\sigma_{\bar{f}}^2 = \frac{1}{n}\sigma_f^2 \approx \frac{1}{n}\frac{n}{n-1}\mathscr{V} = \frac{1}{n-1}\left(\overline{f^2} - \bar{f}^2\right) \tag{7.184}$$

The first step is from Eq. (7.180), namely the relationship between the variance of the sample mean, on the one hand, and the population variance, on the other. The second step employs the unbiased estimator of the population variance, i.e., the sample variance. The third step cancelled the n and plugged in $\mathscr{V}$ from Eq. (7.182). To summarize, we

[44] Remember: $\mathscr{V}$ is an estimator, so we take its expectation to see if it gives the population variance.

[45] Obviously, for large values of n we have $(n-1)/n \approx 1$, so $\mathbb{E}(\mathscr{V})$ is basically equal to σ_f^2.

approximate the population variance σ_f^2 by the sample variance, $n\mathscr{V}/(n-1)$; then, we use that in Eq. (7.180) to produce an estimate for the variance of the sample mean.

You may wish to reread the last few pages at this point, as the—standard—terminology can be somewhat confusing. For example, the "variance of the sample mean" is not the same thing as the "sample variance". To reiterate the main point for the umpteenth time: we have been able to approximate the *population* mean μ_f using only n function evaluations, i.e., *sample* properties; our best estimate is $\overline{f}$, with $\sigma_{\overline{f}}$ quantifying how much we should trust the value of $\overline{f}$ we've produced. Observe that we have expressed our quadrature problem in the language of statistical inference.

7.7.2.3 Practical Prescription

Recall that our goal has been to approximate an integral. We can recast Eq. (7.175) as:

$$\int_a^b f(x)dx = (b-a)\,\mu_f \tag{7.185}$$

The previous subsection studied the estimation of the population mean μ_f via the sample mean $\overline{f}$, Eq. (7.178). In terms of our integral, this gives:

$$\int_a^b f(x)dx \approx (b-a)\,\overline{f} \pm (b-a)\,\sigma_{\overline{f}} \tag{7.186}$$

To interpret this in the usual (Gaussian) sense of $\pm$, we need to assume the population variance is finite. In other words, we employ the *central limit theorem* which tells us that asymptotically the $\overline{f}$ obeys a Gaussian distribution regardless of which $P(x)$ was employed to draw the samples (e.g., a uniform one here); this is derived in problem 6.47 and further explored in problem 6.48. As per page 394, Eq. (7.186) gives us a confidence interval with *long-run* performance guarantees (i.e., as we keep recomputing $\overline{f}$ from scratch).

Using Eq. (7.184) for the standard deviation of the sample mean, we find:

$$\int_a^b f(x)dx \approx (b-a)\overline{f} \pm (b-a)\sqrt{\frac{\overline{f^2}-\overline{f}^2}{n-1}} \tag{7.187}$$

If we also take the opportunity to plug in $\overline{f}$ and $\overline{f^2}$ as in Eq. (7.182), we arrive at the following practical formula for *Monte Carlo integration*:

$$\int_a^b f(x)dx \approx \frac{b-a}{n}\sum_{i=0}^{n-1} f(\mathscr{X}_i) \pm \frac{b-a}{\sqrt{n-1}}\sqrt{\frac{1}{n}\sum_{i=0}^{n-1} f^2(\mathscr{X}_i) - \left[\frac{1}{n}\sum_{i=0}^{n-1} f(\mathscr{X}_i)\right]^2} \tag{7.188}$$

where the $\mathscr{X}_i$'s are chosen uniformly from a to b. When we implement Eq. (7.188) we will see that the $\sqrt{n-1}$ in the denominator slowly but surely decreases the standard deviation of the sample mean: more work pays off. For now, focus on the first term on the right-hand side: this is a sum over terms of the form $f(\mathscr{X}_i)(b-a)/n$. This is similar to the rectangle

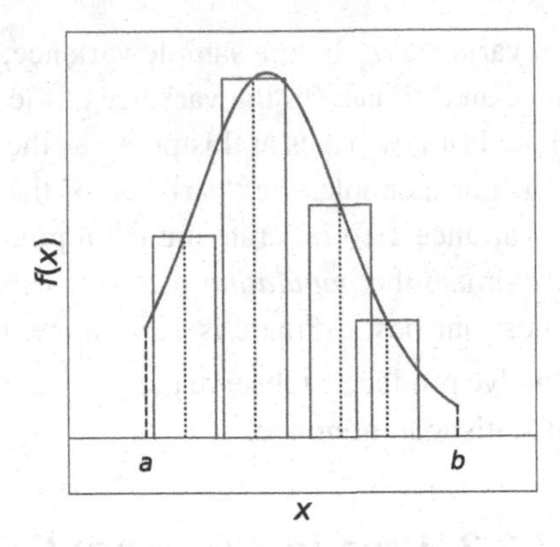

Fig. 7.6 Examples of randomly chosen abscissas, each of which leads to a rectangle

or midpoint methods, Fig. 7.2, in that the total area is evaluated as a sum of rectangle areas. One difference is that this time the width of the rectangles is $(b-a)/n$ instead of $h = (b-a)/(n-1)$; more importantly, the height is now determined by the value of the function at the uniformly distributed $\mathscr{X}_i$'s. This is illustrated in Fig. 7.6.[46]

7.7.3 Monte Carlo beyond the Uniform Distribution

We now turn to non-uniformly distributed random numbers, a subject to which we return in section 7.7.5.2 below, when we introduce the *Metropolis–Hastings algorithm.*

7.7.3.1 Generalizing to Weight Functions

The discussion in the previous subsection started from Eq. (7.174), which gives us the expectation of a function of a continuous random variable:

$$\mathbb{E}[f(\mathscr{X})] = \int_{-\infty}^{+\infty} P(x)f(x)dx \tag{7.189}$$

This time around, we take our probability density function to be $w(x)/(b-a)$ from a to b and zero elsewhere; this $w(x)$ is kept general, except for being positive. This leads to:[47]

$$\mu_f \equiv \mathbb{E}[f(\mathscr{X})] = \frac{1}{b-a}\int_a^b w(x)f(x)dx \tag{7.190}$$

Notice that the integral on the right-hand side is precisely of the form we encountered in Eq. (7.129), in our discussion of general Gaussian quadrature. It should therefore come as no surprise that $w(x)$ is known as a *weight function*. For this new population mean, we can write down the corresponding population variance; this follows from Eq. (7.176) and would lead to a generalization of Eq. (7.177), where this time the integrands would be $[w(x)f(x)]^2$ and $w(x)f(x)$.

[46] The bridge between Monte Carlo integration and *Lebesgue* integration is, alas, beyond the ambit of this work.

[47] Note that $w(x) = 1$ takes us back to Eq. (7.175); also, since $\int_{-\infty}^{+\infty} P(x)dx = 1$, we have that $\int_a^b w(x)dx = b - a$. If we were interested in an improper integral, we would make a different choice here.

We could now introduce a new sample mean, as per Eq. (7.178):

$$\overline{f} = \frac{1}{n}\sum_{i=0}^{n-1} f(\mathscr{X}_i) \tag{7.191}$$

where, crucially, this time the $\mathscr{X}_0, \mathscr{X}_1, \mathscr{X}_2, \ldots, \mathscr{X}_{n-1}$ are IID drawn randomly *from the probability density function* $w(x)/(b-a)$, i.e., *not* from a uniform distribution.[48] Intriguingly, if you go over our derivation on the expectation of the sample mean, Eq. (7.179), you will see that all the steps carry over in exactly the same way. The only thing that's changed is the meaning of the expectation, which this time is given as per Eq. (7.190). Similarly, the derivation on the variance of the sample mean is also identical to Eq. (7.180) and, finally, the same holds for the expectation of $\mathscr{V}$, our proposed estimator for the population variance, Eq. (7.183). As before, the only thing that's changed is the meaning of the expected value. To make a long story short, the entire argument carries over, implying that we have a practical formula for *Monte Carlo integration with a general weight function*:

$$\int_a^b w(x)f(x)dx \approx \frac{b-a}{n}\sum_{i=0}^{n-1} f(\mathscr{X}_i) \pm \frac{b-a}{\sqrt{n-1}}\sqrt{\frac{1}{n}\sum_{i=0}^{n-1} f^2(\mathscr{X}_i) - \left[\frac{1}{n}\sum_{i=0}^{n-1} f(\mathscr{X}_i)\right]^2} \tag{7.192}$$

where this time the $\mathscr{X}_i$'s are chosen from $w(x)$—strictly speaking, from $w(x)/(b-a)$—e.g., if $w(x)$ is exponential, these are exponentially distributed random variables.

7.7.3.2 Inverse Transform Sampling

In the preceding discussion we took it for granted that the random variables $\mathscr{X}_0, \mathscr{X}_1, \mathscr{X}_2, \ldots, \mathscr{X}_{n-1}$ could be drawn from the distribution $w(x)$. You may have wondered how that is possible, given that section 7.7.1 discussed only the generation of uniformly distributed random numbers. We now go over a specific technique that helps you accomplish this task, known as *inverse transform sampling* or, sometimes, simply *inverse sampling*.

The task facing us in Eq. (7.192) can be viewed from two equivalent perspectives: first, as mentioned in the previous paragraph, you can focus on the right-hand side, which involves non-uniformly distributed random numbers. These can be generated starting from uniformly distributed random numbers; in probability theory, this is accomplished via the *transformation of a random variable*, also known as the *derived-distribution approach*. A distinct (yet equivalent) tack is to focus on the left-hand side of Eq. (7.192): it would be nice if we could somehow convert the integrand wf to f, via an appropriate change of variables; in calculus, this is accomplished via *integration by substitution*.[49] If we succeeded in this, then we could apply the Monte Carlo approach of Eq. (7.188) for our simpler integral, making use of *uniformly* distributed random numbers. We will take the latter approach; let us start by integrating the weight function:

[48] As per page 384, for us *distribution* is a synonym for *density function* (not for *cumulative distribution function*).
[49] Also known as *u-substitution* in elementary (and pullback/pushforward in advanced) treatments.

$$g(x) = \int_a^x w(x')dx' \tag{7.193}$$

If you have taken a course on probability, you will recognize this integral over the probability density function as the *cumulative distribution function*.[50] Notice that $g(a) = 0$ and $g(b) = b - a$. Let's see why this helps us "absorb" the $w(x)$ in Eq. (7.192):

$$\int_a^b w(x)f(x)dx = \int_a^b f(x)\frac{dg}{dx}dx = \int_a^b f\left(g^{-1}(g(x))\right)\frac{dg}{dx}dx = \int_0^{b-a} f\left(g^{-1}(u)\right)du \tag{7.194}$$

In the first step we employed the fundamental theorem of calculus, i.e., we differentiated Eq. (7.193). In the second step we thought of Eq. (7.193) as defining $u = g(x)$, inverting which gave us $x = g^{-1}(u)$. In the third step we used integration by substitution:

$$\int_a^b h(g(x))\frac{dg}{dx}dx = \int_{g(a)}^{g(b)} h(u)du \tag{7.195}$$

itself a result of the fundamental theorem of calculus.[51] Our final expression in Eq. (7.194) is an *unweighted* integral, which can be tackled via Eq. (7.188) and uniformly chosen $\mathscr{U}_i$'s:

$$\int_a^b w(x)f(x)dx = \int_0^{b-a} f\left(g^{-1}(u)\right)du \approx \frac{b-a}{n}\sum_{i=0}^{n-1} f\left(g^{-1}(\mathscr{U}_i)\right) \tag{7.196}$$

The $\mathscr{U}_i$'s are uniformly distributed from 0 to $b - a$ and therefore the $\mathscr{X}_i = g^{-1}(\mathscr{U}_i)$ are distributed according to $w(x)$, from a to b;[52] basically, the $\mathscr{X}_i$'s are $w(x)$-aware, via the inverse transform we carried out. In summary, we've managed to employ uniform random variables, together with a transformation, to apply Monte Carlo to a weighted integral.

To see how this all works in practice, let's study a specific example:

$$I = \int_1^3 e^{-x}\sin x dx \tag{7.197}$$

We could treat the entire integrand as one piece here, $f(x) = e^{-x}\sin x$; we could then directly apply Eq. (7.188) and use uniformly distributed random numbers to compute it (recall footnote 27). Clearly, this is wasteful: by choosing your abscissas uniformly, you are not taking into consideration the exponential drop in the magnitude of the integrand.

Thus, another choice is to apply Eq. (7.192) instead of Eq. (7.188); take $w(x) = ce^{-x}$, where the normalization gives $c = 2e^3/(e^2 - 1)$. To get the prefactors to cancel, take $f(x) = \sin x/c$. The inverse-transform method will let us draw samples from the (doubly truncated)

[50] This is $\int_{-\infty}^{x} P(x')dx'$, but our $P(x)$ is 0 below a; note our normalization, involving $w(x)$ and $b - a$.

[51] To say Eq. (7.195) follows from $du = g'(x)dx$ is to confuse derivatives with fractions (*qui sine peccato est*). Physicist-level rigor is usually fine, but can get you in trouble for higher orders or for partial derivatives.

[52] The inverse cumulative distribution function is known as the *quantile function*.

exponential distribution by first evaluating $g(x)$ as per Eq. (7.193):

$$g(x) = c \int_1^x e^{-x'} dx' = c(e^{-1} - e^{-x}) \tag{7.198}$$

and then inverting to give:

$$g^{-1}(u) = -\ln\left(e^{-1} - \frac{u}{c}\right) \tag{7.199}$$

in which case Eq. (7.196) for the integral in Eq. (7.197) becomes:

$$\int_1^3 ce^{-x}\frac{1}{c}\sin x dx = \int_0^2 \frac{1}{c}\sin\left(-\ln\left[e^{-1} - \frac{u}{c}\right]\right)du \approx \frac{2}{n}\sum_{i=0}^{n-1}\frac{1}{c}\sin\left(-\ln\left[e^{-1} - \frac{\mathcal{U}_i}{c}\right]\right) \tag{7.200}$$

where the $\mathcal{U}_i$'s are distributed uniformly from 0 to 2; problem 7.36 asks you to implement this in Python. In short, you have managed to sample from an exponential distribution, even though your input random-number generator was for a uniform distribution.

7.7.3.3 Importance Sampling

You may be thinking that it was easy to handle an integrand of the form $e^{-x}\sin x$, with its clean separation between an exponential enhancement/suppression and everything else. This was basically the integrand itself telling us which weight function to employ. As it turns out, the technique we just introduced is much more useful than we've hinted at so far: even if no weight function is present, you can introduce one yourself.

To see what we mean, let's repeat the previous argument, for the case where the integrand appears to be a single (i.e., unweighted) function:

$$\int_a^b f(x)dx = \int_a^b w(x)\frac{f(x)}{w(x)}dx = \int_0^{b-a}\frac{f\left(g^{-1}(u)\right)}{w\left(g^{-1}(u)\right)}du \approx \frac{b-a}{n}\sum_{i=0}^{n-1}\frac{f\left(g^{-1}(\mathcal{U}_i)\right)}{w\left(g^{-1}(\mathcal{U}_i)\right)} \tag{7.201}$$

In the first equality we multiplied and divided by a positive weight function of our choosing. In the second equality we used the change of variables from Eq. (7.193), in complete analogy to what we saw in Eq. (7.194); the only difference is that this time after we carry out the transform there is still a w left over in the denominator. In the third equality we treated f/w as our ("unweighted") integrand and therefore used $\mathcal{U}_i$'s which are uniformly distributed from 0 to $b - a$. The entire process is known as *importance sampling*.[53]

Let's see why we bothered doing this: if you choose a (monotonic) $w(x)$ that behaves approximately the same way that $f(x)$ does, then your random numbers will be distributed in the most "important" regions,[54] instead of uniformly (recall Fig. 7.6). To put the same idea differently: since our (unweighted) integrand is f/w, the variance will be computed

[53] Use the same trick if your integral is $\int_a^b w(x)f(x)dx$ but you cannot sample from $w(x)$, i.e., write the problem as $\int_a^b h(x)w(x)f(x)/h(x)dx$ for a manageable $h(x)$. Make sure that $h(x)$ has thicker/heavier tails than $w(x)$.

[54] "Be sure not to elongate the edifice beyond all probability, lest we come crashing down" (Lucian, *Charon*, 5).

as per Eq. (7.188), with f/w in the place of f. Since $w(x)$ is chosen to be similar to $f(x)$, we see that f/w will be less varying than f itself was; this will lead to a reduction of the variance, i.e., a better overall estimate of the integral we are trying to compute.

To see this in action, we turn to the example we've been revisiting throughout this chapter, namely the electrostatic potential in Eq. (7.2), which integrates the function:

$$f(x) = \frac{1}{\sqrt{x^2+1}} \tag{7.202}$$

from 0 to 1. While this integrand doesn't consist of two easily separable parts, we can plot it to get a feel for it (left panel of Fig. 7.7). We see that $f(x)$ is decreasing from 1 to 0.7 in our interval; with that in mind, we decide to employ a linear weight function:

$$w(x) = c_0 + c_1 x \tag{7.203}$$

We wish $w(x)$ to roughly track the behavior of $f(x)$ in our interval; one way to do this is to ensure that $f(0)/w(0)$ is equal to $f(1)/w(1)$. This gives one equation relating c_0 and c_1. If we then also impose the normalization condition $\int_0^1 w(x)dx = 1$ we get another relation. Thus, we are able to determine both parameters:

$$c_0 = 4 - 2\sqrt{2}, \qquad c_1 = -6 + 4\sqrt{2} \tag{7.204}$$

It comes as no surprise that c_1 is negative, since $w(x)$ should be a decreasing function in our interval. Our weight function is also plotted in the left panel of Fig. 7.7, together with the ratio f/w: we see that $f(x)/w(x)$ varies between 0.85 and 0.9, which is considerably smaller than the variation of our original integrand. It will therefore not come as a surprise in the following subsection that the computed variance in the importance-sampling case turns out to be much smaller than the variance in the original, uniform-sampling case.

Backing up for a second, what we're interested in doing is applying the importance-sampling prescription in Eq. (7.201). Up to this point, we've only selected the $w(x)$ we'll use. The next step is to carry out the change of variables as per Eq. (7.193):

$$g(x) = \int_0^x (c_0 + c_1 x')dx' = c_0 x + \frac{1}{2}c_1 x^2 \tag{7.205}$$

After that, we need to invert this relation to find $g^{-1}(u)$; since we're dealing with a quadratic, we will have two roots. Of these, we pick the one that leads to x from 0 to 1 for u in $[0, 1]$, since that's the interval we're interested in. We get:

$$g^{-1}(u) = \frac{-c_0 + \sqrt{2c_1 u + c_0^2}}{c_1} \tag{7.206}$$

Having calculated both the weight function $w(x)$ and the inverted $g^{-1}(u)$, we are now ready to apply importance sampling to our problem, as per Eq. (7.201):

$$\int_0^1 \frac{1}{\sqrt{x^2+1}}dx \approx \frac{1}{n}\sum_{i=0}^{n-1} \frac{1}{\sqrt{[g^{-1}(\mathcal{U}_i)]^2+1}\,(c_0 + c_1 g^{-1}(\mathcal{U}_i))} \tag{7.207}$$

where we get the $g^{-1}(\mathcal{U}_i)$ from Eq. (7.206). As before, the $\mathcal{U}_i$'s are uniformly distributed.

Incidentally, you may be wondering why we picked a linear $w(x)$ in Eq. (7.203). As

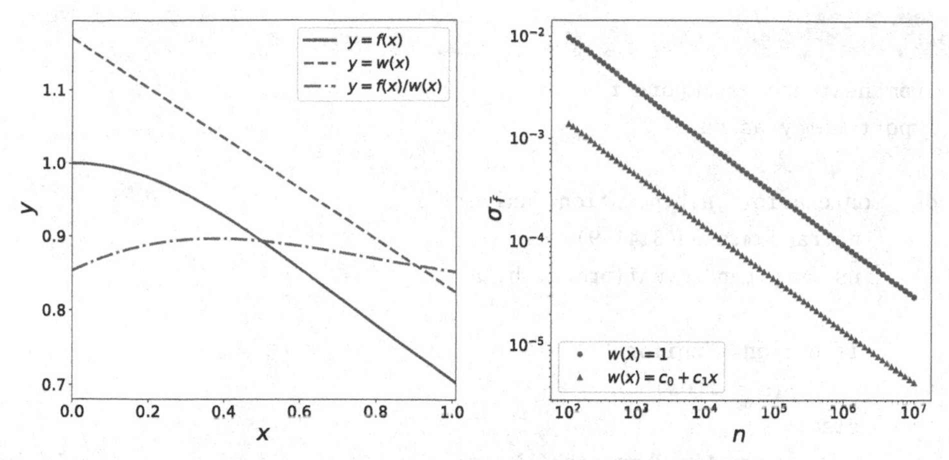

Fig. 7.7 The integrand, the weight, and their ratio (left); results of MC integration (right)

the left panel of Fig. 7.7 clearly shows, our integrand itself is not linear. Thus, you may feel tempted to pick a $w(x)$ that is slightly more complicated (e.g., a quadratic), since this would imply that the variation in $f(x)/w(x)$ is even smaller. This is fine, in principle, but in practice you have to remember that the integral of $w(x)$, which you need to carry out for the change of variables as per Eq. (7.193), must always remain invertible; for example, if your $w(x)$ is quadratic then you will need to invert a cubic. As this example shows, as you introduce further structure into your $w(x)$, the inversion becomes increasingly hard to do analytically; of course, there's nothing holding you from trying to carry out this inversion numerically, but you will have to be careful about picking the right solution, as we saw in Eq. (7.206). To provide a quick preview of coming attractions, we note that this approach does not work that well if you try to generalize it to a multidimensional problem; we will discover a better alternative in section 7.7.5 below.

7.7.4 Implementation

Code 7.5 implements Monte Carlo quadrature for the case of a one-dimensional integral. This is done in `montecarlo()`, which addresses both the case of uniform sampling and that of importance sampling. Crucially, uniform sampling is the default parameter value, which means that you can call this function with `montecarlo(f,a,b,n)`, i.e., with precisely the same interface as our three integrators in `newtoncotes.py` and the one in `gauleg.py`. In other words, this is a drop-in replacement for a general-purpose integration routine. In the function body, we start by seeding the random-number generator and producing an array of n numbers, uniformly distributed from a to b, using `numpy` functionality described in section 7.7.1. With these in hand, the option of uniform sampling is a straightforward application of Eq. (7.188). In the interest of separating our concerns, we have farmed out the evaluation of $\overline{f}$ and $\overline{f^2}$, which we need as per Eq. (7.187), to a separate function, `stats()`.[55]

[55] We could have used `numpy.mean()` and `numpy.std()` but, as usual, we prefer to roll our own.

Code 7.5 montecarlo.py

```
from newtoncotes import f
import numpy as np

def montecarlo(f,a,b,n,option="uniform"):
    np.random.seed(314159)
    us = np.random.uniform(a, b, n)

    if option=="uniform":
        fs = f(us)
    else:
        c0 = 4 - 2*np.sqrt(2)
        c1 = -6 + 4*np.sqrt(2)
        xs = (-c0 + np.sqrt(2*c1*us + c0**2))/c1
        fs = f(xs)/(c0 + c1*xs)

    fbar, err = stats(fs)
    return (b-a)*fbar, (b-a)*err

def stats(fs):
    n = fs.size
    fbar = np.sum(fs)/n
    fsq = np.sum(fs**2)/n
    varfbar = (fsq - fbar**2)/(n - 1)
    return fbar, np.sqrt(varfbar)

if __name__ == '__main__':
    for n in 10**np.arange(2,7):
        avu, erru = montecarlo(f, 0., 1., n)
        avi, erri = montecarlo(f, 0., 1., n, option="is")
        rowf = "{0:7d}  {1:1.9f} {2:1.9f}  {3:1.9f} {4:1.9f}"
        print(rowf.format(n, avu, erru, avi, erri))
```

For the importance-sampling case, we code up the parameters of the linear weight function, Eq. (7.204), then produce an array of $\mathcal{X}_i$'s using $g^{-1}(\mathcal{U}_i)$ from Eq. (7.206), and then we remember to divide $f(g^{-1}(\mathcal{U}_i))$ by $w(g^{-1}(\mathcal{U}_i))$, just like in Eq. (7.201) or Eq. (7.207). As always, we employ numpy functionality to carry out the division f/w for all i's in one line; that's why, as usual, the code contains no indices. The same holds, of course, with respect to stats(), where the squaring and the sums are carried out without needing explicit indices. Incidentally, Eq. (7.201) shows only the sample mean, not its variance; that is taken from our general formula for Monte Carlo integration, Eq. (7.188), with f/w in the place of f. Finally, we note that the uniformly distributed numbers for this case should in general go from 0 to $b - a$, not a to b; thus, our importance sampling option only works because $a = 0$ and $b = 1$. That's fine, though, since the $w(x)$ itself (i.e., both its values at the two endpoints and the normalization) was tuned only to the interval [0, 1].

The main program picks n to be the first several powers of 10 and carries out Monte Carlo integration using either uniform sampling, Eq. (7.188), or importance sampling, Eq. (7.201), with a linear weight function. The output table format is then set up. For comparison, recall that the exact answer for this integral is `0.88137358702`. Running this code produces the following output:

```
    100    0.873135430  0.009827018    0.880184046  0.001397861
   1000    0.878313494  0.003014040    0.880653976  0.000439206
  10000    0.879343920  0.000933506    0.881029489  0.000139055
 100000    0.881289768  0.000292906    0.881400087  0.000043577
1000000    0.881433836  0.000092589    0.881389786  0.000013775
```

First, looking at the uniform-sampling results we observe that they do converge to the right answer as n is increased; the exact answer is always within a few standard deviations of the sample mean (though often it lies within less than one standard deviation). However, it's fair to say that the convergence is quite slow: using a million points, the standard deviation is roughly 10^{-4}. As you may recall, Gauss–Legendre quadrature got all the printed digits right with only 10 points! Of course, this is an unfair comparison, since the true strength of Monte Carlo is in multidimensional integration, where even Gauss–Legendre can't help you, as we saw in section 7.6.4.

We now turn to the importance-sampled results. Overall, we see that the importance sampling has helped reduce the standard deviation by roughly an order of magnitude, which was certainly worthwhile. Note that, in both cases, we are not showing the absolute error (i.e., comparing with the exact answer) but the standard deviation that comes out of the Monte Carlo machinery. Incidentally, the n increases by a factor of 10 as you move to the next row: as a result, we see the standard deviation get reduced by roughly a factor of 3 each time; this is consistent with a scaling of $1/\sqrt{n}$. In order to drive this point home, the right panel of Fig. 7.7 is a log-log plot of the standard deviation of the sample mean for the two cases of uniform sampling and importance sampling; we've used more n points to make the plot look good and that it does: the slope of -0.5 is pretty unmistakable in both cases. Just like we found in our table of values, we again see that importance sampling consistently helps us reduce the standard deviation by an order of magnitude.

7.7.5 Monte Carlo in Many Dimensions

We now turn to a truly relevant application of the Monte Carlo approach: multidimensional integration. As in the one-dimensional case, we start from uniform sampling, but then try to do a better job: we explain how to carry out weighted sampling via the *Metropolis–Hastings algorithm*, which is one of the most successful methods ever.

7.7.5.1 Uniform Sampling

We briefly talked about multidimensional integration in section 7.6.4, see Eq. (7.156):

$$I = \int_a^b \int_a^b f(x, y) dx dy \tag{7.208}$$

This notation can get cumbersome if we need many dimensions, so it makes more sense to employ notation like the one used in section 5.6.1; thus, we bundle together the variables $x_0, x_1, \ldots, x_{d-1}$ into $\mathbf{x}$. We will be dealing with a scalar function of many variables, i.e., $f(\mathbf{x})$ produces a single number when given d variables as input.[56] We emphasize that the number of dimensions (and therefore variables) is d. Our starting point is the same as in section 7.7.2.1, i.e., we define a *population mean*:

$$\mu_f \equiv \mathbb{E}[f(\boldsymbol{\mathscr{X}})] = \frac{1}{V} \int f(\mathbf{x}) d^d x \tag{7.209}$$

where the expectation on the left-hand side is of a function of a random vector $\boldsymbol{\mathscr{X}}$ (i.e., the bold version of $\mathscr{X}$). The multidimensional integration on the right-hand side is over the volume V: this is the multidimensional generalization of writing $P(x) = 1/(b - a)$ in Eq. (7.175).

At this point, there is nothing discrete going on: we simply defined the population mean as the multidimensional integral in Eq. (7.209). We can now step through all the arguments in section 7.7.2, defining the population variance, the sample mean and its variance, testing that the sample mean is a good estimator of the population mean, and so on. The entire argument carries over naturally, leading to the following practical formula for *multidimensional Monte Carlo integration*:

$$\int f(\mathbf{x}) d^d x \approx \frac{V}{\mathfrak{N}} \sum_{i=0}^{\mathfrak{N}-1} f(\boldsymbol{\mathscr{X}}_i) \pm \frac{V}{\sqrt{\mathfrak{N}-1}} \sqrt{\frac{1}{\mathfrak{N}} \sum_{i=0}^{\mathfrak{N}-1} f^2(\boldsymbol{\mathscr{X}}_i) - \left[\frac{1}{\mathfrak{N}} \sum_{i=0}^{\mathfrak{N}-1} f(\boldsymbol{\mathscr{X}}_i)\right]^2} \tag{7.210}$$

which is a d-dimensional generalization of Eq. (7.188). Thus, the $\boldsymbol{\mathscr{X}}_i$'s are chosen uniformly; this is a good time to unpack our notation a little. Each of the $\boldsymbol{\mathscr{X}}_i$'s on the right-hand side is a d-dimensional random sample, whose components are $(\mathscr{X}_i)_0, (\mathscr{X}_i)_1, \ldots, (\mathscr{X}_i)_{d-1}$. In order to keep our notation consistent with that in section 7.6.4, in Eq. (7.210) we use $\mathfrak{N}$ for the total number of samples; this plays the role n played in the one-dimensional case.

As before, we've approximated the expectation of a continuous random variable (i.e., the population mean) with the arithmetic average of many function values taken at (discrete)

[56] So our f here plays the role of ϕ in section 5.6.1. Of course, now we're integrating, not minimizing.

uniformly distributed points. Each $\mathscr{X}_i$ is made up of d uniformly distributed random numbers; crucially, the values of a given $\mathscr{X}_i$ are totally unrelated to those of another sample (i.e., say, $\mathscr{X}_5$ is chosen independently of $\mathscr{X}_4$). It should therefore be straightforward to see how to implement this approach programmatically. Problem 7.38 asks you to apply Eq. (7.210) to a five-dimensional integral and compare with the corresponding five-dimensional Gauss–Legendre computation. The most significant feature of Eq. (7.210) is that the standard deviation of the sample mean goes as $1/\sqrt{\mathfrak{y}-1}$, which is the standard Monte Carlo result that applies regardless of dimensionality:[57] it involves $\mathfrak{y}$ (the number of samples), not d (the dimensionality of each sample). This is why Monte Carlo is to be preferred over conventional quadrature for many-dimensional problems.

7.7.5.2 Weighted Sampling: Metropolis–Hastings Algorithm

As you may recall from section 7.7.3, uniform sampling is not the best we can do. While the simple prescription in Eq. (7.210) does work and does have the pleasant $1/\sqrt{\mathfrak{y}-1}$ scaling, it is far from perfect: since the $\mathscr{X}_i$'s are chosen uniformly this may (and in practice does) imply that many random samples are wasted trying to evaluate $f(\mathscr{X}_i)$ in regions of the d-dimensional space where not much is going on. It would be nice to have a Monte Carlo technique where the random samples, i.e., the $\mathscr{X}_i$'s, are preferentially chosen to be such that they can make a difference: an integrand-aware sampling would be considerably more efficient than uniformly producing random samples.

You may have noticed a pattern in our discussion of Monte Carlo techniques: the main feature (approximate the population mean via the sample mean) is always the same. Thus, we can straightforwardly generalize the argument in section 7.7.3.1 to produce the following practical formula for *multidimensional weighted Monte Carlo integration*:

$$\int w(\mathbf{x})f(\mathbf{x})d^dx \approx \frac{V}{\mathfrak{y}}\sum_{i=0}^{\mathfrak{y}-1} f(\mathscr{X}_i) \pm \frac{V}{\sqrt{\mathfrak{y}-1}}\sqrt{\frac{1}{\mathfrak{y}}\sum_{i=0}^{\mathfrak{y}-1} f^2(\mathscr{X}_i) - \left[\frac{1}{\mathfrak{y}}\sum_{i=0}^{\mathfrak{y}-1} f(\mathscr{X}_i)\right]^2} \tag{7.211}$$

where this time the $\mathscr{X}_i$'s are drawn from $w(\mathbf{x})$, e.g., if $w(\mathbf{x})$ is exponential, these are samples made up of exponentially distributed random numbers. Note that, just like $f(\mathbf{x})$, the weight function $w(\mathbf{x})$ is now a function of d variables. If you're interested in integrating over all of space, simply drop the V term; to see why, picture Eq. (7.190) for the case of integrating from $-\infty$ to $+\infty$: as per Eq. (7.189), you would have $P(x) = w(x)$ and no $(b-a)$ term would appear on the right-hand side.

For the one-dimensional case, when we were faced with Eq. (7.192), we employed the inverse transform sampling method to produce $\mathscr{X}_i$'s which were drawn from the probability density function $w(x)$. In other words, we employed the quantile function to go from uniform samples to weighted samples, Eq. (7.196). When you move on to a multidimensional problem, i.e., producing random samples drawn from $w(\mathbf{x})$, our trivial change of

[57] Alas, sometimes the prefactor may implicitly depend on d, or you may be interested in the *relative* accuracy.

variables would have to be replaced by its generalization, namely the determinant of the Jacobian matrix. To make matters worse, the generalization of the integral in Eq. (7.193) would have to be carried out somehow (analytically?). As if that weren't enough, the inversion (in, say, 100 variables) cannot be carried out analytically; thus, you would have to employ one of the multidimensional root-finders from chapter 5. As you know by now, these approaches are prone to fail if you don't start sufficiently near the root. In a multidimensional problem, you will also likely have many roots and would be faced with the task of automatically selecting the correct one. Putting it all together, this approach is not practical when the number of dimensions is large. While other techniques exist to carry out weighted sampling in one-dimensional problems, they also get in trouble when the dimensionality increases.

Markov Chains

Basically the only thing that works as far as multidimensional weighted sampling is concerned is *Markov chain Monte Carlo* (MCMC).[58] The main new concept involved here is that of a *Markov chain*: imagine you have a sequence of random samples $\mathscr{X}_0, \mathscr{X}_1, \dots, \mathscr{X}_{\mathfrak{y}-1}$ for which a given sample $\mathscr{X}_i$ depends only on the previous one, i.e., on $\mathscr{X}_{i-1}$, but not on any of the earlier ones, i.e., $\mathscr{X}_{i-2}$, $\mathscr{X}_{i-3}$, and so on. In other words, one starts from a random sample $\mathscr{X}_0$, uses that to produce sample $\mathscr{X}_1$, then uses that in its turn to produce sample $\mathscr{X}_2$, and so on, always using the current sample to create the next one. This sequence of samples, $\mathscr{X}_0, \mathscr{X}_1, \dots, \mathscr{X}_{\mathfrak{y}-1}$, is known as a *random walk*.[59] Note that this is quite different from what we were doing in earlier subsections: there the $\mathscr{X}_i$ were independent from one another, whereas now we use a given $\mathscr{X}_{i-1}$ to produce the next one, i.e., $\mathscr{X}_i$: these are *not* IID samples. The reason Markov chains are so useful is that they can be produced such that they asymptotically (i.e., as $\mathfrak{y} \to \infty$) have the distribution we would like them to, which in our case would be $w(\mathbf{x})$. One could therefore do an increasingly better job at computing a d-dimensional integral by continuing the Markov chain for larger values of $\mathfrak{y}$.

Detailed Balance

We wish to produce a Markov chain with an asymptotic distribution of our choosing, which would therefore be the stationary distribution of the chain (if you don't know what that means, keep reading). Thus, we can borrow ideas from the statistical mechanics of systems in equilibrium. A sufficient (but not necessary) condition of evolving toward equilibrium and staying there is the *principle of detailed balance*:

$$w(\mathbf{x})T(\mathbf{x} \to \mathbf{y}) = w(\mathbf{y})T(\mathbf{y} \to \mathbf{x}) \tag{7.212}$$

[58] We will address (discrete-time) Markov chains on a *continuous* state-space. The overwhelming majority of the literature studies concepts like "Harris recurrence", "ergodicity", and so on [124], on a discrete state-space. Some authors distinguish between the terms "equilibrium" and "stationary", but we use them interchangeably.

[59] One could employ an ensemble of independent initial configurations (thereby producing distinct *walkers*) to visualize the distribution of each $\mathscr{X}_i$. Below we mostly stick to a single walker, for the sake of simplicity.

Here $T(\mathbf{x} \to \mathbf{y})$ is the (conditional) probability density that you will move to $\mathbf{y}$ if you start at $\mathbf{x}$; it is often called the *transition probability*.[60] Since we're dealing with a Markov chain, we need to know how to go from one sample to the next; this is precisely what the transition probability will allow us to do, soon. Since $w(\mathbf{x})$ is the probability density of being near $\mathbf{x}$, $w(\mathbf{x})T(\mathbf{x} \to \mathbf{y})$ quantifies how likely it is overall to start at $\mathbf{x}$ and move to $\mathbf{y}$. Similarly, $w(\mathbf{y})T(\mathbf{y} \to \mathbf{x})$ tells us how likely it is to start at $\mathbf{y}$ and move to $\mathbf{x}$. In words, Eq. (7.212) says that it is equally likely that we will go in one direction as in the reverse direction. The principle of detailed balance is sometimes known as the *reversibility condition*, due to the fact that the reverse process would result if everything went backward in time. Intuitively, detailed balance tells us that if you're in equilibrium then effectively not much is changing: you could go somewhere, but you're just as likely to come back.

At this stage, detailed balance is just a condition: we haven't shown how to actually produce a Markov chain that obeys it. Even so, we will now spend some time seeing exactly how the detailed-balance condition can help us accomplish our goal. Instead of thinking about moving from an individual sample to another one, it can be helpful to think in terms of going from one probability density function to another. Assume that $P_{i-1}(\mathbf{x})$ is the distribution of values of the random vector $\mathscr{X}_{i-1}$ and, similarly, $P_i(\mathbf{x})$[61] is the distribution of $\mathscr{X}_i$ (recall footnote 59). Then, $P_i(\mathbf{x})$ satisfies the following recurrence relation:[62]

$$P_i(\mathbf{x}) = P_{i-1}(\mathbf{x}) + \int [P_{i-1}(\mathbf{y})T(\mathbf{y} \to \mathbf{x}) - P_{i-1}(\mathbf{x})T(\mathbf{x} \to \mathbf{y})]\, d^d y \qquad (7.213)$$

In words, what this is saying is that the probability of being near $\mathbf{x}$ at step i is equal to the probability of being near $\mathbf{x}$ at step $i-1$, plus the probability of leaving all other configurations $\mathbf{y}$ and coming to $\mathbf{x}$, minus the probability of leaving $\mathbf{x}$ and going to any other configurations $\mathbf{y}$. We will now put Eq. (7.213) together with the condition of detailed balance, Eq. (7.212), to see what we get.

First, we will show that if Eq. (7.212) holds, then Eq. (7.213) shows that $w(\mathbf{x})$ is a fixed point of the iteration (see section 5.2.3). For $P_{i-1}(\mathbf{x}) = w(\mathbf{x})$, Eq. (7.213) turns into:

$$P_i(\mathbf{x}) = w(\mathbf{x}) + \int [w(\mathbf{y})T(\mathbf{y} \to \mathbf{x}) - w(\mathbf{x})T(\mathbf{x} \to \mathbf{y})]\, d^d y = w(\mathbf{x}) \qquad (7.214)$$

In the last step we noticed that the integrand vanishes, due to the detailed-balance condition. Thus, we have shown that $P_i(\mathbf{x}) = P_{i-1}(\mathbf{x}) = w(\mathbf{x})$, i.e., we are dealing with a stationary distribution. This is saying that if at some point of the iteration our distribution becomes the desired distribution, then it will stay there.

Second, we would like to know that we are actually approaching that stationary distribution:[63] it wouldn't do us much good if a fixed point existed but we could never reach it.

[60] More generally, one often says "probability" instead of "probability density"; this is analogous to classical field theory, where many times people will say "Lagrangian" even when they actually mean "Lagrangian density". Note that $T(\mathbf{x} \to \mathbf{y})$ would be called $T(\mathbf{y}|\mathbf{x})$ using the notation of section 6.6.

[61] Obviously, this $P_i(\mathbf{x})$ has nothing to do with the Legendre polynomial $P_n(x)$.

[62] Looking at the equivalent relation Eq. (7.262) may help you solidify your intuition.

[63] Two sufficient conditions: it should be possible for the Markov chain to (eventually) reach the neighborhood of any $\mathbf{x}$ from any starting $\mathbf{y}$ (*irreducibility*), but the return to $\mathbf{x}$ should not happen periodically (*aperiodicity*).

To see this, divide Eq. (7.213) by the desired distribution, $w(\mathbf{x})$:[64]

$$\frac{P_i(\mathbf{x})}{w(\mathbf{x})} = \frac{P_{i-1}(\mathbf{x})}{w(\mathbf{x})} + \int \left[P_{i-1}(\mathbf{y})\frac{T(\mathbf{y}\to\mathbf{x})}{w(\mathbf{x})} - P_{i-1}(\mathbf{x})\frac{T(\mathbf{x}\to\mathbf{y})}{w(\mathbf{x})} \right] d^d y$$
$$= \frac{P_{i-1}(\mathbf{x})}{w(\mathbf{x})} + \int T(\mathbf{x}\to\mathbf{y}) \left[\frac{P_{i-1}(\mathbf{y})}{w(\mathbf{y})} - \frac{P_{i-1}(\mathbf{x})}{w(\mathbf{x})} \right] d^d y \qquad (7.215)$$

To get to the second line, we first used the detailed balance condition, Eq. (7.212), to replace $T(\mathbf{y}\to\mathbf{x})/w(\mathbf{x})$ by $T(\mathbf{x}\to\mathbf{y})/w(\mathbf{y})$ and then pulled out the common term $T(\mathbf{x}\to\mathbf{y})$. Observe that the two terms inside the square brackets are expressed in the form of "actual distribution" divided by "desired distribution", i.e., P_{i-1}/w. Keep in mind that $T(\mathbf{x}\to\mathbf{y})$ is a conditional probability density and is therefore non-negative. Think about what Eq. (7.215) tells us will happen to a $P_{i-1}(\mathbf{x})/w(\mathbf{x})$ that is near a maximum, i.e., is larger than other ratios, $P_{i-1}(\mathbf{y})/w(\mathbf{y})$: the square bracket will be negative, therefore the entire right-hand side will drive this ratio down. Correspondingly, if $P_{i-1}(\mathbf{x})/w(\mathbf{x})$ is near a minimum, i.e., is smaller than the other ratios, the square bracket will be positive, therefore the entire right-hand side will drive this ratio up. In both cases, the ratio $P_i(\mathbf{x})/w(\mathbf{x})$ will be closer to 1 than $P_{i-1}(\mathbf{x})/w(\mathbf{x})$ was.

While we *still* haven't shown how to produce a Markov chain that obeys detailed balance, our two results in Eq. (7.214) and Eq. (7.215) are that *if* you have a Markov chain that obeys detailed balance then: (a) $w(\mathbf{x})$ *is a stationary distribution*, and (b) $P_i(\mathbf{x})$ *asymptotically approaches that stationary distribution*. In other words, our Markov chain will approach a d-dimensional equilibrium distribution of our choosing. We will now introduce an elegant trick that is able to produce a Markov chain obeying detailed balance.

Metropolis–Hastings Algorithm

The *Metropolis–Hastings algorithm* starts by splitting the transition probability:

$$T(\mathbf{x}\to\mathbf{y}) = \pi(\mathbf{x}\to\mathbf{y})\alpha(\mathbf{x}\to\mathbf{y}) \qquad (7.216)$$

Here $\pi(\mathbf{x}\to\mathbf{y})$ is the probability of making a *proposed* step from $\mathbf{x}$ to $\mathbf{y}$ and $\alpha(\mathbf{x}\to\mathbf{y})$ is the probability of *accepting* that move. Since we are dealing with an *acceptance probability* $\alpha(\mathbf{x}\to\mathbf{y})$, some moves will be accepted (i.e., the system moves from $\mathbf{x}$ to $\mathbf{y}$) and some moves will be rejected (i.e., the system will stay at $\mathbf{x}$); this leads to an aperiodic Markov chain (recall footnote 63). The *proposal probability* $\pi(\mathbf{x}\to\mathbf{y})$ is not unique; the acceptance probability will be chosen in such a way that detailed balance is obeyed.

The Metropolis–Hastings algorithm proceeds by evaluating the following quantity:

$$R(\mathbf{x}\to\mathbf{y}) = \frac{w(\mathbf{y})\,\pi(\mathbf{y}\to\mathbf{x})}{w(\mathbf{x})\,\pi(\mathbf{x}\to\mathbf{y})} \qquad (7.217)$$

[64] Problem 7.39 employs another argument, using differences between these entities (instead of ratios).

which is sometimes known as the *Metropolis–Hastings ratio*. Note that everything on the right-hand side is known: the desired distribution w is of our choosing, as is also true of the proposal distribution π. As a matter of fact, a simpler version of the Metropolis–Hastings algorithm, known as the *Metropolis algorithm* since that's how it was originally put forward, employs a symmetric proposal distribution; when $\pi(\mathbf{x} \to \mathbf{y}) = \pi(\mathbf{y} \to \mathbf{x})$ you can see that the ratio is simply $R(\mathbf{x} \to \mathbf{y}) = w(\mathbf{y})/w(\mathbf{x})$, namely the ratio of the (analytically known) desired weight at the configuration $\mathbf{y}$ and at the configuration $\mathbf{x}$.

The next part of the Metropolis–Hastings algorithm is to use the ratio $R(\mathbf{x} \to \mathbf{y})$ to determine the acceptance probability as follows:

$$\alpha(\mathbf{x} \to \mathbf{y}) = \min\left[1, R(\mathbf{x} \to \mathbf{y})\right] \tag{7.218}$$

We already know that $R(\mathbf{x} \to \mathbf{y})$ is non-negative. What Eq. (7.218) does is to account for the possibility that the ratio $R(\mathbf{x} \to \mathbf{y})$ is larger than 1: in that case, the acceptance probability is taken to be 1 (i.e., the step is guaranteed to be taken). If $R(\mathbf{x} \to \mathbf{y})$ is less than 1, then the proposed step is taken with probability $\alpha(\mathbf{x} \to \mathbf{y})$—i.e., it is not taken with probability $1 - \alpha(\mathbf{x} \to \mathbf{y})$.

As Eq. (7.218) now stands, it may not be totally obvious why we said that the acceptance probability was selected so that we can satisfy detailed balance. It's true that the evaluation of α involves only π and w, both of which are known, but what does this have to do with detailed balance? We will now explicitly answer this question; as you will see, showing this is quite easy, now that you are comfortable with all the concepts involved. Let us start with the left-hand side of the detailed-balance condition, Eq. (7.212):

$$\begin{aligned} w(\mathbf{x})T(\mathbf{x} \to \mathbf{y}) &= w(\mathbf{x})\pi(\mathbf{x} \to \mathbf{y})\alpha(\mathbf{x} \to \mathbf{y}) = w(\mathbf{x})\pi(\mathbf{x} \to \mathbf{y})\frac{w(\mathbf{y})\,\pi(\mathbf{y} \to \mathbf{x})}{w(\mathbf{x})\,\pi(\mathbf{x} \to \mathbf{y})} \\ &= w(\mathbf{y})\pi(\mathbf{y} \to \mathbf{x})\alpha(\mathbf{y} \to \mathbf{x}) = w(\mathbf{y})T(\mathbf{y} \to \mathbf{x}) \end{aligned} \tag{7.219}$$

In the first equality we split the transition probability into its two parts as per Eq. (7.216). In the second equality we assumed that $R(\mathbf{x} \to \mathbf{y}) < 1$, in which case Eq. (7.218) tells us that $\alpha(\mathbf{x} \to \mathbf{y}) = R(\mathbf{x} \to \mathbf{y})$; we also took the opportunity to plug in the ratio as per Eq. (7.217). In the third equality we cancelled what we could and also introduced an $\alpha(\mathbf{y} \to \mathbf{x})$ term: the definition of R in Eq. (7.217) shows that $R(\mathbf{x} \to \mathbf{y})R(\mathbf{y} \to \mathbf{x}) = 1$ so, since we assumed that $R(\mathbf{x} \to \mathbf{y}) < 1$, we will have $R(\mathbf{y} \to \mathbf{x}) > 1$; but in that case, Eq. (7.218) will lead to $\alpha(\mathbf{y} \to \mathbf{x}) = 1$, meaning we were justified in including this term. The fourth equality identified the two parts of the transition probability as per Eq. (7.216). Thus, we have shown that for the Metropolis–Hastings algorithms if $R(\mathbf{x} \to \mathbf{y}) < 1$, the left-hand side of Eq. (7.212) is equal to the right-hand side, i.e., the condition of detailed balance is met. The argument for the case of $R(\mathbf{x} \to \mathbf{y}) > 1$ is precisely analogous, this time starting from the right-hand side and reaching the left.

In summary, we have shown that *the Metropolis–Hastings algorithm satisfies detailed balance*. As we discussed after Eq. (7.215), obeying detailed balance means that $w(\mathbf{x})$ is a stationary distribution and our random walk will be asymptotically approaching that stationary distribution, i.e., our equilibrium distribution will be $w(\mathbf{x})$. Thus, we have managed to draw random samples from a d-dimensional distribution, which was our goal all along. Once again, take a moment to appreciate how different this all is from what we were doing in earlier subsections of this chapter: our $\mathscr{X}_i$'s now form a Markov chain. This means that they are *not statistically independent* from one another (recall page 528). As problem 7.46 shows, some care must be taken in evaluating averages along the random walk; we implement one possible strategy below, but other approaches also exist, e.g., "binning". Even so, it's important to emphasize that the Metropolis–Hastings algorithm has allowed us to draw random samples from a d-dimensional $w(\mathbf{x})$ simply by calculating ratios of w's (and perhaps also of π's) at two configurations each time. There was no need to worry about a change of variables, no numerical multidimensional inversion, and so on. Simply by evaluating known quantities at a current and a proposed configuration, we managed to solve a complicated sampling problem. This is why the Metropolis–Hastings prescription is routinely listed among the most important algorithms of the twentieth century.

How to Implement the Metropolis Algorithm

We now give a practical step-by-step summary of the Metropolis algorithm—which allows us to sample from the d-dimensional distribution $w(\mathbf{x})$—in random-vector notation.

0. Start at a random location, $\mathscr{X}_0$. It shouldn't matter where you start, since the Markov chain will "equilibrate", reaching the same stationary distribution, anyway. You could account for this "burn-in"/warm-up time by discarding some early iterations or you could start from an $\mathscr{X}_0$ that is not unlikely, i.e., pick $\mathscr{X}_0$ such that $w(\mathscr{X}_0)$ is not too small.
1. Take $\mathscr{X}_{i-1}$ as given (this is $\mathscr{X}_0$ the first time around) and produce a uniformly distributed proposed step according to:

$$\mathscr{Y}_i = \mathscr{X}_{i-1} + \theta \times \mathscr{U}_i \tag{7.220}$$

   where $\mathscr{U}_i$ is a d-dimensional vector of uniformly distributed random numbers from -1 to 1; here θ is a number that controls the "step size" and $\mathscr{Y}_i$ is the proposed walker configuration. This step, being uniformly distributed[65] in a multidimensional cube of side 2θ, is *de facto* employing a proposal distribution π that is symmetric;[66] this means that we are actually dealing with the simple Metropolis algorithm (as will also be the case in the Project below; the full Metropolis–Hastings algorithm is explored in problem 7.43). The value of θ is chosen such that roughly 15% to 50% of the proposed steps are accepted.
2. We can combine Eq. (7.217) together with Eq. (7.218). Since we're dealing with a symmetric proposal distribution, the acceptance probability is:

$$\alpha(\mathscr{X}_{i-1} \to \mathscr{Y}_i) = \min\left[1, \frac{w(\mathscr{Y}_i)}{w(\mathscr{X}_{i-1})}\right] \tag{7.221}$$

   Only the ratio $w(\mathscr{Y}_i)/w(\mathscr{X}_{i-1})$ matters: even if you don't know w's normalization, you can still carry out the entire process, since the normalization constant cancels.

[65] This is, obviously, not a unique choice; an even more commonly employed step is normal, i.e., Gaussian.
[66] The proposal probability density when starting at $\mathscr{X}_{i-1}$ is $1/(2\theta)^d$, while the probability density had we started at $\mathscr{Y}_i$ would have been $1/(2\theta)^d$, i.e., the same.

3. With probability $\alpha(\mathscr{X}_{i-1} \to \mathscr{Y}_i)$, set $\mathscr{X}_i = \mathscr{Y}_i$ (the proposed step is accepted), otherwise set $\mathscr{X}_i = \mathscr{X}_{i-1}$ (the proposed step is rejected). In practice, this is done as follows: generate a realization of $\mathscr{Q}_i$, a random variable uniformly distributed from 0 to 1; then set:[67]

$$\mathscr{X}_i = \begin{cases} \mathscr{Y}_i, & \text{if } \mathscr{Q}_i \le \alpha(\mathscr{X}_{i-1} \to \mathscr{Y}_i) \\ \mathscr{X}_{i-1}, & \text{if } \mathscr{Q}_i > \alpha(\mathscr{X}_{i-1} \to \mathscr{Y}_i) \end{cases} \tag{7.222}$$

   Random numbers appear in two distinct roles: first, they allow us to produce the proposed walker configuration, $\mathscr{Y}_i$, as per Eq. (7.220), and, second, they help us decide whether to accept or reject the proposed step as per Eq. (7.222). If $\alpha(\mathscr{X}_{i-1} \to \mathscr{Y}_i) = 1$ the proposed step is accepted regardless of the value of $\mathscr{Q}_i$. Crucially, the next configuration is determined by the magnitude of $\alpha(\mathscr{X}_{i-1} \to \mathscr{Y}_i)$, which in its turn depends on the value of the ratio $w(\mathscr{Y}_i)/w(\mathscr{X}_{i-1})$; thus, two evaluations (one of which is new) of the (analytically known) desired distribution w are enough to select a new sample.

   At this point, you can appreciate our comment about how to pick the value of θ: if θ is tiny, you're taking a small step; since you were most likely already in a region where $w(\mathscr{X}_{i-1})$ was large, it's reasonable to expect that $w(\mathscr{Y}_i)$ will also be large, in which case the step will likely be accepted; but in this case you're not moving very far from where you started, so your sampling is of poor quality. On the other hand, if θ is huge, then you are likely to leave the region of large $w(\mathscr{X}_{i-1})$ and go to a region of small $w(\mathscr{Y}_i)$, in which case the step is likely to be rejected; in this case, too, your sampling is bad, since you are not producing truly new (accepted) samples. By picking θ between these two extremes, you ensure that you can carry out the sampling process efficiently; the value of θ impacts the *acceptance probability* (which, as per Eq. (7.221), refers to a single Metropolis step) and therefore also the *acceptance rate* (how many steps were accepted overall, for the whole calculation). While the lore of computational physics instructs you to aim for an acceptance rate of 50%, statisticians [125] have shown under reasonably general conditions[68] that in many dimensions the optimal acceptance rate is 23.4%. The detailed argument can get complicated, but you shouldn't lose sleep over this: an acceptance rate from 15% to 50% is typically efficient enough.
4. Every n_m steps, make a "measurement", i.e., evaluate $f(\mathscr{X}_i)$. You will need this in order to evaluate the right-hand side of Eq. (7.211); note that, because the Markov chain samples are not statistically independent, we are computing the sample mean (and its variance) using *not* every single sample, but every n_m-th one. This *thinning* is carried out in order to eliminate the correlation between the samples (see also problem 7.46).
5. Increment i by 1 and return to step 1. Terminate the entire process when you've generated enough measurement samples ($\mathfrak{y}$) to get the desired variance of the sample mean.

If this all seems too abstract to you, you'll be pleased to hear that we will apply the Metropolis algorithm to a 12-dimensional integral in the Project below. Another way of becoming more comfortable with this algorithm is to implement it for the case of a one-dimensional integral (as problem 7.40 asks you to do): it's always a good idea to first apply a new algorithm to a problem you've solved before, where you know what to expect.

[67] The cumulative distribution function of the uniform distribution gives $\mathbb{P}(\mathscr{Q}_i \le \alpha(\mathscr{X}_{i-1} \to \mathscr{Y}_i)) = \alpha(\mathscr{X}_{i-1} \to \mathscr{Y}_i)$.
[68] For a Gaussian proposal distribution which, as mentioned in footnote 65, is the typical choice.

7.8 Project: Variational Quantum Monte Carlo

We will now see how the Metropolis algorithm can help us describe *many-particle quantum-mechanical systems*. This is not the first time we have encountered quantum mechanics: the Project in chapter 3 revolved around taking the second derivative of a single-particle wave function, whereas the Project in chapter 4 studied several interacting spins. The latter investigation assumed that there were no orbital degrees of freedom and therefore turned into a matrix eigenvalue problem. In the present section we will make the "opposite" assumption, i.e., we will study spinless particles, leaving the fermionic case for problem 7.51. Note that to fully solve the quantum-mechanical problem for many interacting particles, one would need to solve the *many-particle Schrödinger equation*, an involved task that is at the forefront of modern research; we provide the tools that will help you solve the *single*-particle Schrödinger equation in the next chapter, which studies differential equations. Our goal here will be more humble: we try to produce an *upper bound* on the ground-state energy of a many-particle quantum-mechanical system; this is a task that can be written down in the form of a many-dimensional integral, allowing us to introduce the method known as *variational Monte Carlo* (VMC).

Assume we are dealing with n_p particles, e.g., for $n_p = 4$ these would be the particles labelled 0, 1, 2, and 3, as usual; their positions would be $\mathbf{r}_0$, $\mathbf{r}_1$, $\mathbf{r}_2$, and $\mathbf{r}_3$. Further assume that we are dealing with n_d-dimensional space, e.g., for $n_d = 3$ we have the usual x, y, and z Cartesian components; the components of the position of the 0th particle are therefore $(\mathbf{r}_0)_x$, $(\mathbf{r}_0)_y$, and $(\mathbf{r}_0)_z$. For n_p particles in n_d dimensions, we have to keep track of $n_p n_d$ components in total, e.g., $4 \times 3 = 12$ components for our running example. Employing the notation used in our earlier Monte Carlo discussion, where $\mathbf{x}$ was a d-dimensional vector, we can bundle the $d = n_p n_d$ components into a vector as follows:

$$\mathbf{x} = \Big((\mathbf{r}_0)_x \quad (\mathbf{r}_0)_y \quad (\mathbf{r}_0)_z \quad (\mathbf{r}_1)_x \quad \dots \quad (\mathbf{r}_{n_p-1})_y \quad (\mathbf{r}_{n_p-1})_z \Big)^T \tag{7.223}$$

One could employ an $n_p \times n_d$ matrix $\mathbf{X}$, instead; regardless of how one stores them, the essential point is that there will be $d = n_p n_d$ numbers to keep track of.

7.8.1 Hamiltonian and Wave Function

Our goal will be, in n_d dimensions for n_p particles described by a Hamiltonian $\hat{H}$, to determine the ground-state energy of the system, E_0, reasonably well, using a trial wave function, ψ_T. In general, $\hat{H}$ will be complicated, so the problem will not be amenable to an analytical solution. We will provide only a *variational* answer: our approximation of E_0 will be an upper bound, i.e., will be guaranteed to be larger than or equal to the true ground-state energy; we will accomplish this by evaluating a multidimensional integral using techniques from the present chapter. We will now pose a specific problem: four interacting particles in a three-dimensional anisotropic harmonic oscillator.

7.8.1.1 Hamiltonian

Our chosen four-particle Hamiltonian consists of a (non-relativistic) kinetic energy, a one-body potential energy, and a two-body interaction term:

$$\hat{H} = -\frac{\hbar^2}{2m}\sum_{j=0}^{3}\nabla_j^2 + \frac{1}{2}m\sum_{j=0}^{3}\left\{\omega_x^2[(\mathbf{r}_j)_x]^2 + \omega_y^2[(\mathbf{r}_j)_y]^2 + \omega_z^2[(\mathbf{r}_j)_z]^2\right\} + g\sum_{\substack{j,k=0\\ j<k}}^{3}\exp\left[-q(\mathbf{r}_j - \mathbf{r}_k)^2\right] \tag{7.224}$$

where the particles have equal mass m and ∇_j^2 is the Laplacian taking the second derivative with respect to position $\mathbf{r}_j$. The oscillator strength in each direction is determined by the angular frequencies ω_x, ω_y, and ω_z, i.e., we are allowing for the possibility of an *anisotropic* oscillator (for which the frequencies are unequal); note the unmistakable quadratic dependence on the position components, identifying this as a *harmonic* oscillator. The last term on the right-hand side is the two-body interaction, which we have taken to be of Gaussian form; this interaction is characterized by a *strength* parameter g (q is there to take care of the units, but we will drop it in what follows). Note the $j < k$, which ensures that we're not double-counting: if the 0th particle interacts with the 2nd particle, then you shouldn't separately include a term for the 2nd particle interacting with the 0th particle.

Let's spend some time interpreting this Hamiltonian physically. The particles have kinetic energy: they're free to move around. They also have a harmonic-trapping potential energy: this is the result of each individual particle separately interacting with the external field (i.e., the harmonic oscillator). Finally, the two-body interaction term has to do with how strongly the particles attract or repel each other; we will study the case of positive g, so the particles will be repelling each other (i.e., overall increasing the total energy); the two-body term depends on the distance between each pair of particles. You can also consider the Hamiltonian from the perspective afforded by *perturbation theory*: split it into an analytically known term, on the one hand, and a complicated term, on the other, i.e., $\hat{H} = \hat{H}_0 + \hat{H}_1$; in this case, $\hat{H}_0$ would be kinetic plus oscillator (which is the textbook problem we covered in section 3.5) and $\hat{H}_1$ would be the two-body interaction term. Note that, in what follows, we will investigate general (i.e., not necessarily small) values of g.

7.8.1.2 Trial Wave Function

The split into $\hat{H}_0$ and $\hat{H}_1$ we just mentioned will also guide our choice of the trial wave function. Perhaps we should first explain why this is called a *trial* wave function: the reason is that the full $\hat{H}$ in Eq. (7.224) is quite complicated, so we won't attempt to completely solve the corresponding many-body Schrödinger equation. Instead, we will limit ourselves to employing an *Ansatz*; we will choose a trial wave function, denoted ψ_T, which will hopefully contain some of the physics involved in our problem. As should be clear by this discussion, this is a never-ending process: unless you can prove (via another method) that your wave function is the true ground state (and therefore your estimate of the ground-state energy E_0 is *exact*), you will always be able to go back to the drawing board and design a new, hopefully better, trial wave function; problem 7.52 invites you to do better than we will with our choice below. Crucially, this is not a directionless endeavor: since we will be obeying a *variational* principle, a trial wave function that leads to a lower energy is *better* (i.e., closer to the, unknown, true ground-state energy). This then becomes a practical question: if your new guesses no longer lead to a marked improvement, then you can call it a day.

We won't have to worry about the particles' spin; this could mean either *bosons* or *boltzmannons*, i.e., either indistinguishable particles obeying Bose–Einstein statistics, or distinguishable particles obeying Maxwell–Boltzmann statistics.[69] In that case, we can place each of the four particles into the same single-particle state ϕ:

$$\psi_T(\mathbf{X}) = \phi(\mathbf{r}_0)\phi(\mathbf{r}_1)\phi(\mathbf{r}_2)\phi(\mathbf{r}_3) \tag{7.225}$$

On the left-hand side we are using our Monte-Carlo-esque notation from Eq. (7.223); obviously, our four-particle state is merely the product of four copies of the same single-particle state. We stress that we are *allowed* to place all four particles in the same ϕ.

We pick this ϕ motivated by the split into $\hat{H}_0$ and $\hat{H}_1$: imagine for a moment that $g = 0$, i.e., we didn't have a two-body interaction term; the resulting Hamiltonian is then simply the sum of four single-particle Hamiltonians. That means that our study in section 3.5 carries over; of course, there we were faced with a single particle in a single dimension, so we first have to generalize to the three-dimensional case. The problems for each of the three Cartesian components are totally decoupled, so ϕ can simply be another product, of the wave functions solving those three problems:

$$\phi(\mathbf{r}_0) = \varphi_{n_x}((\mathbf{r}_0)_x)\,\varphi_{n_y}((\mathbf{r}_0)_y)\,\varphi_{n_z}((\mathbf{r}_0)_z) \tag{7.226}$$

where we chose the 0th particle for concreteness. Here the φ_{n_x}, φ_{n_y}, and φ_{n_z} have the form of the one-dimensional solution in Eq. (3.72); be sure to distinguish between ϕ and φ. We reiterate that there are two products involved here: Eq. (7.225) is a product over particle labels and Eq. (7.226) a product over Cartesian components for a given particle label. Note also that each Cartesian component gets its own quantum number; since we are studying the ground-state of the system, we are justified in choosing each of these to be zero, i.e., $n_x = n_y = n_z = 0$. Similarly to what we did in code 3.3, i.e., `psis.py`, we also allow for the possibility of including an extra variational parameter, α, in the exponent of the Gaussian, leading to:

$$\varphi_0\left((\mathbf{r}_0)_x\right) = e^{-\alpha m\omega_x[(\mathbf{r}_0)_x]^2/(2\hbar)} \tag{7.227}$$

and similarly for the y and z components. Note that if you take the two-particle interaction—i.e., g in Eq. (7.224)—to be strong, you can find a minimizing α that is not the "obvious" one ($\alpha = 1$); we will see this explicitly below.

Putting the last three equations together, we are led to a many-particle wave function that is a product of 12 terms: four ϕ's, each of which is a product of three φ's. For our

[69] This is merely the distinction between quantum *mechanics* and quantum *statistics*. Imagine neutrons and protons with different spin projections, or ultracold atoms belonging to distinct species.

specific case, we can further manipulate the product to write it in terms of a sum in the exponent. Note that this form is not set in stone: our steps *motivate* the wave function, but do not derive it: one could envision more complicated choices being made (e.g., instead of a single α, having different variational parameters in each direction—see problem 7.52).

7.8.1.3 Implementation

Before we proceed to develop the variational principle and the VMC method, let's implement what we already have, namely the trial many-particle wave function of Eq. (7.225) and the effect the Hamiltonian of Eq. (7.224) has on it. We now wish to quantify this effect; we recall that in Eq. (3.83) we had defined the local kinetic energy. We generalize that to the following definition of the *local energy*:

$$E_L(\mathbf{x}) = \frac{\hat{H}\psi_T(\mathbf{x})}{\psi_T(\mathbf{x})} \tag{7.228}$$

Since this involves the full Hamiltonian, it is the local *total* energy. It's called local because it depends on the specific "walker" $\mathbf{x}$ you are at; in other words, it's something you can evaluate once you've placed all the particles at given locations, as per Eq. (7.223). In the code, we will split this local energy into two parts, by dividing the full Hamiltonian into kinetic plus potential energy, $\hat{H} = \hat{T} + \hat{V}$; note that this is different from the earlier split $\hat{H} = \hat{H}_0 + \hat{H}_1$, which was designed to capture as much as possible of the interactions in a form that could be analytically handled; our latest split is simply bookkeeping: we separate the kinetic energy (which, for a non-relativistic system, is always going to be the same) from the potential energy (which depends on the specific problem).

Code 7.6 starts by defining `psi()`, a Python function that evaluates the wave function of the many-particle system for a given variational parameter (α), a given set of oscillator frequencies ($\omega_x, \omega_y, \omega_z$), and a given set of particle positions (which bundles together all the $\mathbf{r}_j$'s). This is an implementation of Eq. (7.225), Eq. (7.226), and Eq. (7.227), one plugged into the other. Our example employs broadcasting of a 1d array with a 2d array, which works in our case because the dimensions match: for our example of four particles in three dimensions, `oms` has three components and `X` is a 4×3 matrix. If you are not comfortable with this, you should write out the wave function evaluation the "naive" way. Note that the code is quite general: if we wanted to study, say, 11 particles in two dimensions, `psi()` would work equally well. We start this program with the wave-function implementation, because the next function (`ekin()`, for the local kinetic energy) uses `psi()` without receiving it as a parameter. However, it's crucial to note that `ekin()` is general and will work for a different wave function as soon as you change one or two lines in the body of `psi()`.

The function `ekin()` accepts the same parameters as `psi()`, as well as two other ones (for the step size h and for the value of $\hbar^2/m$) which are given default values. We first extract the numbers of particles and dimensions, based on the position matrix that was passed in: once again, this is all general and would work just as well for other problems. The core

Code 7.6 eloc.py

```
import numpy as np

def psi(al, oms, X):
    rexp = np.sum(oms*X**2)
    return np.exp(-0.5*al*rexp)

def ekin(al, oms, X, h=0.01, hom = 1.):
    npart, ndim = X.shape
    psiold = psi(al, oms, X)
    kin = 0.
    for (j,el), r in np.ndenumerate(X):
        X[j,el] = r + h
        psip = psi(al, oms, X)
        X[j,el] = r - h
        psim = psi(al, oms, X)
        X[j,el] = r
        lapl = (psip + psim - 2.*psiold)/h**2
        kin += -0.5*hom*lapl/psiold
    return kin

def epot(oms, X, strength = 3, m = 1., q = 1.):
    npart, ndim = X.shape
    pot = 0.5*m*np.sum(oms**2*X**2)
    for k in range(1,npart):
        for j in range(k):
            r2 = np.sum((X[j,:] - X[k,:])**2)
            pot += strength*np.exp(-q*r2)
    return pot

if __name__ == '__main__':
    npart, ndim, al = 4, 3, 0.6
    oms = np.arange(1, 1 + ndim)
    X = 1/np.arange(1, 1 + npart*ndim).reshape(npart, ndim)
    print(ekin(al, oms, X), epot(oms, X))
```

of this function is very similar to code 3.4, i.e., `kinetic.py` from chapter 3; in essence, we are implementing a finite-difference approximation to the Laplacians in Eq. (7.224). This time around, in addition to studying a many-particle problem, we are also employing `numpy` arrays instead of Python lists. The naive approach here would have been to set up two loops iterating over, say, j for the particle index and l for the Cartesian-component index. We do this more compactly via `numpy.ndenumerate()`, a generalization of Python's built-in `enumerate()`. Similarly to code 3.4, we shift each coordinate twice corresponding to $\pm h$ (recall Eq. (3.85)): this time we index into our 2d array (our walker) in order to tweak only a single element each time (modifying twice and then restoring). It's important to emphasize that `ekin()` doesn't depend on the details of the wave function; if you take the second derivatives analytically (as problem 7.49 asks you to do) you will avoid these $6n_p + 1$ wave function evaluations (more generally: $2n_d n_p + 1$ evaluations), thereby producing a more efficient but less general code. As a matter of fact, `ekin()` doesn't even depend on details of the Hamiltonian in Eq. (7.224): even if you wanted to study a different problem in the future (say, particles in a periodic box with no onc-body potential and a different two-body potential) the kinetic energy (function) would stay the same.

The function `epot()` is a straightforward implementation, corresponding to the one- and two-body interaction terms in the Hamiltonian of Eq. (7.224). We mentioned above that `ekin()` doesn't depend on the details of the Hamiltonian, since the kinetic energy is always the same; `epot()` *does* depend on the details of the Hamiltonian, but it encapsulates them, i.e., if you wish to use different interaction terms in the future you would have to touch only these lines of code. Just like in `ekin()`, we start out by extracting the number of particles and number of dimensions, meaning that everything that follows will be quite general. The implementation of the one-body term uses broadcasting, just as we saw in our discussion of `psi()`. The implementation of the two-body term imposes the $j < k$ condition in the summation by having the two indices take the values $k = 1, 2, \ldots, n_p - 1$ and $j = 0, 1, \ldots, k - 1$. It uses `numpy` array slicing to select $\mathbf{r}_j$ and $\mathbf{r}_k$, and then carries out the squaring and summing of the components automatically. It's worth noting that we have picked the default value of the strength g to be large; this will allow us, below, to investigate a case that is not an already-solved problem. Note, finally, that `epot()` does not take in α as a parameter; as a matter of fact, it doesn't even call the function `psi()`; from the definition in Eq. (7.228) we see that for local interaction terms the wave function cancels: $\hat{V}\psi_T(\mathbf{x})/\psi_T(\mathbf{x}) = V(\mathbf{x})$.

The main program assigns specific values to our input parameters and then makes up an arbitrary configuration of particle positions, merely as a test case. It then calls the kinetic and potential functions, which evaluate the energies *for that one configuration of particles*. The output of this code is not very important, since there is no deep physical meaning behind the specific choice of where the particles are located. Incidentally, this entire program does not involve *any* random numbers: for a given configuration, we can deterministically evaluate the wave function, the local kinetic energy, and the local potential energy. There is nothing stochastic going on so far: in the following subsection we'll see how to combine these functions together with Monte Carlo integration to approximate the ground-state energy of the system (i.e., not the local energy for a single walker).

If you are looking to build some intuition around the magnitude of the numbers involved, you may wish to turn off the two-body interaction (i.e., set $g = 0$) as a test case: if you also take $\alpha = 1$, then you can analytically calculate the energy of the system. Since there are no two-body interactions in this scenario, we are dealing with four independent particles, so we can focus on a single one. Then, the energy of a single particle in a given dimension, say y, is simply $(n_y + 0.5)\hbar\omega_y$ as per Eq. (3.71); for the specific set of frequencies we chose (given also that $n_x = n_y = n_z = 0$), we therefore expect each particle to contribute an energy of $0.5 + 0.5 \times 2 + 0.5 \times 3 = 3$, leading to a value of 12 in total; we expect to find this *regardless* of the specific positions the particles are placed in (here and below, all energies are given in units of $\hbar\omega_x$). Of course, the function `ekin()` in code 7.6 involves a default step size,[70] `h`, which may lead to a numerical error; for our example walker `X` the code gives `11.9994839168`, which is showing that things are behaving as expected.

7.8.2 Variational Method

Code 7.6 evaluates the wave function and the local kinetic and local potential energies, but doesn't tell you how to pick the particle positions. The $\mathbf{r}_j$'s have to be selected somehow; in the present subsection, we see how to pick them stochastically, in a way that allows us to approximate the ground-state energy.

7.8.2.1 Rayleigh–Ritz Principle

We now go over a very important result from elementary quantum mechanics: this is variously known as the *Rayleigh–Ritz principle* or the *variational principle*. In words, it says that the (normalized) expectation value of the Hamiltonian $\hat{H}$ evaluated with a trial wave function ψ_T is an upper bound on the true ground-state energy E_0. If you've encountered this principle in a course on quantum mechanics, this was most likely in the context of single-particle wave functions; however, as you'll soon see, the derivation and result are fully general, which means that we'll be able to apply this principle also in the context of *many*-particle quantum mechanics.

Take the complete set of orthonormal eigenstates $|u_n\rangle$ of the Hamiltonian $\hat{H}$:

$$\hat{H}|u_n\rangle = E_n|u_n\rangle \tag{7.229}$$

Note that this is "diagonal", since the $|u_n\rangle$ are the exact energy eigenstates. The corresponding eigenvalues are E_n and are ordered such that $E_0 \leq E_1 \leq \ldots$; thus, the ground state has the state vector $|u_0\rangle$ and energy E_0. Of course we don't actually know these: if we did, then we would have already solved a major problem in many-particle quantum mechanics. Even so, we are free to expand our trial wave function (state vector) $|\psi_T\rangle$ in terms of the (unknown) exact eigenstates $|u_n\rangle$:

$$|\psi_T\rangle = \sum_{n=0}^{\infty} \gamma_n |u_n\rangle \tag{7.230}$$

which is a step that is frequently carried out in quantum mechanics courses.[71]

[70] The finite-difference approximation to the Laplacian employs a "step size" h, but we also encountered another "step size" θ in Eq. (7.220); the latter refers to the steps carried out in the d-dimensional space.

[71] Another way of looking at this relation is by taking $|\psi_T\rangle$ and inserting a resolution of the identity, i.e., $|\psi_T\rangle = \sum_{n=0}^{\infty} |u_n\rangle\langle u_n|\psi_T\rangle$, from which we immediately see that $\gamma_n = \langle u_n|\psi_T\rangle$.

We now decide to define the following *variational energy* E_V:[72]

$$E_V = \frac{\langle\psi_T|\hat{H}|\psi_T\rangle}{\langle\psi_T|\psi_T\rangle} \tag{7.231}$$

and then write the kets and bras as per Eq. (7.230):

$$E_V = \frac{\sum_{n=0}^{\infty}\sum_{n'=0}^{\infty}\gamma_n\gamma_{n'}^*\langle u_{n'}|\hat{H}|u_n\rangle}{\sum_{n=0}^{\infty}\sum_{n'=0}^{\infty}\gamma_n\gamma_{n'}^*\langle u_{n'}|u_n\rangle} = \frac{\sum_{n=0}^{\infty}\sum_{n'=0}^{\infty}\gamma_n\gamma_{n'}^*E_n\langle u_{n'}|u_n\rangle}{\sum_{n=0}^{\infty}\sum_{n'=0}^{\infty}\gamma_n\gamma_{n'}^*\langle u_{n'}|u_n\rangle} = \frac{\sum_{n=0}^{\infty}|\gamma_n|^2E_n}{\sum_{n=0}^{\infty}|\gamma_n|^2} \tag{7.232}$$

In the second step we took advantage of the fact that we know what effect $\hat{H}$ has on the energy eigenstates, as per Eq. (7.229). In the third step we employed the orthonormality of the exact eigenstates (in both the numerator and the denominator) to get rid of the sum(s) over n'. We now write $E_n = E_0 + E_n - E_0$ in the numerator to find:

$$E_V = E_0 + \frac{\sum_{n=1}^{\infty}|\gamma_n|^2(E_n - E_0)}{\sum_{n=0}^{\infty}|\gamma_n|^2} \geq E_0 \tag{7.233}$$

where the last step follows from the positivity of $|\gamma_n|^2$ and of $E_n - E_0$; you can take the sum in the numerator to start at $n = 0$ if you wish, but that term would give zero anyway.

In short, we have the foundation of the *Rayleigh–Ritz (or variational) principle*:

$$E_V = \frac{\langle\psi_T|\hat{H}|\psi_T\rangle}{\langle\psi_T|\psi_T\rangle} \geq E_0 \tag{7.234}$$

As advertised, this says that by forming the expectation value of the Hamiltonian with respect to a trial wave function you produce an upper bound on the true ground-state energy. When ψ_T is the exact ground-state wave function, then we reach the equality in this relation (i.e., we get the exact ground-state energy).[73] Of course, since the true ground state is actually unknown, what we do, instead, is write down ψ_T in terms of one or more variational[74] parameters, like α in Eq. (7.227). Thus, $\psi_T = \psi_T(\alpha)$ leads to a variational energy which also depends on the value of α, i.e., $E_V = E_V(\alpha)$. By minimizing $E_V(\alpha)$ we will therefore try to get as close to E_0 as we can. As promised, none of these steps or conclusions have anything to do with single-particle quantum mechanics; they are completely general.

7.8.2.2 Variational Monte Carlo

Armed with the Rayleigh–Ritz principle, we will now see how to practically evaluate the variational E_V; if you have a way of accomplishing that task, then you can also minimize E_V to get a good upper bound on the ground-state energy. Start by taking the variational energy from Eq. (7.234); this time, instead of expanding in terms of the exact eigenstates

[72] The formal similarity with Eq. (4.133), defining the Rayleigh quotient, is pretty unmistakable.
[73] You will explicitly show this in problem 7.45.
[74] We can finally see why this parameter was called "variational" all this while.

like in Eq. (7.230), write out the matrix elements in terms of integrals in coordinate space:[75]

$$E_V = \frac{\int d^d x\, \psi_T^*(\mathbf{x})\hat{H}\psi_T(\mathbf{x})}{\int d^d x\, \psi_T^*(\mathbf{x})\psi_T(\mathbf{x})} = \frac{\int d^d x\, |\psi_T(\mathbf{x})|^2 \frac{\hat{H}\psi_T(\mathbf{x})}{\psi_T(\mathbf{x})}}{\int d^d x\, |\psi_T(\mathbf{x})|^2} = \int d^d x \left(\frac{|\psi_T(\mathbf{x})|^2}{\int d^d x'\, |\psi_T(\mathbf{x}')|^2} \right) \frac{\hat{H}\psi_T(\mathbf{x})}{\psi_T(\mathbf{x})} \tag{7.235}$$

where $\mathbf{x}$ is a d-dimensional entity. In the second step we multiplied and divided by the wave function. In the third step we renamed the dummy integration variable and regrouped.

Our final result is now of a familiar form: it looks exactly like the $\int w(\mathbf{x})f(\mathbf{x})d^d x$ in Eq. (7.211) for multidimensional weighted Monte Carlo integration, if we identify:

$$w(\mathbf{x}) = \frac{|\psi_T(\mathbf{x})|^2}{\int d^d x'\, |\psi_T(\mathbf{x}')|^2}, \qquad f(\mathbf{x}) = \frac{\hat{H}\psi_T(\mathbf{x})}{\psi_T(\mathbf{x})} \tag{7.236}$$

with the latter being none other than the *local energy* of Eq. (7.228). In words, we have mapped the problem of computing the variational energy E_V into a multidimensional integral with the weight being controlled by $|\psi_T(\mathbf{x})|^2$ and the function we are integrating being the local energy $E_L(\mathbf{x})$. This is probably a good time to note that the variational energy E_V does *not* depend on the positions of the particles, since it integrates over all the $\mathbf{x}$'s; this makes it markedly different from the local energy E_L, which is a function of $\mathbf{x}$.

Given our mapping, we now have a prescription in Eq. (7.211) to produce the sample mean and its standard deviation: draw samples from $w(\mathbf{x})$ and use them to evaluate the local energy. The only complication is how to produce d-dimensional variates drawn from $w(\mathbf{x})$; but that, too, is a task we've already accomplished, via the Metropolis algorithm, see Eq. (7.220) and below. As per Eq. (7.221), the acceptance probability is determined by evaluating the ratio of the weight function at the proposed and current walker configurations; from Eq. (7.236) we see that this is simply the ratio of the squares of the moduli of the trial wave functions (the denominators cancel—hence our cavalier attitude to normalization). Using this acceptance probability in the Metropolis algorithm will lead to a random walk that is asymptotically approaching the desired weight distribution. To reiterate:[76]

$$E_V \approx \frac{1}{\mathfrak{N}} \sum_{i=0}^{\mathfrak{N}-1} E_L(\mathcal{X}_i) \pm \frac{1}{\sqrt{\mathfrak{N}-1}} \sqrt{\frac{1}{\mathfrak{N}} \sum_{i=0}^{\mathfrak{N}-1} E_L^2(\mathcal{X}_i) - \left[\frac{1}{\mathfrak{N}} \sum_{i=0}^{\mathfrak{N}-1} E_L(\mathcal{X}_i)\right]^2} \tag{7.237}$$

This is merely Eq. (7.211) cast in the language of our problem; the $\mathcal{X}_i$'s are drawn from $|\psi_T(\mathbf{x})|^2$ via the Metropolis algorithm.

At this point, let us return to step 4 in our suggested implementation of the Metropolis algorithm: this involved making a "measurement", i.e., an evaluation of E_L only every n_m steps, whereas the evaluation of the $|\psi_T|^2$ ratio *has* to be carried out at every step in order to make a decision. In order to see why, imagine making several hundred Metropolis steps using a weight of $|\psi_T|^2$ and evaluating E_L at every single one of them (i.e., with $n_m = 1$):

75 Imagine inserting resolutions of the identity, this time in terms of position eigenstates: $\int d^d x\, |\mathbf{x}\rangle\langle\mathbf{x}|$.

76 As mentioned on page 527, we dropped the volume V because we're interested in integrating over all of space; note that this V has nothing to do with the (local) potential energy, $V(\mathbf{x})$.

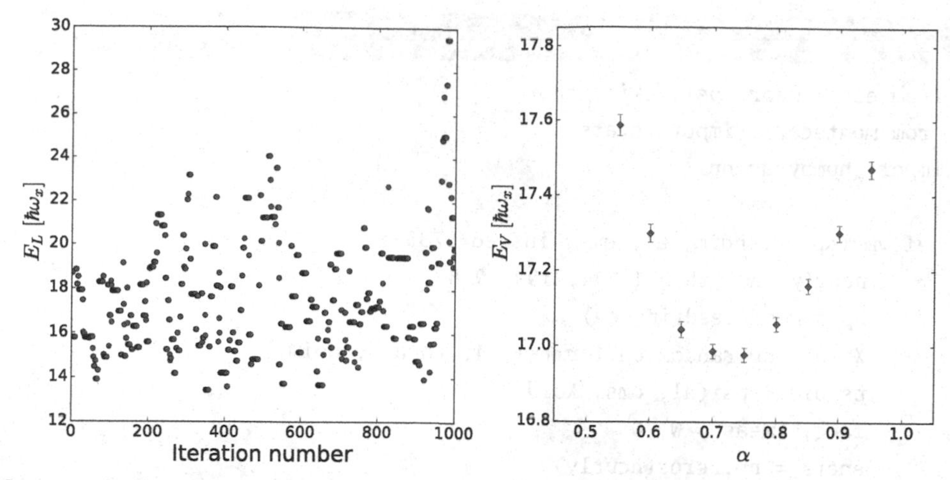

Fig. 7.8 Consecutive local energies for $\alpha = 0.6$ (left), variational minimization in action (right)

the result is shown in the left panel of Fig. 7.8, for the same input parameters as in code 7.6. It is in the nature of our algorithm to start from a given walker configuration and produce a new one by making a small change (a step): as a result, the configurations making up our Markov chain are clearly correlated. This can be seen in the figure from the fact that several of the circles are overlapping, implying that very small steps can and do get accepted; if you make θ smaller, you will encounter even more features of large-scale structure. You can see, especially for the last 50 iterations of the figure, that if your walker gets "stuck" in a region where the energy is unrealistically large, it takes a while to climb down to more reasonable values. The thinning prescription says to only measure the local energy every once in a while, letting the Markov chain de-correlate in the meantime; in the code that we are about to introduce, the E_L's for the 1000 consecutive iterations (being produced by 1000 evaluations of the $|\psi_T|^2$ ratio in the background) shown in the left panel of Fig. 7.8 will be replaced by 10 E_L evaluations. Note that Eq. (7.237) involves $\mathfrak{y}$ evaluations of the local energy: since we are making measurements every n_m steps, this means we'll need $n_m\mathfrak{y}$ Metropolis steps in total. Problem 7.46 studies these issues in more detail.

7.8.2.3 Implementation

Code 7.7 starts out by importing the wave function, the local kinetic and local potential energy functions from code 7.6 and the statistics function from code 7.5; note that the latter is a one-dimensional code, but `stats()` works just as well here, since the evaluation of the sample mean and its standard deviation in Eq. (7.237) is fully analogous to that in Eq. (7.192); our walker is sampling a d-dimensional distribution and, similarly, the local energy is a function of a d-dimensional variable, but at the end of the day we're carrying out the same basic statistical operations (mean and variance of a scalar). Having already built all the necessary infrastructure, our VMC code is quite simple and is placed in a single function, `vmc()`; our earlier point about `stats()` working in both one-dimensional

Code 7.7 vmc.py

```
from eloc import psi, ekin, epot
from montecarlo import stats
import numpy as np

def vmc(npart, ndim, al, oms, inseed=8735):
    ncurly, nm, th = 10**4, 100, 0.8
    np.random.seed(inseed)
    Xold = np.random.uniform(-1, 1, (npart, ndim))
    psiold = psi(al, oms, Xold)
    iacc, imeas = 0, 0
    eners = np.zeros(ncurly)

    for itot in range(nm*ncurly):
        Xnew = Xold + th*np.random.uniform(-1,1,(npart, ndim))
        psinew = psi(al, oms, Xnew)
        psiratio = (psinew/psiold)**2

        if np.random.uniform(0,1) <= psiratio:
            Xold = np.copy(Xnew)
            psiold = psinew
            iacc +=1
        if (itot%nm)==0:
            eners[imeas] = ekin(al,oms,Xold) + epot(oms,Xold)
            imeas += 1

    return iacc/(nm*ncurly), eners

if __name__ == '__main__':
    npart, ndim, al = 4, 3, 0.6
    oms = np.arange(1, 1 + ndim)
    accrate, eners = vmc(npart, ndim, al, oms)
    av, err = stats(eners)
    print(accrate, av, err)
```

and d-dimensional settings has an analogue here: vmc() is written for the d-dimensional case, manipulating walkers that are matrices with $n_p \times n_d$ elements, but it is quite similar to the one-dimensional Metropolis code that you are asked to develop in problem 7.40.

We now turn to a discussion of vmc(). There is no parameter for the particle positions, since `vmc()` will generate the walker configurations itself; thus, the particle number (n_p) and number of dimensions (n_d) have to be explicitly passed in this time. As before, the next two parameters correspond to the variational parameter (α) and the oscillator frequencies ($\omega_x, \omega_y, \omega_z$). The final parameter is the seed that gets the random number sequence going, given a default value here. The first line in the body of `vmc()` sets up three other quantities: the total number of evaluations of the local energy ($\mathfrak{y}$), the number controlling how often these evaluations should be carried out (n_m), and the quantity controlling the size of the proposed step in Eq. (7.220), namely θ; our chosen value for n_m is large enough that we are likely removing the correlations (but perhaps are also being a little wasteful). We then pass the seed into `numpy.random.seed()`: we employ `numpy` (as opposed to standard Python) functionality, as discussed on page 513. We then use `numpy.random.uniform()` to produce the (arbitrary) initial walker configuration, $\mathscr{X}_0$: marvel at how convenient this is, generating an entire matrix of uniformly distributed numbers (with the dimensions we need) in one line of code. Later in the program, `numpy.random.uniform()` is used twice more: first, when making a proposed step; this is what Eq. (7.220) called $\mathscr{U}_i$ which, you may recall, is a d-dimensional entity. Second, we also need a realization of the random variable $\mathscr{Q}_i$ as per Eq. (7.222); for this case we make use of the fact that if you don't pass a third argument to `numpy.random.uniform()` then it returns a single number.

It's important to keep straight the different integers involved here: n_p, n_d, $\mathfrak{y}$, and n_m are kept fixed for a given VMC run, i.e., they are "input" in a way. Knowing $\mathfrak{y}$ ahead of time allows us to set up a NumPy array to keep track of the $E_L(\mathscr{X}_i)$'s that we will be evaluating. In addition to these four constants, our function makes use of three more integers, `itot`, `iacc`, and `imeas`. Of these, `itot` is the index that keeps track of every single Metropolis step (i.e., plays the role of i in Eq. (7.222)); the total number of such steps is $n_m\mathfrak{y}$, as discussed on page 543. Next, we have `iacc` which is an index we use to keep track of how many proposed steps have been accepted so far; this allows us to return the acceptance rate when the function is done, computed via `iacc/(nm*ncurly)`. Finally, we also have `imeas`, an index we use to store our $\mathfrak{y}$ energy measurements: `itot` goes from 0 to $n_m\mathfrak{y} - 1$ and we evaluate the local energy every n_m-th time; when we do so, we increment `imeas` by 1, so we can write the local energy into `eners` the next time we need to make a measurement.

Inside the loop over `itot`, we evaluate the square of the ratio of the trial wave function at the new and old walker configurations: this is what the Metropolis algorithm of Eq. (7.221) requires us to do for a weight function of $|\psi_T(\mathbf{x})|^2$, as mentioned on page 542. Instead of the $\mathscr{X}_{i-1}$, $\mathscr{X}_i$, and $\mathscr{Y}_i$ involved in Eq. (7.221) and Eq. (7.222) we here need only two walker variables, `Xold` and `Xnew`. When a proposed step is accepted, we use `numpy.copy()` to update `Xold`; we could have used a simple assignment, but we wish to instill best practices, lest you get over-confident with re-assigning `numpy` arrays. If the step gets rejected, we don't have to explicitly do anything (i.e., there's no need for an `else:`), since `Xold` simply keeps the value it already had. Finally, when it's time to make a measurement we call our earlier functions `ekin()` and `epot()` and store their sum into `eners`.

It's worth observing that when we evaluate a proposed step in `vmc()`, we move all the particles at once, as per Eq. (7.220); there exist alternative Monte Carlo algorithms that make "local updates" instead, which in our case could translate to moving one particle at a time (before evaluating a new Metropolis ratio). Another scenario is to employ "heat-bath updates" (called a "Gibbs sampler" by statisticians) whereby a few variables are brought to equilibrium, while all the other ones are kept fixed; this is most commonly used in lattice Monte Carlo approaches (like that of problem 7.65).

The main program sets up the input parameters, calls `vmc()` to produce the acceptance rate and array of local energies, and then passes the latter to `stats` to compute the sample mean and its standard deviation. Repeating this exercise for many α's leads to the right panel of Fig. 7.8; we have chosen to tune θ to keep the acceptance rate reasonably stable and also used different seeds for different α's. The variational energy E_V exhibits a clear minimum at $\alpha \approx 0.75$; crucially, $\alpha = 1$ most certainly is *not* the minimum, meaning that our variational parameter has helped us produce a (small) decrease in the energy. If the specific numbers shown in the right panel of Fig. 7.8 don't mean much to you, you may choose to turn off the two-body interaction (i.e., set $g = 0$) and also take $\alpha = 1$ as a test case: the energy is 12, so that's what we expect our VMC run to produce *for every sample*! If you carry out the test explicitly you therefore find a tiny standard deviation.

It's worth taking a moment to appreciate the significance of our achievement: we were dealing with a strongly interacting system of four particles in three dimensions, so this isn't really a problem we would have attempted to solve without the use of a computer;[77] we were able to produce an upper bound to the ground-state energy for this system, by evaluating 12-dimensional integrals via the Metropolis algorithm. Equally important is the fact that we've kept the code sufficiently general to allow us to study in the future different particle numbers, different dimensionalities, and different interactions.

It's hard to avoid noticing that the right panel of Fig. 7.8 is minimizing by checking different values of α "by hand". You could envision using one of the minimization techniques from section 5.5, instead of such a manual process; it's important to realize that in the present case our energy estimates are always associated with a *statistical error*, so minimizing the sample mean is not enough. Problems 7.52 and 7.64 guide you through precisely such automated minimizations, using golden-section search or Powell's method.

Problems

7.1 [$\mathcal{A}$] We will see how to evaluate the following Gaussian integral analytically:

$$I = \int_{-\infty}^{\infty} dx e^{-x^2/2} \tag{7.238}$$

The main trick is to first look at the square of the integral:

$$I^2 = \int_{-\infty}^{\infty} dx e^{-x^2/2} \int_{-\infty}^{\infty} dy e^{-y^2/2} \tag{7.239}$$

where we used a new name for the second integration variable. Switch from Cartesian coordinates, x and y, to polar coordinates, r and θ. It should be straightforward from there. Make sure to take the square root at the end of the entire process.

[77] "Reputation of power, is Power" (Thomas Hobbes, *Leviathan*, part I, chapter X).

7.2 Carry out an error analysis for the midpoint rule (Taylor expand around $x_i + h/2$ and keep terms up to second order). This should include both the one-panel case and the composite case. Then, implement the midpoint rule in Python for the usual case of $f(x) = 1/\sqrt{x^2+1}$ ($a = 0, b = 1$) for $n = 51$ and compare with the analytical answer, as well as with the other methods.

7.3 [$\mathcal{A}$] Take $f(x) = 1$ in Eq. (7.26) and thereby show that, for a Newton–Cotes integration rule, the sum of all the weights is equal to the width of the elementary interval.

7.4 [$\mathcal{A}$] Rederive Simpson's rule for an elementary interval, Eq. (7.45), the way you would if you didn't know about Lagrange interpolation: take $f(x) \approx p(x) = \alpha x^2 + \beta x + \gamma$ and then integrate it from x_i to x_{i+2}. Then, enforce the fact that the quadratic passes through the points $f(x_i)$, $f(x_{i+1})$, and $f(x_{i+2})$. You could now try to solve these three equations for the three unknowns α, β, and γ. Instead, take a linear combination of these three equations that gives the same answer as the result of integrating $p(x)$ earlier in this problem; you have thereby arrived at the two-panel Simpson's rule.

7.5 [$\mathcal{A}$] In the main text we computed the leading error in the one-panel version of the trapezoid rule starting from the Lagrange-interpolation error formula. Now rederive it using a Taylor expansion of $f(x)$ around x_i, keeping terms up to second order. Re-express the first derivative using the forward-difference approximation, Eq. (3.8), starting at the point x_i. Integrate your equation for $f(x)$ from x_i to x_{i+1} and compare the result with the sum of Eq. (7.30) and Eq. (7.35).

7.6 [$\mathcal{A}$] In the main text we derived the error of the trapezoid rule beyond the leading term, Eq. (7.41), by making use of Eq. (7.38). Now write down two Taylor expansions: first, expand $f(x)$ around x_i and, second, expand $f(x)$ around x_{i+1}. Add the two equations together and divide by 2. Integrate from x_i to x_{i+1} and then add together all the one-panel equations. Observe that the even-order derivative terms can be expressed as composite trapezoid-rule approximations to integrals, e.g., the term proportional to h^3 is an approximation to $(h^2/6)\int_a^b f''(x)dx$; note that these integrals can be trivially carried out. To find the error in *those* terms, simply repeat the whole exercise, using $f''(x)$ instead of $f(x)$, and so on.

7.7 [$\mathcal{A}$] In the main text we derived the error in Simpson's rule using a Taylor expansion, see Eq. (7.48) and below. We now examine another derivation for the same quantity, this time starting from the interpolation formula Eq. (6.38). In essence, we need to carry out a more complicated version of the integral in Eq. (7.34); the important difference is that the integrand changes sign and therefore you cannot immediately employ the mean-value theorem (if you could pull out the derivative term, the remaining integral would give zero, which is obviously wrong). Instead, employ:

$$\mathcal{E}_i = \frac{1}{4!} f^{(4)}(\eta) \int_{x_i}^{x_{i+2}} (x - x_i)(x - x_{i+1})^2 (x - x_{i+2}) dx \tag{7.240}$$

The middle term is now squared and the derivative is of one order higher. Carry out the integral and compare with Eq. (7.52).

7.8 We will now discuss a fascinating method [100] which combines contour integration and the trapezoid rule to compute high-order derivatives of analytic functions (!).[78]

[78] Cf. complex steps (problem 3.4), finite differences (problem 3.9), and spectral differentiation (problem 6.30).

(a) Start from the *Cauchy integral formula for differentiation*, which gives us the k-th derivative in terms of a contour integral:

$$f^{(k)}(a) = \frac{k!}{2\pi i} \oint_\gamma \frac{f(z)}{(z-a)^{k+1}} dz \tag{7.241}$$

The contour γ can be taken to be a circle (around the point a) oriented counter-clockwise; parametrizing this as $z - a = re^{2\pi i t}$, rewrite things as:

$$f^{(k)}(a) = \frac{k!}{r^k} \int_0^1 f\left(r\, e^{2\pi i t} + a\right) e^{-2\pi i t k} dt \tag{7.242}$$

(b) Compute $f^5(0)$ for $f(x) = e^x/(\cos^3 x + \sin^3 x)$ by evaluating the integral in Eq. (7.242) with `trapezoid()` from code 7.1 (recalling section 7.6.1). You can take $r = 0.5$; do you understand what happens when you take $r = 1$?

(c) Presumably, when solving the previous part you didn't think too carefully about what the trapezoid method is doing. Now explicitly apply the composite version of the trapezoid rule, Eq. (7.31), to the integral in Eq. (7.242) to show that:

$$f^{(k)}(0) = \frac{k!}{r^k} \frac{1}{n-1} \sum_{j=0}^{n-2} f(r\, e^{2\pi i j/(n-1)})\, e^{-2\pi i k j/(n-1)} \tag{7.243}$$

The n used here is the one employed by the trapezoid rule, so it is at our disposal. We are therefore free to take $n \to n+1$, at which point we recognize that Eq. (7.243) is precisely of the form of the shifted DFT, Eq. (6.97); recompute $f^5(0)$ for our $f(x)$ using `fft()`. While you may use a large value for n, your final answer should be a single number, just like in part (b).

7.9 Generalize Simpson's rule for complex integrands:

(a) Compute the integral given on the left-hand side of Eq. (4.287), for $L = 5$ and a few values of $n - n'$. Compare the answer to the analytically derived one. (This is the same integral that problem 6.25 asked you to compute using the FFT.)

(b) Compute the following integral:

$$I = \int_{-L/2}^{L/2} dx e^{i2\pi(n-n')x/L} |x| \tag{7.244}$$

You are now in a position to study the band structure resulting from arbitrary periodic potentials (by generalizing problems 4.55 and 4.56).

7.10 [$\mathcal{A}$] In the main text we derived the trapezoid and Simpson's rules from Lagrange interpolation; we now continue along that path, for the case of Simpson's 3/8 rule. For $q = 4$ analytically derive the elementary-interval (a) weights from Eq. (7.27), and (b) absolute error, by integrating Eq. (6.38). Compare with table 7.1.

7.11 [$\mathcal{A}$] Repeat the argument of section 7.3.1 (twice) to prove Eq. (7.63) and Eq. (7.64).

7.12 Starting from the elementary-interval expressions for Simpson's 3/8 rule and for Boole's rule given in table 7.1, implement the composite versions of these approaches. Then, generalize the derivation of section 7.3.1 to cover Boole's rule and then modify our code in `adaptive.py` so that it also works for this case.

7.13 In `adaptive.py` we gave a general adaptive integration routine. Note, however, that this was wasteful: when you double the number of panels from N to $N' = 2N$, you don't need to throw away all the function values that you've already computed. Explicitly write out the first few cases (I_1, I_2, I_4, I_8) to convince yourself that:

$$I_{N'} = \frac{1}{2}I_N + h_{N'} \sum_{k=1,3,\ldots}^{N'-1} f(a + kh_{N'}) \tag{7.245}$$

Implement this approach to produce an adaptive trapezoid integrator that is efficient.

7.14 We will study the *Glasser function*, a function that exhibits many oscillations and thereby necessitates an adaptive integrator to compute accurately:

$$G(x) = \int_0^x \sin(t\sin(t))dt \tag{7.246}$$

(a) Use an adaptive Simpson's method to plot $G(x)$ from $x = 0$ to $x = 20$.
(b) Plot the number of integration points n needed at each x value (for an absolute tolerance of 10^{-12}); use a logarithmic scale on the y axis.
(c) Locate the first 20 minima of $G(x)$ after the origin.

7.15 [$\mathscr{A}$] Observe that the second column in the output of `romberg.py` is identical to the first few lines in the output for Simpson's rule from `adaptive.py`. Motivated by this fact, apply Eq. (7.88) to the case of $j = 1$, i.e., for $R_{i,1}$. You immediately see that this is expressed in terms of $R_{i,0}$ and $R_{i-1,0}$. Both of these are from the first column, so they are composite-trapezoid rule results; write them out in terms of function evaluations, use the resulting expressions to do the same for $R_{i,1}$, and then compare the final result with the composite Simpson's rule, Eq. (7.46).

7.16 In problem 7.13 you produced an efficient (i.e., non-wasteful) adaptive integrator. Do the same for Romberg integration, i.e., rewrite `romberg.py` such that you don't throw away earlier function evaluations when you go from N to $N' = 2N$ panels.

7.17 [$\mathscr{A}$] Apply Eq. (7.89) to the case of $n = 1$, i.e., demand that you can exactly integrate monomials up to order $2n - 1 = 1$. Find c_0 and x_0 and compare the resulting formula with the one-panel version of the midpoint rule, Eq. (7.20).

7.18 [$\mathscr{A}$] Show that the monic orthogonal polynomial of degree n is unique. To do so, employ proof by contradiction, i.e., assume that there are two monic orthogonal polynomials, $p_n^A(x)$ and $p_n^B(x)$; their difference is a polynomial of degree $n - 1$ and is therefore orthogonal to both $p_n^A(x)$ and $p_n^B(x)$. Write this out and figure out what it means for the relationship between these two monic polynomials.

7.19 [$\mathscr{A}$] In order to derive the Christoffel–Darboux identity for Legendre polynomials, given in Eq. (7.107), take the recurrence relation in Eq. (2.86) and multiply by $P_j(y)$. Rewrite your equation with x interchanged with y and then subtract these two equations from each other. If you sum the result for j from 0 to n, you find a telescoping series, i.e., the middle terms cancel. You should be left with Eq. (7.107).

7.20 [$\mathscr{A}$] If you compute the weights for high-order Newton–Cotes rules, you will see that some of these are negative. Show that for Gauss–Legendre quadrature the weights are always positive: plug Eq. (6.48) into Eq. (7.97), split the resulting expression into two integrals, and notice that the second one vanishes. The first one is a definite integral of a non-negative function (and is therefore also non-negative).

7.21 For Gauss–Legendre quadrature with $n = 5, 50, 100$, plot the weights vs the abscissas, i.e., the c_k's vs the x_k's that appear in Eq. (7.102).

7.22 [$\mathcal{A}$] As promised at the start of section 7.5.2, we now discuss another way of showing that the nodal abscissas in Gauss–Legendre quadrature are given by roots of Legendre polynomials. Take $f(x)$ to be a polynomial of degree up to $2n - 1$; divide $f(x)$ by $P_n(x)$, the Legendre polynomial of degree n (recall problem 3.11):

$$f(x) = Q_{n-1}(x)P_n(x) + R_{n-1}(x) \tag{7.247}$$

where the quotient and remainder are polynomials of degree up to $n - 1$. Integrate both sides from -1 to $+1$ and justify why one of these integrals vanishes. Express the non-vanishing integral as a sum over weights and function values, as in our defining Eq. (7.7). Now take the x_k's to be the roots of $P_n(x)$ and re-express the sum using Eq. (7.247) evaluated at the x_k's.

7.23 Yet another approach to Gaussian quadrature was mentioned at the start of section 7.5.2: the Golub–Welsch algorithm. In problem 5.20 we saw how to cast computing the roots of a Legendre polynomial in the form of evaluating the eigenvalues of tridiagonal Jacobi matrices; we now see how to get the weights from the eigenvectors. Use the matrix $\mathcal{J}$ to produce the corresponding symmetric Jacobi matrix $\tilde{\mathcal{J}}$:

$$\tilde{\mathcal{J}}_{i,i} = \mathcal{J}_{i,i}, \qquad \tilde{\mathcal{J}}_{i-1,i} = \tilde{\mathcal{J}}_{i,i-1} = \sqrt{\mathcal{J}_{i-1,i}\mathcal{J}_{i,i-1}} \tag{7.248}$$

First check (programmatically) that $\tilde{\mathcal{J}}$ is similar to $\mathcal{J}$, i.e., it has the same eigenvalues. Then, find $\tilde{\mathcal{J}}$'s normalized eigenvectors, $\mathbf{v}_j$, and compute the weights from the formula $c_j = 2[(\mathbf{v}_j)_0]^2$. Here $(\mathbf{v}_j)_0$ is the first component of the eigenvector $\mathbf{v}_j$.

7.24 We will now discuss *Clenshaw–Curtis quadrature* in its simplest version, i.e., without making use of the FFT. This amounts to employing Eq. (7.89) with the Chebyshev nodes of Eq. (6.12), i.e., $x_k = -\cos(\pi k/(n-1))$: observe that the abscissas obey nesting (compare, e.g., $n = 8$ vs $n = 15$). The weights take the form:

$$c_k = \frac{4\alpha_k}{n-1}\sum_{j=0}^{m}\frac{\beta_j}{1-4j^2}\cos\left(\frac{2jk\pi}{n-1}\right) \tag{7.249}$$

where $n = 2m + 1$ or $n = 2m$. Here $\alpha_k = 1/2$ if $k = 0$ or $k = n - 1$ ($\alpha_k = 1$ otherwise) and, similarly, $\beta_j = 1/2$ if $j = 0$ or $j = m$ ($\beta_j = 1$ otherwise). Compare with the output of `gauleg()` for $m = 4, 8, \ldots, 64$ (odd or even n)—don't forget Eq. (7.128).

7.25 In code 3.3, i.e., `psis.py`, we computed the values of Hermite polynomials and their derivatives; then, in problem 5.18 we saw how to compute their roots, by modifying code 5.4, i.e., `legroots.py`. Following the approach outlined in section 7.5.3 one arrives at the following formula for the weights:

$$c_j = \frac{2^{n+1}n!\sqrt{\pi}}{[H_n'(x_j)]^2} \tag{7.250}$$

which is the Gauss–Hermite analogue of Eq. (7.117). Implement a general routine for Gauss–Hermite quadrature; feel free to use `numpy.sort()` and `numpy.argsort()`. The factorials in our equation for the c_j's can cause problems, but our root-solver doesn't work for too large values of n, anyway, so don't go above $n = 20$. Compare the results with what you get from `numpy.polynomial.hermite.hermgauss()`.

7.26 Carry out the change of variables $u = \sqrt{x}$ for the integral:

$$I = \int_0^\infty \frac{x^2 + x}{\sqrt{x}} e^{-x} dx \tag{7.251}$$

and observe that you can compute it using Gauss–Hermite quadrature, if you also notice that the integrand is now even. If you solved problem 7.25, use the function you developed there, otherwise call `numpy.polynomial.hermite.hermgauss()`.

7.27 Compare the trapezoid rule, Simpson's rule, and Gauss–Legendre quadrature for:

$$I_1 = \int_0^{2\pi} e^{\sin(2x)} dx, \quad I_2 = \int_0^{2\pi} \frac{1}{2 + \cos x} dx, \quad I_3 = \int_{-1}^{1} e^{-x^2/2} dx \tag{7.252}$$

Do you understand why the trapezoid rule is not giving terrible results?

7.28 Apply the trapezoid rule to the following integral:

$$I = \int_0^\pi \sqrt{x} \cos x dx \tag{7.253}$$

(a) Print out the answer for increasing values of n, checking to see which digits are no longer changing. Do you understand why this function is slow to converge?

(b) Carry out the change of variables $u = \sqrt{x}$ and then apply the trapezoid rule again. Compare this convergence behavior with that in part (a).

7.29 [$\mathcal{A}$] You will now manipulate the integral I_G from the main text in two distinct ways. First, apply the slicing and then a different transformation for each integral as discussed in Eq. (7.152) and in the following discussion. Second, transform Eq. (7.154) twice, first by taking $v = 1 - u$ and then $w = \sqrt{v}$.

7.30 [$\mathcal{A}$] Manipulate the integral:

$$I = \int_{10^{-6}}^{1} \frac{e^{-x}}{x} dx \tag{7.254}$$

using the techniques of section 7.6.2. Specifically, subtract off the singularity and then use a Taylor expansion where necessary.

7.31 Reproduce both panels of Fig. 7.5. Then, make a pairwise plot using the numbers provided by `random.random()`.

7.32 We now introduce the *equidistribution test* for random-number generators. Split the region from 0 to 1 into, say, $n_b = 10$ bins of equal size. Generate $n = 10^5$ samples in total and count the populations in each bin, n_j, where $j = 0, 1, \ldots, n_b - 1$. We expect to find n/n_b samples in each bin (so 10^4 for our example). Compute the following statistic:

$$\chi^2 = \frac{n_b}{n} \sum_{j=0}^{n_b-1} \left(n_j - \frac{n}{n_b} \right)^2 \tag{7.255}$$

for the three generators used in problem 7.31. A "good" χ^2 is equal to the number of degrees of freedom (as you learned in problem 6.46). Since $\sum_j n_j = n$, the n_b populations are not fully independent, so we actually have $n_b - 1$ degrees of freedom.

7.33 [$\mathcal{A}$] The *Chebyshev inequality* for a general random variable $\mathscr{X}$, whose mean is $\mathbb{E}(\mathscr{X})$, can be expressed as:

$$\mathbb{P}\left(|\mathscr{X} - \mathbb{E}(\mathscr{X})| \geq \epsilon\right) \leq \frac{\mathbb{V}(\mathscr{X})}{\epsilon^2} \tag{7.256}$$

where ϵ is positive and arbitrarily small. In words, this says that the probability of the random variable's value being more than ϵ away from the mean is less than the variance divided by ϵ^2. To show this, apply the definition of the left-hand side, then multiply the integrands by $[x - \mathbb{E}(\mathscr{X})]^2/\epsilon^2$ (why are you allowed to do this?), extend the integration interval from $-\infty$ to $+\infty$, and then employ the definition of the variance from Eq. (C.30).

Having shown Eq. (7.256), one can apply it to our sample mean $\overline{f}$ which is, after all, a random variable. Then, re-express the right-hand side using Eq. (7.180) and take the limit $n \to \infty$; since we're working with a bounded σ_f, the right-hand side goes to 0. Of course, this also means that:

$$\lim_{n\to\infty} \mathbb{P}\left(|\overline{f} - \mu_f| < \epsilon\right) = 1 \tag{7.257}$$

This result is known as the *weak law of large numbers.*

7.34 [$\mathcal{A}$] Given the definition of the estimator $\mathscr{V}$ in Eq. (7.182), show that its expectation is as per Eq. (7.183). To show this, explicitly take the expectation as defined in Eq. (7.174), following steps similar to those in Eq. (7.179). Crucially, when taking the expectation of $[\sum_i f(\mathscr{X}_i)/n]^2$ you have to distinguish between "diagonal" and "off-diagonal" terms and also use the independence of different samples.

7.35 This chapter is focused on numerical integration, but the ability to draw samples from a desired $w(x)$ distribution, which we developed when introducing Monte Carlo techniques (section 7.7) is much more general. In this problem (and in problem 7.41) we will see how to generate "synthetic data" from a desired distribution. The following weight function is an example of the *Cauchy distribution*:[79]

$$w(x) = \frac{1}{10\pi[(x - 2.5)^2 + 0.01]} \tag{7.258}$$

Apply the inverse-transform sampling method from section 7.7.3.2 to generate Cauchy-distributed samples, $\mathscr{X}_i = g^{-1}(\mathscr{U}_i)$.

(a) Produce a histogram of raw counts, i.e., the y axis should be the population of x's in a given narrow "bin". You should use 100 bins of equal width from from -9 to 14 for 10^4 Cauchy-distributed samples. Use the appropriate NumPy/Matplotlib functionality for the visualization.

(b) Make a plot of: (i) the $w(x)$ you started from, Eq. (7.258), and (ii) the raw counts normalized such that the total area in the previously produced histogram is one.

[79] It is also known as the *Lorentz distribution* and the *Breit–Wigner form.* We encountered the latter in problem 6.74, though that was not a probability density function, due to the presence of an overall prefactor.

(c) To understand how important the tails are, generate $n = 10^3, 10^4, \ldots, 10^8$ Cauchy-distributed samples and compute each sample mean (no need to plot anything). For each n, compare the answer with the corresponding sample mean using n normally distributed samples (for $\mu = 2.5$ and $\sigma^2 = 0.01$).

7.36 Produce a plot that looks like the right panel of Fig. 7.7, for the integral of Eq. (7.197), i.e., $I = \int_1^3 e^{-x} \sin x dx$, using uniformly and exponentially distributed numbers.

7.37 [$\mathcal{A}$] We now discuss the *Box–Muller method* of producing normal deviates; this makes a detour through two-dimensional distributions. In the one-dimensional case we carried out the integration in Eq. (7.193); the analogue here is the joint cumulative distribution function:

$$g(x_0, x_1) = \int_{-\infty}^{x_0} \int_{-\infty}^{x_1} w(x'_0, x'_1) dx'_0 dx'_1 = \int_{-\infty}^{x_0} \int_{-\infty}^{x_1} \frac{e^{-(x'_0)^2/2}}{\sqrt{2\pi}} \frac{e^{-(x'_1)^2/2}}{\sqrt{2\pi}} dx'_0 dx'_1 \quad (7.259)$$

As problem 7.1 showed, the Gaussian integral that seems complicated in one dimension becomes near-trivial in two dimensions (in polar coordinates). You'll find that your result for $g(r, \theta)$ depends on r and θ separately, i.e., it is of the form $u_0(r)u_1(\theta)$. Inverting these two functions, you can get $r(u_0)$ and $\theta(u_1)$. Going back to Cartesian coordinates (recalling that if u_0 is a uniform deviate, then so is $1 - u_0$), show that:

$$x_0 = \sqrt{-2 \log u_0}\, \cos(2\pi u_1), \qquad x_1 = \sqrt{-2 \log u_0}\, \sin(2\pi u_1) \quad (7.260)$$

This accomplishes our task of producing two normal deviates (x_0 and x_1) from two uniform deviates (u_0 and u_1).

7.38 Generalize Eq. (7.161) so that you can tackle the following five-dimensional integral:

$$I = \int_{-1}^{1} du\, dv\, dw\, dx\, dy \left(\frac{2 + u + v}{5 + w + x} \right)^y \quad (7.261)$$

using Gauss–Legendre quadrature and $n = 5$–30. Then, apply uniform multidimensional Monte Carlo integration as per Eq. (7.210), with $\mathfrak{y} = 10^7$–10^8. Observe that the $\mathfrak{y}$'s in the two cases are different, but the total runtimes not so much.

7.39 [$\mathcal{A}$] This problem examines the convergence of a Markov chain to the stationary distribution, employing a different argument than in the main text.

(a) Re-express Eq. (7.213) so that it takes the form:

$$P_i(\mathbf{x}) = \int P_{i-1}(\mathbf{y}) T(\mathbf{y} \to \mathbf{x}) d^d y \quad (7.262)$$

Feel free to employ the normalization of the transition probability.

(b) Integrate Eq. (7.212) over $\mathbf{y}$, use the normalization again, and combine your result with Eq. (7.262) to show that $w(\mathbf{x})$ is a stationary distribution.

(c) Use Eq. (7.262) as is and then apply it to equilibrium, i.e., for the special case where $P_i(\mathbf{x}) = P_{i-1}(\mathbf{x}) = w(\mathbf{x})$, to show that:

$$\int |P_i(\mathbf{x}) - w(\mathbf{x})|\, d^d x \leq \int |P_{i-1}(\mathbf{x}) - w(\mathbf{x})|\, d^d x \quad (7.263)$$

Discuss the relevance of the equality here to the irreducibility condition from footnote 63. Conclude that $P_i(\mathbf{x})$ is closer to $w(\mathbf{x})$ than $P_{i-1}(\mathbf{x})$ was.

7.40 The one-dimensional integral $\int_0^\infty e^{-x}x^2 dx$ is of the form $\int_0^\infty w(x)f(x)dx$, for $w(x) = e^{-x}$ and $f(x) = x^2$. Implement the Metropolis algorithm for 10^8 total steps, measuring f every 100 steps. (To keep the x's positive you may interpret $w(x < 0) = 0$, allowing you to simply reject any move that tries to make x negative.) Observing that you can also employ Gauss–Laguerre quadrature, recompute this integral using $n = 2$ together with the function `numpy.polynomial.laguerre.laggauss()`.

7.41 Just like in problem 7.35, we are here interested in sampling from a desired $w(x)$ distribution (without computing any weighted integrals). The following weight function is an example of the *beta distribution*:

$$w(x) = 30x(1-x)^4 \tag{7.264}$$

which is commonly used for random variables that are restricted between 0 and 1. For this case the approach of problem 7.35 (analytical integration and then inversion) is not an option. You could compute the relevant integral numerically; instead, here you are asked to draw samples from $w(x)$ by employing the Metropolis algorithm, thereby obviating any integration whatsoever. (Hint: you have to be a little careful when enforcing $0 < x < 1$.) Produce two figures as in problem 7.35, i.e., a histogram of raw counts and a plot of normalized data.[80]

7.42 We will now try to sample from a *bimodal distribution*. For concreteness, we will study a one-dimensional problem, namely the average of two normal distributions of the form of Eq. (6.169): $(\mathcal{N}(6, 1/2) + \mathcal{N}(4, 1))/2$. You should use the Metropolis algorithm to sample from this probability density, carefully investigating any dependence on the initial position ($\mathcal{X}_0$) and the step size (θ) in Eq. (7.220). Plot $\mathcal{N}(6, 1/2)$, $\mathcal{N}(4, 1)$, their average, as well as normalized histogram counts corresponding to the output of your runs. You will find that, if you're not careful, you sample from only one or the other of the two peaks;[81] provide plots showing this.

7.43 We will tackle the integral of Eq. (7.155):

$$I_H = \int_{-\infty}^{\infty} \frac{e^{-x^2}}{\sqrt{x^2+1}} dx \tag{7.265}$$

(a) Use Gauss–Hermite quadrature, making sure you have enough integration points.

(b) Use the Metropolis algorithm with uniform steps as per Eq. (7.220). In addition to computing the integral, you should also produce a normalized histogram, comparing with $w(x) = e^{-x^2}/\sqrt{\pi}$ (you should also do this for the following parts).

(c) Use the Metropolis algorithm, but this time employ Gaussian steps, with zero mean and θ being the standard deviation. Since these are centered, you are still dealing with the Metropolis algorithm; this is because the quantities in the two proposal-probability exponents will cancel out, since $(\mathcal{X}_{i-1} - \mathcal{Y}_i)^2 = (\mathcal{Y}_i - \mathcal{X}_{i-1})^2$.

[80] The Metropolis algorithm does not care about the normalization factor, so you are actually drawing samples from a function that is *proportional* to Eq. (7.264). That is, in order to produce the second figure, you will be *de facto* comparing the area under the probability density function (which is 1, even you can't do that analytically) with the sum of the areas in the bins.

[81] This Markov chain is nearly reducible (recall footnote 63), i.e., it exhibits slow/poor mixing.

(d) Use the Metropolis algorithm, i.e., an acceptance probability as per Eq. (7.221), with Gaussian steps again, but this time employ a mean of 1 (with θ controlling the standard deviation, as above). You should find a wrong answer for both the integral and the normalized histogram counts.

(e) You should be able to see what went wrong: the proposal-probability distribution in the previous part was non-symmetric, so you should have been using the *Metropolis–Hastings algorithm*, instead. That means that you should use the acceptance probability of Eq. (7.218) where, crucially, the ratio will be coming from Eq. (7.217). Qualitatively, the reason you need to be careful is that this time around the two terms in the exponent do not cancel out, since $(\mathscr{X}_{i-1} - \mathscr{Y}_i - 1)^2 \neq (\mathscr{Y}_i - \mathscr{X}_{i-1} - 1)^2$.

To employ the Metropolis–Hastings algorithm you need to do two distinct things: (i) draw random samples from your distribution of choice to produce the proposed step $\mathscr{Y}_i$, and (ii) include the ratio of proposal probabilities. The latter is trivial, since the proposal probability is of your choosing; the former can be non-trivial, as it implies you have the ability to draw samples from complicated probability distributions (and if you truly could do so, you wouldn't be using this algorithm in the first place).

7.44 [$\mathcal{A}, \mathcal{P}$] Generalize Eq. (7.4) to address the Maxwell–Boltzmann distribution for the speed of a three-dimensional ideal gas. Then, establish its relationship (if any) with the chi-squared distribution of Eq. (6.255).

7.45 [$\mathcal{A}, \mathcal{P}$] We now show that, when the variational energy E_V for the trial wave function ψ_T gives the ground-state energy E_0, then ψ_T *is* the ground-state wave function; imagine slightly perturbing ψ_T such that Eq. (7.231) turns into:

$$E_0 + \Delta E = \frac{(\langle\psi_T| + \langle\Delta\psi_T|)\, \hat{H}\, (|\psi_T\rangle + |\Delta\psi_T\rangle)}{(\langle\psi_T| + \langle\Delta\psi_T|)\, (|\psi_T\rangle + |\Delta\psi_T\rangle)} \tag{7.266}$$

Manipulate this equation by neglecting terms of order $\langle\Delta\psi_T|\Delta\psi_T\rangle$. Do this a second time in order to move all $|\Delta\psi_T\rangle$ terms onto the numerator. Then, set $\Delta E = 0$ to show that $\langle\Delta\psi_T|(\hat{H} - E_0)|\psi_T\rangle = 0$. Discuss what this equation implies on the status of $|\psi_T\rangle$.

7.46 [$\mathcal{P}$] This problem studies the *autocorrelation function*, defined to be:[82]

$$C(k) = \frac{\overline{f f_k} - \overline{f}^2}{\overline{f^2} - \overline{f}^2} \tag{7.267}$$

Here we repeatedly use the arithmetic average, defined in Eq. (7.178). It involves terms we've encountered before, like $\overline{f^2}$ and $\overline{f}^2$ which are used in Eq. (7.182), as well as a new, strange-looking term, $\overline{f f_k}$; this is simply shorthand for:

$$\overline{f f_k} = \frac{1}{\mathfrak{y} - k} \sum_{i=0}^{\mathfrak{y}-k-1} f(\mathscr{X}_i) f(\mathscr{X}_{i+k}) \tag{7.268}$$

In this equation, the value of f at a given sample is multiplied by that of f for the

[82] This brings to mind Eq. (6.271); for the random walk the numerator in Eq. (7.267) is known as the *autocovariance*. One should generalize Eq. (7.184) for the variance of the sample mean to also involve the autocorrelation time; in problem 7.34 this term vanished, because the samples were independent there.

sample that is k steps apart (and such pairwise multiplications are carried out *for the entire chain*); for example, $k = 1$ would connect $f(\mathscr{X}_i)$ with $f(\mathscr{X}_{i+1})$. Of course, given our definition, $C(0) = 1$; $C(1)$ would be smaller, and then at some point $C(k)$ would be small enough not to worry about. The value of k at which this happens (the *correlation length*) tells you how many steps you should skip when thinning, in order to eliminate the correlation between succeeding walker configurations.

For our random walk in `vmc.py`, E_L will play the role of f in the definition above. Note that you need to take consecutive measurements to apply Eq. (7.267), i.e., $n_m = 1$. Your first task is to plot the autocorrelation function for $k = 0, 1, \ldots, 99$ for $\mathfrak{y} = 10^6$. Then, compute the *autocorrelation time*: $T = 1 + 2\sum_{k=0}^{99} C(k)$. Repeat this computation for different step sizes (θ) and plot the autocorrelation time vs the acceptance rate; the minimum gives you the optimal acceptance rate.

7.47 [$\mathcal{P}$] In our discussion of the Metropolis–Hastings algorithm we provided a general formula for the acceptance probability, Eq. (7.218). This, however, is not unique; an alternative choice, known as *Barker's algorithm*, is:

$$\alpha(\mathbf{x} \to \mathbf{y}) = \frac{R(\mathbf{x} \to \mathbf{y})}{1 + R(\mathbf{x} \to \mathbf{y})} \tag{7.269}$$

First, show that this acceptance probability allows you to satisfy detailed balance. Second, update `vmc.py` to use this new criterion.

7.48 [$\mathcal{P}$] Rewrite `vmc.py`, breaking down the contributions to E_V in terms of kinetic, one-body, and two-body energies. Which contribution is the most important one?

7.49 [$\mathcal{P}$] Rewrite the function `ekin()` in `eloc.py` such that you don't use any finite-difference approximation (and therefore also no step size h). This means you first have to analytically take the second derivatives and then code up your expressions; make sure that you (analytically) cancel any terms that do cancel. Compare your output with that of `eloc.py` for different values of h.

7.50 [$\mathcal{P}$] To make this problem more manageable, study $n_p = 4$ and $n_d = 2$. Observe from the first equality in Eq. (7.235) that we have to compute eight-dimensional integrals. However, writing out $\hat{H}_1$ you realize that the sum contains six terms, each of which couples only a single pair of particles (the j-th and k-th particle each time). Thus, the eight-dimensional integral decouples into an "easy" four-dimensional one and a "hard" four-dimensional one. Carrying these out by Gauss–Legendre, you should compare with the output of `vmc.py` (if it helps, choose smaller values of g).

7.51 [$\mathcal{P}$] Generalize `vmc.py` to handle particles that obey Fermi–Dirac statistics (do you expect the energy to increase or decrease?). The many-particle wave function is not a simple product of individual ϕ's as in Eq. (7.225), but has to be antisymmetrized, giving rise to a Slater determinant involving the ϕ's (recall problem 4.57); feel free to use Eq. (4.102) and `ludec.py`. For concreteness, study only the case of four particles in three dimensions. Since you're dealing with fermions, you can place only a single particle in the state $n_x = n_y = n_z = 0$; the next particle goes in $n_x = 1, n_y = n_z = 0$, and so on. When you code these up you can use `hermite()` from `psis.py`, or simply hard-code the needed Hermite polynomials.

7.52 [$\mathcal{P}$] We will produce a better trial wave function ψ_T than the one in `vmc.py`, by employing different variational parameters in each direction, i.e., $\alpha_x, \alpha_y, \alpha_z$. For $\eta = 10^3$ (in order to speed the computations up) you should find smaller values of E_V than in Fig. 7.8 by using: (a) a (somewhat coarse) three-dimensional grid for the three parameters $\alpha_x, \alpha_y, \alpha_z$, and (b) Powell's derivative-free method from section 5.6.5.

7.53 [$\mathcal{P}$] Evaluate the following four integrals that arise in the study of the *simple pendulum* beyond the small-angle approximation:

$$I_1 = \int_0^{\theta_0} \frac{1}{\sqrt{\cos\theta - \cos\theta_0}} d\theta, \qquad I_2 = \int_0^{\pi/2} \frac{1}{\sqrt{1 - A^2 \sin^2\phi}} d\phi$$
$$I_3 = \int_0^{\theta_0} \sqrt{\cos\theta - \cos\theta_0}\, d\theta, \qquad I_4 = \int_0^{\pi/2} \frac{1 - \sin^2\phi}{\sqrt{1 - A^2 \sin^2\phi}} d\phi \tag{7.270}$$

where $A = \sin(\theta_0/2)$ and we take $\theta_0 = \pi/3$. Of these, I_3 and I_4 appear when you use "action-angle coordinates", whereas I_2 and I_4 can be expressed in terms of the *elliptic integrals* of the first and second kind. Compute these integrals using the adaptive trapezoid rule and Gauss–Legendre quadrature for a comparable number of points.

7.54 [$\mathcal{P}$] In quantum statistical mechanics, the total number of particles takes the form:

$$N = A(k_B T)^{\alpha+1} \int_0^\infty dx \frac{x^\alpha}{e^{x-w} \pm 1} \tag{7.271}$$

where $x = \epsilon/(k_B T)$, $w = \mu/(k_B T)$, the plus sign corresponds to a Fermi gas, the minus sign to a Bose gas, and α is an exponent related to whether or not the gas is relativistic ($\alpha = 2$ or $\alpha = 1/2$). Textbooks typically address only the degenerate (vanishing T) or classical (huge T) limits. Here we will focus on intermediate temperatures, doing the integrals numerically. Plot the value of the integral for bosons (w from -30 to 0) and fermions (w from -10 to $+20$) with a logarithmic scale on the y axis; in each plot include results for both α cases. You can use any method you want to compute the integrals, but it's probably easiest to employ Gauss-Legendre quadrature up to some large energy cutoff/right endpoint of integration.

7.55 [$\mathcal{P}$] Here is a wave packet, given in both coordinate and wave-number space:

$$\psi(x) = \frac{\pi^{1/4}}{2\cosh(\sqrt{\pi}x/2)}, \qquad \phi(k) = \frac{\pi^{1/4}}{\sqrt{2}\cosh(\sqrt{\pi}(k-3))} \tag{7.272}$$

Numerically check that the *Heisenberg uncertainty relation*, $\sigma_x \sigma_k \geq 1/2$, is satisfied. A reminder:

$$\sigma_x^2 = \int_{-\infty}^{+\infty} x^2 |\psi(x)|^2 dx - \left(\int_{-\infty}^{+\infty} x|\psi(x)|^2 dx\right)^2,$$
$$\sigma_k^2 = \int_{-\infty}^{+\infty} k^2 |\phi(k)|^2 dk - \left(\int_{-\infty}^{+\infty} k|\phi(k)|^2 dk\right)^2 \tag{7.273}$$

7.56 [$\mathcal{P}$] We now return to the Roche potential of problem 1.17, which we extremized in problem 5.50. This time we do not limit ourselves to the orbital plane (i.e., we will *not* be imposing $z = 0$). We will compute the volume of the Roche lobe around

m_1, for the case of a tiny mass ratio, $q = m_0/m_1 \ll 1$. For infinitesimal q, the equipotential surface of Eq. (1.6) becomes connected at a critical value of Φ:

$$\frac{3}{2} = \frac{1}{\sqrt{x^2 + y^2 + z^2}} + \frac{x^2 + y^2}{2} \tag{7.274}$$

We are faced with an oblate figure of rotation around the z axis, which has poles at $(0, 0, \pm 2/3)$. Thus, the volume of the Roche lobe around m_1 will be proportional to:

$$I = \int_{-2/3}^{2/3} (x^2 + y^2) dz \tag{7.275}$$

Use Eq. (7.274) to express z in terms of $u = x^2 + y^2$, thereby rewriting I completely in terms of u; you have to be careful when re-expressing the integral (both the differential and the limits). Then, employ the trapezoid rule, Simpson's rule, and Gauss–Legendre quadrature to evaluate I. (You first have to eliminate a singularity in your integrand; you can do this either via the substitution $w = \sqrt{3 - u}$ or simply cancelling the appropriate terms in the numerator and denominator.)

7.57 [$\mathcal{P}$] We now study *classical scattering* using the *Yukawa potential*, $V(r) = ce^{-r/\alpha}/r$, where c is a strength parameter and α reflects the range of the interaction. For simplicity, take $c = 1$; in what follows, you should give α the values 1, 10, and 50.

(a) Take β (the impact parameter) to range from 0.01 to 20 and solve the algebraic equation for the turning point r_0 (setting $E = 1$ for simplicity):

$$1 - \frac{\beta^2}{r_0^2} - \frac{V(r_0)}{E} = 0 \tag{7.276}$$

(b) We wish to use the β's and r_0's we produced to compute the scattering angle θ:

$$\theta = \pi - 2\beta \int_{r_0}^{\infty} \frac{dr}{r^2 \sqrt{1 - \beta^2/r^2 - V(r)/E}} \tag{7.277}$$

This is asking for trouble, since you know from part (a) that you will have a singularity at the left endpoint. Carry out the change of variables $u = \sqrt{1 - r_0/r}$ and then use Gauss–Legendre quadrature for the new integral:

$$\theta = \pi - 4\beta \int_0^1 \frac{du}{\sqrt{\beta^2(2 - u^2) + \frac{r_0^2}{Eu^2}\{V(r_0) - V(r_0/[1 - u^2])\}}} \tag{7.278}$$

which has finite endpoints. While the new denominator may look like it leads to numerical issues at *both* endpoints, it is much better behaved: plot the integrands of Eq. (7.277) and Eq. (7.278) for $\alpha = 10$ and $\beta = 5$.

(c) Combine your results with the formula giving the differential *cross section*:

$$\sigma(\theta) = \frac{\beta}{\sin\theta} \frac{1}{|d\theta/d\beta|} \tag{7.279}$$

You can use a central difference to approximate the derivative.[83] Plot your results

[83] In one problem we employed root-finding, a change of variables, quadrature, and finite differences.

together with those coming from the *Rutherford scattering formula*:

$$\sigma(\theta) = \left(\frac{c}{4E}\right)^2 \frac{1}{\sin^4(\theta/2)} \tag{7.280}$$

and discuss qualitative features. Your x axis values (for the θ's) should go from 0 to 0.4; use a logarithmic scale on the y axis (for the σ's).

7.58 [$\mathcal{P}$] In the *Bardeen–Cooper–Schrieffer (BCS) theory of superfluidity* a pairing gap, Δ, arises in the single-particle excitation spectrum. Instead of the single-particle energy dispersion $\hbar^2k^2/(2m) - \mu$ (where we're employing the chemical potential, μ, as an offset), one is faced with:

$$E(k) = \sqrt{\left(\frac{\hbar^2k^2}{2m} - \mu\right)^2 + \Delta^2} \tag{7.281}$$

While the BCS theory was originally introduced at weak coupling, in the twenty-first century it became possible to experimentally probe (using ultracold atoms) the strong-coupling limit of infinite scattering length. For this case (two components in 3d, with an interaction that has zero range), the BCS gap equation takes the form:

$$0 = \int_0^\Lambda \frac{dkk^2}{4\pi^2}\left[\frac{2m}{\hbar^2k^2} - \frac{1}{E(k)}\right] \tag{7.282}$$

where Λ is a theory cutoff, which we will take to be large (but finite). To compute Δ and μ self-consistently, one needs to solve Eq. (7.282) together with the equation:

$$n = \int_0^\Lambda \frac{dkk^2}{2\pi^2}\left[1 - \frac{\frac{\hbar^2k^2}{2m} - \mu}{E(k)}\right] \tag{7.283}$$

where $n = k_F^3/(3\pi^2)$ is the number density (and k_F is the Fermi wave number). Use the multi-dimensional Newton's method to solve these two coupled equations; employ Gauss–Legendre quadrature to carry out the two integrals. At the end of your calculation, you should print out the dimensionless numbers μ/E_F and Δ/E_F, where $E_F = \hbar^2k_F^2/(2m)$ is the Fermi energy. This means that you are free to give convenient values to $\hbar^2/m$ and to k_F. Hint: in contradistinction to the usual textbook case, here the chemical potential μ is most certainly *not* equal to the Fermi energy E_F.

7.59 [$\mathcal{P}$] The static dielectric function for a system made up of non-interacting electrons can be evaluated starting from a simple sum:

$$\mathcal{L}_Q(q) = \frac{1}{2\pi L^3}\sum_{|\mathbf{k}|<Q}\frac{1}{(\mathbf{q}-\mathbf{k})^2} \overset{L\to\infty}{=} \int_0^Q dkk^2 \int_0^\pi d\theta \sin\theta \frac{1}{q^2 + k^2 - 2qk\cos\theta} \tag{7.284}$$

where we are working in three spatial dimensions, Q is the maximum allowable momentum, and in the second step we took the thermodynamic limit. We will now explicitly continue the calculation, giving rise to the *Lindhard function*, and then proceed to study the screening of a static charged impurity embedded in the fermionic gas, as part of which we will encounter what are known as *Friedel oscillations*.

(a) Set $x = \cos\theta$ in Eq. (7.284) and integrate over x using an elementary result on integrals of rational functions. Think of the remaining integral as $\int dk f(k) dg(k)/dk$ and integrate by parts; to do so, you need to evaluate df/dk. The boundary term is easy to handle; you can use partial fractions to simplify the $-\int dk g(k) df(k)/dk$ term (i.e., go from squares to first powers in the denominator). You can then employ the same elementary rational/logarithmic integral as above. If you did all the steps correctly, the final result should be:

$$\mathcal{L}_Q(q) = Q + \frac{Q^2 - q^2}{2q} \ln\left|\frac{Q+q}{Q-q}\right| \tag{7.285}$$

The Lindhard function appears in a number of interrelated physical settings.

(b) Take $Q = 2k_F$ and plot $\mathcal{L}_{2k_F}(q)$ vs q (both in units of k_F). Observe that for $q \ll k_F$ the decrease is not very rapid; also note that something interesting is taking place near $q = 2k_F$. Investigate this by plotting the first derivative of $\mathcal{L}_{2k_F}(q)$ vs q.

(c) In the random-phase approximation (RPA) one uses the Lindhard function we derived in Eq. (7.285) to arrive at the induced screening density change:

$$\Delta n(\mathbf{r}) = \int \frac{d^3q}{(2\pi)^3} e^{i\mathbf{q}\mathbf{r}} \frac{\xi \mathcal{L}_{2k_F}(q)}{q^2 + \xi \mathcal{L}_{2k_F}(q)} \tag{7.286}$$

for a unit positive charge. Write out the exponent and carry out the integral over $x = \cos\theta$, similarly to what you did in part (a) above. Notice that most of the structure of your (fully real) integrand is to be found near $q = 2k_F$; use Gauss–Legendre quadrature three times, for the intervals $(0, 1.5k_F)$, $(1.5k_F, 2.5k_F)$, and $(2.5k_F, 20k_F)$, adding up these contributions at the end. Plot $\Delta n(\mathbf{r})$ vs r, where r is measured in units of $1/k_F$ and $\Delta n(\mathbf{r})$ in units of k_F^3. Show curves for $\xi = 0.5, 0.75, 1$ (in units of k_F), going up to $r = 12$; zoom in to the region beyond $r = 2$ to find the Friedel oscillations, as advertised.

7.60 [$\mathcal{P}$] In statistical mechanics, the Gaussian approximation allows for (simple) fluctuations about the uniform mean field. A nice example, involving the specific heat, allows us to illustrate the concept of *critical exponents*. The two terms that arise in this context can be rescaled in terms of the *correlation length*, ξ:

$$C_A = \xi^{4-n_d} \int_0^{\xi\Lambda} dq \frac{q^{n_d-1}}{(1+q^2)^2}, \qquad C_B = \xi^{2-n_d} \int_0^{\xi\Lambda} dq \frac{q^{n_d-1}}{1+q^2} \tag{7.287}$$

where n_d is the number of spatial dimensions and Λ is a momentum-space cutoff which we will take to be large (but finite). Our goal is to investigate the behavior of C_A and C_B as $\xi \to \infty$ for $n_d = 1, 2, 3, 4, 5$; it is important to separately examine, in each case, the effect of the prefactor and of the integral.

(a) Use Gauss–Laguerre quadrature for each of the two integrals for the case of $\xi \to \infty$. Since there are no exponentials in the integrands, you will have to introduce the appropriate weight term yourself. Check for convergence by increasing the number of integration points, n.

(b) Use Gauss–Legendre quadrature for each of the two integrals for a number of (increasingly large) ξ values.

(c) For the same large ξ values as in part (b), combine the prefactors with the integrals to see when C_A and C_B appear to converge (and when they don't).

7.61 [$\mathcal{P}$] We now return to the zero-dimensional field-theory integral $Z(g)$ of problem 2.27.

(a) Use Gauss–Hermite quadrature for the integral in Eq. (2.130) and compare with the exact answer from Eq. (2.131). Plot the two curves for g from 0.01 to 0.99.

(b) To solidify your understanding of the rather abstract Eq. (2.134), we will now *demand* that its right-hand side be equal to the right-hand side of Eq. (2.130) to actually find a specific expression for $\mathcal{B}_{\mathcal{A}}(g)$. You should analytically determine the Borel transform starting from:

$$\mathcal{B}_{\mathcal{A}}(t,g) \equiv \int_{-\infty}^{+\infty} \delta\left(\frac{1}{2}x^2 + \frac{g}{4!}x^4 - t\right) dx \tag{7.288}$$

(c) Now that you know what the Borel transform looks like, try using Gauss–Laguerre quadrature to compute the integral (for $g = 0.1$) in Eq. (2.134), comparing with the exact answer in Eq. (2.131), or with the result you produced in part (a).

(d) Use Gauss–Legendre quadrature to compute the same integral, i.e., Eq. (2.134) with the analytically known Borel transform of part (b), for $g = 0.1$. You'll first have to use the transformation from Eq. (7.153), followed by Eq. (7.128).

(e) You may be feeling uncomfortable that our Eq. (7.288) gives $\mathcal{B}_{\mathcal{A}}(t,g)$, whereas Eq. (2.134) involved $\mathcal{B}_{\mathcal{A}}(tg)$.[84] If so, you can use the following, generalized, Borel transform:

$$\mathcal{B}_{\mathcal{A}}(g) = \sum_{m=0}^{\infty} \frac{c_m g^m}{\Gamma(m+1/2)}, \qquad \Gamma(m+1/2) = \frac{(2m)!}{4^m m!}\sqrt{\pi} \tag{7.289}$$

which uses the *gamma function* instead of a simple factorial (as in Eq. (2.133)). Then, you can introduce the corresponding Borel sum:

$$B(g) = \int_0^{\infty} t^{-1/2} e^{-t} \mathcal{B}_{\mathcal{A}}(tg)dt, \qquad \mathcal{B}_{\mathcal{A}}(tg) \equiv t^{1/2} \int_{-\infty}^{+\infty} \delta\left(\frac{1}{2}x^2 + \frac{g}{4!}x^4 - t\right) dx \tag{7.290}$$

(i) Repeat the argument we gave after Eq. (2.134), for the case of the new equations above, i.e., show that $B(g)$ from Eq. (7.290) is asymptotically the same as $\mathcal{A}(g)$ after you plug in the expansion of Eq. (7.289); you will need to use the definition of the gamma function from Eq. (6.258), and (ii) analytically evaluate the second integral in Eq. (7.290) and then use Gauss–Laguerre quadrature to compute the first integral in Eq. (7.290), comparing with part (c).

7.62 [$\mathcal{P}$] The following identity is essentially a more elaborate version of the Gaussian integral we encountered at the start of this problem set, Eq. (7.238):

$$\int_{-\infty}^{+\infty} e^{-x^2+Ax} = \sqrt{\pi}e^{A^2/4} \tag{7.291}$$

This identity has many applications in quantum field theory and many-body physics,

[84] Such discomfort is unwarranted: for the argument we made after Eq. (2.134) we need a $\mathcal{B}_{\mathcal{A}}(t,g)$ that becomes a function of only g when we take $t \to t/g$. The Borel transform you found in part (b) certainly fits the bill.

where it is known as the *Hubbard–Stratonovich transformation*. While these applications involve the square of an operator or a matrix in the exponent (instead of a simple scalar, as here), the idea is the same: you can convert a quadratic term into a linear one, at the cost of an extra integration (over an *auxiliary field*).

(a) Analytically show that this identity holds, by completing the square on the left-hand side and then shifting.
(b) Compute this integral (for $A = 1/4$) via uniform Monte Carlo.
(c) Again for $A = 1/4$, recompute the integral using Gauss–Hermite quadrature, as per `numpy.polynomial.hermite.hermgauss()`, or problem 7.25, for $n = 2$ and $n = 3$. (This is known as the *discrete Hubbard–Stratonovich transformation*.)

7.63 [$\mathcal{P}$] This problem addresses the loop integrals that arise in quantum field theory. Specifically, we first discuss a typical term that appears when evaluating Feynman diagrams, after carrying out a Wick rotation. The idea of *dimensional regularization* is to study the general case of d spacetime dimensions (where d is not necessarily an integer) to investigate what type of convergence (if any) you find. Our integral is:

$$I_A = \int \frac{d^d k}{(2\pi)^d} \frac{1}{(\mathbf{k}^2 + m^2)^2} \tag{7.292}$$

If none of the above terms mean anything to you, that's OK: you can treat this as an integration problem.

(a) Observe that the integrand depends only on the magnitude of the d-dimensional vector, so we can employ spherical polar coordinates in d dimensions to find:

$$I_A = \int \frac{d\Omega_d}{(2\pi)^d} \int_0^\infty dk_r \frac{k_r^{d-1}}{(k_r^2 + m^2)^2} \tag{7.293}$$

We will first carry out the angular integral, which will give us the area of a d-dimensional unit sphere (e.g., 4π in three dimensions). To do so, we will (again) make use of the Gaussian integral from Eq. (7.238); knowing it means we actually know how to do the corresponding d-dimensional integral:

$$I_B = \int_{-\infty}^{+\infty} d^d k \, e^{-(k_0^2 + k_1^2 + \ldots + k_{d-1}^2)} = \left(\int_{-\infty}^{+\infty} dx e^{-x^2} \right)^d = (\sqrt{\pi})^d \tag{7.294}$$

You should re-express the term in the first equality in a form similar to that of Eq. (7.293), i.e., separating the radial and angular terms. Then, use the definition of the gamma function from Eq. (6.258) to find the d-dimensional unit-sphere area whose value you were after in Eq. (7.293).
(b) We're more interested in the (one-dimensional) radial integral in Eq. (7.293), call it $I_{A,r}$. Carry out the transformation $u = m^2/(k_r^2 + m^2)$ to cast $I_{A,r}$ as an integral with finite integration limits.
(c) Take $m^2 = 1$ and use Gauss–Legendre quadrature with $n = 200, 500, 1000$ points to plot the value of $I_{A,r}$ vs d (the latter taking on real values from 1 to 3).
(d) Use an increasingly large number of integration points n to try to compute $I_{A,r}$ for $d = 3.98, 3.99, 4.0$. Do you understand what is going on?

(e) Having realized what went wrong, you should now subtract off the singularity, i.e., subtract out (and add back in) a function that shows the same misbehavior as your integrand at the left endpoint, but is also analytically integrable. Interpret things in terms of the convergence or divergence of the integral $I_{A,r}$.

(f) We now turn to a more complicated integral:

$$I_C = \int \frac{d^d k}{(2\pi)^d} \frac{1}{(\mathbf{k}^2 + m^2)[(\mathbf{k} - \mathbf{q})^2 + m^2]} \tag{7.295}$$

On the face of it, this appears to be much harder to tackle than I_A in Eq. (7.292), since the $(\mathbf{k} - \mathbf{q})^2$ term gives rise to the angle between the d-dimensional vectors $\mathbf{k}$ and $\mathbf{q}$: this means that we cannot simply separate out the angular integration on its own, as we did in Eq. (7.293) and you were asked to do for Eq. (7.294). This problem is handled by the *Feynman parametrization*:

$$\frac{1}{VW} = \int_0^1 dx \frac{1}{[(1-x)V + xW]^2} \tag{7.296}$$

This has introduced an extra integration variable, but has also allowed us to combine the two terms in the denominator (summing them, instead of multiplying them together). Apply Eq. (7.296) to I_C in Eq. (7.295) and complete the square. You should then be able to make a change of variables to $\mathbf{k}' = \mathbf{k} - x\mathbf{q}$, which will allow you to employ the trivial angular integration as in earlier parts of this problem. (No need to numerically evaluate anything.)

7.64 [$\mathcal{P}$] In Fig. 7.8 we tweaked the wave-function parameter α "by hand", while in problem 7.52 we carried out a similar study for three parameters (either manually or automatically); in both instances, we directly minimized the variational energy. Here we introduce a distinct approach, involving the minimization of the variance.

(a) Modify `vmc()` such that it also returns the particle configurations at which the energy is measured. Use this new function to store such a set of configurations into a file (for the case of $\alpha_0 = 0.6$).

(b) We will now learn how to find the optimal α by re-using the stored configurations and *minimizing the variance*. Before doing so, let us set up some more scaffolding: write a Python function (of a single variable/parameter) that computes the variance of the local energy for a given α:[85]

$$\mathbb{V}[E_L](\alpha) = \frac{\int d^d x \left|\psi_T^{\alpha_0}(\mathbf{x})\right|^2 \left[E_L^{\alpha}(\mathbf{x}) - E_V^{\alpha}\right]^2}{\int d^d x \left|\psi_T^{\alpha_0}(\mathbf{x})\right|^2} \tag{7.297}$$

where we are explicitly labelling E_L and E_V with α, to emphasize that they give different answers for different choices of α; this means that E_V^{α} is a straightforward generalization of Eq. (7.235):

$$E_V^{\alpha} = \frac{\int d^d x \, |\psi_T^{\alpha_0}(\mathbf{x})|^2 \frac{\hat{H}\psi_T^{\alpha}(\mathbf{x})}{\psi_T^{\alpha}(\mathbf{x})}}{\int d^d x \, |\psi_T^{\alpha_0}(\mathbf{x})|^2} \tag{7.298}$$

[85] This is called the *unreweighted* variance, because it involves $|\psi_T^{\alpha_0}(\mathbf{x})|^2$ without a weight $|\psi_T^{\alpha}(\mathbf{x})|^2/|\psi_T^{\alpha_0}(\mathbf{x})|^2$.

While the local energy and the variational energy are recomputed anew for each value of α, we are using "frozen" particle configurations (i.e., those which we produced in part (a) and we now load in) to approximate the integrals. Print out a table of the values $\mathbb{V}[E_L](\alpha)$ takes on when α goes from 0.1 to 1.5.

(c) Part (b) is printing out a table, i.e., is still doing things manually. Apply the golden-section search technique to $\mathbb{V}[E_L](\alpha)$ to automatically compute the optimal wave-function parameter and the corresponding energy (with the associated statistical error). Note that we arrived at this answer by minimizing the variance, not the energy; the variance has the nice property that its lower bound is zero.[86]

(d) There's a complication: the variance, energy, and error we found are taking it for granted that the particle configurations we produced in part (a) are good enough to describe the system even at a different value of α. To see why this is a problem, compare the energy estimate you printed out in part (c) with that resulting from a new VMC run for the optimal α value.

(e) It should be easy to see what you should do next: you can repeat the entire process, i.e., generate new particle configurations for the optimal α value, use those to minimize the variance, and compare the corresponding energy estimate with a "fresh" VMC run at the latest α. Typically a few such cycles are all that's needed. Try this out for the present problem; you can call it a day when the optimal α stops changing (within some tolerance).

7.65 [$\mathcal{P}$] We will address the *Ising model in two dimensions*. This can be viewed as a simple model of a magnet, using classical degrees of freedom which represent the magnetic moments of atoms in a solid. We place these magnetic moments on a two-dimensional *lattice* and will refer to them below as "spins" (though, again, this will be a classical system). Take the lattice to have dimensions $n \times n$; to each lattice site will correspond a given spin, which we label using the indices of the two spatial directions, i.e., $S_{i,j}$. Each spin can take only two possible values: $S_{i,j} = +1$ or $S_{i,j} = -1$ (spin-up or spin-down, respectively). Similarly to what we did in section 4.5 (which tackled the interaction of many *quantum* spins), one can also use a single label to denote a given lattice site, e.g., μ, in which case the corresponding spin would be given as S_μ. Either way, we'll use $\mathbf{S}$ to denote a given collection of n^2 individual spins, i.e., a specific lattice configuration.

Assume you are dealing with periodic boundary conditions, i.e., the generalization of Fig. 6.8 to two dimensions; it may help you to look at Fig. 8.13, which corresponds to a different physical problem but involves a similar two-dimensional periodic grid. Note how our grid builds in the periodicity, i.e., we don't carry around needless information. In its simplest form, the Hamiltonian for the Ising model is:

$$H(\mathbf{S}) = -J \sum_{\langle \mu\nu \rangle} S_\mu S_\nu \tag{7.299}$$

where J is a strength parameter and $\langle \mu\nu \rangle$ means that we are taking the sum over

[86] Unfortunately, when this global minimum is not attainable (i.e., when your trial wave function is not flexible enough to allow you to reach the true ground-state wave function) it may be that the variance does not exhibit a local minimum near the approximate eigenstate, so you may end up missing the latter [33].

nearest-neighbor spin pairs, i.e., the spin at the lattice site i, j will be interacting with the (four) spins at $i \pm 1, j$ and $i, j \pm 1$; have a look at Fig. 8.13, again. When $J > 0$ the energy is lower if the neighbors point in the same direction, so this is called a ferromagnetic interaction; similarly, $J < 0$ is called an antiferromagnetic interaction. Working in the canonical ensemble, a given configuration **S** will be weighted by:

$$w(\mathbf{S}) = \frac{e^{-H(\mathbf{S})/(k_B T)}}{\sum_{\mathbf{S}} e^{-H(\mathbf{S})/(k_B T)}} \tag{7.300}$$

As you should recognize, the denominator here (a sum over Boltzmann factors) is the partition function, Z. We are interested in evaluating the energy:

$$E = \sum_{\mathbf{S}} w(\mathbf{S}) H(\mathbf{S}) \tag{7.301}$$

and the magnetization:

$$M = \sum_{\mathbf{S}} w(\mathbf{S}) \left(\sum_{\mu} S_{\mu} \right) \tag{7.302}$$

Unlike our Project in section 7.8, this problem is inherently discrete, so we are faced with sums instead of integrals. Even so, all our stochastic machinery carries over. Start from a randomly chosen configuration of spins. You will employ the Metropolis algorithm to make proposed steps which should only "touch" a single spin, $S_\mu = S_{i,j}$, at a time. In that case, the Metropolis ratio of Eq. (7.221) becomes quite simple:

$$\frac{w(\mathbf{S}')}{w(\mathbf{S})} = e^{[-H(\mathbf{S}')+H(\mathbf{S})]/(k_B T)} = e^{-2JS_{i,j}\left[S_{i+1,j}+S_{i-1,j}+S_{i,j+1}+S_{i,j-1}\right]/(k_B T)} \tag{7.303}$$

where the current and the proposed configurations are denoted by **S** and **S′** (respectively) and the second equality (which you should show) assumes that we are only flipping a single spin $S_{i,j}$, i.e., $S'_{i,j} = -S_{i,j}$. Stepping through the entire lattice one site at a time, you will be carrying out multiple "sweeps" of the whole lattice.

(a) Repurpose our (continuum) Metropolis code `vmc.py` to study the Ising model for a 16×16 lattice; take $k_B T = 1$ and $J = 0.3$ and plot the energy per spin vs iteration number. Then, plot the magnetization per spin vs iteration number. (Both quantities are "per spin": this means you should divide by n^2.)

(b) Having learned how to take statistical averages (i.e., whether or not you should throw away some equilibration/burn-in iterations) in the previous part, compute the energy (and the magnetization) per spin for J ranging from 0.1 to 0.9. Hint: you should be on the lookout for interesting behavior around $J \approx 0.45$.

(c) We now include the interaction of each spin with an external magnetic field:

$$H(\mathbf{S}) = -J \sum_{\langle \mu\nu \rangle} S_\mu S_\nu - B \sum_{\mu} S_\mu \tag{7.304}$$

Compute the energy (and the magnetization) per spin for the case of a 16×16 lattice, $k_B T = 1$, $J = 0.3$, and $B = 0.5$. Obviously, you should first analytically examine how to generalize the expression we gave above for the Metropolis ratio.

8 Differential Equations

My hopes are fluttering, I can see neither what has passed nor what is behind.

Sophocles

8.1 Motivation

This is the final chapter in the book, so we will take advantage of several of the tools we have developed so far, most notably: finite differences (chapter 3), linear algebra (chapter 4), root-finding (chapter 5), discrete Fourier transforms (chapter 6), and quadrature rules (chapter 7). Our exposition is cumulative, so you are expected to have studied the preceding material; even so, we provide specific references when we use techniques introduced in earlier chapters, so you can brush up on selected topics if the need arises.

8.1.1 Examples from Physics

Differential equations are at the heart of physics and (together with linear algebra) are the most prominent part of numerical analysis. As usual, here are a few examples:

1. **Stellar structure with relativistic corrections**
   Assume you are given an *equation of state*, $P(\rho)$, relating the pressure P and the mass density ρ for a perfect fluid. We would like to see how to go from the equation of state (which describes microphysics) to the structure of a star (which is a macrophysical consequence). In other words, we wish to determine how the matter is distributed as you start from the center of a star and go all the way out to its edge. In general relativity, the structure of a spherically symmetric static star can be determined by combining the Einstein field equations with an equation describing hydrostatic equilibrium; this gives rise to the following relation, where c is the speed of light:

$$\frac{dP(r)}{dr} = -\frac{Gm(r)\rho(r)}{r^2}\left[1+\frac{P(r)}{\rho(r)c^2}\right]\left[1+\frac{4\pi r^3 P(r)}{m(r)c^2}\right]\left[1-\frac{2Gm(r)}{c^2 r}\right]^{-1} \tag{8.1}$$

   known as the *Tolman–Oppenheimer–Volkoff (TOV) equation*; one way of recovering the Newtonian limit is to take $c^2 \to \infty$. Here, G is Newton's gravitational constant, r is the radial coordinate, and $m(r)$ is the mass inside a sphere of radius r, which satisfies:

$$\frac{dm(r)}{dr} = 4\pi r^2 \rho(r) \tag{8.2}$$

Note that in the last two relations P, ρ, and m are functions of only r; as a result we are dealing with a pair of *ordinary* differential equations. From the equation of state, given $\rho(r)$ you can determine $P(\rho(r))$, meaning that Eq. (8.1) and Eq. (8.2) are two first-order differential equations for $\rho(r)$ and $m(r)$; crucially, these cannot be solved independently of each other, since both of them involve $m(r)$ and $\rho(r)$; we say that these are *simultaneous differential equations* (or *coupled* differential equations).

We start solving these simultaneous differential equations at the center of the star (where we impose $\rho(0) = \rho_c$ and $m(0) = 0$) and then integrate outward. The mass M and radius R of the star are determined by $P(R) = 0$ and $m(R) = M$, i.e., the star ends when the pressure vanishes, at which radius we get the star's entire mass.[1] Mathematically, this is an *initial-value problem,* even though there's no time involved here: "initial" means that we know the starting values and are trying to determine the behavior elsewhere.[2]

2. **Projectile motion with air resistance**

Consider a projectile of mass m in two dimensions. Its motion will obey Newton's second law, i.e., $\mathbf{F} = m\mathbf{a}$; recall that $\mathbf{a} = d^2\mathbf{r}/dt^2$, where $\mathbf{r}$'s components are x and y. In two dimensions we get a pair of second-order differential equations for $x(t)$ and $y(t)$. Our projectile will feel the force of gravity and also air resistance; at large speeds, the latter can be expressed as follows:

$$\mathbf{F}_d = -km\mathbf{v}^2\frac{\mathbf{v}}{|\mathbf{v}|} \tag{8.3}$$

This force is opposite to the direction of motion. The k is a parameter containing what is known as the drag coefficient, though the details are not relevant here: what matters is that the magnitude of the force in Eq. (8.3) is proportional to the square of the velocity. Putting everything together, Newton's second law takes the form:

$$\frac{d^2x}{dt^2} = -k\frac{dx}{dt}\sqrt{\left(\frac{dx}{dt}\right)^2 + \left(\frac{dy}{dt}\right)^2}, \qquad \frac{d^2y}{dt^2} = -g - k\frac{dy}{dt}\sqrt{\left(\frac{dx}{dt}\right)^2 + \left(\frac{dy}{dt}\right)^2} \tag{8.4}$$

where the acceleration due to gravity, g, is present only in the y direction. Note that we have $x = x(t)$ and $y = y(t)$, i.e., both x and y depend only on t, so these are second-order ordinary differential equations; even so, our problem consists of two *simultaneous* ordinary differential equations, since $x(t)$ and $y(t)$ appear in both equations.

As you may recall, to start solving a second-order differential equation you need two pieces of information, the starting value of the function and the starting value of its derivative. For the projectile problem, we are faced with two second-order differential equations, so our input information would have to be the four quantities $x(0)$, $v_x(0)$, $y(0)$, and $v_y(0)$, where $v_x = dx/dt$ and $v_y = dy/dt$. In other words, we need to know the starting positions and starting velocities of the projectile. So far, this is another example of an initial-value problem, like the one we encountered in solving the TOV equation. Here's a twist, though: what if, instead, you were given only $x(0)$, $v_x(0)$, $y(0)$, and $y(\tau)$? These are still four numbers, namely the starting positions, one component of the starting velocity, and one component of the final position. In physical terms: you shoot a

[1] M is the total *gravitational* mass: the proper volume element is *not* $4\pi r^2 dr$ (i.e., Eq. (8.2) is deceptively simple).

[2] The situation is a bit more complicated; see problem 8.51 for the full story.

cannonball at a given point and with a given v_x and you know how much time passes before it hits the ground (τ where $y(\tau) = 0$). This is no longer an initial-value problem, since we don't have all the starting information at our disposal; instead, it is a *boundary-value problem*, where we know a final piece of information, which we'll try to use to extract the missing initial value, in this example $v_y(0)$—see problems 8.30 and 8.31.

3. **Schrödinger equation(s)**
We encountered the Schrödinger equation in section 4.6: at the time, we were studying spins, so we needed its formulation in terms of state vectors. Here, we examine the (simpler) scenario of a single spinless non-relativistic particle of mass m in one spatial dimension, x, similarly to section 3.5. The system is described by its *wave function* $\Psi(x, t)$ which satisfies the *time-dependent Schrödinger equation* in the position basis:

$$i\hbar \frac{\partial \Psi(x,t)}{\partial t} = \left[-\frac{\hbar^2}{2m} \frac{\partial^2}{\partial x^2} + V(x,t) \right] \Psi(x,t) \tag{8.5}$$

The right-hand side is made up of a *kinetic* energy term (the second derivative) and a *potential* energy term. Observe that Eq. (8.5) involves a second-order spatial derivative and a first-order time derivative.[3] As you can see, the time-dependent Schrödinger equation is a linear partial differential equation; it is a dynamical equation, allowing us to determine how the wave function changes in time (i.e., it is an *initial-value problem*, for a partial differential equation this time)—see problem 8.59.
For the special case where the potential is time independent, $V(x)$, Eq. (8.5) is *separable*, giving rise to a trivial time dependence:

$$\Psi(x,t) = \psi(x) e^{-iEt/\hbar} \tag{8.6}$$

with $\psi(x)$ obeying the *time-independent Schrödinger equation* in the position basis:

$$\left[-\frac{\hbar^2}{2m} \frac{d^2}{dx^2} + V(x) \right] \psi(x) = E\psi(x) \tag{8.7}$$

Here the energy E is a real number. Crucially, Eq. (8.7) is no longer a dynamical equation but an *eigenvalue problem*: this equation has acceptable solutions only for specific values of E. In one spatial dimension, the time-independent Schrödinger equation is an *ordinary* differential equation. We had already seen the Schrödinger equation as an eigenproblem in the case of interacting spins in section 4.6, but there we were naturally faced with matrices. In the present case, we have an eigenvalue equation even though we're not using the language of matrices (yet); see, e.g., problem 8.32.

8.1.2 The Problems to Be Solved

For most of this chapter, we will be tackling *ordinary differential equations* (ODEs). The simplest case is when we are trying to solve a single differential equation for a single function, $y(x)$, where x is known as the *independent variable* and y is the *dependent variable*, namely the solution we are after.

One could start by studying equations involving, say, second-order derivatives of y. For

[3] This asymmetry is lifted when you introduce special relativity, giving rise to the *Klein–Gordon equation*.

reasons that will become clear in section 8.2.4 we, instead, focus on the more elementary problem of a first-order derivative. Thus, the first class of equation we will tackle in this chapter will be the following *initial-value problem* (IVP):

$$y'(x) = f(x, y(x)), \qquad y(a) = c \tag{8.8}$$

where $y' = dy/dx$ and $f(x, y)$ is a known function (in general non-linear) which could depend on both x and y,[4] e.g., $f(x, y) = 3x^5 - y^3$. We emphasize that f is known: what we wish to solve for is y. This is called an *initial-value* problem because we know the value of $y(x)$ at a certain point, a (i.e., we know that $y(a) = c$, where c is given), so we start from there and try to integrate up to, say, the point b. We used the word "integrate", bringing to mind the theme of chapter 7. Formally, we can integrate Eq. (8.8) to find:

$$y(x) = c + \int_a^x f(z, y(z))dz \tag{8.9}$$

If we now pick the special case where f does not depend on y (i.e., the opposite of an autonomous ODE) and set $x = b$, then our problem looks similar to Eq. (7.7). Thus, you should not be surprised to find out that terms introduced in the previous chapter (midpoint, trapezoid, Simpson's) will re-appear in the present chapter. That being said, we are here interested in producing not a single number (the value of a definite integral) but a full function, $y(x)$; also, in practice the function f most often *does* depend on y, so the problem we are faced with is more complicated than that of the previous chapter.

For a higher-order ODE, we would have to specify all necessary information at the starting point; for example, in an initial-value problem for a second order ODE we would need to specify the values of both y (the sought-after function) and y' (its derivative) at a:

$$y'' = f(x, y, y'), \qquad y(a) = c, \qquad y'(a) = d \tag{8.10}$$

In this, more general, case we see that f could also depend on y'; in other words, we have isolated the highest-order derivative on the left-hand side. Observe that both $y(a)$ and $y'(a)$ are given at the same point, so this is still an initial-value problem. A much harder problem, which we will also discuss in this chapter, arises when you are given the values of the function at two distinct points, without being provided the starting value of y'. This is known as a *boundary-value problem* (BVP), which we will first tackle in its simplest possible form, namely that of a second-order ODE:

$$y'' = f(x, y, y'), \qquad y(a) = c, \qquad y(b) = d \tag{8.11}$$

Our input data, $y(a) = c$ and $y(b) = d$, are known as *boundary conditions*. While in Eq. (8.10) we always got a unique solution, BVPs are more complicated: they can have no solutions or infinitely many solutions, but we will mainly focus on BVPs that have a single solution.

[4] If f depends on x only through its dependence on $y(x)$—i.e., $f(y)$—we are dealing with an *autonomous* ODE.

An even harder problem arises when f in Eq. (8.11) is generalized to depend not only on x, y, and y', but also on a parameter s; this is an *eigenvalue problem* (EVP):

$$y'' = f(x, y, y'; s), \qquad y(a) = c, \qquad y(b) = d \tag{8.12}$$

This equation only has non-trivial solutions for special values of s. This means that we will need to computationally find the appropriate values of s. Viewed from another perspective, in Eq. (8.10) or Eq. (8.11) we have a single equation which we try to solve, but our newest problem in Eq. (8.12) describes a family of equations (and correspondingly more than one solution). Unsurprisingly, such an eigenvalue problem has close connections both to the matrix eigenvalues we encountered in section 4.4 and to the eigenvalues appearing in the Schrödinger equation, as touched upon in section 8.1.1 (where the role played by s here was played by the eigenenergy E).

All three problems discussed above appeared in the context of *ordinary differential equations*. Of course, in practice one is also faced with *partial differential equations* (PDEs), namely equations where the dependent variable depends on more than one independent variable. Here we will provide only the briefest of introductions to the subject, in section 8.5. Then, in section 8.6 we will tackle *Poisson's equation* in two dimensions:

$$\frac{\partial^2 \phi}{\partial x^2} + \frac{\partial^2 \phi}{\partial y^2} = f(x, y) \tag{8.13}$$

employing a method that is different from what we used for ODEs (though tools that were introduced earlier in the chapter are certainly also applicable to PDEs). Note that here our dependent variable is ϕ, which depends on both x and y, i.e., $\phi(x, y)$. Crucially, this y is *not* a dependent variable as in previous paragraphs.[5]

8.2 Initial-Value Problems

While this chapter discusses a variety of questions, most of it is dedicated to initial-value problems, since the techniques that can be used to tackle these have wider applicability and will re-appear later on. As mentioned above, we will write a general IVP in the form:

$$y'(x) = f(x, y(x)), \qquad y(a) = c \tag{8.14}$$

where $y' = dy/dx$ and $f(x, y)$ is known. We wish to solve this equation for $y(x)$, with x going from a to b.

[5] You may wish to think of this problem as $y(x_0, x_1)$ instead, but it's important not to get confused: x_0 and x_1 would then be independent variables in their own right, not points (e.g., x_0 could be varied from a to b).

In most of this chapter, we will employ a *discretization*, more specifically an equally spaced grid of points, as for Newton–Cotes quadrature, Eq. (7.8). In other words, we will be trying to compute the function $y(x)$ at a set of n grid points x_j, from a to b:

$$x_j = a + jh \tag{8.15}$$

where, as usual, $j = 0, 1, \ldots, n-1$. The step size h is obviously given by:

$$h = \frac{b-a}{n-1} \tag{8.16}$$

We could use different step sizes h in different regions, as touched upon in section 8.2.3.5 and problem 8.11, but we will mostly stick to a fixed h, as per Eq. (8.16), in what follows.

Let us introduce some more notation. At a given grid point, x_j, the exact solution of our ODE will be $y(x_j)$; we will use the symbol y_j to denote the approximate solution, i.e., the value that results from a given discretization scheme.[6] We now turn to a discussion of specific methods that will help us make y_j increasingly closer to $y(x_j)$.

8.2.1 Euler's Method

We will now motivate the simplest possible approach and discuss its error behavior. We'll encounter a scenario where it gets in trouble and then we'll see what we can do about that.

8.2.1.1 Forward Euler

We can motivate what is known as the *forward Euler method* starting from the forward-difference formula. Translating Eq. (3.8) into our present notation gives:

$$y'(x_j) = \frac{y(x_{j+1}) - y(x_j)}{h} - \frac{h}{2}y''(\xi_j) \tag{8.17}$$

where, as usual, ξ_j is a point between x_j and x_{j+1}. Here it is assumed that we are starting at x_j and trying to figure out how to make a step onto the next point, x_{j+1}. Now, we know that $y(x_j)$ is the exact solution at the point x_j, so it must satisfy our ODE, Eq. (8.14):

$$y'(x_j) = f(x_j, y(x_j)) \tag{8.18}$$

We can use this equation to eliminate the first derivative from the left-hand side of Eq. (8.17):

$$y(x_{j+1}) = y(x_j) + hf(x_j, y(x_j)) + \frac{h^2}{2}y''(\xi_j) \tag{8.19}$$

We've made no approximations, which is why we don't see any y's with indices.

It is now time to make an approximation. Assuming h is small, the term proportional to h^2 will be less important, so we can drop it. This leads to the following prescription:

[6] For each technique, the y_j's will satisfy a recurrence relation (often called a *difference equation* in this context).

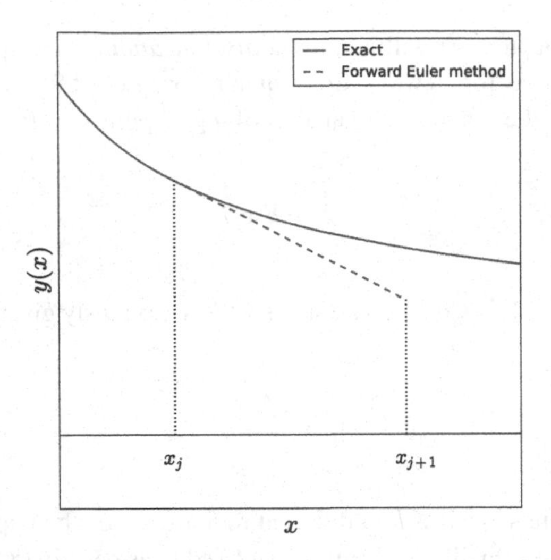

Fig. 8.1 Illustration of a step of the forward Euler method

$$y_{j+1} = y_j + hf(x_j, y_j), \qquad j = 0, 1, \ldots, n-2$$
$$y_0 = c \tag{8.20}$$

Since we started from the forward-difference formula in Eq. (8.17), the resulting method is known as the *forward* Euler method. Note that this is an approximate formula, involving y_j and y_{j+1} instead of $y(x_j)$ and $y(x_{j+1})$. As you have seen, higher-order terms in the Taylor expansion are ignored. We'll see how to do (much) better in later sections.

The method is illustrated in Fig. 8.1: since the right-hand side of Eq. (8.20) contains a term proportional to $f(x_j, y_j)$, and we know from our starting equation, Eq. (8.14), that $f(x_j, y_j)$ is (an approximation to) the slope of the tangent to the exact solution at x_j, the geometrical interpretation is reasonably straightforward. Note that in this figure we assumed that $y(x_j) = y_j$ for the purposes of illustration. This is certainly true when we start (i.e., $y(x_0) = y_0$), but will most likely not continue to be true later on (i.e., the starting point of the dashed line will not lie on top of the exact solution). Similarly, the forward Euler method approximates the slope as $f(x_j, y_j)$, whereas the true slope would have been $f(x_j, y(x_j))$. However, that's inevitable: at a given iteration, we have only y_j at our disposal, i.e., our approximation to the true $y(x_j)$.

8.2.1.2 Local and Global Error

We turn to the error incurred in a single step of the forward Euler method. This is known as the *local truncation error*; you can think of this as the error made when you go from x_0 to x_1, i.e., the difference between $y(x_1)$ and y_1. More generally, we define:

$$t_j = y(x_{j+1}) - y(x_j) - hf(x_j, y(x_j)) \tag{8.21}$$

This is defined in terms of exact quantities ($y(x_j)$, not y_j): this is because we are interested in the error incurred in a single step *assuming* we were starting from the exact solution. For the forward Euler method, we can immediately plug Eq. (8.19) into Eq. (8.21) to find:

$$t_j = \frac{1}{2}h^2 y''(\xi_j) \tag{8.22}$$

Thus, the local discretization error of the forward Euler method is $O(h^2)$. This is reminiscent of the absolute error in the one-panel rectangle formula, as per Eq. (7.17).[7]

Despite the simplicity (and popularity) of the analogy with the rectangle rule, if you wanted to proceed by analogy to the *composite* rule, as per Eq. (7.18), you would be on shaky ground. The reason is that here the different steps are *not* independent from one another: the actual error at the j-th step is *not* given by Eq. (8.22) because, as hinted at above, $y(x_j)$ is actually different from y_j and therefore $f(x_j, y(x_j))$ is also different from $f(x_j, y_j)$, due to the errors made in previous iterations; in other words, the local truncation error doesn't tell the whole story.

Let's be clear about our goal: we wish to determine the *global error* that accumulates when we numerically integrate our differential equation from $x_0 = a$ to $x_{n-1} = b$, i.e.,

$$\mathcal{E} = y(b) - y_{n-1} \tag{8.23}$$

As in the previous chapter, this definition is of the form "exact minus approximate". Working toward this goal, let us introduce some new notation:

$$e_j = y(x_j) - y_j \tag{8.24}$$

where, obviously, $e_{n-1} = \mathcal{E}$. We are consciously not calling this new quantity $\mathcal{E}_j$, in order to emphasize the conceptual differences from what was carried out in the previous chapter. With a view to seeing how the error in one step is related to that in the next step, subtract Eq. (8.20) from Eq. (8.19) to find:

$$e_{j+1} = e_j + h\left[f(x_j, y(x_j)) - f(x_j, y_j)\right] + t_j \tag{8.25}$$

where we also made use of Eq. (8.22). If we take the absolute value and use the triangle inequality (twice), this gives:

$$|e_{j+1}| \leq |e_j| + h\left|\left[f(x_j, y(x_j)) - f(x_j, y_j)\right]\right| + |t_j| \tag{8.26}$$

If we now assume that f satisfies Lipschitz continuity[8] and also that the second derivatives (contained in t_j) are bounded by M, we get:

$$|e_{j+1}| \leq (1 + Lh)|e_j| + \frac{Mh^2}{2} \tag{8.27}$$

where we grouped together the first two terms on the right-hand side.

[7] Some authors define the local truncation error with a factor of h divided out. We work by analogy with the scaling in chapter 7, where the local error is always one order higher than the global error.

[8] Lipschitz continuity means that there exists a constant L such that $|f(x, y) - f(x, \tilde{y})| \leq L\,|y - \tilde{y}|$ holds. This is satisfied if $f(x, y)$ has bounded partial derivatives, as will be the case for us.

For $j = 0$, Eq. (8.27) gives:

$$|e_1| \leq \frac{Mh^2}{2} \tag{8.28}$$

since $e_0 = y(x_0) - y_0 = 0$. Similarly, for $j = 1$ we find:

$$|e_2| \leq (1 + Lh)|e_1| + \frac{Mh^2}{2} \leq [1 + (1 + Lh)] \frac{Mh^2}{2} \tag{8.29}$$

and the next step gives:

$$|e_3| \leq (1 + Lh)|e_2| + \frac{Mh^2}{2} \leq \left[1 + (1 + Lh) + (1 + Lh)^2\right] \frac{Mh^2}{2} \tag{8.30}$$

Repeating this until the end, and summing up the geometric series, leads to:

$$|\mathcal{E}| = |e_{n-1}| \leq \frac{(1 + Lh)^{n-1} - 1}{Lh} \frac{Mh^2}{2} \tag{8.31}$$

where we can cancel the h in the denominator.

A Maclaurin expansion for the exponential, as per Eq. (C.2), gives $e^x = 1 + x + x^2 e^{\xi}/2$. This implies that $e^x \geq 1 + x$ and therefore $(1 + x)^{n-1} \leq e^{(n-1)x}$. Applied to our result:

$$|\mathcal{E}| \leq h \frac{M}{2L} (e^{L(b-a)} - 1) \tag{8.32}$$

where we also used $(n - 1)h = b - a$. Even if you need to re-read the above derivation, the main conclusion is clear: the forward Euler method converges and its global error decreases linearly, $O(h)$. While a $O(h)$ scaling is not great, it does imply that as h goes to 0 the global error also goes to 0 (ignoring considerations arising from roundoff error). We say that this is a *convergent* method. To reiterate, up to this point we were interested in what happened as we kept making h smaller; we now turn to what happens for a fixed h (and find out that we can get into serious trouble).

8.2.1.3 Stability Considerations

We now go over a pathological scenario: a case where the forward Euler method simply fails to converge. Put another way, we'll see an example where a small change in the step size has a huge effect on the global error, considerably beyond what is captured in our $O(h)$ result.

To make the discussion concrete, we will examine the (aptly named) *test equation*:

$$y'(x) = \mu\, y(x), \qquad y(0) = 1 \tag{8.33}$$

where μ is a real constant. This is a very simple autonomous ordinary differential equation (recall footnote 4). Its exact solution can be straightforwardly seen to be $y(x) = e^{\mu x}$.

The steps to follow will be for the test equation only but, as you will find out in problem 8.2, an appropriately generalized derivation can be carried out for any differential equation. Basically, the behavior of solutions to the test equation is pretty representative of more general features, so this equation will re-appear later in the chapter.

Stability of Forward Euler Method

Applying the forward Euler method, Eq. (8.20), to the test equation, Eq. (8.33), leads to $y_{j+1} = y_j + h\mu y_j$, which can be re-expressed in the form:

$$y_{j+1} = (1 + \mu h)y_j \tag{8.34}$$

If we now repeatedly apply this equation from $j = 0$ to $j = n - 2$, we get:

$$y_{n-1} = (1 + \mu h)^{n-1} y_0 = (1 + \mu h)^{n-1} \tag{8.35}$$

where, in the last step, we took advantage of the fact that $y_0 = 1$.

When $\mu > 0$, our result in Eq. (8.35) grows pretty rapidly, but that's OK, since so does the exact solution $y(x) = e^{\mu x}$. We are more interested in the case of $\mu < 0$, for which the exact solution is decaying exponentially. We immediately realize that $(1 + \mu h)^{n-1}$ may or may not decay as n gets larger, depending on the value of μh. Thus, if we wish to ensure that the solution gets smaller from one step to the next, i.e., $|y_{j+1}| < |y_j|$, we see from Eq. (8.34) that we need to demand that:

$$|1 + \mu h| < 1 \tag{8.36}$$

holds. In other words, we have arrived at the following *stability condition*:

$$h < \frac{2}{|\mu|} \tag{8.37}$$

If this is not satisfied, the forward Euler method leads to a numerically unstable solution.

In Fig. 8.2 we show the result of applying the forward Euler method to the test equation $y'(x) = -20y(x)$, i.e., for the case of $\mu = -20$. Based on the above criterion, we expect that a qualitatively new feature emerges when we cross the $h = 0.1$ threshold. This expectation is clearly borne out by the figure. The global error of the forward Euler method is supposed to be $O(h)$, but when we change $n = 12$ to $n = 10$ (i.e., go from $h = 0.091$ to $h = 0.111$) we encounter a dramatic change in behavior. As a matter of fact, the $n = 10$ results oscillate and grow considerably away from the exact solution; clearly, for this step size and μ, the forward Euler method is *unstable*. To reiterate the main point: while accuracy requirements may lead us to expect a given behavior for a given step size, stability requirements have to be taken into account separately. In other words, the forward Euler global error is $O(h)$, *only when the method is stable*. This behavior is rarely acceptable, so we now try to see if we can come up with other techniques that perform better.

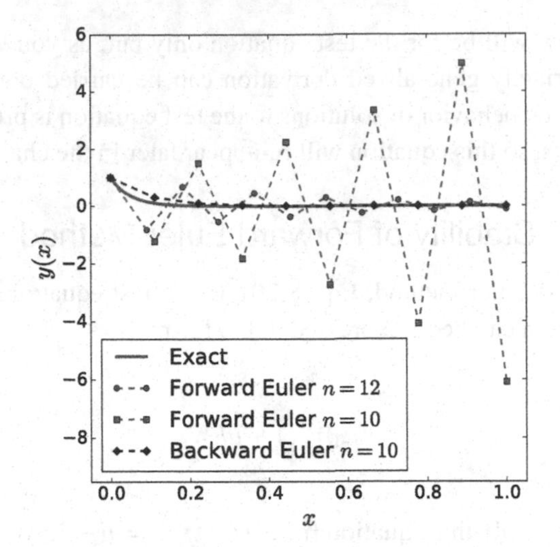

Fig. 8.2 The forward and backward Euler methods applied to the test equation

Backward Euler Method

In Fig. 8.2 we also show a set of results for a mysterious "backward Euler" method, so let us now see what that is. We will introduce this technique at a general level, i.e., for any (first-order) differential equation. After that, we will see how the new method helps us do better for the problem of the test equation.

The forward Euler method is so named because we motivated it starting from the forward-difference approximation to the first derivative. We can do something analogous for the backward-difference approximation; this means we can translate Eq. (3.11) to find:

$$y'(x_{j+1}) = \frac{y(x_{j+1}) - y(x_j)}{h} + \frac{h}{2}y''(\xi_j) \tag{8.38}$$

This certainly seems like an innocuous variation on what we were doing earlier. However, we'll now see that its impact is major. Recall that we are trying to solve our general ODE, Eq. (8.14); since $y(x_{j+1})$ is the exact solution at the point x_{j+1}, we can eliminate the first derivative from the left-hand side of Eq. (8.38), getting:

$$y(x_{j+1}) = y(x_j) + hf(x_{j+1}, y(x_{j+1})) - \frac{h^2}{2}y''(\xi_j) \tag{8.39}$$

This motivates our approximation, which is to drop the term proportional to h^2:

$$\begin{aligned} y_{j+1} &= y_j + hf(x_{j+1}, y_{j+1}), \qquad j = 0, 1, \dots, n-2 \\ y_0 &= c \end{aligned} \tag{8.40}$$

Since this time we started from the backward-difference formula in Eq. (8.38), the resulting method is known as the *backward* Euler method. We observe that the local truncation error (i.e., the $O(h^2)$ term we dropped to go from Eq. (8.39) to Eq. (8.40)) is almost identical to what we were faced with in the forward Euler case, but has the opposite sign.

While Eq. (8.40) appears, at first sight, to be quite similar to Eq. (8.20), in reality it is very different: in Eq. (8.20) we have a way of producing the y_{j+1} on the left-hand side simply by plugging in results from earlier steps, such as y_j; methods that allow you to do this are called *explicit*. However, Eq. (8.40) involves y_{j+1} on both the left-hand side and the right-hand side. Since $f(x, y)$ is in general a non-linear function, we are thereby faced with a root-finding problem of the form:

$$z = y_j + hf(x_{j+1}, z) \tag{8.41}$$

at each iteration! Methods for which the evaluation of y_{j+1} implicitly depends on y_{j+1} itself are called *implicit*. Based on what you know from chapter 5, in the general case such methods require considerably more computational effort per iteration than analogous explicit methods; this immediately raises the question why anyone would bother implementing them.

Stability of Backward Euler Method

To give the conclusion ahead of time: the backward Euler method has much better stability properties, thereby justifying the extra work we need to put in per iteration. This happens to be a general feature: *implicit methods are often better from the perspective of stability*. In the rest of the chapter we will introduce a mixture of explicit and implicit methods: the former are typically faster but, when they get in trouble, implicit methods can save the day.

In order to examine the stability features of the backward Euler method, we turn once again to the test equation of Eq. (8.33). For such a simple right-hand side, it is straightforward not only to write down Eq. (8.40):

$$y_{j+1} = y_j + h\mu y_{j+1} \tag{8.42}$$

but also to analytically solve it for y_{j+1}:

$$y_{j+1} = \frac{1}{1 - \mu h} y_j \tag{8.43}$$

As you go from the j-th step to the $(j + 1)$-th one, this equation shows you that your approximate solution for $\mu > 0$ may grow; however, that's fine, since we know that the exact solution also grows. In reality, we are more interested in the case of $\mu < 0$, for which the forward Euler method could get into serious trouble. In the present case, you can see from Eq. (8.43) that when $\mu < 0$, regardless of the magnitude of $h > 0$, you will always satisfy $|y_{j+1}| < |y_j|$. This is as it should be, since the exact solution will be decaying. In other words, we have shown that, for the test equation with $\mu < 0$, the backward Euler method is *unconditionally stable*, i.e., is stable regardless of the value of μh. We have thereby explained why this implicit method did so well in Fig. 8.2.

8.2.2 Second-Order Runge–Kutta Methods

At a big-picture level, the Euler method (whether explicit or implicit) followed from truncating a Taylor expansion to very low order. Thus, a way to produce increasingly better methods would be to keep more terms in the Taylor expansion. Of course, higher-degree terms are associated with derivatives of higher order, which are generally difficult or expensive to compute. In this and the following section, we will investigate an alternative route: so-called *Runge–Kutta methods* employ function evaluations (i.e., not derivatives) at x_j or x_{j+1} (or at points in between), appropriately combined such that the prescription's Taylor expansion matches the exact solution up to a given order. This may sound too abstract, so let's explicitly carry out a derivation that will lead to a family of second-order methods.

8.2.2.1 Derivation

Before considering the Runge–Kutta prescription (which will allow us to produce an approximate solution), let us examine the Taylor expansion of the exact solution. As advertised, we will explicitly evaluate higher-order terms this time. We have:

$$\begin{aligned} y(x_{j+1}) &= y(x_j) + hy'(x_j) + \frac{h^2}{2}y''(x_j) + \frac{h^3}{6}y'''(x_j) + O(h^4) \\ &= y(x_j) + hf(x_j, y(x_j)) + \frac{h^2}{2}f' + \frac{h^3}{6}f'' + O(h^4) \\ &= y(x_j) + hf(x_j, y(x_j)) + \frac{h^2}{2}\left(\frac{\partial f}{\partial x} + f\frac{\partial f}{\partial y}\right) \\ &\quad + \frac{h^3}{6}\left[\frac{\partial^2 f}{\partial x^2} + 2f\frac{\partial^2 f}{\partial x \partial y} + \frac{\partial f}{\partial x}\frac{\partial f}{\partial y} + f^2\frac{\partial^2 f}{\partial y^2} + f\left(\frac{\partial f}{\partial y}\right)^2\right] + O(h^4) \end{aligned} \tag{8.44}$$

In the first equality we expanded $y(x_j + h)$ around the point x_j. In the second equality we employed the fact that $y(x_j)$ is the exact solution at the point x_j, so it must satisfy our ODE, Eq. (8.14), namely $y'(x_j) = f(x_j, y(x_j))$; since the notation gets messy, we write f', where it is implied that the derivatives are taken at the point $x_j, y(x_j)$. To get to the third equality, we noticed that when evaluating f' we need to account for both the explicit dependence on x and for the fact that y depends on x:

$$f' = \frac{\partial f}{\partial x} + \frac{\partial f}{\partial y}\frac{dy}{dx} = \frac{\partial f}{\partial x} + \frac{\partial f}{\partial y}f \tag{8.45}$$

Similarly, the next order term in the third equality applied this fact again:

$$f'' = \frac{\partial}{\partial x}\left(\frac{\partial f}{\partial x} + \frac{\partial f}{\partial y}f\right) + \frac{\partial}{\partial y}\left(\frac{\partial f}{\partial x} + \frac{\partial f}{\partial y}f\right)\frac{dy}{dx} \tag{8.46}$$

Our final result in Eq. (8.44) is an expansion of the exact solution, explicitly listing all the terms up to order h^3. Any Ansatz, such as those we will introduce below, will have to match these terms order by order, if it is to be correct up to a given order.

We now turn to the *second-order Runge–Kutta* prescription; as its name implies, this will turn out to match Eq. (8.44) up to order h^2—i.e., it will have a local truncation error of $O(h^3)$. Crucially, it will accomplish this without needing to evaluate any derivatives:

$$y_{j+1} = y_j + c_0 hf(x_j, y_j) + c_1 hf\left[x_j + c_2 h, y_j + c_2 hf(x_j, y_j)\right] \tag{8.47}$$

The c_0, c_1, and c_2 will be determined below. This prescription requires two function evaluations in order to carry out a single step: it may look as if there are three function evaluations, but one of them is re-used. To see this, take:

$$\begin{aligned} k_0 &= hf(x_j, y_j) \\ y_{j+1} &= y_j + c_0 k_0 + c_1 hf\left[x_j + c_2 h, y_j + c_2 k_0\right] \end{aligned} \tag{8.48}$$

As advertised, this doesn't involve any derivatives; it *does* involve the evaluation of f at a point other than x_j, y_j: this involves a shifting of both x_j (to $x_j + c_2 h$), as well as a more complicated shifting of y_j (to $y_j + c_2 hf(x_j, y_j)$). However, if c_0, c_1, c_2, x_j, y_j, and f are all known, then the right-hand side can be immediately evaluated; in other words, we are here dealing with an *explicit* Runge–Kutta prescription.

Our plan of action should now be clear: expand the last term in Eq. (8.47), grouping the contributions according to which power of h they are multiplying; then, compare with the exact solution expansion in Eq. (8.44) and see what you can match:

$$\begin{aligned} f\left[x_j + c_2 h, y_j + c_2 hf(x_j, y_j)\right] &= f(x_j, y_j) + c_2 h\frac{\partial f}{\partial x} \\ &+ c_2 hf(x_j, y_j)\frac{\partial f}{\partial y} + \frac{c_2^2 h^2}{2}\frac{\partial^2 f}{\partial x^2} + \frac{c_2^2 h^2}{2} 2f(x_j, y_j)\frac{\partial^2 f}{\partial x \partial y} \\ &+ \frac{c_2^2 h^2}{2} f^2(x_j, y_j)\frac{\partial^2 f}{\partial y^2} + O(h^3) \end{aligned} \tag{8.49}$$

It may be worthwhile at this point to emphasize that in Eq. (8.44) we expanded $y(x_j + h)$, i.e., a function of a single variable. This is different from what we are faced with in the left-hand side of Eq. (8.49), namely f, which is a function of both our independent variable and our dependent variable. Our result on the right-hand side involves all the necessary partial derivatives, similarly to what we did in the last step of Eq. (8.44).

If we now plug Eq. (8.49) into Eq. (8.47) and collect terms, we find:

$$\begin{aligned} y_{j+1} &= y_j + (c_0 + c_1)hf(x_j, y_j) + c_1 c_2 h^2 \left[\frac{\partial f}{\partial x} + f(x_j, y_j)\frac{\partial f}{\partial y}\right] \\ &+ \frac{c_1 c_2^2 h^3}{2}\left[\frac{\partial^2 f}{\partial x^2} + 2f(x_j, y_j)\frac{\partial^2 f}{\partial x \partial y} + f^2(x_j, y_j)\frac{\partial^2 f}{\partial y^2}\right] + O(h^4) \end{aligned} \tag{8.50}$$

Comparing this result with the expansion for the exact solution, Eq. (8.44), we realize that we can match the terms proportional to h and to h^2 if we assume:

$$c_0 + c_1 = 1, \qquad c_1 c_2 = \frac{1}{2} \tag{8.51}$$

Unfortunately, this does not help us with the term that is proportional to h^3. Simply put, the term in brackets in Eq. (8.44) contains more combinations of derivatives than the corresponding term in Eq. (8.50). Thus, since we have agreement up to order h^2, we have shown that the explicit Runge–Kutta prescription in Eq. (8.47) has a local truncation error which is $O(h^3)$, as desired.[9] Since the local error of our Runge–Kutta prescription is $O(h^3)$, you can guess that the global error is one order worse, i.e., $O(h^2)$, similarly to what we saw in section 7.2 for the midpoint and trapezoid quadrature rules (cf. section 8.2.1.2 on the forward Euler method). Hence the appellation *second-order* Runge–Kutta.

8.2.2.2 Explicit Midpoint and Trapezoid Methods

There is more than one way to satisfy Eq. (8.51)—two equations in three unknowns; we will examine a couple of options. The first one obtains when you take the parameters to be:

$$c_0 = 0, \quad c_1 = 1, \quad c_2 = \frac{1}{2} \tag{8.52}$$

For this case, the prescription in Eq. (8.47) takes the form:

$$y_{j+1} = y_j + hf\left[x_j + \frac{h}{2}, y_j + \frac{h}{2}f(x_j, y_j)\right] \tag{8.53}$$

Things may be more transparent if we break up the evaluations into two stages:

$$\begin{aligned} k_0 &= hf(x_j, y_j) \\ y_{j+1} &= y_j + hf\left[x_j + \frac{h}{2}, y_j + \frac{k_0}{2}\right] \end{aligned} \tag{8.54}$$

This is an explicit method, since the right-hand side is already known; it is also a method that employs the midpoint between x_j and x_{j+1}. Observe that the second argument passed to f is $y_j + k_0/2$, namely the (approximation to the) slope of the tangent at the midpoint. As a result, this is known as the *explicit midpoint method.*

Its interpretation can be understood from Fig. 8.3: recall that Fig. 8.1 showed Euler's method employing the slope at the point x_j, y_j. In contradistinction to this, here we are finding the slope at the midpoint and then (parallel transporting to) make a move starting from the point x_j, y_j.[10] As a result, even in this illustration it can be seen that the explicit midpoint method does a better job than Euler's method; of course, you already expected this to be the case, since you know the local (or global) error to be better by one order.

We now turn to our second choice of how to satisfy the conditions in Eq. (8.51):

$$c_0 = \frac{1}{2}, \quad c_1 = \frac{1}{2}, \quad c_2 = 1 \tag{8.55}$$

For this case, the prescription in Eq. (8.47) takes the form:

$$y_{j+1} = y_j + \frac{h}{2}f(x_j, y_j) + \frac{h}{2}f\left[x_j + h, y_j + hf(x_j, y_j)\right] \tag{8.56}$$

[9] This is better by one order compared to the local truncation error of the Euler method(s), which was $O(h^2)$. On the other hand, Euler needs a single function evaluation per step, whereas our new method requires two.

[10] As before, we are assuming here that $y_j = y(x_j)$ for the purposes of illustration.

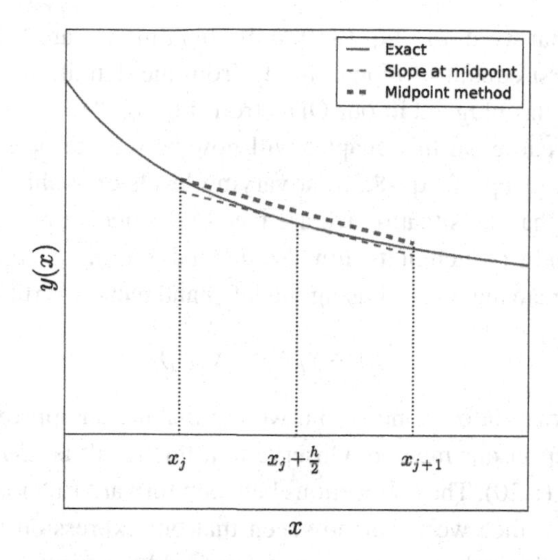

Fig. 8.3 Illustration of a step of the midpoint method

Once again, this is slightly easier to understand if we break up the evaluations in two stages:

$$k_0 = hf(x_j, y_j)$$
$$y_{j+1} = y_j + \frac{h}{2}\left[f(x_j, y_j) + f(x_{j+1}, y_j + k_0)\right] \tag{8.57}$$

This, too, is an explicit method. Given that it involves a prefactor of $h/2$ multiplied by the sum of the function values at the left endpoint and (what appears to be) the right endpoint, you may not be surprised to hear that this is known as the *explicit trapezoid method.*[11] Just like the explicit midpoint method, this is a second-order Runge–Kutta technique; it requires two function evaluations per step and is better than the basic Euler method.

8.2.2.3 Quadrature Motivation

So far, we've mentioned terms from the previous chapter on numerical integration several times; we will now see in a bit more detail how these connections between the two chapters arise. In the process, we will also introduce two more (implicit) methods.

The goal, as before, is to figure out how to make the step from x_j to x_{j+1}, i.e., how to produce y_{j+1}. We have:

$$y(x_{j+1}) - y(x_j) = \int_{x_j}^{x_{j+1}} \frac{dy}{dx}dx = \int_{x_j}^{x_{j+1}} f(x, y(x))dx \tag{8.58}$$

[11] These methods are also known by several other names. For example, the explicit trapezoid method is sometimes called *Heun's method* and the explicit midpoint method is known as the *modified Euler method.*

This is similar to what we did in Eq. (8.9), only this time we are limiting ourselves to a single step h. The first equality follows simply from the definition of a definite integral. The second equality has plugged in our ODE from Eq. (8.14), $y' = f(x, y)$. The different methods introduced earlier in this chapter will now be seen to be a result of evaluating the integral in the last step of Eq. (8.58) at varying levels of sophistication. Of course, it is important to note that the situation we are faced with here is not quite so clean: our f depends on both x and y (which in its turn also depends on x).

First, we approximate the integral using the left-hand rectangle rule of Eq. (7.10):

$$y_{j+1} = y_j + hf(x_j, y_j) \tag{8.59}$$

Notice how $y(x_j)$ turned into y_j and so on: we are making an approximation, in order to produce the next step in our method. Observe that this result is *identical* to the forward Euler method of Eq. (8.20). The connections between forward Euler and the rectangle rule aren't too surprising, since we've already seen that our expression for the local error in the two cases was fully analogous. Of course, we could have just as easily employed the right-hand rectangle rule, which would have led to:

$$y_{j+1} = y_j + hf(x_{j+1}, y_{j+1}) \tag{8.60}$$

Again, this is identical to the backward Euler method of Eq. (8.40). Of the last two methods, the first was explicit and the second implicit, as you know.

Now, we approximate the integral in Eq. (8.58) using the midpoint rule of Eq. (7.20):

$$y_{j+1} = y_j + hf\left[x_j + \frac{h}{2}, y\left(x_j + \frac{h}{2}\right)\right] \tag{8.61}$$

We are now faced with a problem: the last term involves $y(x_j + h/2)$ but that is off-grid. We don't (and will never) know the value of our dependent variable outside our grid points.[12] What we can do is to further approximate $y(x_j + h/2)$ by the average of y_j and y_{j+1}:

$$y_{j+1} = y_j + hf\left(x_j + \frac{h}{2}, \frac{y_j + y_{j+1}}{2}\right) \tag{8.62}$$

This is easily identifiable as an implicit method: it involves y_{j+1} on both the left-hand side and the right-hand side. (Note, however, that our earlier problem is resolved: y_j and y_{j+1} are approximate values *at* the grid points.) The resulting prescription is known as the *implicit midpoint method*. At this point, you could choose to approximate y_{j+1} on the right-hand side by the forward Euler method; if you do that, you will find that you recover the *explicit* midpoint method. If in the previous subsection you had any lingering doubts about why this method was so named, these should have been put to rest by now.

We can also approximate the integral in Eq. (8.58) using the trapezoid rule of Eq. (7.30):

[12] You could take a page out of section 3.3.6.2's book and double your step size: $h \to 2h$ turns Eq. (8.61) into $y_{j+2} = y_j + 2hf(x_{j+1}, y_{j+1})$; similarly to Eq. (8.154) or Eq. (8.172), this needs two points to get going. Confusingly, this multistep method sometimes also goes by the name of the "explicit midpoint method".

$$y_{j+1} = y_j + \frac{h}{2}\left[f(x_j, y_j) + f(x_{j+1}, y_{j+1})\right] \quad (8.63)$$

You're probably getting the hang of it by now: since this is an implicit method that resulted from using the trapezoid rule, it is called the *implicit trapezoid method*. Just like we did in the previous paragraph, we could now approximate y_{j+1} on the right-hand side by the forward Euler method; unsurprisingly, doing that leads to the *explicit* trapezoid method. You may have noticed that we didn't actually prove that the *implicit* midpoint and trapezoid methods are second-order; the previous subsection carried out the Taylor-expansion matching for the *explicit* Runge–Kutta prescription in Eq. (8.47). Problem 8.5 asks you to check the order of the error for the implicit case.

8.2.3 Fourth-Order Runge–Kutta Method

Up to this point, we've introduced six distinct methods: Euler, midpoint, and trapezoid, with each one appearing in an implicit or explicit version. We also saw the connections of these techniques with the corresponding quadrature rules from the previous chapter. Higher-order methods can be arrived at by applying higher-order quadrature rules; unfortunately, the connection betwen the former and the latter stops being as clean. One (unpleasant) option would then be to completely drop the connection with quadrature and revert to a Taylor expansion of a higher-order prescription, generalizing Eq. (8.47), and then matching terms order-by-order. Problem 8.6 guides you toward that goal.

Before we go any farther, let us write down a very important prescription, belonging to the *fourth-order Runge–Kutta method*:

$$\begin{aligned}
k_0 &= hf(x_j, y_j)\\
k_1 &= hf\left(x_j + \frac{h}{2}, y_j + \frac{k_0}{2}\right)\\
k_2 &= hf\left(x_j + \frac{h}{2}, y_j + \frac{k_1}{2}\right)\\
k_3 &= hf\left(x_j + h, y_j + k_2\right)\\
y_{j+1} &= y_j + \frac{1}{6}(k_0 + 2k_1 + 2k_2 + k_3)
\end{aligned} \quad (8.64)$$

This Ansatz is sufficiently widespread that it is also known as *classic Runge–Kutta* or *RK4*. It requires four function evaluations in order to produce y_{j+1} starting from y_j; the specific form it takes tells us that this is an explicit method.

Just like our earlier prescriptions, RK4 does *not* need to evaluate any derivatives. Even though it is more costly than any of the other methods we've seen up to now, it is certainly a worthwhile investment, as RK4 has a local error of $O(h^5)$, which is two orders better than anything we encountered before. There are many other methods on the market, e.g., higher-order Runge–Kutta techniques, multistep methods, Richardson-extrapolation approaches, or techniques tailored to specific types of ODEs; the problem set introduces several of these. For most practical purposes, however, the fourth-order Runge–Kutta method is an efficient and dependable tool which is likely to be your technique of choice.

The broad outlines of a graphic interpretation of Eq. (8.64) are easy to grasp: the prescription involves evaluations at x_j and x_{j+1}, as well as at their midpoint. Similarly, k_0, k_1, k_2, and k_3 are approximations to the slope at the endpoints and the midpoint. Of course, you may be wondering exactly why these k_i's have been chosen to have this specific form (and also end up being combined in the final line in precisely that manner). As mentioned above, a full Taylor-expansion derivation is quite messy, without providing any new insights. However, the essential points can be grasped if we examine two extreme cases for $f(x, y)$: (a) the case where f does not depend on y, and (b) the case where f does not depend on x (an autonomous ODE). We take these up in turn.

8.2.3.1 First Case: Quadrature

As advertised, we first consider the case where f depends only on our independent variable x. Then, Eq. (8.58) takes the form:

$$y(x_{j+1}) - y(x_j) = \int_{x_j}^{x_{j+1}} f(x)dx \tag{8.65}$$

This is now not *similar* to what we had in the previous chapter, but *identical*. We've already seen the rectangle, midpoint, and trapezoid rule approximations to this integral. The next quadrature formula we encountered in the previous chapter was *Simpson's rule*, see Eq. (7.45). Applied to Eq. (8.65), this gives:

$$y_{j+1} = y_j + \frac{h}{6}\left[f(x_j) + 4f\left(x_j + \frac{h}{2}\right) + f(x_{j+1})\right] \tag{8.66}$$

where we took care of the fact that Eq. (7.45) was written down for two panels (i.e., from x_j to x_{j+2}). If we now look at the RK4 prescription of Eq. (8.64), we realize that when f does not depend on y we have $k_1 = k_2 = hf(x_j + h/2)$. Thus, Eq. (8.66) exactly matches the RK4 prescription. We have therefore accomplished (the first part of) what we set out to do: the fourth-order Runge–Kutta method is equivalent to Simpson's rule (for the, simpler, case where f does not depend on y). This is one way of seeing that RK4 has a local error of $O(h^5)$, as shown for Simpson's rule in Eq. (7.52).

Incidentally, the connections between the numerical integration of ODEs, on the one hand, and plain quadrature, on the other, do not stop here. We've focused on Newton–Cotes methods, but one could just as well employ Gauss–Legendre quadrature schemes; as a matter of fact, the midpoint method is a trivial example of this. One could, analogously, introduce fourth-order or sixth-order implicit Gauss–Legendre Runge–Kutta methods. That being said, we won't need to do so, since RK4 is enough for our purposes.

8.2.3.2 Second Case: Autonomous

So far, we've only accomplished half of what we wanted: we've seen what happens when we are faced with $f(x)$. We'll now examine the other extreme, that of an autonomous differential equation, namely one with $f(y)$. To make things concrete, we will study (again) the test equation of Eq. (8.33):

$$y'(x) = \mu y(x) \tag{8.67}$$

Since this is a case where we explicitly know the right-hand side (which also happens to be simple), we can follow the explicit route we opted against in the general case: we can compare the RK4 prescription with a Taylor expansion of the exact solution.

We start by explicitly applying the RK4 prescription of Eq. (8.64) to the test equation:

$$\begin{aligned}
k_0 &= hf(x_j, y_j) = \mu h y_j \\
k_1 &= hf\left(x_j + \frac{h}{2}, y_j + \frac{k_0}{2}\right) = \left[\mu h + \frac{(\mu h)^2}{2}\right] y_j \\
k_2 &= hf\left(x_j + \frac{h}{2}, y_j + \frac{k_1}{2}\right) = \left[\mu h + \frac{(\mu h)^2}{2} + \frac{(\mu h)^3}{4}\right] y_j \\
k_3 &= hf\left(x_j + h, y_j + k_2\right) = \left[\mu h + (\mu h)^2 + \frac{(\mu h)^3}{2} + \frac{(\mu h)^4}{4}\right] y_j \\
y_{j+1} &= y_j + \frac{1}{6}(k_0 + 2k_1 + 2k_2 + k_3) = \left[1 + \mu h + \frac{(\mu h)^2}{2} + \frac{(\mu h)^3}{6} + \frac{(\mu h)^4}{24}\right] y_j
\end{aligned} \tag{8.68}$$

The last step in the first four lines plugs in $f(y) = \mu y$ (and earlier results as they are produced). The last step in the final line grouped all the terms together.

We now recall that the exact solution to the test equation is $y(x) = ce^{\mu x}$. We can apply this at the two grid points of interest:

$$y(x_j) = ce^{\mu x_j}, \quad y(x_{j+1}) = ce^{\mu(x_j+h)}, \quad \frac{y(x_{j+1})}{y(x_j)} = e^{\mu h} \tag{8.69}$$

where in the last equation we formed the ratio, in order to emphasize that the specific starting point does not matter here. As usual, we now Taylor expand $e^{\mu h}$ up to a sufficienty high order:

$$\frac{y(x_{j+1})}{y(x_j)} = 1 + \mu h + \frac{(\mu h)^2}{2} + \frac{(\mu h)^3}{6} + \frac{(\mu h)^4}{24} + \frac{(\mu h)^5}{120} + O(h^6) \tag{8.70}$$

Comparing y_{j+1}/y_j to $y(x_{j+1})/y(x_j)$ we find perfect agreement up to h^4; significantly, when evaluating y_{j+1}/y_j we did *not* have to evaluate any derivatives! The RK4 prescription does not capture the term $(\mu h)^5/120$ (or any higher-order terms), so we see that the local error is $O(h^5)$. This completes the second half of our investigation of the local error of RK4. Since this is $O(h^5)$, we see that the global error of the fourth-order Runge–Kutta prescription of Eq. (8.64) is $O(h^4)$, as expected.

Code 8.1 ivp_one.py

```
import numpy as np

def f(x,y):
    return - (30/(1-x**2)) + ((2*x)/(1-x**2))*y - y**2

def euler(f,a,b,n,yinit):
    h = (b-a)/(n-1)
    xs = a + np.arange(n)*h
    ys = np.zeros(n)
    y = yinit
    for j,x in enumerate(xs):
        ys[j] = y
        y += h*f(x, y)
    return xs, ys

def rk4(f,a,b,n,yinit):
    h = (b-a)/(n-1)
    xs = a + np.arange(n)*h
    ys = np.zeros(n)
    y = yinit
    for j,x in enumerate(xs):
        ys[j] = y
        k0 = h*f(x, y)
        k1 = h*f(x+h/2, y+k0/2)
        k2 = h*f(x+h/2, y+k1/2)
        k3 = h*f(x+h, y+k2)
        y += (k0 + 2*k1 + 2*k2 + k3)/6
    return xs, ys

if __name__ == '__main__':
    a, b, n, yinit = 0.05, 0.49, 12, 19.53
    xs, ys = euler(f,a,b,n,yinit); print(ys)
    xs, ys = rk4(f,a,b,n,yinit); print(ys)
```

8.2.3.3 Implementation

While developing all these ODE integration techniques up to this point, we didn't say anything about whether or not f is linear or non-linear. This is because for the initial-value problem of an ordinary differential equation our methods are so robust that the non-linearity of the equation typically doesn't really matter.[13] To drive this point home we will now solve the following (mildly non-linear) *Riccati equation*:

$$y'(x) = -\frac{30}{1-x^2} + \frac{2x}{1-x^2}y(x) - y^2(x), \qquad y(0.05) = 19.53 \tag{8.71}$$

The right-hand side here depends on both x and y; recall that the non-linearity comes from $y^2(x)$, not from the x dependence. Since this is an initial-value problem, we have pulled an initial value out of a hat, in order to get things going.

We are now ready to solve this equation in Python, using two of the seven numerical-integration techniques introduced so far; the result is code 8.1. If you've been reading this book straight through, your programming maturity will have been developing apace. Thus, our comments on this and the following codes will be less extensive than in earlier chapters.

We start with a function implementing the right-hand side of Eq. (8.71). As usual, the main workhorses of the code are general, i.e., they allow you to solve a different ODE with a different initial value with minimal code modifications. This is obvious in the definition of `euler()`, which implements the forward Euler method of Eq. (8.20) and takes in as parameters the right-hand side of the ODE to be solved, the starting and end points, the number of integration points to be used, as well as the initial value (i.e., the value of the dependent variable at the starting point). The body of the function is standard: we define the step size, set up the grid of x_j's, initialize y_0, and then step through the y_j's as per Eq. (8.20). We try to be Pythonic in the sense of minimizing index use, but even so we require a call to `enumerate()` to help us store the latest y_j in the appropriate slot. The function `rk4()` is a straightforward generalization of `euler()`: this time we are implementing Eq. (8.64) line by line. The main program simply sets up the input parameters and calls the two solvers one after the other.

In Fig. 8.4 we have taken the liberty of plotting the output produced in the main program. We've also included a curve marked as the exact solution: we won't tell you how we produced that for now (but stay tuned). Overall, we see that the fourth-order Runge–Kutta method does an excellent job integrating out from our starting point, despite the fact that we used a very coarse grid, i.e., only 12 points in total. The same cannot quite be said about the forward Euler method: already after the first point, we are visibly distinct from the exact solution. As a matter of fact, for this specific ODE, it is the first and last steps for which Euler does a bad job: these are the regions where the function is most interesting. Since Euler doesn't build in extra information (other than the slope at the left point), it fails to capture complicated behavior. In problem 8.3 you will implement the explicit midpoint and trapezoid methods for the same ODE in order to see if they do any better.

[13] It may appear that this statement is somewhat softened in the following subsection, but even there the cause of trouble is not the non-linearity.

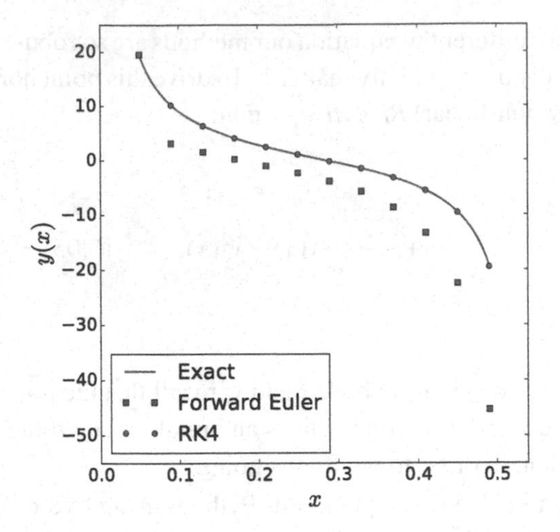

Fig. 8.4 Forward Euler and Runge–Kutta methods for our non-linear initial-value problem

8.2.3.4 Stiffness

As mentioned earlier, the fourth-order Runge–Kutta method is fairly robust and should usually be your go-to solution. As another example of how well it does, we can solve the test equation for which, as we saw in our discussion of Fig. 8.2, the forward Euler method faced stability issues. The left panel of Fig. 8.5 shows the fourth-order Runge–Kutta and the backward Euler methods for $y'(x) = -20y(x)$. Overall, we see that RK4 is certainly decent: for $n = 10$ it does slightly more poorly than the backward Euler method, but qualitatively the behavior is the same for both $n = 10$ and $n = 12$. This doesn't really come as a surprise: RK4 has a local error of $O(h^5)$, so it can "get the most out" of a finite step size h; in other words, it does a dramatically better job than the forward Euler method and that's enough. Based on this panel, you might be tempted to think that our earlier digression on stability and implicit methods was pointless: you can use a high-order *explicit* method without having to worry about complicated root-finding steps.

Of course, this is too good to be true. In earlier chapters (especially in our discussion of linear algebra in chapter 4) we've seen that one must distinguish between the *stability* of a given method and the *ill-conditioning* of a given problem. There are well-conditioned problems which some methods fail to solve (we say they are *unstable*) and other methods manage to solve (we say they are *stable*). On the other hand, there exist ill-conditioned problems, which no method can hope to attack very fruitfully. In between these two extremes, there are mildly ill-conditioned problems which can be hard to solve even if you're employing a generally stable method. The ill-conditioning of a differential equation is known as *stiffness*: a *stiff* ODE makes many methods unstable. Quantifying stiffness in the general case can get quite complicated (but see section 8.2.4.4 for more on this), so let's look at a specific example to see stiffness in action.

Here's an example that will put our earlier RK4-oriented triumphalism to rest:

$$y'(x) = 501e^x - 500y(x), \qquad y(0) = 0 \tag{8.72}$$

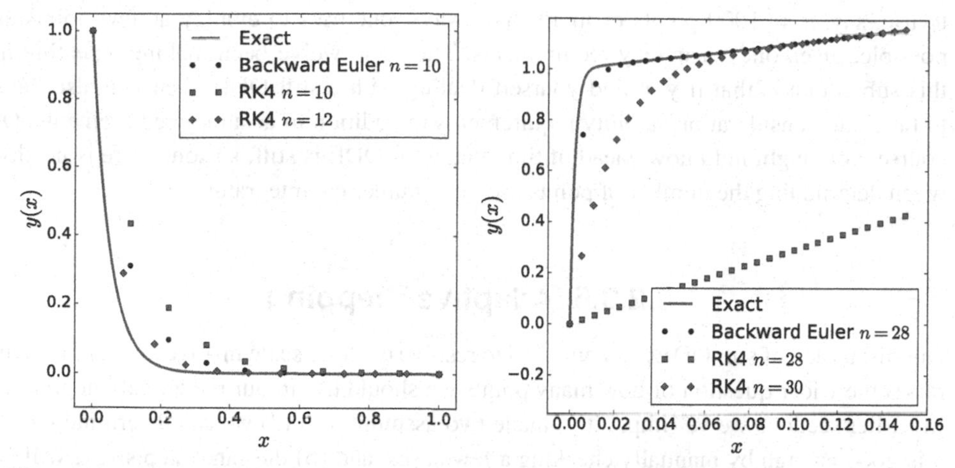

RK4 and backward Euler applied to the test equation (left) and our stiff problem (right) **Fig. 8.5**

As usual, this is a first-order initial-value problem; its exact solution is $y(x) = e^x - e^{-500x}$. The exact solution is already warning us of what could go wrong: one needs to keep track of two very different scales: a growing exponential that has x in the exponent and a decaying exponential that has $-500x$ in the exponent. Roughly speaking, stiff equations will always exhibit this feature of needing to keep track of two different scales at the same time. As you may have imagined, in order to be able to handle both scales, one needs to employ a step size h that is short enough to resolve the smallest of the scales in our problem: of course, that means that the calculation is likely to be pretty expensive.

This discussion is made concrete in the right panel of Fig. 8.5, where we show the backward Euler and RK4 methods applied to Eq. (8.72). We find that, even for a small number of points ($n = 28$), the backward Euler method does a reasonably good job. On the other hand, we find that RK4 behaves strangely: the global error of this Runge–Kutta method is supposed to be $O(h^4)$, but when we change $n = 28$ to $n = 30$ (i.e., go from $h = 0.0056$ to $h = 0.0052$) we, once again, encounter a dramatic change in behavior. In short, RK4 becomes unstable at some point, even though the (implicit) backward Euler method has no trouble handling the same problem. While it is true that if you "throw more points at the problem" RK4 will behave properly (i.e., converge), the figure shows that sometimes even this solid technique gets in trouble.[14] Of course, Eq. (8.72) was hand-picked to emphasize the problems that arise when dealing with stiff ODEs. In problem 8.10 you will encounter another, more innocuous-looking, case; often, such issues arise when you have to deal with two (or more) exponentials or two (or more) oscillatory behaviors. You can sometimes transform your troubles away, i.e., employ an analytical trick *before* you apply a technique such as RK4 (see problem 8.9).

A general point should be made here: in these figures we have been employing a small number of points ($n = 10$ or $n = 30$). In practical applications, you typically use more points; this is fine if the evaluation of your right-hand side f is trivial but, as is sometimes the case, if your ODE itself is the result of another computation, it may be very costly

[14] "God delights in clipping everything that stands out" (Herodotus, *The Histories*, paragraph 7.10ε).

to use, say, $n = 1000$ points or more. As always, one tries to employ as few points as possible, given one's accuracy requirements. The point we've been making/repeating in this subsection is that if you find yourself dealing with a stiff ODE, then you also have to take into consideration stability requirements in addition to accuracy requirements. Of course, you might not know ahead of time that your ODE is stiff, so some care is needed when determining the number of points used in a numerical integrator.

8.2.3.5 Adaptive Stepping

Our discussion of stiff ODEs and our need to resolve the finer scale involved in our problem raises the wider question of how many points we should use in our numerical integration. In earlier sections we've (implicitly) made two assumptions: (a) we can determine which n is good enough by manually checking a few cases, and (b) the same step size h will be employed throughout the integration region. As the reader may have already guessed, both of these assumptions are not tenable in heavy-duty applications: it would be nice to have an automatic routine, which determines on its own how many steps to use; it would be even better if such a routine can sub-divide the problem into "easy" and "hard" regions, distributing the total number of points appopriately. In this section, we remove each of these assumptions in turn, slowly ramping up the complications that have to be introduced into the algorithm. Problem 8.12 asks you to recover our results by implementing such adaptive integration schemes yourself.

Global Adjustment

The simplest tack is completely analogous to our adaptive-integration schemes in section 7.3, where we were doubling the number of panels.[15] To refresh your memory, what we did there was to carry out a calculation employing N panels and another one for $N' = 2N$ panels. We then took advantage of the fact that the error in the composite Simpson's rule is ch_N^4 in the first case and $ch_{N'}^4$ in the second. After a brief derivation, we were able to show that the error in the calculation employing $N' = 2N$ panels can be expressed in terms of the difference between the two estimates (employing N and N' panels, respectively). Crucially, this was a derivation for the *composite* Simpson's rule: in the first case we were using $n = N - 1$ points and in the second case $n' = N' - 1$ points; as per Eq. (7.65), the two numbers are related by $n' = 2n - 1$. Both n and n' were expected to get quite large, depending on the error budget one was dealing with.

Repeating that derivation for the present case (of integrating an IVP from a to b) would entail first carrying out a calculation using n points (where the best approximation to $y(b)$ would be y_{n-1}) and then another calculation using n' points (where the best approximation to $y(b)$ would be $y_{n'-1}$); as before, $n' = 2n - 1$. Observe that y_{n-1} and $y_{n'-1}$ correspond to two distinct calculations. Since the global error for the RK4 method, just like the error in

[15] Step halving is also used in ODE solvers that employ Richardson extrapolation, as you will discover when you solve problem 8.34.

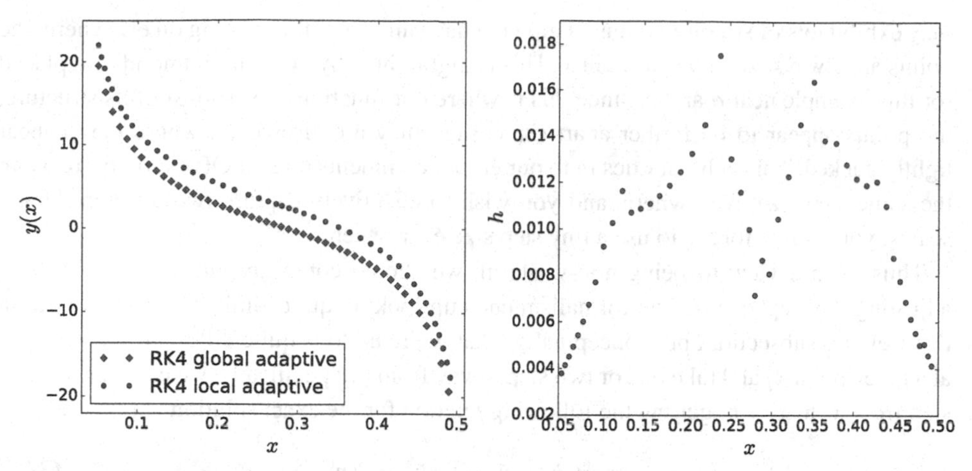

Global and shifted local adaptive RK4 (left) and the step size for the local case (right) Fig. 8.6

the composite Simpson's rule, is $O(h^4)$, the derivation is completely analogous:

$$y(b) = y_{n-1} + ch_N^4 = y_{n'-1} + ch_{N'}^4 \tag{8.73}$$

This, together with $h_N = 2h_{N'}$, can be manipulated to eliminate $h_{N'}$, thereby giving:

$$\mathcal{E} = \frac{1}{15}\left(y_{n'-1} - y_{n-1}\right) \tag{8.74}$$

where the 15 in the denominator is a direct consequence of using RK4. Note that this is the absolute error; you can divide by $y_{n'-1}$ (i.e., the best of the two estimates) if you want to approximate the relative error. Just like for Simpson's rule, we have here been doing calculations with many points which are equally spaced from a to b.

As you will find out when you solve problem 8.12, applying this scheme to our nonlinear Riccati equation from Eq. (8.71) leads to the points shown with diamonds in the left panel of Fig. 8.6. The trend is identical to that in Fig. 8.4, only this time we are not manually choosing a specific value of n: instead, we have pre-set a specific *global* tolerance for our approximation to $y(b)$ and then keep doubling the number of panels until we meet our accuracy goals. The advantage of such an adaptive scheme is that we don't have to worry too much about the details of our problem: the algorithm will keep on using more points until the error tolerance goals are met.

Local Adjustment

As should be immediately obvious from the distribution of our diamond points, the global adaptive scheme can be quite wasteful: it employs equally spaced points and doubles the total number of panels used every time, without regard to the behavior of the solution. In other words, this scheme has no way of accounting for the possibility that our function may exhibit lots of structure in one region but have almost nothing going on elsewhere: the points are always equally spaced in x. This is highlighted by how our diamonds are placed for this example near a and b: since this is where our function shows most of its structure, the points appear to be farther apart (in y) than they are elsewhere, where they appear tightly packed. This behavior ties in to our earlier comments on stiff ODEs: if you're using the same step size everywhere, and you wish to effectively capture two different length scales, you will be forced to use a tiny step size everywhere.[16]

Thus, with a view to being non-wasteful, we should come up with a way of *locally* adjusting the step size h. The formalism ends up looking quite similar to what we did in the previous subsection, but conceptually what we're up to is quite different: we will start at a given point x_j and take one or two steps away from it. Specifically, we can take a single step from x_j to $x_j + h$, getting the following relation for the exact solution $y(x_j + h)$:

$$y(x_j + h) = \tilde{y}_{j+1} + \kappa h^5 + O(h^6) \tag{8.75}$$

Let's take a moment to unpack the notation here: $\tilde{y}_{j+1}$ is the result of starting at the point x_j, y_j, and taking a step as per Eq. (8.64): we're using a tilde to emphasize that this is a single step, but other than that we haven't done anything new; you can think of y_j as $\tilde{y}_j$ if it helps; we use Eq. (8.64) to produce $\tilde{y}_{j+1}$ from $\tilde{y}_j$, implicitly assuming that we knew the starting value exactly. We're explicitly showing the local error for this one step, κh^5, and then capturing all higher-order terms with $O(h^6)$.

Now picture taking two steps instead of one: first, from x_j to $x_j + h/2$ and then from $x_j + h/2$ to $x_j + h$. The error for the first step will look like:

$$y\left(x_j + \frac{h}{2}\right) = \tilde{\tilde{y}}_{j+1/2} + \kappa\left(\frac{h}{2}\right)^5 + O(h^6) \tag{8.76}$$

where we are using double-tildes to emphasize that in this scenario we will be taking two steps to go from x_j to $x_j + h$. In Eq. (8.76) we used Eq. (8.64) to go from y_j to $\tilde{\tilde{y}}_{j+1/2}$. After we take the second step we will have:

$$y\left(x_j + h\right) = \tilde{\tilde{y}}_{j+1} + 2\kappa\left(\frac{h}{2}\right)^5 + O(h^6) \tag{8.77}$$

where now there is an extra factor of 2 in the error, because we didn't actually start from $y(x_j + h/2)$ but from $\tilde{\tilde{y}}_{j+1/2}$; we are assuming that κ stays constant from x_j to $x_j + h$, which is likely true for small steps.

If we now equate Eq. (8.75) and Eq. (8.77), we find:

$$\tilde{\tilde{y}}_{j+1} - \tilde{y}_{j+1} = \frac{15}{16}\kappa h^5 \tag{8.78}$$

which clearly shows that the difference between our two estimates, $\tilde{\tilde{y}}_{j+1} - \tilde{y}_{j+1}$, is proportional to h^5. We can now plug this relation back in to Eq. (8.77) to get:

[16] The exact solution to Eq. (8.72) consists of two exponentials with little overlap, so it's not obvious why one should use the same step size everywhere.

$$y\left(x_j + h\right) = \tilde{\tilde{y}}_{j+1} + \frac{1}{15}\left(\tilde{\tilde{y}}_{j+1} - \tilde{y}_{j+1}\right) + O(h^6) \tag{8.79}$$

This quantifies the error in our best estimate of the value $y\left(x_j + h\right)$. Observe that both Eq. (8.74) and Eq. (8.79) involve the number 15, though the origins are not quite the same: in order to derive the first we dealt with global errors which are $O(h^4)$, whereas for the second we used the local error which is $O(h^5)$. Furthermore, the two function values we are subtracting in Eq. (8.79) both correspond to the point $x_j + h$, whereas in Eq. (8.74) we were dealing with estimates of the function at $x_{n-1} = x_{n'-1} = b$.

We don't have to stop here, though: we can take advantage of the fact that the difference of the two estimates scales as h^5, in order to produce a guess for the *next* step size! In other words, for a given h we get a $\tilde{\tilde{y}}_{j+1} - \tilde{y}_{j+1}$ that is proportional to h^5 as per Eq. (8.78). We can reverse the argument, to ask: which $\bar{h}$ should we use if we want the estimate-difference to be equal to an absolute tolerance, Δ? In equation form:

$$\frac{|\tilde{\tilde{y}}_{j+1} - \tilde{y}_{j+1}|}{\Delta} = \left(\frac{h}{\bar{h}}\right)^5 \tag{8.80}$$

This relationship can be solved for $\bar{h}$; we choose to be conservative and multiply the result of this equation by a *safety factor*, $\alpha < 1$:

$$\bar{h} = \alpha h \left|\frac{\Delta}{\tilde{\tilde{y}}_{j+1} - \tilde{y}_{j+1}}\right|^{0.2} \tag{8.81}$$

The safety factor ensures that we make the estimate even smaller, just in case; typical values used are 0.8 or 0.9. As a secondary precaution, one can choose to employ Eq. (8.81) only if it's telling us to increase the step size by a factor of, say, up to 5: this ensures that we don't have huge jumps in the magnitude of the step size.

Let's interpret what's going on here. In the right-hand side of Eq. (8.79) we've seen a way to quantify the error in $\tilde{\tilde{y}}_{j+1}$. Then, Eq. (8.81) tells us how to use the scaling of Eq. (8.78) in order to guess the next step size: (a) if $\tilde{\tilde{y}}_{j+1} - \tilde{y}_{j+1}$ is larger in magnitude than the desired tolerance Δ, then this equation tells us how to pick a new step size before re-trying the current step (from x_j to $x_j + h$); the step size will be *decreasing*, to allow us to have better chances at succeeding this time around, and (b) if $\tilde{\tilde{y}}_{j+1} - \tilde{y}_{j+1}$ is smaller in magnitude than the desired tolerance Δ, then that means that we've already met the tolerance requirement for the present step and can now turn to the next step (from $x_j + h$ to $x_j + 2h$); in that scenario, Eq. (8.81) tells us by how much we can *increase* the step size for that next step (except when α reduces it slightly). The only thing we've left out is which h to pick for the very first step: the short answer is that it doesn't really matter; if our first guess is too large, the algorithm will keep applying Eq. (8.81) until the tolerance test is successful.

Problem 8.12 asks you to implement this local adaptive scheme using RK4 for the usual non-linear Riccati equation. The results are shown in the left panel of Fig. 8.6: note that the two sets of results (global and local) lie on the same curve, but we have artificially shifted one set in order to help you distinguish it from the other. We have chosen the tolerances (global and local, respectively) such that both approaches give rise to roughly 50 points in total. The most prominent feature is that the circles are *not* equally spaced in x: as advertised, they have been placed according to the behavior of the function we are solving for. As a consequence of this, we see that the large spacing (in y) that was exhibited by the global method near a and b is *not* a feature of the local method: our algorithm has figured out that the step size needs to be small in those regions and therefore placed more points there. You will also notice a larger density of points near the middle of the plot, but more diluteness on either side of the middle. Once again, this is the algorithm itself determining whether it needs a large or small step size, according to the solution's trend.[17] To drive this point home, the right panel of Fig. 8.6 shows the step size h corresponding to each point in x:[18] you can immediately see that this is fully consistent with the statements we made in connection with the left panel. Crucially, for this example, we see a variation of more than a factor of 4: the local adaptive method dedicates fewer points and therefore spends less time in regions where nothing exciting is taking place.

Before concluding this section, we note that we have discussed only the simplest possible approaches to (global and local) adaptive integration; both of these employed RK4, which has been the best method we've seen so far. In problem 8.13 you will encounter the higher-order method RK5; it is possible to use a combination of RK4 and RK5 as a way of quantifying the local error, instead of the single-step and double-step calculations employed above.

8.2.4 Simultaneous Differential Equations

We've spent quite a few pages up to this point discussing initial-value problems, all of which have had the form of Eq. (8.8), namely they have been first-order ordinary differential equations. The motivation behind this is that everything we've learned so far can also be applied to systems of simultaneous first-order differential equations (with many dependent variables) with minimal modification. As a matter of fact, even ODEs of higher order can be cast as such systems; as a result, the techniques we've introduced (such as fourth-order Runge–Kutta) can be very straightforwardly generalized to solve much more complicated initial-value problems. Let us take some time to unpack these statements.

8.2.4.1 Two Equations

In physical problems, it is quite common that one has to solve two ordinary differential equations together: take the independent variable to be x (as before) and the two dependent variables to be $y_0(x)$ and $y_1(x)$; note that there is nothing discretized going on here: $y_0(x)$ and $y_1(x)$ are functions of x which we must solve for. An example of such coupled ODEs is given by the TOV problem which we discussed at the start of this chapter. In the general case, the two ODEs we need to solve are:

[17] "Division is made only to help bring out the meaning of the thing divided" (Dante, *Vita Nuova*, chapter 14).

[18] This is always the *successful* step size, i.e., the one that was able to meet our tolerance requirement.

$$
\begin{aligned}
y_0'(x) &= f_0\left(x, y_0(x), y_1(x)\right) \\
y_1'(x) &= f_1\left(x, y_0(x), y_1(x)\right) \\
y_0(a) &= c_0, \qquad y_1(a) = c_1
\end{aligned}
\tag{8.82}
$$

Note that since we have two dependent variables, $y_0(x)$ and $y_1(x)$, we also have two equations, two right-hand sides, f_0 and f_1, and two known initial-values, $y_0(a)$ and $y_1(a)$.

Crucially, the prototypical problem of a (single) second-order IVP, given in Eq. (8.10):[19]

$$
w'' = f(x, w, w'), \qquad w(a) = c, \qquad w'(a) = d \tag{8.83}
$$

can be re-cast as a set of two simultaneous first-order ODEs. To see that, examine the following two definitions:

$$
\begin{aligned}
y_0(x) &= w(x) \\
y_1(x) &= w'(x) = y_0'(x)
\end{aligned}
\tag{8.84}
$$

Our first new function is equal to our starting $w(x)$ and the second new function is equal to $w(x)$'s derivative. In the last step of the second relation we used the first relation. We can now combine these two equations together with Eq. (8.83) to get:

$$
\begin{aligned}
y_0'(x) &= y_1(x) \\
y_1'(x) &= f\left(x, y_0(x), y_1(x)\right) \\
y_0(a) &= c, \qquad y_1(a) = d
\end{aligned}
\tag{8.85}
$$

The first equation is simply the second relation in Eq. (8.84), written with the derivative on the left-hand side. The second relation is Eq. (8.83) expressed in terms of $y_0(x)$ and $y_1(x)$. Similarly, we've translated the initial values into our new language. All in all, Eq. (8.85) is precisely of the form of Eq. (8.82): observe that f_0 is in this case particularly simple, whereas f_1 is the f contained in the second-order ODE. Thus, we have validated our earlier claim that second-order equations can be seen as simultaneous first-order ones.

8.2.4.2 General Case

Everything we discussed in the previous subsection can be straightforwardly generalized to the case of ν coupled equations (or that of a single ODE of ν-th order). Specifically, Eq. (8.82) takes the form:

$$
\begin{aligned}
y_i'(x) &= f_i\left(x, y_0(x), y_1(x), \ldots, y_{\nu-1}(x)\right) \\
y_i(a) &= c_i, \qquad i = 0, 1, \ldots, \nu - 1
\end{aligned}
\tag{8.86}
$$

This is just begging to be converted into vector notation:

[19] We are calling the dependent variable $w(x)$ for clarity, but we could just as well have called it $y(x)$.

$$\begin{aligned}\mathbf{y}'(x) &= \mathbf{f}\,(x, \mathbf{y}(x))\\ \mathbf{y}(a) &= \mathbf{c}\end{aligned} \tag{8.87}$$

where $\mathbf{y}(x)$ bundles together ν functions, $\mathbf{f}$ stands for ν right-hand sides, and $\mathbf{c}$ is a vector of ν numbers (the initial values). Keep in mind that x, a, and b are *not* vectors but plain scalars. Since a single ODE of ν-th order is just a special case of Eq. (8.87), you should assume that whatever we say below applies to that case, also.

Armed with our vector notation, we are now in a position to start discretizing. Instead of rederiving all the methods for IVPs in ODEs that we introduced in earlier subsections, we will merely observe that they all carry over if you make the obviously necessary modifications. For example, the backward Euler method of Eq. (8.40) now takes the form:

$$\begin{aligned}\mathbf{y}_{j+1} = \mathbf{y}_j + h\,\mathbf{f}(x_{j+1}, \mathbf{y}_{j+1}), \qquad j = 0, 1, \ldots, n-2\\ \mathbf{y}_0 = \mathbf{c}\end{aligned} \tag{8.88}$$

Take a moment to realize that we are no longer dealing with exact solutions (which are functions of x) but with discretizations (which are approximations that exist only at specific grid points). Also, note that $\mathbf{y}_j$ is the approximate value of *all* ν functions at x_j. If you need to spell things out, you will have to employ some extra notation here: for example, $\mathbf{y}_0$ groups together all the function values at the initial point; that means that it contains the components $(y_0)_0$, $(y_1)_0$, and so on. Crucially, n controls how many grid points we use from a to b, whereas ν tells us how many simultaneous equations we need to solve.

Similarly, the forward Euler method of Eq. (8.20) now becomes:

$$\begin{aligned}\mathbf{y}_{j+1} = \mathbf{y}_j + h\,\mathbf{f}(x_j, \mathbf{y}_j), \qquad j = 0, 1, \ldots, n-2\\ \mathbf{y}_0 = \mathbf{c}\end{aligned} \tag{8.89}$$

Our workhorse, the fourth-order Runge–Kutta prescription of Eq. (8.64), turns into:

$$\begin{aligned}\mathbf{k}_0 &= h\,\mathbf{f}(x_j, \mathbf{y}_j)\\ \mathbf{k}_1 &= h\,\mathbf{f}\left(x_j + \frac{h}{2}, \mathbf{y}_j + \frac{\mathbf{k}_0}{2}\right)\\ \mathbf{k}_2 &= h\,\mathbf{f}\left(x_j + \frac{h}{2}, \mathbf{y}_j + \frac{\mathbf{k}_1}{2}\right)\\ \mathbf{k}_3 &= h\,\mathbf{f}\left(x_j + h, \mathbf{y}_j + \mathbf{k}_2\right)\\ \mathbf{y}_{j+1} &= \mathbf{y}_j + \frac{1}{6}(\mathbf{k}_0 + 2\mathbf{k}_1 + 2\mathbf{k}_2 + \mathbf{k}_3)\end{aligned} \tag{8.90}$$

Marvel at how straightforward our generalization is. It should come as no surprise that NumPy's array functionality is perfectly matched to such vector manipulations, a topic we turn to next.

ivp_two.py

Code 8.2

```
import numpy as np

def fs(x,yvals):
    y0, y1 = yvals
    f0 = y1
    f1 = - (30/(1-x**2))*y0 + ((2*x)/(1-x**2))*y1
    return np.array([f0, f1])

def rk4_gen(fs,a,b,n,yinits):
    h = (b-a)/(n-1)
    xs = a + np.arange(n)*h
    ys = np.zeros((n, yinits.size))

    yvals = np.copy(yinits)
    for j,x in enumerate(xs):
        ys[j,:] = yvals
        k0 = h*fs(x, yvals)
        k1 = h*fs(x+h/2, yvals+k0/2)
        k2 = h*fs(x+h/2, yvals+k1/2)
        k3 = h*fs(x+h, yvals+k2)
        yvals += (k0 + 2*k1 + 2*k2 + k3)/6
    return xs, ys

if __name__ == '__main__':
    a, b, n = 0.05, 0.49, 12
    yinits = np.array([0.0926587109375, 1.80962109375])
    xs, ys = rk4_gen(fs,a,b,n,yinits)
    print(ys)
```

8.2.4.3 Implementation

We start this subsection with a confession: our earlier implementation example—the Riccati equation of Eq. (8.71)—was chosen because it can be repurposed to exemplify many distinct questions. We already saw how to solve this non-linear first-order initial-value problem in code 8.1. We now see that that equation can be recast if one makes the following *Riccati transformation*: $y(x) = w'(x)/w(x)$, where $w(x)$ is a new function. Using this transformation, Eq. (8.71) turns into:

$$w''(x) = -\frac{30}{1-x^2}w(x) + \frac{2x}{1-x^2}w'(x)$$
$$w(0.05) = 0.0926587109375, \qquad w'(0.05) = 1.80962109375 \tag{8.91}$$

This is now a *second*-order ODE. Crucially the $y^2(x)$ has dropped out of the problem, meaning that this is now a *linear* ODE! Since it's of second order, we need to supply two initial values: $w(a)$ and $w'(a)$. It's time for a second confession: this is not just any old second-order linear ODE. It's the *Legendre differential equation*, namely the differential equation whose solution gives us Legendre polynomials; given the presence of the number 30 on the right-hand side (which is the result of 6×5), we expect our solution to be $P_5(x)$: this is how we have produced the initial values at $a = 0.05$, namely by using code 2.8, i.e., `legendre.py`.[20]

We can now rewrite our problem in Eq. (8.91) in the form of two coupled first-order ODEs, as per Eq. (8.85):

$$y_0'(x) = y_1(x)$$
$$y_1'(x) = -\frac{30}{1-x^2}y_0(x) + \frac{2x}{1-x^2}y_1(x) \tag{8.92}$$
$$y_0(0.05) = 0.0926587109375, \qquad y_1(0.05) = 1.80962109375$$

This is now in sufficiently general form to allow us to solve it via minimal modifications of our earlier code.

Code 8.2 shows a Python implementation that solves this IVP problem of two simultaneous ODEs. Specifically, the function `fs` expects as input the x_j and a one-dimensional numpy array of two values, representing $\mathbf{y}_j$ at a given x_j. It evaluates the two right-hand sides in Eq. (8.92) and then returns its own array of two values. You may be thinking that this is too formal for a simple problem involving only two ODEs. The reason we've done things this way is because the other function in this program, `rk4_gen()`, is completely general: it is an implementation of Eq. (8.90) that works for the ν-dimensional problem just as well as for our two-dimensional problem here; as mentioned after that equation, NumPy functionality is very convenient, in that the code closely follows the vector relations. The specialization to the case of two equations is carried out only in the input parameter `yinits`, which bundles together the initial values, i.e., $\mathbf{y}_0$. This is completely analogous to what we did in code 5.5, i.e., `multi_newton.py`, when implementing Newton's multidimensional root-finding method.[21] We employ an $n \times 2$ array to store all the $\mathbf{y}_j$'s together; we move to a new row each time we step on to the next x_j.

The main program sets up the integration interval and the number of points, as well as the initial values, calls our numerical integrator, and prints out the 12×2 `ys` array, suppressed here for brevity; if you're solely interested in the w_j's, you can look at the 0th column only. We're not plotting the result here because you already know what it looks like: it's

[20] Almost every chapter in this book has had some relation to Legendre polynomials!

[21] Note also that our implementation is a linear combination of the two previous subsections: `rk4_gen()` applies to the case of ν equations, but `fs()` to the case of two equations.

simply $P_5(x)$ from Fig. 2.6; as a matter of fact, the 1st column can also be benchmarked by comparing to $P_5'(x)$ in the right panel of that figure.

8.2.4.4 Stability and Stiffness for Simultaneous Equations

Having introduced and implemented the more general problem of ν simultaneous first-order ODEs, we now take some time to investigate the concepts of stability and stiffness. You may recall that stability was introduced in the context of Euler's method and stiffness in the context of RK4, but these are much broader concepts: stability has to do with the method one is employing (Euler, RK4, etc.) while stiffness has to do with properties of the problem, which then impacts which methods are stable (and which aren't).[22] We will now examine these two concepts in turn for the general problem of ν simultaneous ODEs.

First, we wish to discuss the idea of *stability* of a given method for the problem of ν simultaneous equations. We introduced stability (of a given method) by studying the test equation, Eq. (8.33). The most obvious way of generalizing this to ν equations is:

$$\mathbf{y}'(x) = \mathbf{A}\mathbf{y}(x), \qquad \mathbf{y}(0) = \mathbf{c} \tag{8.93}$$

where, for us, $\mathbf{A}$ is a $\nu\times\nu$ matrix made up of real numbers. This is the minimal modification to the test equation: it contains a first-order derivative on the left-hand side and the function itself (i.e., the functions themselves) on the right-hand side. Obviously, these are *coupled* ODEs: the first equation contains $y_0'(x)$ on the left-hand side but, as a result of $\mathbf{A}$, involves all the $y_0(x), y_1(x), \ldots, y_{\nu-1}(x)$. Note that this is a *linear* system of ODEs.

We would have preferred it if we had been faced with the problem where the ODE for $y_0'(x)$ only involves $y_0(x)$, the one for $y_1'(x)$ only $y_1(x)$, and so on. In short, we would like to *diagonalize* this problem. We immediately remember Eq. (4.125), which led to the eigendecomposition of a matrix:[23] $\mathbf{A} = \mathbf{V}\mathbf{\Lambda}\mathbf{V}^{-1}$; here $\mathbf{\Lambda}$ is the diagonal "eigenvalue matrix" made up of the eigenvalues λ_i and $\mathbf{V}$ is the "eigenvector matrix", whose columns are the eigenvectors $\mathbf{v}_i$ (have a look at section 4.4 if you need a refresher). If we express $\mathbf{A}$ in this way and then multiply by $\mathbf{V}^{-1}$ on the left, our equation becomes:

$$\mathbf{V}^{-1}\mathbf{y}'(x) = \mathbf{\Lambda}\mathbf{V}^{-1}\mathbf{y}(x), \qquad \mathbf{V}^{-1}\mathbf{y}(0) = \mathbf{V}^{-1}\mathbf{c} \tag{8.94}$$

If we now introduce a new (vector) variable:[24]

$$\mathbf{z}(x) = \mathbf{V}^{-1}\mathbf{y}(x) \tag{8.95}$$

our equation takes the form:

$$\mathbf{z}'(x) = \mathbf{\Lambda}\mathbf{z}(x), \qquad \mathbf{z}(0) = \mathbf{V}^{-1}\mathbf{c} \tag{8.96}$$

[22] A method may be assessed in terms of: (a) *consistency*, i.e., whether it addresses the true problem as $h \to 0$, (b) *stability*, i.e., whether a small change in the input causes a small change in the output, and (c) *convergence*, i.e., whether it can eventually approach the true solution. According to the *Lax equivalence theorem*, a consistent and stable method is convergent. (The converse is obvious: a convergent method is consistent and stable.)

[23] Incidentally, the present stability analysis requires the ability to diagonalize non-symmetric matrices.

[24] This entire argument is very similar to the derivation around Eq. (4.164); see also Eq. (4.206).

or, in index notation:

$$z_i'(x) = \lambda_i z_i(x), \qquad z_i(0) = \left(\mathbf{V}^{-1}\mathbf{c}\right)_i, \quad i = 0, 1, \ldots, \nu - 1 \tag{8.97}$$

We have accomplished what we set out to: $z_i'(x) = \lambda_i z_i(x)$ involves only a single function, so it is of the form of the original test equation, Eq. (8.33). This means that each of these equations can be immediately solved to give the analytical solution $z_i(x) = e^{\lambda_i x} z_i(0)$. We will deal only with real eigenvalues; thus, we see that if all the eigenvalues are *negative*, then all the $z_i(x)$'s will be decaying functions of x (i.e., $z_i(x)$ goes to 0 as x goes to infinity). From Eq. (8.95) we see that $\mathbf{y}(x) = \mathbf{V}\mathbf{z}(x)$, meaning that all the $y_i(x)$'s will also decay away.

Note that up to this point in this subsection we haven't actually studied any specific method, i.e., we were dealing with a system of ODEs, not with a discretization scheme. Let us now investigate a specific technique, namely the forward Euler method of Eq. (8.89); for the system in Eq. (8.93) this gives:

$$\mathbf{y}_{j+1} = \mathbf{y}_j + h\,\mathbf{A}\mathbf{y}_j \tag{8.98}$$

If we now repeatedly apply this equation from $j = 0$ to $j = n - 2$, we get:

$$\mathbf{y}_{n-1} = (\mathcal{I} + h\mathbf{A})^{n-1}\mathbf{y}_0 \tag{8.99}$$

In order to get a decaying solution as $n \to \infty$, we need to impose as a condition that the eigenvalues of the matrix $\mathcal{I} + h\mathbf{A}$ are all strictly less than one in magnitude. We learned this fact in our discussion of the power method, section 4.4.1; at the time this was an unwanted side-effect, but here we *want* the solution to decay. If λ_i are the eigenvalues of $\mathbf{A}$, then the eigenvalues of $\mathcal{I} + h\mathbf{A}$ are simply $1 + h\lambda_i$; take a moment to see that this is true.

In other words, we have reached the requirement:

$$|1 + h\lambda_i| < 1 \tag{8.100}$$

for any i. If we now assume, as per Eq. (4.128), that our eigenvalues are sorted, with λ_0 being the dominant eigenvalue, we can formulate the following *stability condition*:

$$h < \frac{2}{|\lambda_0|} \tag{8.101}$$

If this condition is not satisfied, the forward Euler method will lead to a numerically unstable solution. It's worth pausing to appreciate that the eigenvalues of the matrix $\mathbf{A}$ appeared both in our study of the system of ODEs, Eq. (8.94), and in our analysis of the forward Euler method's stability, Eq. (8.101); in the former case, we saw that a decaying solution corresponds to all the eigenvalues being negative, whereas in the latter case it implied a specific constraint on the step size: it needs to be smaller than $2/|\lambda_0|$.

It is now time to discuss a *stiff* system of ODEs. Our earlier analysis tells us that we will need a step size that is very small if the largest eigenvalue is very large. You can see that if the eigenvalues are several orders of magnitude apart, we will need to step through our system with a tiny step determined by the largest eigenvalue, even though the exact solution might not be severely impacted by that term; otherwise, we risk our method becoming unstable, at which point we don't trust its results at all. Here's a specific example:

$$\mathbf{y}'(x) = \begin{pmatrix} 98 & 198 \\ -99 & -199 \end{pmatrix} \mathbf{y}(x), \qquad \mathbf{y}(0) = \begin{pmatrix} 1 \\ 0 \end{pmatrix} \tag{8.102}$$

The exact solution is:

$$y_0(x) = 2e^{-x} - e^{-100x}, \qquad y_1(x) = -e^{-x} + e^{-100x} \tag{8.103}$$

as you can verify by substitution. We immediately see that the exact solution contains two contributions of very different scales: e^{-100x} and e^{-x}. This is consistent with the fact that the two eigenvalues here are -100 and -1. The stability condition in Eq. (8.101) tells us that we need to take $h < 2/100$ if we want the forward Euler method to be stable: this, despite the fact that e^{-100x} is considerably smaller than e^{-x}, so we wouldn't have expected the presence of e^{-100x} to matter for reasonable values of x. This is a clear-cut case of a stiff system of ODEs: the two eigenvalues are orders of magnitude apart, thereby imposing a stringent stability step size, much beyond what we would have needed for accuracy purposes. As in earlier examples, implicit methods do a better job for such stiff problems; you will see this in action in the problem set (see, e.g., problems 8.16 and 8.50).

8.3 Boundary-Value Problems

As mentioned at the start of section 8.2, this chapter is not limited to initial-value problems; even so, the techniques we introduced earlier will also be helpful in what follows. Given this, the present section, tackling boundary-value problems, as well as the following one, on eigenvalue problems, will be shorter; the emphasis will be on implementation (of already known methods, this time for the new problems) and on introducing new implicit methods.

Our prototypical boundary-value problem will be a second-order ODE:

$$w'' = f(x, w, w'), \qquad w(a) = c, \qquad w(b) = d \tag{8.104}$$

where $w(a) = c$ and $w(b) = d$ are the boundary conditions; observe that we do *not* know $w'(a)$, so this is not an initial-value problem. In other words, we can't just start from $x = a$ and then step through the ODE using Euler, RK4, or any of the earlier methods. The f in Eq. (8.104) could be either linear or non-linear; the latter case is harder, as you will discover in problem 8.21. Similarly, the boundary conditions specified could also be more complicated: what we have in Eq. (8.104), where the function's value is provided at the endpoint, $w(b)$, is known as a *Dirichlet boundary condition*. Problem 8.19 examines the alternative case, where the function-derivative's value is given at the endpoint, $w'(b)$ or $w'(a)$, a scenario known as a *Neumann boundary condition*.

Below, we will discuss two general approaches to solving this BVP: (a) treat it as a

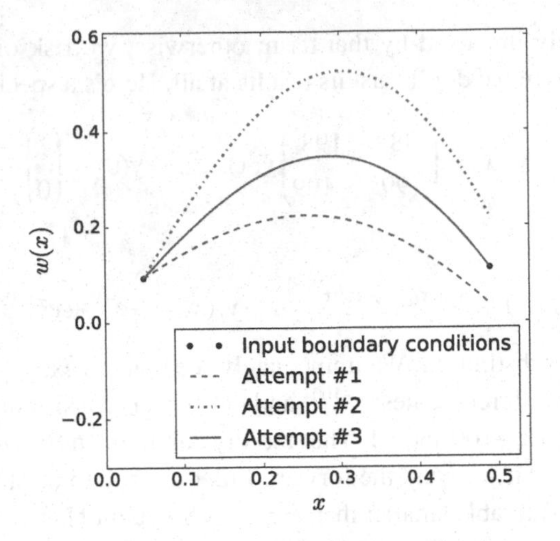

Fig. 8.7 Illustration of the shooting method for a boundary-value problem

more complicated version of an initial-value problem, or (b) face head-on the fact that we know two boundary conditions and set up a method accordingly. We discuss these two approaches in turn.

8.3.1 Shooting Method

We said (correctly) that Eq. (8.104) is not an initial-value problem, because we do not have all the necessary information to start integrating out from a. In other words, we can't blindly apply the material of section 8.2.4 to this problem. The simplest way to approach this problem is: what if we *could* treat this as an initial-value problem? It's true that we don't know $w'(a)$, but why not simply guess it and see if that works? This is illustrated in Fig. 8.7: what we *do* have as input are the values of $w(a)$ and $w(b)$, shown with circles in the figure. We then guess $w'(a)$ and see what happens. Obviously, a random guess is extremely unlikely to be able to satisfy the boundary condition $w(b)$; that's OK, though, because we can try again. In the example of this figure, it's easy to see that the value for $w'(a)$ we first chose was too small; we can simply adjust that to be larger and try again. The second attempt doesn't satisfy the boundary condition $w(b)$ either; we can keep playing this game, though. This time, we reduce the value of $w'(a)$ and try again. Eventually, this process will lead to a value of $w'(a)$ that is "just right", allowing us to go through the point $w(b)$, as desired. (The figure shows the, highly optimistic, scenario where it only took us three tries to get there.) This iterative procedure is obviously analogous to aiming at a target and, as a result, is known as the *shooting method*.

Let us try to recast the above prescription in a more systematic way. Our input is $w(a) = c$ and $w(b) = d$. In the language of our discretization scheme(s), these input conditions can be expressed as $w_0 = c$ and $w_{n-1} = d$. What we do not know is $w'(a)$, but let's assume that its (trial) value is σ, i.e, $w'(a) = \sigma$. A moment's thought will convince you that you can take c and σ, plug them into a numerical integrator, like those in section 8.2.4, and determine the corresponding w_{n-1} value that comes out. In other words, a method like RK4 for this problem transforms the input guess σ into a specific output at the right endpoint, w_{n-1}; mathematically, we can say that our integration method is a function that takes us from σ to w_{n-1}, i.e., $g(\sigma) = w_{n-1}$, where g is a new function describing the effect of solving our ODE with a given σ, from a to b. In this language, since our goal is that $w_{n-1} = d$, we can demand that:

$$g(\sigma) = d \tag{8.105}$$

To reiterate, g is not an analytically known function—like f, which is the right-hand side in Eq. (8.104); instead, it is the result of carrying out a full initial-value problem solution from a to b. Then, what we are faced with in Eq. (8.105) is a *root-finding* problem, like those we spent so much time studying in chapter 5. The function g may be linear or non-linear, but this is irrelevant, since we know already how to solve a general algebraic equation; as a matter of fact, f itself may be linear or non-linear, but this doesn't really change how the shooting method works. To summarize, the shooting method combines initial-value solvers like forward Euler with iterative solvers like Newton's method: for each guess of σ an IVP needs to be solved. This confirms our earlier claim that boundary-value problems are more complicated to solve than initial-value problems.

8.3.1.1 Implementation

A code implementing the shooting method will fuse two elements: an ODE integrator and a root-finder. As already mentioned, this is a general feature, regardless of whether or not the ODE we are faced with is linear. For the sake of concreteness, we will here tackle the Legendre differential equation (again) that, in its latest incarnation, was used to illustrate the solution of two simultaneous differential equations, Eq. (8.91). This time around, instead of providing $w(a)$ and $w'(a)$ like before, our input will be $w(a)$ and $w(b)$:

$$\begin{gathered} w''(x) = -\frac{30}{1-x^2}w(x) + \frac{2x}{1-x^2}w'(x) \\ w(0.05) = 0.0926587109375, \qquad w(0.49) = 0.1117705085875 \end{gathered} \tag{8.106}$$

If you're wondering where we got the value of $w(0.49)$, the answer is that we already know the solution: we know this should be the Legendre polynomial $P_5(x)$, so we can simply use `legendre.py`. Alternatively, we can look at the output of running `ivp_two.py`. Put another way, we already know what $w'(a)$ should be in order to get the $w(0.49)$ shown in Eq. (8.106): we could simply look at the input provided in `ivp_two.py`. Of course, these avenues would not be open to us in general: we would simply be given the ODE and the boundary values and would have to solve for $w'(a)$. In other words, in what follows we ignore all prior knowledge other than what is shown in Eq. (8.106).

Code 8.3 bvp_shoot.py

```
from secant import secant
from ivp_two import fs, rk4_gen
import numpy as np

def shoot(sig):
    a, b, n = 0.05, 0.49, 100
    yinits = np.array([0.0926587109375, sig])
    xs, ys = rk4_gen(fs,a,b,n,yinits)
    wfinal = 0.11177050858750004
    return ys[-1, 0] - wfinal

if __name__ == '__main__':
    wder = secant(shoot,0.,1.)
    print(wder)
```

Code 8.3 shows a Python implementation of the shooting method for our problem. From a practical perspective, we plan to re-use two of our earlier codes: we are faced with a second-order differential equation, so we'll use `rk4_gen()` from `ivp_two.py`; as a matter of fact, since the ODE hasn't changed, we can also re-use `fs()` from the same code. The only thing that will be different this time around is the input: we (pretend to) no longer know both of the elements of `yinits` that `rk4_gen()` expects to receive as input. We will also need a general root-finding function; Newton's method requires information on the derivative, so it's simply easier to reach for the secant method from `secant.py`. Other than importing these three functions, our code doesn't actually do very much: we define a new function, `shoot()`, which plays the role of $g(\sigma) - d$ in Eq. (8.105). It does this by carrying out an RK4 solution for the given σ argument and then subtracting the desired $w(b)$ from the value of w_{n-1} it has produced. Make sure you understand why we wrote `ys[-1, 0]`: the first index corresponds to the discretization of the x_j's and the second index to the equation we are solving, see Eq. (8.85).

The main program simply calls `secant()`, giving it `shoot()` as the function whose root we're after. As you may recall, the secant method requires two initial guesses to get going, so we arbitrarily provide it with two numbers; as you will discover when you run this code, these don't really matter: in any case, this code solves $g(\sigma) = d$ very quickly, giving us the $w'(a)$ that respects our desired boundary condition $w(b)$.

8.3.2 Matrix Approach

Let's do a brief recap: the shooting method approaches the second-order ODE boundary-value problem of Eq. (8.104) by casting it as a set of two simultaneous ODEs, guessing one initial value, and then combining an IVP solver (for simultaneous first-order ODEs) with a root-finder (for algebraic equations in one variable). We now turn to an alternative approach, which tackles our second-order ODE directly: to see what this means, remember that we started our discussion of IVP methods in section 8.2 by discussing the forward and backward Euler methods; these were motivated by finite-difference approximations to the *first* derivative, since that's what was involved in the ODE we were faced with at the time. Here, we will follow a similar route: use a finite-difference approximation for both the *second* and the *first* derivative involved in Eq. (8.104).

To be explicit, we first repeat our problem from Eq. (8.104):

$$w'' = f(x, w, w'), \qquad w(a) = c, \qquad w(b) = d \tag{8.107}$$

We now reach for the central-difference approximation to the first and second derivatives, for points on a grid. That is precisely what we studied in section 3.3.6, so all we need to do is to apply Eq. (3.38) and Eq. (3.39), thereby transforming our problem into:

$$\begin{gathered}\frac{w_{j-1} - 2w_j + w_{j+1}}{h^2} = f\left(x_j, w_j, \frac{w_{j+1} - w_{j-1}}{2h}\right), \qquad j = 1, 2, \ldots, n-2 \\ w_0 = c, \qquad w_{n-1} = d\end{gathered} \tag{8.108}$$

Note that this set of equations involves only the w_j's, i.e., there are no derivatives left. Posing and solving these equations is a process that is completely different from what we were doing for IVPs above: we *cannot* start at the left endpoint and work our way to the right one; instead, we need to solve for *all* of these w_j values at the same time. Another way of putting the same fact: w_{j-1}, w_j, and w_{j+1} appear on both the left-hand side and the right-hand side, bringing to mind the implicit methods we encountered earlier (only this time it's not just y_{j+1} on both sides). These features will become clearer in the following subsection, where we discuss a specific example.

The approach given in Eq. (8.108) is known by many names in the literature, e.g., *finite-difference method* or *equilibrium method*. In order to avoid any confusion, and so as to emphasize its difference from the shooting method, we here keep things general and call it the *matrix approach* to BVPs. Observe that we haven't said much about f so far: if this is linear (as in our example below), then Eq. (8.108) gives rise to a linear system of equations (with a *single* solution), of the type that we spent much of chapter 4 solving (section 4.3); if, on the other hand, f is non-linear, then we would be faced with a system of non-linear equations which, as we learned in section 5.4, is a formidable task, especially if you don't have a decent initial guess for the (entire) solution, as is usually the case. Since we need to solve for all the w_j's simultaneously, you can see that taking, say, n = 100 for the non-linear case is already a non-trivial task. It might help to recall that the shooting method employed a root-finder regardless of the linearity of f; however, that was a root-finding problem in a single variable and therefore reasonably straightforward. If you're wondering why you would need to use such a large n with the matrix approach, recall the truncation error involved in our finite-difference approximations for the first and second derivatives, Eq. (3.38) and Eq. (3.39); in both cases, the error was $O(h^2)$, which is considerably worse than the $O(h^5)$ exhibited by RK4 (see also problem 8.20). On the other hand, the matrix approach satisfies the boundary condition *exactly*, which reflects its better *stability* properties (barring the extreme case where the matrix involved is singular).

8.3.2.1 Matrix Approach Applied to Our Equation

This might all be a tad too abstract for you (e.g., what *matrix* are we referring to?); to shed some light, let us apply Eq. (8.108) to a specific example. To keep things simple, we will study the BVP of the (linear) Legendre differential equation, Eq. (8.106), again. For this case, the right-hand side of Eq. (8.107) takes the form:

$$f(x, w, w') = -\frac{30}{1-x^2}w(x) + \frac{2x}{1-x^2}w'(x) \tag{8.109}$$

We can now plug this equation into Eq. (8.108) to find:

$$\begin{gathered} w_{j-1} - 2w_j + w_{j+1} = -h^2\frac{30}{1-x_j^2}w_j + \frac{hx_j}{1-x_j^2}(w_{j+1} - w_{j-1}), \qquad j = 1, 2, \dots, n-2 \\ w_0 = 0.0926587109375, \qquad w_{n-1} = 0.1117705085875 \end{gathered} \tag{8.110}$$

The equation on the first line involves w_{j-1}, w_j, and w_{j+1} on both the left-hand side and the right-hand side. If we group terms appropriately, we can re-express this problem as:

$$\begin{aligned} w_0 &= 0.0926587109375 \\ \alpha_j w_{j-1} + \beta_j w_j + \gamma_j w_{j+1} &= 0, \qquad j = 1, 2, \dots, n-2 \\ w_{n-1} &= 0.1117705085875 \end{aligned} \tag{8.111}$$

where we've defined:

$$\alpha_j = 1 + \frac{hx_j}{1-x_j^2}, \quad \beta_j = -2 + \frac{30h^2}{1-x_j^2}, \quad \gamma_j = 1 - \frac{hx_j}{1-x_j^2} \tag{8.112}$$

in order to make the notation less cumbersome. Now, Eq. (8.111) can be straightforwardly re-expressed in matrix format:

$$\begin{pmatrix} 1 & 0 & 0 & \dots & 0 & 0 & 0 \\ \alpha_1 & \beta_1 & \gamma_1 & \dots & 0 & 0 & 0 \\ 0 & \alpha_2 & \beta_2 & \dots & 0 & 0 & 0 \\ \vdots & \vdots & \vdots & \ddots & \vdots & \vdots & \vdots \\ 0 & 0 & 0 & \dots & \beta_{n-3} & \gamma_{n-3} & 0 \\ 0 & 0 & 0 & \dots & \alpha_{n-2} & \beta_{n-2} & \gamma_{n-2} \\ 0 & 0 & 0 & \dots & 0 & 0 & 1 \end{pmatrix} \begin{pmatrix} w_0 \\ w_1 \\ w_2 \\ \vdots \\ w_{n-3} \\ w_{n-2} \\ w_{n-1} \end{pmatrix} = \begin{pmatrix} 0.0926587109375 \\ 0 \\ 0 \\ \vdots \\ 0 \\ 0 \\ 0.1117705085875 \end{pmatrix} \tag{8.113}$$

where you can, finally, see why this is called the *matrix* approach. Crucially, Eq. (8.113) is of the form $\mathbf{Ax} = \mathbf{b}$: the n quantities w_j are unknown and everything else in this equation is already known. As noted above, you can't simply step through the elements from left to right (or from right to left). You might benefit from writing out the $n = 6$ case explicitly: w_0 is immediately known; the next row gives an equation relating w_1 to w_2, neither of which

Code 8.4

bvp_matrix.py

```
from gauelim_pivot import gauelim_pivot
import numpy as np

def matsetup(a,b,n):
    h = (b-a)/(n-1)
    xs = a + np.arange(n)*h

    A = np.zeros((n,n))
    np.fill_diagonal(A, -2 + 30*h**2/(1-xs**2))
    A[0,0] = 1; A[-1,-1] = 1
    np.fill_diagonal(A[1:,:], 1 + h*xs[1:]/(1-xs[1:]**2))
    A[-1,-2] = 0
    np.fill_diagonal(A[:,1:], 1 - h*xs/(1-xs**2))
    A[0,1] = 0

    bs = np.zeros(n)
    bs[0] = 0.0926587109375
    bs[-1] = 0.11177050858750004
    return A, bs

def riccati(a,b,n):
    A, bs = matsetup(a, b, n)
    ws = gauelim_pivot(A, bs)
    return ws

if __name__ == '__main__':
    a, b, n = 0.05, 0.49, 400
    ws = riccati(a, b, n); print(ws)
```

is known; the row after that gives an equation involving w_1, w_2, and w_3, but we still have more unknowns than equations. However, if you write out *all* the corresponding equations then you *can* solve them together; after all, this is a linear system of equations. Here we merely wanted to emphasize that the matrix approach to the BVP is conceptually different from how we used finite differences in section 8.2 to produce a given function value from the previous one (remember: we were studying *single-step* techniques).

Our matrix equation, Eq. (8.113), can be solved via, say, Gaussian elimination. Of course, this is a *tridiagonal* coefficient matrix, so applying a general-purpose method like Gaussian elimination is quite wasteful; this is why we introduced the Jacobi iterative method, the Gauss–Seidel method, or an LU method which was tailored to tridiagonal matrices in the main text and the problem set of chapter 4. Here we are not interested in the details of the implementation or its efficiency, but in the big-picture concepts differentiating the shooting method from the matrix approach to BVPs.

8.3.2.2 Implementation

Code 8.4 is a Python implementation of Eq. (8.113). Most of the code is inside a single function, `matsetup()`, which sets up the coefficient matrix and the constant vector. Just like in earlier codes, this is accomplished by slicing the arrays we pass as the first argument to `numpy.fill_diagonal()`. We manually set some special values as needed. The only other function in this code, `riccati()`, calls `matsetup()` and then our earlier Gaussian-elimination-with-pivoting function from chapter 4. This is all quite straightforward. The main program sets up the problem and then calls `riccati()`. The (standard) notation here is potentially confusing: a and b are the (scalar) endpoints, while $\mathbf{A}$ and $\mathbf{b}$ (a matrix and a vector, respectively) define the system of Eq. (8.113). In other words, `a` is not `A` and, similarly, `b` is not `bs`.

Keep in mind that this time around `ws` is a 1d array (as opposed to `ys` in `bvp_shoot.py`, which was a 2d array). This is as it should be, since Eq. (8.113) was set up by discretizing Eq. (8.107) directly, i.e., without having to introduce a new function corresponding to the derivative values. We don't display the output, since it closely matches what we encountered earlier: this is the same Legendre differential equation solution we've been studying for the last few subsections. Of course, a few minor tweaks in how you set up the matrix would allow you to study totally different problems on your own.

8.4 Eigenvalue Problems

We now turn to a specific subclass of boundary-value problems, which is sufficiently distinct that it gets its own section. This is the case where the boundary-value problem's equation involves—in addition to x, $w(x)$, $w'(x)$, and $w''(x)$—a parameter s: this one parameter, known as the *eigenvalue*, will make a world of difference in how this problem is tackled. Our prototypical eigenvalue problem will be a second-order ODE:

$$w'' = f(x, w, w'; s), \qquad w(a) = c, \qquad w(b) = d \tag{8.114}$$

This looks an awful lot like Eq. (8.104); the only difference is the presence of s. This means that we are not faced with a *single* differential equation, but a *family* of ODEs. For given values of s there may be no solution to the corresponding ODE; thus, we will have to compute the "interesting" values of s (i.e., the ones that lead to a solution) in addition to producing an approximation to $w(x)$. The f contained in Eq. (8.114) could be quite complicated, but then the EVP typically becomes intractable. We will here focus on the special case of a linear second-order differential equation such as:

$$w''(x) = \zeta(x)w'(x) + \eta(x)w(x) + \theta(x)\, s\, w(x) \tag{8.115}$$

where the functions $\zeta(x)$, $\eta(x)$, and $\theta(x)$ are known. If we also impose homogeneous boundary conditions, this is (close enough to) a *Sturm–Liouville form.*

We could now proceed to see how to solve this general problem. The obvious candidates are the techniques we introduced in the previous section on BVPs: the shooting method and the matrix approach. Especially in the latter case, the details of the derivation become cumbersome; we opt, instead, to study a specific example of a linear second-order eigenvalue problem, for which we will show all the details explicitly. The problem set guides you to study other physical scenarios, including some drawn from quantum mechanics. Our problem of choice will be the *Mathieu equation*, which appears in the study of string vibrations and a host of other topics in physics:

$$w''(x) = (2q\cos 2x - s)\,w(x), \qquad w(0) = w(2\pi) \tag{8.116}$$

We didn't actually specify the values of the function at the two boundaries; we merely stated that the solution is periodic. You should not be surprised that the solution can be periodic: the equation contains a $\cos 2x$ term, so it is certainly plausible that that "drives" a corresponding periodicity in $w(x)$; we will see below that the solution to this ODE does not *have* to be periodic, in which case it is not a solution to our eigenvalue problem (since it doesn't obey the boundary condition we have imposed).

The s parameter will be very important below: we will need to find the values of s that lead to non-trivial (i.e., non-zero) solutions for $w(x)$. The q parameter determines the strength of the cosine term and will be held fixed (at a finite value) in what follows. Take a moment to appreciate that when $q = 0$ this ODE takes the form $w''(x) + sw(x) = 0$, on which you have spent a good chunk of your undergraduate education. The mere fact that we are now exploring possible values of q and s shows that this problem is qualitatively different from what we were studying earlier in this chapter. Let us see how to solve the Mathieu equation in practice.

8.4.1 Shooting Method

As you may recall, the shooting method amounts to guessing a value of $w'(a)$, solving an IVP problem, and seeing if you managed to hit upon $w(b)$. Obviously, this process can be systematized, as we saw earlier when we employed the secant root-finding method to produce a better guess for $w'(a)$, by finding the root(s) of Eq. (8.105), $g(\sigma) = d$. Unfortunately, this approach doesn't trivially generalize to the problem at hand. To see that, let's look at the Mathieu equation again:

$$w''(x) = (2q\cos 2x - s)\,w(x), \qquad w(0) = w(2\pi) = 0 \tag{8.117}$$

This time around we took $w(0) = w(2\pi) = 0$, i.e., we are searching for odd solutions.[25]

[25] As we'll find out soon enough, the solutions will have a period that is *at most* 2π.

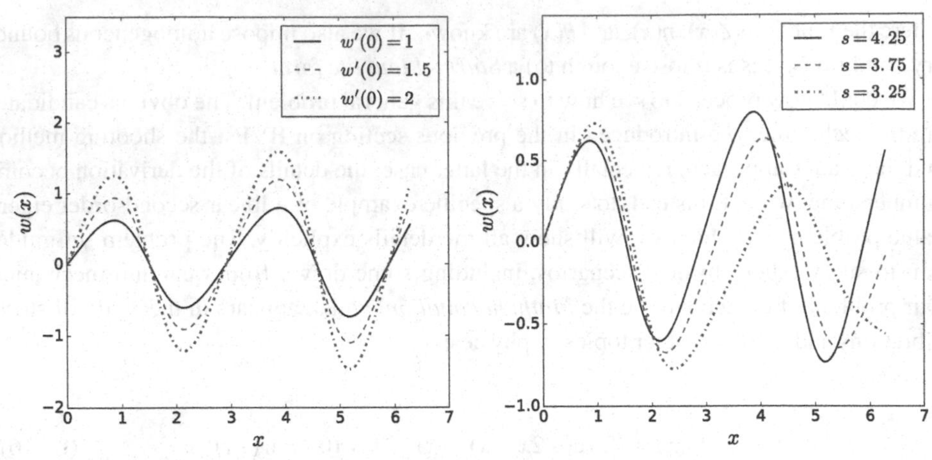

Fig. 8.8 Mathieu solutions for $q = 1.5$: $s = 4.25$ (left) and $w'(0) = 1$ (right)

One problem with applying the shooting method as described above to this equation is that we don't actually know the value of s. But let's assume you do. Even then, you will have trouble using the solution of the IVP to get the derivative at the first endpoint. Let's see why. For this linear ODE, if $w(x)$ is a solution then so is $\alpha w(x)$, where α is an arbitrary constant. Observe that if $w(0) = w(2\pi)$ then $\alpha w(0) = \alpha w(2\pi)$ also holds. But that means that trying out different values of $w'(0)$ won't get us anywhere, since the first derivative of $\alpha w(x)$ at $x = 0$ is $\alpha w'(0)$ and α could be anything. As you may know from a course on quantum mechanics, determining the α in $\alpha w(x)$ is a process known as *normalization*, which has nothing to do with actually satisfying our boundary conditions (since those are satisfied regardless of the value of α). If you haven't studied quantum mechanics yet, there's no need to worry: we are merely stating that you will need an external requirement if you wish to find α.

If you think this is too vague, look at the left panel of Fig. 8.8: we've arbitrarily picked q and s. We see that, for our choice of s, the boundary condition is simply not met (here and below we keep $q = 1.5$ fixed). Then, by changing the value of $w'(0)$ all we're accomplishing is to multiply the entire solution by a constant: if it didn't meet the boundary condition for one initial value $w'(0)$, then no amount of stretching is going to save us. In other words, the value of $w'(0)$ is related to the normalization, but doesn't help us accomplish our goal. The right panel of the same figure shows that it is s which might help us actually satisfy the boundary conditions in Eq. (8.117): this time around we know that the value of $w'(0)$ is irrelevant, so we keep it fixed. We see that as we tune s we get close to satisfying $w(2\pi) = 0$; obviously, we need to come up with a systematic way of doing so.

Our goal can be accomplished by slightly tweaking our earlier argument: instead of using a root-finder to determine $w'(0)$ (a hopeless task), arbitrarily pick $w'(0)$ and then use a root-finder to determine s. Our input boundary conditions $w(0) = w(2\pi) = 0$ can, for our discretization scheme(s), be expressed as $w_0 = w_{n-1} = 0$. As explained, we are free to randomly guess a value of $w'(0) = \sigma$. What we do next is to take our starting w_0 and σ, plug them into a numerical integrator, like those in section 8.2.4, and determine the corresponding w_{n-1} value that comes out. For a randomly picked value of s this approach is fated to fail: for a given s, a method like RK4 transforms the input s into a specific output at the right endpoint, w_{n-1}, without caring about the "desired" value of $w_{n-1} = 0$; mathematically, we can say that our integration method is a function that takes us from s to w_{n-1}, i.e., $g(s) = w_{n-1}$, where g is a new function describing the effect of solving our ODE with a given s, with x going from 0 to 2π. This g is (analogous to, but) different from that appearing in Eq. (8.105): the latest g is a function of the s parameter in our Mathieu equation. In this language, since our goal is that $w_{n-1} = 0$, we can demand that:

$$g(s) = 0 \tag{8.118}$$

As before, g is the result of carrying out a full initial-value problem solution from 0 to 2π. Once again, we can apply a root-finding algorithm to solve this equation; the only difference is that this time we are not searching for a starting derivative value, but for the value of s for which our boundary condition $w(2\pi) = 0$ can be met.

8.4.1.1 Implementation

Implementing the shooting method for eigenvalue problems should be simple enough. After all, we're still combining the same tools as in the BVP case: a numerical integrator for an IVP (like RK4) and a root-finder (like the secant method). Just as in Fig. 8.8, we are setting the cosine strength to $q = 1.5$, but it's totally straightforward to explore other values later. Thinking of how to proceed, our first thought is that code 8.3, i.e., `bvp_shoot.py`, might be a good template. However, things are not so simple. In that program we imported `fs()` and `rk4_gen()` from `ivp_two.py`, which made sense at the time, since we were faced with a well-defined second-order ODE, namely the Legendre differential equation of Eq. (8.106). In the present case, this won't work: even keeping q fixed, we cannot avoid the fact that our differential equation itself depends on the value of the s parameter. As a result, `rk4_gen()` cannot be used, since it assumes the entire differential equation is given (i.e., not parameter dependent).

One could consider using a nested function here, "binding" one parameter to fixed values for selected purposes. Instead, in code 8.5 we follow the quick-and-dirty route, writing new versions of the f and RK4 routines. This still follows the general philosophy of casting a second-order ODE as two first-order ODEs, as per Eq. (8.85), but this time our f and RK4 functions take in an extra parameter corresponding to s. This allows us, in the new version of `shoot()`, to solve for s by determining the value that satisfies Eq. (8.118), i.e., the one that leads to $w_{n-1} = 0$. As advertised, we pass in an arbitrary value of $w'(0)$ which shouldn't (and doesn't) affect anything other than the normalization.

In the main program, we provide a few initial guesses and then print out the corresponding solutions (for the s's). The output of running this code is:

Code 8.5 evp_shoot.py

```
from secant import secant
import numpy as np

def fs(x,yvals,s):
    q = 1.5
    y0, y1 = yvals
    f0 = y1
    f1 = (2*q*np.cos(2*x) - s)*y0
    return np.array([f0, f1])

def rk4_gen_eig(fs,a,b,n,yinits,s):
    h = (b-a)/(n-1)
    xs = a + np.arange(n)*h
    ys = np.zeros((n, yinits.size))

    yvals = np.copy(yinits)
    for j,x in enumerate(xs):
        ys[j,:] = yvals
        k0 = h*fs(x, yvals,s)
        k1 = h*fs(x+h/2, yvals+k0/2,s)
        k2 = h*fs(x+h/2, yvals+k1/2,s)
        k3 = h*fs(x+h, yvals+k2,s)
        yvals += (k0 + 2*k1 + 2*k2 + k3)/6
    return xs, ys

def shoot(s):
    a, b, n = 0, 2*np.pi, 500
    yinits = np.array([0., 5.])

    xs, ys = rk4_gen_eig(fs,a,b,n,yinits,s)
    wfinal = 0.
    return ys[-1, 0] - wfinal

if __name__ == '__main__':
    for sinit in (-0.4, 3.3, 8.5):
        sval = secant(shoot,sinit,sinit+0.5)
        print(sval, end=" ")
```

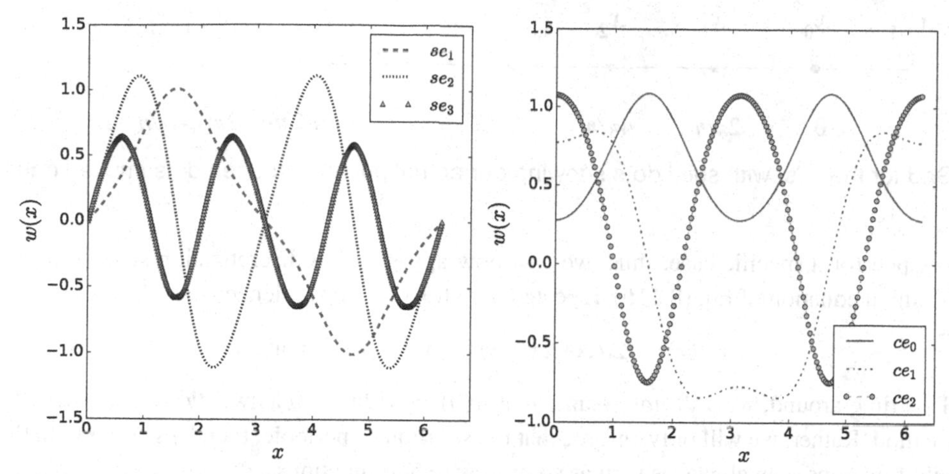

Eigenfunctions for the Mathieu problem: odd (left) and even (right)

Fig. 8.9

```
-0.73326514905   3.81429091649   9.09260876706
```

where we have suppressed the intermediate print-outs from `secant()`. We had implied this before, but now have irrefutable evidence: there is more than one value of s for which we can solve our eigenvalue problem (i.e., there is more than one eigenvalue).

Each eigenvalue (i.e., each value of s for which we can satisfy the boundary conditions non-trivially) corresponds to a solution for $w(x)$; unsurprisingly, these are known as *eigenfunctions*. In the left panel of Fig. 8.9 we are plotting the three eigenfunctions corresponding to the three eigenvalues of the output. We've taken the opportunity to label these eigenfunctions according to their standard names from the theory of Mathieu functions, but this doesn't really matter for our purposes. What *does* matter is that these are periodic functions that look *almost* like sines, but not quite; all three of them are odd functions of x. The difference from a sine is more pronounced in the case of se_1. Note that one of these curves has period of π and the other two a period of 2π; of course, even se_2, which has a period of π, still satisfies the desired boundary condition, $w(0) = w(2\pi) = 0$. This happens to be the solution that the right panel of Fig. 8.8 was moving toward. You should play around with the main program, trying to find the next eigenvalue/eigenfunction pairs.

Intriguingly, we have computed these eigenvalues and eigenfunctions by building on our earlier concepts: no complicated theory needed to be invoked. Even better: our techniques are of much wider applicability than the one problem of the Mathieu equation.

8.4.2 Matrix Approach

We just saw how to tweak the idea of the shooting method developed for the BVP, in order to have it work for the EVP. In the present subsection we'll do something analogous for the matrix approach. In other words, similarly to what we did in Eq. (8.108), we can directly tackle the second-order ODE of Eq. (8.114), replacing the second and first derivatives with the corresponding central-difference approximations. Once again, this is most easily

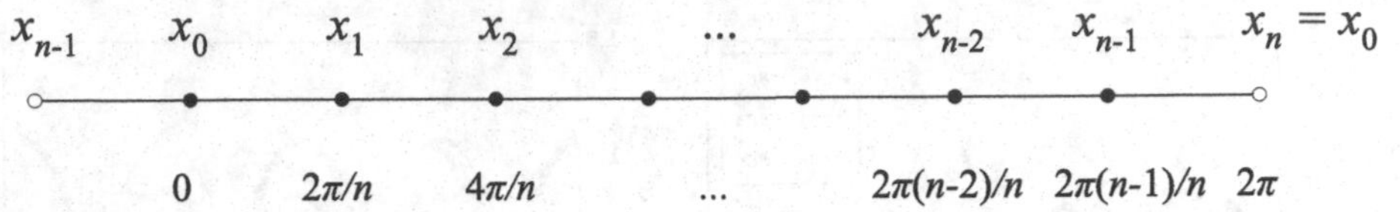

x_{n-1} x_0 x_1 x_2 ... x_{n-2} x_{n-1} $x_n = x_0$

0 $2\pi/n$ $4\pi/n$ ... $2\pi(n-2)/n$ $2\pi(n-1)/n$ 2π

Fig. 8.10 Grid for the x_j's, with solid dots showing our actual points and open dots implied ones

grasped for a specific case; thus, we will now apply such a discretization scheme to the Mathieu equation of Eq. (8.116), repeated here for your convenience:

$$w''(x) = (2q\cos 2x - s)\, w(x), \qquad w(0) = w(2\pi) \tag{8.119}$$

This time around, we will *not* assume that $w(0) = w(2\pi) = 0$ as we did for the shooting method. Rather, we will only enforce that the solution be periodic, thereby getting both the odd functions seen above, as well as some new (even) functions.

We wish to build the periodicity in to our method. To do so, we recall that we had faced a similar situation in chapter 6: specifically, in Eq. (6.83) we set up a grid accordingly:

$$x_j = \frac{2\pi j}{n}, \qquad j = 0, 1, \ldots, n-1 \tag{8.120}$$

Unlike the grid of Eq. (8.15), here we include the left endpoint but not the right endpoint: this is because we know our function is periodic, so we avoid storing needless information. As always, we are still dealing with n points; this is illustrated in Fig. 8.10, which is a repeat of Fig. 6.8. Take a moment to compare to what we did for the shooting method: there we started at the left endpoint, 0, and explicitly integrated all the way up to the right one, 2π, in order to check if our solution has the desired value there. In the present case, we will always satisfy the boundary condition *exactly*. Of course, just as in the case of the BVP, this pleasant feature has to be weighed against the unfortunate detail that the truncation error involved in our finite-difference approximations for the first and second derivatives, Eq. (3.38) and Eq. (3.39), is $O(h^2)$, which is much worse than the $O(h^5)$ exhibited by RK4.

The Mathieu equation involves $w''(x)$ and $w(x)$, but not $w'(x)$: this means we only need to use the central-difference approximation to the *second* derivative, Eq. (3.39). Doing so leads to the following set of equations:

$$\frac{w_{j-1} - 2w_j + w_{j+1}}{h^2} = \left(2q\cos 2x_j - s\right) w_j, \qquad j = 0, 1, \ldots, n-1 \tag{8.121}$$

We can now group terms, to get:

$$w_{j-1} + \alpha_j w_j + w_{j+1} = -h^2 s w_j, \qquad j = 0, 1, \ldots, n-1 \tag{8.122}$$

where we defined:

$$\alpha_j = -2 - 2h^2 q \cos 2x_j \tag{8.123}$$

for ease of reading. Observe that both in Eq. (8.121) and in Eq. (8.122), we made no mention of the boundary terms. If you actually write down Eq. (8.122) for $j = 0$ and

$j = n-1$, you are faced with the pesky terms w_{-1} and w_n which don't actually exist, if you take our grid in Eq. (8.120) seriously. We interpret these in the most natural way possible:

$$w_{-1} \equiv w_{n-1}, \qquad w_n \equiv w_0 \tag{8.124}$$

These definitions are fully consistent with the spirit of Fig. 8.10. With that clarification in mind, we are now in a position to write out the n equations of Eq. (8.122) in matrix form:

$$\begin{pmatrix} \alpha_0 & 1 & 0 & \dots & 0 & 0 & 1 \\ 1 & \alpha_1 & 1 & \dots & 0 & 0 & 0 \\ 0 & 1 & \alpha_2 & \dots & 0 & 0 & 0 \\ \vdots & \vdots & \vdots & \ddots & \vdots & \vdots & \vdots \\ 0 & 0 & 0 & \dots & \alpha_{n-3} & 1 & 0 \\ 0 & 0 & 0 & \dots & 1 & \alpha_{n-2} & 1 \\ 1 & 0 & 0 & \dots & 0 & 1 & \alpha_{n-1} \end{pmatrix} \begin{pmatrix} w_0 \\ w_1 \\ w_2 \\ \vdots \\ w_{n-3} \\ w_{n-2} \\ w_{n-1} \end{pmatrix} = -h^2 s \begin{pmatrix} w_0 \\ w_1 \\ w_2 \\ \vdots \\ w_{n-3} \\ w_{n-2} \\ w_{n-1} \end{pmatrix} \tag{8.125}$$

Unmistakably, this is of the form $\mathbf{Av} = \lambda\mathbf{v}$, eliminating any lingering doubts you may have had about why this is called an *eigenvalue problem*. Seeing the role that s plays in this equation similarly justifies why the s in Eq. (8.116) is called an *eigenvalue*.

Take a moment to compare this equation with our result for the matrix approach as applied to the BVP, Eq. (8.113), which was of the form $\mathbf{Ax} = \mathbf{b}$. There, the right-hand side was known, whereas here the unknown w_j's appear on both the left-hand side and the right-hand side. The $n \times n$ matrix on the left-hand side of Eq. (8.125) is cyclic tridiagonal, similar to the one we ecountered in problem 4.18. You should think about the origin of those two units (at the bottom left and top right). Gratifyingly, our final result allows us to apply the eigenvalue/eigenvector methods we spent much of chapter 4 (section 4.4) developing. As advertised, we are *building in* the periodicity, but not making any assumptions about the value of the solution at the endpoints (i.e., we are not limited to homogeneous boundary conditions). Finally, Eq. (8.125) allows us to produce the *spectrum* of eigenvalues, not just the lowest one (or few); using n points you produce approximations for n eigenvalues. This deserves to be emphasized: in the shooting method you need to zero-in on candidate eigenvalues manually; you may end up getting the same solution more than once, or you might accidentally skip over a given solution (a fact you would try to determine by counting nodes/zeros in the eigenfunctions). In contradistinction to this, the matrix approach gives you estimates for n eigenvalues. In problem 8.35 you will explore if it does equally well for all eigenvalues.

8.4.2.1 Implementation

Code 8.6 is an implementation of the matrix approach to the eigenvalue problem for the Mathieu equation. Just like in code 8.4, most lines are dedicated to a function setting up the coefficient matrix, `matsetup()`. This implements Eq. (8.125), once again using `numpy.fill_diagonal()` and a couple of manual assignments. A separate function, `mathieu()`, calls `matsetup()` and then `qrmet()` from chapter 4 to apply the QR method for

Code 8.6 evp_matrix.py

```
from qrmet import qrmet
import numpy as np

def matsetup(q,n):
    h = 2*np.pi/n
    xs = np.arange(n)*h

    A = np.zeros((n,n))
    np.fill_diagonal(A, -2 - 2*h**2*q*np.cos(2*xs))
    np.fill_diagonal(A[1:,:], 1)
    np.fill_diagonal(A[:,1:], 1)
    A[0,-1] = 1
    A[-1,0] = 1
    return A

def mathieu(q,n):
    A = matsetup(q, n)
    qreigvals = qrmet(A,200)
    h = 2*np.pi/n
    qreigvals = np.sort(-qreigvals/h**2)
    return qreigvals

if __name__ == '__main__':
    q, n = 1.5, 200
    qreigvals = mathieu(q, n)
    print(qreigvals[:6])
```

eigenvalue evaluation. We then divide by $-h^2$ and sort so that we can extract the s's in order. As expected, the quality of the output (for a given finite-difference approximation) gets better as the number of points n increases; of course, given the (lack of) efficiency in our implementation of the QR method, this slows things down, practically speaking. Even so, it's typically worth the wait since, as noted, our output consists of the first n eigenvalues.

The main program sets up the cosine strength, q, and the number of points on our grid, n, and calls `mathieu()`. We limit the output to the first six eigenvalues:

```
[-0.93706036 -0.73350696 2.16553604 3.81267332 4.74538385 9.08581402]
```

Every other number here is (close to) the output of `evp_shoot.py`: we have produced *all* the eigenvalues; half correspond to odd eigenfunctions and half to even. We have plotted the latter in the right panel of Fig. 8.9: two have a period of π and one has a period of 2π; these look somewhat like cosines but, especially in the case of ce_1, they are clearly different. This approach can be used to produce the left panel of Fig. 8.9, as well. Incidentally, Mathieu functions also arise in quantum mechanics (the "quantum pendulum").

8.5 Partial Differential Equations

Having spent the entire chapter up to this point studying *ordinary* differential equations, we now turn to *partial* differential equations; this is a huge subject, deserving of its own book. Here we first provide some general comments about different types of PDEs, then choose a specific one and solve it using elementary techniques.

8.5.1 Examples of PDEs

The classification of PDEs is easier to grasp in the context of a few specific examples:

- A typical example of a *hyperbolic* PDE is the *wave equation* involving $\phi(x, t)$:

$$\frac{\partial^2 \phi}{\partial t^2} - c^2 \frac{\partial^2 \phi}{\partial x^2} = 0 \tag{8.126}$$

  where $c > 0$ and we assumed one spatial dimension. This equation involves a second-order time derivative and a second-order space derivative, which have opposite signs.
- A typical example of a *parabolic* PDE is the *diffusion equation* involving $\phi(x, t)$:

$$\frac{\partial \phi}{\partial t} - \alpha \frac{\partial^2 \phi}{\partial x^2} = 0 \tag{8.127}$$

  where $\alpha > 0$ and we assumed one spatial dimension. This equation involves a first-order time derivative and a second-order space derivative, with opposite signs.
- A typical example of a *dispersive* PDE is the *Schrödinger equation* involving $\Psi(x, t)$:

$$i\hbar \frac{\partial \Psi}{\partial t} = \left[-\frac{\hbar^2}{2m} \frac{\partial^2}{\partial x^2} + V(x, t) \right] \Psi \tag{8.128}$$

  where $m > 0$. Like Eq. (8.127), this involves a first-order time derivative and a second-order space derivative; unlike Eq. (8.127), it contains a pesky i on the left-hand side instead of having opposite signs and the wave function $\Psi(x, t)$ is, in general, *complex*.
- A typical example of an *elliptic* PDE is *Poisson's equation* involving $\phi(x, y)$:

$$\frac{\partial^2 \phi}{\partial x^2} + \frac{\partial^2 \phi}{\partial y^2} = f(x, y) \tag{8.129}$$

  where we assumed two spatial dimensions. Note that this involves two second-order spatial derivatives, which come in with the same sign.

Here's another way of looking at the classification of PDEs: Eq. (8.126), Eq. (8.127), and Eq. (8.128) describe time evolution, which starts at a given point; thus, such dynamical equations correspond to what we called *initial-value problems*. On the other hand,

Eq. (8.129) does not involve time; such a static situation involves specified boundary conditions, thereby giving rise to a *boundary-value problem*. In the interest of being concrete, we will now focus on a couple of the aforementioned scenarios.

8.5.2 A Parabolic PDE: the Diffusion Equation

We first examine a reasonably straightforward PDE problem setup: that of the *diffusion equation* of Eq. (8.127) with simple boundary conditions. We will then discuss how to solve this equation via discretization, emphasizing both the commonalities and the differences from what we encountered in our earlier discussion of ODEs. The problem set elaborates on the techniques we introduce here for the diffusion equation, but also discusses analogous approaches that apply to the wave equation and the time-dependent Schrödinger equation. In the Project concluding this chapter, the only remaining representative PDE (Poisson's equation) is tackled, using a technique that is somewhat different from the approaches introduced up to this point.

8.5.2.1 Problem and Discretization

Before discussing solution techniques, let us see what our problem is. The diffusion equation, which provides a macroscopic description of large numbers of particles undergoing random movement in the context of *Brownian motion*, has the form:

$$\frac{\partial \phi}{\partial t} - \alpha \frac{\partial^2 \phi}{\partial x^2} = 0 \tag{8.130}$$

where α is a diffusion coefficient, here taken to be constant. The same problem arises in the study of heat flow, in which case Eq. (8.130) is known as the *heat equation* and α is the thermal diffusivity of the medium.

Regardless of the physical interpretation, we immediately see that Eq. (8.130) is a *partial* differential equation, since ϕ depends on both space and time, $\phi = \phi(x, t)$—we are assuming a single spatial dimension for simplicity. As noted above, this equation involves a first-order time derivative and a second-order space derivative, with opposite signs. Since this problem involves a time dependence, we see that we are dealing with an initial-value problem: we need to specify an initial condition, i.e., the behavior at $t = 0$, and then try to solve for the behavior at later times. Of course, there is also a space dependence, i.e., we need to provide spatial boundary conditions at all times. Putting it all together, our problem takes the form:

$$\begin{gathered} \frac{\partial \phi(x,t)}{\partial t} = \alpha \frac{\partial^2 \phi(x,t)}{\partial x^2}, \\ \phi(x,0) = f(x), \quad \phi(0,t) = u, \quad \phi(L,t) = v \end{gathered} \tag{8.131}$$

This clearly shows that we are providing (constant) boundary conditions for $\phi(x, t)$, here taken to be at the two positions $x = 0$ and $x = L$ and applied to any time label t; as

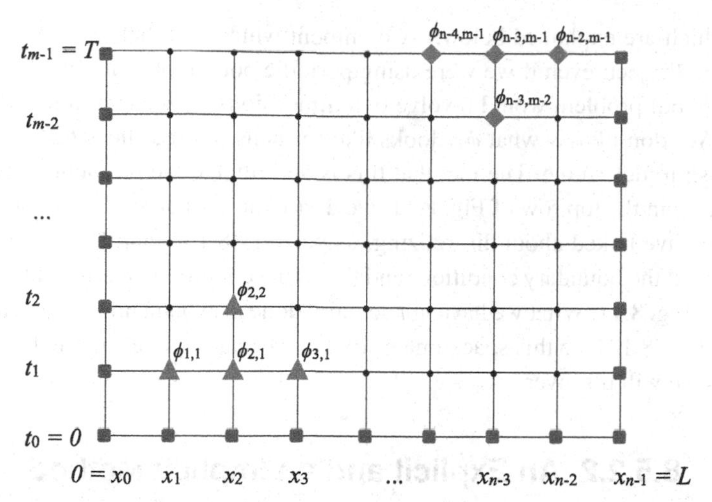

Spacetime grid for diffusion equation with some example discretizations

Fig. 8.11

mentioned on page 601, these are known as *Dirichlet boundary conditions*. It also shows that the initial condition (corresponding to $t = 0$) needs to be a *function* specifying the starting solution over the entire spatial domain.

In the spirit of this chapter (and this book, more widely) we will tackle the diffusion equation by discretizing, i.e., assuming that the positions can only take values on a grid:

$$x_j = j\frac{L}{n-1}, \qquad j = 0, 1, \ldots, n-1 \tag{8.132}$$

This goes from 0 to L, as it should. Note that we are *not* making any assumptions about periodicity: unlike, say, Fig. 8.10, our latest grid is *inclusive* of both endpoints. As you may have expected, we also discretize time:

$$t_k = k\frac{T}{m-1}, \qquad k = 0, 1, \ldots, m-1 \tag{8.133}$$

Again, this goes from 0 to T (some final time), as expected. Observe that we are using L for the largest spatial position and T for the largest time value. Observe also that we are using different indices to keep track of spatial and time discretization, j and k, respectively. Finally, note that we are allowing for the possibility of using a different number of points in x vs t: these are n and m, respectively; this will come in very handy below, when we learn that generally one needs to be more careful with the time evolution.

It may help you to look at Fig. 8.11 when thinking about our problem and its discretization. The small circles correspond to a given space and time value, i.e., x_j and t_k. The squares denote an externally provided solution. For example, the bottom row shows the initial condition using squares: in Eq. (8.131) this is stated as $\phi(x, 0) = f(x)$; in our new discretized language, where $\phi(x, t)$ turns into $\phi_{j,k}$, the bottom row of the figure corresponds to $\phi_{j,0}$. Similarly, the boundary conditions represented by squares on the first and last columns correspond to $\phi_{0,k}$ and $\phi_{n-1,k}$, respectively.[26] (The figure also contains other

[26] Here the (standard) notation is confusing: rows in $\phi_{j,k}$ correspond to columns in the figure (and vice versa).

symbols, which are explained below.) A comment which may help you when you study the following Project: even if we were using periodic boundary conditions—i.e., $u = v$ in Eq. (8.131)—our problem would involve two dimensions (space and time) which are not equivalent. We don't know what $\phi_{j,k}$ looks like for general time slices t_k: that is precisely what we wish to determine. The fact that this is an initial-value problem translates to the lack of squares in the top row of Fig. 8.11: we don't know what we'll be evolving toward.

So far we have talked about discretizing, as per Eq. (8.132) and Eq. (8.133), and seen how to interpret the boundary conditions and the initial condition in terms of the discretized values $\phi_{j,k}$ in Fig. 8.11. What we have *not* actually done is explain how to map the diffusion equation of Eq. (8.131) to this spacetime grid; that's because there's more than one way of doing so, as we will discover.

8.5.2.2 An Explicit and an Implicit Method

We will now see how to discretize the diffusion equation of Eq. (8.131). At a big-picture level, what we'll be doing is to approximate the derivatives in our PDE using finite-difference formulas from chapter 3. A bit more of notation may be helpful at this point: recall that the discretizations employed in space and time are different, as per Eq. (8.132) and Eq. (8.133). The distance between points along a row (i.e., positions) and a column (i.e., time slices) is also correspondingly different, the former being $h_x = L/(n-1)$ and the latter $h_t = T/(m-1)$—similar notation as on page 299. We can now rewrite Eq. (8.131) by employing the forward-difference approximation to the first derivative as per Eq. (3.9), as well as the central-difference approximation to the second derivative as per Eq. (3.39):

$$\frac{\phi_{j,k+1} - \phi_{j,k}}{h_t} = \alpha \frac{\phi_{j+1,k} - 2\phi_{j,k} + \phi_{j-1,k}}{h_x^2} . \tag{8.134}$$

It is important to keep in mind that when you encounter an entity like $\phi_{j,k}$, the first index corresponds to space and the second index to time; this is why the left-hand side (corresponding to a temporal derivative) shifts the second index, whereas the right-hand side (corresponding to a spatial derivative) shifts the first index. In both cases, we have picked $\phi_{j,k}$ as the "point of reference", in the sense that we start at x_j and t_k and make a forward step in time and centered steps in space. You won't be surprised to hear that the technique this gives rise to (presented below) is known as the *forward-time centered-space (FTCS) scheme*. Qualitatively, we have rewritten the PDE in Eq. (8.131) by equating the temporal and spatial derivatives at the time slice t_k (and it just so happened that the temporal derivative necessitated an evaluation at the next time slice, t_{k+1}).

Even at this early stage, it is worth keeping in mind that we are treating time and space differently (which is hardly surprising, since we were faced with a first and second derivative, respectively): the forward-difference approximation on the left-hand side has an approximation error of $O(h_t)$, while the central-difference approximation on the right-hand side has an approximation error of $O(h_x^2)$. (Needless to say, these are errors for a *single step*, i.e., the approximation of a derivative at a single point, not for the full FTCS approach.) Incidentally, you should be able to see why we didn't reach for a better approximation to the temporal derivative, e.g., the second forward-difference of Eq. (3.27): this would involve *three* time slices (t_k, t_{k+1}, and t_{k+2}) thereby making it hard for us to even get going (since we have an initial condition only at a single time slice, $t_0 = 0$).

An example is visualized with triangles in Fig. 8.11: this shows that out of the four elements involved in Eq. (8.134) three lie at the current time slice t_k and only one ($\phi_{j,k+1}$) corresponds to the next time slice t_{k+1}. Thus, we can straightforwardly rewrite Eq. (8.134), solving for the unknown element:

$$\phi_{j,k+1} = \gamma\phi_{j-1,k} + (1 - 2\gamma)\phi_{j,k} + \gamma\phi_{j+1,k} \tag{8.135}$$

where we defined $\gamma = \alpha h_t / h_x^2$ and this applies to $j = 1, 2, \ldots, n-2$ and $k = 0, 1, \ldots, m-2$. Our final equation tells us how to go from k (one time slice) to $k+1$ (the next time slice); applying this to different j's is trivial. Since the right-hand side involves only known quantities, we are here dealing with an *explicit* method, similar to many techniques we encountered in earlier sections in the context of ODEs. Of course, since we are now tackling a PDE, the index structure is more complicated: Eq. (8.135) tells us how to produce a given $\phi_{j,k+1}$ given three of its "ancestors", but we also have to step through different j values to produce results across all positions (for a new time slice). Speaking of which, we could cast Eq. (8.135) in matrix form (similarly to what we did for the matrix approach to the BVP and EVP above), with a column vector on the left-hand side and a matrix multiplication on the right-hand side, $\boldsymbol{\phi}_{k+1} = \mathbf{B}\boldsymbol{\phi}_k$. This would not really constitute progress: it is straightforward to implement Eq. (8.135) in the form in which it has already been given.

At this point, we realize that the finite-difference formulas we employed to go from Eq. (8.131) to Eq. (8.134) were not set in stone; better yet, there is no reason why we should be equating the temporal and spatial derivatives at the time slice t_k, in the first place. We could have just as well decided to equate the derivatives at the time slice t_{k+1}, i.e., we can take $\phi_{j,k+1}$ as our "point of reference". Specifically, this means that we can start at x_j and t_{k+1} and make a *backward* step in time and centered steps in space. You won't be surprised to hear that the technique this gives rise to is known as the *backward-time centered-space (BTCS) scheme*. To be explicit, this means employing the backward-difference approximation to the first derivative as per Eq. (3.12), as well as the central-difference approximation to the second derivative as per Eq. (3.39) to arrive at:

$$\frac{\phi_{j,k+1} - \phi_{j,k}}{h_t} = \alpha \frac{\phi_{j+1,k+1} - 2\phi_{j,k+1} + \phi_{j-1,k+1}}{h_x^2} \tag{8.136}$$

Take a moment to appreciate the fact that the left-hand side resulted from a backward-difference formula and the right-hand side from a central-difference formula (both around $\phi_{j,k+1}$, as advertised). Just like in the FTCS scheme, our formulas have approximation error $O(h_t)$ and $O(h_x^2)$, respectively.

An example is visualized with diamonds in Fig. 8.11: this shows that out of the four elements involved in Eq. (8.136) three lie at the next time slice t_{k+1} and one ($\phi_{j,k}$) corresponds to the current time slice t_k. Thus, you can already see that solving this problem will not be as trivial as for the FTCS scheme above: since this is an initial-value problem, we always know the solution at a given time slice k (starting with $k = 0$) and try to find the solution at the following time slices (starting with $k = 1$). If you're not seeing why our task is non-trivial, you may appreciate our re-organizing the terms to produce:

$$-\gamma\phi_{j-1,k+1} + (1 + 2\gamma)\phi_{j,k+1} - \gamma\phi_{j+1,k+1} = \phi_{j,k} \tag{8.137}$$

where, once again, $\gamma = \alpha h_t/h_x^2$ and this applies to $j = 1, 2, \dots, n-2$ and $k = 0, 1, \dots, m-2$. On the face of it, Eq. (8.137) is one equation involving three unknowns, making us unsure of how to proceed. Put another way, if you rewrite Eq. (8.137) such that you have only $\phi_{j,k+1}$ on the left-hand side, similarly to Eq. (8.135), you will be faced with the inconvenient fact that your right-hand side would involve $\phi_{j-1,k+1}$ and $\phi_{j+1,k+1}$, which you don't actually know. In yet other words, we are here dealing with an *implicit* method.

The way to make progress is to realize that we are not totally helpless: as per Eq. (8.131), we have the boundary conditions $\phi_{0,k+1} = \phi_{0,k} = u$ and $\phi_{n-1,k+1} = \phi_{n-1,k} = v$. This brings to mind what we encountered using the matrix approach to BVP for ODEs in Eq. (8.113): if we write out Eq. (8.137) for all possible j values, we *will* be able to solve the (linear) system of equations that they make up. This is:

$$\begin{pmatrix} 1 & 0 & 0 & \dots & 0 & 0 & 0 \\ -\gamma & 1+2\gamma & -\gamma & \dots & 0 & 0 & 0 \\ 0 & -\gamma & 1+2\gamma & \dots & 0 & 0 & 0 \\ \vdots & \vdots & \vdots & \ddots & \vdots & \vdots & \vdots \\ 0 & 0 & 0 & \dots & 1+2\gamma & -\gamma & 0 \\ 0 & 0 & 0 & \dots & -\gamma & 1+2\gamma & -\gamma \\ 0 & 0 & 0 & \dots & 0 & 0 & 1 \end{pmatrix} \begin{pmatrix} \phi_{0,k+1} \\ \phi_{1,k+1} \\ \phi_{2,k+1} \\ \vdots \\ \phi_{n-3,k+1} \\ \phi_{n-2,k+1} \\ \phi_{n-1,k+1} \end{pmatrix} = \begin{pmatrix} \phi_{0,k} \\ \phi_{1,k} \\ \phi_{2,k} \\ \vdots \\ \phi_{n-3,k} \\ \phi_{n-2,k} \\ \phi_{n-1,k} \end{pmatrix} \tag{8.138}$$

The first row in the coefficient matrix encapsulates the fact that the solution at the left endpoint is both time-independent and known (it is a square in Fig. 8.11). The next row involves three solution elements, but we now understand that one of them is known ($\phi_{0,k+1} = \phi_{0,k} = u$). The following rows involve three elements each, until we get to the bottom of the coefficient matrix where we, again, know the answer from the right endpoint ($\phi_{n-1,k+1} = \phi_{n-1,k} = v$). In all, our problem has taken the form $\mathbf{A}\boldsymbol{\phi}_{k+1} = \boldsymbol{\phi}_k$ i.e., $\mathbf{Ax} = \mathbf{b}$, as advertised. Since here the first and last elements ($\phi_{0,k+1}$ and $\phi_{n-1,k+1}$) are fixed (bringing to mind the Project of section 5.7), we could have moved them to the right-hand side of Eq. (8.137) and set up the linear system only for the elements that actually need to be solved for (from $j = 1$ to $j = n-2$). However, we are being slightly wasteful, explicitly listing the $j = 0$ and $j = n - 1$ elements (which are known from the boundary conditions) both in order to bring out the analogy with our approach to the BVP in Eq. (8.113) and in order to streamline the programming implementation below.

Speaking of the analogy with earlier problems/approaches, it is important not to push these too far: take a moment to appreciate the fact that after solving the entire linear system you get the solution only at the next time slice (i.e., all the $\phi_{j,k+1}$'s): you have to solve many such linear systems if you want to evolve your solution from $t = 0$ all the way to $t = T$. Put another way, you need to solve a problem of the form $\mathbf{Ax} = \mathbf{b}$ for each new time slice. Since the coefficient matrix does not depend on the time, this is more efficiently done if you, e.g., LU-decompose it and then carry out only forward substitution and back substitution at each time slice, as per Eq. (4.97). At this point, you may be wondering: since the BTCS approach appears to employ approximations of the same accuracy as FTCS—$O(h_t)$ and $O(h_x^2)$—is the extra cost involved worth it? We will address this point in more detail below (and in problem 8.27) but, at a qualitative level, you should not be surprised to hear that the BTCS method, being an implicit technique, has better stability properties.

As mentioned earlier, you can cast the FTCS scheme of Eq. (8.135) in matrix form: this would involve a column vector on the left-hand side and a matrix multiplication on the right-hand side. This is to be juxtaposed with the BTCS scheme of Eq. (8.138), which involves a matrix multiplication on the left-hand side and therefore gives rise to a linear system of equations. If you squint (and have a good memory), you can see that this is completely analogous to two techniques we encountered all the way back in chapter 4 when computing eigenvalues: the (direct) power method of Eq. (4.136) involved a matrix multiplication on the right-hand side, whereas the inverse-power method of Eq. (4.143) involved a matrix multiplication on the left-hand side. You may not be surprised to hear that the analogy does not stop there: the direct power method was easy to implement but could not help us tackle the full eigenvalue problem, whereas the inverse-power method was more expensive but also more versatile. Similarly, as you will discover below and in problem 8.27, the FTCS scheme is trivial to implement but can get you in trouble, whereas the BTCS approach is a bit messier (and more costly) but also more robust.

8.5.2.3 Implementation

The implementation of the FTCS scheme, shown in code 8.7, i.e., `diffusion.py`, is quite straightforward: the update of Eq. (8.135) can be carried out in a single line (were it not for the column count), taking advantage of NumPy-array slicing. We simply assume that the array provided as input, corresponding to the initial condition, $\phi_{j,0}$, contains the boundary conditions in its first and last elements ($\phi_{0,0} = u$ and $\phi_{n-1,0} = v$, respectively). Thus, all we need to do is to touch the "middle" elements as per Eq. (8.135). Note that this allows the coding to be way cleaner than in `action.py` (where `xini` and `xfin` had to be treated separately), at the cost of us carrying around an extra two elements which are never modified. Since the update for all j's taking us from k to $k + 1$ is carried out via slicing, our function `ftcs()` contains only a single explicit loop, taking us from $t = 0$ to $t = T$.

Our function `btcs()` is only slightly more complicated. Incidentally, note that both `ftcs()` and `btcs()` take in only a single parameter `fac` standing in for γ, i.e., they don't need to bother with α, h_t, and h_x. The function `btcs()` first sets up the coefficient matrix from Eq. (8.138), similarly to several of our earlier codes. It then LU-decomposes that matrix, in order to avoid having to carry out a full Gaussian elimination for each time-step evolution. Indeed, the only loop in `btcs()`, as in the previous function, takes us from $t = 0$ to $t = T$; it simply carries out a forward substitution and a back substitution as per Eq. (4.97), followed by a variable renaming in preparation for the next iteration.

The main program sets up the parameters and an initial condition $\phi_{j,0}$ of the form:

Code 8.7 diffusion.py

```
from triang import forsub, backsub
from ludec import ludec
import numpy as np

def ftcs(fac, m, polds):
    pnews = np.copy(polds)
    for k in range(m-1):
        pnews[1:-1] = fac*(polds[:-2] + polds[2:] - 2*polds[1:-1])
        pnews[1:-1] += polds[1:-1]
        polds = np.copy(pnews)
    return pnews

def btcs(fac, m, polds):
    n = polds.size
    A = np.zeros((n,n))
    np.fill_diagonal(A, 1.+2*fac)
    A[0,0] = 1; A[-1,-1] = 1
    np.fill_diagonal(A[1:,:], -fac)
    A[-1,-2] = 0
    np.fill_diagonal(A[:,1:], -fac)
    A[0,1] = 0
    LA, UA = ludec(A)

    for k in range(m-1):
        ys = forsub(LA,polds)
        pnews = backsub(UA,ys)
        polds = np.copy(pnews)

    return pnews

if __name__ == '__main__':
    al, L, T, n, m = 0.1, 1., 0.4, 51, 200
    hx = L/(n-1); ht = T/(m-1)
    fac = al*ht/hx**2
    pinits = 30*np.ones(n)
    pinits[:n//2] = 10
    phis = ftcs(fac, m, pinits); print(phis)
    phis = btcs(fac, m, pinits); print(phis)
```

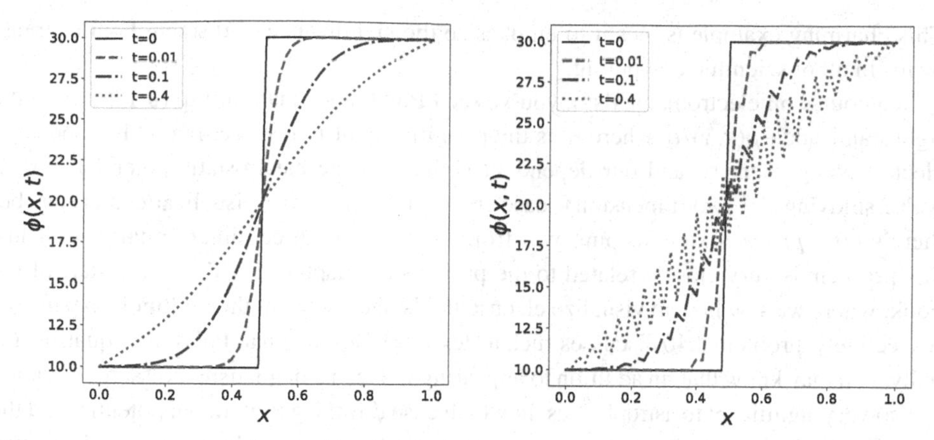

Fig. 8.12 Diffusion-equation solutions with BTCS (left) and FTCS (right)

$$\phi(x,0) = \begin{cases} u, & \text{if } x < L/2 \\ v, & \text{if } x \geq L/2 \end{cases} \tag{8.139}$$

While discontinuous at $x = L/2$, this agrees with the boundary conditions at $x = 0$ and $x = L$ (which hold at all times). Physically, you can think of this scenario as placing two metal rods of different temperature next to each other; the free end of each rod is kept at a fixed temperature (same as the starting temperature), but the rods are free to exchange heat, which will flow and thereby eliminate the discontinuity in the temperature profile. The BTCS results are shown in the left panel of Fig. 8.12, where we see "snapshots" of the time evolution. To be clear, T and m have fixed values, those shown in our program, but we are also plotting results at intermediate time slices. Perhaps unsurprisingly, given that we started with a discontinuous initial condition, at early times $\phi_{j,k}$ exhibits a pretty sudden switch from one "temperature" to another. As the time increases, this is smoothened out completely. The right panel of Fig. 8.12 shows analogous snapshots for the FTCS scheme. It exhibits some rather unpleasant oscillations, which are clearly an artifact of this explicit method, i.e., they are signatures of its impending instability. To be fair, this plot would look quite different if we had picked a much larger value of m (i.e, the FTCS approach would give essentially the same results as the BTCS scheme then). The reasons behind this, having to do with the relative magnitudes of h_t and $h_x^2/(2\alpha)$ are explored in the problem set, where you will also learn about an improved scheme, the *Crank–Nicolson method* (see problem 8.26), which behaves even better than the BTCS approach.

8.6 Project: Poisson's Equation in Two Dimensions

The approach(es) to the diffusion equation in section 8.5.2 may have left you feeling uninspired: they are simple extensions of what we did for ODEs in earlier sections. To spice things up, we will now tackle Poisson's equation using techniques introduced in chapter 6.

This charming example is meant to emphasize the significance of discrete Fourier transforms in all of scientific computing.

In a course on electromagnetism, you've seen Poisson's equation, Eq. (8.129), with the right-hand side $-\rho(x, y)/\epsilon_0$ where ϵ_0 is the permittivity of free space, $\rho(x, y)$ is a specified electric charge density, and our dependent variable is the electrostatic potential $\phi(x, y)$; we're studying the two-dimensional case in order to make the visualizations easier, but there's no *a priori* reason keeping you from tackling the three-dimensional case. Thus, this problem is very closely related to the projects of chapters 1 and 2 at the start of the book, where we saw how to visualize electric fields and carry out the multipole expansion, respectively; problem 8.40 discusses such a "textbook" approach to Poisson's equation. Finally, you may know that, in addition to appearing in electromagnetism, Poisson's equation is also very significant to astrophysics, in which case ϕ is the gravitational potential and the right-hand side takes the form $-4\pi G\rho(x, y)$—again in two spatial dimensions—with $\rho(x, y)$ being the mass density and G Newton's gravitational constant. Regardless of the physical application, mathematically speaking our problem is a constant-coefficient elliptic equation and we'll also specialize it further, studying only *periodic* boundary conditions.

8.6.1 Problem and Discretization

In order to be concrete, we will tackle the following specific problem:

$$\begin{gathered}\frac{\partial^2 \phi}{\partial x^2} + \frac{\partial^2 \phi}{\partial y^2} = f(x, y), \qquad f(x, y) = \cos(3x + 4y) - \cos(5x - 2y) \\ \phi(x, 0) = \phi(x, 2\pi), \qquad \phi(0, y) = \phi(2\pi, y)\end{gathered} \tag{8.140}$$

As advertised, this is a periodic problem, both in its boundary conditions and in its forcing term. You may be thinking that you can solve this problem analytically, in which case you may not see the point of using a computer; but what if you were dealing with Mathieu functions instead of cosines? The computational tools we'll introduce below will apply even if the right-hand side of Poisson's equation is not an analytically known function, as you will discover when you tackle problem 8.42.

We are now ready to discretize, placing our points on a grid:[27]

$$x_p = \frac{2\pi p}{n}, \qquad p = 0, 1, \ldots, n-1 \tag{8.141}$$

where we are building in the fact that our problem is periodic—just like in Eq. (8.120)—the last point is at $2\pi(n-1)/n$, not at 2π, because we already know that all quantities have the same value at 2π as at 0. Since we're dealing with a two-dimensional problem, we also need to discretize the y's and therefore employ another index:

$$y_q = \frac{2\pi q}{n}, \qquad q = 0, 1, \ldots, n-1 \tag{8.142}$$

[27] We introduce new symbols to emphasize that both dimensions are spatial, unlike the j and k of Eq. (8.134).

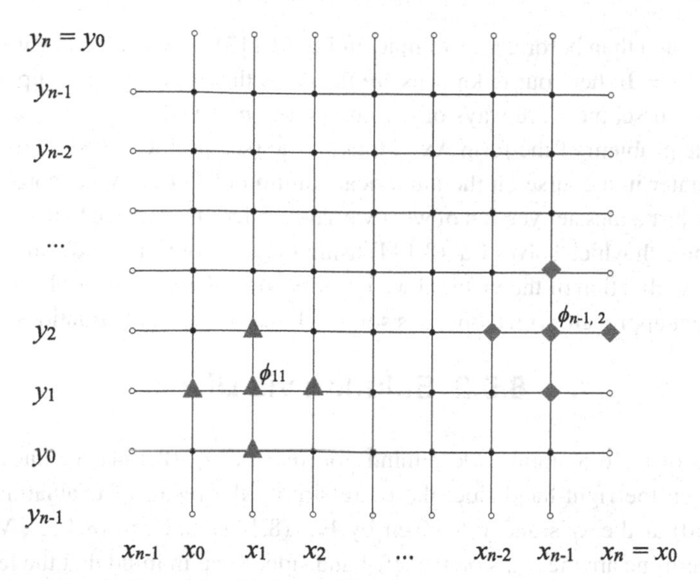

Two-dimensional periodic grid, with examples

Fig. 8.13

We have chosen to use the same discretization scheme for both coordinates: they both go from 0 to 2π, involving n points for each dimension. This is simply because that's what we'll need in what follows, but you will go beyond this in problem 8.39.

Similarly to what we did for the BVP and EVP, we now apply the central-difference approximation to the second-derivative (once for the x's and once for the y's) of Eq. (3.39):

$$\frac{\phi_{p+1,q} - 2\phi_{p,q} + \phi_{p-1,q}}{h^2} + \frac{\phi_{p,q+1} - 2\phi_{p,q} + \phi_{p,q-1}}{h^2} = f_{pq}, \qquad p, q = 0, 1, \ldots, n-1 \quad (8.143)$$

which can be slightly manipulated into the form:

$$\phi_{p+1,q} + \phi_{p-1,q} + \phi_{p,q+1} + \phi_{p,q-1} - 4\phi_{pq} = h^2 f_{pq} \quad (8.144)$$

Again, p and q are spatial indices, as per Eq. (8.141) and Eq. (8.142). Our two-dimensional grid is illustrated in Fig. 8.13 (cf. Fig. 8.10); as before, solid circles show actual points and open circles are implied points. Each point ϕ_{pq} enters the discretized Poisson's equation, Eq. (8.144), together with its four nearest neighbors[28] (cf. the FTCS and BTCS schemes in Fig. 8.11). The figure also shows a couple of examples, highlighting the fact that our setup works equally well if ϕ_{pq} is somewhere in the middle of the grid, or right at its edge. This is not always the case: very often in the study of PDEs the implementation of the boundary conditions has to be handled separately (and can be a headache). Here, with physics in mind, we study a periodic setting, tailoring our solution to the problem at hand.

You can solve Eq. (8.144) with linear algebra machinery, as we did earlier: it is nothing other than a finite-difference version of a PDE. However, notice that the problem here is

[28] This is sometimes called a *five-point stencil*, though we generally prefer the term "discretization scheme".

more complicated than before: for example, in Eq. (8.113) we were faced with an equation of the form $\mathbf{Ax} = \mathbf{b}$; here our unknowns are the ϕ_{pq}'s that together make up a *matrix*, not a vector! Of course, there are ways of organizing our n^2 equations in such a way that we can solve one problem of the form $\mathbf{Ax} = \mathbf{b}$ at a time (see problem 8.41); this is what you would encounter in a course on the numerical solution of PDEs. At its core, this is not a new method, just a messier version of what we saw earlier. Instead, we here opt for a totally different approach which solves Eq. (8.144) using the fast Fourier transform; in addition to being a nice application of the material we introduced in chapter 6, this also happens to be a very efficient approach, which obviates solving linear systems of equations altogether.

8.6.2 Solution via DFT

For the sake of (re)orientation, we remind you that in Eq. (8.144) we know the values of the f_{pq}'s on the right-hand side: these are simply the result of evaluating the $f(x, y)$ in Eq. (8.140) at the x_p's and y_q's given by Eq. (8.141) and Eq. (8.142). What we are trying to determine are the ϕ_{pq}'s on the left-hand side. Keep in mind that the left-hand side in Eq. (8.144) is basically the Laplacian operator in discretized coordinates. As you may recall from Fourier analysis—see Eq. (6.4)—derivatives turn into multiplications in Fourier (wave number) space. That means that we can be hopeful the complicated left-hand side in Eq. (8.144) will be dramatically simplified if we turn to the (two-dimensional) discrete Fourier transform; we haven't seen this before, but it is a straightforward generalization of the one-dimensional case. At a big-picture level, we are about to trade the p, q, which are spatial indices, with k, l, which are going to be Fourier indices. As mentioned, we are willing to go to Fourier space in order to simplify the effect of the derivative(s).

Generalizing Eq. (6.99) to the two-dimensional case allows us to write the inverse discrete Fourier transform of the discretized ϕ as follows:

$$\phi_{pq} = \frac{1}{n^2}\sum_{k=0}^{n-1}\sum_{l=0}^{n-1}\tilde{\phi}_{kl}e^{2\pi ikp/n}e^{2\pi ilq/n}, \qquad p, q = 0, 1 \ldots, n-1 \tag{8.145}$$

where we assumed we are using the same number of points, n, for each of the two dimensions in the problem, as above. We can do the same thing for the function appearing on the right-hand side of Poisson's equation, namely the discretized f:

$$f_{pq} = \frac{1}{n^2}\sum_{k=0}^{n-1}\sum_{l=0}^{n-1}\tilde{f}_{kl}e^{2\pi ikp/n}e^{2\pi ilq/n}, \qquad p, q = 0, 1, \ldots, n-1 \tag{8.146}$$

Plugging these two relations into the equation we wish to solve, Eq. (8.144), we get:

$$\frac{1}{n^2}\sum_{k=0}^{n-1}\sum_{l=0}^{n-1}e^{2\pi ikp/n}e^{2\pi ilq/n}\left[\tilde{\phi}_{kl}\left(e^{2\pi ik/n} + e^{-2\pi ik/n} + e^{2\pi il/n} + e^{-2\pi il/n} - 4\right) - h^2\tilde{f}_{kl}\right] = 0 \tag{8.147}$$

At this point, we can use the orthogonality of complex exponentials in the discrete case, Eq. (6.92), to eliminate the sums. Explicitly, we multiply by $e^{-2\pi ik'p/n}$ and then sum over p and, similarly, multiply by $e^{-2\pi il'q/n}$ and then sum over q. If we also replace the four exponentials inside the parentheses with two cosines, we are led to:

$$\tilde{\phi}_{kl} = \frac{1}{2} \frac{h^2 \tilde{f}_{kl}}{\cos(2\pi k/n) + \cos(2\pi l/n) - 2} \tag{8.148}$$

Thus, we can produce the $\tilde{\phi}_{kl}$ by taking the $\tilde{f}_{kl}$ and plugging them into Eq. (8.148).

We're almost done: since we've already computed $\tilde{\phi}_{kl}$, all that's left is for us to use the inverse DFT in Eq. (8.145) to evaluate the real-space ϕ_{pq}. That was our goal all along and we reached it without having to explicitly solve a linear system of n^2 coupled equations as per Eq. (8.144). The one thing we skipped over is that Eq. (8.148) involves $\tilde{f}_{kl}$ on the right-hand side, not the f_{pq} which is actually our input. That's OK, though, because we can simply use the following definition of the *direct* DFT in the two-dimensional case, which is a generalization of Eq. (6.97):

$$\tilde{f}_{kl} = \sum_{p=0}^{n-1} \sum_{q=0}^{n-1} f_{pq} e^{-2\pi ikp/n} e^{-2\pi ilq/n}, \qquad k, l = 0, 1, \ldots, n-1 \tag{8.149}$$

Our strategy is: (a) take f_{pq}, (b) use it to get $\tilde{f}_{kl}$ from the direct DFT in Eq. (8.149), (c) use that to evaluate $\tilde{\phi}_{kl}$ from Eq. (8.148), and (d) take the inverse DFT in Eq. (8.145) to get ϕ_{pq}.

8.6.3 Implementation

Thinking about how to implement the above approach to solving Poisson's equation, we realize that we don't actually have an implementation of a two-dimensional FFT (or even DFT, for that matter) available.[29] This is not a major obstacle: we will simply use the one-dimensional FFT implementation of code 6.4, i.e., `fft.py`, to build up a two-dimensional version. In order to see how we could do that, we return to Eq. (8.149) and group the terms in a more suggestive manner:

$$\tilde{f}_{kl} = \sum_{p=0}^{n-1} \left[\sum_{q=0}^{n-1} f_{pq} e^{-2\pi ilq/n} \right] e^{-2\pi ikp/n} \tag{8.150}$$

As the brackets emphasize, it is possible to first take the one-dimensional DFT of each row (i.e., f_{pq} with p held fixed each time), and then take the one-dimensional DFT of each column of the result. Having applied the one-dimensional DFT algorithm $2n$ times, we have accomplished what we set out to do.

The only thing left is to figure out how to implement the two-dimensional *inverse* DFT when we have an implementation of the two-dimensional direct DFT at our disposal (as per the previous paragraph). We saw how to code up the one-dimensional inverse DFT in problem 6.30; the idea generalizes straightforwardly to the two-dimensional case. Take the complex conjugate of the inverse DFT in Eq. (8.146):

$$f^*_{pq} = \frac{1}{n^2} \sum_{k=0}^{n-1} \sum_{l=0}^{n-1} \tilde{f}^*_{kl} e^{-2\pi ikp/n} e^{-2\pi ilq/n}, \qquad p, q = 0, 1, \ldots, n-1 \tag{8.151}$$

[29] Of course, numpy has one but, as usual, it's more educational to write our own. Incidentally, in this section we use DFT and FFT interchangeably: the latter is the efficient implementation of the former, as always.

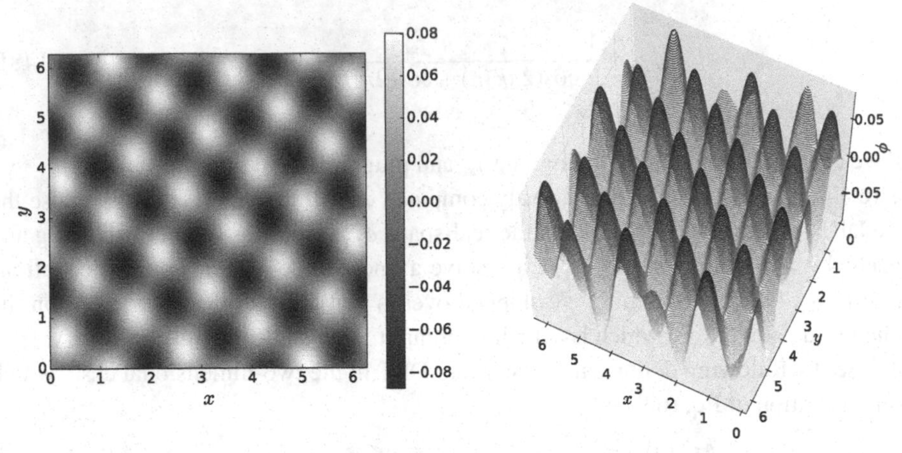

Fig. 8.14 Poisson-equation solution for a two-dimensional periodic boundary-value problem

Thus, the direct DFT of $\tilde{f}^*_{kl}$, divided by n^2, is simply f^*_{pq}. But that means that, in order to produce the inverse DFT, we can simply: (a) take the complex conjugate of $\tilde{f}_{kl}$, (b) carry out the direct DFT of that, (c) divide by n^2, and (d) take the complex conjugate at the end.

Code 8.8 starts by importing the one-dimensional FFT functionality of `fft.py`. It then uses the outlined procedure to define two functions that carry out the direct and inverse two-dimensional FFT. We use NumPy's `astype()` to copy the input array (casting to complex numbers) and then idiomatically step through the columns as per page 20. Another function is defined to provide the $f(x, y)$ from the right-hand side of Poisson's equation.

The physics is contained in the longest function in this program, called `poisson()`. After defining a one-dimensional grid of points, this function calls `numpy.meshgrid()` to produce two coordinate matrices: these are created in such a way that taking the `[q,p]` element from `Xs` and the `[q,p]` element from `Ys` gives you the pair (x_p, y_q).[30] The rows of `Xs` repeat the x_p's and the columns of `Ys` repeat the y_q's. In short, `numpy.meshgrid()` has allowed us to produce coordinate matrices, which we pass into `func()` to produce the $\{f_{pq}\}$ matrix in a single line of code. The next line calls our brand-new two-dimensional FFT function to create the $\{\tilde{f}_{kl}\}$ matrix. We then do something analogous for the Fourier-space indices, i.e., we create two coordinate matrices for the k's and l's, which we use to carry out the division in Eq. (8.148) most simply. There is a slight complication involved here: the element `[0,0]` leads to division by zero, which gives rise to a `RuntimeWarning`. We're not really worried about this, however: this is simply the DC component; our boundary conditions are periodic, as per Eq. (8.140), so we are only able to determine the potential ϕ_{pq} up to an overall constant. Thus, after our division we set this overall offset to obey our chosen normalization. As per our earlier strategy, we then take the inverse DFT in Eq. (8.145) to calculate ϕ_{pq} from $\tilde{\phi}_{kl}$ and then we're done.

The main program calls our function `poisson()`, passing in $n = 128$: as you may recall from chapter 6, our implementation of the (one-dimensional) FFT algorithm kept halving,

[30] If you expected (x_q, y_p), then you should look into the (optional) `indexing` argument; see also page 24.

poisson.py

Code 8.8

```
from fft import fft
import numpy as np

def fft2(A):
    B = A.astype(complex)
    for i, row in enumerate(A):
        B[i,:] = fft(row)
    for j, col in enumerate(B.T):
        B[:,j] = fft(col)
    return B

def inversefft2(A):
    n2 = A.size
    newA = fft2(np.conjugate(A))/n2
    return np.conjugate(newA)

def func(x,y):
    return np.cos(3*x+4*y) - np.cos(5*x-2*y)

def poisson(a,b,n):
    h = (b-a)/(n-1)
    xs = a + np.arange(n)*h
    Xs, Ys = np.meshgrid(xs,xs)

    F = func(Xs, Ys)
    Ftil = fft2(F)

    ks = np.arange(n)
    Kxs, Kys = np.meshgrid(ks,ks)
    Denom = np.cos(2*np.pi*Kxs/n) + np.cos(2*np.pi*Kys/n) - 2
    Phitil = 0.5*Ftil*h**2/Denom
    Phitil[0,0] = 0

    Phi = np.real(inversefft2(Phitil))
    return Phi

if __name__ == '__main__':
    Phi = poisson(0, 2*np.pi, 128); print(Phi)
```

so it relied on n being a power of 2. This routine is here used to carry out the two-dimensional direct and inverse FFT, which in our case gives rise to matrices of dimension 128×128. (You might want to try larger values of n and compare the timing with that resulting from the "slow" DFT implementation.) In Fig. 8.14 we are attempting to visualize the $\{\phi_{pq}\}$, which is our final result. We are showing two different ways of plotting the value of the electrostatic potential. Unsurprisingly, the solution is periodic; the coefficients in the arguments of the cosines in the $f(x, y)$ term have led to rapid modulations. In all, we managed to solve a two-dimensional boundary-value problem with a minimum of fuss (and a code that fits on a single page).

Problems

8.1 [$\mathcal{A}$] It is standard to assume that a (linear) second-order ODE doesn't have a first-order derivative term. To see why this is legitimate, start with $y''(x) + p(x)y'(x) + q(x) = 0$ and make the substitution:

$$z(x) = y(x)e^{\frac{1}{2}\int p(x)dx} \tag{8.152}$$

Show that you get the ODE $z''(x) + r(x)z(x) = 0$, which contains no first-order term.

8.2 [$\mathcal{A}$] In the main text we investigated the stability of the Euler method(s) using the test equation, see Eq. (8.33). To generalize this, start from Eq. (8.25), which relates e_{j+1} to e_j. Multiply and divide the term in square brackets by e_j and then interpret the fraction appropriately. One of the terms on the right-hand side should be e_j times a factor that is linear in h; explain what this means.

8.3 Reproduce Fig. 8.4, augmented to also include results for the explicit midpoint and explicit trapezoid methods, Eq. (8.54) and Eq. (8.57).

8.4 Implement the backward Euler method for the Riccati equation, Eq (8.71). You could use a root-finder or simply solve the quadratic equation "by hand"; as part of this process, you need to pick one of the two roots (justify your choice).

8.5 [$\mathcal{A}$] Show that the implicit trapezoid method, Eq. (8.63), has local error $O(h^3)$. To do so, you can re-use the first equality in Eq. (8.44). The main trick involved is that, just like $y'(x_j) = f(x_j, y(x_j))$ holds, we can also take $y'(x_{j+1}) = f(x_{j+1}, y(x_{j+1}))$ and then Taylor expand $y'(x_{j+1})$ around x_j.

8.6 [$\mathcal{A}$] In the main text we explicitly derived the prescription for second-order Runge–Kutta methods, see section 8.2.2. When we studied the fourth-order Runge–Kutta method, we provided motivation from quadrature and then carried out an explicit derivation for the test equation. We now turn to *third*-order Runge–Kutta methods; to make things manageable, focus on an autonomous ODE, $y' = f(y)$, which is a specific case, but still more general than the single example of the test equation. The prescription is:

$$k_0 = hf(y_j), \qquad k_1 = hf\left(y_j + c_0k_0\right), \qquad k_2 = hf\left(y_j + c_1k_0 + c_2k_1\right)$$
$$y_{j+1} = y_j + d_0k_0 + d_1k_1 + d_2k_2 \tag{8.153}$$

Your task is, by carrying out a number of Taylor expansions as before, to derive the relations these c and d parameters need to obey.

8.7 Implement the following (explicit) two-step *Adams–Bashforth method*:

$$y_{j+1} = y_j + \frac{h}{2}\left[3f(x_j, y_j) - f(x_{j-1}, y_{j-1})\right] \tag{8.154}$$

for the Riccati equation, Eq (8.71). Obviously, you'll need two y values to get going (y_0 and y_1); since a first-order IVP corresponds to only a single input y value (y_0), we'll need to produce an estimate of y_1 before we can start applying this method; use the forward Euler method to do that.

8.8 Reproduce the two plots shown in Fig. 8.5. Note that you will need to tailor the backward Euler routine to the specific equation you are tackling each time.

8.9 Tackle the problem in Eq. (8.72) by setting $u(x) = y(x) + e^{-500x}$. As you will discover when you implement this yourself, this trick eliminates RK4's troubles, even though the ODE is unchanged. Do you understand why?

8.10 Use RK4 to solve the following problem:

$$y'(x) = 100\,(\sin x - y), \qquad y(0) = 0 \tag{8.155}$$

from $x = 0$ to $x = 8$, with $n = 285$ and $n = 290$ points. Do you understand why the two results are so dramatically different?

8.11 Generalize code 8.1 to use Chebyshev points. Then, employ Lagrange interpolation in between the points and produce a new version of Fig. 8.4.

8.12 Implement global and local adaptive stepping in RK4, to reproduce the panels in Fig. 8.6.

8.13 The *Runge–Kutta–Fehlberg method* (RK5), with local error $O(h^6)$, follows the steps:

$$\begin{aligned}
k_0 &= hf(x_j, y_j)\\
k_1 &= hf\left(x_j + \frac{h}{4}, y_j + \frac{k_0}{4}\right)\\
k_2 &= hf\left(x_j + \frac{3}{8}h, y_j + \frac{3}{32}k_0 + \frac{9}{32}k_1\right)\\
k_3 &= hf\left(x_j + \frac{12}{13}h, y_j + \frac{1932}{2197}k_0 - \frac{7200}{2197}k_1 + \frac{7296}{2197}k_2\right)\\
k_4 &= hf\left(x_j + h, y_j + \frac{439}{216}k_0 - 8k_1 + \frac{3680}{513}k_2 - \frac{845}{4104}k_3\right)\\
k_5 &= hf\left(x_j + \frac{h}{2}, y_j - \frac{8}{27}k_0 + 2k_1 - \frac{3544}{2565}k_2 + \frac{1859}{4104}k_3 - \frac{11}{40}k_4\right)\\
y_{j+1} &= y_j + \frac{16}{135}k_0 + \frac{6656}{12825}k_2 + \frac{28561}{56430}k_3 - \frac{9}{50}k_4 + \frac{2}{55}k_5
\end{aligned} \tag{8.156}$$

The reason this approach is popular is that one can embed a fourth-order (Runge–Kutta) technique in it—with local error $O(h^5)$—which we denote by $\tilde{y}_{j+1}$; subtracting the two estimates gives us an expression for the error:

$$y_{j+1} - \tilde{y}_{j+1} = \frac{1}{360}k_0 - \frac{128}{4275}k_2 - \frac{2197}{75240}k_3 + \frac{1}{50}k_4 + \frac{2}{55}k_5 \tag{8.157}$$

This doesn't require the step-halving that we employed for RK4, see Eq. (8.76) and below. Implement both the technique itself and its error estimate programmatically for the Riccati equation, Eq (8.71), and compare with RK4 for fixed $n = 7$.

8.14 Use RK4 to solve the following *Lotka–Volterra equations*, used to describe the populations of a predator and a prey species:

$$y_0'(x) = 0.1y_0(x) - 0.01y_0(x)y_1(x), \qquad y_1'(x) = -0.5y_1(x) + 0.01y_0(x)y_1(x),$$
$$y_0(0) = 60, \qquad y_1(0) = 20 \tag{8.158}$$

Plot $y_0(x)$ and $y_1(x)$, integrating up to $x = 80$.

8.15 Solve the following IVP using RK4:

$$w''(x) = -w(x) - \frac{1}{x}w'(x), \qquad w(0) = 1, \qquad w'(0) = 0 \tag{8.159}$$

integrating up to $x = 0.5$. Applying `rk4_gen()` gets you in trouble due to the $1/x$ term at $x = 0$. Apply L'Hôpital's rule to that term, getting another contribution of $w''(x)$ at the origin; code up `fs()` so that it takes into account both possibilities. (This is the *Bessel equation* for the *Bessel function* $J_0(x)$, so you may wish to benchmark against `scipy.special.jv()`.)

8.16 First, repeat the analysis in Eq. (8.98) and below, this time for the case of the backward Euler method. Then, apply both the forward and the backward Euler methods for the system of ODEs in Eq. (8.102), integrating up to $x = 0.1$; start with $n = 1000$ to make sure you implemented both techniques correctly, and then repeat the exercise for $n = 6$.

8.17 [$\mathcal{A}$] Generalize the analysis of the backward Euler method from problem 8.16 for the case of an autonomous, possibly non-linear, system of equations:

$$\mathbf{y}_{j+1} = \mathbf{y}_j + h\mathbf{f}(\mathbf{y}_{j+1}) \tag{8.160}$$

Linearize this system by Taylor expanding $\mathbf{f}(\mathbf{y}_{j+1})$ around $\mathbf{y}_j$, just like in Eq. (5.76) when introducing Newton's method for root-finding in many dimensions. Formally solve the resulting equation for $\mathbf{y}_{j+1}$ and discuss what condition needs to be satisfied in order for you to do so in a dependable manner.

8.18 Apply the shooting method to the following fourth-order BVP:

$$w''''(x) = -13w''(x) - 36w(x), \quad w(0) = 0, \quad w'(0) = -3, \quad w(\pi) = 2, \quad w'(\pi) = -9 \tag{8.161}$$

Note that we are missing two pieces of information at the starting point, $w''(0)$ and $w'''(0)$, so you will have to use the multidimensional Newton's method. Specifically, make sure to define one function for the four right-hand sides of our system of ODEs (after you rewrite our fourth-order ODE) and another function for the two conditions we wish to satisfy ($w(\pi) - 2 = 0$ and $w'(\pi) + 9 = 0$).

8.19 We will now see how to handle *Neumann boundary conditions* in the matrix approach to the BVP. Specifically, take the following problem:

$$w''(x) = -\frac{30}{1-x^2}w(x) + \frac{2x}{1-x^2}w'(x)$$
$$w'(0.05) = 1.80962109375, \qquad w(0.49) = 0.1117705085875 \tag{8.162}$$

which is a variation on Eq. (8.106). Our finite-difference scheme leads to equations like those in Eq. (8.111), with the first one being replaced by:

$$\frac{w_1 - w_{-1}}{2h} = 1.80962109375 \tag{8.163}$$

Here the solution is *not* required to be periodic, so you *cannot* assume $w_{-1} = w_{n-1}$ holds, as in Eq. (8.124): w_{-1} is simply outside the solution domain. Apply the discretized ODE, Eq. (8.110), for the case of $j = 0$; combining this with Eq. (8.163) allows you to eliminate w_{-1}. Implement the solution to this BVP programmatically.

8.20 [$\mathcal{A}$] After introducing the matrix approach to the BVP in Eq. (8.108), we discussed the error stemming from the finite-difference approximations employed. If you consider the problem more carefully, you will realize that the error in approximating the derivatives is not necessarily the same as the error in approximating $w(x_j)$ by w_j (which quantifies how well or how poorly we are satisfying the ODE we are trying to solve). Expand $w(x_{j+1})$ and $w(x_{j-1})$, both around x_j, to explicitly investigate how good an approximation w_j is, for the case of a linear BVP.

8.21 Solve the following non-linear BVP:

$$w''(x) = -e^{w(x)}, \qquad w(0) = 0, \qquad w(1) = 0 \tag{8.164}$$

using both the shooting method and the matrix approach. Note that you'll need a root-finder for both cases (one-dimensional and multidimensional, respectively). Try out different initial guesses to find *two* solutions to this problem. For the matrix approach, have your initial guess vector be either zero or $20x - 20x^2$; you could evaluate the Jacobian numerically (as in `multi_newton.py`) or analytically.

8.22 You should now tackle the BVP:

$$w''(x) = 5\sinh[5w(x)], \qquad w(0) = 0, \qquad w(1) = 1 \tag{8.165}$$

(a) Start from a simple modification of the shooting method in `bvp_shoot.py` for $n = 50$ points. The secant method gets into trouble, so update the code such that you use the bisection method, instead.

(b) Having solved part (a), you realize that not much is happening in the solution up to $x = 0.8$ or so. This gives rise to the following idea, known as *shooting to a fitting point*: instead of shooting from the left endpoint and trying to match the value of the function at the right endpoint, you can carry out two independent RK4 integrations, one from a outward (using $n_l = 20$ points) and another one from b inward (using $n_l = 30$ points). The advantage of this approach is that you will be (by construction) satisfying the boundary conditions at both a and b.
This involves *two* unknown quantities, namely the starting derivatives for each integration—$w'(a)$ and $w'(b)$. To determine them, you need to solve *two* equations: this is naturally done by matching the solutions for $w(x)$ and $w'(x)$ at the fitting point (also known as a *joint*), which you should take to be at $x = 0.8$. Use the multidimensional Newton's method to solve these two simultaneous equations. Hint: be sure not to confuse the `fs()` that are the right-hand sides of your ODEs with the `fs()` that you need to pass in to Newton's method.

8.23 You will now solve the following BVP:

$$w''(x) = \left(1 - \frac{4}{x + 0.5}\right) w(x), \qquad w(0) = 1, \qquad w(+\infty) = 0 \tag{8.166}$$

(a) Establish the solution the "naive" way, i.e., by taking the right endpoint of integration to be large but finite, e.g., $b = 20$. Apply both the shooting method and the matrix approach to this problem and make sure the two solutions agree.

(b) We will now use a joint, just like in problem 8.22, but this time we (won't be employing the shooting method but we) will be using the matrix approach. From the solution in the previous part of the present problem, we notice that $w(x)$ is oscillatory up to roughly $x = 3.5$ and then decays after that, so we decide to place the joint at $x = 3.5$. Our integration outward (starting at $a = 0$) will employ n_l points and will look similar to what you did in part (a). Unlike what we had in Eq. (8.111), w_{n_l-1} in the present case is not at our disposal, since it corresponds to the value of the function at the joint, which we don't actually know.
Turn to the interval $(3.5, +\infty)$: the change of variables $u = 4/(x + 0.5)$ allows us to transform onto the finite interval $(0, 1)$. Our differential equation is now:

$$u^2 W''(u) + 2uW'(u) - \frac{16}{u^2}(1 - u^2)W(u) = 0 \tag{8.167}$$

and, since we were integrating the original ODE inward, our starting point is $W_0 = 0$. Discretize this equation, similarly to Eq. (8.111), using n_r points and noting, as above, that we don't actually know the value of W_{n_r-1}.
In order to produce the two missing equations we will now, just like in problem 8.22, impose the continuity of the function and derivative values at the joint. The former is $w(3.5) = W(1)$, which translates to $w_{n_l-1} = W_{n_r-1}$. For the latter, you should rewrite

$$\left.\frac{dW}{du}\right|_{u=1} = \left.\frac{dw}{dx}\right|_{x=3.5} \left.\frac{dx}{du}\right|_{u=1} \tag{8.168}$$

in terms of the second forward-difference approximation of Eq. (3.27) and the corresponding second backward difference. In total, you should have $(n_l - 1) + (n_r - 1) + 2 = n_l + n_r$ equations and as many unknowns.
Implement this approach in Python. Be sure to set up the solution vector as consisting first of all the w_i's and then the W_i's *in reverse order* (so that the two values at the middle will end up being identical). Obviously, in order to show the final solution on the same plot you will have to employ $x = -0.5 + 4/u$.

8.24 Experiment with the shooting method for an eigenvalue problem. Specifically, reproduce Fig. 8.8 which shows that it is the eigenvalue, and not the starting derivative, which allows us to satisfy the boundary condition.

8.25 Reproduce our plots of the Mathieu functions in Fig. 8.9 using `evp_shoot.py` for the left panel and `evp_matrix.py` for the right panel.

8.26 Average Eq. (8.134) and Eq. (8.136), developed when introducing the FTCS and BTCS scheme, respectively. Show that this leads to the scheme:

$$-\gamma\phi_{j-1,k+1} + (1 + 2\gamma)\phi_{j,k+1} - \gamma\phi_{j+1,k+1} = \gamma\phi_{j-1,k} + (1 - 2\gamma)\phi_{j,k} + \gamma\phi_{j+1,k} \tag{8.169}$$

where, this time, $\gamma = \alpha h_t/(2h_x^2)$ and this equation applies to $j = 1, 2, \ldots, n-2$ and $k = 0, 1, \ldots, m-2$. This is known as the *Crank–Nicolson method*. Implement it for the problem of `diffusion.py`. Produce a plot like Fig. 8.12; when debugging, you can try doubling m: the results should be essentially unchanged (in contradistinction to what happens for BTCS and, even more so, for FTCS). Hint: it may help you to cast Eq. (8.169) in the form of $\mathbf{A}\boldsymbol{\phi}_{k+1} = \mathbf{B}\boldsymbol{\phi}_k$.

8.27 [$\mathcal{A}$] We now guide you toward a *von Neumann stability analysis* of the techniques used to solve the diffusion equation of Eq. (8.130). In essence, this means writing down a generic *eigenmode*:

$$u_{j,k} = [\xi(q)]^k \exp[iqjh_x] \tag{8.170}$$

and demanding that the (complex) $\xi(q)$ satisfy $|\xi(q)| < 1$. Plug Eq. (8.170) into the FTCS Eq. (8.135), the BTCS Eq. (8.137), and the Crank–Nicolson Eq. (8.169). You should find that the BTCS and Crank–Nicolson methods are unconditionally stable, whereas FTCS is stable only when $h_t < h_x^2/(2\alpha)$.

8.28 [$\mathcal{P}$] In *molecular dynamics* simulations (see problem 8.58), the *velocity Verlet algorithm*[31] tackles the second-order ODE $w'' = f(x, w, w')$ by splitting it into two first-order ODEs, employing the definitions in Eq. (8.84), as usual. However, instead of bundling $y_0(x)$ and $y_1(x)$ into a vector $\mathbf{y}(x)$ and stepping through each in parallel like we did in the main text, the velocity Verlet method employs different steps for each of the two functions (see also problem 8.48):

$$\begin{aligned} k &= (y_1)_j + \frac{h}{2} f[x_j, (y_0)_j, (y_1)_j] \\ (y_0)_{j+1} = (y_0)_j + hk, \qquad (y_1)_{j+1} &= k + \frac{h}{2} f[x_{j+1}, (y_0)_{j+1}, k] \end{aligned} \tag{8.171}$$

Implement this explicit method and compare its output with RK4 for the Legendre differential equation, Eq. (8.91), for $n = 6$ and $n = 100$.

8.29 [$\mathcal{P}$] To solve the time-independent Schrödinger equation, one often employs *Numerov's method*. For $w''(x) = f(x)w(x)$—see problem 8.1—the Numerov algorithm (is more accurate than RK4 and) produces the next value of the solution by:

$$w_{j+1} = \frac{\left[2 + \frac{5}{6}h^2 f(x_j)\right] w_j - \left[1 - \frac{1}{12}h^2 f(x_{j-1})\right] w_{j-1}}{1 - \frac{1}{12}h^2 f(x_{j+1})} \tag{8.172}$$

This discretizes the $w(x)$ directly, i.e., in contradistinction to what we did for other IVP methods, here we do not define $y_0(x)$ and $y_1(x)$. Solve the problem:

$$w''(x) = \frac{5}{1+x^2} w(x), \qquad w(0) = 1, \qquad w'(0) = 0 \tag{8.173}$$

using both Numerov's method and RK4. Observe that Numerov's method requires

[31] A similar idea goes under the name of the *leapfrog method* in the study of partial differential equations. Somewhat surprisingly, its application to the diffusion equation of Eq. (8.130)—a centered-time centered-space (CTCS) scheme known as the Richardson method—is unconditionally *unstable*. This, despite the fact that its time-derivative approximation error is $O(h_t^2)$, i.e., better than, say, that of the FTCS scheme (cf. problem 8.27).

two w values to get going (it is a multistep method). However, in addition to $w_0 = w(0)$, you also know $w'(0)$, so you can use it to approximate w_1 (cf. problem 8.7).

8.30 [$\mathcal{P}$] Derive Eq. (8.4) for the problem of the two-dimensional projectile in the presence of air resistance. Implement this programmatically, using `rk4_gen()` and plot the trajectory—i.e., $y(t)$ as a function of $x(t)$. To be concrete, take $k = 1, g = 9.81, x(0) = 1, v_x(0) = 2, y(0) = 5$, and $v_y(0) = 7.808$; integrate from $t = 0$ up to $t = 2.5$.

8.31 [$\mathcal{P}$] Solve Eq. (8.4) for the problem of the two-dimensional projectile in the presence of air resistance, this time as a BVP, i.e., with the input being $x(0) = 2, v_x(0) = 3, y(0) = 4$, and $y(2.5) = 0$, compute the needed $v_y(0)$. While benchmarking, try out the starting values given in problem 8.30 (where you know how large $v_y(0)$ turns out to be).

8.32 [$\mathcal{P}$] Use the shooting method for the quantum-mechanical eigenvalue problem of the *infinite square well* (which is different from the periodic box we encountered in section 3.5.1.2 and the finite well of section 5.1.1). Specifically, take the potential to be zero for $-a < x < a$ and infinite at the boundaries; compute the first six eigenvalues and compare with:

$$E_n = \frac{\hbar^2\pi^2}{8ma^2}n^2 \tag{8.174}$$

which you learned in a course on quantum mechanics. Take $\hbar^2/m = 1$ and $a = 0.5$.

8.33 [$\mathcal{P}$] Use code 8.6 to tackle the *quantum harmonic oscillator*, for which the "potential" term in Eq. (8.116) is $0.5qx^2$ instead of $2q\cos 2x$. Tackle two finite-difference discretization schemes: first, Eq. (8.121) which has error $O(h^2)$ and, second, the approximation which has error $O(h^4)$. The eigenfunctions are "periodic" in that they are expected to die off at large distances. Compare the answer to the (analytically known) eigenvalues of Eq. (3.71). You may use `numpy.linalg.eigvals()`.

8.34 [$\mathcal{P}$] For the first six eigenvalues of the harmonic oscillator of problem 8.33, apply *Richardson extrapolation*: for a given finite-difference approximation to the second derivative, say, that of Eq. (8.121), carry out a calculation for step size h and another one for $h/2$ and use Eq. (3.47); compare with the "unextrapolated" values.

8.35 [$\mathcal{P}$] If you take $q = 0$ in code 8.6, i.e., `evp_matrix.py`, you get the non-interacting particle in a box (of length 2π), see section 3.5.1. From Eq. (3.82) the energy is analytically known to be (proportional to) n^2; thus (except for the 0th eigenvalue), the energies will come in pairs, e.g., $n = \pm 3$. For the first 50 distinct eigenvalues, plot: (a) the exact eigenvalue, and (b) the numerical answer for matrix-approach discretizations using $n = 100, 200, 300, 400, 500$ vs the cardinal number of the eigenvalue; feel free to use `numpy.linalg.eigvals()`. Do you notice a pattern?

8.36 [$\mathcal{A}, \mathcal{P}$] In quantum mechanics, the Mathieu equation takes the form:

$$-\frac{\hbar^2}{2m}\psi''(x) + 2q\cos 2x\,\psi(x) = E\psi(x) \tag{8.175}$$

In the spirit of the Project in section 7.8, we can view the Hamiltonian on the left-hand side as $\hat{H} = \hat{H}_0 + \hat{H}_1$. We will now see how to tackle this problem in perturbation theory. In problem 8.35 we took $q \to 0$: this leads to the non-interacting particle-in-box, with the eigenfunctions and eigenenergies being given by Eq. (3.80) and

Eq. (3.82), respectively; let us denote the state vector of the non-interacting problem by $|n^{(0)}\rangle$ and the corresponding energy by $E_n^{(0)}$. Show that the first-order correction to the energy, $E_n^{(1)} = \langle n^{(0)}|\hat{H}_1|n^{(0)}\rangle$, vanishes. Then, derive a simple analytical expression for the second-order correction to the energy, starting from:

$$E_n^{(2)} = \sum_{k\neq n} \frac{|\langle n^{(0)}|\hat{H}_1|k^{(0)}\rangle|^2}{E_n^{(0)} - E_k^{(0)}} \tag{8.176}$$

8.37 [$\mathcal{P}$] Compute the first six eigenvalues of the Mathieu equation (for $q = 1.5$ with $n = 150$) using a higher-order approximation for the second derivative. That is, instead of Eq. (8.121), which has error $O(h^2)$, use an approximation which has error $O(h^8)$.

8.38 [$\mathcal{P}$] In the main text (implicitly) and in earlier problems (explicitly), we have been treating the Schrödinger equation as a differential equation, working in the position basis. We now learn that it can be cast and solved as an *integral equation*, by working in the momentum basis. First, show that, by using the resolution of the identity in one dimension, $\int dk|k\rangle\langle k|$, the Schrödinger equation in Eq. (4.276) takes the form:

$$\frac{k^2}{2m}\psi(k) + \int dk' V(k,k')\psi(k') = E\psi(k) \tag{8.177}$$

Use quadrature and show that this equation is equivalent to:

$$\sum_j H_{ij}\psi_j = E\psi_j \tag{8.178}$$

This is the eigenvalue problem $\mathbf{H}\boldsymbol{\psi} = E\boldsymbol{\psi}$, see Eq. (4.226). Finally, compute the ground-state energy for $V(k,k') = -e^{-(k-k')^2/4}/(2\sqrt{\pi})$; use 100 Gauss–Legendre points from -10 to $+10$. While benchmarking, use the matrix approach to also solve the corresponding differential equation in coordinate space, for $V(x) = -e^{-x^2}$.

8.39 [$\mathcal{P}$] Compute $\phi(x,y)$ for Eq. (8.140), allowing for different discretizations in the x and y coordinates. Specifically, reproduce Fig. 8.14 for $n_x = 64$ and $n_y = 128$.

8.40 [$\mathcal{P}$] In a course on electromagnetism you've seen how to solve Poisson's equation via the use of *Green's functions*; for the two-dimensional case this takes the form:

$$\phi(x,y) = \int \frac{f(x',y')}{|\mathbf{r}-\mathbf{r}'|} d^2r' \tag{8.179}$$

Plot the electric potential for the problem of Eq. (8.140) by carrying out this two-dimensional integral numerically, say, via Gauss–Legendre quadrature; note that you need to compute such an integral *for each* value of x and y. As if that wasn't bad enough, you now have to worry about the denominator value(s) being close to zero.

8.41 [$\mathcal{P}$] Tackle the problem of Eq. (8.140) by setting up Eq. (8.144) as a linear algebra problem and solving it. Take $n = 4$ and assume that $\phi = 7$ at the boundary. This leads to nine equations for the nine unknown $\phi_{p,q}$'s, for the interior points. Start from $p = q = 1$: some of the terms in Eq. (8.144) are known from the boundary condition, so they can be moved to the right-hand side. Explicitly write out the other eight equations, grouping everything into the form $\mathbf{Ax} = \mathbf{b}$, where $\mathbf{x}$ bundles the nine interior $\phi_{p,q}$'s column-wise: that means it contains $\phi_{1,1}, \phi_{2,1}, \phi_{3,1}, \phi_{1,2}, \phi_{2,2}, \ldots$.

8.42 [$\mathcal{P}$] Solve a problem that is similar to Eq. (8.140), but this time for a driving term which is $f(x, y) = ce_1(3x+4y) - ce_1(5x-2y)$, where ce_1 is the second even Mathieu function for the case of $q = 1.5$; make sure to normalize this Mathieu function in such a way that $ce_1(0) = 1$. You want to be able to evaluate the right-hand side of Poisson's equation for any possible argument, so implement ce_1 as a combination of an eigenvalue-problem (to produce the function on a grid) and an interpolation via FFT (to allow you to compute its value for points off the grid).

8.43 [$\mathcal{P}$] Let us revisit the (non-linear) pendulum, first introduced in problem 1.16, along with the phase plane which helped us physically interpret what was going on. If you did not solve that problem when working through chapter 1, you should do so now.

(a) In order to facilitate debugging and physical understanding, let us start from the same scenario we encountered in problem 1.16. Since we can now tackle differential equations directly, we turn to the equation of motion corresponding to Eq. (1.5):

$$\ddot{\theta}(t) + \frac{g}{l}\sin\theta(t) = 0 \tag{8.180}$$

where, as usual, the double-dot denotes a second time derivative. Take $l = 1$ m, $g = 9.8$ m/s^2, $\theta(0) = 0$; give $\dot{\theta}(0)$ a range of values from -10 to $+10$ and use RK4 to integrate from $a = 0$ to a b that you think is sufficient. To determine how many $\dot{\theta}(0)$ values to use as well which specific b to employ, you should take your results and overlay them on the $\dot{\theta}$ vs θ plot from problem 1.16. This will also allow you to debug your new code (or the old one, for that matter).

(b) Now turn to the *damped pendulum*, which is more realistic in that it can capture the effects of friction/losses:

$$\ddot{\theta}(t) + \frac{g}{l}\sin\theta(t) + \tau\dot{\theta}(t) = 0 \tag{8.181}$$

For concreteness, take $\tau = 0.5$; this is an example of *subcritical damping* (also known as an *underdamped pendulum*). As in the previous part of this problem, you should use RK4 to integrate your ODE; this time use a b that is $3-4$ times larger. You should find that the open curves now have decreasing amplitudes; as a result, just because the pendulum could swing over once, it does not mean that it will be able to do so twice (or thrice, and so on). Compare this to what you found for the case of the undamped pendulum.

8.44 [$\mathcal{P}$] We turn to the *deflection of light* in general relativity, one of the early signatures that Newtonian gravity is not the full story. Specifically, in a Schwarzschild spacetime, picture a photon starting out with $\phi = 0$ at some large distance and passing by a star of mass M. The shape of a photon worldline $r(\phi)$ is given by the ODE:

$$\frac{d^2u}{d\phi^2} = 3GMu^2 - u \tag{8.182}$$

where $u = 1/r$. For $M = 0$ the solution is simply $u(\phi) = u_c \sin\phi$; this is consistent with the initial conditions being (in appropriate units) $u(0) = 10^{-7}$ and $u'(0) = 10^{-5}$

(i.e., $u_c = 10^{-5}$). Compare that answer with the solution to the full problem for $GM = 45\,000$, integrating up to $\phi = \pi$.

8.45 [$\mathcal{P}$] We return to the textbook setting of a single particle in a (one-dimensional) finite square well (mentioned in section 5.1.1) this time viewed as a (differential-equation) eigenvalue problem—the time-independent Schrödinger equation, Eq. (8.7). Recall that the potential is V_0 for $x < -a$ and $x > a$ and is 0 inside the well, i.e., for $-a < x < a$. As in problem 5.48, we take $V_0 = 20$, $a = 1$, and $\hbar^2/m = 1$; once again, you will need to find all the (even) bound-state solutions supported by this well. Use the shooting method (along with RK4) to tackle this problem. Hint: you will need to be careful with the limits of the integration, especially for the most weakly bound state.

8.46 [$\mathcal{P}$] In the *restricted gravitational three-body problem* one of the three masses is taken to be negligible in comparison to the other two; the latter are on a circular orbit, which is not disturbed by the presence of the third body. The problem involving the two non-negligible masses is independent and taken to be solved (giving rise to a circular orbit): this is precisely the Roche model that we encountered in problems 1.17, 5.50, and 7.56. Thus, the restricted three-body problem is focused on describing the motion of the small body in the presence of the gravitational fields of the two large bodies. In the earlier problems we studied binary stars; another physical example could be the Earth (of mass M), the Moon (of mass m), and a spacecraft (of negligible mass); take all three bodies to be in the same plane.

It is convenient to switch to a coordinate frame that eliminates the circular orbit of the two large bodies, i.e., in which these two bodies are fixed; this is accomplished by employing a co-rotating/synodic frame. In the co-rotating frame, the differential equations for the $x(t)$ and $y(t)$ characterizing the small body are:

$$\begin{aligned}\frac{d^2x}{dt^2} &= x + 2\frac{dy}{dt} - w\frac{x+u}{[(x+u)^2+y^2]^{3/2}} - u\frac{x-w}{[(x-w)^2+y^2]^{3/2}}\\ \frac{d^2y}{dt^2} &= y - 2\frac{dx}{dt} - w\frac{y}{[(x+u)^2+y^2]^{3/2}} - u\frac{y}{[(x-w)^2+y^2]^{3/2}}\end{aligned} \tag{8.183}$$

where $u = m/(m+M)$ and $w = 1 - u$. For the specific case where the two large bodies are the Earth and the Moon, we have $M/m = 80.45$. Our equations assume that the Earth is fixed at the origin and the Moon is fixed at the point $(1\ 0)^T$. You should employ the following initial conditions:

$$x(0) = 0.994, v_x(0) = 0, y(0) = 0, v_y(0) = -2.00158510637908 \tag{8.184}$$

where $v_x = dx/dt$ and $v_y = dy/dt$ (like in the two-dimensional projectile). These initial conditions give rise to a special periodic solution, the *Arenstorf* orbit, which was instrumental in the planning of the Apollo missions to the Moon.

Use RK4 and plot the trajectory—i.e., $y(t)$ as a function of $x(t)$—of the small body. Study the dependence on the point up to which you will be integrating, b; you should stop when you've traced a full period. You should also investigate the dependence of the trajectory on the number of integration points, n, you use.

8.47 [$\mathcal{P}$] In quantum optics, a standard problem involves the study of a two-level atom interacting with a classical field. One way of quantitatively describing such a system is via the *optical Bloch equations*:

$$\frac{d}{dt}\begin{pmatrix}\rho_{ee}\\ \tilde{\rho}_{eg}\\ \tilde{\rho}_{ge}\\ \rho_{gg}\end{pmatrix} = \begin{pmatrix}-\Gamma & +i\Omega/2 & -i\Omega/2 & 0\\ +i\Omega/2 & 0 & -(\gamma-i\Delta) & -i\Omega/2\\ -i\Omega/2 & 0 & -(\gamma+i\Delta) & +i\Omega/2\\ +\Gamma & -i\Omega/2 & +i\Omega/2 & 0\end{pmatrix}\begin{pmatrix}\rho_{ee}\\ \tilde{\rho}_{eg}\\ \tilde{\rho}_{ge}\\ \rho_{gg}\end{pmatrix} \tag{8.185}$$

$$\rho_{ee}(0) = 1, \qquad \tilde{\rho}_{eg}(0) = 0, \qquad \tilde{\rho}_{ge}(0) = 0, \qquad \rho_{gg}(0) = 0$$

where the ρ's are density-matrix elements, corresponding to excited-state and ground-state populations as well as the quantum coherences between the two levels. (The $\tilde{\rho}$'s are given in a different frame and are known as the "slowly varying coherences".) The coefficients, too, have a physical interpretation: Ω is the Rabi frequency, Δ is the detuning, Γ is the longitudinal decay rate, and γ is the transverse decay rate. For concreteness, take: $\Omega = 2$, $\Delta = 0.3$, $\Gamma = 0.2$, and $\gamma = 0.15$. Note that we have also provided a possible set of initial conditions. Computationally, we are dealing with an IVP for an autonomous system of four coupled ODEs; the only complication in comparison to what we saw in the main text is that you will need to generalize your RK4 function such that it works with complex numbers. It's up to you whether you address all four equations, or you use the relations:

$$\rho_{ee} + \rho_{gg} = 1, \qquad \tilde{\rho}_{eg} = \tilde{\rho}^*_{ge} \tag{8.186}$$

to eliminate two of the ODEs. (If you do work with four equations, you should check to make sure that the relations in Eq. (8.186) are always satisfied.) Experiment with your RK4 function to determine how many integration points, n, you need to use. In order to determine the right endpoint of the integration, you can compare with the steady-state solutions:

$$\rho_{ee}(t \to \infty) = \frac{\Omega^2}{2\gamma\Gamma}\frac{1}{1+\frac{\Delta^2}{\gamma^2}+\frac{\Omega^2}{\gamma\Gamma}}, \qquad \tilde{\rho}_{eg}(t\to\infty) = -\frac{i\Omega}{2\gamma}\frac{1+\frac{i\Delta}{\gamma}}{1+\frac{\Delta^2}{\gamma^2}+\frac{\Omega^2}{\gamma\Gamma}} \tag{8.187}$$

Plot (the modulus of) each of the four density-matrix elements as a function of time, thereby showing that you have reached the steady state.

8.48 [$\mathcal{P}$] We now go over another example of *symplectic integration*: such techniques are often applied to periodic problems, because they satisfy certain conditions (e.g., decent energy conservation) even when the specific technique employed is not a high-order one.[32] To illustrate what's at stake, we will address a very simple ODE, the (classical) simple harmonic oscillator:

$$\frac{d^2x(t)}{dt^2} + \omega^2 x^2(t) = 0 \tag{8.188}$$

For concreteness, take $\omega = 2$, $x(0) = 3$, and $v(0) = 0$, for which the analytical solution is $x(t) = 3\cos(2t)$; we will be integrating from 0 to 5, using 200 integration points.

[32] The velocity Verlet algorithm of Eq. (8.171) also has such properties (actually, even better ones).

(a) Plot the solutions coming from the explicit Euler method as well as RK4. Also show the analytic solution, as well as the solution corresponding to the technique:

$$v_{j+1} = v_j - h\omega^2 x_j, \qquad x_{j+1} = x_j + hv_{j+1} \tag{8.189}$$

As you may have gathered, v here is the time-derivative of the position. This is known as the *Euler–Cromer method* (or the *semi-implicit Euler method*); note that you need to evaluate the equations in the order shown.

(b) Plot the quantity $E = (v^2 + \omega^2 x^2)/2$ as a function of time, for the four cases of analytic, explicit Euler, RK4, and Euler–Cromer. Zoom in to see the detailed trend of the RK4 behavior and qualitatively compare it to the Euler–Cromer one.

8.49 [$\mathcal{P}$] This problem addresses the *stability of the outer solar system* in the context of Hamiltonian mechanics, i.e., the trajectories of five small-ish masses m_{1-5} (those of Jupiter, Saturn, Uranus, Neptune, and Pluto, respectively) in the presence of one large mass m_0, i.e., the Sun. All masses will be measured relative to the solar mass; the latter is itself taken to be slightly larger than unity, to account for the inner planets. We will take the masses, initial positions, and initial velocities of all six objects as given; these are provided in a data file at `www.numphyspy.org` for your convenience. We will also be measuring time in Earth days, distances in astronomical units, and the gravitational constant will be $G = 2.95912208286 \times 10^{-4}$. The Hamiltonian of the six-body system is:

$$H = \sum_{k=0}^{5} \frac{\mathbf{p}_k^T \mathbf{p}_k}{2m_k} - G \sum_{k=1}^{5} \sum_{j=0}^{k-1} \frac{m_j m_k}{\|\mathbf{q}_j - \mathbf{q}_k\|} \tag{8.190}$$

where $\mathbf{p}_k$ is the momentum of the k-th mass (made up of its x, y, z components) and, similarly, $\mathbf{q}_k$ is the three-dimensional position of the k-th mass. The denominator in the second term involves the Euclidean norm of Eq. (4.23), as elsewhere.

(a) Show that Hamilton's equations take the form:

$$\dot{\mathbf{q}}_k = \frac{\mathbf{p}_k}{m_k}, \qquad \dot{\mathbf{p}}_k = G \sum_{j=0, j\neq k}^{5} \frac{m_j m_k (\mathbf{q}_j - \mathbf{q}_k)}{\|\mathbf{q}_j - \mathbf{q}_k\|^3} \tag{8.191}$$

for the Hamiltonian in Eq. (8.190). As usual, a dot here denotes a time derivative.

(b) Taken together, Eq. (8.191) constitute 36 coupled first-order ODEs. Set this IVP problem up programmatically and solve it using the explicit Euler and RK4 methods for a step size of $h = 5$ days, integrating for 40,000 steps. Note that you will have to figure out how to plot three-dimensional trajectories in order to visualize the output of your code. Discuss detailed features of your results, e.g., whether or not Saturn stays on the original plane, and so on.

(c) If you solved problem 8.48, produce another plot using the Euler–Cromer method. You now have to deal with more equations and, crucially, to distinguish between momentum and velocity.

8.50 [$\mathcal{P}$] In the study of the kinetics of autocatalytic reactions, a well-known problem is that of the *Robertson reaction network*:

$$\begin{aligned} y_0'(x) &= -0.04y_0(x) + 10^4 y_1(x)y_2(x), \\ y_1'(x) &= 0.04y_0(x) - 10^4 y_1(x)y_2(x) - 3\times 10^7 y_1^2(x), \\ y_2'(x) &= 3\times 10^7 y_1^2(x), \\ y_0(0) &= 1, \qquad y_1(0) = 0, \qquad y_2(0) = 0 \end{aligned} \tag{8.192}$$

At first sight, this is simply an IVP for an autonomous system of three (non-linear) coupled ordinary differential equations. Upon closer inspection, we realize that the three distinct coefficients on the right-hand side are of wildly varying magnitudes. As a result, a generalization of the (linear) analysis we carried out in section 8.2.4.4 would show that the eigenvalues involved here are also of considerably different magnitudes, and hence that we are faced with a *stiff* system of ODEs. This means that the solutions will also be characterized by different magnitudes (and will be interesting in different regions). Specifically, $y_1(x)$ will turn out to exhibit a very quick initial transient (and its overall size will remain small), whereas $y_0(x)$ and $y_2(x)$ take considerably longer to decay and grow (respectively).

(a) Use RK4 to integrate this system up to $b = 5$. Be sure to use enough integration points, n, such that you can trust the final answers. Plot all three functions, multiplying $y_1(x)$ by an appropriate prefactor that allows its behavior to show up on the same scale as for the other two functions.

(b) With the RK4 results under your belt, you are free to turn to other techniques. In the main text we stressed the fact that stiff systems are frequently tackled using implicit methods. Start from the backward Euler method, namely the non-linear system of algebraic equations of Eq. (8.88), giving you $\mathbf{y}_{j+1}$; such a non-linear system needs to be solved at each iteration. Use the multidimensional Newton's method to do this; it is always good to use analytical derivatives when possible, so code up a separate function for the Jacobian. As is usual in Newton's method, you will need an initial guess for the solution; the easiest thing to do is to employ $\mathbf{y}_j$ (each time) as the best available estimate for $\mathbf{y}_{j+1}$. Hint: given how we set up the problem, you will need to write a function called, say, `fs()` that you will then pass in to Newton's method. This `fs()`, however, will *not* correspond to the `fs()` in `ivp_two.py`—or part (a) of the present problem, for that matter. You are now using a multidimensional root-finder, so your `fs()` will involve the entire equation(s) in Eq. (8.88), with all the terms moved to the same side.

(c) Having implemented an implicit technique for a non-linear stiff system of ODEs in part (b), it shouldn't be too hard to repeat the exercise for the implicit-trapezoid method. First, generalize Eq. (8.63) to the case of ν ODEs, and then write out the non-linear algebraic system that corresponds to the implicit-trapezoid method applied to Eq. (8.192). Take the needed derivatives in order to set up the Jacobian analytically. Implement this and produce a plot that shows results for both the implicit-Euler and the implicit-trapezoid methods.

8.51 [$\mathcal{P}$] We now return to the problem of determining *stellar structure* by solving the Tolman–Oppenheimer–Volkoff (TOV) equation(s). After Eq. (8.2) we suggested that one way to proceed is to use the equation of state, giving us $P(\rho)$, and convert Eq. (8.1) such that it involves a derivative of $\rho(r)$. To be explicit, that meant using:

$$\frac{dP(r)}{dr} = \left(\frac{dP(\rho)}{d\rho}\right)\left(\frac{d\rho}{dr}\right) \tag{8.193}$$

and then keeping only $d\rho/dr$ on the left-hand side of the equation. This is the natural approach to take if you're employing an equation of state in the form $P(\rho)$. Here we explore an alternative scenario, where you have been given the equation of state in the form $\rho(P)$. (Obviously, in trivial cases one can simply invert to go from one formulation to the other.) Thus, we will end up with versions of Eq. (8.1) and Eq. (8.2) in which both the left-hand sides and the right-hand sides involve only $P(r)$ and $m(r)$.

(a) Instead of getting lost in a sea of units and factors of c^2, it is convenient to make things dimensionless as follows:

$$m = M_\odot \overline{m}, \quad r = R_S \overline{r}, \quad P = \gamma \overline{P} \tag{8.194}$$

where $M_\odot = 1.989 \times 10^{30}$ kg is the mass of our Sun, $R_S = 2GM_\odot/c^2 = 2.954$ km is the corresponding Schwarzschild radius, and γ will be chosen below to make our lives easier. Crucially, $\overline{m}$, $\overline{r}$, and $\overline{P}$ are dimensionless. Think about the units of P, γ, and ρ and find a way to produce a dimensionless mass density, $\overline{\rho}$, by re-using γ (possibly appropriately rescaled).

Analytically show that when they are rewritten in terms of the above dimensionless quantities, Eq. (8.1) and Eq. (8.2) take the form:

$$\begin{aligned}\frac{d\overline{P}}{d\overline{r}} &= -\frac{\overline{m}\,\overline{\rho}}{2}\left[1+\frac{\overline{P}}{\overline{\rho}}\right]\left[1+\frac{4\pi\overline{r}^3\overline{P}}{\overline{m}}\frac{\gamma R_S^3}{M_\odot c^2}\right]\left[\frac{1}{\overline{r}^2-\overline{m}\,\overline{r}}\right]\\ \frac{d\overline{m}}{d\overline{r}} &= 4\pi\overline{r}^2\overline{\rho}\frac{\gamma R_S^3}{M_\odot c^2}\end{aligned} \tag{8.195}$$

Notice that both equations involve $\gamma R_S^3/(M_\odot c^2)$: if you now take $\gamma = M_\odot c^2/R_S^3$ (as you can), that term goes away and both equations look particularly simple.

(b) We now prepare to solve our ODEs numerically. Since there is a numerical singularity at $r = 0$, you should start the integration of the TOV equations at a small (but finite) r_Δ. Analytically integrate Eq. (8.2) from 0 to r_Δ to find:

$$m(r_\Delta) = \frac{4}{3}\pi r_\Delta^3 \rho_c \tag{8.196}$$

which can, in turn, be expressed in terms of the dimensionless quantities. The only missing piece of the puzzle is the equation of state. Here we employ the following phenomenological form:

$$\overline{\rho} = 0.871\overline{P}^{3/5} + 2.867\overline{P} \tag{8.197}$$

which has been tailored to capture both non-relativistic and relativistic microphysics (in the first and second terms, respectively) for a neutron star. It's probably also wise to programmatically account for the, numerical, possibility of negative pressure; make sure your code produces a tiny positive $\overline{\rho}$ in that case.

Guess a value of the dimensionless central pressure $\overline{P}_c = 0.01$, use it to get $\overline{\rho}_c$, and from there compute $\overline{m}(\overline{r}_\Delta)$; then, use RK4 to solve Eq. (8.195). Observe that you don't actually need to know the numerical values of $M_\odot$, R_S, and γ to do this: you will simply find the total radius $\overline{R}$ and total mass of the star $\overline{M}$ (determined when the pressure goes to zero) as dimensionless quantitities.

Note a further complication: in the language of RK4, we don't know ahead of time up to where we wish to integrate, i.e., we don't know which b to pass in. You should employ two different strategies: (i) guess a large b value and keep only that portion of the solution that corresponds to positive pressures that are larger than some threshold, and (ii) treat this as a BVP, where you know $\overline{P}(0) = \overline{P}_c$ and $\overline{P}(\overline{R}) = 0$; use a root-finder to determine $\overline{R}$. In both approaches, your end goal is to compute the mass $\overline{M}$ (in units of the solar mass, $M_\odot$) and the radius R (in km).

(c) Repeat the above calculation for a whole range of initial $\overline{P}_c$ values, thereby producing a *mass–radius plot* with the total radius on the x axis and the total mass on the y axis. Discuss the plot's qualitative features. Hint: you should be careful when choosing the number of integration points, n.

8.52 [$\mathcal{P}$] In scalar self-interacting field theory, there exist non-trivial solutions of the (Euclidean) field equations for which the action is finite. These *instanton solutions* satisfy the following ordinary differential equation:

$$\phi''(r) = -\frac{n_d - 1}{r}\phi'(r) + \phi(r) - \phi^3(r) \tag{8.198}$$

where n_d is the number of spatial dimensions ($n_d = 1, 2, 3$). This could be reinterpreted as describing the mechanical motion of a unit-mass particle in the double-well potential $V(\phi) = -\phi^2/2 + \phi^4/4$ (and r re-interpreted as the time).

Regardless of the interpretation, our task here will be to use RK4 to numerically solve Eq. (8.198) from the origin ($a = 0$) out to infinity ($b = +\infty$), imposing regularity at the origin ($\phi'(0) = 0$) and a starting $\phi(0)$ that will be discussed below. While we haven't dealt with an infinite right endpoint much, this is straightforward to handle: you can either carry out a transformation like those introduced in section 7.6 (or in problem 8.23) to make the interval finite or, for this simple case, you can take the right endpoint of integration to be large but finite, e.g., $b = 12$.

(a) For $n_d = 1$ compare your solution to the analytical answer $\phi(r) = \sqrt{2}/\cosh r$. You can analytically drop the $n_d - 1$ term from the right-hand side of Eq. (8.198), or employ the prescription introduced in the next part of this problem. Here you are free to "guess" the $\phi(0)$, based on the exact answer we have provided.

(b) We now turn to $n_d = 2$. You can immediately see that RK4 gets into trouble when starting out (for the cases of $n_d = 2$ and $n_d = 3$), since our ODE has a singularity at $r = 0$. This is easily handled by taking the starting point to be tiny but finite, say, $a = 10^{-12}$. Use starting values $\phi(0) = 2.0, 2.5, 2.20620086465$ and interpret the behavior of the three solutions in relation to the afore-mentioned double-well potential. Note that if your last curve does not stay flat as r approaches b then you need to use more integration points (i.e., a larger value of n). Thus, one of your findings will be that for a "magic" value of $\phi(0)$ you will obey the boundary condition $\phi(+\infty) = 0$.

(c) Next, turn to $n_d = 3$ and do an analogous study, starting with $\phi(0) = 4.0$ and 4.5; this time we are not providing the intermediate "magic" value, so you will have to determine that yourself. In short, you are faced with a BVP, but this time around you know the value of the starting derivative $\phi'(0) = 0$ and need to determine $\phi(0)$. You will have to combine RK4 with a root-finding algorithm; the secant method behaves too wildly for the present purposes, so you should use the bisection method, instead.

(d) Now that you have produced a general framework, you are able to find more solutions. As you have probably gathered already, for most values of $\phi(0)$ the solution will *not* obey $\phi(+\infty) = 0$. However, there will be other "magic" values that will allow you to satisfy this boundary condition. You should compute another such solution for each of $n_d = 2$ and $n_d = 3$.

8.53 [$\mathcal{P}$] The *Cornell potential* (linear plus Coulomb) was introduced in the study of *quarkonium*, which is a model for heavy-quark mesons. Since the quark and antiquark involved are heavy, it is reasonable to use the non-relativistic Schrödinger equation, which becomes (roughly):

$$\frac{d^2u(r)}{dr^2} = \left(r - \frac{1}{r} - E\right)u(r) \tag{8.199}$$

This looks similar to the instanton problem of Eq. (8.198), in terms of the singularity and the fact that we need to integrate from $r = 0$ out to infinity but, crucially, here we are dealing with an EVP (there is an energy eigenvalue on the right-hand side). Demand that $u(r)$ go to zero (or to a small value) at both a and b; these can be picked as in problem 8.52. Use the shooting method to compute the first five eigenvalues, also plotting the corresponding eigenfunctions.

8.54 [$\mathcal{P}$] The *theory of S-wave scattering* is typically covered near the end of a course on quantum mechanics; far from being an "advanced" topic, it is absolutely crucial to the connection of subatomic theory with experiment. Courses usually limit themselves to analytically solvable scenarios (e.g., hard-sphere scattering); this may leave students unsure how to generalize things. In what follows, we will discuss a spherically symmetric short-range potential, $V(r)$; in the center-of-momentum frame, employing the relative coordinate r, the relevant differential equation takes the form:

$$-\frac{\hbar^2}{2\mu}\frac{d^2u(r)}{dr^2} + V(r)u(r) = Eu(r) \tag{8.200}$$

where $\mu = m_0m_1/(m_0 + m_1)$ is the reduced mass (and the scattering is between two particles of mass m_0 and m_1). Here $u(r)$ is intimately connected to the (radial part of

the) scattering wave function $\psi(r) = u(r)/r$. It is conventional to express the energy in terms of the relative wave number, $E = \hbar^2 k^2/(2\mu)$.
Given what we discussed on page 568, you may be thinking that this is an eigenvalue problem for which only discrete energies are allowed. However, what distinguishes scattering from bound-state problems is precisely the fact that the present problem has a solution for any (positive) value of E. Similarly, while the initial value $u(0) = 0$ (ensuring regularity at the origin) is something that you encounter for both scattering and bound-state problems, the situation regarding the boundary condition at some distance R (taken to be outside the range of the potential) is different: for bound-state problems the wave function rapidly goes to zero when $V(r) = 0$ whereas, in the present case, when the potential vanishes the wave function becomes a linear combination of the non-interacting solutions, which can be expressed as:

$$u(R) = A \sin(kR + \delta_0) \tag{8.201}$$

Here A is a normalization constant and $\delta_0 = \delta_0(k)$ is the S-wave *phase shift*. Practically speaking, the situation we are now faced with is quite different from what we had in section 8.4: there we guessed the starting derivative (since we didn't care about the normalization) and combined an IVP-solver with a root-finder to compute the eigenvalue. Here, any E (or equivalently k) will work, so it's not clear how one should go about matching the numerical solution of Eq. (8.200) with the analytically determined Eq. (8.201), since we still don't know the phase shift δ_0 in Eq. (8.201). To put it differently: our task is precisely the determination of the phase shift δ_0 for a given (input) wave number k.
The way forward is to realize that we will be tackling the IVP in Eq. (8.200) using RK4, by setting up the second-order differential equation as two coupled first-order differential equations. This means that when we arrive at $r = R$ we will have (numerical) access not only to $u(R)$ but also to $u'(R)$. On the analytical front, differentiating Eq. (8.201) immediately gives:[33]

$$u'(R) = Ak \cos(kR + \delta_0) \tag{8.202}$$

Combining Eq. (8.201) and Eq. (8.202) we get:

$$k\frac{u(R)}{u'(R)} = \tan(kR + \delta_0) \tag{8.203}$$

Our strategy is now clear: we will numerically solve Eq. (8.200), with $u(0) = 0$ and an immaterial guess for $u'(0)$, and then plug the values of $u(R)$ and $u'(R)$ into Eq. (8.203) to compute δ_0.[34] Let us specialize to (low-energy) *neutron–neutron scattering*. This means we can take:

$$V(r) = -c\frac{\hbar^2}{2\mu}\frac{\beta^2}{\cosh^2(\beta r)} \tag{8.204}$$

This potential is of the (modified) Pöschl–Teller form; you can use $c = 1.814164$ and

[33] This is a bit sloppy. More properly, we first differentiate $u(r)$ with respect to r and then set $r = R$.
[34] In the literature, this idea is known as the matching of the (inverse) logarithmic derivative, since $(\ln u)' = u'/u$. Note that the normalization constant cancels when you form the quotient in Eq. (8.203).

$\beta = 0.79959$ fm^{-1}. Since the potential range is a few fm, you can take $R = 15$ fm. Plot the phase shift δ_0 (measured in degrees) for $k = 0.01, 0.02, \ldots, 2.0$ fm^{-1}; make sure to use the fact that the phase shift will here always be positive to shift the output of your inverse-tangent function (thereby ensuring that it looks continuous).

8.55 [$\mathcal{P}$] In problem 8.54 we solved the IVP in Eq. (8.200), for the potential in Eq. (8.204), and determined the phase shift using Eq. (8.203). This was made (much) easier by the fact that we were given the values of c and β in Eq. (8.204): these had been tuned to correspond to the strength and range of the neutron–neutron interaction. Our present task is to carry out such a tuning process ourselves, i.e., to determine the potential parameters c and β that correspond to the desired S-wave phase shift(s). This is a necessary step in connecting nuclear forces (and, in turn, *ab initio* nuclear many-body theory) to experiment. To keep things interesting, we now turn to the case of ($^1S_0$) *neutron–proton scattering*, for which we have extracted the following from a partial-wave analysis [136]:

$$\delta_0(0.0245522951103708) = 0.51934174301976,$$
$$\delta_0(0.034722188732473) = 0.67081991599790 \tag{8.205}$$

where the phase shifts are given in radians. This is a non-linear problem with two unknowns (c and β) and two equations (Eq. (8.205)). Of course, to set up these two equations you need to solve the IVP problem of Eq. (8.200) and compute the phase shift by matching the logarithmic derivative as per Eq. (8.203).
First, combine the multidimensional Newton's method with RK4 to find c and β; after you have done this, you should plot the phase shift δ_0 (measured in degrees) for $k = 0.01, 0.02, \ldots, 2.0$ fm^{-1} and compare to what you had found in problem 8.54 for the neutron–neutron case. Hint: if you're feeling overwhelmed, first try computing the neutron–neutron c and β, for which you already know the answer, by using as input two phase shifts you get from the (simpler) code you developed for problem 8.54.

8.56 [$\mathcal{P}$] Problems 8.54 and 8.55 took the two-nucleon interaction to be of the modified Pöschl–Teller form of Eq. (8.204). There is nothing special about this form: nuclear forces are not uniquely determined by experimental data. Thus, the present problem will focus on the *shape-independent approximation*, which relates the phase shift to the scattering length (a_0) and the effective range (r_e) as follows:

$$k \cot \delta_0 = -\frac{1}{a_0} + \frac{1}{2} r_e k^2 \tag{8.206}$$

Terms that are of higher degree in k have been dropped, hence the name of the approximation. Qualitatively, you may think of the scattering length as encapsulating the strength of an interaction and the effective range as its width/spatial extent.

(a) For the neutron–neutron case ($c = 1.814164$ and $\beta = 0.79959$ fm^{-1}) produce a table of δ_0 vs k (for $k = 0.01, \ldots, 0.05$ fm^{-1}) by solving the IVP and matching the logarithmic derivative, as in problem 8.54. Then, carry out linear least-squares fitting in Eq. (8.206), with the model $c_0 + c_1 x^2$, to determine a_0 and r_e.

(b) Still for the neutron–neutron case, we will now try to recover (something close to) the same answers for a_0 and r_e by taking advantage of the $k \to 0$ limit of

Eq. (8.206); practically speaking, this means that we will be considering the low-k behavior. First, we return to the non-interacting/asymptotic solution, which we wrote down in Eq. (8.201) at a large distance R. We are now interested in its behavior for any r, so we use another symbol to denote this solution, $y(r)$, and also take the opportunity to "normalize" it:

$$y(r) = \frac{\sin(kr + \delta_0)}{\sin \delta_0} \tag{8.207}$$

Note that this obeys an inhomogeneous boundary condition ($y(0) = 1$)[35] in contradistinction to our interacting solution $u(r)$. Compute a_0 and r_e from:

$$a_0 = -\lim_{k\to 0} \frac{\tan \delta_0}{k}, \qquad r_e = 2\int_0^{\infty} dr \lim_{k\to 0} \left[y^2(r) - u^2(r)\right] \tag{8.208}$$

by taking $k = 0.001$ fm^{-1}. Note that you will have to explicitly rescale $u(r)$ such that it matches $y(r)$ in the asymptotic regime (e.g., at R). It's probably best to plot $y(r)$ and $u(r)$ together before trying to compute the integral that will give you r_e.

(c) We now return to the neutron–proton case; we will, again, try to compute c and β. This time, instead of taking two phase shifts that have been extracted from experiment, like we did in Eq. (8.205), we will employ as input:

$$a_0 = -23.749 \text{ fm}, \qquad r_e = 2.81 \text{ fm} \tag{8.209}$$

Plug these two numbers into the shape-independent approximation of Eq. (8.206) for $k = 0.01, \ldots, 0.05$ fm^{-1} and produce a table of the "true" δ_0 values vs k. These will now be your raw input to carry out a *non-linear* least-squares fit with c and β being your two undetermined parameters (i.e., we have five data points and two parameters). In problem 8.55, we had two equations and two unknowns, so we could solve that problem exactly; we are now faced with a fitting problem giving rise to an overdetermined system of *non-linear* equations; this is precisely why we introduced the Gauss–Newton method in section 6.7.1. This time around we don't know how to (or certainly don't wish to) produce the derivatives (which are needed for the Jacobian) analytically, so you should turn to a finite-difference approximation. After you've used the Gauss–Newton method to find the optimal c and β values, repeat the (*linear* least-squares) calculation you carried out in part (a) of this problem and compare the output a_0 and r_e with your input values in Eq. (8.209).

8.57 [$\mathcal{P}$] While two neutrons (in vacuum) do not form a bound state and, similarly, neither do two protons, a system composed of one neutron and one proton does form a bound nucleus, the *deuteron*. This motivates the introduction of isotopic spin T (also known as isospin) and the conclusion that this lightest nucleus (i.e., the deuteron) is that two-nucleon state which has $T = 0$ (isospin-singlet). It has been inferred that the deuteron has even parity, implying that its orbital angular momentum L is even. Since both the neutron and the proton are spin-half particles, the spin part of the deuteron wave

[35] Yet the $k \to 0$ Schrödinger equation for $y(r)/r$, involving $\partial^2/\partial r^2$ and $(2/r)\partial/\partial r$, is satisfied without a hitch.

function can be either $S = 0$ (spin-singlet) or $S = 1$ (spin-triplet)—recall Fig. 4.1. Given that the orbital part is symmetric and the isospin part antisymmetric, we see that, in order to ensure that the full wave function is antisymmetric with respect to the exchange of two nucleons, the spin part has to be symmetric, i.e., $S = 1$. The total angular momentum of the deuteron has been observed to be $J = 1$. The angular-momentum addition rule of Eq. (4.278), applied to $\mathbf{J} = \mathbf{L} + \mathbf{S}$, together with the positivity/evenness of the parity, leads to the conclusion that the deuteron is characterized by $L = 0$ and $L = 2$. Thus, in spectroscopic notation the deuteron is made up of a combination of the ${}^3S_1$ and ${}^3D_1$ states.

Since a central interaction $V(r)$ cannot produce such a mixture of ${}^3S_1$ and ${}^3D_1$ states, a new ingredient is needed, going beyond the simple potential we encountered for S-wave scattering in Eq. (8.200). While the modern theory of nuclear forces involves many different operators, we here limit ourselves to the simplest non-trivial case, that where our potential is a sum of a central force and a *tensor force*:

$$V = V_C(r) + V_T(r)\hat{S}_{12} = -47\frac{\exp(r/1.18)}{r/1.18} - 24\frac{\exp(r/1.7)}{r/1.7}\hat{S}_{12} \tag{8.210}$$

where $\hat{S}_{12}$ is the tensor operator, the strength of the potential is measured in MeV, and the distance r is measured in fm. This is, of course, only a specific choice (that of employing two Yukawa potential wells), but it will serve us in good stead below. Just like in Eq. (8.200), the radial part of the wave function in the ${}^3S_1$ state is $u(r)/r$ and, similarly, we denote the radial part of the wave function in the ${}^3D_1$ state by $w(r)/r$. Writing out the full Schrödinger equation (including the centrifugal term, as well as the effect of the tensor operator on the angular parts) leads to the following two coupled second-order ODEs:

$$\begin{aligned} -\frac{\hbar^2}{m}\frac{d^2u(r)}{dr^2} + V_C(r)u(r) - Eu(r) &= -\sqrt{8}V_T(r)w(r) \\ -\frac{\hbar^2}{m}\left(\frac{d^2w(r)}{dr^2} - \frac{6w(r)}{r^2}\right) + [V_C(r) - 2V_T(r)]\,w(r) - Ew(r) &= -\sqrt{8}V_T(r)u(r) \\ u(0) = w(0) = 0, \qquad u(+\infty) = w(+\infty) &= 0 \end{aligned} \tag{8.211}$$

where we also took the opportunity to write down the boundary conditions; you can use $\hbar^2/m = 41.47208$ MeV fm^2.

(a) Plot the potentials $V_C(r)$ and $V_T(r)$ in Eq. (8.210). Interpret what you see.
(b) Note that in Eq. (8.211) we are faced with two differential equations, but only one eigenvalue; just like in most of QM (but unlike problems 8.54, 8.55, and 8.56 on scattering), here E needs to be determined via a combination of an IVP-solver and a root-finder. There is a numerical singularity at $r = 0$ but, as usual, you can simply start the numerical integration at a small-but-finite a; to get things started, assume that $u(r)$ goes like r near $r = 0$ and $w(r)$ goes like r^3 (and pick the derivative values out of a hat, as usual). Try to implement a combination of RK4 and the secant method for this problem, integrating up to $b = 10$; this problem seems to be a straightforward application of material we developed in

the main text. As you will soon find out, you have only one parameter at your disposal, E, but you need to find a way of satisfying *two* boundary conditions ($u(+\infty) = w(+\infty) = 0$) and this turns out to be not-so-trivial.

(c) Since it is hard to start with $u(0) = w(0) = 0$ and manage to satisfy $u(+\infty) = w(+\infty) = 0$, a natural idea is to use the modification of the shooting method which we called shooting to a fitting point in problem 8.22: carry out two independent integrations, (i) one starting near $r = 0$ and going up to some intermediate point, say $R = b/10$, and (ii) one starting at (the large) b and integrating backward to the same intermediate point R. (As before, the advantage of this approach is that you will be, by construction, satisfying the boundary conditions at both a and b.) Implement this to see what happens.

(d) Unfortunately, shooting to a fitting point for this problem manages to satisfy the boundary conditions, but is not able to find a solution where E is negative; remember that we are looking for a bound state. The way to make progress involves a new idea: since we are faced with *two* coupled second-order equations, Eq. (8.211) will have *two* linearly independent solutions that are regular at the origin; for any negative value of E, we call them $u_A(r), w_A(r)$ and $u_B(r), w_B(r)$. You can produce these by integrating outward, starting from two linearly independent sets of initial conditions at a (i.e., first use what you set up in the previous part of this problem and then tweak it, making sure not to multiply every component by the same number, as that would lead to linear dependence). Similarly, for any negative value of E, Eq. (8.211) will have *two* linearly independent solutions that have the characteristic decaying asymptotic behavior of a bound state; let's call them $u_C(r), w_C(r)$ and $u_D(r), w_D(r)$. You can produce these by integrating inward, starting from two linearly independent sets of initial conditions at b. Form linear combinations of the outward and inward solutions:

$$\begin{aligned} u_{\text{out}}(r) &= Au_A(r) + Bu_B(r), \qquad & w_{\text{out}}(r) &= Aw_A(r) + Bw_B(r), \\ u_{\text{in}}(r) &= Cu_C(r) + Du_D(r), \qquad & w_{\text{in}}(r) &= Cw_C(r) + Dw_D(r) \end{aligned} \tag{8.212}$$

We will now choose the coefficients A, B, C, and D such that (for any negative value of E) the following conditions are satisfied at $r = R$:

$$u_{\text{out}}(R) = u_{\text{in}}(R) = u_R, \quad w_{\text{out}}(R) = w_{\text{in}}(R), \quad w'_{\text{out}}(R) = w'_{\text{in}}(R) \tag{8.213}$$

where u_R is some constant you pick (e.g. 3). Analytically set up the four equations in Eq. (8.213), then use the numerical-solution values at $r = R$ to determine A, B, C, and D. Step back for a second to understand what we've accomplished: we now have a way of making the outward and inward solutions match at $r = R$ *for any* negative value of E. Of course, a random guess for E will *not* lead to a zero value for $u'_{\text{out}}(R) - u'_{\text{in}}(R)$ since that was the only condition that we did not impose in Eq. (8.213). The way forward is now obvious: use a rootfinder to find that value of E that leads to $u'_{\text{out}}(R) - u'_{\text{in}}(R)$ being zero (within some tolerance). The absolute value of that E will be the binding energy of the deuteron (which has no bound excited states).

(e) Having computed E, plot $u_{\text{out}}(r)$ and $w_{\text{out}}(r)$ from a to R on the same figure as

$u_{\text{in}}(r)$ and $w_{\text{in}}(r)$ from b to R. Marvel at how cleanly everything matches. Either before plotting or after, normalize $u(r)$ and $w(r)$ such that:

$$\int_0^\infty dr\left[u^2(r) + w^2(r)\right] = 1 \tag{8.214}$$

Using the normalized wave functions, compute:

$$\begin{aligned} P_D &= \int_0^\infty drw^2(r) \\ r_d &= \frac{1}{4}\int_0^\infty dr\, r^2\left[u^2(r) + w^2(r)\right] \\ Q_d &= \frac{1}{20}\int_0^\infty dr\, r^2\, w(r)\left[\sqrt{8}u(r) - w(r)\right] \end{aligned} \tag{8.215}$$

i.e., the D-state probability, root-mean-square radius, and quadrupole moment.

8.58 [$\mathcal{P}$] A mainstay of modern computational physics is the approach known as *molecular dynamics*, which is commonly used to study the properties of materials. In its simplest form, molecular dynamics amounts to solving the classical equations of motion (i.e., Newton's second law applied to each particle) for a collection of n_p particles, placed in periodic boundary conditions and experiencing pairwise interactions. In equation form, we have:[36]

$$\frac{d^2x_i(t)}{dt^2} = \frac{F_i(t)}{m}, \qquad i = 0, 1, \ldots, n_p - 1 \tag{8.216}$$

Here x_i is the position of the i-th particle and F_i is the force acting on that particle (coming from all the other particles); the notation $F_i(t)$ is shorthand for the fact that the force depends on all the x_i's (and these, in turn, are time-dependent). We also assumed that we are dealing with equal-mass particles. We will study liquid argon, for which the simple *Lennard–Jones potential* does a good job:

$$V(r_{ij}) = 4\epsilon\left[\left(\frac{\sigma}{r_{ij}}\right)^{12} - \left(\frac{\sigma}{r_{ij}}\right)^6\right] \tag{8.217}$$

where r_{ij} is the distance between two particles; in the language of Eq. (8.216), this is simply the distance between particles i and j, i.e., $r_{ij} = |x_i - x_j|$. Observe that this potential has a repulsive core at short distances (due to the $1/r_{ij}^{12}$ term) and a well at intermediate and large distances (due to the $-1/r_{ij}^6$ term). To connect Eq. (8.216), in which the force on the i-th particle shows up, and Eq. (8.217), containing the potential for the interaction between the i-th and j-th particles, we take the derivative of the total potential energy to get the force:

$$F_i = -\sum_{\substack{j=0\\ j\neq i}}^{n_p-1}\frac{\partial V(r_{ij})}{\partial x_i} = -\sum_{\substack{j=0\\ j\neq i}}^{n_p-1}\frac{\partial V(r_{ij})}{\partial r_{ij}}\frac{\partial r_{ij}}{\partial x_i} = -\sum_{\substack{j=0\\ j\neq i}}^{n_p-1}\frac{48\epsilon}{r_{ij}^2}\left[\left(\frac{\sigma}{r_{ij}}\right)^{12} - \frac{1}{2}\left(\frac{\sigma}{r_{ij}}\right)^6\right](x_i - x_j) \tag{8.218}$$

[36] For simplicity, we study a one-dimensional problem, but the equations given here (and the wider molecular-dynamics approach) carry over with trivial modifications for a two- or three-dimensional system.

At its most basic level, molecular dynamics for liquid argon involves solving the coupled second-order differential equations in Eq. (8.216) with the right-hand sides being given as per Eq. (8.218). This could be done using, say, the RK4 method, but it is much wiser to employ a technique that has been tailored to the problem at hand: we will employ the velocity Verlet algorithm of problem 8.28. Updating Eq. (8.171) to the present notation, this becomes:

$$\begin{aligned} x_i(t+h) &= x_i(t) + hv_i(t) + h^2 F_i(t)/(2m) \\ v_i(t+h) &= v_i(t) + h\left[F_i(t) + F_i(t+h)\right]/(2m) \end{aligned} \tag{8.219}$$

where we are denoting the current and next slice using the time arguments t and $t + h$. Notice that different updates are employed for the positions and for the velocities; crucially, we need to compute the new positions in order to evaluate the new forces, $F_i(t+h)$, on the right-hand side of the second equation. Building a molecular-dynamics code involves several distinct steps; in what follows we split each part of the calculation into a separate task, to make things more manageable. For concreteness, study $n_p = 20$ particles.

(a) The physical values of the mass and Lennard–Jones parameters are $m \approx 6.7 \times 10^{-26}$ kg, $\epsilon \approx 1.6 \times 10^{-21}$ J, and $\sigma \approx 3.4 \times 10^{-10}$ m, respectively.[37] Pick a timestep of $h = 2 \times 10^{-17}$ s and find what that corresponds to in units of the time scale (which is that unique combination of m, ϵ, and σ that is measured in seconds). To simplify our code, we will use the above values for m, ϵ, and σ as the scales of mass, energy, and length; practically speaking, this means that we will set $m = 1$, $\epsilon = 1$, and $\sigma = 1$.

(b) Write a Python function that takes care of initialization, i.e., sets up the starting ($t = 0$) values of the positions and velocities. The initial positions should be uniformly distributed from 0 to $L = 100$. To arrive at the initial velocities you should, first, draw n_p numbers from a Gaussian distribution; then, compute the average velocity $\bar{v}$ (i.e., $\sum_i v_i/n_p$) and subtract it from each v_i (thereby ensuring that the total velocity/momentum is zero).

(c) Experiments take place at a given temperature T_0; unfortunately, the initialization in the previous part will *not* lead to velocities that correspond to this desired temperature. This is handled by carrying out a simple *rescaling* of the velocities, $v_i \to \lambda v_i$, and demanding that the rescaled velocities correspond to T_0:

$$\frac{1}{2} m\lambda^2 \sum_{i=0}^{n_p-1} v_i^2 = (n_p - 1)\frac{1}{2} k_B T_0 \tag{8.220}$$

where the number of degrees of freedom on the right-hand-side is $n_p - 1$; we know that the total velocity is zero, as per part (b). Write a function that takes in the v_i's and T_0 and carries out this rescaling; set $k_B T_0 = 200$.

(d) Write a Python function that carries out the updates, i.e., one integration step of our equations of motion via the velocity Verlet method of Eq. (8.219). Do yourself a favor and write a separate function for the computation of the forces as per

[37] This is an *effective interaction*, i.e., we would employ different values for a pair of argon atoms in vacuum.

Eq. (8.218). You should be proactive and also make arrangements for the case where the distance (appearing in the denominator in the Lennard–Jones potential/force) underflows. (Both of these functions will have to be slightly modified in the following part of the problem.)

(e) We turn to the *periodic boundary conditions*, according to which our collection of particles is envisioned as being in a "main" box of side L that is surrounded by infinitely many copies of itself (i.e., boxes containing the same configuration of particles). The periodicity impacts both the positions and the separations. First, when applying the velocity Verlet algorithm, as per part (d), it is conceivable that some positions (which started out as being from 0 to $L = 100$) may exit the box. The solution is to translate them back into the box (using Python's modulo operator); crucially, this should be done *after* computing the velocities. Second, in the computation of the forces (but also for other quantities, see below) we are faced with the problem that a given particle i is interacting not only with another particle j, but also with all the images of j (i.e., the copies of j in the other boxes); we are, obviously, not interested in computing infinitely many terms. The solution is to employ the *minimum-image convention*, according to which i interacts only with that j particle which is nearest to i; that closest j particle could be inside our main box (if $r_{ij} < L/2$) or in one of the neighboring boxes. Implement this solution using `np.rint()`.

(f) Carry out a full molecular-dynamics run, calling the functions you prepared in previous parts: start by initializing, rescaling, and then carrying out several integration steps in sequence. Carry out 1500 "burn-in" (i.e., preparatory) steps, during which the system is still out of equilibrium; be sure to carry out a rescaling of the velocities every 10 time steps. Then, make another 3500 "equilibrium" (i.e., production) steps, during which you don't rescale the velocities. You should plot the kinetic energy ($m \sum_i v_i^2/2$), potential energy ($\sum_{i \neq j} V(r_{ij})/2$), and total energy at each time step. You should find that, in the production steps, the total energy is constant to at least 11 significant figures.

8.59 [$\mathcal{P}$] In problem 3.24 we addressed the time-independent setting of scattering from a (simple-step) potential barrier; in problem 1.18 we introduced the Kronig–Penney model, corresponding to an infinite array of barriers. We now turn to the time evolution of a wave packet impinging on a (rectangular) *potential barrier*:

$$V(x) = \begin{cases} 0, & x < a \text{ or } x > b \\ V_0, & a < x < b \end{cases} \tag{8.221}$$

(a) The Crank–Nicolson method of Eq. (8.169) involved taking the average of the equations for FTCS and BTCS. The left-hand side remained unchanged, given by, e.g., the forward-difference approximation to the first derivative. The right-hand side involved the average of central-difference approximations (to the second derivative) centered at time slices t_k and t_{k+1}. Apply the same approach to the time-dependent Schrödinger equation of Eq. (8.5); the only complication involves the potential term: for a time-independent potential, this gives rise to

$V_j(\Psi_{j,k} + \Psi_{j,k+1})/2$. Set $\hbar = m = 1$ and show that this approach leads to:

$$-\gamma\Psi_{j-1,k+1} + (Z_j - i)\Psi_{j,k+1} - \gamma\Psi_{j+1,k+1} = \gamma\Psi_{j-1,k} - (Z_j + i)\Psi_{j,k} + \gamma\Psi_{j+1,k} \quad (8.222)$$

where $\gamma = h_t/(4h_x^2)$, $Z_j = 2\gamma + h_t V_j/2$, and this applies to $j = 1, 2, \ldots, n-2$ and $k = 0, 1, \ldots, m-2$; as before, time goes from 0 to T and space from 0 to L.

(b) The boundary conditions are that the wave function vanishes asymptotically, i.e., $\Psi_{0,k} = \Psi_{n-1,k} = 0$. Take the initial condition to be a Gaussian wave packet:

$$\Psi(x, 0) = \exp\left[-\frac{1}{2}\left(\frac{x-\mu}{\sigma}\right)^2\right] e^{ikx} \quad (8.223)$$

which we didn't bother normalizing (the Schrödinger equation is linear). Take $L = 100$, $T = 5$, $V_0 = 50$, $a = 48$, $b = 52$, $\mu = 30$, $\sigma = 1$, $k = 10$, $n = 4001$ and $m = 1001$. Plot the modulus squared of your solution for a few time slices, also showing the barrier (appropriately scaled down) for comparison. Be careful when handling complex numbers, most notably when initializing arrays; use `lu_factor()` and `lu_solve()` from `scipy.linalg`. Interpret the results.

8.60 [$\mathcal{P}$] Use the central-difference approximation to the second derivative of Eq. (3.39), twice, to show that the *wave equation* of Eq. (8.126) turns into:

$$\phi_{j,k+1} = \gamma\phi_{j-1,k} + 2(1-\gamma)\phi_{j,k} + \gamma\phi_{j+1,k} - \phi_{j,k-1} \quad (8.224)$$

where, this time, $\gamma = (ch_t/h_x)^2$ and this equation applies to $j = 1, 2, \ldots, n-2$ and $k = 0, 1, \ldots, m-2$. This is different from what were faced with in section 8.5.2 because, unlike the diffusion equation, the wave equation has a second-order time derivative; as a result, the right-hand side of Eq. (8.224) involves results at two time slices (t_k and t_{k-1}), making it analogous to the Adams–Bashforth method of Eq. (8.154) or Numerov's method of Eq. (8.172). Note that, once you get going, everything on the right-hand side of Eq. (8.224) is known, so you are dealing with an *explicit* method. Implement this approach for $L = 100$, $T = 20$, $c = 10$, $n = 2000$, $m = 8001$, and an initial wave packet of the form:

$$\phi(x, 0) = \frac{\sin(x - L/2)}{x - L/2} \quad (8.225)$$

which also provides us with the boundary conditions at $x = 0$ and $x = L$. Since the wave equation is second-order in time, we also need an initial condition for the velocities. Take the initial velocities to be zero and use the forward-difference approximation to the first derivative of Eq. (3.9) to conclude that you can get the solution process going with $\boldsymbol{\phi}_1 = \boldsymbol{\phi}_0$. Plot your solution for a few time slices. Hint: if you decide to implement Eq. (8.224) in the form of $\boldsymbol{\phi}_{k+1} = \mathbf{A}\boldsymbol{\phi}_k - \boldsymbol{\phi}_{k-1}$, you need to be careful with the boundary conditions (so you don't zero them out).

The parts I understood are excellent as, I suppose, are also the parts I didn't understand; it's just that one needs to be an expert swimmer.

Socrates (on Heraclitus' treatise)

Appendix A **Installation and Setup** A

But they gulp down the first puzzle, as though assuming that it is trivial.

Aristotle

Python packages The packages we touch upon can be divided into two categories:

- *Required:* Python 3 and its standard library; Matplotlib; NumPy.
- *Mentioned, but not required:* Jupyter Notebook; SciPy; SymPy; JAX.

The matrix-related material uses Python's infix `@` operator instead of `numpy.dot()`; thus, for a seamless experience you should be using Python 3.5 or a later version. Incidentally, if you're still using Python 2 but would like to follow along, you should read the relevant section in our online tutorial (though you really should switch to Python 3).

Installation The easiest way to download and install the needed packages is via the *Anaconda Distribution*. Actually, this also includes functionality that we don't use, like `pandas`, `TensorFlow`, or `Numba`; if you keep doing numerical work with Python, you are likely to run into many of Anaconda's packages in the future. Practically speaking, you should head on over to `www.anaconda.com/distribution` and click on "Download". There are packaged versions for Windows, macOS, or Linux. If you are asked to choose between Python 2.n and Python 3.n, choose the latter. Follow the instructions after that and all should be well. Another option is to use the *Enthought Deployment Manager*. Finally, you could try to install the reference implementation, *CPython*, and the other needed libraries "by hand", but if this is your first foray into the Python world it's probably best to stick to a distribution, which takes care of all the technical details for you.

Running The simplest way of using Python is to launch the Python interpreter; in addition to interactive sessions, you can also use a text editor (like `vim` or `emacs`) to save code into a file, which you then run. Another popular option is the *Spyder* integrated development environment (IDE), which allows you to edit, debug, and profile your programs. A third option is to use *Jupyter Notebook*, which involves interactive documents that combine code, text, equations, and plots all in one place (displayed in your browser). If you used the Anaconda distribution, then you already have both Spyder and Jupyter Notebook installed on your system; in that case, you can work through our Python and NumPy tutorials interactively (otherwise, you can read the html versions). Depending on your taste, you may prefer another programming environment, e.g., Colab or JupyterLab. You can access the tutorials and the codes discussed in the main text at the companion website `www.numphyspy.org` or at the publisher's website.

B Appendix B Number Representations

Such things should not remain unknown,
for it is upon them that knowledge of other things depends.

Thomas Aquinas

We now discuss floating-point numbers more carefully than in chapter 2. As you may recall, computers use *binary digits or bits*: bits can take on only two possible values, by convention 0 or 1; compare with decimal digits, which can take on 10 different values, from 0 to 9. Bits are grouped to form *bytes*: 1 byte ≡ 1 B ≡ 8 bits ≡ 8 b.

B.1 Integers

The number of bytes needed to store variables of different type varies (by implementation, language, etc.), but nowadays it is common that characters use 1 B, integers use 4 B, and long integers use 8 B; trying to store an integer that's larger than that leads to overflow. This doesn't actually happen in Python, because Python employs arbitrary-precision integers.

For simplicity, let us start with an example for the case of 4 bits:

$$1100_2 = 1 \times 2^3 + 1 \times 2^2 + 0 \times 2^1 + 0 \times 2^0 = 12_{10} \tag{B.1}$$

We use subscripts to denote binary and decimal numbers; the equality sign is here used loosely. We're interested only in *unsigned integers*, where all bits are used to represent powers of 2. Applying the logic of this equation, you can also show that, e.g., $0011_2 = 3_{10}$. We see that, using 4 bits, the biggest integer we can represent is:

$$1111_2 = 1 \times 2^3 + 1 \times 2^2 + 1 \times 2^1 + 1 \times 2^0 = 15_{10} \tag{B.2}$$

Note that $2^4 - 1 = 15$. More generally, with n bits we can represent integers in the range $[0, 2^n)$, i.e., the largest integer possible is $2^n - 1$. Here's another example with $n = 6$ bits:

$$100101_2 = 1 \times 2^5 + 0 \times 2^4 + 0 \times 2^3 + 1 \times 2^2 + 0 \times 2^1 + 1 \times 2^0 = 37_{10} \tag{B.3}$$

Explicitly check that the maximum integer representable with 6 bits is $2^6 - 1 = 63$.

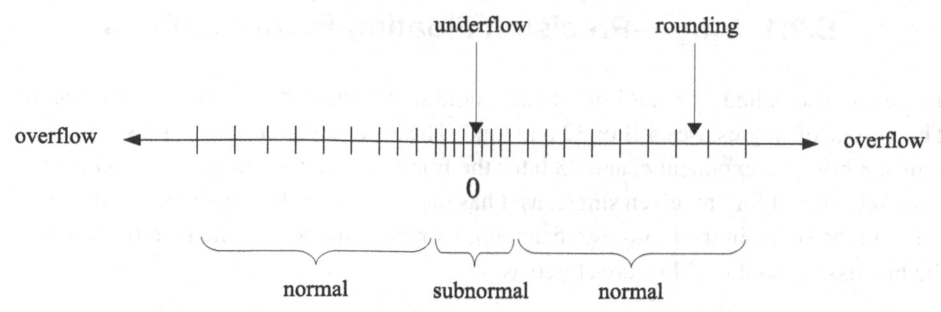

Fig. B.1 Illustration of exactly representable floating-point numbers

B.2 Real Numbers

Several ways of storing real numbers on a computer exist. The simplest is known as *fixed-point representation*, but this is not commonly used in computational science. Most often, real numbers on a computer employ *floating-point representation*: $\pm$ mantissa $\times$ 10^{exponent}. For example, the speed of light in scientific notation is $+2.997\,924\,58 \times 10^8$ m/s. The mantissa here has nine significant figures.

Computers store only a finite number of bits, so cannot store exactly all possible real numbers. As a result, there are "only" finitely many exact representations/machine numbers; this refers to finitely many decimal numbers that can be stored exactly using a floating-point representation. There are three ways of losing precision, as shown qualitatively in Fig. B.1 (which you also encountered as Fig. 2.2): *underflow* for very small numbers, *overflow* for very large numbers, and *rounding* for decimal numbers whose value falls between two exactly representable numbers.

In the past, there existed a polyphony of conventions (in programming languages and operating systems) on how to treat floating-point representations; nowadays, the Institute of Electrical and Electronics Engineers has established a Standard for Floating-Point Arithmetic (known as *IEEE 754*), which is widely implemented. As usual, we go over only selected features.[1] The general convention is:

$$x_{\text{floating point}} = (-1)^s \times 1.f \times 2^{e-bias} \tag{B.4}$$

where: (a) s is the *sign bit*, (b) $1.f$ follows the convention of starting with a 1 (called a *hidden bit*) and then having the *fraction* of the *mantissa*,[2] and (c) the (fixed-value) *bias* allows us to use an *exponent* e that's always positive, i.e., it is an unsigned integer (but then $e - bias$ can range from negative to positive values). We need to store s, f, and e, following the above convention. We go over two cases: singles and doubles.

[1] We do not discuss the alternative concept of universal numbers, *unums*, and their latest incarnation as *posits*. This mention of them should be enough for you if you want to delve into this topic further.

[2] What we colloquially refer to as the mantissa is more properly called the *significand*.

B.2.1 Single-Precision Floating-Point Numbers

These are also called "singles" or "floats"; note that *Python floats are actually doubles.*[3] The storage of singles uses 4 B or 32 b in total. These are distributed as follows: 1 b for the sign s, 8 b for the exponent e, and 23 b for the fraction of the mantissa f (the *bias* doesn't have to be stored for any given single, as it has the same value for all singles.) Specifically, we store the sign s in the most-significant bit, then the exponent e, and then the fraction of the mantissa f, so the 32 bits are stored as:

s	e	f
31	30...23	22...0

(B.5)

If $s = 0$ we have $(-1)^s = +1$, whereas if $s = 1$ we have $(-1)^s = -1$.

Exponent The e is always positive (unsigned) and stored in 8 bits so, given that $2^8 - 1 = 255$, the exponent values go from $0000\,0000_2 = 0_{10}$ to $1111\,1111_2 = 255_{10}$. Thus, we have $0 \le e \le 255$. In reality, $e = 0$ and $e = 255$ are special cases, so we have $1 \le e \le 254$. The bias for all singles is $bias = 127$ so the actual exponent in 2^{e-bias} takes the values $-126 \le e - 127 \le 127$. In other words, numbers with $1 \le e \le 254$ are called *normal* and are represented as per Eq. (B.4):

$$x_{\text{normal single}} = (-1)^s \times 1.f \times 2^{e-127} \tag{B.6}$$

Now for the special cases:

- $e = 255$ with $f = 0$: this is either $+\infty$ (for $s = 0$) or $-\infty$ (for $s = 1$). In NumPy, these are the `inf` and `NINF` constants, respectively.
- $e = 255$ with $f \neq 0$: this is NaN (not a number). In NumPy, this is the `nan` constant.

We return to the remaining special case, $e = 0$, below.

Mantissa We have at our disposal 23 bits for the fraction f of the mantissa. Denoting them by $m_{22}, m_{21}, m_{20}, \ldots, m_1, m_0$, they are combined together as follows:

$$\text{normal single mantissa} = 1.f = 1 + m_{22} \times 2^{-1} + m_{21} \times 2^{-2} + \cdots + m_0 \times 2^{-23} \tag{B.7}$$

This expression only includes powers of inverse 2, i.e., it cannot be used to represent numbers that cannot be expressed as a sum of powers of inverse 2. You'll do an example of a "middle-of-the-road" floating-point single-precision number in problem B.2.

Largest normal single For now, let us examine the largest possible positive normal single. The largest possible f is all 1s, so:[4]

$$\begin{aligned} \text{normal single mantissa}\Big|_{max} &= 1.1111\,1111\,1111\,1111\,1111\,111 \\ &= 1 + 0.5 + 0.25 + \cdots \approx 2 \end{aligned} \tag{B.8}$$

[3] Core Python has no singles, but NumPy allows you to control the data type in finer detail, see section 1.6.
[4] We stop explicitly denoting numbers as binary or decimal; the distinction should be clear from the context.

On the other hand, we already know that the largest possible normal exponent e is 254 (since 255 is a special case, as seen above). In binary, this is 1111 1110. Putting it all together, the largest normal single is stored as:

$$x_{\text{normal single}}\Big|_{max} = 0\ 1111\ 1110\ 1111\ 1111\ 1111\ 1111\ 1111\ 111 \tag{B.9}$$

This starts with a positive sign bit, then moves on to the largest normal exponent, and then on to the largest possible fraction of the mantissa. Its value is:

$$(-1)^s \times 1.f \times 2^{e-127} \approx +1 \times 2 \times 2^{127} = 2^{128} \approx 3.4 \times 10^{38} \tag{B.10}$$

Subnormal singles We left the special case $e = 0$ for last. This is used to represent numbers that are smaller than the smallest normal number, as per Fig. B.1. These are called *subnormal* numbers and are represented as follows:

$$x_{\text{subnormal single}} = (-1)^s \times 0.f \times 2^{-126} \tag{B.11}$$

Comparing this with Eq. (B.6) we observe, first, that the sign convention is the same. Second, the subnormal mantissa convention is slightly different: it's written $0.f$, so the bits in the fraction of the mantissa for subnormals are interpreted as:

$$\text{subnormal single mantissa} = 0.f = m_{22} \times 2^{-1} + m_{21} \times 2^{-2} + \cdots + m_0 \times 2^{-23} \tag{B.12}$$

where there is no leading 1 (since there is a leading 0). Third, the exponent term is 2^{-126}, which is different from what we would get from Eq. (B.6) if we plugged in $e = 0$. The vertical lines in Fig. B.1 are not equidistant from each other: this is most easily seen in the case of subnormal numbers, which are more closely spaced than normal numbers.

Incidentally, note that if all the bits in f are 0 we are left with the number zero: since we still have a sign bit, s, this gives rise to a *signed zero*, meaning that +0 and −0 are two different ways of representing the same real number. In NumPy, these are the `PZERO` and `NZERO` constants, respectively.

Smallest subnormal single The smallest possible (non-zero) positive subnormal single is also the smallest possible positive single-precision floating-point number (since subnormals are smaller than normals). The smallest possible positive subnormal single has only a single 1 in the mantissa:

$$\text{subnormal single mantissa}\Big|_{min} = 0.0000\ 0000\ 0000\ 0000\ 0000\ 001 = 2^{-23} \tag{B.13}$$

On the other hand, we already know that the exponent is $e = 0$, since we're dealing with a subnormal. Putting it all together:

$$x_{\text{subnormal single}}\Big|_{min} = 0\ 0000\ 0000\ 0000\ 0000\ 0000\ 0000\ 0000\ 001 \tag{B.14}$$

This starts with a positive sign bit, then moves on to an exponent that is zero, and then on to the smallest possible fraction of the mantissa. Its value is:

$$(-1)^s \times 0.f \times 2^{-126} = +1 \times 2^{-23} \times 2^{-126} = 2^{-149} \approx 1.4 \times 10^{-45} \tag{B.15}$$

Summary Singles can represent:

$$\pm 1.4 \times 10^{-45} \leftrightarrow \pm 3.4 \times 10^{38} \tag{B.16}$$

This refers to the ability to store very large or very small numbers; most of this ability is found in the term corresponding to the exponent. If we try to represent a number that's larger than $2^{128} \approx 3.4\times10^{38}$ we get *overflow* (up to infinity). Similarly, if we try to represent a number that's smaller than $2^{-149} \approx 1.4 \times 10^{-45}$ we get *underflow* (down to 0).

Being able to represent 1.4×10^{-45} does *not* mean that we are able to store 45 significant figures in a single. The number of significant figures (and the related concept of *precision*) is found in the mantissa. For singles, the precision is 1 part in $2^{23} \approx 1.2 \times 10^{-7}$, which amounts to six or seven decimal digits.

B.2.2 Double-Precision Floating-Point Numbers

These are also called "doubles"; keep in mind that Python floats are actually doubles. You will probe these more deeply in problem B.4 – here we just summarize several facts. Their storage uses 8 B or 64 b in total. These are distributed as follows: 1 b for the sign s, 11 b for the exponent e, and 52 b for the fraction of the mantissa f.[5] Specifically, we store the sign s in the most-significant bit, then the exponent e, and then the fraction of the mantissa f. In other words, similarly to what we did for singles, the 64 bits of a double are stored as follows:

s	e	f
63	62...52	51...0

(B.17)

Exponent The e is always positive (unsigned) and stored in 11 bits so, given that $2^{11} - 1 = 2047$, the exponent goes from $0000\ 0000\ 000_2 = 0_{10}$ to $1111\ 1111\ 111_2 = 2047_{10}$. Thus, we have $0 \leq e \leq 2047$. As you may have guessed, $e = 0$ and $e = 2047$ are special cases, so we have $1 \leq e \leq 2046$. The bias for all doubles is $bias = 1023$ so the actual exponent in 2^{e-bias} takes the values $-1022 \leq e - 1023 \leq 1023$. In other words, numbers with $1 \leq e \leq 2046$ are called *normal* and are represented as per Eq. (B.4):

$$x_{\text{normal double}} = (-1)^s \times 1.f \times 2^{e-1023} \tag{B.18}$$

Mantissa The fraction of the mantissa is interpreted analogously to what we did for singles. There are 52 bits, $m_{51}, m_{50}, m_{49}, \ldots, m_1, m_0$, which appear in:

$$\text{normal double mantissa} = 1.f = 1 + m_{51} \times 2^{-1} + m_{50} \times 2^{-2} + \cdots + m_0 \times 2^{-52} \tag{B.19}$$

for normal doubles. We do not provide more details here because everything works by analogy to the case of singles.

[5] Again, the *bias* doesn't have to be stored, as it has the same value for all doubles.

Summary Doubles can represent:

$$\pm 4.9 \times 10^{-324} \leftrightarrow \pm 1.8 \times 10^{308} \tag{B.20}$$

This refers to the ability to store very large or very small numbers. As mentioned above, most of this ability is found in the term corresponding to the exponent. If we try to represent a number that's larger than 1.8×10^{308} we get *overflow*. Similarly, if we try to represent a number that's smaller than 4.9×10^{-324} we get *underflow*.

Again, being able to represent 4.9×10^{-324} does *not* mean that we are able to store 324 significant figures in a double. The number of significant figures (and the related concept of *precision*) is found in the mantissa. For doubles, the precision is 1 part in $2^{52} \approx 2.2 \times 10^{-16}$, which amounts to 15 or 16 decimal digits. Observe that both the digits of precision and the range in Eq. (B.20) are much better than the corresponding results for singles.

Problems

B.1 We saw how to convert binary integers to decimal, for example:
$0101_2 \rightarrow$ `0 × 2`^{3} `+ 1 × 2`^{2} `+ 0 × 2`^{1} `+ 1 × 2`^{0} $= 5_{10}$
By analogy to this, write a program that breaks decimal numbers down, e.g.:
$5192_{10} \rightarrow$ `5 × 10`^{3} `+ 1 × 10`^{2} `+ 9 × 10`^{1} `+ 2 × 10`^{0}
Do this for any four-digit integer the user inputs. (Use * instead of ×.) Can you generalize this to work for an integer of any number of digits?

B.2 [$\mathcal{A}$] Convert the following single-precision floating point numbers from binary to decimal:

(a) `0 0000 0011 0111 1011 0000 0000 0000 000`
(b) `1 0101 0101 0111 0110 0000 0000 0001 011`

B.3 [$\mathcal{A}$] Find the smallest possible (non-zero) *normal* single-precision floating-point number.

B.4 [$\mathcal{A}$] Repeat our singles argument for doubles and thereby explicitly show Eq. (B.20).

B.5 This problem studies underflow and overflow in Python in more detail.

(a) Modify the code in section 2.3.2 to investigate underflow for floating-point numbers in Python. In order to make your output manageable, make sure you start from a number that is sufficiently small.
(b) Now investigate whether or not overflow occurs for integers in Python. Do yourself a favor and start not from 1, but from a very large positive number and increment from there. You should use `sys.maxsize` to get the ballpark of near where you should start checking things. You should use `type()` to see if the type changes below `sys.maxsize` and above it.

C Appendix C Math Background

To me it's all the same where I begin, for I shall go back there again.

Parmenides

Given the finitude of this book, we cannot cover everything about everything; there are some prerequisites that will help you benefit from our exposition. We assume you've already taken courses on calculus, linear algebra, and differential equations; some familiarity with probability and statistics (e.g., that gained in a lab course) is also necessary. Here we provide the briefest of summaries on some essential topics, in order to refresh your memory; this is intended as reference (i.e., not pedagogical) material.

C.1 Taylor Series

A *Taylor expansion* of a real function $f(x)$ about the point x_0 is the infinite series:

$$f(x) = f(x_0) + (x - x_0)f'(x_0) + \frac{(x - x_0)^2}{2!}f''(x_0) + \frac{(x - x_0)^3}{3!}f'''(x_0) + \frac{(x - x_0)^4}{4!}f^{(4)}(x_0) + \cdots \tag{C.1}$$

If $x_0 = 0$, this is known as a *Maclaurin series*. Except for a few counter-examples in section 7.6, we will always assume that these derivatives exist. We typically express the Taylor series as a *partial sum* (i.e., with a finite number of terms) plus the *Lagrange remainder*:

$$\begin{aligned} f(x) = f(x_0) + (x - x_0)f'(x_0) + \frac{(x - x_0)^2}{2!}f''(x_0) + \frac{(x - x_0)^3}{3!}f'''(x_0) + \cdots \\ + \frac{(x - x_0)^{n-1}}{(n-1)!}f^{(n-1)}(x_0) + \frac{(x - x_0)^n}{n!}f^{(n)}(\xi) \end{aligned} \tag{C.2}$$

where ξ is some point between x_0 and x.

We will often encounter this series in different forms, e.g., expanding $f(x+q)$ around the point x: you should be able to apply Eq. (C.1) to such scenarios. To help you get oriented, let's see this Taylor series for the case where we only keep the first few terms:

$$f(x + q) = f(x) + qf'(x) + \frac{q^2}{2}f''(x) + \frac{q^3}{6}f'''(x) + \frac{q^4}{24}f^{(4)}(\xi) \tag{C.3}$$

If we have no further information on $f^{(4)}(\xi)$, this can be compactly expressed by saying that the remainder is of fourth order, also known as $O(q^4)$. This is a way of listing only the most crucial dependence, dropping constants, prefactors, etc.

We will also encounter the case of a scalar function of many variables, i.e., $\phi(\mathbf{x})$, where

$\mathbf{x}$ bundles together the variables $x_0, x_1, \ldots, x_{n-1}$ but ϕ produces scalar values. We can then employ a multivariate Taylor expansion, which takes the form:

$$\phi(\mathbf{x}+\mathbf{q}) = \phi(\mathbf{x}) + \sum_{j=0}^{n-1} \frac{\partial \phi}{\partial x_j} q_j + \frac{1}{2} \sum_{i,j=0}^{n-1} \frac{\partial \phi}{\partial x_i \partial x_j} q_i q_j + \cdots \tag{C.4}$$

where $\mathbf{q}$ bundles together the steps $q_0, q_1, \ldots, q_{n-1}$. This can be recast as:

$$\phi(\mathbf{x}+\mathbf{q}) = \phi(\mathbf{x}) + (\nabla\phi(\mathbf{x}))^T \mathbf{q} + \frac{1}{2}\mathbf{q}^T \mathbf{H}(\mathbf{x})\mathbf{q} + \cdots \tag{C.5}$$

where $\nabla\phi(\mathbf{x})$ is the *gradient vector* and $\mathbf{H}(\mathbf{x})$ is the *Hessian matrix*. This is more compact, but it assumes you know what the bold symbols mean, a topic we now turn to.

C.2 Matrix Terminology

Here we establish the matrix-related notation and terminology heavily used from chapter 4 and onward. Crucially, in order to be consistent with Python's 0-indexing, we will employ notation that goes from 0 to $n-1$; this is different from most (all?) books on linear algebra; even so, having both the equations and the code employ the same notation seems to be an advantage, helping one avoid frustrating "off-by-one" errors.

Definitions A matrix is a rectangular array of numbers. We will be mainly dealing with *square matrices*, i.e., matrices of dimensions $n \times n$ (in fact, we will often refer to these simply as *matrices*). Here's a 3×3 example:

$$\mathbf{A} = \begin{pmatrix} A_{00} & A_{01} & A_{02} \\ A_{10} & A_{11} & A_{12} \\ A_{20} & A_{21} & A_{22} \end{pmatrix} \tag{C.6}$$

We denote the whole square matrix $\mathbf{A}$ (in bold) and the individual elements A_{ij} (where each of the indices goes from 0 to $n-1$). Note that the first index denotes the row and the second one the column. Since this is a 3×3 matrix, it has three rows and three columns.

In this book, we use bold uppercase symbols to denote matrices. You may also encounter the notation $\{A_{ij}\}$, where a matrix element is surrounded by curly braces; here the braces mean that you should imagine i and j taking on all their possible values (and as a result, you get all possible matrix elements). In other words, this notation implies $\mathbf{A} = \{A_{ij}\}$.

We will also be using *column vectors*, which have a single column, e.g.:

$$\mathbf{x} = \begin{pmatrix} x_0 \\ x_1 \\ x_2 \end{pmatrix} \tag{C.7}$$

Here we denote the whole column vector $\mathbf{x}$ (in bold) and the individual elements x_i. Viewed as a matrix, this has dimensions 3×1, i.e., three rows and one column. Similarly, one could define a *row vector*, which has a single row, for example:

$$\mathbf{y} = \begin{pmatrix} y_0 & y_1 & y_2 \end{pmatrix} \tag{C.8}$$

We denote the whole row vector $\mathbf{y}$ (in bold) and the individual elements y_i. This matrix has dimensions 1×3, i.e., one row and three columns.

Operations There are several mathematical operations one can carry out using matrices. For example, multiplication by a scalar can be expressed as either $\mathbf{B} = \kappa\mathbf{A}$ or $B_{ij} = \kappa A_{ij}$. The most interesting operations are *matrix–vector multiplication*:

$$\mathbf{y} = \mathbf{A}\mathbf{x}, \qquad y_i = \sum_{j=0}^{n-1} A_{ij} x_j \tag{C.9}$$

and (two)-*matrix multiplication*:

$$\mathbf{C} = \mathbf{A}\mathbf{B}, \qquad C_{ij} = \sum_{k=0}^{n-1} A_{ik} B_{kj} \tag{C.10}$$

You could also take the *transpose* of a matrix:

$$\mathbf{B} = \mathbf{A}^T, \qquad B_{ij} = A_{ji} \tag{C.11}$$

A further definition: the *trace* of a square matrix $\mathbf{A}$ is the sum of the diagonal elements.

Finally, note that, if $\mathbf{x}$ is a 5×1 column vector, it is easier to display its transpose:

$$\mathbf{x}^T = \begin{pmatrix} x_0 & x_1 & x_2 & x_3 & x_4 \end{pmatrix} \tag{C.12}$$

since it fits on one line. Our bold lowercase symbols will (almost) always be column vectors, so we'll often be using the transpose to save space on the page.

Special matrices There are several special matrices that you already know: diagonal matrices, the identity matrix, triangular matrices, symmetric matrices, real matrices, Hermitian matrices, and so on. Make sure you remember what these definitions mean; for example, symmetric means $\mathbf{A} = \mathbf{A}^T$ or $A_{ij} = A_{ji}$. In numerical linear algebra an important class consists of *sparse* matrices, for which most matrix elements are zero. Equally important are *tridiagonal* matrices, for which the only non-zero elements are on the main diagonal and on the two diagonals next to the main diagonal (more generally, *banded* matrices have non-zero elements on one or more diagonals).

Finally, a *diagonally dominant* matrix is one where each diagonal element is larger than or equal to (in absolute value) the sum of the magnitudes of all other elements on the same row. This might be easier to grasp by writing out an example:

$$\mathbf{A} = \begin{pmatrix} -3 & 2 & -7 \\ -9 & 1 & 6 \\ 1 & -5 & -2 \end{pmatrix}, \qquad \mathbf{B} = \begin{pmatrix} -9 & 1 & 6 \\ 1 & -5 & -2 \\ -3 & 2 & -7 \end{pmatrix} \tag{C.13}$$

A is not diagonally dominant, but **B** is (though we simply re-arranged the rows).

Determinant For a square matrix **A**, one can evaluate a number known as the *determinant*,[1] denoted by det(**A**) or |**A**|. It's easiest to start with the 2×2 case:

$$|\mathbf{A}| = \begin{vmatrix} A_{00} & A_{01} \\ A_{10} & A_{11} \end{vmatrix} = A_{00}A_{11} - A_{01}A_{10} \tag{C.14}$$

This is nothing other than a sum of products of matrix elements. For the $n \times n$ case:

$$|\mathbf{A}| = \sum_{i_0, i_1, \ldots, i_{n-1}=0}^{n-1} (-1)^k A_{0,i_0} A_{1,i_1} \ldots A_{(n-1),i_{n-1}} \tag{C.15}$$

This sum is over all the $n!$ permutations of degree n and k is the number of interchanges needed to put the i_j indices in the order $0, 1, 2, \ldots, n-1$. Qualitatively, we have a product of n elements with the appropriate sign and are dealing with $n!$ such products.

This is not the most efficient way of evaluating determinants. For a 10×10 matrix, we will need to sum $10! = 3\,628\,800$ products, each of which involves nine multiplications (since you need nine multiplications to multiply 10 elements together). In total, this requires $9 \times 3\,628\,800 = 32\,659\,200$ multiplications and 3 628 799 additions/subtractions (since we were faced with 3 628 800 products to be added together). We would also need to keep track of the interchanges in order to make sure we are using the correct sign. In practice, one, instead, relies on the fact that the determinant of a triangular matrix is the product of the diagonal elements:

$$|\mathbf{A}| = \prod_{i-0}^{n-1} A_{ii} \tag{C.16}$$

In chapter 4 we see how to transform a matrix without changing the determinant; thus, we will be able to compute the determinant of a non-triangular matrix simply by evaluating the determinant of a "corresponding" triangular matrix.

Inverse Often, we can define the *inverse* of a matrix **A**, denoted by $\mathbf{A}^{-1}$, as follows:

$$\mathbf{A}^{-1}\mathbf{A} = \mathbf{A}\mathbf{A}^{-1} = \mathcal{I} \tag{C.17}$$

where $\mathcal{I}$ is the identity matrix having the same dimensions as **A** (and $\mathbf{A}^{-1}$). Sometimes (rarely), we are not able to define the inverse: this happens when the determinant of the matrix is 0; we say the matrix is *singular*: such a matrix is made up of linearly dependent rows (or columns).

We are now in a position to introduce two more definitions of "special matrices". First, an *orthogonal matrix* has a transpose that is equal to the inverse: $\sum_{k=0}^{n-1} A_{ik}A_{jk} = \delta_{ij}$ or $\mathbf{A}^{-1} = \mathbf{A}^T$. Second, a *unitary matrix* has a Hermitian conjugate that is equal to the inverse: $\sum_{k=0}^{n-1} A_{ik}A^*_{jk} = \delta_{ij}$ or $\mathbf{A}^{-1} = \mathbf{A}^\dagger$. There are several other important cases (e.g., skew-symmetric: $\mathbf{A}^T = -\mathbf{A}$, skew-Hermitian: $\mathbf{A}^\dagger = -\mathbf{A}$), but what we have here is enough for our purposes.

[1] Geometrically, the signed volume of the parallelepiped determined by the columns of the matrix.

C.3 Probability

We now summarize basic results from probability theory, which will come in handy in section 6.6 (on linear regression) and section 7.7 (on stochastic integration).

C.3.1 Discrete Random Variables

Consider a *discrete* random variable $\mathscr{X}$: its possible values are x_i, each one appearing with the corresponding probability $\mathbb{P}_i^{\mathscr{X}}$. Observe that we are using a small script-like symbol for the random variable and an ordinary lowercase symbol for its possible values.

Mean and variance The *expectation* of this random variable (also known as the *mean* or *expected value*) is simply:

$$\mathbb{E}(\mathscr{X}) = \sum_i \mathbb{P}_i^{\mathscr{X}} x_i \tag{C.18}$$

sometimes denoted by μ or $\mu_{\mathscr{X}}$. One can take the expected value of other quantities, for example the random variable $\mathscr{X}^2$. This is called the second moment of $\mathscr{X}$ and is simply:

$$\mathbb{E}(\mathscr{X}^2) = \sum_i \mathbb{P}_i^{\mathscr{X}} x_i^2 \tag{C.19}$$

This helps us calculate another useful quantity, known as the *variance*, $\mathbb{V}(\mathscr{X})$. The variance is the expectation of the random variable $[\mathscr{X} - \mathbb{E}(\mathscr{X})]^2$:

$$\mathbb{V}(\mathscr{X}) = \mathbb{E}\left\{[\mathscr{X} - \mathbb{E}(\mathscr{X})]^2\right\} \tag{C.20}$$

also denoted by σ^2 or $\sigma^2_{\mathscr{X}}$. A simple calculation gives:

$$\begin{aligned}\mathbb{V}(\mathscr{X}) &= \sum_i \mathbb{P}_i^{\mathscr{X}} [x_i - \mathbb{E}(\mathscr{X})]^2 = \sum_i \mathbb{P}_i^{\mathscr{X}} x_i^2 - 2\sum_i \mathbb{P}_i^{\mathscr{X}} x_i \mathbb{E}(\mathscr{X}) + \sum_i \mathbb{P}_i^{\mathscr{X}} \mathbb{E}(\mathscr{X})^2 \\ &= \mathbb{E}(\mathscr{X}^2) - 2\mathbb{E}(\mathscr{X})^2 + \mathbb{E}(\mathscr{X})^2 = \mathbb{E}(\mathscr{X}^2) - \mathbb{E}(\mathscr{X})^2\end{aligned} \tag{C.21}$$

In the first equality we applied the definition of an expected value. In the second equality we expanded out the square. In the third equality we identified a couple of expected values and used $\sum_i \mathbb{P}_i^{\mathscr{X}} = 1$. In the fourth equality we collected terms. This final result is very often the expression used to evaluate the variance. Another concept that is often used is that of the *standard deviation*, namely the square root of the variance: $\sqrt{\mathbb{V}(\mathscr{X})}$ or σ or $\sigma_{\mathscr{X}}$.

Note that if $\mathscr{X}$ is a random variable then $f(\mathscr{X})$ is also a random variable. Its expectation is given by the *law of the unconscious statistician*:

$$\mathbb{E}[f(\mathscr{X})] = \sum_i \mathbb{P}_i^{\mathscr{X}} f(x_i) \tag{C.22}$$

and, similarly, its variance is:

$$\mathbb{V}[f(\mathscr{X})] = \mathbb{E}\left\{(f(\mathscr{X}) - \mathbb{E}[f(\mathscr{X})])^2\right\} = \mathbb{E}[f^2(\mathscr{X})] - \mathbb{E}[f(\mathscr{X})]^2 \tag{C.23}$$

where the second step follows from a derivation analogous to that above.

Properties of the mean and variance We now turn to the problem of two random variables, $\mathscr{X}$ and $\mathscr{Y}$. Specifically, we take the expectation of a linear combination of the two random variables:

$$\mathbb{E}(\lambda_0\mathscr{X} + \lambda_1\mathscr{Y}) = \sum_{i,j} \mathbb{P}^{\mathscr{X}\mathscr{Y}}_{ij}(\lambda_0 x_i + \lambda_1 y_j) = \lambda_0 \sum_{i,j} \mathbb{P}^{\mathscr{X}\mathscr{Y}}_{ij} x_i + \lambda_1 \sum_{i,j} \mathbb{P}^{\mathscr{X}\mathscr{Y}}_{ij} y_j$$
$$= \lambda_0 \sum_i \mathbb{P}^{\mathscr{X}}_i x_i + \lambda_1 \sum_j \mathbb{P}^{\mathscr{Y}}_j y_j = \lambda_0\mathbb{E}(\mathscr{X}) + \lambda_1\mathbb{E}(\mathscr{Y}) \tag{C.24}$$

In the first equality we employed the *joint probabilities* $\mathbb{P}^{\mathscr{X}\mathscr{Y}}_{ij}$; this is the bivariate version of the law of the unconscious statistician applied to $f(\mathscr{X}, \mathscr{Y}) = \lambda_0\mathscr{X} + \lambda_1\mathscr{Y}$. In the third equality we evaluated sums like $\sum_j \mathbb{P}^{\mathscr{X}\mathscr{Y}}_{ij} = \mathbb{P}^{\mathscr{X}}_i$, leading to the *marginal* distributions. We call our final result the *addition rule* for expectations of random variables: the expectation of a linear combination of random variables is the same linear combination of the expectations of the random variables. This trivially generalizes to more random variables.

We now examine the variance of a linear combination of random variables:

$$\mathbb{V}(\lambda_0\mathscr{X} + \lambda_1\mathscr{Y}) = \mathbb{E}[(\lambda_0\mathscr{X} + \lambda_1\mathscr{Y})^2] - \mathbb{E}(\lambda_0\mathscr{X} + \lambda_1\mathscr{Y})^2$$
$$= \mathbb{E}(\lambda_0^2\mathscr{X}^2 + \lambda_1^2\mathscr{Y}^2 + 2\lambda_0\lambda_1\mathscr{X}\mathscr{Y}) - [\lambda_0\mathbb{E}(\mathscr{X}) + \lambda_1\mathbb{E}(\mathscr{Y})]^2$$
$$= \lambda_0^2\mathbb{E}(\mathscr{X}^2) + \lambda_1^2\mathbb{E}(\mathscr{Y}^2) + 2\lambda_0\lambda_1\mathbb{E}(\mathscr{X}\mathscr{Y}) - \lambda_0^2\mathbb{E}(\mathscr{X})^2 - \lambda_1^2\mathbb{E}(\mathscr{Y})^2 - 2\lambda_0\lambda_1\mathbb{E}(\mathscr{X})\mathbb{E}(\mathscr{Y})$$
$$= \lambda_0^2\mathbb{V}(\mathscr{X}) + \lambda_1^2\mathbb{V}(\mathscr{Y}) + 2\lambda_0\lambda_1[\mathbb{E}(\mathscr{X}\mathscr{Y}) - \mathbb{E}(\mathscr{X})\mathbb{E}(\mathscr{Y})]$$
$$= \lambda_0^2\mathbb{V}(\mathscr{X}) + \lambda_1^2\mathbb{V}(\mathscr{Y}) + 2\lambda_0\lambda_1\text{cov}(\mathscr{X}, \mathscr{Y}) \tag{C.25}$$

In the first line we applied the generalization of Eq. (C.21). In the second line we expanded out the square (in the first term) and used the addition rule that we established in the previous derivation (in the second term). In the third line we used the same addition rule again (in the first term) and expanded out the square (in the second term). In the fourth line we grouped terms (first and fourth, second and fifth, third and sixth, respectively). In the fifth line we introduced the *covariance*, $\text{cov}(\mathscr{X}, \mathscr{Y}) = \mathbb{E}(\mathscr{X}\mathscr{Y}) - \mathbb{E}(\mathscr{X})\mathbb{E}(\mathscr{Y})$, which measures the degree of independence of two random variables $\mathscr{X}$ and $\mathscr{Y}$.

Let's make our statement on the interpretation of the covariance more concrete. If $\mathscr{X}$ and $\mathscr{Y}$ are two independent random variables, then $\mathbb{P}^{\mathscr{X}\mathscr{Y}}_{ij} = \mathbb{P}^{\mathscr{X}}_i\mathbb{P}^{\mathscr{Y}}_j$, so the $\mathbb{E}(\mathscr{X}\mathscr{Y})$ that appears on the right-hand side in the definition of the covariance can be calculated as follows:

$$\mathbb{E}(\mathscr{X}\mathscr{Y}) = \sum_{i,j} \mathbb{P}^{\mathscr{X}\mathscr{Y}}_{ij} x_i y_j = \sum_i \mathbb{P}^{\mathscr{X}}_i x_i \sum_j \mathbb{P}^{\mathscr{Y}}_j y_j = \mathbb{E}(\mathscr{X})\mathbb{E}(\mathscr{Y}) \tag{C.26}$$

implying that *for two independent random variables the covariance vanishes.*[2]

We now specialize the main result in our second derivation, Eq. (C.25), to the case of two independent random variables (and therefore vanishing covariance), to find:

$$\mathbb{V}(\lambda_0\mathscr{X} + \lambda_1\mathscr{Y}) = \lambda_0^2\mathbb{V}(\mathscr{X}) + \lambda_1^2\mathbb{V}(\mathscr{Y}) \tag{C.27}$$

In words, we find that for independent random variables the variance of a linear combination of the variables is a new linear combination of the variances of the random variables we started with: the coefficients on the right-hand side turn out to be squared. We will

[2] The converse is not true: you can have two random variables that are not independent but give zero covariance.

refer to this as the addition rule for the variances of random variables. This, too, trivially generalizes to more random variables.

C.3.2 Continuous Random Variables

For the case of a *continuous* random variable $\mathscr{X}$ we are faced with a *probability density function*, $P(x)$. While this is basically analogous to the $\mathbb{P}_i^{\mathscr{X}}$ of the discrete case, the possible values x are now continuous, so we need to integrate $P(x)$ to get an actual probability $\mathbb{P}$. We typically assume that the probability density function is normalized, i.e.:[3]

$$\int_{-\infty}^{+\infty} P(x)dx = 1 \tag{C.28}$$

holds. This time around, the definition of the *expectation* is:

$$\mathbb{E}(\mathscr{X}) = \int_{-\infty}^{+\infty} P(x)xdx \tag{C.29}$$

sometimes denoted by μ or $\mu_{\mathscr{X}}$. Similarly, we have for the *variance*:

$$\begin{aligned} \mathbb{V}(\mathscr{X}) &= \mathbb{E}\left\{[\mathscr{X} - \mathbb{E}(\mathscr{X})]^2\right\} = \int_{-\infty}^{+\infty} P(x)[x - \mathbb{E}(\mathscr{X})]^2 dx \\ &= \mathbb{E}(\mathscr{X}^2) - \mathbb{E}(\mathscr{X})^2 = \int_{-\infty}^{+\infty} P(x)x^2 dx - \left(\int_{-\infty}^{+\infty} P(x)xdx\right)^2 \end{aligned} \tag{C.30}$$

also denoted by σ^2 or $\sigma_{\mathscr{X}}^2$. Everything here is analogous to the discrete-variable case.

Just like in the previous subsection, if $\mathscr{X}$ is a random variable then $f(\mathscr{X})$ is also a random variable; its expectation is given by the law of the unconscious statistician (also called the *rule of the lazy statistician*):

$$\mathbb{E}[f(\mathscr{X})] = \int_{-\infty}^{+\infty} P(x)f(x)dx \tag{C.31}$$

and, similarly, its variance is:

$$\mathbb{V}[f(\mathscr{X})] = \mathbb{E}\left\{(f(\mathscr{X}) - \mathbb{E}[f(\mathscr{X})])^2\right\} = \mathbb{E}[f^2(\mathscr{X})] - \mathbb{E}[f(\mathscr{X})]^2 \tag{C.32}$$

Both steps are identical to Eq. (C.23) but the meaning of the expectation is now different.

In complete analogy to the discrete-variable case, we can take linear combinations of random variables. For the expectation we find:

$$\mathbb{E}(\lambda_0 \mathscr{X} + \lambda_1 \mathscr{Y}) = \lambda_0 \mathbb{E}(\mathscr{X}) + \lambda_1 \mathbb{E}(\mathscr{Y}) \tag{C.33}$$

and for the variance, for independent random variables, we get:

$$\mathbb{V}\left(\lambda_0 \mathscr{X} + \lambda_1 \mathscr{Y}\right) = \lambda_0^2 \mathbb{V}(\mathscr{X}) + \lambda_1^2 \mathbb{V}(\mathscr{Y}) \tag{C.34}$$

Observe that these two results are identical to Eq. (C.24) and to Eq. (C.27), respectively, but once again the expectation should now be interpreted as an integral instead of a sum. In other words, the addition rule for expectations or variances of random variables is the same, regardless of whether the variables are discrete or continuous.

[3] Given the intended audience, we do not discuss things like measure theory, σ-fields, and so on.

Bibliography

To find out if a door is bolted, we must first push up against it.

Michel de Montaigne

[1] Abramowitz, M., and Stegun, I. A. (eds). 1965. *Handbook of Mathematical Functions*. Dover.

[2] Acton, F. S. 1990. *Numerical Methods That Work*. Mathematical Association of America.

[3] Acton, F. S. 1996. *Real Computing Made Real*. Princeton University Press.

[4] Allen, M. B. III, and Isaacson, E. L. 2019. *Numerical Analysis for Applied Science*. Second edn. John Wiley & Sons.

[5] Allen, M. P, and Tildesley, D. J. 2019. *Computer Simulation of Liquids*. Second edn. Oxford University Press.

[6] Andrae, R., Schulze-Hartung, T., and Melchior, P. 2010. Dos and don'ts of reduced chi-squared. *arXiv*, 1012.3754.

[7] Antoniou, A., and Lu, W.-S. 2007. *Practical Optimization*. Springer.

[8] Arfken, G. B., and Weber, H. J. 2005. *Mathematical Methods for Physicists*. Sixth edn. Elsevier.

[9] Ascher, U. M., and Greif, C. 2011. *A First Course in Numerical Methods*. Society for Industrial and Applied Mathematics.

[10] Audi, G., Wapstra, A. H., and Thibault, C. 2003. The Ame2003 atomic mass evaluation. *Nucl Phys A*, **729**, 337.

[11] Bailey, D. H., and Swarztrauber, P. N. 1994. A fast method for the numerical evaluation of continuous Fourier and Laplace transforms. *SISC*, **15**, 1105.

[12] Bajorski, P. 2012. *Statistics for Imaging, Optics, and Photonics*. John Wiley & Sons.

[13] Barlow, R. J. 1989. *Statistics*. John Wiley & Sons.

[14] Baym, G. 1969. *Lectures on Quantum Mechanics*. Westview Press.

[15] Becca, F., and Sorella, S. 2017. *Quantum Monte Carlo Approaches for Correlated Systems*. Cambridge University Press.

[16] Bender, C. A., and Orszag, S. A. 1978. *Advanced Mathematical Methods for Scientists and Engineers*. McGraw-Hill.

[17] Bernardo, J. M., and Smith, A. F. M. 2000. *Bayesian Theory*. John Wiley & Sons.

[18] Berrut, J.-P., and Trefethen, L. N. 2004. Barycentric Lagrange interpolation. *SIAM Review*, **46**, 501.

[19] Beu, T. A. 2015. *Introduction to Numerical Programming*. CRC Press.

[20] Bevington, P. R., and Robinson, D. K. 2003. *Data Reduction and Error Analysis for the Physical Sciences*. Third edn. McGraw-Hill.

[21] Bishop, C. M. 2006. *Pattern Recognition and Machine Learning*. Springer.

[22] Blatt, J. M., and Weisskopf, V. F. 1979. *Theoretical Nuclear Physics*. Springer.

[23] Boudreau, J. F., and Swanson, E. S. 2018. *Applied Computational Physics*. Oxford University Press.

[24] Box, G. E. P., and Luceño, A. 1997. *Statistical Control*. John Wiley & Sons.

[25] Brent, R. P. 1973. *Algorithms for Minimization without Derivatives*. Prentice-Hall.

[26] Broyden, C. G. 1965. A class of methods for solving nonlinear simultaneous equations. *Math Comp*, **19**, 577.

[27] Burden, R. L., Faires, J. D., and Burden, A. M. 2016. *Numerical Analysis*. Tenth edn. Cengage Learning.

[28] Byron, F. W. Jr, and Fuller, R. W. 1992. *Mathematics of Classical and Quantum Physics*. Dover.

[29] Casella, G., and Berger, R. L. 2002. *Statistical Inference*. Second edn. Duxbury.

[30] Ceder, N. 2018. *The Quick Python Book*. Third edn. Manning Publications.

[31] Conte, S. D., and de Boor, C. 1980. *Elementary Numerical Analysis*. Third edn. McGraw-Hill.

[32] Creutz, M. 1983. *Quarks, Gluons and Lattices*. Cambridge University Press.

[33] Cuzzocrea, A. et al. 2020. Variational principles in quantum Monte Carlo. *J Chem Theory Comput*, **16**, 4203.

[34] Daley, A. J. 2014. Quantum trajectories and open many-body quantum systems. *Adv Phys*, **63**, 77.

[35] Dalhquist, G., and Björck, Å. 1974. *Numerical Methods*. Prentice-Hall.

[36] Davio, M. 1981. Kronecker products and shuffle algebra. *IEEE Trans Comp*, **C-30**, 116.

[37] Davis, P. J., and Rabinowitz, P. 1984. *Methods of Numerical Integration*. Second edn. Academic Press.

[38] Deisenroth, M. P., Faisal, A. A., and Ong, C. S. 2020. *Mathematics for Machine Learning*. Cambridge University Press.

[39] Delves, L. M., and Mohamed, J. L. 1985. *Computational Methods for Integral Equations*. Cambridge University Press.

[40] Demmel, J. W. 1997. *Applied Numerical Linear Algebra*. Society for Industrial and Applied Mathematics.

[41] Downey, A. B. 2016. *Think Python*. Second edn. O'Reilly.

[42] Duncan, A. 2012. *The Conceptual Foundations of Quantum Field Theory*. Oxford University Press.

[43] Dunn, W. L., and Shultis, J. K. 2012. *Exploring Monte Carlo Methods*. Elsevier.

[44] Efron, B., and Tibshirani, R. J. 1993. *An Introduction to the Bootstrap*. Springer.

[45] Fetter, A. L., and Walecka, J. D. 1980. *Theoretical Mechanics of Particles and Continua*. McGraw-Hill.

[46] Feynman, R. P., Leighton, R. B., and Sands, M. L. 2010. *The Feynman Lectures on Physics, Vol. 3*. New millennium edn. Addison-Wesley.

[47] Fletcher, R. 1987. *Practical Methods of Optimization*. Second edn. John Wiley & Sons.

[48] Flügge, S. 1994. *Practical Quantum Mechanics*. Springer.

[49] Franklin, J. 2013. *Computational Methods for Physics*. Cambridge University Press.

[50] Freeman, W. T., and Adelson, E. H. 1991. The design and use of steerable filters. *IEEE Trans Pattern Anal Mach Intell*, **13**, 891.

[51] Gander, W., Gander, M. J., and Kwok, F. 2014. *Scientific Computing*. Springer.

[52] Gil, A., Segura, J., and Temme, N. M. 2007. *Numerical Methods for Special Functions*. Society for Industrial and Applied Mathematics.

[53] Gilks, W. R., Richardson, S., and Spiegelhalter, D. J. (eds). 1996. *Markov Chain Monte Carlo in Practice*. CRC Press.

[54] Giorgini, S., Pitaevskii, L. P., and Stringari, S. 2008. Theory of ultracold atomic Fermi gases. *Rev Mod Phys*, **80**, 1215.

[55] Goldberg, D. 1991. What every computer scientist should know about floating-point arithmetic. *ACM Comp Surv*, **23**, 5.

[56] Goldenfeld, N. 2019. *Lectures on Phase Transitions and the Renormalization Group*. CRC Press.

[57] Goldstein, H., Poole, C., and Safko, J. 2001. *Classical Mechanics*. Third edn. Addison-Wesley.

[58] Golub, G. H., and Van Loan, C. F. 1996. *Matrix Computations*. Third edn. Johns Hopkins University Press.

[59] Goodfellow, I., Bengio, Y., and Courville, A. 2016. *Deep Learning*. MIT Press.

[60] Greenbaum, A., and Chartier, T. P. 2012. *Numerical Methods*. Princeton University Press.

[61] Greene, W. H. 2018. *Econometric Analysis*. Eighth edn. Pearson.

[62] Griffiths, D. J. 2008. *Introduction to Elementary Particles*. Second edn. Wiley-VCH.

[63] Griffiths, D. J. 2017. *Introduction to Electrodynamics*. Fourth edn. Cambridge University Press.

[64] Hammarling, S. 2005. An introduction to the quality of computed solutions. Pages 43–76 of: Einarsson, B. (ed), *Accuracy and Reliability in Scientific Computing*. Society for Industrial and Applied Mathematics.

[65] Hamming, R. W. 1973. *Numerical Methods for Scientists and Engineers*. Second edn. McGraw-Hill.

[66] Hamming, R. W. 2012. *Introduction to Applied Numerical Analysis*. Dover.

[67] Heath, M. T. 2002. *Scientific Computing*. Second edn. McGraw-Hill.

[68] Higham, N. J. 2002. *Accuracy and Stability of Numerical Algorithms*. Second edn. Society for Industrial and Applied Mathematics.

[69] Hildebrand, F. D. 1974. *Introduction to Numerical Analysis*. Second edn. McGraw-Hill.

[70] Hilditch, R. W. 2001. *An Introduction to Close Binary Stars*. Cambridge University Press.

[71] Hill, C. 2020. *Learning Scientific Programming with Python*. Second edn. Cambridge University Press.

[72] Hoffman, J. D. 2001. *Numerical Methods for Engineers and Scientists*. Second edn. Marcel Dekker.

[73] Isaacson, E., and Keller, H. B. 1994. *Analysis of Numerical Methods*. Dover.

[74] Iserles, A. 2008. *A First Course in the Numerical Analysis of Differential Equations*. Second edn. Cambridge University Press.

[75] Izaac, J., and Wang, J. 2018. *Computational Quantum Mechanics*. Springer.

[76] Jackson, J. D. 1999. *Classical Electrodynamics*. Third edn. John Wiley & Sons.

[77] Jammer, M. 1966. *The Conceptual Development of Quantum Mechanics*. McGraw-Hill.

[78] Jaynes, E. T. 2003. *Probability Theory*. Cambridge University Press.

[79] Kahan, W. 1966. Numerical linear algebra. *Can Math Bul*, **7**, 757.

[80] Kahan, W. 1981. Why do we need a floating-point arithmetic standard? *Technical Report, UC Berkeley*.

[81] Kahan, W. 2005. How futile are mindless assessments of roundoff in floating-point computation? *Hous Sympos XVI*.

[82] Kalos, M. H., and Whitlock, P. A. 2008. *Monte Carlo Methods*. Second edn. Wiley-VCH.

[83] Kendall, M. G., and Stuart, A. 1961. *The Advanced Theory of Statistics, Vol. 2*. Hafner Publishing Company.

[84] Kernighan, B. W., and Pike, R. 1999. *The Practice of Programming*. Addison-Wesley.

[85] Kernighan, B. W., and Plauger, P. J. 1976. *Software Tools*. Addison-Wesley.

[86] Kittel, C. 2005. *Introduction to Solid State Physics*. Eighth edn. John Wiley & Sons.

[87] Kiusalaas, J. 2013. *Numerical Methods in Engineering with Python 3*. Cambridge University Press.

[88] Klainerman, S. 2008. Partial differential equations. Pages 455–483 of: Gowers, T. (ed), *The Princeton Companion to Mathematics*. Princeton University Press.

[89] Knuth, D. E. 1998. *The Art of Computer Programming, Vol. 2*. Third edn. Addison-Wesley.

[90] Koonin, S. E., and Meredith, D. C. 1990. *Computational Physics*. Addison-Wesley.

[91] Krylov, V. I. 2005. *Approximate Calculation of Integrals*. Dover.

[92] Lacava, F. 2016. *Classical Electrodynamics*. Springer.

[93] Landau, R., and Páez, M. 2018. *Computational Problems for Physics*. CRC Press.

[94] Langer, J. S., and Vosko, S. H. 1960. The shielding of a fixed charge in a high-density electron gas. *J Phys Chem Sol*, **12**, 196.

[95] Le Cam, L. 1990. Maximum likelihood: an introduction. *IS Review*, **58**, 153.

[96] Liboff, R. L. 2002. *Introductory Quantum Mechanics*. Fourth edn. Pearson.

[97] Longair, M. 2013. *Quantum Concepts in Physics*. Cambridge University Press.

[98] Lovitch, L., and Rosati, S. 1965. Direct numerical integration of the two-nucleon Schrödinger equation with tensor forces. *Phys Rev*, **140**, B877.

[99] Lummer, O., and Pringsheim, E. 1897. Die Strahlung eines "schwarzen" Körpers zwischen 100 and 1300 C. *Ann Phys Chem*, **299**, 395.

[100] Lyness, J. N., and Moller, C. B. 1967. Numerical differentiation of analytic functions. *SIAM J Numer Anal*, **4**, 202.

[101] Lyons, L. 2013. Bayes and frequentism: a particle physicist's perspective. *Contemp Phys*, **54**, 1.
[102] Mariño, M. 2015. *Instantons and Large N.* Cambridge University Press.
[103] McConnell, S. 2004. *Code Complete.* Second edn. Microsoft Press.
[104] McKinney, W. 2022. *Python for Data Analysis.* Third edn. O'Reilly.
[105] Merzbacher, M. 1970. *Quantum Mechanics.* Second edn. John Wiley & Sons.
[106] Meyn, S., and Tweedie, R. L. 2009. *Markov Chains and Stochastic Stability.* Second edn. Cambridge University Press.
[107] Millikan, R. A. 1916. A Direct photoelectric determination of Planck's "h". *Phys Rev*, **7**, 355.
[108] Newman, M. 2012. *Computational Physics.* Revised edn. CreateSpace.
[109] Nocedal, J., and Wright, S. J. 2006. *Numerical Optimization.* Second edn. Springer.
[110] Oliveira, S., and Stewart, D. 2006. *Writing Scientific Software.* Cambridge University Press.
[111] Overton, M. L. 2001. *Numerical Computing with IEEE Floating Point Arithmetic.* Society for Industrial and Applied Mathematics.
[112] Pang, T. 2006. *An Introduction to Computational Physics.* Second edn. Cambridge University Press.
[113] Pavelich, R. L., and Marsiglio, F. 2015. The Kronig-Penney model extended to arbitrary potentials via numerical matrix mechanics. *Am J Phys*, **83**, 773.
[114] Peskin, M. E., and Schroeder, D. V. 2018. *An Introduction to Quantum Field Theory.* CRC Press.
[115] Pethick, C. J., and Smith, H. 2008. *Bose–Einstein Condensation in Dilute Gases.* Second edn. Cambridge University Press.
[116] Pines, D., and Nozières, P. 1989. *The Theory of Quantum Liquids, Vol. 1.* Westview Press.
[117] Poisson, E., and Will, C. M. 2014. *Gravity.* Cambridge University Press.
[118] Press, W. H., Teukolsky, S. A., Vetterling, W. T., and Flannery, B. P. 1992. *Numerical Recipes in Fortran.* Second edn. Cambridge University Press.
[119] Ralston, A., and Rabinowitz, P. 1978. *A First Course in Numerical Analysis.* Second edn. McGraw-Hill.
[120] Ramalho, L. 2022. *Fluent Python.* Second edn. O'Reilly.
[121] Ree, F. H., and Holt, A. C. 1973. Thermodynamic properties of the alkali-halide crystals. *Phys Rev B*, **8**, 826.
[122] Richardson, O. W., and Compton, K. T. 1912. The photoelectric effect. *Phil Mag*, **24**, 575.
[123] Ridders, C. 1979. A new algorithm for computing a single root of a real continuous function. *IEEE Trans Circ Syst*, **26**, 979.
[124] Robert, C. P., and Casella, G. 2004. *Monte Carlo Statistical Methods.* Second edn. Springer.
[125] Roberts, G. O., Gelman, A., and Gilks, W. R. 1997. Weak convergence and optimal scaling of random walk Metropolis algorithms. *Ann Appl Prob*, **7**, 110.
[126] Rogers, S., and Girolami, M. 2017. *A First Course in Machine Learning.* Second edn. CRC Press.

[127] Schutz, B. F. 2022. *A First Course in General Relativity*. Third edn. Cambridge University Press.

[128] Scopatz, A., and Huff, K. D. 2015. *Effective Computation in Physics*. O'Reilly.

[129] Shankar, R. 1994. *Principles of Quantum Mechanics*. Second edn. Plenum Press.

[130] Shapiro, S. L., and Teukolsky, S. A. 2004. *Black Holes, White Dwarfs, and Neutron Stars*. Wiley-VCH.

[131] Širca, S., and Horvat, M. 2018. *Computational Methods in Physics*. Second edn. Springer.

[132] Slatkin, B. 2019. *Effective Python*. Second edn. Addison-Wesley.

[133] Stewart, G. W. 1973. *Introduction to Matrix Computations*. Academic Press.

[134] Stewart, G. W., and Sun, J. 1990. *Matrix Perturbation Theory*. Academic Press.

[135] Stoer, J., and Bulirsch, R. 1993. *Introduction to Numerical Analysis*. Second edn. Springer.

[136] Stoks, V. G. J. et al. 1993. Partial-wave analysis of all nucleon-nucleon scattering data below 350 MeV. *Phys Rev C*, **48**, 792.

[137] Stowe, K. 2007. *An Introduction to Thermodynamics and Statistical Mechanics*. Second edn. Cambridge University Press.

[138] Strang, G. 2005. *Linear Algebra and Its Applications*. Fourth edn. Brooks/Cole.

[139] Szebehely, V. 1967. *Theory of Orbits*. Academic Press.

[140] Szegö, G. 1975. *Orthogonal Polynomials*. Fourth edn. American Mathematical Society.

[141] Taylor, J. R. 1972. *Scattering Theory*. John Wiley & Sons.

[142] Theodoridis, S. 2020. *Machine Learning*. Second edn. Academic Press.

[143] Thijssen, J. M. 2007. *Computational Physics*. Second edn. Cambridge University Press.

[144] Thornton, S. T., and Marion, J. B. 2004. *Classical Dynamics of Particles and Systems*. Fifth edn. Brooks/Cole.

[145] Tierney, L. 1994. Markov chains for exploring posterior distributions. *Ann Stat*, **22**, 1701.

[146] Toussaint, D. 1989. Introduction to algorithms for Monte Carlo simulations and their application to QCD. *Comput Phys Commun*, **56**, 69.

[147] Trefethen, L. N. 2019. *Approximation Theory and Approximation Practice*. Extended edn. Society for Industrial and Applied Mathematics.

[148] Trefethen, L. N., and Bau, D. III. 1997. *Numerical Linear Algebra*. Society for Industrial and Applied Mathematics.

[149] Tucker, W. 2011. *Validated Numerics*. Princeton University Press.

[150] van der Vaart, A. W. 1998. *Asymptotic Statistics*. Cambridge University Press.

[151] Wasserman, L. 2004. *All of Statistics*. Springer.

[152] Weinberg, S. 2020. *Lectures on Astrophysics*. Cambridge University Press.

[153] Wilkinson, J. H. 1963. *Rounding Errors in Algebraic Processes*. Prentice-Hall.

[154] Wilkinson, J. H. 1965. *The Algebraic Eigenvalue Problem*. Oxford University Press.

[155] Williams, B. 1978. *Descartes: The Project of Pure Inquiry*. Penguin.

[156] Zinn-Justin, J. 2021. *Quantum Field Theory and Critical Phenomena*. Fifth edn. Oxford University Press.

Index

Though the lastnamed locality was not easily getatable.

James Joyce